THERMODYNAMICS

THERMODYNAMICS

GILBERT NEWTON LEWIS

MERLE RANDALL

Revised by

KENNETH S. PITZER

Professor of Chemistry
University of California, Berkeley

LEO BREWER

Professor of Chemistry
University of California, Berkeley

SECOND EDITION

McGRAW-HILL BOOK COMPANY, INC.

New York   Toronto   London

1961

THERMODYNAMICS

III

37622

THE MAPLE PRESS COMPANY, YORK, PA.

PREFACE

"There are ancient cathedrals which, apart from their consecrated purpose, inspire solemnity and awe. Even the curious visitor speaks of serious things, with hushed voice, and as each whisper reverberates through the vaulted nave, the returning echo seems to bear a message of mystery. The labor of generations of architects and artisans has been forgotten, the scaffolding erected for their toil has long since been removed, their mistakes have been erased, or have become hidden by the dust of centuries. Seeing only the perfection of the completed whole, we are impressed as by some superhuman agency. But sometimes we enter such an edifice that is still partly under construction; then the sound of hammers, the reek of tobacco, the trivial jests bandied from workman to workman, enable us to realize that these great structures are but the result of giving to ordinary human effort a direction and a purpose.

"Science has its cathedrals, built by the efforts of a few architects and of many workers. In these loftier monuments of scientific thought a tradition has arisen whereby the friendly usages of colloquial speech give way to a certain severity and formality. While this may sometimes promote precise thinking, it more often results in the intimidation of the neophyte. Therefore we have attempted, while conducting the reader through the classic edifice of thermodynamics, into the workshops where construction is now in progress, to temper the customary severity of the science in so far as is compatible with clarity of thought. But since it is improbable that we have been successful in this endeavor, to more than a limited extent, we shall take this opportunity of conversing very informally with the reader concerning our book and its purpose.

"There are several kinds of audience to which a book on thermodynamics might be addressed. There is the beginner who, in order that he may decide whether the subject will meet his needs or arouse his interest, asks what thermodynamics is and what sorts of problems in physics, chemistry and engineering can be solved by its aid; there is the reader who looks for the philosophical implications of such concepts as energy and entropy; above all there is the investigator who, attacking problems of pure or applied science, seeks the specific thermodynamic methods which are applicable to his problem and the data requisite for its solution.   Perhaps we have been over-ambitious in attempting, within the confines of a single volume, to meet all these demands—to lead the beginner through the intricacies of thermodynamic theory and to guide the experienced investigator to the extreme limits now set by existing methods and data."

It was with these comments that Gilbert Newton Lewis and Merle Randall introduced to the world their book on thermodynamics.   The success of this book was so great and its popularity so prolonged that we were not easily persuaded to undertake the task of revision.   It seemed at first more appropriate to leave the masterpiece unchanged as a monument to its authors and their era in the development of chemical thermodynamics.   Then we realized that the original edition would remain as that monument whatever we did, but that those portions of the masterpiece which are still timely could best be made available to the present generation by combining them with new discussion of the recent advances.

In the first edition the statistical aspect of thermodynamic phenomena was emphasized even though at that time a quantitative statistical treatment was available for only a few phenomena.   Subsequently a major advance has been made in the development of statistical methods for the calculation of thermodynamic properties from spectroscopic and other microscopic molecular data.   Both the statistical viewpoint and the precise equations are now an integral part of modern thermodynamics and have been so presented.   There are, of course, basic postulates in microscopic terms upon which the science of statistical mechanics may be built, and these postulates yield also the equations of statistical thermodynamics.   Indeed the best empirical evidence for the truth of statistical mechanics lies in the agreement of calculated and observed thermodynamic properties.   But in this book it has seemed preferable to develop these statistical equations simply and directly from thermodynamic laws.

Other major advances of recent decades have concerned systems with variables in addition to pressure, temperature, and composition.   The treatment of the thermodynamics of surfaces and of systems in gravitational, centrifugal, electric, or magnetic fields has been greatly expanded in this edition.   Also thermodynamic methods have been applied to systems where the separation of the reversible phenomena from heat flow or other irreversible effects is not self-evident.

Another major change which has occurred since the publication of the first edition is the extent to which thermodynamics has become an integral part of the first course in physical chemistry.   Thus, because of this preparation, many readers will be able to pass more quickly over some of the earlier chapters and come sooner to those chapters dealing with the rapidly expanding knowledge near the present frontier.

Just as Lewis and Randall wrote of the first edition, "Our work is not a textbook in the ordinary sense of the term. A textbook is a sort of table d'hôte to which any one may sit down and satisfy his hunger for information, with no thought of the complex agricultural processes which gave rise to the raw materials, nor of the mills which converted the raw materials into foodstuffs, nor of the arts of cookery responsible for the well-prepared meal which is set before him. It has not been our desire to offer such a repast to the reader. Our book is designed rather as an introduction to research, and as a guide to anyone who wishes to use thermodynamics in productive work. We have wished to permit every statement made during the course of the book to be traced either to the fundamental postulates of thermodynamics or to experimental work which is described in the literature and to which copious references have been furnished.

"In spite of this departure from the textbook tradition—and perhaps even because of this departure—we trust that this volume will be found serviceable in advanced chemical courses. For the benefit of the student, whether he is a member of a class, or is attempting by himself to master the fundamentals of thermodynamics, we have introduced numerous exercises during the earlier part of the book. These exercises will suggest many others, and it is only by repeated application of theory to concrete example that a real grasp of thermodynamics can be obtained."

One of the main objectives of the first edition was the collection, for the practical use of the chemist and the chemical engineer, of data needed for calculations of chemical equilibria. In recent years the number of accurate thermodynamic data has increased to such an extent that many volumes would be required to cover the available data as comprehensively as in the first edition. However, an attempt has been made to include the accurately known data for the more common substances as a convenient basis for equilibrium calculations.

At times one wishes to treat a problem with the greatest possible accuracy, and at other times one seeks an answer of lesser accuracy by the quickest and easiest method. Throughout the book we have given the methods for the exact solution of problems in terms of the thermodynamic laws and the experimental data. These methods avoid any further assumptions except possibly some fully established limiting law such as the ideal-gas law for the limit of zero pressure. Ordinarily one may use in calculations either graphical methods or analytical methods. With graphical methods the magnitude of experimental error is usually obvious, but with analytical methods special care must be taken not only to ensure that the empirical equations retain the full accuracy of the data but also to estimate the possible error in the final result. Both methods are illustrated many times. In addition generalizations of reasonable accuracy are very useful if they permit rapid calculation or estimation of the desired quantity. Our selection of the particular systems to be presented has naturally been influenced by our personal experience. We believe the acentric-factor system for real gases, the specific-ion-interaction equations of Bronsted and Guggenheim for electrolyte solutions, and our other choices are as accurate and convenient as any other systems now available. In some cases there are, however, other systems of comparable merit, and in these instances we were forced to make an arbitrary choice.

The choice of names and symbols for certain thermodynamic functions is a troublesome matter. In Chapter 13 we retained the text and footnotes of the first edition, which explained the choices made at that time. We also comment on the current situation and indicate a possible course of action which would eliminate the conflicting usage of the same name or symbol for different functions. Since Gibbs, the inventor of the functions under discussion, did not name them and since his symbols are not extensively used at present, there is no strong historical precedent. The system used in the first edition has been adopted for the major tabulations of chemical thermodynamic data. While we are aware of the conflicting usage in some other countries or other areas of science and would welcome an international notation if it is free from conflict with the major current systems, we believe that partial or premature changes will only increase confusion in the end. Hence, pending international agreement on a fully satisfactory system, we believe it best to retain the present scheme, which is in a sense also a recognition of the leadership of G. N. Lewis in the formulation of thermodynamics for accurate and convenient chemical application and in systematic accumulation of numerical data for important substances.

We are greatly indebted to colleagues both here at the University of California and elsewhere for their advice and help. The many suggestions of Professor W. F. Giauque have been especially valuable. Professor George Scatchard reviewed the entire manuscript, and his comments led to many important improvements. Very valuable suggestions were received from Doctors John Chipman, L. S. Darken, Harold Friedman, H. S. Harned, L. G. Hepler, H. S. Johnston, George Jura, K. K. Kelley, Norman Phillips, W. J. Ramsey, R. Schuhmann, Jr., and C. H. Shomate. Our thanks go to all these colleagues as well as to many students who have also made useful suggestions. Both of us wish to express special appreciation for the inspiration and the wise counsel we received from Professor Wendell M. Latimer from our first meeting until

his untimely death.   We thank Mrs. Virginia Shirley for preparation
of the index and Mrs. Betty North for her painstaking help with the
proofs.   Finally we wish to acknowledge with thanks the support from
the U.S. Atomic Energy Commission and from the American Petroleum
Institute for our research activities, which have contributed directly or
indirectly to this work.

Kenneth S. Pitzer
Leo Brewer

Let this book be dedicated to the chemists of the newer generation, who will not wish to reject all inferences from conjecture or surmise, but who will not care to speculate concerning that which may be surely known. The fascination of a growing science lies in the work of the pioneers at the very borderland of the unknown, but to reach this frontier one must pass over well-traveled roads; of these one of the safest and surest is the broad highway of thermodynamics.

CONTENTS

1

THE SCOPE OF THERMODYNAMICS

Aside from the logical and mathematical sciences, there are three great branches of natural science which stand apart by reason of the variety of far-reaching deductions drawn from a small number of primary postulates. They are mechanics, electromagnetics, and thermodynamics.

These sciences are monuments to the power of the human mind; and their intensive study is amply repaid by the aesthetic and intellectual satisfaction derived from a recognition of the order and simplicity which have been discovered among the most complex of natural phenomena. Also much of the material progress of the past century has sprung from the development of mechanical and electrical engineering and from the application of thermodynamics to the steam engine and other power-generating apparatus.

When it was first discovered that heat and work could be transformed one into the other and the laws governing such transformation were embodied in the science of thermodynamics, it was the primary function of this science to increase efficiency in the design and the use of engines for the production of work. Although this function has grown steadily in importance, it is now overshadowed by numerous applications of thermodynamics to physics and especially to chemistry. Here the methods of thermodynamics have brought quantitative precision in place of the old, vague ideas of chemical affinity, and thus chemistry has made the greatest advance toward the status of an exact science since the early chemists Lavoisier, Richter, and Dalton laid the foundations of stoichiometry. Even in the superchemistry of today, in which processes of stellar evolution, of radioactivity, and of intense electric discharge lead to the consideration of phenomena of a very different order from those of traditional chemistry, sometimes involving the dissolution of the atom itself, thermodynamics affords an unerring guide to the investigator.

1

THE POWER AND THE LIMITATIONS OF THERMODYNAMICS

Our book might be introduced by the very words used by Le Chatelier[1] several generations ago: "These investigations of a rather theoretical sort are capable of much more immediate practical application than one would be inclined to believe.   Indeed the phenomena of chemical equilibrium play a capital role in all operations of industrial chemistry."

Le Chatelier writes further:

It is known that in the blast furnace the reduction of iron oxide is produced by carbon monoxide, according to the reaction

$$Fe_2O_3 + 3CO = 2Fe + 3CO_2$$

but the gas leaving the chimney contains a considerable proportion of carbon monoxide, which thus carries away an important quantity of unutilized heat. Because this incomplete reaction was thought to be due to an insufficiently prolonged contact between carbon monoxide and the iron ore, the dimensions of the furnaces have been increased.   In England they have been made as high as thirty meters.   But the proportion of carbon monoxide escaping has not diminished, thus demonstrating, by an experiment costing several hundred thousand francs, that the reduction of iron oxide by carbon monoxide is a limited reaction.   Acquaintance with the laws of chemical equilibrium would have permitted the same conclusion to be reached more rapidly and far more economically.

Fortunately, the importance of thermodynamics in designing chemical and metallurgical processes is now generally appreciated in the major industries, although there are still many particular instances of neglect.

Before speaking more in detail of that which can be accomplished through the aid of thermodynamics, we must from the outset recognize its limitations.   In mechanics it is possible to foretell by simple laws the minimum expenditure of work by which a certain operation may be effected; but unless we know what frictional resistance may be encountered, we cannot predict how much work will actually be required.   So thermodynamics tells us the minimum amount of work necessary for a certain process, but the amount which will actually be used will depend upon many circumstances.   Likewise thermodynamics shows us whether a certain reaction may proceed and what maximum yield may be obtained but gives no information as to the time required.   The rate of a reaction is determined by factors which have hitherto eluded any exact analysis. The rate of a reaction is influenced by many factors in addition to those which determine the equilibrium yield.   Considerable progress has been made in the development of a general theory of reaction rates; indeed thermodynamics is one of the foundation stones of the theory.   But

[1] H. Le Chatelier, *Ann. mines*, (8)**13,** 157 (1888).

other basic postulates are required in addition to thermodynamics before a theory of chemical kinetics is possible.

Although subject to these limitations, thermodynamics is an instrument of great power and universality. It shows the engineer the maximum amount of work which a given quantity of fuel can produce in a given type of steam engine; it shows that more work is available when the fuel is burned in an explosion engine and, finally, that still more could be utilized if we were to solve the technical difficulties in obtaining "electricity direct from coal." The maximum efficiency of a turbine or of a refrigerator or of a weapon of ordnance is a subject for thermodynamic calculation.

To the manufacturing chemist thermodynamics gives information concerning the stability of his substances, the yields which he may hope to attain, the methods of avoiding undesirable substances, the optimum range of temperature and pressure, the proper choice of solvent, the limitations of methods of fractional distillation and crystallization.

To the analytical chemist it offers the means of predicting the limits of possible error, of avoiding side reactions, of choosing the concentrations best suited to his work.

Examples illustrating these applications will be discussed at various points in the book. Also the special treatises in these other areas now make extensive use of thermodynamics and thus illustrate the applications more fully than is possible here. However, even those already familiar with the applications will find it rewarding to examine carefully the foundations of thermodynamics and to note the numerous interlocking relationships which may yield additional material of interest.

THE MODERN STAGE OF THERMODYNAMICS

The whole foundation of classical thermodynamics was laid before the middle of the nineteenth century. The work of Black, Rumford, Hess, Carnot, Mayer, Joule, Clausius, Kelvin, and Helmholtz established the basic principles of the theory of energy.

Next came the task of building up from these cardinal principles a great body of thermodynamic theorems. This was the work of many men, among whom may be mentioned van't Hoff, and especially J. Willard Gibbs, whose great monograph on "The Equilibrium of Heterogeneous Substances"[1] has proved a rich and still unexhausted mine of thermodynamic material.

The third stage of thermodynamic development, in which we now find ourselves, is characterized by the design of more specific thermodynamic methods and their application to particular chemical processes, together

[1] J. Willard Gibbs, *Trans. Conn. Acad. Sci.*, **3**, 228 (1876).

with a systematic accumulation and utilization of the data of thermodynamic chemistry.

While this work was begun much earlier in isolated investigations by Sainte-Claire Deville[1] and by Horstmann,[2] it is only since the beginning of the present century that chemists have made general use of thermodynamics. The first systematic study of all the thermodynamic data necessary for the calculation of the free-energy changes in a group of important reactions was published in Germany by Haber. This book, "Thermodynamik der technischen Gas Reaktionen,"[3] is a model of accuracy and of critical insight. Also in Germany, Nernst and his associates made remarkable contributions to both theory and practice, while in Denmark Bronsted pursued a most valuable series of investigations.

It was, however, particularly the senior author of the first edition of this book, Gilbert N. Lewis, together with his many colleagues and students, who accomplished the major part of this third stage of development. In numerous studies they showed how to apply thermodynamics to various chemical systems, and in an extensive array of experimental investigations they accumulated the necessary data for the common chemical substances. Tables of free energies and of electrode potentials took their place in basic chemical handbooks alongside the table of atomic weights.

In addition to the extensive application of classical thermodynamics to chemical problems in the first half of the present century, quantum mechanics was developed, and our understanding of molecular properties progressed greatly. The third law of thermodynamics, which depends in essence on the quantization of molecular energies, was developed, tested, and applied most fruitfully to chemical problems. Also statistical methods were perfected which make it possible to calculate macroscopic thermodynamic properties from the molecular energy levels observed spectroscopically. These developments are an integral part of modern thermodynamics.

Statistical mechanics constitutes in large part a parallel science to thermodynamics. Frequently the fundamental structure of statistical mechanics is presented in terms of abstract postulates and relatively complex mathematics which seem remote from the laws of thermodynamics. However, Giauque and others have developed the important

[1] Sainte-Claire Deville, Leçons sur la dissociation, "Leçons de chimie," 1864–1865, p. 255, Librairie Hachette, Paris, 1866.

[2] August Horstmann, *Ber. deut. chem. Ges.*, **1**, 137 (1869); *Ann. Chem. Pharm.*, **170**, 192 (1873).

[3] Fritz Haber, "Thermodynamik der technischen Gas Reaktionen," R. Oldenbourg-Verlag, Munich, 1905 (trans. by A. B. Lamb, Longmans, Green & Co., Inc., New York, 1908).

equations of statistical thermodynamics on the basis of thermodynamic postulates, and we shall follow this course as the one appropriate to a book on thermodynamics. The statistical viewpoint gives great insight into the thermodynamic behavior of chemical systems. The concept of entropy, otherwise strange and abstract, is statistically related to familiar ideas.

2

DEFINITIONS; THE CONCEPT OF EQUILIBRIUM

As a science grows more exact, it becomes possible to employ more extensively the accurate and concise methods and notation of mathematics. At the same time it becomes desirable, and indeed necessary, to use words in a more precise sense. For example, if we are to speak, in the course of this work, of a pure substance, or of a homogeneous substance, these words must convey as nearly as possible the same meaning to writer and to reader.

Unfortunately it is seldom possible to satisfy this need by means of formal definitions, partly because the most fundamental concepts are the least definable, partly because of the inadequacy of language itself, but more particularly because we often wish to distinguish between things which differ rather in degree than in kind. Frequently, therefore, our definitions serve to divide for our convenience a continuous field into more or less arbitrary regions—as a map of Europe shows roughly the main ethnographic and cultural divisions, although the actual boundaries are often determined by chance or by political expediency.

The distinction between a solid and a liquid is a useful one, but no one would attempt to fix the exact temperature at which sealing wax or glass passes from the solid to the liquid state. Any attempt to make the distinction precise makes it the more arbitrary.

CLASSIFICATION OF SUBSTANCES

Whatever part of the objective world is the subject of thermodynamic discourse is customarily called a system. Sometimes it is desirable to use this term in a more definite sense, implying a spatial content. If we make an enclosure by means of physical walls, or if we imagine such an enclosure made by a mathematical surface, such an enclosing surface serves as the boundary of the system, which then comprises everything of thermodynamic interest contained within that boundary. The boundary or surface that defines the physical extent of the system can be distorted—as when the system is expanding—or it may have a fixed

6

position.   The choice of the boundary in many problems is often arbitrary, and judicious choice of the boundary of a system may often simplify the treatment of changes taking place.   Thus, for example, a crystal or any chosen cubic centimeter of that crystal may be chosen as the system. A thermodynamic system may contain no substance at all, in the ordinary sense, but may consist of radiant energy or an electric or magnetic field. Usually, however, a system comprises a substance, which may be homogeneous or heterogeneous.

Systems which can exchange energy with the surroundings but which cannot transfer matter across the boundaries are known as *closed systems.* Open systems can exchange both energy and matter with the surroundings.   Our discussion in the first 13 chapters of this book will be restricted to closed systems, and we shall not consider open systems until we are prepared to treat as variables the concentrations of the components of a solution.

Homogeneous Systems.   At the outset we meet the difficulty of exact definition or classification if we attempt to make this traditional distinction between homogeneous and heterogeneous systems.   Nevertheless it will be convenient to define a homogeneous system as one whose properties are the same in all parts, or at least which vary continuously from point to point, a system, in other words, in which there are no apparent surfaces of discontinuity.

This definition would include among homogeneous systems some which are of little importance to us in thermodynamic work.   Thus a tube of water through which copper sulfate is diffusing in such a manner that the concentration varies gradually from one end of the tube to the other gives, according to our definition, a homogeneous system.   But it is one that is not susceptible to simple treatment by the ordinary methods of thermodynamics.   On the other hand, a long vertical tube containing a solution of copper sulfate will, under the influence of gravity, finally reach a condition in which there is a definite and constant concentration gradient, governed by simple thermodynamic laws.   These are, however, unusual cases, and ordinarily, by a homogeneous system, we shall mean one whose properties, as judged by our ordinary criteria, are the same throughout.

Heterogeneous Systems.   A heterogeneous system consists of two or more distinct homogeneous regions.   Thus benzene and water or ice and water form heterogeneous systems.   The homogeneous regions, which are called *phases*, appear to be separated from one another by surfaces of discontinuity.   (In our usual thermodynamic work it is immaterial whether we have one piece of ice or several pieces of ice in contact with a mass of water.   In such cases it is commonly considered that there are only two phases present, the ice phase and the water phase.)

The boundaries between phases are not surfaces in a strict mathematical sense but are very thin regions, in which the properties change with great abruptness from the properties of the one homogeneous phase to those of the other.   The effect of surface area on the total thermodynamic properties is so small that it may be neglected unless the subdivision is exceedingly fine.   But when we study surface tension, adsorption, and kindred phenomena, these surface regions become of great importance.

Indeed the classification of systems according to homogeneity or heterogeneity seemed formerly more precise than it does now, when so much attention is being devoted to the interesting types of substances known as colloids.   When, therefore, we speak of a heterogeneous system, without further qualification, it must be understood that the system is one in which each homogeneous region is large.

Components and Solutions.   Any homogeneous system, whether solid, liquid, or gaseous, is called a solution if its composition is variable. Thus air, brine, glass, and a mixed crystal of alum and chrome alum are all called solutions.   The components are the substances of fixed composition which can be mixed in varying amounts to form the solution. For thermodynamic purposes, the choice of components of a system is often arbitrary and depends upon the range of conditions for the problem under consideration.   A system consisting of a single element of fixed isotopic composition is usually considered to be a single-component system, but it is necessary to consider each isotope as an independent component if their amounts are being varied independently and if the properties being studied vary with isotopic content within the accuracy of the measurements.

If the elements are combined in a definite ratio as a compound, so that they are not varied independently, the compound is considered as a single component.   A mixture of O_2 and N_2 is a two-component system if the amount of each element can be varied independently, but a system consisting of only NO molecules, or produced from only NO molecules, is considered a one-component system.   A system produced by adding NO, O_2, and N_2 molecules independently of one another is a three-component system at low temperatures.   However, at high temperatures where these three molecular species are in rapid equilibrium with one another and their relative amounts are fixed by the equilibrium constant for the dissociation of NO, the system is only a two-component system, since it can be produced by the addition of only O_2 and N_2.   If the system is restricted to compositions corresponding to equal amounts of nitrogen and oxygen, the high-temperature system can be considered a one-component NO system.   Likewise, a mixture of 10 hydrocarbons would be a 10-component system at low temperatures and a two-component system at very high temperatures where the molecular composition is not at one's disposal but is fixed by the equilibrium constants and the amounts of carbon and hydrogen added.   A system consisting of HF gas contains many different molecular species.   However, they are all in rapid equilibrium with one another under ordinary conditions, and all species can be derived from any one. Thus the system is a one-component system, but the choice of which species, for example, HF, H_2F_2, etc., to use for its designation is arbitrary.

The choice of the molecular species to be considered as components in a solution is not even restricted to those species actually known to be present in appreciable concentration. Thus we can consider the components of a sodium chloride aqueous solution to be H_2O and NaCl even though we know that there is no appreciable concentration of NaCl molecules in dilute solutions. We can consider solid solutions of composition in the range $FeO_{1.05-1.19}$ to be formed from the components FeO and Fe_3O_4 even though the composition FeO has no stable existence as a solid phase. We could equally well have chosen FeO and Fe_2O_3, Fe and Fe_3O_4, Fe and O_2, or any pair of arbitrarily chosen compositions between Fe and O_2 as the two components of the solution, depending upon which pair proves to be most convenient for the calculations in mind. Thus Darken and Gurry[1] chose as the components of this system $FeO_{1.0477}$ and oxygen.

Solids, Liquids, and Gases; Crystals and Noncrystals. The ancient categories represented by earth, water, and air have persisted in a simple classification of substances into solids, liquids, and gases. While this useful classification may ordinarily be employed without fear of ambiguity, there are, as we have already pointed out, some substances which are unquestionably solid, like glass, but which, when heated, pass by imperceptible gradations into typical liquids. Also, since the pioneer investigation of the critical state by Andrews,[2] it has been known that a liquid may be changed to a gas by a process in which the substance remains as a pure phase from beginning to end, without the appearance of discontinuity at any stage.

A more fundamental distinction at present seems to be the one between crystalline and noncrystalline states; as yet no substance has succeeded in passing by a continuous process from one of these states to the other. In contrast to the critical point for liquid-gas systems, no similar point corresponding to the solid-liquid transformation has been observed, and there is no clear indication that such a point exists. Crystalline substances, although usually solid, range from hard, rigid substances like diamond, through soft crystals like rubidium, to the extremely fluid crystals.[3,4]

---

[1] L. S. Darken and R. W. Gurry, *J. Am. Chem. Soc.*, **67**, 1398–1412 (1945).

[2] T. Andrews, *Trans. Roy. Soc. (London)*, **159**, 575 (1869).

[3] F. C. Frank, *Discussions Faraday Soc.*, **25**, 19 (1958), and other papers of this discussion.

[4] C. W. Oseen, *Trans. Faraday Soc.*, **29**, 883 (1933), and other papers of the 1933 discussion presented in vol. 29, pt. 9, 1933.

STATES AND PROPERTIES

If it were possible to know all the details of the internal constitution of a system, in other words, if it were possible to find the distribution, the arrangement, and the modes of motion of all the ultimate particles of which it is composed, this great body of information would serve to define what may be called the microscopic state of the system and this microscopic state would determine in all minutiae the properties of the system.

We possess no such knowledge, and in thermodynamic considerations we adopt the converse method. The *state* of a system (macroscopic state) is determined by its *properties* just in so far as these properties can be investigated directly or indirectly by experiment. We may therefore regard the state of a substance as adequately described when all its properties which are of interest in a thermodynamic treatment are fixed with a definiteness commensurate with the accuracy of our experimental methods. Let us quote from Gibbs[1] in this connection:

> So when gases of different kinds are mixed, if we ask what changes in external bodies are necessary to bring the system to its original state, we do not mean a state in which each particle shall occupy more or less exactly the same position as at some previous epoch, but only a state which shall be undistinguishable from the previous one in its sensible properties. It is to states of systems thus incompletely defined that the problems of thermodynamics relate.

The *properties* of a substance describe its present state and do not give a record of its previous history.

It is an obvious but highly important corollary of this definition that, when a system is considered in two different states, the difference in volume or in any other property between the two states *depends solely upon those states themselves, and not upon the manner in which the system may pass from one state to the other.*

Extensive and Intensive Properties. Most of the properties which we measure quantitatively may be divided into two classes. If we consider two identical systems, let us say two kilogram weights of brass or two exactly similar balloons of hydrogen, the volume, or the internal energy, or the mass of the two is double that of each one. Such properties are called *extensive.*

On the other hand, the temperature of the two identical objects is the same as that of either one, and this is also true of the pressure and the density. Properties of this type are called *intensive.* Some intensive

[1] J. W. Gibbs, *Trans. Conn. Acad. Sci.*, **3**, 228 (1876); "Collected Works of J. Willard Gibbs," vol. I, "Thermodynamics," Longmans, Green & Co., Inc., New York, 1928.

properties are derived from the extensive properties; thus, while mass and volume are both extensive, the density, which is mass per unit volume, and the specific volume, which is volume per unit mass, are intensive properties. These intensive properties are the ones which describe the specific characteristics of a substance in a given state, for they are independent of the amount of substance considered. Indeed in common usage it is only these intensive properties which are meant when the properties of a substance are being described.

Interaction of Systems. Many of the words such as work, heat, energy, etc., that are important in thermodynamics often have connotations from their use in nonscientific discussions that conflict with their meaning as used here. To clarify the meaning of these words, their use will be illustrated by describing the behavior of simple interacting systems. To avoid unnecessary complications, the systems chosen first for discussion will be highly idealized to limit the types of interactions that can take place simultaneously and will be restricted to systems with no temperature changes due to frictional effects. From this initial idealized discussion we shall later generalize when we discuss the first law of thermodynamics.

When two systems come into contact, they will generally interact as shown by changes in their properties. Can we predict the extent of interaction or the extent of change of the properties of the systems from knowledge of the initial states of the systems? Or knowing how much the properties of one system have changed, can we state how much the other system must have changed? Let us apply these questions to the interaction resulting from the collision of two systems. When their relative velocity changes as the result of the interaction, we say that each system is exerting a force on the other that is defined as the rate of change of linear momentum (the product of mass and linear velocity). The units of force, the cgs dyne and the mks newton, correspond to rates of change of linear momentum of 1 g cm/sec² and 1 kg m/sec², respectively. From Newton's third law of mechanics, the two forces are equal in magnitude though opposite in sign or direction. If there are no other forces or interactions between the two systems, the laws of mechanics require, in addition to conservation of momentum, that the changes in relative velocities of the two systems be given by the equation $m_1 \Delta(v_1)^2 = m_2 \Delta(v_2)^2$ or that the quantity $\frac{1}{2}mv^2$, called the kinetic energy, be conserved for the two systems as a whole. The unit of kinetic energy in the cgs system, the erg, corresponds to the increase in kinetic energy of a system when a force of 1 dyne acts on the system while it moves 1 cm. In the mks system, the unit of energy is the joule, or 10^7 ergs. We may paraphrase our description of the interaction by saying that the force acting on a given system has accelerated the system to a greater velocity and thus has increased its kinetic energy. In thermodynamic terms, we would say that the first system did an amount of work on the second system equal to the increase in kinetic energy of the second system.

For this simple interaction, we can readily answer the questions that we have posed above. Knowledge of the properties mass and velocity of each system fixes the kinetic energy of each system, which is thus a property itself of each system. From the change of kinetic energy of one system, we can calculate the change in kinetic energy of the other interacting system and thus the change in the velocity of the other system.

The velocity of a given system is defined not absolutely but rather in terms of a particular coordinate system or frame of reference. The fundamental laws of motion and of thermodynamics are valid for various frames of reference, although the equations may become unnecessarily complex. These complications need not disturb us at this point since we shall have no need to change the frame of reference in our problems.

For such a simple interaction, the increase in kinetic energy of one system or the work done on that system is just equal to the decrease in kinetic energy of the second system. If there are other interactions, the kinetic energy is not conserved. Thus, for a system moving upward in the earth's gravitational field, we observe a decrease in the velocity of the system. We describe this as due to a gravitational interaction between the earth and the system, with the gravitational force given by the rate of change of linear momentum of the system. As the velocity of the system gradually decreases to zero with respect to the earth, the kinetic energy of the system is decreased without any corresponding increase in the kinetic energy of any other system. However, when the system has fallen back to its initial height, it will have regained its initial kinetic energy if there have been no other interactions with the system. The equation $\frac{1}{2}mv^2 + mgh = $ constant represents the variation of kinetic energy of such a system with height where mg is the gravitational force in dynes acting on the system. At the earth's surface $g = 980.665$ dynes/g. The quantity mgh is called the gravitational potential energy of the system relative to the earth. Thus the movement of a system upward a height h against a gravitational force mg results in an increase in the potential energy of the system by the amount mgh in ergs if m is expressed in grams and h in centimeters. The kinetic energy of a system is a property of the system, i.e., a function of the velocity of the system; the potential energy is a property which is a function of the system's position. The increase in potential energy is just equal to the decrease in kinetic energy if there are no interactions other than the gravitational interaction and the sum of the kinetic and potential energies is conserved.

The gravitational potential energy of a system is a property of that system, but it cannot be given an absolute value. Its value depends upon the reference point for measurement of height. However, it is the change in kinetic energy or the change in potential energy that is of interest in describing interactions of systems, and the changes will not depend upon the origin selected for the coordinate of height.

Interactions between systems can result not only in forces that change the kinetic and gravitational energies but in electrical, magnetic, and other types of forces. Through electric motors and other similar devices, it is always possible to measure the work done on an interacting system by having the system ultimately lift a weight as a result of returning to its initial state.

For each type of interaction we shall usually associate some intensive property which will be different initially for the two systems and which will have become equal for both systems when further interaction can no longer take place. For example, the interaction of two systems with different pressures will result in forces causing compression of the system of lower pressure and expansion of the system of higher pressure until the pressures are balanced. Likewise, systems of different electrical potential will interact until the potentials are equalized. The differences of these potentials or similar intensive properties are the cause of the interaction between systems, but the work done or the magnitude of the change in state of the interacting systems depends not only upon the potential difference but also upon the size or amount of material of the system. As an example, the amount of electrical work required to increase the electrical potential of a hollow sphere depends upon the size of the sphere. The electrical capacity of a system is proportional to the increase in energy for a unit increase of electrical potential. For many other interactions, a similar extensive property can be defined as the energy change associated with a unit change in potential. These capacities are generally complex functions of the potentials.

In summary, for every type of interaction between systems, there will be a characteristic intensive property, such as pressure or a potential, for each system. The difference between this property of the systems will be the cause or driving force of the interaction, and no interaction takes place with equal values of the potential or driving force. In addition, for each type of interaction, there will be a characteristic extensive property of each system which together with the corresponding intensive property fixes an energy for each system. Each characteristic energy is a state function. For a given change in the state of a system, the change in the characteristic energy will be a fixed quantity. When a system has been changed from one state to another state corresponding to a given change in a characteristic energy, e.g., electrical energy, we say that an amount of work, e.g., electrical work, has been done on the system equal to the increase in the electrical energy of the system.

Thermal interactions are likewise associated with the intensive property temperature and the extensive property heat capacity. Thermal interactions are of special importance to thermodynamics, and we shall discuss the definition and application of these thermal properties in considerable detail in Chapter 4.

Reproducibility of States. We have tacitly assumed in the preceding paragraphs that a pure substance always exists in one of a few well-defined forms, so that if a few conditions are fixed, all the properties are determined. Indeed this is true for so large a number of substances that unless otherwise stated it will be taken for granted.

Until recently only three forms of water were known: vapor, liquid, and ice. The properties of a given amount of pure water vapor can be completely determined by external conditions. Thus, if the temperature and the pressure are fixed, two equal quantities of water vapor will, by any experimental test, be found identical in all respects. The same is true of liquid water, and it is probably nearly true for ice, as well as for the various other forms of solid water.

On the other hand, certain metals, even when pure, vary greatly according to their previous treatment, and two samples are not identical although all external conditions are the same. In such cases a substance, instead of appearing only in a few well-defined states, may assume any one of an infinite number of states, depending upon its mode of preparation and its mechanical or thermal treatment. Cases of this sort deserve more careful consideration than they usually receive. Thus there is no doubt that many measurements of the electrode potentials of metals have been deprived of value because of the lack of definition of the surface conditions in the electrodes.

In a perfect crystal the atoms are supposed to be arranged in a perfectly definite order. At a given temperature and pressure, we should thus expect the properties to be unambiguously determined. It is, however, doubtful whether there are many actual crystalline substances in which the conditions are so simple.

Let us consider common ice. When pure water is frozen, long crystals first traverse the mass; these are then connected by shorter crystals, until finally a mesh is produced in which the last remaining drops of liquid may not be free to form the same perfect crystals as were produced at the beginning. It is conceivable, therefore, that the material formed at the end of the process has somewhat different properties from those of the more ideal crystals produced at the beginning. Such a difference might exhibit itself, for example, by a slightly lower melting point.

In the case of a typical liquid also we may expect the properties to be definitely determined by external conditions, not because of any ordered arrangement of the particles, but rather because their mobility permits a complete randomness of arrangement, so that with large numbers of molecules the average properties of the mass are constant.

However, in the case of substances of high viscosity, the mobility of the particles is so small that they do not readily assume the positions of symmetry in a space lattice which are characteristic of the perfect crystal or the random arrangement which is characteristic of the mobile fluid. In such a case the particles may remain for long periods of time in strained positions which are determined by their previous treatment or by the fortuitous circumstances of their original assemblage. As examples we may cite, on the one hand, a drawn wire of hard metal or a piece of unevenly cooled glass; on the other, such materials as are obtained when a metal is deposited by electrical spattering.[1] Here each particle may lie as it strikes without later rearrangement.

It is evident that, if we are to treat quantitatively and numerically the properties of substances, the state of a substance must be described with great particularity, unless we can assume that its properties are completely determined by the external conditions. When the properties are so determined, the state of a given amount of a substance can ordinarily be fixed merely by stating the temperature and pressure and, in the case of solutions, the composition. Only in special cases shall we consider the properties of substances as dependent upon the degree of subdivision of the phases, upon gravitational, electric, magnetic or centrifugal fields, or upon other external influences which have only a minute effect upon the system.

EQUILIBRIUM AND REACTION SPEED

We have developed in some detail the ideas of the preceding section because they lead us directly to the idea of *equilibrium,* and in all thermodynamics there is no concept more fundamental than this.

If a substance like water is under fixed external conditions which completely determine its state, its properties do not change with time. It is said to be in a state of rest. If the external conditions are momentarily altered, the water returns immediately thereafter to its original state and properties.

If, instead of water, a substance like soft tar is employed, the same thing happens, but more slowly. If the tar is subjected to some temporary distortion or to some unevenness of pressure, it slowly yields, or flows, until the former state of rest is once more established.

When a system is in such a state that after any slight temporary disturbance of external conditions it returns rapidly or slowly to the initial

[1] I. Langmuir, *J. Am. Chem. Soc.,* **38,** 2221 (1916).

state, this state is said to be one of equilibrium.   *A state of equilibrium is a state of rest.*

Even crystalline substances of the softer sort fail to retain for long any condition differing from the characteristic state of equilibrium. Thus, for example, we may account for the extraordinary reproducibility of the electrode potentials of soft metals, such as sodium or lead, as compared with metals like iron or nickel.   Even in the case of substances of great viscosity or rigidity it seems reasonable to suppose that they also behave in a similar manner, although the changes may be imperceptible because of their slowness.

Any change in the properties of a system is called a *process*, and if the process is one which is roughly termed chemical, it is sometimes called a *reaction*, but we shall employ these terms almost interchangeably.   The idea which we have developed regarding the restoration of equilibrium after a mechanical disturbance we may extend to cases in which chemical reactions are involved.

If we dissolve methyl acetate in water, hydrolysis will set in and the properties of the system will change until a definite state is reached, which is fixed by conditions such as temperature and pressure.   In addition to the original substances, methyl alcohol and acetic acid will be found in solution in fixed amounts.   We have again a state of equilibrium. If the system is temporarily disturbed, for example by raising the temperature for a short time and then bringing it back to the original value, the solution will once more return to the same state.

Slow establishment of equilibrium, after mechanical disturbance, we have attributed to such factors as viscosity.   When the rate of a chemical reaction is involved, the time required to establish an equilibrium depends upon factors which we may suspect to be often fundamentally analogous to viscosity, but which are still obscure.

We shall therefore consider not only that every state of equilibrium is a state of rest but that *every state of absolute rest is a state of equilibrium*, and therefore that every system which has not reached a state of equilibrium is changing continuously toward such a state with greater or less speed although possibly at an immeasurably slow speed.

Stable Systems.   Frequently one speaks of *stable* systems, and much confusion of thought has resulted from the use of this term for two different ideas which are separable, and indeed must be separated if any clarity is to be obtained in the application of thermodynamics to chemistry.   In common usage a system is said to be stable when it undergoes no apparent changes.   Now a system which is apparently in a stationary state may be so because it has reached one of the states of equilibrium from which it has no tendency to depart, no matter how great its mobility; or it may be because processes occurring within it are so slow as to be imperceptible, even though the system may be far from a true state of equilibrium. It is only systems of the first kind, which are really in a state of equilibrium, which we shall call stable in any thermodynamic sense. Systems of the second kind may be called *inert* or *unreactive*.

A mixture of oxygen and hydrogen might be kept for a long time without the formation of any measurable quantity of water, but the system is inert, and not thermodynamically stable, as shown by the fact that any one of a number of catalytic substances causes a rapid formation of water. Such a catalyzer merely increases the rate of attainment of equilibrium. In the absence of such catalyzer, and at room temperature, the rate of the reaction is entirely too slow to be measurable. Nevertheless we can make an approximate calculation of that rate by actually measuring it at a number of high temperatures and employing the method of extrapolation.

Partial Equilibrium. Of the various possible processes which may occur within a system, some may take place with extreme slowness, others with great rapidity. Hence we may speak of equilibrium with respect to the latter processes before the system has reached equilibrium with respect to all the possible processes. Thus, in a system of oxygen, hydrogen, and water, the two gases dissolve rapidly until the water is saturated, and we may say that the system is in equilibrium with respect to the process of solution. It is far from equilibrium with respect to the reaction by which water is formed from oxygen and hydrogen—a process in which the speed is of a far different order of magnitude. As another example we may consider nitrogen tetroxide, which dissociates rapidly until a state of equilibrium is soon reached between N_2O_4 and NO_2. But each of these substances is really extremely unstable with respect to elementary oxygen and nitrogen, although, without catalysts, the process of decomposition into these elements is an extraordinarily slow one.

Degrees of Stability. A stone lying in a hollow upon a hillside is considered to be in a stable position, although, if pushed over the edge, it will roll to a position of greater stability at the bottom. So in thermodynamics a system may be in a state of rest and, if slightly disturbed, may revert to this same state of rest, but if largely disturbed it may proceed toward some entirely new condition of equilibrium. Thus liquid water, a degree or two below the freezing point, reaches a state of equilibrium to which it will return after a slight disturbance. Any large disturbance, however, may cause it to seek a new condition of equilibrium in the more stable form of ice.

In practice we often assume the existence of several such equilibrium states toward which a system may tend, all these states being stable but representing higher or lower degrees of stability. From a theoretical standpoint it might be doubted whether there is any condition of real equilibrium, with respect to every conceivable process, except the one which represents the most stable state. This, however, is not a question which need concern us greatly, nor is it one which we could discuss adequately at this point, without largely anticipating what we shall later have to say regarding the statistical view of thermodynamics.

Equilibrium as a Macroscopic State.   Even here it is desirable to emphasize that by a state of rest, or equilibrium, we mean a state in which the properties of a system, as experimentally measured, would suffer no further observable change even after the lapse of an indefinite period of time.   It is not intimated that the individual particles are unchanging.   Thus, when sulfuric acid is heated in a closed vessel, a condition is ultimately reached in which definite amounts of the liquid sulfuric acid and of the gases sulfuric acid, sulfur trioxide, and water have been produced.   These amounts, as determined by any of our quantitative methods, then remain constant.   This is what we call the state of equilibrium.   If, however, we were in a position to follow the paths of the individual molecules, we should perceive the wildest chaos: molecules of the liquid evaporating, some molecules of vapor entering the liquid phase, others dissociating into molecules of water and sulfur trioxide, and these in turn constantly combining.   The absolute number of molecules of each of these species varies from instant to instant, but these variations are so small compared with the total numbers that they would be imperceptible even if the accuracy of our analytical processes were increased a billionfold.

3

CONVENTIONS AND MATHEMATICAL METHODS

In an extended application of thermodynamics to chemistry, involving, as it does, a large number of arithmetical computations, it is essential that a definite notation be established and that certain conventions be laid down and adhered to rigorously. Such conventions usually have no theoretical significance, but their practical value can hardly be over-estimated. Some of these conventions, for example, will be made solely to prevent that confusion of sign which is so fatal to accuracy and speed in numerical calculations.

It is a necessary consequence of the haphazard growth of science that certain terms are used in various senses at different times and by different authors. Thus the heat of reaction ordinarily means the heat evolved in a chemical reaction, while the heat of vaporization and the heat of fusion mean the heat absorbed in these processes. The so-called equilib-rium constant of one author may be the reciprocal, or the square root, of that used by another. There is frequently not the slightest a priori reason for preferring one definition to another, and yet, as it is important that persons of one locality decide by artificial convention whether to pass, on meeting, to the right or to the left, it is likewise important that scientists should endeavor to establish uniform usage of scientific terms. When there is no urgent reason for the choice of one convention rather than another, we shall employ that one which seems best to conform to permanent international usage.

The Mole. The gram and the kilogram are the units of mass in the cgs and mks systems, respectively, and are commonly employed as the units of quantity of material. However, when we are dealing with chemical reactions, it is far more convenient to employ the mole or the equivalent for such a unit. For general purposes the mole is the better unit, since an equivalent of a substance may have different meanings according to the kind of reaction into which the substance enters. Thus one equivalent of permanganic acid has a variable significance, according as we consider the power to neutralize a base or to act as an oxidizing agent in acid or in alkaline solution.

19

If M is the molecular weight of the substance, a mole is defined as M grams. This unit is by no means as free from ambiguity as the gram. The atomic weights are subject to revision, but most are now very accurately known, and in case of doubt the value assumed may be stated. The more common ambiguity concerns the formula of the molecules actually present in the system. In many cases the composition of these ultimate molecules is not known, and the formula employed represents an estimate or is merely the simplest expression of the stoichiometric proportions.

In general we regard the mole as identical with what has also been called the formula weight. Therefore the mole is not defined unless the chemical formula is established by universal usage or is definitely stated. In this book we shall choose the chemical formula with regard more to convenience than to consistency. For gases the actual molal weight has frequently been experimentally determined, and in the case of liquids we shall, for the most part, use the same formula as in the gaseous state. In the case of solids we shall sometimes do the same thing. Thus for solid halogens we write I_2, Br_2, etc. On the other hand, we shall use the formula S for solid sulfur, although it is known that in both the rhombic and monoclinic forms the structural unit is the molecule S_8.

Molal Properties. We have defined an extensive property as one whose quantitative measure is proportional to the amount of substance taken. Thus volume is an extensive property, but the volume per mole of any substance is an intensive property. If we denote the volume in general by V, we may denote the volume per mole, or the molal volume,[1] by v. Morever, if Y is any extensive property, y will denote the molal value of Y and is an intensive property.

CHEMICAL SYMBOLS AND EQUATIONS

Our chemical symbols will frequently be used to indicate not only the substance under consideration but also a definite quantity of that substance. Thus HCl denotes 1 mole of hydrogen chloride, and $\frac{1}{2}O_2$ denotes $\frac{1}{2}$ mole of oxygen. Thus, when we write for water at 4°C, v = 18.02 cc,

---

[1] In general, molal quantities will be denoted by small Roman capitals, as in the examples above, v and y. With few exceptions, most properties are used as intensive quantities and only in a few general discussions will they be used as extensive quantities. Thus the difficulty of distinguishing between small and large capitals on the blackboard or in notes does not appear to be a serious problem. No special symbol will be used for specific quantities as it is invariably more convenient to discuss and present data on a mole rather than a gram basis. Specific quantities tabulated in tables can be simply headed "Specific volume," for example. If it is desired to use a symbol, V/M can be used for specific volume, where M is the molecular weight.

we could state the same thing by the expression $V(H_2O) = 18.02$, or $V(\frac{1}{2}H_2O) = 9.01$ or $v_{H_2O} = 18.02$ when we are definitely speaking of molal quantities but we wish to indicate the formula to ensure that there is no question of the molecular weight used.   When there are several polymers of a molecule, the use of v_{CH_2O} or $v_{(CH_2O)_4}$ clearly indicates the molecular weight we have used.   It is generally wise to add the formula as a subscript unless there is no question about the formulas being used. When that is so, the subscript position can be used for temperature, for example, V_{298}, or other designations. $C_V(Pb) = 6$, $C_{V,Pb} = 6$, or Pb; $c_V = 6$ cal/deg are alternative ways of expressing "The heat capacity at constant volume of 1 mole of Pb is 6 cal/deg."

Thus the formulas in a chemical equation may indicate not only the substances but the amounts of those substances involved.   They do not, however, indicate the particular condition of each substance.   It will often be necessary to amplify the chemical equation by specifying the exact state of each substance, whether it is a solid, liquid, or gas or is in solution in some solvent at a certain concentration.   The pressure on each substance must also be given, and the temperature, although the latter will ordinarily be stated not in the equation itself but in the context.   In the case of certain substances, it will be necessary to add further specifications in order to make absolutely unambiguous the exact character of the substance.   In most cases the specifications which we have mentioned will suffice.

We may therefore append to each formula appearing in a chemical equation an abbreviated statement of the physical state.   Thus HCl(g, 2 atm) will indicate 1 mole of hydrogen chloride gas at two atmospheres pressure.   HCl(s) and HCl(l) will indicate solid and liquid, respectively, while HCl(aq, 0.01 M) means hydrochloric acid in aqueous solution at a concentration of one-hundredth molal.   Sometimes, for example in dealing with approximate heats of reaction, it is not necessary to specify the exact concentration; hence HCl(aq) will indicate hydrochloric acid in more or less dilute aqueous solution.

In general the pressure will be assumed to be one atmosphere unless otherwise specifically indicated, and the distinguishing marks (l), (s), (g), (aq) may sometimes be omitted when no ambiguity is likely.   Thus the plain symbol H_2 will indicate hydrogen gas at unit pressure.

Then, if we write

$$Ag(s) + \tfrac{1}{2}Cl_2(g, 10 \text{ atm}) = AgCl(s)$$

we are considering a process whereby 1 mole of solid silver at unit pressure and $\frac{1}{2}$ mole of chlorine gas at 10 atm disappear and 1 mole of solid silver chloride is produced.

It should perhaps be emphasized that there is some difference between this thermodynamic significance of a chemical equation and the one which is usually understood in chemistry. Ordinarily, when the equation is written, there is some implication as to the mechanism of the reaction. Here we are not interested in such a mechanism. We take 1 mole of silver metal, we draw the chlorine gas from a cylinder containing ½ mole at a pressure of 10 atm, then by any process, direct or indirect, these are brought into combination, and we finally have 1 mole of silver chloride, at the same temperature (unless otherwise stated) as the original silver and chlorine. We have expressed by our equation merely what existed at the beginning and what exists at the end.

We shall use similar equations for processes which are not ordinarily represented by chemical equations. Thus

$$H_2O(l) = H_2O(g)$$

will indicate the formation of 1 mole of water vapor from 1 mole of liquid water, both at atmospheric pressure. Similarly

$$H_2O(g) = H_2O(g, 0.01 \text{ atm})$$

indicates any process whereby 1 mole of water vapor has been changed from a pressure of 1 atm to a pressure of one-hundredth of an atmosphere.

If any quantity, such as the volume, is determined by the state of a system, in other words, if it is a *property* of the system, then when the system changes from one state to another it will be convenient to designate the increase[1] in that quantity by the symbol Δ. Thus, if we consider the fusion of ice at atmospheric pressure and at 0°C, or approximately 273° abs (273°K), we write

$$H_2O(s) = H_2O(l) \qquad \Delta V_{273} = -2 \text{ cc}$$

The system in the first state consists of 1 mole of ice, the volume of which is about 20 cc; the system in the second state consists of 1 mole of water, of which the volume is about 18 cc. We could equally well write,

$$H_2O(l) = H_2O(s) \qquad \Delta V_{273} = 2 \text{ cc}$$

In general, when a system passes from state A to state B, $\Delta V = V_B - V_A$.

As we proceed we shall find it necessary to make numerous other conventions which we cannot discuss here without anticipating ideas which will be developed in subsequent chapters. In the meantime we may turn our attention briefly to the most important of all our conventions, those of mathematics.

---

[1] This symbol Δ may be used to denote any such increment, whether finite or infinitesimal. It is made to correspond in sign with the symbol d used only for infinitesimal change. Both dx and Δx are to be read as "the increase in x," although this increase may be negative.

THE LANGUAGE OF MATHEMATICS

It is said that, during his long membership in the Yale faculty, Willard Gibbs made but one speech, and that of the shortest. After a prolonged discussion of the relative merits of language and mathematics as elementary disciplines, he rose to remark, "Mathematics is a language." However, such a language, usually acquired late in life, must not be used unnecessarily in place of our mother tongue if we wish to avoid an appearance of affectation.

Mathematics offers a wonderful shorthand for the precise formulation of well-standardized ideas. On the other hand, the expressions of mathematics are lacking in humor, which is to say that they are no suitable medium for those finer shades of thought which are often necessary in the exposition of ideas which are on the way toward standardization. The formal severity of a mathematical treatment has its disadvantages. Indeed in our opinion absolute mathematical rigor is a sort of *ignis fatuus*, which must not serve as a guide to the scientific investigator, although we do not claim that its pursuit, with proper safeguards, may not offer a very wholesome exercise.

In this book we have not consciously sacrificed any desirable elements of mathematical rigor. If we have the appearance of doing so, it is because we feel the great need of a visualization of the numerous problems before us, and because this end seems best to be attained by mitigating rather than accentuating the formality of mathematical analysis. It is a dangerous thing to use any kind of mathematical equation unless we keep its meaning before us and are able to express this meaning without the symbolism which mathematics affords.

The amount of calculus really essential for a compact statement of the equations of thermodynamics is very small. Since we are to deal with numerous properties which depend upon several variables, it will be necessary to use repeatedly the methods of partial differentiation. As a review of the more important equations of partial differentiation, and as a practice in the translation of equations into common language, we shall discuss briefly the dependence of a variable property, such as the volume, upon such other variables as temperature and pressure.

PROPERTIES WHICH DEPEND UPON TWO OR MORE VARIABLES

In the preceding chapter we have said that the state of a system is defined by the properties of a system. Now, since there is no end to the number of properties of a system, it might seem that it would be necessary to ascertain an infinite number of facts before the state of a system could be definitely fixed. As a matter of fact, however, this is far from being the case. If we are dealing with a mobile system, which readily comes to equilibrium with its environment, we find that, when a very few properties are fixed, all its properties are fixed.

Indeed in the ordinary case we find that, for a given amount of a pure substance, the fixing of almost any pair of properties fixes all the others. Thus, if we consider a liquid like benzene and impose the condition that it shall have a certain viscosity and a certain refractive index, there will be one temperature, one pressure, one density, one dielectric constant compatible with these conditions.

In mathematical parlance the properties are called variables, and we say that we are dealing with a case of two independent variables. We may arbitrarily choose two of the properties and call them *the* independent variables, and then the other properties are called dependent variables.

Most frequently it will be convenient to call pressure and temperature, P and T, the independent variables. Thus, when we are dealing with a given amount of some substance, we say that the volume V is a function of P and T. Any equation giving the relation between these three variables for a certain substance is called an equation of state of that substance.

When P is given a certain value and is maintained constant at that value, V becomes a function of the single variable T and we may express the relationship by means of a curve with axes of V and T. The slope of this curve at any point is called the partial differential coefficient of V with respect to T, is denoted by $\partial V/\partial T$, and is read as the rate of change of volume with temperature alone. If we wish to indicate explicitly the constancy of pressure, we write $(\partial V/\partial T)_P$.

Similarly, if T is kept constant, we obtain, at each value of T, a V-P curve, which (because of the constancy of temperature) is known as an *isotherm*. The slope of this curve at any point is expressed by $\partial V/\partial P$, or $(\partial V/\partial P)_T$.

We shall find it convenient to call these two partial differential coefficients the volume-temperature coefficient, or simply the V-T coefficient, and the volume-pressure coefficient, or V-P coefficient. These are extensive quantities, which depend upon the amount of the substance present, and must be distinguished from the ordinary coefficient of thermal expansion and compressibility, which are intensive quantities (namely, the above coefficients divided by the volume).

The whole relationship between V, P, and T can be expressed geometrically only by means of a surface. Thus, if we have perpendicular T and P axes in a horizontal plane and a vertical axis of V, the height of the surface at any point gives the volume corresponding to the temperature and pressure at that point.

If the volume is determined at a number of temperatures and pressures, it is not difficult, by means of beads on a wire frame, to plot the surface from the experimental data, much as we plot a curve when the property is a function of a single variable. Or if we have a number of results at each of several temperatures, the several isotherms may be drawn, cut out, and mounted vertically at the proper distance from one another so as to give a very satisfactory idea of the surface. Ordinarily, for a qualitative idea of such a surface it suffices to give in a single plane a number of such isotherms, as illustrated in Fig. 16-3.

TYPICAL EQUATIONS OF PARTIAL DIFFERENTIATION

If V depends upon the two variables P and T, the change in V, when we pass from T and P, to $T + dT$ and $P + dP$, is given by the basic equation of partial differentiation,

$$dV = \left(\frac{\partial V}{\partial T}\right)_P dT + \left(\frac{\partial V}{\partial P}\right)_T dP \quad \checkmark \tag{3-1}$$

It states that the change in V is equal to the rate of change of V with T alone, multiplied by the change in T, plus the rate of change with P alone, multiplied by the change in P.

This equation is illustrated in Fig. 3-1, where a and c are two infinitesimally distant points on a V-T-P surface, and $abcd$ is an infinitesimal ele-

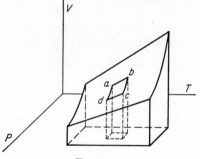

Fig. 3-1

ment of this surface obtained by planes parallel to the V-T plane and the V-P plane. Then $dV = V_c - V_a = (V_b - V_a) + (V_c - V_b)$. The slope of the line ab is the V-T coefficient, $(\partial V/\partial T)_P$, and $V_b - V_a$ is equal to the slope multiplied by dT. Likewise the slope of bc is the V-P coefficient, and $V_c - V_b$ is $(\partial V/\partial P)_T dP$. For such an infinitesimal change it is immaterial which operation is considered first, and we find the same terms if we pass from a to d and thence to c. One may also note that $(\partial V/\partial P)_T = 1/(\partial P/\partial V)_T$.

There are special forms of the general equation (3-1) which are frequently useful. Thus, if we impose the condition that V is constant, i.e., if we move along a contour line of the surface,

$$\left(\frac{\partial V}{\partial T}\right)_P dT + \left(\frac{\partial V}{\partial P}\right)_T dP = 0 \tag{3-2}$$

and expressing the constancy of V in the equation itself,

$$\left(\frac{\partial P}{\partial T}\right)_V = - \frac{(\partial V/\partial T)_P}{(\partial V/\partial P)_T} \tag{3-3}$$

Again, if we have some other dependent variable, i.e., some other quantity which, like the volume, depends only upon the temperature and pressure, let us say the energy E, we may impose the condition that E is constant and obtain from (3-1) equations of the form

$$\left(\frac{\partial V}{\partial T}\right)_E = \left(\frac{\partial V}{\partial T}\right)_P + \left(\frac{\partial V}{\partial P}\right)_T \left(\frac{\partial P}{\partial T}\right)_E \tag{3-4}$$

This equation states that, when we proceed upon the surface along a line of constant energy, the change in V corresponding to a given infinitesimal change in T is the sum of two terms, namely, the change in V, which would be caused by this same change in T alone, and the change in V, caused by such a change in P as is necessary to keep the energy constant. It is also often convenient to introduce derivatives involving dependent variables by means of expressions such as

$$\left(\frac{\partial V}{\partial P}\right)_T = \left(\frac{\partial V}{\partial E}\right)_T \left(\frac{\partial E}{\partial P}\right)_T$$

Finally there is a familiar equation involving second derivatives which we must frequently employ, namely,

$$\frac{\partial}{\partial P}\left(\frac{\partial V}{\partial T}\right)_P = \frac{\partial^2 V}{\partial P\, \partial T} = \frac{\partial}{\partial T}\left(\frac{\partial V}{\partial P}\right)_T \tag{3-5}$$

According to this equation, the rate of change with P of the V-T coefficient is equal to the rate of change with T of the V-P coefficient. The intermediate member of the equation is merely a shorthand method of expressing either of the others.

When a property depends upon three or more independent variables, the geometrical method of interpreting the relations is no longer available but the equations assume a similar form. Thus, if the volume of a given quantity of material depends not only upon temperature and pressure but also upon one or more other independent variables, such as the intensity of an electric field, we write

$$dV = \frac{\partial V}{\partial T} dT + \frac{\partial V}{\partial P} dP + \frac{\partial V}{\partial X} dX + \cdots \tag{3-6}$$

Here also subscripts may be employed to show the independent variables which remain constant during the differentiation.   Thus for $\partial V/\partial T$ we mean $(\partial V/\partial T)_{P,X}$, ...

Perfect Differentials.   We shall have occasion to consider expressions of the type

$$\delta Z = L(x,y)\ dx + M(x,y)\ dy \tag{3-7}$$

where δZ represents an infinitesimal quantity and $L(x,y)$ and $M(x,y)$ are functions of both independent variables as indicated.   This type of expression may or may not be the total differential of a function such as Eq. (3-1).   If there is such a function $Z(x,y)$, then

$$L(x,y) = \left(\frac{\partial Z}{\partial x}\right)_{y} \quad \text{and} \quad M(x,y) = \left(\frac{\partial Z}{\partial y}\right)_{x}$$

and
$$\frac{\partial L}{\partial y} = \frac{\partial^2 Z}{\partial x\ \partial y} = \frac{\partial M}{\partial x} \tag{3-8}$$

The equality $\partial L/\partial y = \partial M/\partial x$ is a necessary and sufficient condition that an expression of the type of Eq. (3-7) is a perfect differential of a function Z.   Suppose we wish to integrate the total change ΔZ from the state x_1, y_1 to the state x_2, y_2.   If Eq. (3-8) holds, then the differential may be integrated and $\Delta Z = Z(x_2,y_2) - Z(x_1,y_1)$.   However, if Eq. (3-8) does not hold, then it is impossible to integrate Eq. (3-7) unless the path from x_1, y_1 to x_2, y_2 is specified.   Also the result will depend on the path chosen, and it is not possible to regard this quantity as a property of the system.

In order to help keep this distinction in mind, we shall use the symbol d (as in dZ) for a perfect differential of a function and the lower-case delta δ (as in δQ) for other infinitesimal quantities which are not perfect differentials.

PROBLEMS

3-1. The relation between the pressure, the temperature, and the volume of 1 mole of hydrogen gas may be expressed over a limited range by the equation of state $P = RT/(v - B)$, where B is independent of P but is a function of T.

(a) Find $(\partial v/\partial T)_P$ and $(\partial v/\partial P)_T$, and express dv as a function of T and P as in Eq. (3-1).

(b) Apply Eq. (3-7) to confirm that dv is a perfect differential.

(c) Confirm Eq. (3-3), using the above equation of state.

3-2. The volume of a circular cylinder is given by the expression $V = \pi r^2 h$.   Find $(\partial V/\partial r)_h$ and $(\partial V/\partial h)_r$.   Write out the complete differential dV, and verify that it is a perfect differential.

4

THE FIRST LAW OF THERMODYNAMICS
AND THE CONCEPT OF ENERGY

The concept of conservation of energy seems intuitively correct to most of us. Yet there is no a priori reason to expect conservation of energy, and some of the current theories of cosmology even claim that energy may not be conserved under all conditions. What then is the basis of the first law of thermodynamics? We have already noted that, whenever an interaction between systems is limited to a single type of interaction and therefore an exchange of a single type of energy, we always find that the loss of energy of one of the systems is just equal to the gain of energy of the other system. Even when there may be a number of mechanical (nonthermal) interactions taking place such that one type of energy is lost by one of the systems and a different type gained by the other system, we find a proportionality between the energy lost by one system and the energy gained by the other. If the various types of mechanical energies are measured in the same terms, as in terms of lifting a mass in the earth's gravitational field, then we find that the net gain of mechanical energy by one system, e.g., the work done on that system, is equal to the net loss of mechanical energy by the other system if thermal changes have been avoided.

For example, in the simple mechanics of rigid elastic bodies there is a quantity, formerly called *vis viva*, which is measured by the sum of the terms $\frac{1}{2}mv^2$, and which remains constant no matter how the system may change through collisions of its component parts. If, however, the system includes such a force as is said to act at a distance, the *vis viva*, or kinetic energy as we now call it, no longer remains constant. Thus, if an object is thrown upward against the earth's gravitational force, its kinetic energy diminishes and at a certain point the object comes momentarily to rest. If it is arrested at this point, the whole of the kinetic energy has disappeared. However, we are accustomed to say that as the object loses in kinetic energy it gains in latent, or potential, energy by an equal amount, and this idea is justified by the fact that, if the object is

28

allowed to drop once again to its original position, it regains the whole of the kinetic energy which it lost in rising.

If in such a mechanical system there are inelastic collisions, or if frictional processes are at work, there may be a net loss of mechanical energy; in other words, the sum of the kinetic and potential energies may diminish. At the end of the eighteenth century when Count Rumford was observing the boring of cannon in the Munich arsenal, he noticed that the mechanical energy expended was roughly measured by the amount of heat produced. This idea, developed by Mayer and by Joule, led to the first determinations of the mechanical equivalent of heat. Whenever a system is subjected to a cyclic process whose net result is merely the conversion of work to heat with the system returning to its initial state, it is always found that the heat produced is strictly proportional to the work done. Therefore the units of heat and work can be so chosen that the amount of heat produced is always equal to the amount of mechanical energy lost. This discovery led to the consideration of heat as a manifestation of thermal energy and to the enunciation of the broad principle which we know as the law of the *conservation of energy*, or the first law of thermodynamics.

As far back as 1762 that remarkably accurate thinker and investigator, Joseph Black, showed, in studying heat alone, that it was necessary to introduce a concept of latent heat (analogous to potential energy). Here again the concept of latent energy is justified by the fact that the amount of heat required to melt 1 g of ice is equal to the heat evolved when 1 g of water freezes.

When the phenomena of electricity became better understood, it was necessary to define electrical energy and the brilliant investigations of Maxwell made it possible to follow the course of electrical energy, not only through material bodies, but through space which is empty of all else but an electromagnetic field.

So, as science has progressed, it has been necessary to invent other forms of energy, and indeed an unfriendly critic might claim, with some reason, that the law of conservation of energy is true because we make it true by assuming the existence of forms of energy for which there is no other justification than the desire to retain energy as a conservative quantity. This is indeed true in a certain sense, as shown by the explanations which have been given for the enormous, and at first sight apparently limitless, energy emitted by radium. But a study of this very case has shown the power and the value of the conservation law in the classification and the comprehension of new phenomena. Also, when the accounting of the energy changes during β decay of radioactive nuclei did not balance, the existence of the neutrino was postulated to account for the missing energy. Because of our confidence in the law of conservation of energy, we felt sure that this elusive particle would eventually be found although it required 30 years before the experimental verification of the neutrino and therefore of the conservation of energy in β-decay processes could be obtained.   Thus we feel certain that, for every new type of interaction that is discovered, it will be possible to find a characteristic energy function of each system that will allow prediction of the change in one system as a result of a change in the other system.   Nevertheless, in spite of our confidence in this law, we must still demand its experimental confirmation when new circumstances arise, and the possibility of its violation under conditions far removed from our present experiments must always be kept open.

THE INTERNAL ENERGY OF A SYSTEM

The observed proportionality of heat and work for any cyclic process requires that the energy contained within a system, or its internal energy, be a *state function*, or a *property* of the system in the technical sense in which this term has been used in Chapter 2.   The increase in such energy when a system changes from state A to state B is independent of the way in which the change is brought about.   It is simply the difference between the final and the initial energy,

$$\Delta E = E_B - E_A$$

where ΔE for this system must, by the conservation law, be equal and opposite in sign to the total ΔE for all other systems involved as given by summing the heat and work.

In a particular process, for example, when we write

$$Cu + \tfrac{1}{2}O_2 = CuO \qquad \Delta E = -37{,}000 \text{ cal}$$

we mean that, at a given temperature, 37,000 cal of energy is evolved when 1 mole of cupric oxide is formed from its elements, or that 1 mole of cupric oxide contains that much less energy than 1 mole of copper and $\tfrac{1}{2}$ mole of oxygen.

In this example, when we write

$$E_B = E(CuO) \qquad \text{and} \qquad E_A = E(\tfrac{1}{2}O_2) + E(Cu)$$

we are able to measure $E_B - E_A$ but we do not claim to know the value of either of these quantities alone.   Nor are we interested in the kinds of energy which make up the total internal energy of a substance like copper oxide.   It presumably consists in the kinetic energy of the moving particles, in the energy of the electric fields emanating from the charged particles of which the atoms are composed, and in other forms of energy to which names are now assigned or may be assigned in the future.   But of all this, thermodynamics takes no cognizance.

Energy and Mass.   While the amount of energy given to or taken from the environment in a case like this is easy to measure, the determination of the total energy of any one material system has, until recently, seemed beyond the range of human possibility.   For practical purposes it is still necessary to regard as undetermined the total energy which a given system possesses.   It is, however, of much theoretical interest to note that the great discovery of Einstein,[1] embodied in the principle of relativity, shows us that every gain or loss of energy by a system is accompanied by a corresponding and proportional gain or loss in mass,[2] and therefore presumably that the total energy of any system is measured merely by its mass.

In other words, mass and energy are different measures of the same thing, expressed in different units, and the law of conservation of energy is but another form of the law of conservation of mass.   These units are indeed very different in magnitude, differing by the square of the velocity of light, so that 1 g is equal to 9×10^{20} ergs.   Even the largest amounts of energy evolved in ordinary chemical reactions produce changes in mass that are below the limits of detectability by means of the balance.

On the other hand, the mass changes associated with the great quantities of energy involved in nuclear reactions and transformations are large enough to be weighed.   Thus, when energy is emitted, there is a diminution in the internal energy of the substances concerned and the sum of the atomic weights of the products is measurably smaller than the atomic weight of the parent substance or substances.   For example, if 1 g atom of helium weighs 0.029 g less than 4 g atoms of hydrogen, we may calculate

$$4H = He \qquad \Delta E = -2.6 \times 10^{19} \text{ ergs}$$

This enormous energy of nearly 1 million million cal/mole of helium produced is believed to be the principal energy source of the sun.

PRESSURE AND TEMPERATURE

There are two intensive properties, pressure and temperature, which play an important role in thermodynamics, since they largely affect, and often completely determine, the state of a system.   Pressure is too familiar an idea to require definition; it has the dimensions of force per

[1] A. Einstein, *Ann. Physik*, **18,** 639 (1905).

[2] For a simple demonstration of this proposition, independent of the principle of relativity, see G. N. Lewis, *Phil. Mag.*, **16,** 705 (1909); *Science*, **30,** 84 (1909).

unit area, and therefore pressure times volume has the dimensions of energy.

The concept of temperature is a little more subtle.   When one system loses energy to another by thermal conduction or by the emission of radiant energy, there is said to be a flow of heat, or a *thermal flow*.   The consideration of such cases leads immediately to the concept of temperature, which may be qualitatively defined as follows: If there can be no thermal flow from one body to another, the two bodies are at the same temperature, but if one can lose energy to the other by thermal flow, the temperature of the former is the greater.   This establishment of a qualitative temperature scale is obviously more than a definition.   It involves a fundamental principle, to which we have already given preliminary expression in the discussion of equilibrium, but which we are not yet ready to put in a general and final form.   For spontaneous thermal flow, this principle requires that if A lose energy to B, B cannot lose it to A; if A lose to B, and B lose to C, C cannot lose to A.

Expressed briefly, our qualitative laws of temperature are: If $T_A = T_B$ and $T_B = T_C$, $T_A = T_C$; if $T_A > T_B$ and $T_B \geqq T_C$, $T_A > T_C$.   As in our general discussion of equilibrium, it must be understood that we are dealing with net gains or losses in energy.   We do not mean that no thermal energy passes from a cold body to a hot, but only that the amount so transferred is always less than that simultaneously transferred from hot to cold.

When we have established the qualitative laws of temperature, we still have a wide freedom of choice in fixing the quantitative scale.   Indeed temperature, as ordinarily measured, or its square or its logarithm would equally satisfy these qualitative requirements.

Since the volumes of most things change appreciably with the temperature and since volumes are easily measured, it early proved convenient to correlate the temperature with the volume of some chosen thermometric substance.   It is obvious that this substance could not be chosen altogether at random, for if water were taken, then, on account of the existence of a maximum density, there would be two temperatures corresponding to one volume.   But, even in the case of substances whose volume always increases with the temperature, one choice may be more convenient than another.

If thermometers made of mercury and alcohol, with linear scales, are made to agree at two temperatures, they will not agree at some intermediate temperature; but if two gases such as hydrogen and air are employed, the agreement will be nearly complete over a wide temperature range.   Hydrogen and air do not behave exactly alike at atmospheric pressure, but the behavior of any two gases becomes more nearly the same the lower the pressure.   We therefore adopt, as our ideal thermometric substance, any gas at very low pressure. Strictly speaking, we measure the PV product of a fixed amount of any gas in contact with the system at a series of pressures and then extrapolate the PV product to zero pressure. This extrapolated value of PV is then directly proportional to the absolute temperature.

We must still define in some manner the size of the unit of temperature. The centigrade scale has the values 0 and 100° for the freezing and boiling points of water. The corresponding absolute scale was defined to have a 100° interval between the ice and steam points. Lord Kelvin pointed out in 1848 that this was a clumsy definition and that one should simply define the temperature of a single fixed point. In a sense the absolute zero is the second fixed point, but there is no need to make any measurement at this point. Giauque[1] renewed this proposal in 1939, and it has recently been adopted by the International Unions of Physics and Chemistry. It is now agreed that the triple point of water where water, ice, and water vapor are in equilibrium shall be 273.1600°K (where K is for Kelvin). (The triple point is more accurately reproducible than the conventional ice point, which is the freezing point in air, because variations from pressure or amount of dissolved air are avoided.) Thus the temperature of a system is given by the expression

$$T = 273.1600 \, \frac{\lim\limits_{P \to 0} (PV)_T}{\lim\limits_{P \to 0} (PV)_{t.p.}} \tag{4-1}$$

The value 273.16 was selected to make the new degree the same size as that of the centigrade scale. Since the ordinary ice point is 0.010° lower than the triple point of water, we say that 0°C = 273.150°K. The temperature 25°C = 298.15°K has been adopted as a standard in the range of room temperatures at which thermodynamic data will be obtained and tabulated. (Since this is a temperature which we shall very frequently use, we shall designate it commonly, for convenience, 298°K, bearing in mind that in our calculations we are to use the more precise value.)

In Chapter 7 we define the thermodynamic scale of temperature, which fortunately proves to be identical with the ideal-gas scale which we have just considered. In that place we shall be able to show that the point which we have defined as the absolute zero of temperature is not an arbitrary point, brought into prominence by the properties of any one substance or class of substances. It is in fact the limit of any rational thermodynamic scale, a true zero where all thermal energy[2] would cease, but as unattainable as the other limit of our scale, the infinite temperature.

[1] W. F. Giauque, *Nature*, **143**, 623 (1939).

[2] The zero-point energy of systems in their lowest quantum state remains, of course, at 0°K.

HEAT AND WORK

There are two terms, *heat* and *work*, that have played an important part in the development of thermodynamics, but their use has often brought an element of vagueness into a science which is capable of the greatest precision. We shall discuss particular limitations on their definition a little later. For our present purpose we may say that when a system gains energy by thermal radiation or conduction as a result of a temperature differential it is absorbing a positive quantity of heat and that when it gains energy by other methods, for example, by the operation of external mechanical forces, a positive quantity of work is being done on the system (or a negative quantity of work is being done by the system).[1]

The amount of heat transferred can be measured in terms of amount of ice melted or frozen in an ice calorimeter, or in any similar calorimeter, with the volume change due to the melting or freezing of a substance at its melting point used as a measure of the heat transfer.

We have previously defined the amount of work done by a system undergoing a change in state in terms of lifting a mass in a gravitational field. Consider a specific example of a gas in a vertical cylinder which is kept from expanding into the upper evacuated space by a piston of mass m and area a which is locked in place by a stop. If the piston is released, the maximum amount of work that the gas could perform for a given expansion is not generally obtained because of frictional effects. If the expansion process is arranged so that no other work can be done except for the lifting of the piston in the earth's gravitational field and all other energy transfers are heat transfers, the work done by the gas on the piston when the gas raises the piston by a height Δh is $mg \, \Delta h$. The external pressure exerted on the gas by the piston is given by $P_{ex} = mg/a$. Thus the work done on the gas is

$$w = -P_{ex} a \, \Delta h = -P_{ex} \, \Delta V \qquad (4\text{-}2)$$

where ΔV is the volume increase of the gas. This is a general result, and the compression work done on any substance is given by the external pressure exerted on the substance times the volume decrease. If the external pressure is not constant, we obtain for a differential element of compression or expansion $\delta w = -P_{ex} \, dV$, or

$$w = \int_{V_A}^{V_B} -P_{ex} \, dV \qquad (4\text{-}3)$$

As dV is positive for expansion and negative for compression, the sign of

[1] Many books on thermodynamics define w with the opposite sign as the work done by the system, which is convenient if the primary emphasis is on heat engines. The definition with the same sign for both q and w seems preferable for a more general thermodynamic treatise.

w is opposite to the sign of dV if work done on the system is taken to be positive.

For a given expansion, it can be seen that the work performed depends upon the external pressure resisting the expansion, and in the instance of expansion into a vacuum where P_{ex} is zero, no work at all is done. In order that an expansion may occur, the internal pressure must, of course, be greater than the external, but this difference may be made as small as is desired, and as a limit we may consider the case in which the internal pressure is equal to the external pressure or differs from it by a negligible amount. This will later be termed a reversible process. For such processes, Eq. (4-3) is valid with P_{ex} replaced by P, the pressure of the system.

According to the law of the conservation of energy, any system in a given condition contains a definite quantity of energy, and when this system undergoes change, any gain or loss in its internal energy is equal to the loss or gain in the energy of surrounding systems. If no matter is transferred in any physical or chemical process, the increase in energy of a given system is therefore equal to q, the heat absorbed from the surroundings, plus w, the work absorbed from the surroundings. If E_A represents the initial energy content of the system and E_B the final energy content, then

$$E_B - E_A = \Delta E = q + w \tag{4-4}$$

We shall use δq and δw to represent infinitesimal quantities of heat and work. Thus for an infinitesimal change we write

$$dE = \delta q + \delta w \tag{4-5}$$

The use of δ instead of d will remind us that q and w are not properties but are defined only for a given path of change.

The choice of boundaries of a system is of great importance for evaluation of heat and work. For some choices of boundaries, heat and work will be found to be unspecifiable. For example, the mechanical and thermal interactions at a boundary where frictional interaction is taking place are so interconnected that one cannot, in general, specify q and w separately. By selecting the boundaries of the system such that the frictional interactions are all within the system rather than at a boundary, one may then specify the thermal and mechanical energies appearing or disappearing in the surrounding systems and therefore specify q and w. Even when q and w cannot be specified, ΔE must still have a definite value for a given change in state, as the energy of a system is a state function and does not depend upon the means of producing the change in state. It is fixed if the initial and final states of the system are fixed. Thus, if ΔE has been measured for a given change in state carried out by a process for which both q and w can be fixed, ΔE is the same for the given change in state even when carried out by a process for which q and w cannot be separately fixed. It is this characteristic of the energy content of a system, e.g., being a state function, that makes it of such value. We shall be able to assign energy values to systems without any regard to the specific processes used for the measurements. It might be also noted that dE, the change in E of the system along a differential element of a path between two states, need not always be defined even though ΔE for the over-all change in state is defined. Thus, if the change in states takes place along a path for which the state of the system is not completely and reproducibly determined by external conditions, due to eddies, gradients, strains, etc., E is not defined along the path as a state function in terms of the usual parameters of the equation of state. These limitations will not handicap us in consideration of chemical systems since we shall normally be interested only in the equilibrium states of a system, which are uniquely fixed by the external parameters of the system.[1]

The application of the law of conservation of energy is merely an accounting process. One adds up all the increases in energy of the system due to all the nonthermal interactions, and this sum is the total work done on the system. One then adds up the total energy increase due to thermal interactions, and this represents the total heat absorbed by the system. The need for separate accounting of thermal and nonthermal energies will be clear when the second law of thermodynamics is discussed. It is not feasible to assign absolute values to energies. One merely measures the change in energy that takes place as a result of a change in the state of a system. If one can measure q, the heat absorbed by the system, and w, the work done on the system, one can then, through the first law of thermodynamics, know the change in the energy of the system, E.

Many attempts have been made to build a perpetual-motion machine for which $q + w$ would not be zero for a cyclic process. Such a machine would produce more thermal energy, for example, than the amount of mechanical energy consumed, and $q + w$ would be greater than zero. No such machine has ever been demonstrated to produce energy. Thus, on the basis of all the accumulated experience on earth, we must accept the existence of a state function, energy, that remains constant for every isolated system and therefore must be conserved for all interacting systems.

The values of q and w depend upon the way in which the process is carried out, and, in general, neither is uniquely determined by the initial and final states of the system. However, their sum is determined, so that if either q or w is fixed by the conditions under which the process occurs, the other is also fixed. Where the work done by the system is the work of expansion against an external pressure, the expansion may be carried out in such a manner that no heat enters or leaves the system. We then say that the process is *adiabatic*, and we may write

$$\Delta E = w = - \int_{V_A}^{V_B} P_{ex} \, dV \tag{4-6}$$

[1] The reader is referred to P. W. Bridgman, "The Logic of Modern Physics," The Macmillan Company, New York, 1927; P. W. Bridgman, "The Nature of Thermodynamics," Harvard University Press, Cambridge, Mass., 1941; and J. H. Keenan and A. H. Shapiro, *Mech. Eng.*, **69**, 915–921 (1947), for a more detailed discussion of heat, work, and the first law of thermodynamics.

On the other hand, if the process is such that no work is done by the surroundings (as in the case of a chemical reaction taking place in a constant volume calorimeter), then $\Delta E = q$.

The Heat Content or Enthalpy. In most calorimetric measurements the pressure and not the volume is kept constant, and the pressure of the system is fixed by the external pressure. No work other than that due to volume change is involved. If the increase in volume during the reaction is ΔV, then

$$\Delta E = q - P\,\Delta V = q - P(V_B - V_A) \tag{4-7}$$
or
$$(E_B + PV_B) - (E_A + PV_A) = q \tag{4-8}$$

and the heat absorbed during the reaction is again a quantity which obviously depends only upon the initial and final states.

We may define a new quantity by the equation

$$H = E + PV \qquad \Delta H = (E_B + P_B V_B) - (E_A + P_A V_A) \tag{4-9}$$

Like E, the property H is one to which we do not attempt to give an absolute value, although changes in H are readily measured. Thus, in any constant-pressure calorimeter,

$$\Delta H = q \tag{4-10}$$

There are two common names for the H function, *heat content* and *enthalpy*. The former is very descriptive of the relationship that we just derived, but it may be misleading when H is used in connection with a process not at constant pressure where ΔH is not equal to q. In general we shall use both names but shall avoid the use of the name heat content when processes involving pressure changes are under discussion.

As an example, we may consider a simple process such as the vaporization of water, in a calorimeter at constant pressure:

$$H_2O(l, 1\ atm) = H_2O(g, 1\ atm)$$
Here
$$\Delta H = H_B - H_A = (E_B + PV_B) - (E_A + PV_A) = q$$

where E_A is the energy, V_A the volume, and H_A the heat content of 1 mole of liquid water; E_B, etc., are the corresponding quantities for 1 mole of the vapor, and q is the heat of vaporization as ordinarily measured (see Problem 4-2).

While we have thus called attention to the equality of ΔH and q in a constant-pressure calorimeter, it must not be supposed that the utility of the quantity H is in any way limited to such a case. Indeed the importance of H in dealing with flow processes will become apparent later. The change in enthalpy is determined only by the initial and final states, and not by the particular manner in which the process is carried out. Thus if we have a process involving substances at different pressures, for example,

$$H_2O(l, 1 \text{ atm}) = H_2O(g, 0.01 \text{ atm})$$

we have ΔH as the difference between

$$H_B = E_B + P_B V_B \qquad \text{and} \qquad H_A = E_A + P_A V_A$$

HEAT CAPACITY

When we impart heat to a system and thereby raise its temperature, the average heat capacity between the initial and final temperatures is defined as $q/\Delta T$ and the limit of this ratio, as q is made indefinitely small, or, in other words, as the second temperature is brought indefinitely near to the first, is called the actual heat capacity at that temperature.

The amount of heat required to produce a given rise in temperature will, however, depend upon the circumstances under which the system is heated. In particular we must consider separately the cases in which the system is heated at constant volume and at constant pressure. Thus we speak of the heat capacity at constant volume C_V, and of the heat capacity at constant pressure C_P.

In the first case, the heat absorbed is equal to the increase in internal energy, for when a system is heated at constant volume there is no work done. Therefore,

$$C_V = \left(\frac{\partial E}{\partial T}\right)_V \qquad (4\text{-}11)$$

Such an equation as this is general and applies to any system, homogeneous or heterogeneous. Indeed, if we are to determine experimentally the heat capacity of a system, it is unnecessary to know what the system contains; it might be some pure liquid, or it might be a mixture of ice and brine.

Even when we are dealing with a pure phase, its internal energy may increase with the temperature in a variety of ways. The translational energy of its molecules may alone be increased, as in the ideal monatomic gas; the molecules may acquire energy of rotation or of oscillation, as in the case of a polyatomic gas; the potential energy of the molecules may be larger owing to their increasing separation, as in the case of liquid mercury; or there may even be an increase of energy owing to a dissociation (or to some other chemical reaction) which occurs when the temperature is raised (as in the case of iodine vapor at high temperatures, and presumably in the case of liquid water at room temperature). All such changes together make up the increase in internal energy, and for the purposes of thermodynamics the particular nature of the transformations of energy within the system need not be specified.

If a system changes in volume during heating, the heat absorbed differs from the increase in internal energy by the work which is done, or

$q = dE + P \, dV$. In particular, if any system is heated infinitesimally at constant pressure,

$$C_P = \left(\frac{\partial E}{\partial T}\right)_P + P \left(\frac{\partial V}{\partial T}\right)_P \qquad \checkmark \qquad (4\text{-}12)$$

When the system is thus heated from one temperature to another, at the same pressure, the increase in its energy is the same as it would be if the system were heated from the first temperature to the second, at constant volume, and then brought at constant temperature to the original pressure. This may be expressed in mathematical form [see Eq. (3-4)].

$$\left(\frac{\partial E}{\partial T}\right)_P = \left(\frac{\partial E}{\partial T}\right)_V + \left(\frac{\partial E}{\partial V}\right)_T \left(\frac{\partial V}{\partial T}\right)_P \qquad \backslash \qquad (4\text{-}13)$$

Combining Eqs. (4-11), (4-12), and (4-13), we obtain an important equation for the difference between the heat capacity at constant pressure and the heat capacity at constant volume,

$$C_P = C_V + \left[P + \left(\frac{\partial E}{\partial V}\right)_T \right]\left(\frac{\partial V}{\partial T}\right)_P \qquad / \qquad (4\text{-}14)$$

Finally, by a slight modification in Eq. (4-12), we may see at once the extremely simple relation between C_P and the heat content. For if P is constant,

$$P\frac{\partial V}{\partial T} = \frac{\partial(PV)}{\partial T}$$

and

$$C_P = \left[\frac{\partial(E + PV)}{\partial T} \right]_P = \left(\frac{\partial H}{\partial T}\right)_P \qquad (4\text{-}15)$$

In other words, the heat capacity, at constant pressure, is equal to the rate of change of heat content with temperature, just as, at constant volume, the heat capacity is the energy-temperature coefficient.

We have pointed out, in discussing the properties of substances, that instead of taking pressure and temperature as our independent variables, we might take another pair such as temperature and volume. There seems, however, to be some a priori reason for selecting P and T, two quantities which are similar to one another in their general characteristics. Certainly this choice is justified on practical grounds. Thus the temperature coefficients of various properties are far more frequently measured at constant pressure than at constant volume. It is for this reason that H and C_P will be more frequently used than E and C_V. In fact, as we proceed we shall find that, besides H and C_P, there are a number of properties which come into prominence in the P-T system, while other closely related properties would be brought into prominence in the V-T system. It will further be noted that our equations assume the simplest form when they include only properties which are characteristic of only one of these systems. Thus, for example $(\partial E/\partial V)_T$ and $(\partial H/\partial P)_T$ will be more frequently met than $(\partial E/\partial P)_T$ or $(\partial H/\partial V)_T$.

SUMMARY OF THE ENERGY CONCEPT

This chapter has been devoted to a development of the concepts of heat, work, and energy. As these concepts are somewhat complex, it is worthwhile at this point to summarize the steps in the development of the concept of energy. In Chapter 3, we first discussed the concept of specific energies associated with specific types of interactions between systems, such as kinetic energy, gravitational potential energy, electrical energy, etc. The value of such concepts is limited in that interactions of a simple type can be approached only as a limit, and likewise the application of conservation of energy to a specific type of energy is of value only as a limit.

The next step in the generalization of the concept of energy was the recognition of the equivalence of different forms of energy such as kinetic, electrical, and other energies in that they are interconvertible. The transfer of any type of energy other than thermal energy from one system to another can be measured on an equivalent basis such as through the lifting of a weight, and the loss of such energy by a system is designated as work done by that system and the gain of such energy is designated as work done on the system. When thermal interactions are excluded as a limit, the conservation principle may be applied to the work transfers or to the nonthermal energies.

The recognition of the distinctness of thermal energy and the lack of complete interconvertibility between thermal and nonthermal energies requires that thermal energies be considered separately from other types of energy. The experimental observations of Rumford, Mayer, and Joule demonstrate that, in a cyclic process which returns a system to its initial state, the heat, or thermal energy, generated is always proportional to the work, or nonthermal, energy consumed, or that total energy is conserved. This establishes the existence of a state function, energy, which is the sum of all the specific energies including the internal energy of substances present. The fact that energy (total energy) is conserved or that it is a state function is not restricted to a limiting situation, as were the previous statements of the conservation of specific energies or of the sum of nonthermal energies, but is a generalization which has been experimentally verified under all circumstances which have been tested.

The next development in the concept of energy of a system is its identification with the mass of a system by Einstein. Although in principle the measurement of mass yields a measure of the energy of a system, in practice one is usually interested in energy changes corresponding to such small mass changes that one must usually measure energy changes through measurements of heat and work transfers. However, the state-function character of energy fixes it as a function of the other properties of a system, and, for well-studied systems, energy changes can be designated without measurement of heat and work for every change of the system. Thus, although energy changes are usually determined through measurement of heat and work transfers, the concept of energy as a state function gives it more general application than heat or work.

UNITS OF ENERGY AND CORRELATED UNITS[1,2]

It must be made clear at the outset that our general equations assume the employment of some self-consistent set of units. On the other hand, in our numerical calculations we shall employ practical units which are not always mutually consistent. In such calculations it is therefore of the utmost importance at all times to bear in mind the units which are being employed and to know the exact numerical relationship between these units.

The basic cgs unit of energy is the erg. The joule $= 10^7$ ergs is the basic mks unit. For practical reasons we shall adopt neither of these, but rather the calorie, as the unit of energy, just as we shall adopt the atmosphere instead of the cgs unit of pressure. This is due to the fact that chemists have become accustomed to using the calorie as the unit of energy and it has proved difficult to gain widespread use of the rational unit, the joule. Since energy is now usually measured in terms of electrical units based on the absolute ohm and ampere from which absolute joules can be measured, energy measurements are actually determined in terms of absolute joules, which are then converted to calories by the defined relationship[3]

$$1 \text{ cal} = 4.1840 \text{ abs joules} \qquad (4\text{-}16)$$

Although the present definition of the calorie is independent of the property of any material, the definition has been so chosen that it represents the same unit of energy

[1] The fundamental constants given here are those evaluated by E. R. Cohen, J. W. M. DuMond, T. W. Layton, and J. S. Rollett, *Revs. Modern Phys.*, **27**, 363–380 (1955), and E. R. Cohen, K. M. Crowe, and J. W. M. DuMond, "The Fundamental Constants of Physics," Interscience Publishers, Inc., New York, 1957.

Substantially the same values were obtained in the independent evaluation of J. A. Bearden and J. S. Thomsen, *Nuovo cimento*, suppl. **5**, 267–360 (1957). The results of both evaluations are based on the physical scale of atomic weights which assigns O^{16} an atomic weight of exactly 16. The chemical scale assigns an atomic weight of 16 to oxygen of average isotopic composition. Owing to variations in isotopic composition, the ratio of the physical to chemical scales may vary as much as 10 parts per million. The values given here have been converted to the chemical scale by taking the ratio of the physical to chemical scale as 1.000272.

[2] Both cgs and mks systems of units will be used, and the conversion factor will be mentioned if it is not obvious. Logically, the calorie is unnecessary since the joule would serve and is the absolute mks unit. The convenience of various simple numerical factors such as R and the heat capacity of water in calories, however, makes the retention of the calorie desirable. The thermochemical table given in kilojoules in the "International Critical Tables" proved to be unpopular, and more recent tabulations have used calories or kilocalories. The units, electron volt and reciprocal centimeter, are also convenient measures of energy in atomic and molecular experiments and are retained in addition to the erg and joule.

[3] H. F. Stimson, *Am. J. Phys.*, **23**, 614–622 (1955).

as the calorie that was formerly defined in terms of the international joule and earlier in terms of the heat capacity of water at 15°C. Thus the energies reported in terms of the calorie before 1947, when the present definition was chosen, and those reported even earlier based on the heat capacity of water at about 15°C require no modification.

Another unit of energy which is of occasional service is the cubic centimeter atmosphere. If 1 atm is defined as the pressure of 760 mm of mercury under standard conditions,

$$1 \text{ atm} = 1,013,250 \text{ dynes/cm}^2 \tag{4-17}$$

and
$$1 \text{ cc atm} = 0.1013250 \text{ joule} \tag{4-18}$$

or
$$1 \text{ cc atm} = 0.0242179 \text{ cal} \tag{4-19}$$

In the next chapter we shall discuss the perfect gas whose equation of state is $Pv = RT$, an equation which is approached by all gases at low pressures. Accurate measurements of the limiting Pv of gases at low pressures yield[1] 22414.6 ± 0.4 cc atm for Pv at 0.00°C (273.15°K) and therefore

$$R = 82.0597 \text{ cc atm/deg mole} \tag{4-20}$$

It will be frequently convenient, however, to express this constant in calories per degree. For this purpose we write

$$R = 1.98726 \text{ cal/deg mole} \tag{4-21}$$

Likewise, we write

$$R = 8.31470 \text{ joules/deg mole} \tag{4-22}$$

In order to obtain energy data from measurements of electromotive force (emf), it is necessary to know the Faraday equivalent, i.e., the number of coulombs of electricity associated with 1 g equiv of any ion,

$$\mathfrak{F} = 96,493.5 \text{ coulombs/equiv} \tag{4-23}$$

If ε is the emf of a cell, then the electrical work done by that cell is $96,493.5 \times \varepsilon$ volt coulombs/equiv, or $23,062.3\varepsilon$ cal/equiv. In other words,

$$\mathfrak{F} = 23,062.3 \text{ cal/volt equiv} \tag{4-24}$$

Planck's constant h, Avogadro's number N_0, and Boltzmann's constant $k = R/N_0$ are three additional useful constants.

$$h = 6.6252 \times 10^{-27} \text{ erg sec} \tag{4-25}$$
$$N_0 = 6.0232 \times 10^{23} \text{ per mole} \tag{4-26}$$
$$k = 1.38045 \times 10^{-16} \text{ erg/deg} \tag{4-27}$$

The probable errors for the fundamental constants have now been reduced to 20 to 40 parts per million. This magnitude of error is negligible for almost all purposes.

For convenience in our future calculations, we have collected the assembled constants, together with several other numerical factors which are of frequent utility, in Appendix 10.

PROBLEMS

4-1. Prove that

$$C_P = C_V + \left[V - \left(\frac{\partial H}{\partial P} \right)_T \right] \left(\frac{\partial P}{\partial T} \right)_V \tag{4-28}$$

[1] F. D. Rossini, H. L. Johnston, F. T. Gucker, L. Pauling, and G. S. Vinal, *J. Am. Chem. Soc.*, **74**, 2699 (1952).

✓ **4-2.** (*a*) 18.02 g of liquid water is enclosed under a frictionless weightless piston at 100°C and 1 atm pressure. The pressure above the piston is lowered slightly below 1 atm and the water allowed to vaporize isothermally until all vaporized. For this process, $q = 9730$ cal. The specific volume of water at 100°C is 1.043 cc, and the specific volume of steam is 1677 cc at 100°C and 1 atm. Calculate the work w attending this vaporization and ΔE and ΔH for the process. (*b*) Find the values of ΔE, ΔH, and q for the process where the piston is just removed and the water is allowed to vaporize freely isothermally into an evacuated space of such volume that the pressure has built up to 1 atm with all the water vaporized. (*c*) Calculate the work in calories attending the fusion of 18.02 g of ice at 0°C and 1 atm. The density of ice is 0.917 g/cc, ✓ and the density of water is 1.000 g/cc. For the fusion process, $q = 1436$ cal/mole. Calculate ΔE and ΔH.

✓**4-3.** The current from a common lead storage cell operating at 2 volts is used for heating purposes. How much heat would be developed per gram of lead consumed in the cell? (This exercise is introduced merely to show one of the many types of elementary problems involving the conversion of units and the employment of numerical factors. The reader who is not practiced in this kind of numerical manipulation should set himself numerous problems of this character.)

5

APPLICATION OF THE FIRST LAW
TO A PURE HOMOGENEOUS SUBSTANCE

In the very general treatment of the last chapter we obtained equations which are valid for any sort of system. It will be well to illustrate the use of these equations by showing their application to the limited case of a single phase of a single component.[1] We have chosen as subjects of such illustration, first, the perfect gas, which has played so large a part in the development of physical science, and, second, the phenomenon of free expansion, which has not only been of great theoretical service in the determination of the absolute scale of temperature and in the choice of suitable equations of state for imperfect gases, but has also been of great practical importance in the liquefaction of air and other gases.

In dealing with a single component, it will frequently be useful to employ molal quantities. Thus, from the general equations of the preceding chapter, we may write such more special equations as

$$\mathrm{H} = \mathrm{E} + P\mathrm{v} \tag{5-1}$$

$$\left(\frac{\partial \mathrm{E}}{\partial T}\right)_V = \mathrm{c}_V \qquad \left(\frac{\partial \mathrm{H}}{\partial T}\right)_P = \mathrm{c}_P \tag{5-2}$$

THE PERFECT GAS

It often happens in science that a law which at first is supposed to be universally valid is found after more thorough investigation of particular systems to be subject to correction. It may indeed be found that no individual system obeys the law rigorously but nevertheless that numerous systems may be made to approach an ideal limiting state in which the law would be exact. Such limiting laws are frequently of great value. Thus, in the theory of radiation, we speak of the laws of reflection from a

---

[1] The discussion in this chapter is not really limited to single-component systems but applies to any solution, provided that its composition is fixed; nor would the equations for free expansion be modified in any essential manner if we were dealing with a heterogeneous system such as a fluid emulsion.

44

perfect reflector and the laws of emission from a perfect blackbody, although the perfect reflector and the perfect blackbody are not found among actual substances.

Such an ideal limiting law is the *gas law*. The perfect gas, or the ideal gas, is an invented substance, defined by certain properties which are not possessed by any actual substance but which are supposed to be approached by every actual gas as its pressure is indefinitely diminished. We may state, then, that the perfect gas is a substance which fulfills the two following conditions: (1) that its energy is a function of the temperature alone, or, in other words, that[1]

$$\left(\frac{\partial E}{\partial V}\right)_T = 0 \qquad (5\text{-}3)*$$

(2) that when its temperature, pressure, and volume are changed, these obey the relation

$$PV = nRT \qquad (5\text{-}4)*$$

where n is the number of moles, or

$$Pv = RT \qquad (5\text{-}5)*$$

Work and Heat of Isothermal Reversible Expansion. At constant temperature, for example, if the containing vessel is immersed in a thermostat, the internal energy of a perfect gas must by definition remain constant. Under such circumstances, therefore, when the gas does work by expansion, the conservation law requires that an equivalent amount of heat must be absorbed from the environment (the thermostat). That is, $-w = q$. In general, $-\delta w = P_{ex}\, dV$ for expansion against the external resisting pressure P_{ex}. If the expansion occurs slowly and without friction, so that the internal pressure is kept equal to the external pressure, then

$$\delta q = -\delta w = P\, dV$$

or, for a finite reversible change,

$$q = -w = \int_{V_A}^{V_B} P\, dV$$

Hence, substituting for P from the gas law, and integrating,

$$q = -w = nRT \ln \frac{V_B}{V_A} = nRT \ln \frac{P_A}{P_B} \qquad (5\text{-}6)*$$

In employing this equation for numerical calculation we use $R = 82.06$ if we wish to measure q and w in cubic centimeter atmospheres or

---

[1] Equations such as these, which are not true for actual substances but which may be used as approximation formulas, will be marked with an asterisk.

$R = 1.9873$ if the result is to be in calories. Thus let us consider the work required to compress reversibly 1 mole of a perfect gas from 1 to 200 atm at 25°C or 298.15°K. We have, per mole,

$$w = -1.9873 \times 298.15 \times 2.3026 \log \tfrac{1}{200} = 3139 \text{ cal}$$

and 3139 cal is also the heat evolved in this reversible isothermal compression.

Work of Reversible Adiabatic Expansion. When 1 mole of a perfect gas expands, not isothermally but adiabatically (in other words, when no heat is transferred to or from the vessel), then $q = 0$ and $\delta w = d\text{E}$. In this process the temperature will change. Since by Eq. (5-3)* the internal energy, at constant temperature, does not depend upon the volume, E must be a function of T alone. It will therefore be evident, for a perfect gas, that, not only for a change at constant volume, but for any change, $d\text{E} = c_V \, dT$. For a reversible process with $P_{ex} = P$, the pressure of the gas, $\delta w = -P \, dV$. Hence we may write

$$-P \, d\text{v} = c_V \, dT$$

or, substituting from the gas law,

$$\frac{RT}{\text{v}} \, d\text{v} = -c_V \, dT$$

or
$$R \, d \ln \text{v} = -c_V \, d \ln T \qquad (5\text{-}7)^*$$

This equation cannot be integrated without knowing more about c_V. In the special case in which c_V is constant, the integration is simple and gives

$$T\text{v}^{R/c_V} = \text{constant} \qquad (5\text{-}8)^*$$

The Difference between the Heat Capacity at Constant Pressure and at Constant Volume. For a perfect gas Eq. (4-14) assumes a very simple form. Since $(\partial E / \partial V)_T = 0$ and $(\partial \text{v} / \partial T)_P = R/P$,

$$c_P - c_V = R \qquad (5\text{-}9)^*$$

Also c_P and c_V are independent of volume (see Problem 5-6).

THE HEAT OF FREE EXPANSION

When a substance under pressure expands against a smaller external pressure, as, for example, when the valve of a cylinder containing gas at high pressure is opened, the process is known as free expansion in contrast to the reversible expansions discussed in the preceding two paragraphs. In the case of any actual substance this free expansion is associated with thermal changes. A perfect gas, when it changes volume at constant temperature, suffers no change in energy or heat content, but an actual gas does.

If we look at the matter from the standpoint not of thermodynamics but of molecular mechanics, we may attribute this change largely[1] to change in the potential energy of the molecules, due to their approach or separation. Thus the heat absorbed in the irreversible, or free, expansion of a compressed gas may be regarded as a residuum of the latent heat of vaporization.

An experiment of great significance in this connection was first performed in an English brewery by Joule and Thomson (Lord Kelvin). Joule had already attempted to show, and did show within the limits of experimental error, that $(\partial E/\partial V)_T = 0$ for the common gases. He allowed the gases to expand, without doing work, from one vessel to another exhausted vessel, both being immersed in a calorimeter. There resulted no measurable change in the temperature of the calorimeter.

It was evident, however, that small energy changes could not be detected in this manner, and the new experiment of Joule and Thomson[2] was therefore undertaken. This experiment consisted in allowing a gas to escape through a porous plug from a high pressure on one side to a low pressure on the other, the resistance of the plug being great enough to ensure a nearly constant pressure in the incoming and also in the outgoing gas. Thermometers placed on either side of the plug showed a temperature difference $T_2 - T_1$ depending upon the pressure difference $P_2 - P_1$. The apparatus was constructed of such poor thermal conductors that no appreciable amount of heat could pass into or out of the system.

At ordinary temperatures and pressures, all gases, except hydrogen and helium, show a cooling effect in such free expansion. This amounts in the case of carbon dioxide to more than 1° for a difference of pressure of 1 atm, while the cooling for air, which approaches more nearly to the perfect gas, is only about one-fifth as great. However, this effect in air is large enough to be of practical importance, and the most common apparatus for the production of liquid air makes use of this phenomenon. Thus, if a certain portion of compressed gas undergoes free expansion and the cooling effect is used to precool another portion, that portion upon expansion will fall to a still lower temperature. By continuing this process a certain fraction of the original compressed gas can be liquefied.

The experiments of Joule and Thomson and of others who have used

---

[1] Of course, the changes in the heat content are due in part to changes in the product PV, but even in regard to the internal energy it would be incorrect to say that its changes are due solely to changes in potential energy. For, contrary to earlier belief, the kinetic energy of molecules is not always the same at the same temperature for a substance in different states. This is because of the deviation at low temperatures of the exact quantum-mechanical properties from those expected from classical mechanics.

[2] J. P. Joule and W. Thomson, *Proc. Roy. Soc. (London)*, **143**, 357 (1853).

the same method show that the cooling produced by a given pressure drop is nearly independent of the pressure (although it is noticeably smaller at the highest pressures which have been studied). On the other hand, it increases rapidly with diminishing temperature. Thus a difference of 200 atm causes a lowering of 45° in air initially at 0°C and of 100° if the air is initially at −90°C. Indeed hydrogen and helium, which at ordinary temperatures are not cooled but are heated by free expansion, behave like other gases at low temperatures. If, therefore, hydrogen is precooled by liquid air, it can be further cooled and eventually liquefied by its own free expansion.

The temperature at which a gas is neither heated nor cooled by free expansion is called its inversion temperature. Probably all gases which at ordinary temperatures are cooled by free expansion will show such an inversion at higher temperatures.

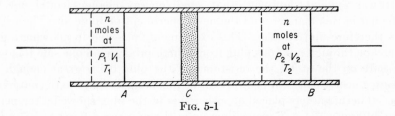

Fig. 5-1

In order to understand clearly the theory of the Joule-Thomson experiment, let us consider the schematic representation in Fig. 5-1. The gas is passing through the porous plug C from left to right. The pistons A and B are moved at such a rate as to keep each of the pressures P_1 and P_2 constant. When 1 mole of gas has passed through the plug, the work P_1v_1 will have been done upon the system and the work P_2v_2 will have been done by the system. The net work done upon the system is $w = P_1v_1 - P_2v_2$. Assuming now that the apparatus is constructed of such good thermal insulators that the process is adiabatic, then $q = 0$, and $E_2 - E_1 = P_1v_1 - P_2v_2$. Hence, by Eq. (4-9), $\Delta H = 0$, and the process is one occurring at constant H.

The ratio $(T_2 - T_1)/(P_2 - P_1)$ having been obtained at several values of $P_2 - P_1$, we may find its value in the limiting case as $P_2 - P_1$ approaches zero. This value is called the Joule-Thomson coefficient μ. It is formally defined by the equation

$$\mu = \left(\frac{\partial T}{\partial P}\right)_H \tag{5-10}$$

Now this coefficient is related to others which we have already employed [see Eq. (3-3)] by the equation

$$\left(\frac{\partial T}{\partial P}\right)_H = -\frac{(\partial H/\partial P)_T}{(\partial H/\partial T)_P}$$

or, by Eqs. (5-10) and (4-15),

$$\left(\frac{\partial H}{\partial P}\right)_T = -\mu C_P \qquad (5\text{-}11)$$

where $(\partial H/\partial P)_T$ shows the heat which would be absorbed, per unit difference in pressure, if the experiment were carried out, not in a thermally insulated tube, but in a calorimeter at constant temperature.

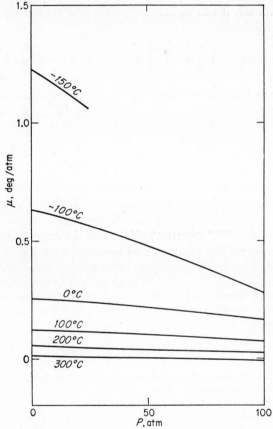

Fig. 5-2. The Joule-Thomson coefficient of nitrogen gas. At the lowest temperature, −150°C, nitrogen liquefies; hence the curve for the gas terminates at the vapor pressure.

The equations of this section are in no way restricted to gaseous substances, and the experiment of Joule and Thomson can be made with any fluid substance. However, few such experiments have been carried out with liquids.

Figure 5-2 shows the results of Roebuck and Osterberg's measurements[1]

[1] J. R. Roebuck and H. Osterberg, *Phys. Rev.*, **48**, 450 (1935).

of the Joule-Thomson coefficient of nitrogen. It is evident that μ approaches a finite value at zero pressure, and this result is characteristic of all gases. The experimental evidence shows also that C_P approaches a finite value at zero pressure. From these conclusions we may prove that real gases follow the equation $(\partial E/\partial V)_T = 0$ in the limit of zero pressure.

$E = H - PV$

$$\left(\frac{\partial E}{\partial V}\right)_T = \left(\frac{\partial H}{\partial V}\right)_T - P - V\left(\frac{\partial P}{\partial V}\right)_T = \left(\frac{\partial P}{\partial V}\right)_T\left[\left(\frac{\partial H}{\partial P}\right)_T - V\right] - P$$

But we have shown that $(\partial H/\partial P)_T = -\mu C_P$; hence

$$\left(\frac{\partial E}{\partial V}\right)_T = -\left(\frac{\partial P}{\partial V}\right)_T(\mu C_P + V) - P \qquad (5\text{-}12)$$

However, in the limit of zero pressure, Eq. (5-4)* applies, and

$$\left(\frac{\partial P}{\partial V}\right)_T = -\frac{nRT}{V^2} = -\frac{P}{V}$$

Hence $$\lim_{P\to 0}\left(\frac{\partial E}{\partial V}\right)_T = \lim_{P\to 0}\frac{P\mu C_P}{V} = 0 \qquad (5\text{-}13)*$$

We recall that temperature was defined in Chapter 4 in a fashion that makes Eq. (5-4)* true in the limit at zero pressure. Since we have now shown that Eq. (5-3)* is also true in this limit, we may conclude that real gases do conform in the limit of zero pressure with our criteria of the perfect gas.

Some other properties of real gases show more complex behavior at low pressures. To illustrate this, we take the equation of state which represents accurately the volume of gases in the low-pressure region.

$$P\text{v} = RT + \text{в}P \qquad (5\text{-}14)*$$

Here в is a function of temperature but not of pressure or volume and is different for each gas. Typically в is negative at low temperatures, is zero at what is called the Boyle point, and is positive at high temperatures.

If we take the limit $P \to 0$, we see that the product $P\text{v}$ approaches the ideal-gas value RT but that the difference between the volume and its ideal value, $\text{v} - RT/P$ approaches the nonzero value в. Similarly we have seen that $(\partial E/\partial V)_T$ becomes zero but $(\partial H/\partial P)_T$ has a finite value $-\mu C_P$ at zero pressure.

Also, in principle a gas such as N_2 may be expected to dissociate into atoms at extremely low pressures. Since this dissociation does not occur at any pressure of actual measurements, there is no practical difficulty in defining the properties of gases such as N_2 in the ideal-gas state. In cases such as nitrogen tetroxide or formic acid, where dissociation does occur at pressures of interest, one may still define the properties of the dissociated species as an ideal gas by working at low pressure and making correction for the associated molecules present.

PROBLEMS

5-1. Show that for a perfect gas $(\partial H/\partial V)_T = 0$.

/ **5-2.** All gases at relatively low pressures approach the equation of state $Pv = RT + {\rm B}P$, where ${\rm B}$ is independent of P or v. Derive an expression for the difference $(\partial H/\partial v)_T - (\partial {\rm E}/\partial v)_T$ for such a gas.

/ **5-3.** Assume the same gas as in Problem 5-2. Obtain the expression for the heat and work of reversible isothermal expansion, and calculate numerically the heat absorbed in expanding 1 mole at 25°C from 250 to 25,000 cc, assuming that ${\rm B} = 20$ cc/mole and that E is independent of V.

5-4. Show that Eq. (5-8)*, for adiabatic change, can be put into the equivalent form

$$PV^{c_P/c_V} = \text{constant} \qquad\qquad (5\text{-}15)^*$$

5-5. Derive an equation valid at low but finite pressures for $(\partial {\rm E}/\partial v)_T$ in terms of P, v, ${\rm B}$, μ, and c_P, using Eq. (5-14)* instead of Eq. (5-4)*. Show that the limit at zero pressure obtained from the simpler equation (5-4)* is correct.

5-6. Show that for an ideal gas

$$\left(\frac{\partial c_V}{\partial v}\right)_T = 0 \qquad\qquad (5\text{-}16)^*$$

$$\left(\frac{\partial c_P}{\partial v}\right)_T = 0 \qquad\qquad (5\text{-}17)^*$$

✓**5-7.** Let us see in a particular case how large a fraction of a compressed gas may be liquefied by a process of free expansion. In so far as the process can be made adiabatic, the enthalpy of the liquefied gas, together with the enthalpy of the residual gas which has not been liquefied, must be equal to the enthalpy of the original compressed gas. In a good liquefier, the heat interchange is so effective that the residual gas issues at a temperature very little lower than that of the compressed gas which enters, and by improving the interchanger this difference can be made indefinitely small. Let us assume then that, of 1 mole of oxygen, at 298°K and at 200 atm, entering the liquefier, the fraction x is converted into liquid oxygen at 1 atm and at 90°K (the boiling point of the liquid), and that the fraction $1 - x$ issues from the liquefier at 1 atm and at 298°K. Then we may write the equation $O_2(g, 200 \text{ atm}, 298°K) = xO_2(l, 1 \text{ atm}, 90°K) + (1 - x)O_2(g, 1 \text{ atm}, 298°K); \Delta H = 0$.

Or if one looks at the matter in another way, when the whole mass of the gas is cooled by free expansion from 200 to 1 atm, it is found by the Joule-Thomson experiment that the drop in temperature is about 50°. Thus the enthalpy at 200 atm and 298°K is the same as the enthalpy at 1 atm and 248°K. Hence we may write for the three states (all at atmospheric pressure)

$$O_2(g, 248°K) = xO_2(l, 90°K) + (1 - x)O_2(g, 298°K) \qquad \Delta H = 0$$

Assuming c_P for the gas to be constant and equal to 6.95 and taking the heat of vaporization per mole as 1629 cal, show that

$$x[1629 + 6.95(248 - 90)] = (1 - x)[6.95(298 - 248)]$$

or that about 11 per cent of the gas is liquefied.

5-8. Assuming Eq. (5-6)* (which produces an error in this calculation of only about 3 per cent), show that, if we compress oxygen to 200 atm at 25°C and use an efficient liquefier, it will require about 28,000 cal of mechanical work for each mole of liquid oxygen produced.

5-9. Show from Eq. (4-28) that for any substance

$$C_V = C_P \left[1 - \mu \left(\frac{\partial P}{\partial T} \right)_V \right] - V \left(\frac{\partial P}{\partial T} \right)_V \qquad (5\text{-}18)$$

How is this equation simplified at the inversion temperature?

5-10. Hoxton[1] has made a careful study of the Joule-Thomson coefficient for air over a limited range of temperature and pressure. His results are summarized in the empirical formula

$$\mu = -0.1975 + \frac{138}{T} - \frac{319P}{T^2} \qquad \text{deg/atm}$$

Show that, at 60°C and at small pressures, $\mu = 0.217$ and $(\partial \mu / \partial T)_P = -0.00124$.

5-11. The differentiation of Eq. (5-11) with temperature leads to an interesting and useful formula. Show that

$$\left(\frac{\partial C_P}{\partial P} \right)_T = -\mu \left(\frac{\partial C_P}{\partial T} \right)_P - C_P \left(\frac{\partial \mu}{\partial T} \right)_P \qquad (5\text{-}19)$$

Few experiments have been made upon the change of heat capacity with pressure, but Holborn and Jakob[2] have made a careful study of the change of C_P with pressure for air at 60°. They found, for 1 g of air at 60°, $C_P = 0.241$ cal/deg and, at small pressures, $(\partial C_P / \partial P)_T = 0.000286$ cal/deg atm. At this same temperature $(\partial C_P / \partial T)_P$ is 0.000035 cal/deg. Using these values and the value of μ from the previous problems, calculate $(\partial \mu / \partial T)_P$, and note how nearly it agrees with the experimental value given in the previous problem.

5-12. It may have been noticed that the equation of Hoxton in Problem 5-10 makes the Joule-Thomson coefficient approach a finite value at zero pressure. Indeed all our experimental evidence indicates that μ [and therefore $(\partial H / \partial P)_T$] does not approach zero as the gas is indefinitely expanded. We have already seen that $(\partial H / \partial V)_T$ is zero at zero pressure. Show analytically that, for a given amount of gas, the ratio of $\partial H / \partial P$ to $\partial H / \partial V$ approaches infinity as the pressure is diminished.

[1] *Phys. Rev.*, **13**, 438 (1919).
[2] *Z. Ver. deut. Ing.*, **58**, 1429 (1914).

6

HEAT CAPACITIES OF PURE SUBSTANCES
AND HEATS OF REACTION

Heat capacities are not only of great practical use but also of real theoretical interest in many cases. We shall see that the change with temperature of the heat of a chemical reaction is determined by the heat capacities of the reactants and products. Also the heat capacity tells us much about the thermal motion of atoms and molecules.

HEAT-CAPACITY VALUES

Direct experimental measurement is the primary source of heat-capacity values, but theoretical calculations based upon the detailed properties of atoms, molecules, or crystals now contribute reliable heat-capacity values in favorable cases. We shall postpone until Chapter 27 the detailed presentation of the theory because of its relatively complex nature and shall merely indicate its scope and general character at this point.

Heat capacity arises because the individual particles comprising a substance are able to move and thus take up kinetic energy and usually also potential energy as the temperature rises. In many substances the electrons do not absorb energy until very high temperatures are reached. Thus most heat-capacity effects are interpreted in terms of the motion of entire atoms. However, in metals and some other substances thermal energy is absorbed by the electrons at low temperatures.

Solid Elements at Room Temperature. In 1819 Dulong and Petit[1] announced the empirical rule that the heat capacity per gram atom is the same for all solid elements. Later this discovery was somewhat discredited when it was found that the specific heats at constant pressure always increase with temperature and sometimes very rapidly, so that at low temperatures, and indeed at room temperature for a number of light elements, the heat capacity proves to be much lower than the value of

---

[1] P. L. Dulong and A. T. Petit, *Ann. chim. phys.*, **10**, 395 (1819).

Dulong and Petit, while at higher temperatures the heat capacities rise somewhat above this value.

Meanwhile, however, Boltzmann[1] had shown that the rule of Dulong and Petit could be directly deduced from the classical kinetic theory and that the constant of Dulong and Petit should be equal to

$$3R = 5.96 \text{ cal/deg}$$

Boltzmann's theory assumes that the atoms in the crystal vibrate as harmonic oscillators, which in turn requires that the interatomic forces follow Hooke's law, i.e., that the restoring force is proportional to the displacement from equilibrium position. We now know that Hooke's law represents only an approximation to true interatomic forces and that the rule of Dulong and Petit is likewise only an approximation to true solid heat capacities even at high temperatures.

Lewis[2] suggested that the Boltzmann theory would be much more nearly correct for C_V than for C_P, whereas most experimental measurements are of the latter quantity. He calculated C_V for a number of elements by use of the thermodynamic equation (which we shall prove in Chapter 10)

$$c_P - c_V = \frac{\alpha^2 \mathrm{v} T}{\kappa} \tag{6-1}$$

where $\alpha = (1/V)(\partial V/\partial T)_P$ is the coefficient of thermal expansion, $\kappa = -(1/V)(\partial V/\partial P)_T$ is the compressibility, v is the molal volume, and T is the absolute temperature. (In ordinary units this equation gives $c_P - c_V$ in cc atm/deg, which may be converted to calories per degree through multiplication by 0.02422.) Lewis found that the values of c_V did conform more closely than values of c_P to the theoretical value of $3R$.

Recent measurements of high precision show that there are several sorts of phenomena which cause c_V to deviate slightly from the value $3R$, but at the same time these results confirm the conclusion that the heat capacity due to atomic vibration is very closely approximated by the value $3R$. Since each atom may vibrate in the crystal in each of three mutually perpendicular directions, we may take R as the heat-capacity contribution at high temperatures per vibrational motion or degree of freedom.

Solid Elements at Low Temperatures. The change of specific heats with temperature had been recognized earlier, but it was called sharply to the attention of scientists by the work of Dewar[3] at very low temperatures. He showed that between the temperatures of liquid

[1] L. Boltzmann, *Sitzb. kgl. Akad. Wiss. Wien*, **63:2**, 679 (1871).

[2] G. N. Lewis, *J. Am. Chem. Soc.*, **29**, 1165, 1516 (1907).

[3] J. Dewar, *Proc. Roy. Soc. (London)*, (A)**74**, 122 (1904).

hydrogen and liquid air the average heat capacity of diamond is less than 1 per cent of the Dulong and Petit value.

Einstein[1] was the first to see the close connection between the diminution of heat capacities at low temperatures and the departure from the classical theory of heat in another of its branches, namely, the one which deals with the spectrum of radiant energy from a hot body. Thus the decreased heat capacity of solids at low temperatures became one of the first confirmations of the proposal of Planck[2] that energy existed in discrete quanta. Einstein derived the equation for the heat capacity of a system of vibrating atoms where there is but a single frequency of oscillation. His result, for a single atom and one direction of oscillation, is

$$C = \frac{ku^2 e^u}{(e^u - 1)^2} \quad \text{with } u = \frac{h\nu}{kT} \tag{6-2}$$

where k = gas constant per molecule (Boltzmann's constant)

h = Planck's constant

ν = frequency of oscillator

In Chapter 27 a derivation is given for Eq. (6-2). At high temperatures this expression approaches the value k. Since a crystal of 1 g atom contains N_0 atoms and each atom has three directions of vibration, we see that Einstein's theory gave the same result as Boltzmann's at high temperatures, that is, $c = 3N_0 k = 3R$.

It was soon found that the heat capacity of solid elements did not conform quantitatively to the Einstein theory. Debye[3] and, independently, Born and von Kármán[4] showed that the theory of crystal vibration yielded not a single frequency but a broad spectrum of frequencies. Debye proposed an approximate expression for this distribution of frequencies and integrated the Einstein expression over the distribution. Debye's result may be expressed in the form[5]

$$c_V = f_D\left(\frac{T}{\theta_D}\right) \tag{6-3}*$$

where θ_D is a parameter for each substance which has the dimension of temperature and

$$f_D(x) = 9Rx^3 \int_0^x \frac{u^4 e^u \, du}{(e^u - 1)^2}$$

The Debye function is complex, but it has been calculated and is tabulated in Appendix 5 in the form $c_V/3R$, which is $f_D/3R$.

[1] A. Einstein, *Ann. Physik*, (4)**22**, 180 (1907).

[2] M. Planck, *Ann. Physik*, (4)**4**, 553 (1901); see also *ibid.*, **1**, 99 (1900).

[3] P. Debye, *Ann. Physik*, (4)**39**, 789 (1912).

[4] M. Born and T. von Kármán, *Physik. Z.*, **13**, 294 (1912); **14**, 15 (1913).

[5] See footnote to Eq. (5-3)* for the significance of the asterisk in these equations.

One important feature of the Debye formula, which arises from the form of the equation,[1] is the property that c_V is a general function of $\log T - \log \theta_D$. Consequently, if we plot c_V for various elements as a function of $\log T$, we should obtain curves of the same shape but displaced from one another horizontally by the differences in $\log \theta_D$ for the

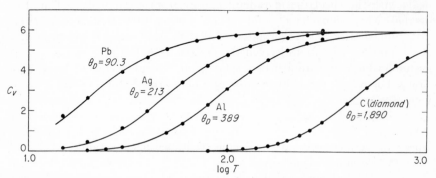

FIG. 6-1. The heat capacity, in cal/deg mole, for several solid elements. The curves are from the Debye function with the θ_D values given.

various substances. Figure 6-1 presents such a plot for several solid elements.

At low temperatures the Debye function assumes the simple form

$$c_V \cong \frac{12\pi^4 R}{5}\left(\frac{T}{\theta_D}\right)^3 = \text{constant} \times T^3 \qquad (6\text{-}4)^*$$

This result has been verified approximately for many substances, and it serves as a very useful formula for the extrapolation of the heat capacity to temperatures below the range of experimental measurement.

The quantity θ_D is proportional to the frequency of atomic vibration (more precisely $\theta_D = h\nu_D/k$, where ν_D is the maximum frequency in the Debye distribution). Consequently, θ_D increases with the rigidity of the interatomic binding force, and it decreases with increasing atomic mass.[2] These trends are illustrated in Table 6-1.

The Born and von Kármán theory is potentially more exact than the Debye theory, but it must be applied individually to each substance, and it requires detailed information about interatomic forces which is not commonly available. Advances[3] are currently being made in the application of the Born and von Kármán theory to particular crystals, and it is

---

[1] Incidentally, the Einstein equation also is of the form of Eq. (6-3)* but with a different function and different θ's.

[2] Strictly speaking, the frequency and hence θ_D are proportional to the square root of the ratio of the interatomic force constant to the atomic mass.

[3] For review see J. de Launay, *Solid State Phys.*, **2**, 219 (1956).

TABLE 6-1. THE HEAT CAPACITY, CAL/DEG, AND DEBYE θ FOR SEVERAL ELEMENTS

	Pb	Au	Ag	Cu	Fe	Al	Diamond
θ_D, °K........	90.3	169	213	313	417	389	1890
c_P (25°K).....	3.36	1.23	0.75	0.23	0.10	0.11 ✓	0.00
c_P (100°K)....	5.84	5.10	4.82	3.85	2.88	3.12 ✓	0.06
c_P (298°K)....	6.41	6.03	6.09	5.85	5.97	5.82	1.45

to be hoped that really precise theoretical heat capacities of solids may become available in the near future.

Complex Solids. In 1864 Kopp[1] proposed that the heat capacity of a solid compound is equal to the sum of the heat capacities of its constituent elements. While this rule is only approximately valid at low temperatures, it becomes quite reliable at temperatures where the rule of Dulong and Petit applies to the elements. Thus, if there are x atoms in the formula, the heat capacity per mole at high temperatures is $3xR$. Just as in the case of solid elements, the theory applies more closely to c_V than to c_P.

There are several types of complication which can arise in solids at high temperatures. One possibility that exists in all metals is the contribution of electrons to the heat capacity. Although electrons are free to move through the metal to conduct electricity, their contribution to the heat capacity is so small that it was unnoticed for many years. This peculiar situation is now explained by the quantum theory of a degenerate Fermi-Dirac gas. This theory predicts that at low temperatures the electronic heat capacity will rise with the first power of the absolute temperature.

$$c_{el} = \gamma T \qquad (6\text{-}5)*$$

The proportionality constant γ is a property of each metal. It is interesting to note that the electronic heat capacity can best be detected at either high or very low temperatures. At high temperatures [although Eq. (6-5)* is no longer strictly valid], one finds a small linear increment over the theoretical value of $3R$ for c_V, due to anharmonicity of lattice vibrations as well as the electronic effect, while at very low temperatures the normal atomic vibrational heat capacity drops off more rapidly (as T^3) than the electronic term. Table 6-2 indicates the relative magnitude of the electronic term at 2 and at 30°K.

There are a number of other phenomena which yield heat-capacity contributions in certain solids. Examples include the change from an ordered to a disordered arrangement of certain atoms or the beginning of

[1] H. Kopp, *Ann. Chem. Pharm. Suppl.*, **3**, 1, 289 (1864).

TABLE 6-2. ELECTRONIC CONTRIBUTION TO HEAT CAPACITY, CAL/DEG

	$\gamma \times 10^4$	At 2°K		At 30°K	
		C_{el}	C_{total}	C_{el}	C_{total}
Cu.........	1.80	0.00036	0.00045	0.005	0.412
Al.........	3.46	0.00070	0.00075	0.010	0.203

rotation of a group of atoms.   If any atom or ion in the solid has a magnetic moment, then heat-capacity effects can arise from the reorientation of these magnetic moments.   Similarly, special effects occasionally arise from reorientation of molecules with electric-dipole moments.   The last two types of effects will be discussed in a later chapter when consideration is given to electric and magnetic fields.

At low temperatures the heat capacity of complex solids falls to zero in a manner similar to that for monatomic solids.   The curves do not necessarily conform quantitatively to any single formula, however.   In a few simple cases such as that of NaCl, the heat capacity is roughly approximated by the Debye theory for a crystal with twice the number of atoms, but even in as simple a substance as AgI the heat capacity deviates widely from this formula.   Indeed, graphite, which has a peculiar arrangement of atoms in layers, has a heat capacity very different from that given by the Debye curve.

While Kopp's rule will always yield a rough estimate for the heat capacity of a complex solid, generally one must depend on direct experimental measurements for precise values.   Also, when using Kopp's rule for estimating heat capacities, one should select the most similar compounds for which data are available.   For example, if one wished an estimate for Rb_2SO_4, one might choose data for K_2SO_4 and then estimate the difference caused by the substitution of Rb for K from the values for $RbClO_3$ and $KClO_3$, which happen to be the only pair of rubidium and potassium salts for which data are available.   At 298.15°K one then finds

$$c_P(Rb_2SO_4) \cong c_P(K_2SO_4) + 2c_P(RbClO_3) - 2c_P(KClO_3)$$
$$= 31.08 + (2 \times 24.66) - (2 \times 23.96) = 32.48 \text{ cal/deg}$$

Monatomic Ideal Gases.   It was in 1867 that Naumann[1] predicted that a monatomic ideal gas would possess no thermal energy except that of translation.   From elementary kinetic theory one then obtains the result $c_V = \frac{3}{2}R$ or from Eq. (5-9)* $c_P = \frac{5}{2}R$.   This prediction has been abundantly verified by measurements on various monatomic gases over

[1] A. Naumann, *Ann. Chem. Pharm.*, **142**, 265 (1867).

wide temperature ranges. For atoms with low-lying electronic levels, there will be an additional contribution to the heat capacity.

Diatomic Ideal Gases. In addition to the translational motion, diatomic molecules can rotate about two mutually perpendicular axes. These rotational motions are similar to translational motions in that there are no restraining forces. Kinetic theory predicts the same contribution to the heat capacity per coordinate (degree of freedom) for rotation as for translation, namely, $\frac{1}{2}R$. Thus rotation and translation will contribute $\frac{5}{2}R$ toward c_V of a diatomic gas. Since each atom has three possible directions of motion, there must be six degrees of freedom in the diatomic molecule. In addition to the five already mentioned there is the motion of the two atoms toward or away from one another. This motion is necessarily subject to restraining forces; hence it is a vibration.

If we assume Hooke's law of force to apply to this vibration, we obtain exactly the Einstein formula, Eq. (6-2), for the heat capacity of vibration. In this case Einstein's formula is exact because each molecule has the same frequency of vibration unaffected by any interaction with other molecules. The heat capacity at constant volume of a diatomic ideal gas is

$$c_V = R\left[\frac{5}{2} + \frac{u^2 e^u}{(e^u - 1)^2}\right] \qquad (6\text{-}6)*$$
$$u = \frac{h\nu}{kT}$$

where ν is the vibration frequency of the particular molecule. Figure 6-2 shows the heat capacities of several diatomic gases. The frequency can be found from appropriate spectral measurements. Just as was the case in solids, molecules with heavy atoms have low vibration frequencies and therefore have substantial vibrational heat capacities at relatively low temperatures.

Equation (6-6)* gives accurate values of the heat capacity of diatomic gases at ordinary temperatures, but it is not absolutely exact because the interatomic force does not follow Hooke's law exactly. Also the centrifugal force of rotation has a small effect on the vibrational motion. Even these small corrections are known from the spectroscopic data in many cases, and it is then possible to calculate the heat capacity with extreme precision.

Polyatomic Ideal Gases. The same principles that were discussed for diatomic molecules apply to larger molecules and allow calculation of the heat capacity whenever all the vibration frequencies can be obtained from the spectrum or from other sources. If all the atoms lie along a straight line, as is the case for CO_2, N_2O, and a few other molecules, there are still two rotational degrees of freedom. Then there will be $3n - 5$

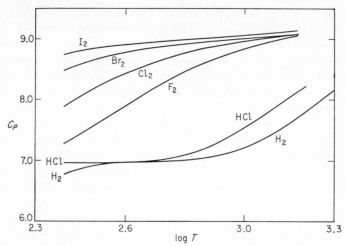

Fɪɢ. 6-2. The heat capacity, in cal/deg mole, of several diatomic gases, showing the vibrational contribution.

vibrations for a molecule with n atoms. If the atoms do not lie in a straight line, there are three axes around which rotation can occur and $3n - 6$ vibrations. The formulas for the heat capacities in the two cases are as follows:

Linear molecules

$$c_V = R\left[\frac{5}{2} + \sum_{i=1}^{3n-5} f_E(u_i)\right] \tag{6-7}*$$

Nonlinear molecules

$$c_V = R\left[3 + \sum_{i=1}^{3n-6} f_E(u_i)\right] \tag{6-8}*$$

$$f_E(u) = \frac{u^2 e^u}{(e^u - 1)^2} \qquad u_i = \frac{h\nu_i}{kT}$$

The frequencies ν_i are, in general, different for each vibration in a molecule but in special cases the same frequency may occur two or three times.

It is beyond the scope of this discussion to give a full treatment of molecular vibrations and of molecular spectroscopy. Nevertheless, it is from these sources that vibration frequencies must be obtained for heat-capacity calculations. We shall merely add that accurate and reliable heat capacities have been calculated by these methods for many polyatomic ideal gases. Figure 6-3 shows the heat capacities of several hydrocarbon gases.

The same sort of minor corrections from failure of Hooke's law and from centrifugal stretching apply to polyatomic molecules as to diatomic molecules.   In addition there are new possibilities of peculiar types of internal motion in certain molecules.   For example, in ethane the two methyl groups may rotate with respect to one another about the C—C axis.   There are forces affecting this rotation; consequently, it is essentially a torsional vibration at low temperatures, but at high temperatures it approximates a rotation.   Our purpose at this point is merely to draw

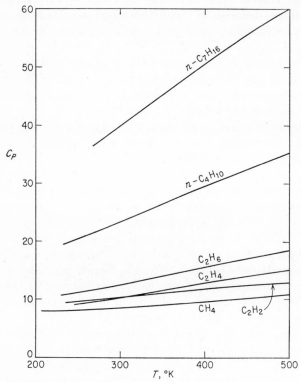

Fig. 6-3. The heat capacity, in cal/deg mole, of several polyatomic hydrocarbon gases.   Note the large vibrational contribution for heptane.

attention to the possibility of such special motions; further discussion of them is given in Chapter 27.

Hydrogen at Low Temperatures.   In our discussion of the heat capacity of gases up to this point we have not considered the possibility that the rotational contribution might decrease at low temperatures in a manner similar to that for vibration.   Actually quantum theory tells us that rotational motion is quantized, and the energy separation of the levels has been measured spectroscopically in many cases.   The separation of the energy levels is usually so small, however, that the decrease in heat capacity is predicted to lie far below the boiling point. Only in the case of hydrogen[1] does this loss of rotational heat capacity occur at a temperature of practical interest.

In principle the translational motion of molecules confined within a limited space is also quantized, but the energy-level spacing is much smaller than for rotation. Even in the case of helium gas at about 4°K the translational heat capacity is still $\frac{3}{2}R$.

Nonideal Gases. Gas molecules do exert forces upon one another, but these forces decrease so rapidly with increasing interatomic distance that it is a useful approximation to ignore them at low pressures. The ideal-gas law expresses gas behavior in the absence of intermolecular force effects. Real gases approach this behavior as $P \rightarrow 0$ and usually follow the ideal-gas law quite closely even at 1 atm. Likewise the ideal-gas heat capacities are those approached by real gases as $P \rightarrow 0$ and are frequently close to the values for 1 atm.

The effect of pressure on the heat capacity of a gas can be calculated from the second temperature derivative of the volume at constant pressure. This will be discussed in detail in Chapter 16. Conversely, direct measurements of the change of heat capacity with pressure offer a very stringent test of the correctness of an equation of state.

It is convenient both theoretically and practically to separate the gas imperfection effect from the heat capacity of the ideal gas. These quantities are usually given by separate equations or tables and are then combined to yield the heat capacity at a given temperature and pressure.

Heat Capacity of Liquids. The liquid state is a much more difficult subject for theoretical calculation than either the solid or the gas. Consequently, we must depend primarily on direct experimental measurement for heat-capacity values for liquids.

The heat capacity of a liquid is usually larger than that for the vapor at the same temperature, and this difference is nearly the same for similar types of liquid but is much smaller for metals than for molecular liquids. The heat capacity of the liquid at the melting point is usually larger than that of the solid for molecular liquids, but the difference is very small for metals. Table 6-3 gives heat-capacity values for a variety of substances to illustrate these trends.

Experimental Measurement of Heat Capacity. The details of calorimetric design and technique constitute an extensive and important subject but one beyond the scope of this book. We shall merely outline the principal methods.

The most direct method is that using an adiabatic calorimeter.[2] The sample container is placed in an evacuated chamber and surrounded by a

[1] A. Eucken, *Sitzber. kgl. preuss. Akad. Wiss.*, **1912**, 141.

[2] H. M. Huffman, *Chem. Rev.*, **40**, 1 (1947), presents details of such a calorimeter.

TABLE 6-3. HEAT CAPACITIES, in CAL/DEG MOLE, FOR LIQUIDS COMPARED
WITH VALUES FOR THE VAPOR OR THE SOLID
Part I. Comparison with Vapor

	T, °K	c_P (liquid)	c_P (vapor)	Difference
Hg...............	500	6.6	5.0	1.6
Zn................	1000	7.5	5.0	2.5
Xe...............	165	10.7	5.0	5.7
CO...............	80	14.5	6.9	7.6
Cl$_2$...............	240	15.7	7.8	7.9
C$_2$H$_6$..............	100	16.4	8.5	7.9
	180	17.3	9.9	7.4
SO$_2$...............	270	20.7	9.3	11.4
CH$_3$CCl$_3$..........	300	34.5	22.4	12.1

Part II. Comparison with Solid

	T, °K	c_P (liquid)	c_P (solid)	Difference
Hg...............	234	6.8	6.8	0.0
Sn................	505	7.3	7.3	0.0
Zn................	693	7.5	7.1	0.4
Cu...............	1357	7.5	7.4	0.1
CO...............	68	14.4	12.4	2.0
Cl$_2$...............	172	16.0	13.3	2.7
SO$_2$...............	198	21.0	16.5	4.5
C$_3$H$_8$..............	85	20.2	12.6	7.6
(CH$_3$)$_2$S..........	175	26.9	20.0	6.9

shield or shields which are maintained at temperatures adjusted to eliminate heat transfer as nearly as possible. An electrical heater is used to introduce a small, accurately measured energy increase, and a sensitive thermometer (usually an electrical resistance thermometer or thermocouple) measures the small temperature rise. Separate measurements yield the heat capacity of the empty container, which must be subtracted. This method is used for solids or liquids at relatively low temperatures (below about 500°K). At higher temperatures the rate of heat transfer from radiation alone becomes so large that it is difficult to keep the shield temperature regulated with the necessary precision to avoid spurious heat transfer.

Below room temperature where heat exchange by radiation is not too rapid, precise measurements may be made by using a shield at constant temperature.[1] Correction must be made for the heat loss or gain during

---

[1] W. F. Giauque and C. J. Egan, *J. Chem. Phys.*, **5**, 45 (1937), and earlier references cited in their paper describe such a calorimeter that has been widely used.

each measurement, but this may be determined accurately under these conditions.

Above room temperature accurate heat-capacity measurements on solids or liquids have been made mostly by the dropping method (or method of mixtures).[1]   The sample is brought to an accurately determined temperature in a furnace and then is dropped suddenly into a calorimeter at room temperature, or an ice calorimeter[2] (in which the heat is measured by the volume change of the ice melted).   Thus one actually measures the total change in heat content above room temperature rather than the differential heat capacity.   The heat-content data may be fitted to an empirical equation, which then may be differentiated to yield c_P if desired; however, for many thermodynamic purposes it is the heat content that is really needed.   Also, if the substance melts or undergoes a transition, this heat is also included and must be appropriately treated.

The measurement of the heat capacity of gases is particularly difficult because the large molal volume of a gas necessarily implies a large surface across which spurious heat transfer can occur.   Among the more successful methods for gases is the flow method.[3]   A steady rate of gas flow is established, and the initial gas temperature is measured.   Then a steady heat input is generated by an electrical heater and the final temperature increase measured to yield the heat capacity.   The velocity of sound in a gas and certain other indirect measurements have also yielded heat-capacity values through thermodynamic relationships which will be discussed in Chapter 10.

The heat capacity of gases at high temperatures has been obtained by measuring the maximum temperature attained in an explosion together with the total heat released by the chemical change.   Dilution of the explosive mixture by unreactive gases allows their heat capacity to be determined also.   The temperatures attained in explosions are frequently so high that some molecules are dissociated into atoms or smaller molecules while excited electronic states are produced in other cases. Thus the explosion method yields checks on these sorts of data also when they cannot be obtained by more accurate methods.

Heat Capacity of Volatile Liquids and Solids.   The heat capacity of a volatile solid or liquid is usually measured in a calorimeter which is nearly full of the condensed phase but contains also some vapor.   A temperature rise causes the vapor pressure to increase, and a correction must be made for material vaporized.[4]   After this correction the result is the

[1] K. K. Kelley, *U.S. Bur. Mines Bulls.* 476, p. 1, 1949; 584, p. 2, 1960, gives a list of references that describe various high-temperature calorimetric techniques.

[2] D. C. Ginnings and R. J. Corruccini, *J. Research Natl. Bur. Standards*, **38,** 583 (1947).

[3] G. Waddington, S. S. Todd, and H. M. Huffman, *J. Am. Chem. Soc.*, **69,** 22 (1947).

[4] H. J. Hoge, *J. Research Natl. Bur. Standards*, **36,** 111 (1946).

heat capacity under the saturated vapor pressure C_{sat}.   Almost all heat-capacity values in the literature for volatile liquids and solids are actually values of C_{sat} even though many are reported as C_P.   The difference is usually trivial, unless the vapor pressure exceeds 1 atm, but may become important under high vapor pressures.

Let us now examine the relationship between C_P and C_{sat} and also that between these quantities and the change of enthalpy with temperature. From the basic definitions of energy and heat content we may write

$$dH = dE + d(PV) = \delta q + \delta w + P\,dV + V\,dP$$

and, since $\delta w = -P\,dV$,

$$C_{\text{sat}} = \lim_{\delta T \to 0} \left(\frac{\delta q}{\delta T}\right)_{\text{sat}} = \left(\frac{\partial H}{\partial T}\right)_{\text{sat}} - V\left(\frac{\partial P}{\partial T}\right)_{\text{sat}} \tag{6-9}$$

Also from Eqs. (3-4) and (4-15)

$$\left(\frac{\partial H}{\partial T}\right)_{\text{sat}} = C_P + \left(\frac{\partial H}{\partial P}\right)_T \left(\frac{\partial P}{\partial T}\right)_{\text{sat}} \tag{6-10}$$

Combination of the last two equations yields

$$C_{\text{sat}} = C_P + \left[\left(\frac{\partial H}{\partial P}\right)_T - V\right]\left(\frac{\partial P}{\partial T}\right)_{\text{sat}} \tag{6-11}$$

and, anticipating Eq. (10-24) to be derived from the second law of thermodynamics,

$$C_P - C_{\text{sat}} = T\left(\frac{\partial V}{\partial T}\right)_P \left(\frac{\partial P}{\partial T}\right)_{\text{sat}} \tag{6-12}$$

Thus the vapor-pressure equation and the coefficient of expansion yield the information necessary to calculate $C_P - C_{\text{sat}}$.   For example, the data for water yield $c_P - c_{\text{sat}} = 0.0046$ cal/deg mole at 373°K, or about 0.03 per cent of the total heat capacity.   At lower temperatures $(\partial P/\partial T)_{\text{sat}}$ drops off rapidly, and the difference between C_P and C_{sat} becomes imperceptible.   Conversely, at temperatures where the vapor pressure is many atmospheres the difference in the two heat capacities may be large.

The difference $C_P - C_{\text{sat}}$ would be large for the saturated vapor, but there is no occasion to measure or use C_{sat} for the vapor.   The enthalpies of both the saturated liquid and vapor are used frequently, and their variation with temperature is given by Eq. (6-9) or (6-10).   The latter equation based upon C_P is much more convenient for the vapor, but either may be used for the liquid.

Representation of Heat-capacity Data.   Heat-capacity and heat-content results are commonly presented by tabulation and by empirical equations.   In either case one should present a quantity closely related to that measured in order to avoid errors from extraneous sources.   Thus, results from the dropping method are best given as $H_T - H_{298.15}$.

In formulating empirical equations it is desirable to use the same form for various substances, since this allows the addition or subtraction of corresponding terms in dealing with the change of heat capacity for a chemical reaction.   Also certain terms are found to be more convenient than others for the eventual use of these equations in free-energy calculations.

The three-term equation

$$c_P = a + bT + cT^{-2} \qquad (6\text{-}13)*$$

has been found to be very convenient for most work above room temperature; equations of this type for a number of common substances are given in Table 6-4.   The range of validity, in each case, should be care-

TABLE 6-4. EQUATIONS FOR HEAT CAPACITY, CAL/DEG MOLE

All monatomic gases with no appreciable electronic excitation:
$$c_P^\circ = 4.97$$

Other gases (applicable 298 to 2000°K):

S	$c_P^\circ = 5.26 - 0.10 \times 10^{-3}T + 0.36 \times 10^5 T^{-2}$
H_2	$c_P^\circ = 6.52 + 0.78 \times 10^{-3}T + 0.12 \times 10^5 T^{-2}$
O_2	$c_P^\circ = 7.16 + 1.00 \times 10^{-3}T - 0.40 \times 10^5 T^{-2}$
N_2	$c_P^\circ = 6.83 + 0.90 \times 10^{-3}T - 0.12 \times 10^5 T^{-2}$
S_2	$c_P^\circ = 8.72 + 0.16 \times 10^{-3}T - 0.90 \times 10^5 T^{-2}$
CO	$c_P^\circ = 6.79 + 0.98 \times 10^{-3}T - 0.11 \times 10^5 T^{-2}$
F_2	$c_P^\circ = 8.26 + 0.60 \times 10^{-3}T - 0.84 \times 10^5 T^{-2}$
Cl_2	$c_P^\circ = 8.85 + 0.16 \times 10^{-3}T - 0.68 \times 10^5 T^{-2}$
Br_2	$c_P^\circ = 8.92 + 0.12 \times 10^{-3}T - 0.30 \times 10^5 T^{-2}$
I_2	$c_P^\circ = 8.94 + 0.14 \times 10^{-3}T - 0.17 \times 10^5 T^{-2}$
CO_2	$c_P^\circ = 10.57 + 2.10 \times 10^{-3}T - 2.06 \times 10^5 T^{-2}$
H_2O	$c_P^\circ = 7.30 + 2.46 \times 10^{-3}T$
H_2S	$c_P^\circ = 7.81 + 2.96 \times 10^{-3}T - 0.46 \times 10^5 T^{-2}$
NH_3	$c_P^\circ = 7.11 + 6.00 \times 10^{-3}T - 0.37 \times 10^5 T^{-2}$
CH_4	$c_P^\circ = 5.65 + 11.44 \times 10^{-3}T - 0.46 \times 10^5 T^{-2}$
TeF_6	$c_P^\circ = 35.33 + 1.62 \times 10^{-3}T - 7.00 \times 10^5 T^{-2}$

Liquids applicable from melting point to boiling point:

I_2	$c_P^\circ = 19.20$
H_2O	$c_P^\circ = 18.04$
NaCl	$c_P^\circ = 16.0$
$C_{10}H_8$	$c_P^\circ = 19.0 + 97.4 \times 10^{-3}T$

Solids applicable from 298°K to melting point or 2000°K:

C(graphite)	$c_P^\circ = 4.03 + 1.14 \times 10^{-3}T - 2.04 \times 10^5 T^{-2}$
Al	$c_P^\circ = 4.94 + 2.96 \times 10^{-3}T$
Cu	$c_P^\circ = 5.41 + 1.50 \times 10^{-3}T$
Pb	$c_P^\circ = 5.29 + 2.80 \times 10^{-3}T + 0.23 \times 10^5 T^{-2}$
I_2	$c_P^\circ = 9.59 + 11.90 \times 10^{-3}T$
NaCl	$c_P^\circ = 10.98 + 3.90 \times 10^{-3}T$
$C_{10}H_8$	$c_P^\circ = -27.7 + 224 \times 10^{-3}T$

fully noted; extrapolation outside this range can lead to serious error in some cases.

K. K. Kelley[1] has given excellent summaries of heat-capacity data for inorganic substances at both low and high temperatures, from which the data of Table 6-4 were obtained. Stull and Sinke[2] give data for the elements from 298 to 3000°K. Extensive tables of such data for hydrocarbons have been presented by Research Project 44 of the American Petroleum Institute.[3] It is to be hoped that tables will become available soon for organic compounds other than hydrocarbons.

Standard States. As heat capacities of most substances do not vary rapidly with pressure, it is convenient to tabulate heat capacities of solids and liquids only at 1 atm pressure. It is also convenient to take advantage of the fact that the heat capacity of a perfect gas is independent of pressure and that real gases approach the perfect-gas state at zero pressure. Consequently, heat capacities of gases are tabulated for the perfect-gas state. The variation of the heat capacity of a gas as the pressure is increased will be discussed in Chapter 16, and the same equation holds for liquids or solids. A superscript is applied, for example, C_P°, to indicate that the conventional standard state is being used. The superscript is used with the corresponding meaning for heat content.

HEAT OF REACTION

There is now a large body of information about the heat effects of chemical reactions. Many of these measurements were made by J. Thomsen, M. Berthelot, and others in late decades of the nineteenth century and are of all grades of accuracy. Some of Thomsen's measurements, in particular, are still the best available. This body of thermochemical data is of major importance to modern chemistry, and we wish now to discuss certain methods of handling this information.

It has been customary to give thermochemical results by such a shorthand expression as the following:

$$Hg(l) + \tfrac{1}{2}Cl_2(g) = \tfrac{1}{2}Hg_2Cl_2(s) + 31,660 \text{ cal}$$

The chemical symbols thus represent not only the nature and the stoichiometrical amounts of the substances involved but also the respective heat contents of these substances. However, it seems unwise to place so heavy a burden upon a single symbol, and we shall express the same result

---

[1] K. K. Kelley, *U.S. Bur. Mines Bulls.* 476, 1949; 584, 1960; and 477, 1950.

[2] D. R. Stull and G. C. Sinke, Thermodynamic Properties of the Elements, *Advances in Chem. Ser.*, No. 18, 1956.

[3] F. D. Rossini et al., "Selected Values of Physical and Thermodynamic Properties of Hydrocarbons and Related Compounds," Carnegie Press, Pittsburgh, 1953.

more explicitly in the form

(a) $Hg(l) + \frac{1}{2}Cl_2(g) = \frac{1}{2}Hg_2Cl_2(s)$ $\Delta H^\circ_{298} = -31{,}660$ cal

In our previous chapters, conforming with universal practice, we have called the heat of vaporization or of fusion the heat absorbed in the process of vaporization or fusion. In the case of the heat of solution usage varies, but here also we have meant the heat absorbed in the process of solution. To be consistent, therefore, we should, in the case of an ordinary chemical reaction, denote by the heat of reaction the heat absorbed in the process. This, however, is contrary to common custom. In order to avoid confusion in this regard, we shall always express reaction heats by giving the value of ΔH, which without ambiguity gives the heat absorbed in the given reaction.

Combination of Thermochemical Equations.   If we reverse the above reaction (a), we write

(b) $\frac{1}{2}Hg_2Cl_2(s) = Hg(l) + \frac{1}{2}Cl_2(g)$ $\Delta H^\circ_{298} = 31{,}660$ cal

In fact such equations may be treated much like algebraic equations, and if we write some other equations, such as

(c) $\frac{1}{2}H_2(g) + \frac{1}{2}Cl_2(g) = HCl(g)$ $\Delta H^\circ_{298} = -22{,}063$ cal

then, by a method which is familiar, we may add (b) to (c), or subtract (a) from (c), and find

(d) $\frac{1}{2}H_2(g) + \frac{1}{2}Hg_2Cl_2(s) = Hg(l) + HCl(g)$
$$\Delta H^\circ = 31{,}660 - 22{,}063$$
$$= 9597 \text{ cal}$$

It is by means of such combinations of existing thermochemical equations that we are able to determine the heats of reaction in many cases which have not been the subject of direct experimental investigation, and indeed in cases where such a determination is impracticable. It would be difficult to measure directly the heat of the reaction

$$2C(\text{graphite}) + 2H_2(g) = C_2H_4(g)$$

but we may burn ethylene and find

(e) $C_2H_4(g) + 3O_2(g) = 2CO_2(g) + 2H_2O(l)$ $\Delta H^\circ_{298} = -337{,}230$ cal

or

(f) $2CO_2(g) + 2H_2O(l) = C_2H_4(g) + 3O_2(g)$ $\Delta H^\circ_{298} = 337{,}230$ cal

Likewise we may burn graphite and hydrogen and find

(g) $2C(\text{graphite}) + 2O_2(g) = 2CO_2(g)$ $\Delta H^\circ_{298} = -188{,}104$ cal
(h) $2H_2(g) + O_2(g) = 2H_2O(l)$ $\Delta H^\circ_{298} = -136{,}634$ cal

Then, by directly adding (f), (g), and (h), we have,

(i) $2C(\text{graphite}) + 2H_2(g) = C_2H_4(g)$ $\Delta H^{\circ}_{298} = 12{,}500 \text{ cal}$

This determination involves the difference between very large numbers, in which even a small percentage error means a large absolute error. Indeed it cannot be too strongly emphasized that almost invariably it is the absolute and not the percentage error which is of significance in our thermochemical calculations.   Thus in the above illustration an error of 100 cal in any one of the determinations (f), (g), or (h) would produce the same error of 100 cal in the final result.

Heat of Formation and Standard States.   It is highly convenient, especially for purposes of concise tabulation, to know the heats of reaction when various substances are formed from their elements.   Such data yield the heats of any other reactions of these substances.   It is therefore desirable to choose some one *standard state of reference* for each element.   For this purpose, we shall, at all temperatures, take the element in that form which is the most stable or more common at the temperature under consideration.   For a gaseous element, the *standard state* is the perfect-gas state or the limit for real gas as the pressure approaches zero. For solids or liquids, we shall take the element at a pressure of 1 atm. Thus at room temperature liquid mercury, solid iodine, rhombic sulfur, and graphite, all under a pressure of 1 atm, and gaseous oxygen in the low-pressure perfect-gas state will be considered to be in their *standard states* for purposes of tabulating heat contents.   We speak, then, of the ΔH of formation of a given substance, or, even more tersely, of its ΔH, meaning the increase in heat content in the reaction by which 1 mole of that substance is prepared from its elements in their standard states. Thus the ΔH_{298} for $\frac{1}{2}Hg_2Cl_2(s)$, for $HCl(g)$, for $C_2H_4(g)$ are seen from (a), (c), and (i) to be $-31{,}660$, $-22{,}063$, and $12{,}500$ cal, respectively.

We may use the same phraseology even for the elements themselves when they are not in their standard states.   Thus for the heat of transition[1] between rhombic and monoclinic sulfur at $368.5°K$

(j) $S(\text{rhombic}) = S(\text{monoclinic})$ $\Delta H_{368.5} = 96 \text{ cal}$

Hence we may say that for monoclinic sulfur $\Delta H_{368.5} = 96$ cal.

In addition to specifying *standard states* for the elements, it is convenient also to specify *standard states* in a similar manner for other substances.   As we have previously noted, heat capacities which are tabulated for the perfect-gas state or for the solid or liquid under a pressure of 1 atm are identified by a superscript °, for example, C°_P, and enthalpies and heats of reactions involving standard states are designated similarly, for example, $H^{\circ}_{298} - H^{\circ}_{273}$ and ΔH°_{298}.

[1] E. D. West, *J. Am. Chem. Soc.*, **81**, 29 (1959).

As an illustration of the use of heats of formation, we may consider the production of potassium chlorate from potassium chloride and oxygen. From the literature we find for $KClO_3(s)$, $\Delta H_{298}^{\circ} = -93,500$ cal; for $KCl(s)$, $\Delta H_{298}^{\circ} = -104,175$ cal. Hence,

$$(k) \quad KCl(s) + \tfrac{3}{2}O_2(g) = KClO_3(s) \qquad \begin{aligned} \Delta H_{298}^{\circ} &= -93,500 + 104,175 \\ &= 10,675 \text{ cal} \end{aligned}$$

As all the substances are in their standard states, all enthalpies are indicated with a <u>superscript</u> °. This convention avoids a great deal of specification concerning the states of the elements and compounds when heats of formation are given. Thus, $KCl(s)$, $\Delta H_{298}^{\circ} = -93,500$ cal corresponds to

$$K(s, 1 \text{ atm}) + \tfrac{1}{2}Cl_2(\text{perfect gas}) = KCl(s, 1 \text{ atm})$$
$$\Delta H^{\circ} = -93,500 \text{ cal at } 298.15^{\circ}K$$

and $S(\text{monoclinic})$, $\Delta H_{368.5}^{\circ} = 96$ cal corresponds to

$$S(\text{rhombic}, 1 \text{ atm}) = S(\text{monoclinic}, 1 \text{ atm})$$
$$\Delta H^{\circ} = 96 \text{ cal at } 368.5^{\circ}K$$

The heat of formation of the form of the element which is taken as the standard reference state is, of course, zero. If a heat of formation is given for nonstandard states, this should always be indicated in detail, as standard states will always be assumed if there is no indication to the contrary.

Indirect Determination of Heat of Reaction at Constant Pressure. In writing a thermochemical equation it is not necessary that all the reagents and the products should be at the same pressure. In most cases, however, we employ each substance in a reaction at atmospheric pressure, and then ΔH is simply the heat measured in the constant-pressure calorimeter. Sometimes, however, a constant-volume calorimeter is used, especially in determining heat of combustion. Here we measure not ΔH but ΔE. The difference between these two quantities is ordinarily quite negligible, except where gases are involved. Most combustions are carried out in a sealed bomb with a high pressure of oxygen to ensure complete combustion. Corrections must be made for the difference in enthalpy between the gases at high pressure and the gases in the low-pressure standard states. These enthalpy differences will be discussed in Chapter 16. There are a number of detailed corrections that must be applied to results obtained from bomb calorimetry. The detailed procedures for use of bomb calorimeters and the details of the various corrections are given by Rossini.[1]

---

[1] F. D. Rossini (ed.), "Experimental Thermochemistry," prepared under the auspices of the International Union of Pure and Applied Chemistry, Interscience Publishers, Inc., New York, 1956.

In addition to the direct calorimetric method of determination of heats of reaction, there are several indirect methods. From the change of an equilibrium constant with the temperature, and especially from the temperature coefficient of emf in a galvanic cell, it is frequently possible to determine a heat of reaction with an accuracy comparable to that in direct calorimetry. We shall discuss these methods in subsequent chapters.

EFFECT OF TEMPERATURE UPON HEAT OF REACTION

Even though the heat of a reaction is known at some one temperature, it is frequently necessary to know the heat of the same reaction at some other temperature. To make a direct calorimetric determination at each desired temperature would require a prohibitive expenditure of effort. But fortunately the equations which we have derived from the law of conservation of energy enable us, in the simplest manner, to calculate the change in a heat of reaction with the temperature.

We know that ΔH is the difference between the sum of the heat contents of the products and the sum of the heat contents of the reagents and that each molal heat content varies with the temperature according to the equation $(\partial H/\partial T)_P = c_P$. By combination of such equations we obtain the Kirchhoff[1] formula

$$\left(\frac{\partial \Delta H}{\partial T}\right)_P = \Delta C_P \qquad (6\text{-}14)$$

where ΔC_P is the sum of the heat capacities of the products, less the corresponding sum for the reactants; in other words, it is the total increase in heat capacity resulting from the reaction. In interpreting such a coefficient as $\partial \Delta H/\partial T$, we regard ΔH as a symbol which represents a single quantity.

If ΔH_{T_1} and ΔH_{T_2} represent the heat of a given isothermal reaction at two different temperatures T_1 and T_2, then by integration of Eq. (6-14)

$$\Delta H_{T_2} - \Delta H_{T_1} = \int_{T_1}^{T_2} \Delta C_P \, dT \qquad (6\text{-}15)$$

Over a small range of temperature, or in general when ΔC_P can be regarded as a constant, this equation takes the form

$$\Delta H_2 - \Delta H_1 = \Delta C_P(T_2 - T_1) \qquad (6\text{-}16)^*$$

Let us consider the freezing of water, first starting with water at 0°C and obtaining ice at 0°C, and then starting with supercooled water at −2°C and obtaining ice at −2°C. We can easily determine the amount of heat

[1] G. Kirchhoff, *Ann. Physik*, (2)**103**, 177 (1858).

set free in the latter process.   Writing the equation

$$H_2O(l) = H_2O(s)$$

ΔH at 0°C has been found to be -79.7 cal/g, or -1436 cal/mole.   The molal heat capacities of water and ice are, respectively, about 18 and 9 cal/deg.   Thus $\Delta c_P = -9$ cal/deg, and ΔH is diminishing by 9 cal for each degree rise in temperature.   Hence, at -2°C,

$$\Delta H = -1436 + (-9)(-2) = -1418 \text{ cal}$$

When the variation of heat capacities is so rapid, or the temperature interval is so great, that ΔC_P cannot be regarded as constant, we may integrate Eq. (6-14) by some graphical method or by an analytical method in which we represent the heat capacities by empirical equations. Ordinarily we assume an algebraic expression for each heat capacity, as in Eq. (6-13)*,

$$C_P = a + bT + cT^{-2} \qquad (6\text{-}13)*$$

or with even more terms, where the number of terms employed will depend upon the range of temperature which has been studied, upon the variability of the heat capacity, and upon the accuracy of the experimental data.

Now if we are dealing with a chemical reaction involving substances for each of which the heat capacity has been studied over the same range of temperature and for each of which we have obtained such an algebraic expression, then we may write

$$\Delta C_P = \Delta a + \Delta bT + \Delta cT^{-2} \qquad (6\text{-}17)*$$

where Δa is the algebraic sum of all the values of a, etc.   Finally, substituting in Eq. (6-14) and integrating, we find

$$\Delta H = \Delta H_I + \Delta aT + \tfrac{1}{2}\,\Delta bT^2 - \Delta cT^{-1} \qquad (6\text{-}18)*$$

where ΔH_I is a constant of integration.

As an illustration of the mode of application of these equations we may consider the formation of water vapor from its elements according to the equation

$$H_2(g) + \tfrac{1}{2}O_2(g) = H_2O(g)$$

As we have seen in Table 6-4, the heat capacities of these three gases are known over the wide range of temperature from 298 to 2000°K.   By combining the molal heat capacities for $H_2O(g)$ and $H_2(g)$ with the heat capacity of $\tfrac{1}{2}$ mole of $O_2(g)$, we obtain

$$\Delta C_P = -2.80 + 1.18 \times 10^{-3}T + 0.08 \times 10^5T^{-2}$$

Hence from Eq. (6-18)*

$$\Delta H = \Delta H_I - 2.80 + 0.59 \times 10^{-3}T^2 - 0.08 \times 10^5T^{-1}$$

Now, at 298.15°K, the experimental value for the heat of the reaction is $\Delta H_{298} = -57{,}798$ cal. Substituting this value with $T = 298.15$ into the above equation yields $\Delta H_I = -56{,}992$ cal. Hence we may write

$$\Delta H = -56{,}992 - 2.80T + 0.59 \times 10^{-3}T^2 - 0.08 \times 10^5T^{-1}$$

If now we wish the value of ΔH at any other temperature, say, at 1000°K, we may substitute $T = 1000$ and find $\Delta H_{1000} = -59{,}210$ cal.

This illustration will suffice to show the method of employing our equations of ΔC_P and of ΔH, although we shall find many a case in which the effect of temperature upon the heat of reaction is far more pronounced. In fact it might be asked why it is necessary to employ so elaborate an equation when, at 1000°K, the whole value of the last term is not much larger than the probable error of most ΔH values. This term, however, is appreciable over a considerable portion of the temperature range, but this is not the whole answer. Indeed, we may assert, as a result of practical experience, that it is not only convenient but necessary, having once determined upon a fundamental equation for some substance, such as the equation for C_P, to continue to use the same equation in all cases in which this substance is involved, even though the choice of the original equation is arbitrary and some other empirical equation may accord equally with the experimental data. Otherwise, in such addition and subtraction of equations as we have already illustrated, discrepancies appear and may be magnified to a surprising degree.

PROBLEMS

6-1. The specific heat of benzene at constant pressure is 0.425 cal/deg at 25°C. Calculate c_P° and c_V° for 1 mole of benzene whose volume is 88.8 cc and for which $\alpha = 0.00124$ and $\kappa = 0.000098$. (Use consistent units.)

6-2. Employ Eq. (6-6) to calculate the value of c_V° for $Br_2(g)$ at 600°K. The vibration frequency is 322 cm^{-1}; hence $\nu = 322c$ sec^{-1}, where c is the velocity of light, 2.99×10^{10} cm/sec.

6-3. Calculate c_V° for aluminum at 25° and 100°K from Eq. (6-4)* and the value 389° for θ_D. Compare with the observed heat capacities in Table 6-1. What do you conclude about the range of applicability of Eq. (6-4)*?

6-4. Integrate the heat-capacity equation for graphite in Table 6-4 to yield $H_T^\circ - H_{298}^\circ$, and compare the results so calculated with the experimental values of 2820 cal at 1000°K and 8530 cal at 2000°K.

6-5. Estimate the value of c_P° for rhodium at 300°K. Give your reasoning.

6-6. Consider the reaction $CO(g) + \tfrac{1}{2}O_2(g) = CO_2(g)$, $\Delta H_{298}^\circ = -67{,}636$ cal. Use the heat-capacity equations of Table 6-4 to find the complete equation for ΔH° as a function of temperature.

6-7. Calculate ΔH_{1000}°, using the equation derived in Problem 6-6. Compare this value with ΔH°_{1000} calculated from heat contents tabulated by Kelley.[1]   What is the significance of the difference?

6-8. By the conservation law, show that, if a reaction occurs at one temperature and the products are then raised to a higher temperature, the change in enthalpy is the same as when the reactants are first heated to the higher temperature and then allowed to react at that temperature.   Assume that carbon (graphite) is burned in pure oxygen to give carbon dioxide; assume that the materials start at 25°C and that the heat developed is used to heat a furnace at 1000°C, at which temperature the carbon dioxide escapes.   Find the maximum amount of energy which can be given to the furnace from the combustion of 1 kg of carbon.

6-9. In the combustion of liquid acetic acid to form gaseous carbon dioxide and liquid water, $\Delta H^{\circ}_{298} = -209.0$ kcal/mole.[†]   With the aid of Eqs. (*g*) and (*h*) find ΔH°_{298} of formation of acetic acid.

6-10. For $C_3O_2(g) + \frac{1}{2}O_2(g) = 3CO(g)$, ΔH°_{298} has been estimated as -70 kcal. The values of $H^{\circ}_T - H^{\circ}_{298}$ in kilocalories per mole of CO are as follows: 2500°K, 17.94; 3000°K, 22.36; 3500°K, 26.83; 4000°K, 31.33.   Calculate the maximum temperature when C_3O_2 gas at 25°C is burned with an equivalent amount of O_2 gas to produce CO at 1 atm.

[1] K. K. Kelley, *U.S. Bur. Mines Bulls.* 476, 1949; 584, 1960.

[†] F. W. Evans and H. A. Skinner, *Trans. Faraday Soc.*, **55,** 260 (1959).

7

THE SECOND LAW OF THERMODYNAMICS
AND THE CONCEPT OF ENTROPY

After the extremely practical considerations in the preceding chapters, we now turn a concept of which neither the practical significance nor the theoretical import can be fully comprehended without a brief excursion into the fundamental philosophy of science.

Clausius summed up the findings of thermodynamics in the statement, "Die Energie der Welt ist konstant; die Entropie der Welt strebt einem Maximum zu," and it was this quotation which headed the great memoir of Gibbs on "The Equilibrium of Heterogeneous Substances." What is this entropy, which such masters have placed in a position of coordinate importance with energy, but which has proved a bugbear to so many a student of thermodynamics?

The first law of thermodynamics, or the law of conservation of energy, was universally accepted almost as soon as it was stated, not because the experimental evidence in its favor was at that time overwhelming, but rather because it appeared reasonable and in accord with human intuition. The concept of the permanence of things is one which is possessed by all. It has even been extended from the material to the spiritual world. The idea that, even if objects are destroyed, their substance is in some way preserved has been handed down to us by the ancients, and in modern science the utility of such a mode of thought has been fully appreciated. The recognition of the conservation of carbon permits us to follow, at least in thought, the course of this element when coal is burned and the resulting carbon dioxide is absorbed by living plants, whence the carbon passes through an unending series of complex transformations.

The second law of thermodynamics, which is known also as the law of the dissipation or degradation of energy, or the law of the increase of entropy, was developed almost simultaneously with the first law through the fundamental work of Carnot, Clausius, and Kelvin. But it met with a different fate, for it seemed in no recognizable way to accord with existing thought and prejudice. The various laws of conservation had been foreshadowed long before their acceptance into the body of scientific thought. The second law came as a new thing, alien to traditional thought, with far-reaching implications in general cosmology.

Because the second law seemed alien to the intuition, and even abhorrent to the philosophy of the time, many attempts were made to find exceptions to this law and thus to disprove its universal validity. But such attempts have served rather to convince the incredulous and to establish the second law of thermodynamics as one of the foundations of modern science. In this process we have become reconciled to its philosophical implications or have learned to interpret them to our satisfaction; we have learned its limitations, or, better, we have learned to state the law in such a form that these limitations appear no longer to exist; and especially we have learned its correlation with other familiar concepts, so that now it no longer stands as a thing apart, but rather as a natural consequence of long-familiar ideas.

Preliminary Statement of the Second Law: The Actual, or Irreversible, Process. The second law of thermodynamics may be stated in a great variety of ways. We shall reserve until later our attempt to offer a statement of this law which is free from every limitation and shall confine ourselves for the present to a discussion of the law sufficient to display its character and content.

As a matter of fact in an early chapter we announced the essential feature of the second law when we stated that every system left to itself changes, rapidly or slowly, in such a way as to approach a definite final state of rest. This state of rest (defined in a statistical way) we also called the state of equilibrium. Now, since it is a universal postulate of all natural science that a system, under given circumstances, will behave in one and only one way, it is a corollary that no system, except through the influence of external agencies, will change in the opposite direction, i.e., away from the state of equilibrium.

Many types of processes leading toward equilibrium are familiarly known. The diffusion of material from a concentrated solution into a dilute solution, leading toward a condition of uniform concentration; the passage of heat from a hot body to a cold, leading to uniformity of temperature; the oxidation of organic substances by the atmosphere; the running down of a clock; the easing of strains in a plastic solid; the self-demagnetization of a magnet are all processes which illustrate the kind of change that occurs spontaneously in nature. Sometimes, as in the motion of the planets, this approach to a final state is extremely slow, but their mechanical energy is gradually being converted into heat by unceasing tidal action.

These processes and all other natural processes are alike in one respect, that they are bringing the various systems toward the condition of ultimate equilibrium or rest, and we may think of these systems as thereby losing in some measure their capacity for spontaneous change.

It is not coal but the combustion of coal which causes a steam engine to operate. As a rule, one system affects other systems in consequence of changes which are going on within it. Hence a system far removed from its condition of equilibrium is the one chosen if we wish to harness its processes for the doing of useful work. A system isolated from all others will always maintain a constant amount of energy, and therefore, if the second law of thermodynamics is spoken of as the law of the dissipation of energy, no loss in energy is meant, but rather a loss in the availability of energy for external purposes. It seems better therefore not to speak of the dissipation or degradation of energy but rather to speak of the degradation of the system as a whole. For in many cases, such as the diffusion of one gas into another, the process does not essentially involve an energy change.

Before proceeding to a more exact characterization of the second law, let us make sure that there is no misunderstanding of its qualitative significance. When we say that heat naturally passes from a hot to a cold body, we mean that, in the absence of other complicating processes, this is the process which inevitably occurs. It is true that by means of a refrigerating machine we may further cool a cold body by transferring heat from it to its warmer surroundings, but here we are in the presence of another dissipative process proceeding in the engine itself. If we include the engine within our system, the whole is moving always toward the condition of equilibrium. A system already in thermal equilibrium may develop large differences of temperature through the occurrence of some chemical reaction, but all such phenomena are but eddies in the general unidirectional flow toward a final state of rest.

The essential content of the second law might be given by the statement that when any actual process occurs it is impossible to invent a means of restoring *every* system concerned to its original condition. Therefore, in a technical sense, any actual process is said to be *irreversible.*

The Ideal, or Reversible, Process. When we speak of an actual process as being always irreversible, we have in mind a distinction between such a process and an ideal process which, although never occurring in nature, is nevertheless imaginable. Such an ideal process, which we shall call *reversible,* is one in which all friction, electrical resistance, or other such sources of dissipation are eliminated. It is to be regarded as a limit of actually realizable processes.

Let us imagine a process so conducted that at every stage an infinitesimal change in the external conditions would cause a reversal in the direction of the process, or, in other words, that every step is characterized by a state of balance. Evidently a system which has undergone such a process can be restored to its initial state without more than infinitesimal changes in external systems. It is in this sense that such an imaginary process is called reversible.

To illustrate, we may consider a system comprising water and water vapor contained in a cylinder with a movable piston. Now in practice the piston cannot be made free from friction, but we have no reason to believe that such friction may not be diminished indefinitely, and therefore we may regard the ideal frictionless piston as a limit which may be approached as nearly as is desired. After a constant temperature is established throughout the system, and when the external pressure upon the piston equals the vapor pressure of water, the system is in equilibrium with respect to internal and external agencies. If, then, the external pressure is increased by any amount, however small, the piston will move inward and the vapor will condense. If the external pressure upon the piston be diminished by any amount, however small, the piston will move outward and the liquid will vaporize. In other words, the work required to condense 1 mole of vapor differs by an infinitesimal amount from the work furnished by the vaporization of 1 mole of liquid.

If this cylinder of water vapor is kept in contact with a thermostat, the pressure required to maintain equilibrium is constant as long as both liquid and vapor are present. Otherwise the vapor pressure will change as the piston is moved, and if our process is to be reversible, the external pressure must vary so that at any stage it differs by no more than an infinitesimal amount from the internal pressure.

An excellent example of an actual process which is very nearly reversible is furnished when the emf of a galvanic cell is measured by means of a sensitive potentiometer. Here the driving force of the cell itself is so nicely balanced against an external emf that in favorable cases a current may be made to flow in one direction or the other by external changes of 0.000001 volt.

Again we may consider a case in which we deal, not with the balance of mechanical or electrical forces, but with a thermal equilibrium. If two bodies differ in temperature only by an infinitesimal amount, the transfer of heat from one to the other is likewise a reversible process, for evidently it would be possible to restore the system to its original condition without causing more than an infinitesimal change in external systems.

A QUANTITATIVE MEASURE OF DEGRADATION

In viewing the reversible process as the limit toward which actual processes may be made to approach indefinitely, it is implied that processes differ from one another in their degree of irreversibility. It is of the utmost importance to establish a quantitative measure of this degree of irreversibility, or this degree of degradation.

When we wish to measure a quantity such as length, we first choose a standard, say a bar of platinum kept at the International Bureau of Weights and Measures, and next we adopt a method of comparing the length of other objects with the length of this standard object. So in defining the degree of irreversibility of a process we shall choose some standard irreversible process and then define the method whereby others may be compared quantitatively with it.

Two familiar types of spontaneous or irreversible processes are (1) the transfer of heat from a higher to a lower temperature and (2) the conversion of work to heat. We have previously seen that the work done on a system can be measured in terms of the lifting of a mass in the earth's gravitational field while the heat transfer can be measured by an ice calorimeter or some similar fusion calorimeter.   We may also use other devices.   For example, a spring may be calibrated by allowing it to lift a mass and determining the height to which the mass is lifted for a given degree of compression of the spring.   Thus it is clear that we have a number of devices at our disposal that can measure the conversion of work to heat during any spontaneous process by allowing a mechanical device to do work on the system to return it reversibly to its initial state and by using some type of calorimeter to measure the resulting heat. We shall choose a standard system composed of a weight and pulley system and a heat reservoir.   In employing this standard system in conjunction with other systems, we are going to use the weight as a source of work and the reservoir as a source or as a sink of heat.   It will therefore be desirable to choose them so that the weight will undergo no thermal change and the reservoir will do no work during the processes we are about to consider.

If the weight is allowed to fall and by some frictional process gives up a part of its energy to the reservoir in the form of heat, the process is an irreversible one and, without the intervention of some external agency, the reverse process, whereby the weight would be lifted again at the expense of the heat of the reservoir, is impossible.

As the weight is gradually lowered, its potential energy being transferred into thermal energy in the reservoir, we might measure the extent of this irreversible process by a pointer and scale attached to the weight to indicate its height or by the amount of heat given to the reservoir. We shall in fact take as the measure of the extent of this standard universal process a quantity which is proportional to the energy exchange, but not equal to it, for it is necessary to our purpose to consider also the *temperature* of the reservoir.

To make this clear we may consider a weight and two separate reservoirs, one at the temperature T_h and one at the lower temperature T_c. If the weight is lowered and a certain amount of heat is given to the reservoir at T_h, and if then this same amount of heat is allowed to flow to the other reservoir at T_c, this latter is also an irreversible process.   The net result is the same as if the heat developed by the falling of the weight were given at once to the cold reservoir at T_c.   Now the sum of the degradation in two successive irreversible processes must be greater than that in either one alone; otherwise our definition would not be quantitative.   Therefore, if we are to have a genuine scale of irreversibility, the transfer of energy from the weight to the hot reservoir at T_h must be regarded as a less irreversible process than the transfer of the same amount of energy from the weight to the cold reservoir at T_c.

It will therefore be expedient to define the extent of irreversibility of our standard process by making it equal, not to q, but to q/θ, where q is the heat received by the reservoir and θ is some quantity which qualitatively satisfies our definition of temperature. Moreover, when the function θ is determined, it completes the quantitative definition of the degree of degradation. We are going to prove later in this chapter that θ may be completely identified with the absolute temperature which we have already defined by means of the perfect gas. Lord Kelvin called θ the thermodynamic temperature, and it is interesting to note that it provides a temperature scale which is entirely independent of the properties of any single substance or class of substances.

The Entropy of the Weight-Reservoir System. So far we have not given a name to our measure of the irreversibility of the standard process. The value of q/θ, when this process occurs, we shall call the increase in *entropy* of the weight-reservoir system. If we denote its entropy at the beginning by S_A and at the end by S_B, we write as our definition,

$$S_B - S_A = \frac{q}{\theta} \quad \checkmark \qquad (7\text{-}1)$$

Our present definition of entropy will be found identical with the definition originally given by Clausius. We have, however, departed radically from the traditional method of presenting this idea, for we have desired to emphasize the fact that the concept of entropy, as a quantity which is always increasing or is being continuously produced in all natural phenomena, is based upon our recognition of the unidirectional flow of all systems toward the final state of equilibrium. In the ordinary definition of entropy the attention is focused upon the reversible process and not upon the irreversible process, the existence of which necessitates the entropy concept. For this reason we have based our definition immediately upon an irreversible process and shall now employ the reversible process only as a means of comparing the degree of degradation, or the extent of production of entropy, during two irreversible processes.

Comparison of Any Irreversible Process with the Standard Irreversible Process. If, in any system, an irreversible process occurs, it is possible, with sufficient ingenuity, to devise a mechanism by which in actuality, or at least in thought, every part of the system may be restored to its original condition at the expense of a degradation in the standard system.

For example, let the system in question be a mixture of oxygen and hydrogen at the temperature of the standard reservoir, and let the irreversible process consist in the combination of these elements to form water. Then by means of an electric generator, operated by the falling weight, the water can be dissociated by electrolysis, and the various parts of the system can be brought to their original temperature by contact with the large standard reservoir. At the end only the weight-reservoir system has suffered degradation.

Of all methods of restoring to its original condition a system in which some process has occurred, there must be at least one which produces the smallest change in the weight-reservoir. Such a method will be one which consists in a reversible process and therefore causes no *further* degradation. In a reversible process there is no entropy production, and thus there is no change in total entropy. Entropy transfers may take place during a reversible process, but no entropy production will result. A cycle carried out by means of reversible processes can restore all the systems involved to their original states.

When a process has occurred in any system, we may define the increase in entropy of that system as equal to the minimum increase in entropy of the weight-reservoir system necessary to restore the system to its original state. In other words, it is the increase in entropy of the weight-reservoir system when the restoration occurs reversibly, since the entropy transferred from the system is equal to the entropy increase of the weight-reservoir.

The Entropy Change in the Free Expansion of a Perfect Gas. As an illustration of the method of calculating the increase of entropy in a simple irreversible process, let us consider a perfect gas enclosed in a flask of volume V_A. Let this flask be attached by a stopcock to an evacuated flask, such that the volume of the two flasks together is V_B. If these flasks are isolated from other systems and the stopcock is opened, the gas will distribute itself uniformly between them. Since the flasks are isolated, the energy does not change during expansion, and since by our previous definition of a perfect gas, the temperature and the internal energy uniquely determine one another, the temperature is the same after the expansion as before.

In order to measure the production of entropy during this process, we shall now restore the system to its initial state by means of a standard weight and a standard reservoir of the same temperature as the gas, namely, T. By keeping the gas in thermal contact with the reservoir it may be compressed isothermally by means of the weight. The work done by the weight and the heat absorbed by the reservoir, if the compression is carried on in a reversible manner, are given by Eq. (5-6)*,

$$q = -w = nRT \ln \frac{V_B}{V_A}$$

Now q/θ is the increase in the entropy of the weight-reservoir system during the restoration step, and $-q/\theta$ must be the increase of entropy of the gas during this reversible step. During the free expansion of the gas, the entropy of the gas must have increased by q/θ. If we write S_A as the entropy of the gas before expansion and S_B as the entropy of the gas after expansion, we find

$$S_B - S_A = nR \left(\frac{T}{\theta}\right) \ln \frac{V_B}{V_A} \qquad (7\text{-}2)^*$$

Table 7-1 reviews the entropy changes in the successive steps that we have been considering. We see that there has been an over-all entropy production of $nR(T/\theta) \ln (V_B/V_A)$. As the gas has been returned to its initial state, there is no net entropy change in the gas and the total entropy increase is found in the standard reservoir. As no net entropy increase took place during the reversible compression when entropy was merely transferred from the gas to the reservoir, the irreversible process that produced entropy took place during the free expansion of the gas.

TABLE 7-1. ENTROPY CHANGES DURING FREE EXPANSION OF A PERFECT GAS

	ΔS of gas	ΔS of weight-reservoir system	Entropy production; total ΔS
Irreversible expansion of perfect gas into vacuum..	$+nR\left(\frac{T}{\theta}\right)\ln\frac{V_B}{V_A}$	0	$+nR\left(\frac{T}{\theta}\right)\ln\frac{V_B}{V_A}$
Reversible compression.....	$-nR\left(\frac{T}{\theta}\right)\ln\frac{V_B}{V_A}$	$+nR\left(\frac{T}{\theta}\right)\ln\frac{V_B}{V_A}$	0
Total for both steps........	0	$+nR\left(\frac{T}{\theta}\right)\ln\frac{V_B}{V_A}$	$+nR\left(\frac{T}{\theta}\right)\ln\frac{V_B}{V_A}$

Entropy as an Extensive Property. In expressing the entropy change during an irreversible process as the difference between the entropy at the end and the entropy at the beginning, we have implied that entropy is a *property* and therefore that the entropy change depends solely upon the initial and final states. Indeed this follows directly from our definition, for, by whatever irreversible path we proceed from state A to state B, the minimum degradation of the weight-reservoir system necessary for the return from state B to state A is the same. It is true that we have not shown how to obtain the absolute value of S_B or S_A, but only their difference, nor shall we need to discuss this question until a later chapter. In the meantime we shall regard the entropy, like the energy and heat content, as a quantity of which the absolute magnitude is undetermined.

Moreover, entropy is an extensive property, for we may consider two systems which are just alike and each of which undergoes the same infinitesimal irreversible process; evidently the change in the standard weight-reservoir system necessary for their restoration is twice as great as it would be for one of them alone.

Since entropy is extensive, we may regard the entropy of a system as equal to the sum of the entropies of its parts. It is therefore important to ascertain how to determine the localization of entropies in the various parts of a system. Owing to the special properties of the standard weight-reservoir system which we assumed at the outset, it will be convenient to postulate that in any operation of the weight-reservoir system the entropy changes occur in the reservoir alone, so that, if the standard reservoir gains heat from any source by the amount q, the reservoir changes in entropy by q/θ. ✓

An Important Criterion for Reversible Processes. We have seen that the total entropy change in a reversible process is zero. For such a process entropy is merely transferred with no entropy production. It follows that in such a process the entropy change in any system must be equal and opposite in sign to the entropy change in all other systems involved. In order to study this case further, let us consider the energy changes which occur in a reversible process between some system and the standard weight-reservoir system. For the sake of simplicity, we shall choose an infinitesimal process. Letting S denote the entropy of the system and S_{st} that of the standard weight-reservoir system, the condition of reversibility, since the total entropy must remain constant, is

$$dS = -dS_{st} \qquad (7\text{-}3)$$

Bearing in mind the condition that the process is to be reversible, it is possible to conduct it so that the system and the standard weight merely exchange mechanical energy and the system and the reservoir merely exchange heat. There must, moreover, be a state of balance between the mechanical forces exerted by the system and by the weight; and the temperature of the system and the reservoir must not differ more than infinitesimally from one another. The total energy gained by the system is equal to the total energy lost by the weight-reservoir system, and, owing to the state of balance, the work done by the weight must equal the work done upon the system. Therefore, by the conservation law, the heat lost by the system is equal to the heat gained by the reservoir.

Algebraically speaking, if the heats absorbed by system and reservoir are δq and δq_{st}, then $\delta q = -\delta q_{st}$ or, since the temperatures are the same, $\delta q/\theta = -\delta q_{st}/\theta$. But by definition of the standard weight-reservoir system, $\delta q_{st}/\theta = dS_{st}$; and therefore, by Eq. (7-3),

$$dS = \frac{\delta q_{\text{rev}}}{\theta} \qquad (7\text{-}4)$$

where we have written δq_{rev} to emphasize that it is heat absorbed reversibly which determines the entropy change of the system.

The Thermodynamic Temperature. At this point let us determine the relationship between the thermodynamic temperature θ and that of the perfect gas T. We have shown that entropy is a property; i.e., its change between a given pair of states is independent of the path followed. Thus the differential of the entropy must be a perfect differential and must satisfy Eq. (3-8), which requires for an expression such as

$$dZ = L\,dx + M\,dy$$

that the functions L and M fulfill the condition

$$\frac{\partial L}{\partial y} = \frac{\partial M}{\partial x} \qquad (3\text{-}8)$$

The properties of a perfect gas were defined by Eqs. (5-3)* and (5-4)*. The heat absorbed in a given change is

$$\delta q = dE - \delta w \qquad (4\text{-}5)$$

and for reversible expansion of a perfect gas this becomes

$$\delta q_{\text{rev}} = C_V\,dT + \frac{nRT}{V}\,dV \qquad (7\text{-}5)^*$$

Although we know that δq is not a perfect differential, it is interesting to verify the point. The energy of a perfect gas depends only on the temperature and not on the volume; hence

$$\frac{\partial C_V}{\partial V} = \frac{\partial^2 E}{\partial V\,\partial T} = 0$$

But
$$\frac{\partial}{\partial T}\left(\frac{nRT}{V}\right) = \frac{nR}{V} \neq 0$$

which confirms that δq_{rev} is not a perfect differential. Division of δq_{rev} by θ yields the differential of the entropy of a perfect gas,

$$dS = \frac{C_V}{\theta}\,dT + \frac{nRT}{V\theta}\,dV \qquad (7\text{-}6)^*$$

The criterion of a perfect differential gives

$$\frac{\partial}{\partial V}\left(\frac{C_V}{\theta}\right) = \frac{\partial}{\partial T}\left(\frac{nRT}{V\theta}\right) \qquad (7\text{-}7)^*$$

But we have already indicated that the left side is zero because the energy of a perfect gas does not depend on the volume. Also n, R, and V have no temperature derivative; consequently,

$$\left(\frac{nR}{V}\right)\left[\frac{\partial}{\partial T}\left(\frac{T}{\theta}\right)\right] = 0 \qquad (7\text{-}8)^*$$

Since θ and T both have the qualitative properties of a temperature, their ratio must be constant to satisfy Eq. (7-8)*. But the size of the degree is arbitrary; consequently, we may define this ratio to be unity and make the thermodynamic temperature identical to the perfect-gas temperature.

Now Eq. (7-2)* for the entropy change on isothermal expansion of a perfect gas may be simplified,

$$S_B - S_A = nR \ln \frac{V_B}{V_A} = nR \ln \frac{P_A}{P_B} \qquad (7\text{-}9)^*$$

General Equation for Entropy. Thus we write for the entropy

$$dS = \frac{\delta q_{rev}}{T} \qquad (7\text{-}10)$$

From this important equation, we conclude that in any *reversible* process the increase in entropy of any system, or part of a system, is equal to the heat which it absorbs, divided by the absolute temperature. Indeed it is this fundamental equation which Clausius used for his original definition of entropy. We note that entropy has the same dimensions as heat capacity and may be expressed in calories per degree. This result does not depend in any way upon the specific weight-reservoir system that we have used for measuring the degree of degradation of the system under examination. It is clear that any source of mechanical and thermal energies that can operate reversibly upon the system to return it to its initial state would yield the same results.

A more general equation is

$$dS = \frac{\delta q}{T} + dS_{irr} \qquad (7\text{-}11)$$

where dS, the increase in entropy of the system, is given by $\delta q / T$, the entropy transferred from the surroundings, plus dS_{irr}, the entropy produced as a result of irreversible processes within the system. For a reversible process $dS_{irr} = 0$, and Eq. (7-11) reduces to (7-10).

Let us consider the entropy increase dS_{irr} caused by a few simple irreversible processes. These examples will illustrate the possible use of Eq. (7-11) with this term included.

Consider first the flow of an amount of heat δq from a body at temperature T_1 to a body at a lower temperature T_2. If we imagine reversible transfer of heat at each temperature, the two entropy terms are readily computed and the net entropy increase is

$$dS_{irr} = \delta q \left(\frac{1}{T_2} - \frac{1}{T_1} \right) = \frac{\delta q(T_1 - T_2)}{T_1 T_2} \qquad (7\text{-}12)$$

The result for the degradation of some form of work to heat follows directly from our discussion of the weight and heat reservoir. If an amount of work δw is degraded to heat at temperature T, the increase in entropy is

$$dS_{\text{irr}} = \frac{\delta w}{T} \qquad (7\text{-}13)$$

As a third example let us take the free expansion of n moles of ideal gas from volume V to volume $V + dV$. Let us first obtain the result from Eq. (7-13) by noting the possible work $P\,dV$ which was degraded instead to heat.

$$dS_{\text{irr}} = \frac{P\,dV}{T} = nR\,d\ln V \qquad (7\text{-}14)^*$$

We may check this result by comparison with Eq. (7-9)*. The free expansion of a perfect gas produces no temperature change in the gas. We assume no heat transfer to or from the surroundings, and hence the only entropy change is the entropy increase of the gas itself in an isothermal expansion. Integration of Eq. (7-14)* should then lead to Eq. (7-9)*, as indeed it does.

In summary we may note that entropy has been defined such that in any irreversible process the total entropy of all systems concerned is increased, whereas in a reversible process the total increase in entropy of all systems is zero. The entropy increase of a system undergoing an irreversible process is the sum of entropy transferred from other systems and entropy produced by the irreversible processes within the system. For a reversible process, which involves only entropy transfers, the increase in the entropy of any individual system, or part of a system, is equal to the heat which it absorbs divided by its absolute temperature. The definition of entropy together with the definition of the thermodynamic temperature scale makes entropy an extensive property of a system. It is important to see clearly that the idea of entropy is necessitated by the existence of irreversible processes; it is only for the purpose of convenient measurement of entropy changes that we have discussed reversible processes here.

PROBLEMS

7-1. Calculate the net entropy increase of all parts of the system, which measures the total irreversibility, when 1 kg of ice of 0°C drops 1 m into an ice-water bath.

7-2. A galvanic cell thermostated at 20°C which produces a reversible emf of 1.07 volts is short-circuited until 1000 coulombs of electricity has passed. What is the net entropy increase of the cell and thermostat if it is assumed that the chemical changes in the cell are the same as would have occurred on reversible discharge by the same amount? What additional information would be needed to divide this entropy increase between the cell and the thermostat?

7-3. A flask initially containing benzene at its freezing point of 5.5°C is brought into contact with an ice-water bath until 1 mole of benzene has frozen. The heat of fusion of benzene is 30.3 cal/g. Calculate the decrease in the entropy of the benzene; the net increase for the combined system.

8

ENTROPY AND PROBABILITY

The second law of thermodynamics not only is a principle of wide-reaching scope and application, but also is one which has never failed to satisfy the severest test of experiment. The numerous quantitative relations derived from this law have been subjected to more and more accurate experimental investigation without detection of the slightest inaccuracy. Nevertheless, if we submit the second law to a rigorous logical test, we are forced to admit that, *as it is ordinarily stated*, it cannot be universally true.

It was Maxwell who first showed the consequences of admitting the possible existence of a being who could observe and discriminate between the individual molecules. This creature, usually known as Maxwell's demon, was supposed to stand at the gateway between two enclosures containing the same gas at the same original temperature. If now he were able, by opening and shutting the gate at will, to permit only rapidly moving molecules to enter one enclosure and only slowly moving molecules to enter the other, the ultimate result would be that the temperature would increase in one enclosure and would decrease in the other. Or, again, we could assume the enclosures filled with air and the demon operating the gate to permit only oxygen molecules to pass in one direction and only nitrogen molecules in the other, so that ultimately the oxygen and nitrogen would be completely separated. Each of these changes is in a direction opposite to that in which a change normally occurs, and each is therefore associated with a *diminution* in entropy.

Of course even in this hypothetical case one might maintain the law of entropy increase by asserting an increase of entropy within the demon, more than sufficient to compensate for the decrease in question. Before conceding this point it might be well to know something more of the demon's metabolism.

While in Maxwell's time it seemed necessary to ascribe demoniacal powers to a being capable of observing molecular motions, we now recognize that the Brownian movement, which is readily observable under the microscope, is in reality thermal motion of large molecules. It would therefore seem possible, by an extraordinarily delicate mechanism in the hands of a careful experimenter, to obtain minute departures from the second law, as ordinarily stated.   But here also we should depend upon a conscious choice exercised by the experimenter.

It would carry us altogether too far from our subject to discuss in detail the degree to which vital processes can be explained by the chemistry and physics of inanimate matter.   Many chemical reactions involved in metabolism, photosynthesis, and other biological processes have been shown to follow the principles of physical chemistry, including the second law of thermodynamics.   However, it is impossible at this point to be sure whether or not other reactions may occur or have at some time occurred in living systems that involve a decrease in entropy.

It is particularly interesting to inquire how living organisms came to have optically active substances.   One can understand means entirely consistent with the second law which will perpetuate optical activity, but it is difficult to see how optical activity could originate without some fluctuation involving a decrease in entropy.

Even if no decrease in entropy is ever found to have occurred in living systems, we must consider the implications of the concepts of statistical probability before we are in a position to understand and to make a final statement of the second law.

Chance.   Sometimes when a phenomenon is so complex as to elude direct analysis, whether it concern the life and death of a human being or the toss of a coin, it is possible to apply methods which are called statistical.   Thus tables and formulas have been developed for predicting human mortality and for predicting the results of various games of chance, and such methods are applied with the highest degree of success.   It is true that in a given community the "expectation of life" may be largely and permanently increased by sanitary improvements, but if a great many individual cases be taken promiscuously from different localities at different times, the mean duration of life, or the average deviation from this mean, becomes more and more nearly constant the greater the number of cases so chosen.

Likewise it is conceivable that a person might become so expert in tossing a coin as to bring heads or tails at will, but if we eliminate the possibility of conscious choice on the part of the player, the ratio of heads to tails approaches a constant value as the number of throws increases. If the two sides of the coin are mechanically alike, and if a number of players are chosen, sufficient to eliminate the effect of habits formed by individual players, this constant ratio becomes equal to unity.   We then say that the chance of turning a head in a random throw is $\frac{1}{2}$.

Now it is characteristic of such a mathematical chance that, if we know the chance of each of several independent events, the chance that all will occur together is the product of the individual chances. Thus, if a coin be thrown 3 times, the chance that it will be a head every time is $(\frac{1}{2})^3 = \frac{1}{8}$. Similarly, if 3 coins are tossed simultaneously, the chance that 3 heads will appear is $\frac{1}{8}$. When we say that the chance of a head in a single throw is $\frac{1}{2}$, we do not mean that 5 heads will necessarily appear in 10 throws; this will be precisely true on the *average*. As the number of throws increases, the chance that the ratio of heads to tails will differ by any specified amount from unity will approach zero, or, in other words, the chance that the ratio will lie within the set limits approaches unity. In the theory of probability a chance equal to unity represents complete certainty.

Let us consider two equal boxes joined by an opening and 3 white and 3 black balls placed within and shaken in so random a manner that any one ball is as often in one box as in the other. Then we may say that the chance of finding a specified ball in box A is $\frac{1}{2}$. What is the chance that after an indefinite shaking we shall find the 3 black balls in box A and the 3 white balls in box B? The chance of each ball being in a specified box is $\frac{1}{2}$; therefore the chance of finding the given arrangement is $(\frac{1}{2})^6$. This system suggests an analogy to a physical system containing molecules of two different gases, each molecule being driven hither and thither in a random manner, i.e., in a manner so complicated as to elude analysis.

An even simpler case is one in which we consider N_0 identical balls shaken in the boxes in a random way, as before. The chance that all the N_0 balls will be in a specified box at a given time is $(\frac{1}{2})^{N_0}$. Likewise, if in a pair of similar flasks connected by a stopcock we have N_0 molecules of a certain gas, then if the stopcock is closed at a certain instant the chance that all the molecules will be in one specified flask is $(\frac{1}{2})^{N_0}$. Thus if $N_0 = 20$, the chance in question is about 1 in a million, and this chance obviously diminishes enormously as we proceed to the large number of molecules such as we deal with in practice. The most recent determinations of the number of molecules in 1 mole give 6.02×10^{23}; in dealing with numbers so vast, the laws of chance lead inexorably to results of an accuracy far exceeding that which is possible even in the most refined physical measurements.

Thus, in the case before us, if 1 mole of gas is distributed between the two flasks, the very randomness of the molecular motions makes it logically certain that minute temporary changes in concentration will from time to time occur. Nevertheless the relative deviations from complete uniformity of distribution between the two flasks must be so exceedingly small that it seems inconceivable that they could ever be detected experimentally. In other words, the chance that, within the limits of accuracy of our own observation, the gas will be equally distributed between the two flasks is, to all intents and purposes, unity. Expressing this mathematical chance or probability by the symbol $\mathcal{P}$, we can write, as a very close approximation, $\mathcal{P} = 1$. On the other hand we have found that the probability of finding all the molecules in one flask is almost zero, namely, $\mathcal{P} = (\frac{1}{2})^{N_0}$, where N_0 is now the number of molecules in 1 mole.

When therefore the gas is at first enclosed in one of the flasks and the stopcock is then opened to allow it to distribute itself between the two flasks, it is legitimate to say that immediately after opening the cock the system passes from a state of very small probability to a state of very large probability, i.e., from $\mathcal{P}_A = (\frac{1}{2})^{N_0}$ to $\mathcal{P}_B = 1$.

In order to obtain a relation which we are about to exhibit, let us in a purely arbitrary manner define a new quantity σ by the equation

$$\sigma = \frac{R}{N_0} \ln \mathcal{P} \qquad \sigma_B - \sigma_A = \frac{R}{N_0} \ln \frac{\mathcal{P}_B}{\mathcal{P}_A} \tag{8-1}$$

where R is the gas constant. Using the above values,

$$\sigma_B - \sigma_A = \frac{R}{N_0} \ln 2^{N_0} = R \ln 2$$

If, instead of using two flasks of equal size, we had allowed the mole of gas to expand from any volume v_A to any other volume v_B, we should have found by precisely similar reasoning

$$\frac{\mathcal{P}_B}{\mathcal{P}_A} = \left(\frac{v_B}{v_A}\right)^{N_0}$$

and
$$\sigma_B - \sigma_A = R \ln \frac{v_B}{v_A} \tag{8-2}$$

This equation is of very great interest since we have obtained in Eq. (8-2) an identical expression for the change in entropy in the expansion of an ideal gas, namely,

$$s_B - s_A = R \ln \frac{v_B}{v_A}$$

Hence in this simple case we find a very simple relation between the entropy and the logarithm of the probability, namely,

$$s_B - s_A = \frac{R}{N_0} (\ln \mathcal{P}_B - \ln \mathcal{P}_A) \tag{8-3}$$

Next let us consider two identical bodies in contact with the same thermostat. If we admit that each individual molecule is sometimes losing and sometimes gaining energy in a purely random manner, there will be in general a difference between the energies of the two bodies, and it is conceivable that by waiting a very long time we might, for an instant, find the system in a state in which this difference would be perceptible.

However, the chance of finding a measurably uneven distribution at any given instant would be exceedingly small.   If, therefore, two bodies at different temperatures are brought into thermal contact, the probability of their maintaining a measurable difference in temperature is entirely negligible.   Indeed we know by experience that in such a case energy always flows from the hot body to the cold until no sensible difference remains.   Again we may look upon this irreversible process as a change in the whole system from a highly improbable to a highly probable condition.

As a further illustration, let us consider a body which is in motion.   In addition to the chaotic motion of its molecules, which depends upon the temperature, each molecule will on the average have a component of velocity equal in direction and magnitude to the velocity of the body as a whole.   In the absence of friction this state of motion will continue indefinitely, but if through frictional processes opportunity is given to the individual molecules to acquire perfectly random motion, the body as a whole will soon come to rest and the average molecular velocity in one direction will be just as great as in another.   In this spontaneous irreversible process the energy of translational motion is converted into the energy of chaotic motion, which we call heat.   The chance of a certain velocity on the part of one molecule will be just as great in one direction as in another, but the chance that all molecules will happen to acquire a component velocity in a single direction so as to set the whole body in motion as before is extremely minute.

Indeed, if a body, initially at rest, is in contact with a reservoir of heat, it must from time to time, owing to the very randomness of the motions which are imparted to its molecules by the molecules of the reservoir, acquire a minute velocity as a whole, first in one direction and then in another.   If the number of molecules in the body is very large, such effects are entirely imperceptible, but in bodies of microscopic, and especially of ultramicroscopic, size these random jostlings become discernible and give rise to the phenomenon known as the Brownian movement.

The distinction between the energy of ordered motion and the energy of unordered motion is precisely the distinction which we have already attempted to make between energy classified as work and energy classified as heat.   Our present view of the relation between entropy and probability we owe largely to the work of Boltzmann, who, however, himself ascribed the fundamental idea to Gibbs, quoting,[1] "The impossibility of an uncompensated decrease of entropy seems to be reduced to an improbability."

It would carry us too far if we should attempt to analyze more fully this idea that the increase in the entropy of a system through processes

[1] L. Boltzmann, "Vorlesungen über Gastheorie," Johann Ambrosius Barth Verlag, Leipzig, 1912.

of degradation merely means a constant change to states of higher and higher probability.   The mere recognition that such a relationship exists suffices to give a new and larger conception of the meaning of an irreversible process and the significance of the second law of thermodynamics.

If we regard every irreversible process as one in which the system is seeking a condition of higher probability, we cannot say that it is inevitable that the system will pass from a certain state to a certain other state. If the system is one involving a few molecules, we can only assert that on the average certain things will happen.   But as we consider systems containing more and more molecules, we come nearer and nearer to complete certainty that a system left to itself will approach a condition of unit probability with respect to the various processes which are possible in that system.   This final condition is the one which we know as equilibrium.   In other words, the system approaches a thermodynamic or macroscopic state, which represents a great group of microscopic states that are not experimentally distinguishable from one another.   With an infinite number of molecules, or with any number of molecules taken at an infinite number of different times, the probability that the macroscopic state of the system will lie within this group is infinitely greater than the probability that it will lie outside of that group.

Leaving out of consideration systems, if such there be, which possess that element of selection or choice that may be a characteristic of animate things, we are now in a position to state the second law of thermodynamics in its most general form: *Every system which is left to itself will, on the average, change toward a condition of maximum probability.*   This law, which is true for *average* changes in any system, is also true for *any* changes in a system of many molecules.

We have thought it advisable to present in an elementary way the ideas touched upon in this chapter in order to give a more vivid picture of the nature of an irreversible process and a deeper insight into the meaning of entropy.   The formal application of statistical methods to physicochemical problems has yielded many valuable results, but these constitute the sister science of statistical mechanics or statistical thermodynamics.   We shall continue our main development in terms of macroscopic properties of matter without formal use of the entropy-probability relationship.   Nevertheless, we shall always tacitly assume that we are dealing with statistical ideas.   Also, we shall make use of statistical methods to calculate thermodynamic properties in some cases where the properties of the individual molecules are known in sufficient detail.

The statistical viewpoint is important when one is interpreting extremely small solubilities or vapor pressures.   Indeed figures are sometimes obtained which are so small as to seem ridiculous to the uninitiated. And yet such figures, when properly interpreted, have as definite a significance, and often as high an accuracy, as others which are capable of direct measurement.   Thus by several methods it has been shown that if 0.004 mole of silver cyanide is dissolved in 1 liter of 3 M potassium cyanide the concentration of silver ion is about 16×10^{-23} mole/liter.   This means that the number of actual molecules of silver ion is about 100 per liter, or one-tenth of a molecule per cubic centimeter.   How then, since the molecules are assumed indivisible, can we say that in such a cubic centimeter there is any concentration of silver ion at all?   We mean simply that, while the complex ions are being dissociated and others are being formed, there will be at any one instant 100 molecules of silver ion per liter *on the average*, or the chance of there being 1 in any given cubic centimeter is 1 in 10.

We shall show in the course of our work that if a suitable catalyzer is placed in saturated water vapor there is at every temperature some dissociation into hydrogen and oxygen and that at 25°C the partial pressure of the hydrogen is 2.50×10^{-28} atm, which is equivalent to the pressure exerted by a single molecule in a space of about 1 million liters.   Yet this value has a precise significance and is certainly known within a few per cent.

One of the most striking results of this character is obtained if we calculate the vapor pressure of tungsten at 100° from experiments at very high temperatures.   The result, 10^{-105} atm, would mean that the concentration of tungsten vapor would be less than one molecule in a space equivalent to the known sidereal universe.   Such a calculation need not alarm us.   Allowing for the possibilities of experimental uncertainty, we may utilize such a calculated vapor pressure in our thermodynamic work with the same sense of security as we use the vapor pressure of water.

9

HEAT ENGINES, HEAT PUMPS, CYCLES, AND FLOW PROCESSES

The Heat Engine. Any system which is not in equilibrium can be made to do useful work. However, in every irreversible process there is always some waste of opportunity in this regard. Let us consider the flow of heat between two reservoirs at different temperatures. Instead of allowing heat to flow directly from one to the other, we may obtain work by means of a steam engine, or a hot-air engine, or any one of the various inventions which are known generically as heat engines. These are characterized by operating in such manner that they themselves undergo no permanent change but do work at the expense of a part of the energy taken from a hot reservoir, while the rest of the energy passes into a cold reservoir. The operation of the heat engine is achieved by subjecting the working substance of the engine to a sequence of processes which form a cycle.

The ratio between the work done and the heat taken from the hot reservoir is known as the conversion factor, or the thermal efficiency, of the engine. The problem of determining the maximum value of this ratio is the one which occupied Carnot[1] in the great monograph which laid the foundations of the second law of thermodynamics.

Every actual heat engine is inefficient because of friction or imperfect design; but even if all sources of degradation are eliminated, it is evident that no engine could be constructed to give a conversion factor of 100 per cent. For if all the heat taken from the hot reservoir were converted into work, the cold reservoir might be removed altogether. The energy taken from the hot reservoir would then be converted into mechanical work without any degradation in other systems. This work could, by friction, be returned as heat to the reservoir, and we should thus find an irreversible process bringing the whole system back to its original state, which would violate the second law of thermodynamics.

[1] S. Carnot, "Reflections sur la puissance motrice du feu," Chez Bachelier Libraire, Paris, 1824.

In order to obtain the maximum possible work from a heat engine, it would be necessary to eliminate friction, to prevent direct flow of heat from hot to cold portions of the system, and to maintain a state of balance with respect to the mechanical forces.   In other words, the process must be reversible.   Under given conditions, therefore, the maximum conversion factor is that of a heat engine which operates reversibly in all its stages, and if we find the thermal efficiency of such an engine, we know the limit which may be approached by any actual engine as its design and construction are improved.

If a heat engine operates reversibly and passes through a whole number of complete cycles, so that it is in the same state at the end of the operation as at the beginning, it will itself suffer no change of entropy.   Hence all the entropy changes are in the rest of the system, and these must sum up to zero in a reversible process.   These entropy changes are immediately obtained from Eq. (7-10).   If q_h is the heat *taken from* the hot reservoir at T_h and q_c is the heat *given* to the cold reservoir at T_c, then the increase of entropy of the hot reservoir is $-q_h/T_h$, and that of the cold reservoir is q_c/T_c.   Equating the sum to zero,

$$- \frac{q_h}{T_h} + \frac{q_c}{T_c} = 0 \tag{9-1}$$

By the conservation law, the net work done by the engine, w, is given by

$$w = q_h - q_c \tag{9-2}$$

and combining these two equations, we find

$$\frac{w}{q_h} = \frac{T_h - T_c}{T_h} \tag{9-3}$$

This important equation gives the conversion factor, or thermal efficiency, of a perfectly efficient engine operating between any two temperatures T_h and T_c.   Any actual engine operating between these temperatures has a lower efficiency, but one which may approach Eq. (9-3) as a limit.   Thus for a steam engine with a condenser at 27°C or 300°K and with a boiler at 327°C or 600°K the maximum work obtainable is equal to one-half the heat taken from the boiler.

The reader should note carefully the method used to obtain Eq. (9-3) because it is applicable to all types of more complex heat-engine and heat-pump problems.   In the first step all entropy changes are calculated and their sum set to zero, in accordance with the second law, since the process is assumed to be reversible.   In the second step all energy changes are expressed, and their sum also is set to zero by the first law.   The work term does not appear in the first equation, since it involves no entropy, but it does appear in the second equation.   These two equations are then solved for the unknown quantities.

The Heat Pump.   By reversing a heat engine it is possible, through the expenditure of work, to transport heat from a cold to a hot reservoir. This is the method employed in a refrigerating machine. The performance of a heat pump for refrigerating purposes is expressed as the coefficient of performance, q_c/w, where w is now the net work done on the working material and q_c is the heat withdrawn from the cold reservoir.

$$\frac{q_c}{w} = \frac{T_c}{T_h - T_c} \tag{9-4}$$

Let us calculate, as an example, the minimum amount of work required to convert 1 kg of water at 0°C into ice at 0°C by an engine operating in a room at 30°C.   In the operation heat will be given up to the room in an amount equal to the heat absorbed from the water, together with the work done on the working material.   If the heat pump is reversible, we employ Eq. (9-4).   The coefficient of performance is $273/30$, q_c is 79,710 cal, and w is $79,710 \times 30/273 = 8760$ cal.   (It will be observed that this is not the work required to convert 1 kg of water at 30° into ice at 0°.   A problem of this more complex type will be considered in Chapter 11; see Problem 11-3.)

The theory of the heat engine leads to an interesting consequence which possesses theoretical, and is beginning to acquire practical, interest. In some localities buildings are heated by electricity, the electricity passing through some form of heater whose resistance converts electrical into thermal energy.   At first sight it would appear, from the law of conservation of energy, that the maximum heating effect would be produced when a certain amount of electrical energy is completely converted into thermal energy.   But this is very far from the truth.

If a heat pump were constructed with the inside of the building serving as the hot reservoir and the outdoor air as the cold reservoir, and if by means of a motor the electrical energy were used to operate this pump so that the heat would be taken from without and given up inside the building, the amount of heating thus produced would, in the limiting case of ideal efficiency, be given by Eq. (9-3), where w represents the electrical energy expended, q_h the heating effect in the building, and T_h and T_c the temperatures within and without.   The amount of heating thus produced per unit of electrical energy expended, or the coefficient of performance of a heat pump for heating purposes, would, in the limiting case of ideal efficiency, be given by

$$\frac{q_h}{w} = \frac{T_h}{T_h - T_c} \tag{9-5}$$

If the internal temperature T_h were 18°C or 291°K and the external temperature T_c were 0°C or 273°K, the coefficient of performance at the maximum would be $291/18$, or more than 16.

The widespread use of heat pumps for heating purposes has been handicapped by the high capital-investment cost and high maintenance costs. However, the development of cheap and dependable refrigerating equipment has now made the heat pump quite feasible for even relatively small heating applications, including individual household heating.

Electrons can be used as the working material of heat engines or heat pumps. An electric current flowing through a junction of one conductor with another absorbs or evolves heat depending on the direction of current flow. This effect, the Peltier heat, is related by thermodynamic principles to the Seebeck effect, which is the potential of the complete thermocouple (see Chapter 28). Semiconductors show larger thermoelectric effects than metals; consequently fewer junctions are needed for a given function. To date, these devices have attained only a fraction of the thermal efficiency, or coefficient of performance, that would be expected from Eqs. (9-3), (9-4), or (9-5) because of direct thermal conduction from hot to cold junctions. Nevertheless, the lack of moving parts and simplicity of design make even presently available thermoelectric devices of considerable interest for small power or refrigerating applications, particularly at remote sites.

Carnot Cycle. The Carnot cycle is a reversible cycle which has played an important historical role in the development of the concept of entropy. A Carnot cycle, which is illustrated in Fig. 9-1, consists of an isothermal expansion, an adiabatic expansion, an isothermal compression, and, finally, an adiabatic compression to return the working material to its original state. The net result of such a cycle is the transfer of heat from a high-temperature reservoir to a low-temperature reservoir with partial conversion into work. For a perfect-gas working material, one can readily calculate that the net work obtained by carrying out the Carnot cycle is equal to $RT_h \times \ln(V_2/V_1) + RT_c \ln(V_4/V_3)$. The two adiabatic processes absorb no

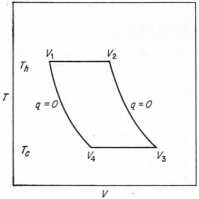

FIG. 9-1. Carnot cycle.

heat, and their work terms cancel. Thus, the heat absorbed from the hot reservoir is given by $q_h = RT_h \ln(V_2/V_1)$. From Eq. (5-8)* one can relate the temperatures and volumes for the adiabatic steps, obtaining $V_3/V_2 = (T_h/T_c)^{C_v/R}$ and $V_4/V_1 = (T_h/T_c)^{C_v/R}$. From these results we see that $V_1/V_2 = V_4/V_3$ so that the net work reduces to $R(T_h - T_c) \times \ln(V_2/V_1)$ and the work obtained divided by the heat absorbed from the hot reservoir reduces to $w/q_h = (T_h - T_c)/T_h$ in agreement with Eq. (9-3). The Carnot cycle can be taken in the reverse direction and can thus operate as a heat pump, and one can confirm the results of Eq. (9-4). Historically,[1-3] the Carnot cycle was used to develop a statement of the second law of thermodynamics and to develop the concept of entropy. Starting with the experimental observation that it is impossible to transfer heat from a low temperature to a higher temperature without the application of work by some outside system, it is possible to show that a function, entropy, as defined in Eq. (7-10) does exist and that dS is an exact differential for any substance carried through a Carnot cycle or any other reversible cycle and thus that entropy is a state function.

An alternate procedure[4-8] for the development of the second law and the concept of entropy is that of Born and Carathéodory, which does not depend upon heat engines but recognizes the relationship between the behavior of real systems and the properties of Pfaffian differential equations. In this treatment, the existence of a state function, entropy, is established by demonstrating the existence of an integrating factor which can convert the differential δq to a perfect differential dS. The integrating factor is defined as the thermodynamic temperature, which can be shown, as has been done in Chapter 7, to be identical with the perfect-gas temperature. These alternate procedures,[9] including the one presented in Chapter 7, are entirely equivalent. But these different approaches are useful for examining the concept of entropy from different points of view.

FLOW PROCESSES

The increasing importance of flow processes in recent years makes it desirable to present briefly the essential features of thermodynamics under these conditions. Examples include flow through jet and rocket engine nozzles as well as flow through all sorts of industrial equipment. The Joule-Thomson experiment is also a flow process, and in Chapter 5 it

[1] R. J. E. Clausius, "The Mechanical Theory of Heat" (trans. by Hirst), Van Voorst, London, 1867.

[2] H. Poincaré, "Thermodynamique," Carré, Paris, 1892.

[3] J. H. Keenan, "Thermodynamics," John Wiley & Sons, Inc., New York, 1941.

[4] M. Born, *Physik. Z.*, **22**, 218, 249, 282 (1921).

[5] S. Chandrasekhar, "An Introduction to the Study of Stellar Structures," University of Chicago Press, Chicago, 1939.

[6] H. A. Buckdahl, *Am. J. Phys.*, **17**, 41, 44, 212 (1949).

[7] M. Born, "Natural Philosophy of Cause and Chance," Oxford University Press, New York, 1949.

[8] R. Eisenschitz, *Sci. Progr.*, **43**, 246 (1955).

[9] M. Planck, "Treatise on Thermodynamics," 3d English ed. (trans. by A. Ogg), Longmans, Green & Co., Inc., New York, 1927 (reprinted by Dover Publications, New York, 1945).

was shown that the work P_1V_1 is done on the fluid flowing into the region of interest, while the work P_2V_2 is done by the fluid leaving. Thus the fluid in any flow system absorbs an amount of energy $-\Delta(PV)$ from the surroundings. The first-law equation may be written in the form

$$\Delta E = q + w = q - \Delta(PV) + w'$$

where w' is the total of all work absorbed except pressure-volume work associated with the flow. In view of the definition of enthalpy, this becomes

$$\Delta H = q + w' \tag{9-6}$$

and since both q and w' were zero in the Joule-Thomson experiment, ΔH is therefore zero.

In the Joule-Thomson experiment the flow velocities are low, and we have neglected the kinetic energy of flow, but in other situations this may be important. The familiar equation of mechanics gives the value, which is $mu^2/2$ per mass m of fluid, where u is the linear velocity. Also, if the fluid changes level, there is a change in gravitational potential energy. It is inconvenient to include gravitational energy on an absolute basis, but its change is $mg(h_2 - h_1)$, where h is the vertical height.

While the internal energy of a substance may be defined to include the kinetic and gravitational energies,[1] it is customary in dealing with flow problems to exclude these terms from E and to include them separately in energy-balance equations. The symbols E, H, etc., thus represent the energy content, enthalpy, etc., of the substance when stationary and at the base height. A general equation for the first law of thermodynamics as applied to a mass m in a flow system is

$$\Delta H + m\frac{\Delta u^2}{2} + mg\,\Delta h = q + w' \tag{9-7}$$

where q is the heat absorbed by the system and w' is work absorbed other than pressure-volume work.

One result which may be readily obtained from Eq. (9-7) is the mechanical work required to operate an adiabatic gas compressor. The gas flows into the compressor at a low pressure and is discharged at a higher pressure. Frequently it is a good approximation to assume that the gas is compressed adiabatically; that is, $q = 0$. Also, if Δh and Δu^2 are negligible, Eq. (9-7) yields the simple result

$$w' = \Delta H$$

While one can analyze any specific compression cycle and obtain this same result, thermodynamics yields it on a perfectly general basis without regard for the type of machine used.

---

[1] Indeed in Chapter 30 the gravitational energy is included in the energy of a substance.

Another interesting application is flow through a jet engine or rocket nozzle. In good approximation Δh, q, and w' are zero in this case, but Δu^2 is large. Equation (9-7) yields at once $\frac{1}{2} \Delta u^2 = -\Delta H/m$. Usually the gas velocity in the combustion chamber is small; hence the exit velocity u_e is given by the expression

$$\frac{1}{2} u_e^2 = -\frac{\Delta H}{m} \qquad \qquad (9\text{-}8)$$

where ΔH is the enthalpy change from the combustion chamber to the nozzle exit.

The second law is also applicable to flow systems. If no heat is transferred and irreversible effects are negligible, the condition $\Delta S = 0$ may be applied. In the next chapter the effect of temperature and pressure on the entropy of a fluid will be considered, and the isentropic pressure-temperature path may then be derived. Thus, if the initial pressure and temperature and the final pressure are fixed for an adiabatic and reversible flow process, the isentropic condition determines the final temperature. Hence the enthalpy change ΔH is determined and may be used in Eq. (9-7) or (9-8) to determine the other quantities of interest.

In some flow systems such as jet or rocket engine nozzles there may be chemical changes occurring in the fluid. If such changes maintain chemical equilibrium, then no irreversibility is introduced. Also, if a given chemical reaction does not occur at all, then it contributes no irreversible entropy increase. However, if reaction occurs but lags significantly behind equilibrium, then an irreversible entropy increase does occur and must be included in the final entropy of the fluid.

We may also obtain the Bernoulli equation, which was first discovered from reasoning based upon mechanical work long before the development of thermodynamics. We consider an incompressible fluid in frictionless adiabatic flow; hence the internal energy of the fluid is constant, and $\Delta H = V \Delta P$, while $q = 0$. Also we assume that no work is transferred except for that due to the volume of fluid flow; that is, $w' = 0$. Now Eq. (9-7) becomes

$$m \frac{\Delta u^2}{2} + mg \, \Delta h + V \, \Delta P = 0 \qquad \qquad (9\text{-}9)$$

The most interesting feature of this equation is the pressure decrease which accompanies an increase in velocity when $\Delta h = 0$.

In the chapters to follow we shall not usually consider flow processes explicitly. But it is hoped that this brief discussion will indicate the additional features which flow introduces into a problem and will assist those readers who wish to apply to flow systems the thermodynamic relations to be presented in later chapters. We note also that certain types of flow systems are discussed extensively in texts[1] directed primarily to chemical engineers and that the flow problems related to jet propulsion are considered in detail in a recent book by Penner.[2] A very general consideration of hydrodynamic phenomena in fluids is given by Hirschfelder, Curtiss, and Bird.[3]

PROBLEMS

9-1. What is the maximum conversion factor of a steam engine operating with a condenser at 30°C and a boiler at 200°C?

9-2. A refrigerating machine operating in a room at 30°C is employed to maintain a cold storage tank at −10°C. What is the minimum amount of work required to withdraw 1000 cal from the tank?

9-3. A thermally powered refrigerator receives heat from a flame at 400°C and withdraws heat from a cold box at −40°C. If the machine is reversible and all heat is discharged at 30°C, calculate the ratio of the heat required at 400°C to that withdrawn at −40°.

9-4. Assume that the total cost of operation of a heat pump is four times the theoretical power cost for perfect efficiency, whereas the cost of direct electrical heating is just the power cost. If room temperature is 27°C, at what outdoor temperature would two systems yield equal total cost?

9-5. Calculate the work required per mole to compress an ideal monatomic gas at 1 atm and 300°K to 10 atm in an adiabatic reversible flow system. Note that the temperature change on reversible adiabatic compression of an ideal gas of constant heat capacity can be obtained from Eq. (5-8)*.

9-6. What is the exit velocity of perfect monatomic helium gas which is expanding from 3000°K and 50 atm through a jet nozzle to 1 atm? Assume reversible adiabatic flow with negligible initial velocity.

✓9-7. It has been suggested that a rocket be propelled by vaporizing lithium in a stream of helium, heating the gas mixture to a temperature of 3000°K and 100 atm in a nuclear reactor, and then expanding through the rocket nozzle.

The enthalpy increase upon heating Li liquid at 1 atm from 500 to 1600°K is 5465 cal/mole. At 1600°K, $\Delta H°$ of vaporization of lithium is 35,260 cal/mole. For He at all temperatures and for Li(g) up to 2000°K, $c_P° = \frac{5}{2}R$. Owing to electronic excitation, $H°_{3000} - H°_{2000} = 5067$ cal/mole for Li(g).

Assume that Li and He gases are perfect gases, and calculate the maximum exit velocity of an equimolal mixture of Li and He if expanded adiabatically from 3000°K to an exit temperature of 500°K with all the Li exhausted as liquid drops in the He gas.

It is not known whether the rate of condensation is rapid enough to condense the lithium during the time of expansion through the nozzle. What would be the maximum exit velocity for an exit temperature of 500°K if all the lithium remained in the supercooled gaseous state?

[1] See, for example, J. M. Smith and H. C. Van Ness, "Introduction to Chemical Engineering Thermodynamics," 2d ed., McGraw-Hill Book Company, Inc., New York, 1959; or O. A. Hougen, K. M. Watson, and R. A. Ragatz, "Chemical Process Principles, II, Thermodynamics," 2d ed., John Wiley & Sons, Inc., New York, 1959.

[2] S. S. Penner, "Chemistry Problems in Jet Propulsion," Pergamon Press, New York, 1957.

[3] J. O. Hirschfelder, C. F. Curtiss, and R. B. Bird, "Molecular Theory of Gases and Liquids," chap. 11, John Wiley & Sons, Inc., New York, 1954.

10

THE RELATIONSHIP OF ENTROPY
TO OTHER STATE FUNCTIONS

When heat flows between two systems of different temperature, the process is more nearly reversible the smaller the temperature difference. (When the two temperatures differ only infinitesimally, there is, of course, a net entropy increase during thermal flow but in mathematical parlance this net entropy increase is an infinitesimal of higher order than the entropy transferred, q/T, and therefore is negligible in comparison with q/T. We then say that the process is reversible.)

If a system, with heat capacity C, absorbs heat from a reservoir of infinitesimally higher temperature and thus rises through the temperature interval dT, so that

$$\delta q = C \, dT \quad \checkmark \tag{10-1}$$

then, by Eq. (7-4),

$$dS = C \frac{dT}{T} \quad \checkmark \tag{10-2}$$

Indeed, the entropy change in the system alone is the same regardless of the temperature from which the heat flows, since the entropy is a *property*. This equation, therefore, is true for any infinitesimal rise of temperature in the system, no matter how it is produced.

Ordinarily the heating is done either at constant volume or at constant pressure. Then

$$dS = C_V \frac{dT}{T} = C_V \, d\ln T \qquad \left(\frac{\partial S}{\partial T}\right)_V = \frac{C_V}{T} \tag{10-3}$$

or
$$dS = C_P \frac{dT}{T} = C_P \, d\ln T \qquad \left(\frac{\partial S}{\partial T}\right)_P = \frac{C_P}{T} \tag{10-4}$$

We shall need presently two equations which are obtained from (10-3) and (10-4) by differentiating the former with respect to volume and the latter with respect to pressure. Remembering the definitions of C_V and

C_P, we find

$$\frac{\partial^2 S}{\partial T \, \partial V} = \frac{1}{T}\left(\frac{\partial C_V}{\partial V}\right)_T = \frac{1}{T}\frac{\partial^2 E}{\partial T \, \partial V} \tag{10-5}$$

$$\frac{\partial^2 S}{\partial T \, \partial P} = \frac{1}{T}\left(\frac{\partial C_P}{\partial P}\right)_T = \frac{1}{T}\frac{\partial^2 H}{\partial T \, \partial P} \tag{10-6}$$

(margin handwriting: $C_V = \left(\frac{\partial E}{\partial T}\right)_V$ $C_P = \left(\frac{\partial H}{\partial T}\right)_P$)

Change of Entropy with Volume and with Pressure.   We may obtain some very useful thermodynamic equations by considering a change in state brought about by a process for which the only possible work is expansion work against the external restraining pressure.   If the process is reversible, from Eq. (7-10), the heat absorbed by the system is $T \, dS$, while the work absorbed by the system is $-P \, dV$.   Thus by the first law of thermodynamics

$$dE = T \, dS - P \, dV \tag{10-7}$$

If $P_{\text{system}} > P_{ex}$, the process is irreversible.   There is entropy production, and $\delta w = -P_{ex} \, dV = -P \, dV + T \, dS_{\text{irr}}$.   Also, from Eq. (7-11), $\delta q = T \, dS - T \, dS_{\text{irr}}$, where dS_{irr} is the same as in the previous term if there is no other source of irreversibility such as incomplete chemical equilibrium within the system.   The combination of these two equations yields again Eq. (10-7), which is now seen to be general for any closed system in internal equilibrium.

In addition, similar equations can be obtained for some of the derived thermodynamic quantities.   Thus from the definition of enthalpy given by Eq. (4-8), we obtain

$$dH = dE + P \, dV + V \, dP = T \, dS + V \, dP \tag{10-8}$$

We shall point out the importance of two additional derived thermodynamic quantities, namely, the *work content* defined by the equation

$$A = E - TS \tag{10-9}$$

(margin handwriting: $dA = dE - T\,ds - s\,dT$)

and the *free energy* defined by the equation

$$F = H - TS = E + PV - TS \tag{10-10}$$

(margin handwriting: $dF = + P\,dV + V\,dP - T\,ds - s\,dT$)

From these definitions and Eq. (10-7) two additional equations are derived which are presented together with Eqs. (10-7) and (10-8).

$$\begin{cases} dE = T \, dS - P \, dV & (10\text{-}7) \\ dH = T \, dS + V \, dP & (10\text{-}8) \\ dA = -S \, dT - P \, dV & (10\text{-}11) \\ dF = -S \, dT + V \, dP & (10\text{-}12) \end{cases}$$

This set of equations is very useful in deriving relationships among the various thermodynamic quantities.   Thus, if we wish to relate the variation of entropy with volume to other thermodynamic quantities, we can take advantage of the fact that all the quantities in the above equations are state functions and therefore have perfect differentials. From the basic equation of partial differentiation, Eq. (3-1), and from Eq. (10-11), we note that

$$\left(\frac{\partial A}{\partial T}\right)_V = -S \qquad \left(\frac{\partial A}{\partial V}\right)_T = -P \qquad (10\text{-}13)$$

and from Eq. (3-7)

$$\frac{\partial^2 A}{\partial V\,\partial T} = -\left(\frac{\partial S}{\partial V}\right)_T = \frac{\partial^2 A}{\partial T\,\partial V} = -\left(\frac{\partial P}{\partial T}\right)_V$$

or

$$\left(\frac{\partial S}{\partial V}\right)_T = \left(\frac{\partial P}{\partial T}\right)_V \qquad (10\text{-}14)$$

Likewise, from Eq. (10-12), we obtain

$$\left(\frac{\partial F}{\partial T}\right)_P = -S = \frac{F-H}{T} \qquad (10\text{-}15)$$

$$\left(\frac{\partial F}{\partial P}\right)_T = V \qquad (10\text{-}16)$$

and

$$\left(\frac{\partial S}{\partial P}\right)_T = -\left(\frac{\partial V}{\partial T}\right)_P \qquad (10\text{-}17)$$

These equations allow us to calculate changes of entropy upon change of volume or pressure if the coefficients of expansion and compressibility are known.

Equations (10-7) to (10-17) deal with the relations between the thermodynamic properties of a system in internal equilibrium. We must now deal with systems undergoing spontaneous, or irreversible, changes.

Criteria of Equilibrium and Spontaneous Change for Closed System. Let us consider a problem of quite general interest. For a given change in state of a system, we wish to know whether or not there is any possible process by which this change in state can take place spontaneously; or if it will not take place spontaneously, we wish to know the amount of work that will be required to bring about the change in state. We shall apply the second law of thermodynamics, which requires that there be an over-all increase in entropy or a production of entropy during any actual process. Thus $dS_{\text{system}} + dS_{sur} = dS_{\text{irr}} \geq 0$. The sum of the entropy increases in the system and in the surroundings is equal to the entropy production, which will always be a positive quantity, but which can be zero in the limit of a reversible process.

If the boundary of the closed system be drawn to include any regions of entropy production, the entropy change of the surroundings will be due merely to entropy transfer and $dS_{\text{system}} + \delta q_{sur}/T_{sur} = dS_{\text{irr}} \geq 0$.

The subscripts *system* and *sur* are used to designate properties of the system and of the surroundings or to designate heat or work absorbed by the system or by the surroundings, respectively.   By substitution of the first-law statements, $\delta q_{sur} + \delta w_{sur} = dE_{sur}$ and $dE_{sur} + dE_{system} = 0$, one obtains $dS_{system} + (-dE_{system} - \delta w_{sur})/T_{sur} = dS_{irr}$.   Finally,

$$-\delta w_{sur} = dE - T_{sur}\, dS + T_{sur}\, dS_{irr} \qquad (10\text{-}18)$$

where one can drop the subscript system as it is clear that E and S refer to the system under discussion.   This result can be used to determine the minimum amount of work to be done by the surroundings to bring about a given change of state by setting the entropy production equal to zero. For any actual process, there will be entropy production, and a larger amount of work will have to be done.

For an isothermal process with the surroundings in thermal equilibrium with the system, $T = T_{sur}$ and $-\delta w_{sur} = \delta w$; thus

$$\delta w = dE - T\, dS + T\, dS_{irr}$$

and from the definition of A, one recognizes that for an isothermal process

$$\delta w = dA + T\, dS_{irr} \qquad (10\text{-}19)$$

For a reversible isothermal process with no entropy production, the amount of work that must be done on the system is equal to the increase in A, the work content.   For any actual isothermal process, a greater amount of work must be done.   Or if the work content decreases, the work done by the system is less than the decrease in work content, $-\Delta A$, but the maximum work that can be done by the system undergoing a given change in state is equal to $-\Delta A$ for the limit of a reversible process. Thus it is clear why A is named the *work content* and is given the symbol A after the German word *Arbeit*.

Associated with any process with a volume change, there will be work of expansion due to expansion of either the system or the surroundings. Thus, if one wishes to determine the maximum amount of electrical work, for example, that can be done by a system, one has to subtract the expansion work from the work-content decrease.   In general the expansion work done by a system is given by $P\, dV$ if the restraining resisting pressure differs only infinitesimally from the pressure of the system.

For isobaric processes $d(PV) = P\, dV$.   For processes involving transport of material from a reservoir of constant pressure P_1 to a reservoir of constant pressure P_2, the expansion work done is also $\Delta(PV)$.   We have already encountered such a process in the Joule-Thompson experiment described in Chapter 5.   We shall encounter other examples of such processes.   For an isothermal process of this type the net work other than expansion work is given by

$$\delta w' = \delta w + d(PV) = dA + d(PV) + T \, dS_{irr}$$
or $$\delta w' = dF + T \, dS_{irr} \tag{10-20}$$

Thus the minimum amount of work other than expansion work required to bring about a given isothermal change of state is given by the increase in free energy. For any actual process, more work would be required. Likewise, the maximum amount of work other than expansion work that can be done by the system is given by the decrease in free energy, and an actual process yields less work.

The chemist is often interested in processes for which no work other than expansion work is done. For such isothermal processes,

$$0 = dF + T \, dS_{irr} \tag{10-21}$$

Such a process can take place spontaneously with resultant entropy production only if the free energy decreases. If the change in state corresponds to no change in free energy, no change can take place spontaneously and the two states will be in thermodynamic equilibrium. In Chapter 13 we shall return to a more detailed examination of free energy.

Equilibrium between a Substance and Its Vapor. Let us apply Eq. (10-14) to a system composed of a substance and its vapor, the two being in equilibrium with one another at the vapor pressure p, which is moreover equal to the applied external pressure P. In this case $\partial S / \partial V$ is the same as $\Delta S / \Delta V$, where ΔS is the increase in entropy and ΔV the increase in volume when an amount of the substance vaporizes. As we are dealing with an equilibrium, and therefore with a reversible process, $\Delta S = q/T$ and q, which we may also write as ΔH, is the ordinary heat of vaporization. Moreover, since the vapor pressure does not depend upon the volume of the system, we may omit the restriction of constancy of volume and thus Eq. (10-14) becomes

$$\frac{dP}{dT} = \frac{\Delta H}{T \, \Delta V} \tag{10-22}$$

This is the famous equation which Clapeyron[1] obtained essentially in this form in 1834. It was the first physicochemical application of what we now call the second law of thermodynamics. This equation is also valid for solid-solid and solid-liquid equilibria.

Thermodynamic Equations of State. From Eq. (10-7) we obtain $(\partial E / \partial V)_T = T(\partial S / \partial V)_T - P$, and substituting Eq. (10-14) yields

$$P = T \left(\frac{\partial P}{\partial T} \right)_V - \left(\frac{\partial E}{\partial V} \right)_T \tag{10-23}$$

[1] E. Clapeyron, *J. école polytech. (Paris)*, 14(23), 153 (1834).

Likewise, from Eqs. (10-8) and (10-17),

$$V = T\left(\frac{\partial V}{\partial T}\right)_P + \left(\frac{\partial H}{\partial P}\right)_T \tag{10-24}$$

A relation between pressure, temperature, and volume is called an equation of state. If either $(\partial E/\partial V)_T$ or $(\partial H/\partial P)_T$ has been determined by experiment, we have in Eq. (10-23) or in Eq. (10-24) a thermodynamic requirement to which any empirical equation of state must conform.

These equations have proved very useful in interpreting thermometric measurements made with gas thermometers. In the case of an actual gas like hydrogen or air, the volume at constant pressure is not proportional to the temperature. If, however, we have measured the Joule-Thomson effect, we may calculate $(\partial H/\partial P)_T$ from Eq. (5-11) and make the needed correction.

Difference between Heat Capacity at Constant Pressure and at Constant Volume. By the first law of thermodynamics we obtained, in Eq. (4-14), a formula for the difference between C_P and C_V, namely,

$$C_P - C_V = \left[P + \left(\frac{\partial E}{\partial V}\right)_T\right]\left(\frac{\partial V}{\partial T}\right)_P$$

This was as far as it was possible to go with the first law of thermodynamics alone, but now we may substitute from Eq. (10-23), which was derived from the second law, and thus obtain the more useful formula

$$C_P - C_V = T\left(\frac{\partial P}{\partial T}\right)_V\left(\frac{\partial V}{\partial T}\right)_P \tag{10-25}$$

Since we do not ordinarily measure the pressure-temperature coefficient at constant volume, we may write from Eq. (3-3)

$$\left(\frac{\partial P}{\partial T}\right)_V = -\frac{(\partial V/\partial T)_P}{(\partial V/\partial P)_T} \tag{10-26}$$

Hence
$$C_P - C_V = -\frac{T(\partial V/\partial T)_P^2}{(\partial V/\partial P)_T} \tag{10-27}$$

If instead of the volume-temperature coefficient and the volume-pressure coefficient we wish to employ the so-called coefficients of thermal expansion and compressibility, $\alpha = (1/V)(\partial V/\partial T)_P$ and $\kappa = -(1/V)(\partial V/\partial P)_T$, then

$$C_P - C_V = \frac{\alpha^2 VT}{\kappa} \tag{10-28}$$

This is a valuable equation, for while most of the general laws of specific heats relate to the heat capacity at constant volume, this is rarely determined directly but must be calculated from the measured values of C_P. Some of the results of such a calculation have already been discussed in Chapter 6.

REVERSIBLE ADIABATIC CHANGES

In connection with the Joule–Thomson effect we have studied a process which was adiabatic. That is to say, it was a process in which no heat entered or left the system which was under investigation. That process, however, was irreversible. We may now study a process which is adiabatic and at the same time reversible. If we compress or expand a substance in such a manner that no sensible pressure gradients develop within the system, and if we prevent by adequate thermal insulation any transfer of heat to or from the surroundings, the process is reversible and adiabatic. But in any reversible change a substance suffers no entropy change unless it gains or loses heat. Therefore in a reversible adiabatic change $\Delta S = 0$, and such a process is sometimes spoken of as *isentropic*.

In such an isentropic compression there will ordinarily be a change in temperature, and by measuring the ratio between this temperature change and the pressure change we obtain $(\partial T/\partial P)_S$. By following the method of Eq. (3-3) and employing also Eqs. (10-4) and (10-17), we find

$$\left(\frac{\partial T}{\partial P}\right)_S = -\frac{(\partial S/\partial P)_T}{(\partial S/\partial T)_P} = \frac{T}{C_P}\left(\frac{\partial V}{\partial T}\right)_P \qquad (10\text{-}29)$$

By the corresponding method we obtain

$$\left(\frac{\partial T}{\partial V}\right)_S = -\frac{(\partial S/\partial V)_T}{(\partial S/\partial T)_V} = -\frac{T}{C_V}\left(\frac{\partial P}{\partial T}\right)_V \qquad (10\text{-}30)$$

Indirect Methods of Measuring the Heat Capacity of Gases. There are two methods whereby Eqs. (10-29) and (10-30) yield the heat capacity of a gas from other measurements. In each case the P-V-T behavior of the gas must be known. While assumption of the perfect-gas law yields approximate values of $(\partial V/\partial T)_P$ or $(\partial P/\partial T)_V$ for any gas at low pressure, accurate work requires knowledge of the equation of state for that gas (see Chapter 16).

In the Lummer–Pringsheim[1] method the value of $(\partial T/\partial P)_S$ is measured directly by a thermometer of very low heat capacity suspended in the center of a large volume of gas. Kistiakowsky and Rice[2] refined earlier techniques and obtained excellent results. They used platinum wire of 0.0075 mm diameter as a thermometer and a 12-liter gas volume. It

[1] O. Lummer and E. Pringsheim, *Ann. Physik*, (3)**64**, 555 (1898).
[2] G. B. Kistiakowsky and W. W. Rice, *J. Chem. Phys.*, **7**, 281 (1939).

proved to be necessary to correct for direct heat exchange by radiation from the thermometer to the wall of the vessel.

Let us assume the simple equation of state which is valid for low pressure in most cases,

$$v = \frac{RT}{P} + B \qquad (10\text{-}31)^*$$

where the second virial coefficient B is a separate function of temperature for each substance. Then

$$\left(\frac{\partial v}{\partial T}\right)_P = \frac{R}{P} + \frac{dB}{dT}$$

and

$$c_P \frac{dT}{T} = \left(\frac{R}{P} + \frac{dB}{dT}\right) dP \qquad \text{at constant } S$$

which, if we take c_P independent of T and P for small changes, integrates to

$$c_P = R \frac{\ln (P_i/P_f)}{\ln (T_i/T_f)} + \left(\frac{dB}{dT}\right) \frac{P_i - P_f}{\ln (T_i/T_f)} \qquad (10\text{-}32)^*$$

where P_i, P_f are the initial and final pressures, respectively, and T_i, T_f are the corresponding temperatures. For carbon dioxide at 300.06°K and approximately 1 atm mean pressure, Kistiakowsky and Rice found the two terms in Eq. (10-32)* to be 8.868 and 0.108, respectively, yielding $c_P = 8.976$ cal/deg mole. This result agrees very well with the value calculated from the spectroscopic energy levels of carbon dioxide and corrected for gas imperfection at 1 atm.

In the second method the velocity of sound is measured. When sound waves of moderate intensity pass through a gas, the local changes approximate closely reversible adiabatic expansions and contractions. If the sound frequency is low, thermal equilibrium is attained with respect to the various rotational and vibrational degrees of freedom and the normal heat capacity is measured. With very high frequencies, however, this intramolecular thermal equilibrium may not be attained, and one finds anomalous absorption of the sound energy or anomalously small heat capacities.

The theory of sound propagation yields for the velocity U the equation[1]

$$U^2 = -\frac{v^2}{M}\left(\frac{\partial P}{\partial v}\right)_S \qquad (10\text{-}33)$$

By use of the method of Eq. (3-4) one obtains

$$\left(\frac{\partial P}{\partial V}\right)_S = \left(\frac{\partial P}{\partial V}\right)_T + \left(\frac{\partial P}{\partial T}\right)_V \left(\frac{\partial T}{\partial V}\right)_S \qquad (10\text{-}34)$$

[1] See, for example, H. Lamb, "Hydrodynamics," 6th ed., p. 476, Cambridge University Press, New York, 1932.

whereupon the substitution of Eqs. (10-34) and (10-30) into (10-33) yields

$$U^2 = - \frac{\mathrm{v}^2}{M} \left\{ \left(\frac{\partial P}{\partial \mathrm{v}}\right)_T - \left(\frac{\partial P}{\partial T}\right)_v^2 \frac{T}{c_V} \right\}$$

or
$$c_V = \frac{\mathrm{v}^2 T (\partial P / \partial T)_V^2}{M U^2 + \mathrm{v}^2 (\partial P / \partial \mathrm{v})_T} \tag{10-35}$$

If we substitute the equation of state for low pressures, Eq. (10-31)*, then

$$c_V = \frac{R \left(1 + \dfrac{T}{\mathrm{v} - \mathrm{B}} \dfrac{d\mathrm{B}}{dT} \right)^2}{\dfrac{M U^2}{RT} \left(\dfrac{\mathrm{v} - \mathrm{B}}{\mathrm{v}} \right)^2 - 1} \tag{10-36}*$$

The simplication for the perfect-gas law is readily obtained by setting $\mathrm{B} = 0$ and $d\mathrm{B}/dT = 0$.

Telfair and Pielemeier[1] discuss apparatus appropriate for velocity of sound measurement together with the problems of eliminating absorption and dispersion corrections in order to obtain reliable heat capacities.

RADIATION AND STEFAN'S LAW

Our respect for the second law of thermodynamics continually grows with the increasing number and diversity of phenomena to which this law is applicable. Indeed the laws of thermodynamics are not merely laws of material systems. There are systems which contain no substance, as the term is commonly used, and yet which are subject to our general thermodynamic equations. Thus we may employ these equations to demonstrate an important principle in the theory of radiation.

A completely enclosed space which is empty of all ordinary matter and is traversed by radiant energy we may call a *hollow*, in the same technical sense as the Germans speak of a *Hohlraum*. On account of the finite velocity of radiant energy, which is the velocity of light, a hollow will contain at any instant a definite amount of energy in transit between the walls. When the surrounding wall is at the same temperature throughout, the hollow will come to a state of equilibrium and the radiant energy which it contains may be shown by simple thermodynamic reasoning to be independent of the materials of which the walls are composed and of the shape of the enclosure.

Thus suppose that a hollow with given walls has come to equilibrium and by some sort of sliding partition we could put in a wall of a different character. If a new equilibrium were now established, there would be a spontaneous process associated with an increase in entropy. If now the original wall were restored, another spontane-

[1] D. Telfair and W. H. Pielemeier, *Rev. Sci. Instr.*, **13**, 122 (1942).

ous process would have to occur bringing the system back to equilibrium, with a further increase of entropy.   But the entropy of the hollow must now be the same as at the beginning, and since the changes of entropy attending the sliding of the partitions may be made negligible, it is evident that the spontaneous changes under consideration could not occur and that the state of equilibrium is independent of the nature of the walls.   Likewise it may be shown that the shape is inessential, for the hollow might be divided into small cells and these cells rearranged spatially with negligible changes in entropy.

The energy therefore will be proportional to the volume of the hollow, and it may readily be shown that it is proportional also to the rate of emission of energy from unit surface of a perfect radiator (blackbody) at the temperature in question, which rate depends upon the temperature alone.

Such a hollow at a given temperature constitutes a simple thermodynamic system.   If it is brought in contact with a heat reservoir of the same temperature and its volume is in some way increased or diminished, heat will be taken from or given to the reservoir.   The case is entirely analogous to a mixture of liquid and vapor enclosed in a cylinder with a moving piston.   Such a system will absorb heat from a reservoir if the piston is pulled out.

When the volume of the hollow is increased in a reversible way and it absorbs heat from the reservoir, it thereby increases in entropy.   Hence $(\partial S/\partial V)_T$ is a positive quantity; but we have already found the universal thermodynamic equation (10-14),

$$\left(\frac{\partial S}{\partial V}\right)_T = \left(\frac{\partial P}{\partial T}\right)_V$$

Now it would be impossible to interpret this equation without assuming (1) that there is a pressure exerted upon the walls of a hollow which is not due to any particular construction of the walls but must rather be attributed to the radiation itself and (2) that this pressure increases with the temperature.   These consequences were first pointed out by Bartoli.[1]

In fact Maxwell[2] predicted from his electromagnetic theory of light that radiant energy would exert a pressure upon a body receiving the radiation, and his equation has been qualitatively and quantitatively verified by the experiments of Lebedew[3] and of Nichols and Hull.[4]   As a consequence of this equation the pressure upon the walls of a hollow is

[1] A. Bartoli, "Sopra i movimenti prodotti dalla luce e dal calore," Le Monnier, Florence, 1876.

[2] J. C. Maxwell, "Treatise on Electricity and Magnetism," vol. II, p. 391, Oxford University Press, New York, 1873.

[3] P. Lebedew, *Ann. Physik*, (4)**6**, 433 (1901); *J. Russ. Phys. Chem. Soc.*, **33**, 53–75 (1901).

[4] E. F. Nichols and G. F. Hull, *Phys. Rev.*, **17**, 26, 91 (1903).

given by the simple formula

$$P = \frac{1}{3}\frac{E}{V} \tag{10-37}$$

where E is the energy of the hollow and V is its volume. At constant temperature, as the volume of a hollow is increased from zero,

$$\frac{E}{V} = \left(\frac{\partial E}{\partial V}\right)_T \tag{10-38}$$

Now by Eq. (10-23)

$$T\left(\frac{\partial P}{\partial T}\right)_V = P + \left(\frac{\partial E}{\partial V}\right)_T$$

and by Eqs. (10-37) and (10-38)

$$T\left(\frac{\partial P}{\partial T}\right)_V = 4P \tag{10-39}$$

By integrating at constant volume we thus find

$$P = \text{constant} \times T^4 \tag{10-40}$$

Hence we see that the energy per unit volume and also the rate of radiation from a perfect radiator (blackbody) are proportional to the fourth power of the temperature.

This important relation was first obtained by Stefan[1] as a purely empirical equation. It was Boltzmann[2] who showed it to be an exact consequence of the principles of electromagnetics and thermodynamics.[3]

Use of the Planck Equation for Fixing High-temperature Scale. In Chapter 4, we noted the use of the gas thermometer, and we showed in Chapter 7 that it yields a temperature identical with the thermodynamic temperature. Thus, over a large part of the temperature range, the gas thermometer is the primary means of establishing the thermodynamic temperature scale. At very low temperatures and at high temperatures, the gas thermometer is not suitable, and other thermometers must be devised. Although the gas thermometer has been used up to 1600°C, 1000°C appears to be the limit for accurate temperature determinations. Even at the present time, different gas-thermometer determinations of the gold point, the melting point of gold, differ by as much as a degree. At higher temperatures, one must use the intensity of

---

[1] J. Stefan, *Sitzb. Akad. Wiss. Wien.*, **79**(2), 391 (1879).

[2] L. Boltzmann, *Ann. Physik*, (2)**22**, 291, 616 (1884).

[3] For a further study of the application of thermodynamics to radiation, especially the extremely important deductions of Wilhelm Wien, the reader is referred to the treatise by Max Planck on the "Theory of Radiation," Johann Ambrosius Barth Verlag, Leipzig, 1913.

emission of radiation from a blackbody hollow together with the Planck radiation law or the Stefan-Boltzman law, Eq. (10-40).   The Planck equation

$$J_\lambda = \frac{c_1}{\lambda^5} \frac{1}{e^{c_2/\lambda T} - 1} \quad \checkmark$$

gives the emission from a blackbody as a function of wavelength and temperature.   With $c_1 = 2\pi hc^2 = 3.741 \times 10^{-7}$ erg cm^2/sec and

$$c_2 = \frac{hc}{k} = 1.4388 \text{ cm deg}$$

J_λ is given in ergs/cm^2 sec for a unit wavelength interval in centimeters.

From this equation, one may determine temperature from the ratio of emission at two wavelengths.   The temperature obtained by this procedure is often called a "color" temperature, particularly when applied to a nonblackbody whose intensity of emission is less than that of a blackbody by a factor called the emissivity.   If the emissivity varies with wavelength, the "color" temperature will differ from the true temperature.

The more common use of the Planck equation involves intensity measurements of emission at a given wavelength with use of a blackbody hollow.   Emission at the gold point is taken as the primary reference with which intensity of emission of hotter objects is compared.   The use of techniques such as dielectric interference filters to define accurately the wavelength and photoelectric detectors considerably improves the accuracy of temperature determinations.   However, the specification of thermodynamic temperature at high temperatures, particularly above 2000°C, is subject to error from an uncertainty of about 1° in the thermodynamic temperature at the gold point.   Also c_2, the second radiation constant of the Planck equation, is uncertain by about 0.1 per cent.   The probable error[1] due to these two sources of error amounts to 7° at 3000° and 12° at 10,000°.   The problem of reducing the error due to these two factors is being actively pursued at a number of laboratories around the world.

As more accurate experiments are carried out, the thermodynamic temperatures for specific points shift.   As it would be very confusing to have the temperature scale used in the published literature change with each new determination of the gold point, for example, a working temperature scale, called the International Temperature Scale, has been set up by the International General Conference on Weights and Measures.   This scale consists of a list of standard fixed points for which temperatures were assigned by agreement.   For example, the steam point was chosen as 373.15°K, the gold point as 1336.15°K, etc.   Temperatures reported in published papers are those based on the International Temperature Scale unless otherwise indicated.   The International Scale was chosen to be as close to the thermodynamic scale as possible, but as more accurate determinations of the thermodynamic scale have become available, differences between the International Temperature Scale and the thermodynamic scale have developed.   These differences may be reconciled by revision of the International Scale at very long intervals, as in 1948, when minor revisions of the 1927 scale were agreed upon.   However, it is advisable to keep the working temperature scale as stable as possible to ensure no ambiguity concerning reported temperatures.   For example, when a temperature of 2505 ± 1°C is reported, it is to be assumed that it is based on the International Temperature Scale of 1948, which means the use of

[1] D. R. Lovejoy, *Can. J. Phys.*, **36**, 1397–1408 (1958).

1336.15°K for the gold point and the use of Planck's equation with $c_2 = 1.438$ cm deg. If one wishes to know the value of this temperature on the thermodynamic temperature scale, one would use a gold-point value around 1338°K and $c_2 = 1.439$ cm deg according to the best evaluation of available data. Thus the reported temperature of 2502 ± 1°C on the International Temperature Scale of 1948 would be corrected to 2507 ± 1°C for use in thermodynamic calculations. The use of a dual temperature system may seem awkward, but the existence of a stable working temperature scale is essential for comparison of measurements of different workers. Except for the most accurate measurements, the difference between the two scales can normally be neglected.

A detailed summary of the standard fixed points of the International Temperature Scale is given by Stimson.[1] This is also reviewed along with details concerning the various temperature measuring instruments in the books "Temperature: Its Measurement and Control in Science and Industry."[2]

PROBLEMS

10-1. For which of the following changes of state is Eq. (10-7) applicable?

I. A sample of NO_2 gas is expanded slowly so that the equilibrium is maintained with respect to $NO + \frac{1}{2}O_2$.

II. A sample of NO_2 gas is expanded at a rate such that the dissociation to $NO + \frac{1}{2}O_2$ lags behind the equilibrium composition.

III. A sample of SO_3 gas is expanded under conditions (absence of catalyst) such that there is no dissociation into SO_2 and O_2.

IV. A sample of water is frozen isothermally at $-10°C$.

10-2. The vapor pressure of liquid ammonia is 7.68 atm at 17°C and is increasing at the rate of 0.25 atm/deg. The specific volumes of vapor and liquid are, respectively, 165 and 2 cc. Calculate the heat of vaporization per gram of ammonia, and compare with the more precisely measured value of 296 cal.

10-3. A gram of ammonia liquid at 7.68 atm and 17°C is vaporized isothermally into an evacuated space to a final volume of 165 cc. Calculate q, ΔE, ΔH, ΔS, and ΔF for this process, using the data of Problem 10-2.

10-4. Consider a substance under two different pressures, at the same temperature. Let the pressure and volume in the first state be P_A and V_A and in the second P_B and V_B. Show that

$$F_B - F_A = \int_A^B V\, dP \tag{10-41}$$

$$A_B - A_A = -\int_A^B P\, dV \tag{10-42}$$

From Eq. (10-10)

$$(F_B - F_A) - (A_B - A_A) = P_B V_B - P_A V_A \tag{10-43}$$

Show that this equation follows also from Eqs. (10-41) and (10-42) and from the principles of integral calculus. Furthermore, show that Eq. (10-43) follows from (10-41) and (10-42) if we consider the area corresponding to the four terms in (10-43) when we use a P-V diagram.

[1] H. F. Stimson, The International Temperature Scale of 1948, *J. Research Natl. Bur. Standards,* **42,** 209–217 (1949).

[2] Vol. 1, 1949, vol. 2, 1955, Reinhold Publishing Corporation, New York. Report of symposia held under the auspices of the American Institute of Physics.

10-5. Show that for a perfect gas, in an isothermal change,

$$F_B - F_A = RT \ln \frac{P_B}{P_A} \qquad (10\text{-}44)^*$$

10-6. For 1 mole of benzene at $T = 298°$, assume $v = 88.8 \, cc$; $\alpha = (1/V)(\partial V/\partial T)_P = 0.00124$; $c_P = 33.2 \, cal/deg$. Bringing these data to consistent units, show that the rise in the temperature of benzene, when its pressure is suddenly increased by 1 atm is about 0.024°.

10-7. Lummer and Pringsheim[1] measured the specific heats of gases by suddenly lowering the pressure in a large balloon of gas and measuring the change in internal temperature by means of a very fine platinum resistance thermometer. The equation they employed is

$$\frac{T_1}{T_2} = \left(\frac{P_1}{P_2}\right)^{(c_P - c_V)/c_P}$$

where c_P/c_V is a constant. Show that this equation may be derived from Eq. (10-29) if we assume a perfect gas.

10-8. The constant of proportionality in Eq. (10-40) can be derived from Planck's radiation theory and is $\frac{8}{45}\pi k(k/hc)^3$ or $2.5215 \times 10^{-15} \, erg/cm^3 \, deg^4$. Derive equations for E and S per unit volume as a function of temperature, and calculate C_V per cubic centimeter for a hollow at 10,000°K.

10-9. Use Eq. (10-18) to show that, for a system kept at constant volume, the minimum amount of work required to transfer an amount of heat, q, to the system is given by

$$\frac{w}{q} = \frac{T_{\text{system}} - T_{\text{sur}}}{T_{\text{system}}}$$

Compare this result with Eq. (9-5).

10-10. Calculate the rate of change with pressure of the freezing point of water. Take the density of ice as 0.917, that of water as 1.000, and the heat of fusion as 80 cal/g. All these values apply to water under 1 atm pressure. What additional data would be required to calculate the freezing point of water at 10,000 atm pressure?

10-11. Calculate the percentage deviation of the volume of isopentane saturated vapor at 298°K from the ideal-gas volume. Use the heat of vaporization and the Clapeyron equation. $\Delta H_{\text{vap}} = 5.878 \, kcal/mole$, density of liquid $= 0.6146$, both at 298°K; and the vapor pressure in millimeters of mercury is given by

$$\log p = 6.7897 - \frac{1020.0}{T - 40.0}$$

[1] O. Lummer and E. Pringsheim, *Ann. Physik*, (3)**64**, 582 (1898).

$$\left(\frac{\partial E}{\partial V}\right)_T = \frac{E}{V} = 3P$$
$$= \text{constant} \times 3 \times T$$

$$T\left(\frac{\partial S}{\partial V}\right)_T = P + 3P = 4P$$

$$\left(\frac{\partial S}{\partial V}\right)_T = \frac{4P}{T} = \cdots$$

11

THE NUMERICAL CALCULATION OF ENTROPY

Unquestionably the idea of entropy appears at first sight a little abstruse. However, it is the universal tool of thermodynamics, and one of great power. Like other tools it cannot be successfully handled without some theoretical knowledge of its mode of operation and some practice in its use. In the preceding chapters we have endeavored to make clear its theoretical significance; and in the present chapter we shall make the concept more concrete by showing how entropy may be handled numerically.

As a starting point we may recall that in any *reversible* process a system, or any part of a system, undergoes an increase of entropy just in so far as it absorbs heat from the surroundings, resulting in an equal decrease of the entropy of the surroundings, and that the increase in entropy is equal to the heat so absorbed divided by the absolute temperature,

$$dS = \frac{\delta q_{\text{rev}}}{T} \qquad (11\text{-}1)$$

The Entropy Change in Fusion. Let us use this equation to calculate the change of entropy when a substance changes from one phase to another at constant temperature and under conditions of equilibrium. We shall consider the fusion of 1 mole of solid mercury at its melting point, which (at atmospheric pressure) is 234.29°K. At the melting point the two phases are in equilibrium. That is to say, there is a state of balance such that if the external temperature is raised by an infinitesimal amount the solid will melt and if it is diminished by an infinitesimal amount the liquid will freeze. So also, at constant temperature, if the pressure is lowered or raised by any amount, the process will occur in the one direction or the other. Hence the process of fusion at the melting point is a reversible one.

In the case of a pure substance like mercury the temperature remains constant during fusion, and we have from Eq. (11-1)

$$\Delta S = \frac{\Delta H}{T} \qquad (11\text{-}2)$$

116

If ΔH is the heat of fusion of 1 mole, namely, 548.6 cal, and T is 234.29°, we may write

$$Hg(s) = Hg(l) \qquad \Delta S_{234.29} = {}^{548.6}\!/_{234.29} = 2.342 \text{ cal/deg}$$

Since in this process 1 mole of solid has disappeared and 1 mole of liquid has appeared, we may say that the molal entropy of liquid mercury is greater than the molal entropy of solid mercury by 2.342 cal/deg at the melting point. Or

$$s_{234.29}(Hg, l) - s_{234.29}(Hg, s) = 2.342 \text{ cal/deg}$$

It must be borne in mind that such a calculation is based upon the fact that we have a state of equilibrium in which every process is reversible. If, on the other hand, we consider the difference in entropy between ice at $-10°C$ and supercooled water at the same temperature, that difference cannot be obtained by dividing the difference in heat content by the absolute temperature, 263°K. We shall revert to this point in Problem 11-1.

The Entropy Change in Vaporization. In vaporization the change of entropy is usually much larger than in fusion. Ether at its boiling point, 307.7°K, absorbs 6500 cal/mole by evaporation. Hence we write for the vaporization,

$$(C_2H_5)_2O(l, 1 \text{ atm}) = (C_2H_5)_2O(g, 1 \text{ atm})$$
$$\Delta S_{307.7} = {}^{6500}\!/_{307.7} = 21.1 \text{ cal/deg}$$

We might equally well have written the equation for condensation,

$$(C_2H_5)_2O(g, 1 \text{ atm}) = (C_2H_5)_2O(l, 1 \text{ atm}) \qquad \Delta S_{307.7} = -21.1 \text{ cal/deg}$$

Benzene at its boiling point, 353.25°K, absorbs 7353 cal/mole in evaporation. Hence we write

$$C_6H_6(l, 1 \text{ atm}) = C_6H_6(g, 1 \text{ atm}) \qquad \Delta S_{353.25} = {}^{7353}\!/_{353.25}$$
$$= 20.81 \text{ cal/deg}$$

Similarly, for cyclohexane, boiling at 353.89°K, and toluene, boiling at 383.77°K, the corresponding values of Δs of vaporization at the two boiling points are 20.30 and 20.81 cal/deg.

It will be noted that the four values are nearly equal. An empirical principle, known as *Trouton's rule*,[1] states that the entropy increase per mole is the same for all so-called normal, or nonpolar, liquids at their boiling points. The constant of Trouton's rule is usually given as about 21 cal/deg.

---

[1] F. Trouton, *Phil. Mag.*, (5)**18**, 54 (1884).

This rough but useful rule, when applied to liquids with a wide range of boiling points, shows a marked trend in the average value of the constant with the temperature, so that the "constant" is about 50 per cent greater for liquids boiling in the neighborhood of 1000°C than it is for low-boiling liquids like oxygen and nitrogen. Several attempts have been made to restate the rule in such a way as to obviate this trend, the most simple being that of Hildebrand.[1]   His rule states that the "entropy of vaporization" is the same for different liquids, not at the boiling points, which are the temperatures where the several liquids have unit vapor pressure, but rather at temperatures where the liquids give the same vapor concentration.   Choosing arbitrarily temperatures at which the concentration of vapor is 0.005 mole/liter, he finds the entropy of vaporization to be 27.6 cal/deg mole for N_2 and O_2, 27.4 for benzene, 27.6 for bromonaphthalene, 26.2 for mercury, and 26.4 for zinc.   However, the values for highly polar substances are substantially larger: 32.4 for ammonia, 32.0 for water, and 33.4 for ethyl alcohol.

It is now apparent that a number of factors influence the precise value of the entropy change on vaporization.   Consequently, no simple rule can be expected to yield exact predictions.

All these processes which we have been considering are reversible.   Let us now consider a similar irreversible process, for example, the evaporation of superheated water at 110°C to steam at 1 atm and 110°C.   The heat of vaporization at this temperature is 9596 cal/mole, so that if steam is formed from water at the equilibrium pressure, namely, 1.414 atm, the entropy of vaporization is readily obtained,

$$H_2O(l, 1.414 \text{ atm}) = H_2O(g, 1.414 \text{ atm})$$
$$\Delta s_{383} = {}^{9596}\!/_{383.15} = 25.04 \text{ cal/deg}$$

Now, if we assume that water is a perfect gas, then we may find the change in entropy of the vapor between 1.414 and 1.000 atm by Eq. (7-9)*, namely,

$$H_2O(g, 1.414 \text{ atm}) = H_2O(g, 1.000 \text{ atm})$$
$$\Delta s_{383} = R \ln ({}^{1.414}\!/_{1.000}) = 0.688 \text{ cal/deg}$$

The difference in entropy of liquid water between 1.414 and 1.000 atm is so small that it may be ignored.

$$H_2O(l, 1.000 \text{ atm}) = H_2O(l, 1.414 \text{ atm}) \qquad \Delta s_{383} = 0.000 \text{ cal/deg}$$

Then we may add our three equations and obtain

$$H_2O(l, 1.000 \text{ atm}) = H_2O(g, 1.000 \text{ atm}) \qquad \Delta s_{383} = 25.73 \text{ cal/deg}$$

It is interesting to note the net increase in the entropy of system and surroundings for this spontaneous process.   The total heat absorbed from a reservoir at 383°K is just the change in heat content for the system since the actual irreversible process occurs at constant pressure and no work is done except that involving change in volume.   Also $\Delta H = 0$ for the

[1] J. H. Hildebrand, *J. Am. Chem. Soc.*, **37**, 970 (1915).

expansion of a perfect gas.   Consequently, we have

$$H_2O(l, \text{ 1 atm}) = H_2O(g, \text{ 1 atm}) \qquad \Delta H_{383} = 9596 \text{ cal}$$

The entropy change in the heat reservoir is just q/T, or

$$-({}^{9596}\!\!\big/_{383.15}) = -25.04$$

and the net entropy increase for the irreversible process is

$$25.73 - 25.04 = +0.69 \text{ cal/deg}$$

The Entropy Change in Expansion.   If in the previous illustration we had desired neither to ignore the change in the entropy of liquid water with the pressure nor to regard the vapor as a perfect gas, our calculation could have been made formally more precise.   It would suffice to know values of the V-T coefficient for liquid and vapor.   Thus from Eq. (10-17)

$$\int dS = -\int \left(\frac{\partial V}{\partial T}\right)_P dP \tag{11-3}$$

For example, let us take the v-T coefficient of 1 mole of liquid water at 110° as constant and equal to 0.015 cc/deg.   The change of entropy when the pressure changes 0.414 atm is 0.0062 cc atm/deg, or 0.00015 cal/deg. In other words,

$$H_2O(l, \text{ 1.414 atm}) = H_2O(l, \text{ 1.000 atm}) \qquad \Delta s_{383} = 0.00015 \text{ cal/deg}$$

and this quantity is evidently less than the experimental errors in the other quantities involved in the above calculation.

We may also compare our entropy value for the expansion of a perfect gas with the value obtained from Eq. (11-3) together with an equation which represents the actual measured volume of steam in this region of temperature and pressure.   At moderate pressures the molal volume steam is reported[1] to be given by the equation

$$v = \frac{RT}{P} + 34.0 - \frac{47{,}590}{T} 10^{8087/T^2} \text{ cc} \qquad (11\text{-}4)^*$$

We differentiate and substitute $T = 383.15$ to obtain

$$\left(\frac{\partial v}{\partial T}\right)_P = \frac{R}{P} + 0.46 \text{ cc/deg}$$

and then substitute in Eq. (11-3).

$$\int ds = R \ln {}^{1.414}\!\!\big/_{1.000} + 0.46(1.414 - 1.000) \text{ cc atm/deg}$$
$$= 0.688 + 0.0046 = 0.693 \text{ cal/deg}$$

[1] J. H. Keenan and F. G. Keyes, "Thermodynamic Properties of Steam," p. 15, John Wiley & Sons, Inc., New York, 1936.

The first term is just that given by the perfect-gas law, and the second term is therefore the correction for gas imperfection. While the correction is small, we see that it is larger than the total entropy change of the liquid which we calculated in the preceding paragraph.

The Entropy Change in a Chemical Reaction. When we consider a chemical reaction, such as the reaction of mercury and chlorine to form mercurous chloride, the change in entropy resulting from the reaction is not to be calculated from the heat of reaction, since the process is a highly irreversible one. However, if we construct a galvanic cell in which this reaction is the one which occurs, and if the emf of this cell is exactly balanced by an external emf, so that slight changes in the latter will cause the current to pass in one direction or the other through the cell, the reaction is made reversible.

The heat absorbed by the cell, when operating under these conditions, is the quantity which, divided by the temperature, gives the increase in entropy during the reaction. This reversible heat of reaction $T \Delta S$ is entirely different in magnitude, and often in sign, from the ordinary calorimetric heat of reaction ΔH.

It was the recognition of these facts by Willard Gibbs which permitted him to place upon a sound foundation the thermodynamics of the galvanic cell. We shall have in our further work frequent occasion to study the reversible galvanic cell, as well as other methods of investigating chemical reactions occurring under reversible conditions.

Change of Entropy with Temperature. We have seen in Eqs. (10-3) and (10-4) how the entropy of a substance changes with the temperature. The equations read

$$\left(\frac{\partial S}{\partial T}\right)_V = \frac{C_V}{T} \qquad \left(\frac{\partial S}{\partial T}\right)_P = \frac{C_P}{T}$$

Thus, if a substance is heated at constant pressure, the change of entropy is given by the equation

$$\int dS = \int \frac{C_P}{T} dT = \int C_P \, d \ln T = 2.303 \int C_P \, d \log T \qquad (11\text{-}5)$$

If, therefore, C_P is known at various temperatures, we may perform the integration by analytical or graphical methods and find the change in entropy of a substance between two temperatures.

The simplest case is the one in which C_P is constant, when we find, between the temperatures T and T'',

$$S' - S = C_P \ln \frac{T'}{T} \qquad (11\text{-}6)^*$$

For liquid Hg in the small temperature range between $T' = 298.15°K$ and the freezing point $T = 234.29°K$, we may regard c_P as approximately constant and equal to 6.75.   Hence

$$s' - s = 6.75 \ln {}^{298.15}\!\!/_{234.29} = 1.63 \text{ cal/deg}$$

We may next consider the case in which c_P is given by a simple empirical equation.   If we employ the equation in Table 6-4 for the heat capacity of oxygen between 298.15 and 1000°K, we obtain by integration of Eq. (11-5)

$$s_{1000} - s_{298.15} = 7.16 \ln {}^{1000}\!\!/_{298.15} + 1.00 \times 10^{-3}(1000 - 298.15)$$
$$+ 0.20 \times 10^{5}(1000^{-2} - 298.15^{-2}) = 9.18 \text{ cal/deg}$$

The analytical method becomes more cumbersome the greater the number of terms in our empirical expression for c_P.   On the other hand, the graphical method of integration can readily be applied to a curve of any degree of complexity.

It is evident from Eq. (11-5) that if we plot c_P against the common logarithm of T, as we have done in Chapter 6, the area under the curve between two points, multiplied by 2.303, gives immediately the difference in entropy between the two points.

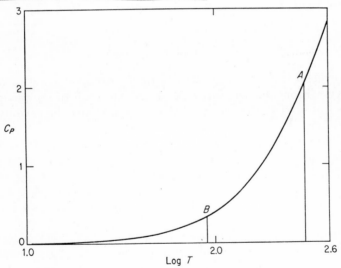

FIG. 11-1. Heat capacity of graphite, in cal/deg mole.

We may illustrate this procedure by means of Fig. 11-1, which shows the atomic heat capacity of graphite[1] plotted against log T. If we wish to determine the difference in entropy between 298°K (log $T = 2.47$), and 90°K (log $T = 1.95$), we merely find the area under the curve between A and B, multiply by 2.303, and obtain $s_{298} - s_{90}$.

[1] W. DeSorbo and W. W. Tyler, *J. Chem. Phys.*, **21**, 1660 (1953).

We note from Eq. (11-5) that we may obtain the entropy from a plot of c_P/T versus T instead of c_P versus log T.   Figure 11-2 shows a plot of c_P/T against T for graphite.   The area under the curve between A and B yields directly the value of $s_{298} - s_{90}$.

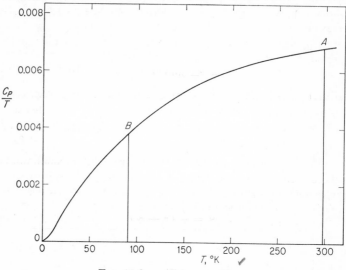

FIG. 11-2. c_P/T for graphite.

It is evident from Fig. 11-2 that the total entropy above 0°K is a finite quantity.   From the total area under the curve up to the point A one obtains $s_{298} - s_0 = 1.37$ cal/deg, where s_0 represents the entropy at the absolute zero.

Absolute Value of Entropy.   In all our discussions of entropy we have shown how to determine, not the entropy of any particular system, but only the change in entropy which accompanies a change in the system.   In this respect our definition of S places it in the same category as the quantities E and H, of which the absolute values are as yet undetermined.   As with these quantities, we might assign an arbitrary value to the entropy of some one substance in a given state, but this could not be done for all substances.   Thus, if we assign an arbitrary value to the entropy of ice at 0°C, the entropy of water is then determined, since we know the entropy change in fusion.   So also if we choose arbitrary values for the entropies of mercury and chlorine, we could not arbitrarily choose a value for mercurous chloride, since we have shown how the change in entropy in the formation of mercurous chloride may be determined.

However, we might at least fix arbitrarily the entropy of each element, in some one state, and this we shall do by taking the entropy of each element in its solid state, or in some one solid state, as zero at the absolute zero of temperature.   Thus in the case of graphite, which we considered above, we may write $s_0 = 0$, and therefore $s_{298} = 1.37$ cal/deg.   The reason for this procedure will be more fully explained in the next chapter.

Extrapolation of Heat-capacity Data to 0°K.   Experimental heat-capacity measurements do not extend completely to 0°K; hence an extrapolation is necessary to obtain the absolute entropy as defined in the preceding section.   From the discussion in Chapter 6 we see that at extremely low temperatures the heat capacity of a normal solid may be expressed as

$$c_P = aT^3 + \gamma T \qquad (11\text{-}7)^*$$

where a and γ are the constants, respectively, in the Debye term for atomic vibration and the term for metallic electrons.   There are several sorts of anomalous effects such as superconductivity or changes in orientation of atomic magnetic moments which may cause deviations from Eq. (11-7)* even at very low temperatures, but we shall not consider these effects here.   Over the range of validity of Eq. (11-7)* and if $S = 0$ at 0°K, the entropy at T is

$$s = \frac{a}{3} T^3 + \gamma T \qquad (11\text{-}8)^*$$

We see that the entropy is less than the heat capacity at T and in case γ may be neglected, $S \cong \frac{1}{3}C_P$.

In our calculation for graphite the experimental data extended to 12.92°K where $c_P = 0.0076$ cal/deg.   The total entropy below this temperature is less than the uncertainty of experimental measurement at higher temperatures; hence even a crude extrapolation will suffice.   While it is always desirable to extend measurements to a sufficiently low temperature to make the extrapolated entropy negligible, this is not always feasible in a given laboratory and it is then necessary to consider the extrapolation carefully.   In the past it has been common to use the Debye equation even above the region where heat capacity is proportional to T^3, but recent measurements indicate that this is not very satisfactory. Indeed the initial deviation from the T^3 law is usually in the opposite direction from that of the Debye function.   This is shown in Fig. 11-3, which includes a sample Debye curve and the experimental data for KCl and Na_2SO_4.

A graph of c_P/T versus T^2 is very convenient since the function of Eq. (11-7)* becomes a straight line with slope a and intercept γ on the $T^2 = 0$ axis.   Figures 11-3 to 11-5 present data for a number of substances on this basis, and in all cases the curves approach the behavior predicted by Eq. (11-7)* at the lowest temperatures.   Also the intercept is at the origin ($\gamma = 0$) except for metals.

In our opinion a plot of the heat-capacity data on the c_P/T versus T^2

basis along with the available data extending to lower temperatures for similar substances provides the best basis for extrapolation to 0°K. The already substantial and rapidly growing body of data extending to very low values of c_P makes this a feasible procedure. One then draws an extrapolated curve similar to the observed curves for similar substances

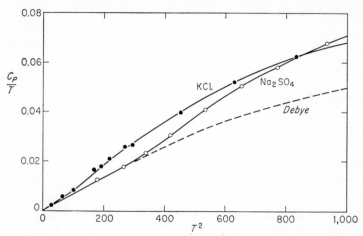

FIG. 11-3. Heat-capacity curves for KCl and Na_2SO_4, together with a Debye curve fitted to the latter substance in the T^3 region. [*KCl data from W. H. Keesom and C. W. Clark, Physica,* **2**, 698 (1935); *J. C. Southard and R. A. Nelson, J. Am. Chem. Soc.,* **55**, 4865 (1933). Na_2SO_4 *data from K. S. Pitzer and L. V. Coulter, J. Am. Chem. Soc.,* **60**, 1310 (1938).]

and in a manner consistent with Eq. (11-7)* in the limit at 0°K. The two curves for $(CH_2)_3S$ in Fig. 11-4 illustrate the probable limits within which the correct extrapolation should lie. Finally the extrapolated values of c_P may be transferred to a plot of c_P/T versus T for the calculation of the entropy by graphical integration.

Entropy of Liquid and Gaseous Elements. To illustrate the simultaneous use of the methods of this chapter, let us calculate the entropy of liquid mercury at 25°C from the entropy of solid mercury and from the changes in entropy when the solid is melted and the liquid is then brought to 25°C. We may first calculate the entropy of the solid at the melting point, namely, 234.29°K. Graphical integration of the data of Smith and Wolcott[1] and of Busey and Giauque[2] yields

$$s_{234.29} = 14.201 \text{ cal/deg}$$

Since the data of Smith and Wolcott extend to 1.2°K where

$$c_P = 0.0028 \text{ cal/deg}$$

the extrapolation to 0°K involves no appreciable uncertainty.

[1] P. L. Smith and N. M. Wolcott, *Phil. Mag.,* (8)**1**, 854 (1956).

[2] R. H. Busey and W. F. Giauque, *J. Am. Chem. Soc.,* **75**, 806 (1953).

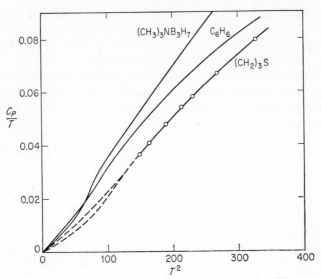

FIG. 11-4. Heat-capacity curves for three molecular crystals.   The pair of dashed lines for $(CH_2)_3S$ indicate the range of reasonable extrapolation curves.   [*J. E. Ahlberg et al., J. Chem. Phys.*, **5**, 539 (1937), *and G. D. Oliver et al., J. Am. Chem. Soc.*, **70**, 1502 (1948), *for* C_6H_6; *D. W. Scott et al., ibid.*, **75**, 2795 (1953), *for* $(CH_2)_3S$; *N. E. Levitin et al., ibid.*, **81**, 3547 (1959) *for* $(CH_3)_3NB_3H_7$.]

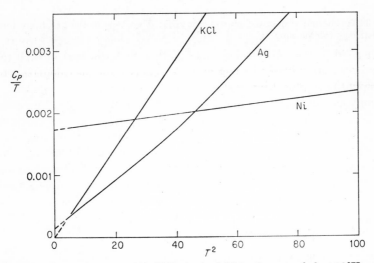

FIG. 11-5. Heat-capacity curves for KCl, Ag, and Ni in the range below 10°K.   Note that the metals have nonzero intercepts for c_P/T at 0°K.   [*KCl data from W. H. Keesom and C. W. Clark, Physica*, **2**, 698 (1935).   *Ag data from W. S. Corak et al., Phys. Rev.*, **98**, 1699 (1955); *W. H. Keesom and J. A. Kok, Proc. Acad. Sci. Amsterdam*, **35**, 301 (1932).   *Ni data from W. H. Keesom and C. W. Clark, Physica*, **2**, 513 (1935).]

Now we have previously calculated the entropy change in the fusion of mercury to be 2.342, and we have also calculated the entropy change in heating liquid mercury from the melting point to 25°C to be 1.63, whence we finally obtain, for Hg(l), $s_{298} = 18.17$ cal/deg.

The same method may be employed in other cases which, however, are frequently more complicated. For example, the element oxygen has three different solid modifications in addition to the liquid and gaseous states. Giauque and Johnston[1] have measured the heat capacity of all

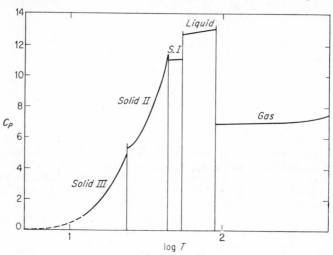

FIG. 11-6. The molal heat capacity of oxygen. The areas in the graph are terms in the entropy calculation.

three solid forms and of the liquid together with the heat of each transition between solid modifications, the heat of fusion, and the heat of vaporization. The heat-capacity curves are shown in Fig. 11-6. The entropy calculation is summarized in Table 11-1.

TABLE 11-1. THE ENTROPY OF OXYGEN

0–14°K, extrapolation....................	0.54
14–23.66°K, solid III, graphical...........	1.500
Transition, $22.42/23.66$...................	0.948
23.66–43.76°K, solid II, graphical.........	4.661
Transition, $177.6/43.76$...................	4.058
43.76–54.39°K, solid I, graphical...........	2.397
Fusion, $106.3/54.39$.....................	1.954
54.39–90.13°K, liquid, graphical...........	6.462
Vaporization $1628.8/90.13$................	18.07
Total, $s_{90.13}$ (gas, 1 atm), cal/deg..........	40.59

Each component of this calculation is of a type which has been discussed earlier in the chapter. The entropy of oxygen at higher tempera-

[1] W. F. Giauque and H. L. Johnston, *J. Am. Chem. Soc.*, **51**, 2300 (1929).

tures may be readily calculated from the known heat capacity of the gas. The value $c_P = \frac{7}{2}R$ may be used for low temperatures, and an equation given in Chapter 6 is valid from 298 to 2000°K.   These heat-capacity values are for the ideal gas.   At low temperatures the heat capacity of the real gas at 1 atm is slightly higher than that of the ideal gas, but we shall postpone discussion of this minor correction until Chapter 16.

We may point out in this connection that thermodynamics imposes certain conditions regarding the form of such heat-capacity curves and the heats of transition. The difference in entropy between solid I at a certain temperature and solid III at another temperature is entirely independent of the path by which we pass from one to the other.   It might have happened that solid II failed to appear, in which case the solid I and solid III curves would have been continued to the transition point III-I. Our calculation of the entropy of oxygen gas would then have contained different terms, but except for possible experimental error the result would have been the same.

PROBLEMS

11-1. Calculate the changes in entropy when 1 mole of water at $-10°C$ is heated to $0°C$, is frozen at $0°C$, and the solid is cooled to $-10°C$.   The sum is equal to Δs_{263} in the *irreversible* isothermal process, $H_2O(l) = H_2O(s)$.   Take c_P as constant for both water and ice, namely, 18 and 9, respectively; and take $\Delta H_{273} = -1436$ cal.

11-2. Taking $c_P = 6.96$ for gaseous oxygen and the entropy of vaporization from Table 11-1, show that for $O_2(g, 298.15°) = O_2(l, 90.13°)$, $\Delta s = -26.4$ cal/deg.

11-3. Calculate the minimum amount of work required to convert 1 mole of oxygen gas at 298.15°K and 1 atm to the liquid state at 90.13°K in a reversible process in which heat is transferred to a heat reservoir at 298.15°.   Note that the entropy decrease of the oxygen was calculated in Problem 11-2 and that the entropy increase of the heat reservoir must be equal so that the net entropy change will be zero.   The first law may then be used to calculate the work.

11-4. Use the data[1] given below to calculate $s_{100} - s_0$ for Na_2SO_4 by graphical integration.   Use the T^3 law to extrapolate c_P to 0°K.

T........	13.74	16.25	20.43	27.73	41.11	52.72	68.15	82.96	95.71
c_P, cal/deg.	0.171	0.286	0.626	1.615	4.346	7.032	10.48	13.28	15.33

11-5. Use the equation in Table 6-4 to calculate $s_{1000} - s_{298}$ for ammonia gas.

11-6. With the aid of Eq. (11-4)* calculate the change in entropy of 1 mole of water vapor at 473°K for compression from 1 to 10 atm pressure.

11-7. Given $\Delta S = 7.2$ cal/mole deg at 300°K for the isothermal change, monatomic gas ($P = 30$ atm) = monatomic gas ($P = 1$ atm), calculate the final temperature if this gas expands adiabatically and reversibly from 300°K and 30 atm to 1 atm. Assume ideal behavior at 1 atm.

11-8. Calculate the net increase in entropy of the system and surroundings when 1 kg of supercooled water is crystallized isothermally at $-20°C$ and 1 atm.   Take $c_P^\circ = 1$ cal/deg g for water and 0.5 for ice and $\Delta H_f^\circ = 80$ cal/g at 273°K.   How much work, other than expansion work, could have been obtained from the isothermal freezing of the supercooled water by setting up a reversible process?

[1] K. S. Pitzer and L. V. Coulter, *J. Am. Chem. Soc.*, **60**, 1310 (1938).

12

THE THIRD LAW OF THERMODYNAMICS

During the nineteenth century the first and second laws of thermodynamics were formulated and applied to many scientific problems. It was not until the first decade of the twentieth century that investigations at very low temperatures had progressed sufficiently to lay the basis for the generalization which constitutes the third law of thermodynamics. Nevertheless, the third law has now satisfied numerous and searching experimental tests and has provided the basis for a large fraction of the numerical data on chemical reactions. It seems appropriate, therefore, to include the third law as an integral portion of the basic foundation of our subject.

We have now seen that heat capacities of solids decrease very rapidly to zero as the absolute zero of temperature is approached. Also we recall the explanation of Einstein that the internal energy of a solid is quantized so that only finite quanta of energy can be absorbed. Thus, as the solid is cooled to $0°K$, all its constituent particles fall into their lowest quantum states; indeed we may say that the entire crystal is in its lowest quantum state. This is evidently an especially simple state, which we may take as an absolute unit of probability.

In Chapter 8 we saw that the second law of thermodynamics was the consequence of probability. The equation for the relationship of entropy difference to probability ratio was found to be

$$s_B - s_A = \frac{R}{N_0} \ln \frac{\mathcal{P}_B}{\mathcal{P}_A} \qquad (8\text{-}3)$$

If we adopt a new probability scale with $\mathcal{P} = 1$ and $S = 0$ for each element in crystalline form at $0°K$, we find the entropy of all substances to be positive at all temperatures above $0°K$. The probability in these other states is thus greater than 1, which is in agreement with the general picture that the substance has its particles distributed over a multitude of quantum states. We shall use on this absolute scale the symbol W for probabilities, which are more appropriately called *multiplicities*.

Now let us consider an element, such as sulfur, which has more than one crystalline form.   If we define $S = 0$ at $0°K$ for rhombic sulfur, we may determine experimentally the entropy of monoclinic sulfur at $0°K$. This experiment requires that we supercool the monoclinic sulfur below the temperature of its transition to rhombic at $368.5°K$, but it is well known that this is possible.   Eastman and McGavock[1] measured with great care the heat capacity of rhombic sulfur and of supercooled monoclinic sulfur in the same calorimeter from 13 to $365°K$.   West[2] measured the heat capacity of rhombic sulfur from $298°K$ to the transition and the heat of transition to monoclinic.   Their results are given in Table 12-1.

TABLE 12-1. ENTROPY OF RHOMBIC AND MONOCLINIC SULFUR

From integration of $c_P \, d \ln T$ for rhombic sulfur:

$s_{368.5}$(rhombic) $-$ s_0(rhombic) $= 8.810 \pm 0.05$
Transition, $\Delta s = (^{96.0}\!/_{368.5})$ $= 0.261 \pm 0.002$
$s_{368.5}$(monoclinic) $-$ s_0(rhombic) $= 9.07 \ \ \pm 0.05$

From integration of $c_P \, d \ln T$ for monoclinic sulfur:

$s_{368.5}$(monoclinic) $-$ s_0(monoclinic) $= 9.04 \pm 0.10$ cal/deg

We see from Table 12-1 that the total entropy difference between s(monoclinic, $368.5°K$) and s(monoclinic, $0°K$) is the same, within experimental error, as the entropy difference between s(monoclinic, $368.5°K$) and s(rhombic, $0°K$).   In other words, the entropy of monoclinic sulfur is zero at $0°K$, since we have already taken the entropy of rhombic sulfur to be zero at that temperature.

TABLE 12-2. ENTROPY OF PHOSPHINE, CAL/DEG MOLE

	Form stable above 49.43°K		Form stable below 49.43°K	
0–15°K, Debye extrapolation...	....	0.495	....	0.338
Graphical integration..........	15–30.29°K	2.185	15–49.43°K	4.041
Transitions....................	$^{19.6}\!/_{30.29}$	0.647	$^{185.7}\!/_{49.43}$	3.757
Graphical integration..........	30.29–49.43	4.800		
Entropy at 49.43°K of form stable above 49.43°K........	....	8.13	....	8.14

The data of Stephenson and Giauque[3] on phosphine yield an even more precise check of the type just given.   Phosphine has several crystalline modifications; the calculation is summarized in Table 12-2.   We find again that two different crystalline modifications of phosphine have very exactly the same entropy at $0°K$.

[1] E. D. Eastman and W. C. McGavock, *J. Am. Chem. Soc.*, **59**, 145 (1937).
[2] E. D. West, *J. Am. Chem. Soc.*, **81**, 29 (1959).
[3] C. C. Stephenson and W. F. Giauque, *J. Chem. Phys.*, **5**, 149 (1937).

We might provide additional examples of this type as well as a large number of more general examples where the entropy of formation of a compound from its elements is measured and the entropy of the compound is found to decrease to zero at 0°K. This sort of information, in the less precise form then available, led Nernst[1] and Planck[2] to attempt the statement of a general principle. Their statements constituted essentially the third law as we know it but contained minor defects which we need not discuss. The first fully satisfactory statement of the third law was that of Lewis and Randall in the first edition of the present book.[3]

If the entropy of each element in some crystalline state be taken as zero at the absolute zero of temperature, every substance has a finite positive entropy; but at the absolute zero of temperature the entropy may become zero, and does so become in the case of perfect crystalline substances.

We shall see, in later chapters, the great utility of this principle in the prediction of chemical equilibrium constants and therefore of the direction of chemical reactions. Before proceeding to that development, however, we discuss a few matters immediately related to the third law.

Absolute Entropy and Multiplicity. Now that we have an absolute basis for entropy, we have also an absolute basis for probability. Instead of the difference in entropy and the ratio of probabilities which appeared in Eq. (8-3), we write the equation,

$$s = \frac{R}{N_0} \ln W \tag{12-1}$$

which is commonly called the Planck-Boltzmann equation. It is beyond the scope of our present discussion to define thermodynamic multiplicity W with full precision. We may indicate, however, that it is the number of quantum states for the entire system which are accessible under the conditions of energy, volume, etc., which are applicable.

With the simultaneous development of quantum mechanics and molecular spectroscopy it became possible to calculate the entropy of ideal gases from detailed information about their energy levels. This very fruitful source of entropy values will be the subject of Chapter 27, but in the meantime we shall not hesitate to use the resulting entropy values for particular substances.

Thermodynamic Properties at the Absolute Zero. The third law clearly implies that the entropy increase from 0°K to any temperature T' must be finite, i.e., the integral $\int_0^{T'} (C/T) \, dT$ must be finite. Thus the

---

[1] W. Nernst, *Nachr. kgl. Ges. Wiss. Göttingen Math.-physik. Kl.*, **1906**, 1.

[2] M. Planck, "Thermodynamik," 3d ed., p. 279, Veit and Co., Leipzig, 1911.

[3] G. N. Lewis and M. Randall, "Thermodynamics and the Free Energy of Chemical Substances," 1st ed., p. 448, McGraw-Hill Book Company, Inc., New York, 1923.

heat capacity C must approach zero with decreasing temperature. The ratio C/T does not necessarily remain finite (for example, the case $C = \text{constant} \times T^{\frac{1}{2}}$ yields a finite entropy but an infinite value of C/T at $T = 0$); however, all known crystalline substances have heat capacities which yield finite limits of C/T. For the atomic vibrations in a crystal the Debye theory gives C proportional to T^3 or C/T proportional to T^2. The electronic heat capacity in metals is found to follow the first power of T, but this still leaves C/T finite as $T \to 0$.

In Chapter 10 we obtained the equations

$$\left(\frac{\partial S}{\partial P}\right)_T = -\left(\frac{\partial V}{\partial T}\right)_P \tag{10-17}$$

$$\left(\frac{\partial S}{\partial V}\right)_T = \left(\frac{\partial P}{\partial T}\right)_V = -\frac{(\partial V/\partial T)_P}{(\partial V/\partial P)_T} \tag{10-14}$$

Since we have postulated that the entropy of a perfect crystalline solid is zero at $0°K$ regardless of the pressure or volume, the quantities on the left side of these two equations must be zero. We conclude then

$$\lim_{T \to 0}\left(\frac{\partial V}{\partial T}\right)_P = 0 \tag{12-2}$$

and

$$\lim_{T \to 0}\left(\frac{\partial P}{\partial T}\right)_V = 0 \tag{12-3}$$

The coefficient of thermal expansion has been measured for a number of substances, and it has always been found to approach zero in accordance with Eq. (12-2). Indeed it appears to approach zero asymptotically with zero slope.

Imperfect Crystals, Glasses, Solid Solutions. In the statement of the third law of thermodynamics the term *perfect crystalline substances* was used. Let us now consider the limitations which this term introduces. Any state in which the arrangement of atoms is disordered will have a greater probability than a perfectly ordered arrangement. Therefore we shall expect that, if any type of disorder remains at $0°K$, the entropy will remain larger than zero.

The atomic arrangement in a glassy solid is essentially similar to that in a liquid, which is known to be less regular than the structure of a crystal. Thus we find it very reasonable that the entropy of a glassy solid fails to approach zero at $0°K$. The classic investigation of this question is the study of glycerine by Gibson and Giauque.[1] They found that the entropy of supercooled liquid, i.e., glassy glycerine at $0°K$, was

---

[1] G. E. Gibson and W. F. Giauque, *J. Am. Chem. Soc.*, **45**, 93 (1923); see also F. Simon and E. Lange, *Z. Physik*, **38**, 227 (1926).

5.6 ± 0.1 cal/deg mole greater than that of crystalline glycerine.   Thus we see clearly that a glassy material does not approach zero entropy at the absolute zero.

In solid solutions different atoms or molecules are distributed randomly over certain sites in the crystal.   If this randomness is still present, we expect the solid solution to have greater than zero entropy at 0°K. Eastman and Milner[1] found this to be true for a solid solution of AgCl and AgBr.   The entropy of formation of their solid solution from pure AgCl and AgBr was 1.12 ± 0.1 cal/deg mole at 298°K.   Since the heat capacity of the solution did not differ significantly from that of its pure constituents anywhere in the range from 15 to 298°K, the entropy of mixing was still present at 15°K and presumably would be present at 0°K.

Mixtures of isotopes form solid solutions.   Thus chlorine at 0°K consists of a solid solution of Cl^{35}-Cl^{35}, Cl^{35}-Cl^{37}, and Cl^{37}-Cl^{37} molecules. Because of the close similarity of chemical properties of these isotopes, we expect that the two types of chlorine atoms are distributed with complete randomness.   Consequently, in a strict sense the entropy of this material is not zero at 0°K.   For most purposes, however, it is convenient to ignore this entropy associated with the mixing of isotopes because the same entropy is present in any other substance in which the element may occur.   Thus it is customary to ignore the entropy of mixing of isotopes unless the change in distribution of isotopes is under particular study.

Studies of hyperfine structure of spectral lines have shown that some nuclei have angular momentum or nuclear spin.   The energy differences associated with the various possible orientations of nuclear spin are very small.   Hydrogen is the only substance in which the equilibrium orientation of nuclear spin differs appreciably from the random orientation at temperatures of practical chemical interest.   These properties of hydrogen are considered in Chapter 35.   For other substances (and under certain conditions also for hydrogen) the entropy of random orientation of nuclear spins remains undiminished when the entropy is extrapolated in the usual manner to 0°K.

Since the entropy associated with the random orientation of a particular set of nuclei with spin is always the same regardless of the chemical form in which the atoms occur, one may ignore nuclear spin for all practical chemical calculations.

The atoms or ions of a crystal sometimes have a net magnetic moment from unpaired electron spin or net orbital angular momentum or both.   The effects of such magnetic moments are, in principle, similar to those from nuclear spin, but the atomic magnetic moments are roughly 1000 times larger and hence interact with one another more strongly.   Thus in many cases the atomic magnetic moments have ordered themselves into some regular pattern at low temperatures, and no complications arise in relation to the third law.   In some cases, however, this ordering process is incomplete even at a temperature of 1°K, and there is a large heat capacity associated with the magnetic effects.   Such systems have very interesting properties, which are discussed in Chapter 31.

---

[1] E. D. Eastman and R. T. Milner, *J. Chem. Phys.*, **1**, 444 (1933).

In addition to the rather straightforward examples of glasses and solid solutions (and the trivial case of nuclear spin) there are a few cases where the crystals of a pure substance fail to attain perfect order at $0°K$. The presently known examples are CO, N_2O, NO (actually N_2O_2 in the solid), the 1-olefins with more than 10 carbon atoms, H_2O (also D_2O) and $Na_2SO_4 \cdot 10H_2O$. In each of these cases there are different geometrical orientations of the molecule which may be expected to have almost exactly the same energy. For example, the carbon monoxide molecule has practically zero dipole moment and therefore may easily form a crystal with random end-for-end orientation.

The entropy of carbon monoxide as calculated from the spectroscopic data for the molecule is 47.20 cal/deg mole at $298.15°K$, and this value is confirmed by chemical-equilibrium studies. However, Clayton and Giauque[1] found from heat-capacity measurements that the entropy of solid carbon monoxide at $0°K$ is only 46.2 cal/deg mole less than that at $298.15°K$. Thus an entropy of 1.0 cal/deg mole remains at $0°K$. If the molecules were randomly oriented in an end-for-end fashion thus:

$$\begin{array}{ccccccc} CO & OC & OC & CO & OC & CO & CO \\ & CO & OC & CO & CO & OC & OC \\ OC & OC & CO & OC & CO & CO & \end{array}$$

the multiplicity would be 2 for each molecule or 2^{N_0} for N_0 molecules in a crystal. Thus

$$s = \frac{R}{N_0} \ln W = \frac{R}{N_0} \ln 2^{N_0} = R \ln 2$$
$$= 1.4 \text{ cal/deg mole}$$

The actual residual entropy is a little smaller, which indicates that the molecular orientation is not quite fully random.

The situation for N_2O(NNO), or a long-chain 1-olefin, is entirely analogous to CO. Apparently the shorter 1-olefin molecules attain perfect order in their crystals, but those longer than decene have end-for-end disorder.[2] The case of NO is similar since we know that it exists as a dimeric N_2O_2 unit in the crystal.[3] The N_2O_2 molecule is rectangular and presumably has the pairs of similar atoms in the diagonally opposite corners. Thus there are two possible orientations in the crystal for the dimeric unit.

[1] J. O. Clayton and W. F. Giauque, *J. Am. Chem. Soc.*, **54**, 2610 (1932).

[2] J. P. McCullough et al., *J. Phys. Chem.*, **61**, 289 (1957).

[3] W. J. Dulmage, E. A. Myers, and W. N. Lipscomb, *J. Chem. Phys.*, **19**, 1432 (1951); *Acta Cryst.*, **6**, 760 (1953).

$$N = O \qquad O = N$$

$$\cdot \quad \cdot \qquad \qquad \cdot \quad \cdot$$

$$\cdot \quad \cdot \quad \text{and} \quad \cdot \quad \cdot$$

$$\cdot \quad \cdot \qquad \qquad \cdot \quad \cdot$$

$$O = N \qquad N = O$$

The observed[1] residual entropy is 1.5 cal/deg for N_2O_2, which is close to the value of $R \ln 2$ expected for this model.

The disorder remaining in ice at the absolute zero is similar in principle but follows a more complex pattern. The model is due to Pauling,[2] while the thermodynamic data are from Giauque and Stout.[3] The residual entropy is 0.8 cal/deg mole. Long and Kemp[4] found the same residual entropy in heavy ice. The residual entropy of $R \ln 2$ in $Na_2SO_4 \cdot 10H_2O$ is explained on the same basis as that in ice.

Pauling's model assumes hydrogen bonding between water molecules but that the proton on any O—O linkage is unsymmetrically located. Each oxygen is known to have four nearest neighbor oxygens, and Pauling assumes that the protons are close to the central oxygen along two links and farther away in the other two. Thus water molecules may be said to exist still in the solid. If no further restriction is placed on the proton locations, Pauling calculates that there are $(\frac{3}{2})^{N_0}$ arrangements for N_0 water molecules. Hence $s = R \ln \frac{3}{2} = 0.806$ cal/deg mole.

The hydrated crystal $Na_2SO_4 \cdot 10H_2O$ shows an entropy discrepancy of approximately $R \ln 2$ per mole[5] which is clearly explained by the crystal structure.[6] One four-membered hydrogen-bonded ring is formed per 10 water molecules. If the protons are unsymmetrically located in the hydrogen bonds, there are two sets of proton positions shown by diagrams A and B which retain water-molecule units with two hydrogens close to

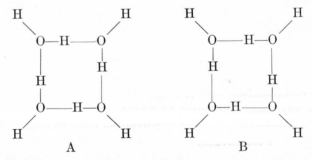

A   B

[1] H. L. Johnston and W. F. Giauque, *J. Am. Chem. Soc.*, **51**, 3194 (1929).
[2] L. Pauling, *J. Am. Chem. Soc.*, **57**, 2680 (1935).
[3] W. F. Giauque and W. Stout, *J. Am. Chem. Soc.*, **58**, 1144 (1936).
[4] E. A. Long and J. D. Kemp, *J. Am. Chem. Soc.*, **58**, 1829 (1936).
[5] K. S. Pitzer and L. V. Coulter, *J. Am. Chem. Soc.*, **60**, 1310 (1938).
[6] D. H. Templeton, private communication.

each oxygen. If the various rings in the crystal have structures A and B at random, then the calculated entropy of disorder is $R \ln 2$. The isomorphous crystal $Na_2CrO_4 \cdot 10H_2O$ has the same ring structure and presumably would have the same residual entropy although the thermal data are not available.

While the possibility of residual entropy from crystal imperfection must always be kept in mind in using the third law, this does not detract from its utility as much as one might think. In most cases one can readily determine that there is no appreciable probability of such a difficulty. Thus, for molecules such as $CO_2(O—C—O)$, CH_4, BF_3, SF_6, their symmetrical structure makes random orientation practically impossible. Even in cases such as $O—C—S$ and the 1-olefins from propene through 1-decene, where random orientation appears probable, the experimental data show that zero entropy is attained at the absolute zero.

Like any other scientific principle, the third law of thermodynamics must be applied with due consideration for the conditions of its applicability. However, it is of immense importance as a source of entropy values for chemical substances since it is always possible to cool a substance to a temperature of 1 to 15°K, but it is not always possible to obtain a chemical reaction at equilibrium which involves a particular substance.

Practical Basis for Absolute Entropies. Absolute entropies as measured and tabulated for practical chemical application are those obtained by experimental measurement down to temperatures in the range 1 to 15°K and extrapolation to 0°K according to the theory of the heat capacity of solids, e.g., the Debye theory. We have noted that the third law holds true on this basis; i.e., the entropies of compounds are found to be zero at 0°K if the entropies of the elements are taken as zero at 0°K. On this practical basis the standard entropy at T' of a solid which is perfectly crystalline at 0°K is given by the integral

$$S_{T'}^{\circ} = \int_0^{T'} \frac{C_P^{\circ}}{T} \, dT \tag{12-4}$$

where the extrapolation to 0°K is carried out as discussed above. The entropy of any transitions must be added separately. For liquids and gasses the entropies of fusion and vaporization must be added as exemplified in Table 11-1 for oxygen.

Two effects have been noted, however, which indicate that randomness still exists in this practical zero state. Isotopes are still randomly mixed and nuclear spins are still randomly oriented at temperatures of 1 to 15°K, and the extrapolation process does not consider these effects. Consequently the entropies of mixing of isotopes and of random orientation of nuclear spins are still present in the state which is taken as the practical standard of zero entropy.   There may be other properties of matter, now unknown, which also contribute entropy in this practical standard state. These problems need not disturb us since these entropy contributions cancel in the calculation of ΔS for the changes which arise ordinarily.

The Entropies of Particular Substances.   The heat capacity has been measured down to very low temperatures for most of the chemical elements and for a substantial number of compounds.   From these data the entropy values have been calculated by the method described in the previous chapter.   It is convenient to tabulate the entropy for a substance in a standard state as was done in Chapter 6 for heat capacities and enthalpies.   For solids and liquids, the standard state is taken at 1 atm as for heat capacities and enthalpies.   Since the entropy of a gas does not approach a constant value as the pressure is reduced, one cannot choose a low-pressure perfect-gas standard state for the tabulation of the entropy of a gas.   The standard state for the tabulation of the entropy of a gas will be the *hypothetical perfect gas at a designated pressure of* 1 *atm.* As for heat capacities and enthalpies, standard-state entropies are also indicated by a superscript, for example, $S°$.

To obtain the standard entropy, we first note the entropy difference between the entropy of a perfect gas at some low pressure P and at 1 atm by Eq. (7-9)*,

$$s° = s_P + R \ln \frac{P}{1} \qquad (12\text{-}5)*$$

If one now takes the limit $P \to 0$, the real gas approaches the ideal-gas state and the quantity $s_P + R \ln P$ will approach a constant independent of pressure.   Thus one has the equation for the standard entropy,

$$s° = \lim_{P \to 0} (s_P + R \ln P) \qquad (12\text{-}6)$$

Convenient methods of dealing with gas imperfection effects are discussed in Chapter 16.

Table 12-3 gives a brief list of entropy values at 298.15°K ($= 25$°C); a more extensive table is given in Appendix 7.   Note that the values of $-(\text{F}° - \text{H}°_{298})/T$ which are tabulated in Appendix 7 at 298.15°K are values of $s°_{298}$, or the entropy at 298.15°K.   In the case of some gases the values calculated by the method to be described in Chapter 27 are more precise than those obtained by the method just considered.   The values in Table 12-3 are those given by the most precise method in each case.   Kelley[1] has tabulated values for the elements and inorganic compounds of $s°_T - s°_{298}$ at 100° intervals to the maximum temperature for which data are available.   These entropy increments may be combined with the $s°_{298}$ values from Table 12-3 or Appendix 7 to obtain entropy values at elevated temperatures.

[1] K. K. Kelley, *U.S. Bur. Mines Bull.* 584, 1960.

TABLE 12-3. THE ENTROPIES OF SOME COMMON SUBSTANCES AT 298.15°K,
CAL/DEG MOLE†

Solids:		Liquids:	
Ag	10.20	Br₂	36.4
AgCl	23.00	HNO₃	37.19
AgBr	25.60	H₂O	16.73
AgI	27.6	Hg	18.17
Al	6.77	Gases:	
As	8.4	CH₄	44.47
C(gr)	1.37	CO	47.20
Ca	9.95	CO₂	51.08
CaO	9.5	Cl₂	53.29
Cd	12.37	F₂	48.49
Cu	7.97	H₂	31.21
Fe	6.49	HCl	44.64
I₂	27.76	H₂S	49.13
S(rh)	7.62	N₂	45.77
Si	4.51	NO	50.34
SiO₂(q)	10.00	O₂	49.01

† See K. K. Kelley, *U.S. Bur. Mines Bull.* 477, 1950, and revision by K. K. Kelley and E. G. King, *U.S. Bur. Mines Bull.* 592, in press, for a critical summary of entropy data and references to original sources.

13

THE FREE ENERGY

We have seen that any actual, or irreversible, process is characterized by an increase in the total entropy of all systems concerned. Therefore a system is subject to spontaneous change if there is any conceivable process for which $\Sigma\, dS > 0$, where the sum covers all the systems affected. On the other hand, a state of equilibrium is one in which every possible infinitesimal process is reversible, or one in which the total entropy remains constant. It is therefore a necessary condition for equilibrium that, for any process, $\Sigma\, dS = 0$.

This is the most general criterion of equilibrium which thermodynamics offers. Indeed for many purposes it is too general. It is not always as easy to study the change in entropy of all the systems which may be affected by a given process as to focus our attention upon some one system. It is for this reason that numerous "thermodynamic functions" have been invented which are less fundamental and less general than the entropy but which are of more practical convenience in the study of some concrete problems. In Chapter 10 we defined the quantities A and F.

$$A = E - TS \qquad\qquad (10\text{-}9)$$
$$F = H - TS \qquad\qquad (10\text{-}10)$$

These quantities are properties of a system, determined by its state and not by its history, since they are defined by means of quantities which themselves are all properties. Moreover, if we consider two identical systems, since T and P are intensive properties, while E, H, and S are extensive, the value of each term will be twice as great for the two systems as for either one, and A and F are extensive properties.

Maximum Work. Let us again reexamine the relationship between the quantity A and the maximum work which can be done by a closed system undergoing a change in state. For a change from one state to another at the same temperature,

$$\Delta A = \Delta E - T\, \Delta S \qquad\qquad (13\text{-}1)$$

138

Now, if the process is also reversible, the last term is the heat absorbed. Hence, by the first law of thermodynamics, the total work done by the system in the reversible process is equal to $-\Delta A = A_A - A_B$.

In any other process leading from the same initial to the same final condition ΔA will be the same, and if the second process is also a reversible one, the work performed by the system will also be $-\Delta A$. In other words, all reversible isothermal processes, leading from the same initial to the same final states, do the same work on the system, $w = \Delta A$. In any actual isothermal process the work performed by the system is less, owing to friction and other sources of degradation, but we may regard $-\Delta A$ as the limit of the work which can be performed by an isothermal process, as its efficiency is indefinitely improved.

As a concrete example, let us consider a process in which zinc acts upon aqueous sulfuric acid to form aqueous zinc sulfate and hydrogen, under constant atmospheric pressure, and in a thermostat. Evidently this process can occur in such a way as to perform no work except the small amount done against the atmosphere by the evolved hydrogen. This, in fact, is just what will occur if impure zinc is added directly to the acid. This process is highly irreversible. On the other hand, if we place in the thermostat the same substances arranged as a galvanic cell, with zinc as one electrode and another electrode of hydrogen in contact with a platinized electrode, and if these two electrodes are connected to a motor or other electrical system in such a way as to utilize the electrical energy which is now available, an amount of work will be done which will depend upon the efficiency of our arrangements.

The maximum work would be obtained if at every instant the external electrical system were arranged to exert so large a counter emf that, when infinitesimally increased, it would force the current in the opposite direction, thus causing hydrogen to be consumed and zinc to precipitate. The process would then be reversible and the total work equal to $A_A - A_B$.

In both the processes which we have described, the system passes from the same initial to the same final states. Therefore $-\Delta A$ is the same in both cases. This difference $-\Delta A$ is, in the reversible process, the work which the system has performed, while in the irreversible process it is the maximum work which *might* have been performed.

As a corollary we may note that, while $T \Delta S$ is also the same for both processes, it represents in the reversible process the amount of heat actually absorbed from the thermostat, while in the irreversible process it merely shows the maximum amount of heat which might have been absorbed; for ΔE is the same in both cases, and therefore the heat absorbed will be greater, the greater the amount of work performed.

For an isothermal process at constant volume which is not harnessed to do any work, we have, from Eq. (10-19), $0 = dA + T \, dS_{\text{irr}}$. Such a process can take place spontaneously with resultant entropy production only if the work content decreases. The condition for thermodynamic equilibrium is $dS_{irr} = 0$, and therefore, with respect to a process occurring at constant temperature and volume, the state of equilibrium is defined by the equation

$$dA = 0 \quad \checkmark \tag{13-2}$$

The Net Work in an Isothermal Process at Constant Pressure. For the electrical cell that we have just considered, the total work done is the sum of the electrical work and the work done against the constant pressure of the atmosphere. The latter quantity of work, which is equal to $P \, \Delta V$, must always be done when a process occurs at constant pressure whether this process be reversible or irreversible. Thus the maximum electrical work that could be done by the system undergoing a given change in state is less than the total maximum work, $-\Delta A$, by the expansion work. In Chapter 10, the total work done by a system less the expansion work was designated by the symbol $-w'$, which we call the net work performed by the system. We shall recall from Eq. (10-20) that w' can be related to the free energy

$$-w' = -\Delta F = F_A - F_B \tag{13-3}$$

Thus, in any process occurring at constant temperature and pressure, $F_A - F_B$ represents the maximum of work in addition to expansion work which can be obtained from a given process. It is for this reason that F is known as the *free energy*.

The quantities F and A, or, better, the molal or partial molal values of these quantities, are sometimes called thermodynamic potentials on account of certain rough analogies to mechanics, which we need not stress. While the value of such functions was pointed out by Massieu,[1] their great utility in the interpretation of the most diverse physico-chemical phenomena was first fully demonstrated in the comprehensive work of Gibbs.[2] Students of Gibbs will observe that the quantities F and A are his functions ζ and ψ. (We may note also that our H is Gibbs's χ and our E his ϵ.)

It was Helmholtz[3] who introduced the term *free energy*, and it is to be remarked that he applied this name to A rather than to F. Later, however, since ΔA and ΔF do not largely differ numerically, unfortunate confusion has arisen in the literature between these two quantities. Several

[1] M. F. Massieu, *Compt. rend.*, **69**, 858, 1057 (1869). He employed $-F/T$ and $-A/T$, the former of which has more recently been used by M. Planck, "Thermodynamik," Veit and Co., Leipzig, 1897.

[2] J. W. Gibbs, *Trans. Conn. Acad. Arts. Sci.*, **2**, 309, 382 (1873); **3**, 108, 343 (1875).

[3] H. von Helmholtz, *Sitzber. kgl. preuss. Akad. Wiss.*, **1**, 22 (1882).

writers[1] who have defined the free energy as A have really used the function F.   It has therefore seemed best to retain this useful expression for the latter function, which has in practice by far the greater importance. European textbooks commonly use the symbol F and the name *free energy* or *Helmholtz free energy* for the function $A = E - TS$.   These books also use the symbol G and names such as *Gibbs function, Gibbs free energy*, or *free enthalpy* for the function $F = H - TS$.   However, virtually all major tabulations of chemical thermodynamic data use the symbol F for the $H - TS$ function, and these tabulations will be the dominant source of thermodynamic data for many years to come.   Thus the use of the symbol F for the work-content function can only lead to serious confusion of the type already noted and should be discouraged.[2]

It can be seen that F bears the same relation to A that H bears to E. We have previously remarked that the choice of pressure and temperature as the chief variables in thermodynamics, together with the fact that most experiments are carried on at constant temperature or pressure, makes quantities like H, C_P, and F more generally useful than the corresponding quantities E, C_V, and A.

The Driving Force of a Chemical Reaction, and a Test for Equilibrium.   The difference between a process conducted reversibly, and the same process conducted with different degrees of irreversibility, is well illustrated by the study of the type of galvanic cell which is called a storage battery.   The minimum electrical work required to charge the battery from one state to another is equal to the maximum work obtainable in discharging from the second state to the first, and both these are equal numerically to the change of free energy within the cell.   Stating this same thing in another way, the minimum emf required for charging and the maximum emf obtainable in discharge are both equal to the reversible emf.

[1] Thus W. Nernst, "Theoretische Chemie," 6th ed., p. 730, Enke, Stuttgart, 1909, writes on the same page two equations, one connecting ΔA and equilibrium constant and the other ΔA and emf.   Both these equations are quite erroneous, but similar equations in which ΔF is substituted for ΔA are correct.   Even F. Haber in his extraordinarily careful work, "The Thermodynamics of Technical Gas Reactions" (trans. by A. B. Lamb), Longmans, Green & Co., Inc., New York, 1908, occasionally confuses these quantities, thus causing a numerical error of 300 cal on page 177 and of 700 cal on page 320.

[2] A reasonable resolution of the problem is the abandonment of the use of the name free energy and the symbol F.   If this were done, the name work content with the symbol A could be used for the function $E - TS$ and the name free enthalpy or Gibbs energy with the symbol G for the function $H - TS$.   This would avoid conflicting usage of the same name or symbol for different functions.   We hope that an agreement to adopt such a nonconflicting system can be reached at an international level in the near future.   In the meantime, however, it seems best to retain consistency with the major tables of chemical thermodynamic data.

In attempting to measure such a value we find that the charging and discharging emfs are nearer together the smaller the current. With modern methods of measuring potentials, it is sometimes possible to balance a cell so nicely that a change of one-millionth volt will charge or discharge the cell, as shown by the deflection of a galvanometer to right or left. However, in order to approach so near to reversibility, it is necessary to use a galvanometer which is sensitive to a current of the order of 10^{-9} amp. At this rate it would take 1 million years to charge a small storage battery.

In practice reversibility is sacrificed to save time, and if we examine the so-called charge and discharge curves of a battery, we find that the emf obtained in discharging the battery at the normal rate is smaller by several tenths of a volt than the reversible emf. That used in charging is correspondingly higher. The more rapid the charging or discharging is, the greater is the loss in energy efficiency. This is somewhat analogous to the mechanical principle that frictional losses increase with velocity.

In practice different processes differ greatly in degree of irreversibility but it is a universal rule that, if any isothermal process is to occur with finite velocity, it is necessary that $\Delta F < w'$. That is, the net work done on the system will have to be greater than the free-energy increase of the system, or the net work done by the system will have to be less than the free-energy decrease of the system.

Systems Which Are Subject to No External Forces except a Constant Pressure. In the preceding discussion we have considered a chemical process which is in some way harnessed for the production of useful work. We may now turn to the far more common case of a reaction which runs freely, like the combustion of fuel or the action of an acid upon a metal. In other words, let us consider a system which is subjected to no external forces except a constant pressure exerted by the environment. In such cases $w' = 0$, and it follows from what we have said above that no actual isothermal process is possible unless $\Delta F < 0$.

Therefore, if we know the value of ΔF for any isothermal reaction, and if this value is positive, then we know that the reaction, in the direction indicated, is thermodynamically impossible. If, on the other hand, the value is negative, the process is one which can occur, and does occur, although perhaps with no measurable speed.

When water is formed from its elements at 25°C, there is a very large loss of free energy, which may be represented by the equation

$$H_2(g, 1 \text{ atm}) + \tfrac{1}{2}O_2(g, 1 \text{ atm}) = H_2O(l) \qquad \Delta F_{298} = -56{,}690 \text{ cal}$$

The reaction can therefore proceed spontaneously, and although ordinarily it takes place at no measurable rate, the presence of a catalyzer like spongy platinum permits it to proceed readily.

We may next consider the union of oxygen and nitrogen to form nitric oxide. We shall find for this reaction the equation

$$N_2 + O_2 = 2NO \qquad \Delta F_{298} = 43{,}200 \text{ cal}$$

The large positive value of ΔF shows not only that the reaction has no tendency to proceed in the direction indicated but also that nitric oxide, from a thermodynamic standpoint, is an extremely unstable substance.

The fact that nitric oxide apparently does not dissociate into its elements under ordinary conditions must be ascribed to an extremely inert or unreactive character which it is beyond the scope of thermodynamics to predict or to explain. In an earlier chapter we have spoken of the ambiguous manner in which the term *stability* is used. In a thermodynamic sense a system is stable when no process can occur with a diminution in free energy. By a rough analogy to mechanics we may think of the quantity $-\Delta F$ as the driving force of a reaction, while the factors which retard a possible process may be likened to friction.

A Necessary Condition for Equilibrium. We have seen that for any isothermal reversible process $dF = \delta w'$, and since it is a condition for equilibrium that every infinitesimal process be reversible, this equation furnishes also a test of equilibrium. The condition that entropy be merely transferred and not produced in a reversible process has its counterpart for free energy. In any reversible process at constant T and P, free energy can be transferred from the system to the surroundings but is not destroyed. Thus, for a reversible process with work being done on the surroundings, the decrease in free energy of the system is just equal to the increase of the free energy of the surroundings (the work stored in the surroundings). When an irreversible isothermal isobaric process takes place, resulting in entropy production, there will be a corresponding free-energy destruction. But if we are dealing with a system which can perform no net work—in other words, if our system is subject to no external forces except a constant pressure—then $w' = 0$ and no free energy can be transferred. Thus, with respect to any process occurring at constant temperature and pressure, the state of equilibrium is defined by the equation

$$dF = 0 \tag{13-4}$$

This seems at first sight to be a criterion of very limited scope until we realize that, if no reaction at constant temperature and pressure is thermodynamically possible, no reaction whatever can occur. For suppose that some spontaneous process could occur in such a way as to produce inequalities of temperature and pressure within the system or to produce a difference between the temperature and pressure of the system and the temperature and pressure of the environment; then this process could be followed by another obviously spontaneous process, consisting in the equalization of pressure and temperature. But these two processes together would be the equivalent of a spontaneous process occurring at constant temperature and pressure.

It is therefore a general criterion of equilibrium, with respect to *every possible change*, that the free energy remain unchanged in any infinitesimal process occurring at constant temperature and pressure.

PROBLEMS

13-1. Consider the process of fusion of ice at 1 atm, namely, $H_2O(s) = H_2O(l)$; what is the sign of ΔF at $-10°C$, at $0°C$, and at $10°C$?

13-2. What is the change in the free energy (calories) of 1 mole of water at $25°C$ when the pressure is increased by 1 atm?

13-3. The free energy of formation from the elements at 1 atm of 1 mole of water vapor at a pressure of 0.001 atm at $25°C$ is $\Delta F_{298} = -58,730$ cal. By comparison with the free energy of formation of liquid water given in the text, calculate whether the equilibrium vapor pressure of water is greater or less than 0.001 atm at $25°C$. Assume that water vapor is a perfect gas, and calculate the pressure of water vapor that would be in equilibrium with liquid water at $25°C$.

14

THE CHEMICAL POTENTIAL AND ESCAPING TENDENCY; EQUILIBRIUM BETWEEN TWO OR MORE PURE PHASES; THE FUGACITY

In the science of mechanics there are two kinds of methods of ascertaining whether or not a system is in equilibrium. One consists in determining the effect of some infinitesimal displacement upon the properties of the system as a whole; the other depends upon a more localized view of the forces operating between different portions of the system. Thus, if a system is at rest, it is a sufficient condition for equilibrium either that the potential energy of the system is at a minimum with respect to any possible infinitesimal displacement or that the resultant of the forces acting on every part of the system is zero.

The former of these is analogous to those thermodynamic criteria of equilibrium which we have hitherto stated. For many purposes, however, it is desirable to possess also in thermodynamics a more intimate criterion of physicochemical equilibrium, such as we have already obtained for thermal equilibrium.

In order that a system be in thermal equilibrium, it is a necessary and sufficient condition that any infinitesimal flow of heat within the system be attended by no change of total entropy. However, this criterion is far less convenient than the one comprised in the simple statement that thermal equilibrium exists when the temperature is the same in all parts of the system. The entropy criterion directs our attention to changes in the whole of a complicated system. The temperature criterion permits us to focus our attention upon the individual parts.

The value of the temperature concept lies in the fact that two bodies have the same temperature if they both have the same temperature as a third and that we have a corresponding law of temperature inequalities. The temperature scale is therefore unambiguous. If we choose, we may imagine everything to have a certain tendency to lose heat, or we may say that heat has a tendency to escape from every system. Temperature is then a measure of this escaping tendency of heat; for if one body has a higher temperature than another, then the escaping tendency of heat can be said to be greater in that body and, whatever mechanism may be used to establish thermal contact, heat will flow from that body to the other.

Now all that we have said regarding the distribution of thermal energy is true also of the distribution of a material substance throughout any system. If three phases of a substance are coexistent, let us say ice, water, and water vapor, all at constant temperature, it would be impossible to find a state of affairs in which at the same time ice spontaneously goes over into water, water into water vapor, and water vapor into ice. For if this were the case, we might have a complete cycle leading to a system identical with the original system and therefore possessing the same entropy. But spontaneous processes without increase of entropy do not exist.

It is evident therefore that if a substance X is distributed through some system, we may speak of the escaping tendency of X in each part, or in each phase, of the system. The condition of equilibrium will be that the escaping tendency of each substance is constant throughout the system. The escaping tendency will thus obey the same laws of equality and inequality which we have found to hold for temperature.

As an illustration, we may state that the escaping tendency of water is the same for liquid and solid at the freezing point, while at lower temperatures the escaping tendency is greater for the liquid than for the solid. The escaping tendency of sodium chloride in solution is more than, equal to, or less than that of solid sodium chloride, according as the solution is supersaturated, saturated, or unsaturated. If water and ether are shaken together to form two liquid phases in equilibrium, the escaping tendency of the water, and also that of the ether, will be the same in both phases. Moreover, if there is a vapor phase present, the escaping tendency of each substance is the same in that phase as in the other two phases. Lest there be any misunderstanding, it should be made clear that we are not comparing the escaping tendency of one substance with that of another; we are merely comparing the escaping tendency of a given substance in one phase with that of the same substance in another phase.

The value of this conception is not limited to the simpler systems of thermodynamics. If a column of some solution is placed in a gravitational field, equilibrium will not be established until the escaping tendency of each substance present is constant throughout the field; if we consider the surface layer of a soap solution, the concentration of soap is very different in this surface layer and in the interior of the liquid but the escaping tendency of the soap must be the same in both places.

In adopting the concept of escaping tendency we are abandoning a view of the system as a whole in order to view more intimately the individual substance.   We seek some property which will measure quantitatively the escaping tendency of each material substance, as temperature measures the escaping tendency of heat.

Molal Free Energy as a Measure of the Escaping Tendency. Let us consider once more the equilibrium existing between ice and water at 0°C and at a pressure of 1 atm.   If an infinitesimal amount of ice be melted at constant temperature and pressure, then, from Eq. (13-4), we see that the condition of equilibrium is

$$d\mathrm{F} = 0$$

Since equilibrium is maintained as long as both phases are present, we may also write for any finite amount melted

$$\Delta\mathrm{F} = 0$$

If we write the chemical equation,

$$H_2O(s,\ 1\ \text{atm}) = H_2O(l,\ 1\ \text{atm})$$

then $\Delta\mathrm{F}$ measures the increase in free energy when 1 mole of ice is converted into 1 mole of water.   If the molal free energy of water is F_A in the solid state and F_B in the liquid state,

$$\Delta\mathrm{F} = F_B - F_A$$

Therefore at the melting point, under atmospheric pressure,

$$F_B = F_A$$

At higher temperatures the process in the direction indicated is a spontaneous one,

$$\Delta\mathrm{F} < 0 \qquad F_B < F_A$$

At lower temperatures, on the other hand, the process is thermodynamically impossible,

$$\Delta\mathrm{F} > 0 \qquad F_B > F_A$$

By similar reasoning we see that in general no system can be in equilibrium unless the molal free energy of each substance[1] involved is the same in every part of the system.   If the molal free energy of any substance is greater in one part of the system than in another, that substance will pass from the former to the latter place.

In order to avoid any misunderstanding, we must repeat a remark made in Chapter 3 regarding the meaning of the mole.   If we had stated that a condition of equilibrium requires that the free energy of 1 g or of any given number of grams of a substance is the same in the several phases, such a statement would be unquestionable.   But

[1] In accordance with usage determined in Chapter 2, the term *substance*, as here employed, may mean a single molecular species, such as H_2O, or it may mean something immediately obtainable from a single molecular species, such as water.

when we state that the molal free energy is the same in all phases, we imply that the same formula is used for the substance in each phase. This is unfortunately not always the case. Iodine vapor at high temperatures dissociates in part into the monatomic form, so that we write $I_2(g) = 2I(g)$, and the condition of equilibrium is that $\Delta F = 0$, or that the molal free energy of $I_2(g)$ is equal to twice that of $I(g)$. In the case of sulfur there are various molecular species known or suspected. Thus we have $S(g)$, $S_2(g)$, $S_4(g)$, $S_6(g)$, $S_8(g)$, and various liquid and solid forms. Both rhombic and monoclinic sulfur form crystal lattices in which the same unaltered molecule S_8 is the primary unit. Nevertheless it is frequently preferable to use the simple formula S. These special cases need not, however, distract our attention from the points which we are now considering.

Molal Free Energy as the Chemical Potential. We are familiar with the use of potentials in mechanical and electrical systems. Each potential has associated with it a capacity quantity, and the work done on the system is given by the product of the capacity factor times the change in potential. Thus for gravitational potential gh the mass is the capacity factor, and the work is $mg(h_2 - h_1)$. For the electrostatic system the charge q is the capacity, and we designate the electrostatic potential Φ. Then the work required to move the charge q from a location of potential Φ_1 to one of potential Φ_2 is $q(\Phi_2 - \Phi_1)$.

With chemical systems we can similarly define capacity and potential quantities. If we choose the mole as the capacity factor and deal with systems at constant temperature and pressure, then the *molal free energy is clearly the chemical potential.* This conclusion follows directly from the discussion of Chapters 10 and 13, where the change in free energy was shown to be equal to work done on the system at constant T and P (and excluding PV work against the constant-pressure atmosphere). Thus, to transfer n_A moles of substance A from a state of molal free energy F_1 to a state of molal free energy F_2, we must do work $n_A(F_2 - F_1)$. For example, suppose that we have two large tanks of an ideal gas at pressures P_1 and P_2, respectively, and that we wish to transfer n_A moles from tank 1 to tank 2. The work of isothermal compression is $n_A RT \ln (P_2/P_1)$, which is just $n_A(F_2 - F_1)$ since $F_2 - F_1 = RT \ln (P_2/P_1)$.

If the chemical substance is free to move from one place to another, then it will move spontaneously to the state of lower chemical potential. At equilibrium the chemical potential must be constant throughout the entire system for each substance that is free to move.

The concept of a chemical potential becomes even more useful when we begin to deal with solutions. Also its utility is not limited to conditions of constant temperature and pressure. However, we shall confine our attention at this point to single component systems at constant T and P. The symbol μ is frequently used for the chemical potential, and we shall employ it particularly whenever other than constant-temperature and -pressure conditions are involved.

While the chemical potential is an excellent quantitative measure of escaping tendency, we shall not restrict ourselves to its use on all occasions, for there are other measures which will prove more convenient for some purposes. But the usefulness of the present definition will be seen when we now proceed to investigate more systematically than we have hitherto the equilibrium between two or more pure phases.

THE EFFECT OF PRESSURE AND TEMPERATURE UPON SIMPLE TYPES OF EQUILIBRIUM

If two phases of a pure substance are in equilibrium with one another and the pressure is increased, the molal free energy of the substance will be increased in each phase, and to a greater extent in the phase of larger molal volume [Eq. (10-16)]. That phase will therefore disappear. Thus at 1 atm ice and water are in equilibrium at 0°C; if the pressure exerted upon the two phases is now increased, the chemical potential of the ice, which is more voluminous, is increased more than the chemical potential of the water and the former will disappear if the temperature remains constant. So also, if the two phases are in equilibrium and the temperature is changed, the two molal free energies will change differently in accordance with Eq. (10-15) and equilibrium will no longer exist.

We may, however, change both temperature and pressure in such a way as to keep the two chemical potentials equal to one another. This is the condition for the maintenance of equilibrium, which now we need only translate into mathematical form.

Change of "Equilibrium Pressure" with Temperature. If we have two phases of the same pure substance, in equilibrium, and if F_A, V_A, S_A, H_A and F_B, V_B, S_B, H_B are the molal free energies, molal volumes, etc., in the two phases, our condition for equilibrium is

$$F_A = F_B \tag{14-1}$$

Moreover, if any change occurs and equilibrium is to be maintained, it is necessary that

$$d F_A = d F_B \tag{14-2}$$

But the two states are completely determined by the two variables P and T. Hence we have [see Eq. (3-1)]

$$d F_A = \left(\frac{\partial F_A}{\partial P}\right)_T dP + \left(\frac{\partial F_A}{\partial T}\right)_P dT \qquad d F_B = \left(\frac{\partial F_B}{\partial P}\right)_T dP + \left(\frac{\partial F_B}{\partial T}\right)_P dT \tag{14-3}$$

By combining this with Eqs. (14-1) and (14-2) and substituting the values of the differential coefficients from Eqs. (10-15) and (10-16), we find (since $\Delta F = 0$)

$$\boxed{\frac{dP}{dT} = \frac{s_B - s_A}{v_B - v_A} = \frac{\Delta s}{\Delta v} = \frac{\Delta H}{T \, \Delta v}}$$ (14-4)

This equation shows how the equilibrium pressure must change with the temperature in any two-phase system such as vapor and liquid, liquid and solid, or two solid forms like rhombic and monoclinic sulfur.

In case we are dealing with vapor pressure it is convenient to denote this equilibrium pressure by p, and Eq. (14-4) is evidently identical with Eq. (10-22), the Clapeyron equation, which we obtained before by using the fundamental equations of entropy. Indeed in this simple case that method seems a little less cumbrous, but the method that we have just described is of such general serviceability, and will be so frequently employed, that its use will become almost automatic.

Two Phases under Different Pressures at Constant Temperature. Strictly speaking, a system cannot be in complete equilibrium unless the pressure is constant throughout (except in so far as pressure differences may be produced by such an influence as a gravitational field). If every actual substance has a finite fluidity, there must be a constant, although perhaps imperceptible, flow toward a state of constant pressure. But as it rarely happens that we are fortunate enough to find a system which is in equilibrium with respect to every possible change, we are content to find the conditions of equilibrium with respect to certain changes, without inquiring whether there may be other processes which are slowly proceeding. Thus, if we are studying the equilibrium between liquid benzene and its vapor, we are not at all disturbed if we are told that benzene is essentially unstable and tends to go spontaneously into carbon and hydrogen.

So it will be profitable to consider the distribution of a substance between two phases of unlike pressure, provided that a change in such distribution is rapid compared with the process of equalization of pressure throughout the system. Attention was first drawn to such systems by the discovery of the so-called semipermeable membrane, which in turn led to the discovery of osmotic pressure. We shall here consider a membrane of somewhat similar properties, which is chosen, not to permit the passage of a single substance in the presence of other substances, but rather to permit the passage of a substance in one state while not transmitting pressure.

In Fig. 14-1, A will represent a piston pressing upon a mass of liquid mercury B, while E is another piston exerting pressure upon pure mercury vapor in D. The diaphragm C is of fine porous porcelain. Owing to the phenomenon of capillarity, an excess pressure can be exerted by the piston A without forcing the liquid mercury through the pores of the diaphragm.

On the other hand, the vapor of mercury may pass readily through these pores, thus permitting a transfer of material between B and D.

If the system comes to equilibrium at constant temperature, there will be a certain pressure in the vapor phase and another pressure in the liquid phase. Now, if the pressure upon one phase is increased, the escaping tendency from that phase becomes greater. In order to maintain equilibrium, the pressure upon the other phase must then be raised until the escaping tendency from that phase is increased by an equal amount.

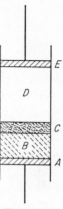

Fig. 14-1

Let us proceed, in the general case, to calculate the ratio of the pressure increments upon two phases of the same pure substance, necessary to maintain equilibrium at constant temperature. Let P_A, v_A, and F_A, P_B, v_B, and F_B be the pressures, molal volumes, and molal free energies in the two phases in a condition of equilibrium. Then, if the pressure upon the first phase is increased by the arbitrary amount dP_A, let us find the change of pressure, dP_B, upon the other phase which will just maintain equilibrium, or, in other words, which will make $dF_A = dF_B$. By Eq. (10-16),

$$dF_A = v_A\, dP_A \qquad dF_B = v_B\, dP_B$$

Hence, for equilibrium,

$$\frac{dP_A}{dP_B} = \frac{v_B}{v_A} \qquad \text{or} \qquad \left(\frac{\partial P_A}{\partial P_B}\right)_T = \frac{v_B}{v_A} \qquad (14\text{-}5)$$

The two changes of pressure must therefore be inversely proportional to the two molal volumes.[1]

Two Phases under Different Pressures, and the Temperature of the Whole System Variable. When the system which we have just been considering is subjected to changes of temperature, there are three independent variables, P_A, P_B, and T, and our criterion of equilibrium merely gives us one relation between the variation in these quantities,

$$dF_A = \left(\frac{\partial F_A}{\partial P_A}\right) dP_A + \left(\frac{\partial F_A}{\partial T}\right) dT = dF_B = \left(\frac{\partial F_B}{\partial P_B}\right) dP_B + \left(\frac{\partial F_B}{\partial T}\right) dT$$

$$(14\text{-}6)$$

and substituting again the values of the various partial differential coefficients,

$$v_B\, dP_B - v_A\, dP_A = (s_B - s_A)\, dT = \frac{\Delta H}{T}\, dT \qquad (14\text{-}7)$$

[1] This equation was first obtained, for the special case of liquid and vapor, by J. H. Poynting, *Phil. Mag.*, (4)**12**, 32 (1881); for the general case, by H. Le Châtelier, *Z. physik. Chem.*, **9**, 335 (1892).

The only case of much interest here is the one in which the pressure of one phase remains constant. Then the pressure on the other phase is determined at each temperature. So in Eq. (14-7) let $dP_A = 0$. Then

$$\left(\frac{\partial P_B}{\partial T}\right)_{P_A} = \frac{\Delta H}{T v_B} \tag{14-8}$$

Thus, if we are considering a system in which the first phase is liquid mercury and the second is mercury vapor, each phase being under a different pressure as in Fig. 14-1, Eq. (14-8) shows the temperature coefficient of vapor pressure, not when the liquid is under the variable vapor pressure as in Eq. (14-4), but when the pressure upon the liquid is constant.[1]

Systems in Which the Temperature Is Not the Same in All Parts. The study of equilibria when different parts of the system are at different temperatures is difficult, owing to the fact that the flow of heat is usually rapid compared with processes involving the transfer of a material substance. When a solution is placed in a tube the two ends of which are maintained at different temperatures, the system reaches a constant state in which the composition varies at the two ends. This is the Soret phenomenon.[2] Since the heat flow occurs at a rate comparable with the rate of transfer of material, we cannot treat this effect in the simple fashion adopted for cases of pressure difference. Also we would not use the free-energy function since it is designed for the study of isothermal processes. Nonisothermal systems will be considered further in Chapter 28.

THE PHASE RULE

When the state of a system cannot be completely determined until at least r data are given, we say that it possesses r degrees of freedom. This is only another way of saying that the state of the system depends upon r independent variables. The state of a single phase of a pure substance depends usually upon only two variables, let us say temperature and pressure. If we impose the further condition that two such pure phases are to exist together in equilibrium, the number of degrees of freedom is reduced to one. With three coexisting phases we have what is known as a nonvariant system. Thus, if water and water vapor are to be coexistent, we may arbitrarily fix the pressure or the temperature, but not both, while the condition that ice, water, and water vapor exist together completely determines the state of the system at a so-called triple point.

[1] This equation is due to G. N. Lewis, *Proc. Am. Acad.*, **37**, 49 (1901); *Z. physik. Chem.*, **38**, 205 (1901).

[2] C. Soret, *Ann. chim. phys.*, (5)**22**, 293 (1881).

There may be other variables besides temperature and pressure which are requisite to determine the state of a phase. Thus we may have to consider the presence of an electric or of a magnetic field or the size of particles. Especially if the phase in question is a solution, the number of degrees of freedom is increased by one for each component beyond the first.

Whatever the number of variables, it still remains true that the number of degrees of freedom of a system as a whole is equal to the number of variables requisite to determine the state of the individual phases, less the number of phases in the system, beyond the first. This is the celebrated phase rule of Gibbs. *[phase rule]*

This simple principle has been the starting point for the development of a large field of exact but qualitative thermodynamic study, into which it will be impossible for us to enter far, since we are presenting the science of thermodynamics with the primary purpose of making it readily applicable to quantitative and numerical calculations. We shall return from time to time to the problem of equilibria between phases. Let us now turn to another useful measure of escaping tendency.

THE FUGACITY

When we were speaking of the qualitative laws of temperature, we saw that any number of quantitative temperature scales might be erected which would conform to these qualitative laws. Thus we might have defined the temperature scale by means of the air thermometer instead of the perfect-gas thermometer; or if T is the reading of the latter scale, we might have defined temperature as T^2 or $\ln T$. It has, however, proved a convenience to adopt one and only one temperature scale.

In the treatment of the analogous concept of escaping tendency, our methods are not so completely standardized. Although we have already seen that the molal free energy or chemical potential measures the escaping tendency, there are certain respects in which this function is awkward. For instance, the molal free energy of a gas approaches an infinite negative value as the pressure approaches zero, and we shall see that this kind of inconvenience enters even more seriously in the study of the partial molal free energy in solutions.

For such reasons another scale of measurement of the escaping tendency is sometimes to be preferred, and we shall not hesitate to employ, side by side with the molal free energy, a second measure of the escaping tendency, which is called the *fugacity*.[1]

The vapor pressure has frequently been used in a qualitative way as a

[1] G. N. Lewis, *Proc. Am. Acad.*, **37,** 49 (1901); *Z. physik. Chem.*, **38,** 205 (1901).

measure of escaping tendency.   Thus the vapor pressure of ice is equal to that of water at the melting point, but less than that of water at all lower temperatures.   So also, if we have two forms of sulfur, at a given temperature, the one which is more stable must have the lower vapor pressure.

Indeed we could with entire correctness use the vapor pressure also as a quantitative measure of escaping tendency, and this would be a very satisfactory procedure if every vapor behaved as a perfect gas.   In the fugacity we are going to define a measure of escaping tendency which bears to the vapor pressure a relation analogous to the relation between the perfect-gas thermometer and a thermometer of some actual gas. The fugacity will be equal to the vapor pressure when the vapor is a perfect gas, and in general it may be regarded as an "ideal" or "corrected" vapor pressure.

We have found in Eq. (10-44)* the difference in free energy of a perfect gas between two pressures, at the same temperature, namely,

$$F_B - F_A = RT \ln \frac{P_B}{P_A} \qquad (14\text{-}9)^*$$

We wish the same functional form to apply exactly for the difference in fugacity between states A and B, thus:

$$\ln f_B = \frac{F_B}{RT} + \left(\ln f_A - \frac{F_A}{RT} \right) \qquad (14\text{-}10)$$

But in the limit of a gas at low pressure we wish to make $f = P$; hence *we define fugacity by the equation*

$$\ln f_B = \frac{F_B}{RT} + \lim_{P_A \to 0} \left(\ln P_A - \frac{F_A}{RT} \right) \qquad (14\text{-}11)$$

The second term on the right is a constant at a given temperature, and one obtains the simple differential form

$$d \ln f = \frac{dF}{RT} \quad \cdot \text{ at constant } T \qquad (14\text{-}12)$$

This definition of the fugacity applies not only to real gases but to liquids and solids as well.   For if we admit that every substance at a finite temperature has a finite vapor pressure, then if the pressure upon a substance is decreased without limit the substance will eventually vaporize and with further diminution in pressure the vapor will approach nearer and nearer to the condition of a perfect gas.   Fugacity will be regarded as having the same dimensions as pressure, and the unit of fugacity will be the atmosphere.

Change of Fugacity with Pressure. We have seen by Eq. (10-16) that

$$\left(\frac{\partial F}{\partial P}\right)_T = v$$

and that by Eq. (14-12)

$$\left(\frac{\partial \ln f}{\partial F}\right)_T = \frac{1}{RT} \tag{14-13}$$

whence

$$\left(\frac{\partial \ln f}{\partial P}\right)_T = \frac{v}{RT} \tag{14-14}$$

By integrating this equation at constant temperature, we may ascertain the fugacity of a substance at any pressure if it is known at some other pressure and if v is known as a function of P. Let us illustrate this procedure at this point by a calculation of the fugacity of a liquid. In Chapter 16 we shall discuss the application of this method to imperfect gases.

Calculation of Fugacity of Liquids or Solids. Whenever two phases are in equilibrium, the fugacity of the substance in both phases is the same. If, therefore, a solid or liquid phase is in equilibrium with vapor, and if we do not desire the highest accuracy, or if the vapor, as nearly as we can measure, obeys the gas law, we take the vapor pressure as a measure of the fugacity. If the fugacity in the vapor phase cannot be taken equal to the pressure, it may be obtained by methods to be discussed in Chapter 16.

Let us now calculate the fugacity of water at 1000 atm and 25°C. At this temperature water vapor is very nearly a perfect gas, and we shall make little error in taking its vapor pressure of 23.75 mm Hg or 0.0312 atm as a measure of the fugacity of the liquid at low pressure. The density of water at 25°C is 0.99704; hence its molal volume 18.070 cc at low pressure. The compressibility of water varies with pressure but it will be a reasonable approximation to take it constant at 45×10^{-6} atm^{-1}. Thus the molal volume, as a function of pressure, is

$$v = 18.07(1 - 45 \times 10^{-6}P) = 18.07 - 8.1 \times 10^{-4}P$$

and the fugacity is

$$RT \ln \frac{f}{0.0312} = \int_{0.03}^{1000} v\, dP$$
$$= 18.07(1000 - 0.03) - 4.0 \times 10^{-4}(1000^2 - 0.03^2)$$
$$= 17{,}670 \text{ cc atm} = 428 \text{ cal}$$
$$f = 0.0312 e^{428/RT} = 0.0643 \text{ atm}$$

Since the fugacity is defined with reference to the gaseous state at low pressure, it is the molal weight in that state which we must use consistently in our fugacity equations. Thus liquid nitrogen dioxide presumably is composed principally of N_2O_4 molecules, but the vapor dissociates at low pressure to NO_2 molecules. Consequently one would adopt the mole of NO_2 as the basis and use the volume of 46 g in all calculations of fugacity. In the case of a substance such as chlorine, however, the dissociation from Cl_2 to $2Cl$ occurs only at very high temperatures or extremely low pressures. At ordinary temperatures we have no difficulty in obtaining the fugacity of chlorine on the basis of a perfect gas of Cl_2 molecules. Thus the usual basis of fugacity of chlorine is the diatomic molecule, but at high temperatures an ambiguity would arise, and it would be necessary to specify whether the basis chosen were Cl or Cl_2.

Change of Fugacity with Temperature. We have seen that at least theoretically any substance may be converted isothermally into a vapor and that this vapor may be made to approach the perfect gas, without limit, as the pressure is indefinitely diminished. Suppose now that we have a substance in a given state and compare its molal free energy F with the molal free energy F* in the vapor state, at some very low pressure and at the same temperature. We are to consider the increase in free energy when 1 mole of the given substance is converted into the highly attenuated vapor. By Eq. (14-10)

$$F^* - F = RT \ln \frac{f^*}{f} \qquad (14\text{-}15)$$

Differentiating this equation with respect to temperature, while the pressure upon each of the two states remains constant,

$$\left(\frac{\partial F^*}{\partial T}\right)_P - \left(\frac{\partial F}{\partial T}\right)_P = R \ln \frac{f^*}{f} + RT\left(\frac{\partial \ln f^*}{\partial T}\right)_P - RT\left(\frac{\partial \ln f}{\partial T}\right)_P \qquad (14\text{-}16)$$

In the gas at very low pressure the fugacity is equal to the pressure. Hence f^* does not change with the temperature at constant pressure, and the next to the last term disappears. Also by Eq. (14-15),

$$R \ln\left(\frac{f^*}{f}\right) = \frac{F^* - F}{T}$$

whence

$$\left(\frac{\partial F^*}{\partial T}\right)_P - \left(\frac{\partial F}{\partial T}\right)_P = \frac{F^*}{T} - \frac{F}{T} - RT\left(\frac{\partial \ln f}{\partial T}\right)_P \qquad (14\text{-}17)$$

Now, employing Eq. (10-15), we obtain the simple equation

$$\left(\frac{\partial \ln f}{\partial T}\right)_P = \frac{H^* - H}{RT^2} \qquad (14\text{-}18)$$

The important quantity H* — H has been called the ideal heat of vaporization. It is the increase in heat content when the substance escapes into a vacuum.

PROBLEMS

✓**14-1.** It is sometimes convenient to regard the vapor pressure of a substance as one of the properties of that substance. If, at a given temperature and pressure, p is the vapor pressure, v is the molal volume of the substance, and ΔH is its heat of vaporization per mole, show that, by assuming the equation of the perfect gas for the vapor, we may obtain the two approximate equations

$$\left(\frac{\partial \ln p}{\partial P}\right)_T = \frac{v}{RT} \tag{14-19}*$$

$$\left(\frac{\partial \ln p}{\partial T}\right)_P = \frac{H^* - H}{RT^2} \tag{14-20}*$$

✓**14-2.** Find approximately the change in vapor pressure which would be produced if the pressure on the liquid benzene alone could be increased by 1 atm at its boiling point, 80°K. Assume the molal volume to be 90 cc.

✓**14-3.** In deriving Eq. (14-5) for the change of vapor pressure upon change of the pressure on the condensed phase at constant temperature, the equilibrium condition $dF = 0$ was used although the change of state was not carried out at constant pressure. Was this justified?

✓**14-4.** The fugacity of liquid water at 25°C is approximately 0.0313 atm. Take the ideal heat of vaporization as 10,450 cal/mole, and calculate the fugacity at 27°C.

15

STANDARD FREE-ENERGY FUNCTIONS AND EQUILIBRIUM CALCULATIONS

To devise various methods of calculating the change of free energy in chemical reactions is a task which will occupy us throughout the greater part of this book. Already, however, we have become acquainted with one or two of the important methods. Thus we know that, whenever a condition of equilibrium is reached in a chemical reaction, the free-energy change of the reaction is zero. For example, there is a transition point between rhombic and monoclinic sulfur at 1 atm and 95.4°C = 368.5°K. Hence we may write

$$S(\text{rhombic}) = S(\text{monoclinic}) \qquad \Delta F_{368.5} = 0$$

So also we have seen that $-\Delta F$ for a reaction which occurs in a galvanic cell is determined by the maximum electrical work which that cell is capable of performing. A further study here of such a cell will be profitable.

Free Energy and the Electromotive Force of a Galvanic Cell. Let us consider a cell composed of a lead electrode in contact with solid lead chloride, a mercury electrode in contact with mercurous chloride, and a solution of potassium chloride as electrolyte. When electrical contact is established between the two electrodes, a current will pass through the cell so that metallic lead is used up, metallic mercury is precipitated, and at the same time the lead chloride increases and the mercurous chloride diminishes in amount according to the chemical equation,

$$Pb(s) + Hg_2Cl_2(s) = PbCl_2(s) + 2Hg(l)$$

In such a case, by Eq. (10-20)

$$\Delta F = w'$$

where $-w'$ is the electrical work capable of being obtained under conditions of maximum efficiency. Under such conditions the counter emf of the storage battery, motor, or other apparatus upon which the electrical work is done must differ only infinitesimally from the maximum or reversible emf of the cell. This emf ε, multiplied by the amount of electricity flowing through the cell, measures the maximum output of electrical work. If $\mathfrak{F}$ is the Faraday equivalent and n is the number of such equivalents passing through the cell when the above reaction occurs, we may write,

$$\Delta F = w' = -n\mathfrak{F}\mathcal{E} \tag{15-1}$$

where $\mathcal{E}$ is positive if the reaction as written is a spontaneous one.

In accordance with the chemical equation which we have written, $n = 2$ and

$$\Delta F = -2 \times 96{,}493.5\mathcal{E} \text{ volt coulombs}$$

Or, by Eq. (4-24),

$$\Delta F = -2 \times 23{,}062.3\mathcal{E} \text{ cal}$$

The measured value of the emf at 25°C is 0.5359 volt, whence

$$\text{Pb(s)} + \text{Hg}_2\text{Cl}_2\text{(s)} = \text{PbCl}_2\text{(s)} + 2\text{Hg(l)} \qquad \Delta F_{298} = -24{,}718 \text{ cal}$$

This is the change of free energy which results from the above reaction, whether the reaction occurs in the cell or lead is added to mercurous chloride and the process takes place in an irreversible way. Any other reversible cell in which this reaction, and only this reaction, occurs can equally well be used to measure the free-energy change, and since n would be the same, $\mathcal{E}$ would be the same. Thus the cell that we have described must give the same emf independently of the particular chloride used as electrolyte, and of its concentration, and of the solvent, provided always that when the current passes through the cell no other process occurs than the one stated. If, for example, we used a very dilute solution of potassium chloride as an electrolyte, the solubility of the two chlorides no longer being negligible, the cell process would not correspond exactly to this chemical equation and the emf force would be slightly different.

Free Energy and Heat of Reaction. In the early days of the first law of thermodynamics, before the second law was fully understood, it was assumed as a matter of course that the most efficient utilization of a chemical reaction for the production of work would consist in converting all the heat of that reaction into work. In other words, the quantity $-\Delta H$ was assumed to represent the limiting quantity of work which could be obtained under the conditions of maximum efficiency. The work of Thomsen, Berthelot, and others has given us a great mass of thermochemical data of all grades of accuracy. The hope of these investigators that the results of their labors would give a direct measure of chemical affinity has proved to be a vain one. Nevertheless, their data are of great utility in the calculations which lead to the true measure of chemical affinity and will be often employed in our later calculations.

However, we have just seen that it is not $-\Delta H$ but $-\Delta F$ which measures this maximum capacity for performing useful work. And these two quantities are not equal unless the entropy of the system in question is the same at the beginning and end of the isothermal reaction under consideration. This is shown by applying Eq. (10-10) to an isothermal process, which gives

$$\Delta F - \Delta H = -T \, \Delta S \qquad (15\text{-}2)$$

According to the sign of ΔS, the work obtainable in a given reversible process may be greater or less than the heat of the reaction.

It is true that, according to the first law, the external work performed must be equal to the loss in heat content of a system, unless some heat is given to or taken from the surroundings, but this is precisely the point first clearly seen by Gibbs. When an isothermal reaction runs reversibly, $T \, \Delta S$ is the heat absorbed from the surroundings, and if this is positive, the work done will be even greater than the heat of reaction.

In the specific reaction which we have just been considering,

$$\Delta H = -22{,}530 \text{ cal}$$

hence, if 1 mole of lead reacts irreversibly with mercurous chloride, as in the calorimeter, 22,530 cal is given up. But we have seen that

$$\Delta F = -24{,}718 \text{ cal}$$

so that, in the reversible isothermal process, the work done is greater than the heat of reaction. Therefore, when this galvanic cell operates reversibly in a thermostat, heat is not given to, but taken from, the thermostat, to the extent of 2188 cal for each mole of lead consumed.

Let us consider another case, namely, a cell with on one side an electrode of mercury and mercurous chloride, on the other side chlorine gas in contact with an electrode of platinum-iridium, and between the two electrodes an electrolyte solution of some chloride. Here again the emf will be independent of the electrolyte, provided that in the electrolyte the solubility of chlorine and of mercurous chloride is negligible, so that only the following reaction occurs in the cell:

$$\text{Hg(l)} + \tfrac{1}{2}\text{Cl}_2(\text{g, 1 atm}) = \tfrac{1}{2}\text{Hg}_2\text{Cl}_2(\text{s})$$

one equivalent

The emf of such a cell is found to be 1.0919 volts, and the reaction as written involves one equivalent; hence

$$\Delta F_{298} = -23{,}062.3 \times 1.0919 = -25{,}182 \text{ cal}$$

The enthalpy of this reaction has also been determined, and the value found is $\Delta H_{298} = -31{,}660$ cal. Here then is a case where, even in the reversible reaction, the work produced is smaller than the heat of reaction, and the cell will give up heat to the surroundings.

In many cases it has been difficult to find an efficient method for utilizing the diminution of free energy for the production of useful work. The most important of all work-producing chemical reactions is the combustion of carbon.   Yet when coal is burned under the boiler of a steam engine, at best the work obtained is less than half the maximum work calculated from the free energy.   To obtain perfect efficiency, it would be necessary to contrive a process in which carbon and oxygen at a given temperature would be consumed, carbon dioxide would be produced at the same temperature, and every step of the process would be reversible, so that, by reversing the engine, carbon dioxide would be dissociated into carbon and oxygen.

This might be done, as in the examples already given, if we could devise a galvanic cell with reversible electrodes of carbon and oxygen. But hitherto all such attempts to obtain "electricity direct from coal" have failed.   It is possible, however, by methods which we shall develop later, to ascertain the theoretical amount of work obtainable in such a process, and we shall find[1]

$$C(s) + O_2(g) = CO_2(g) \qquad \Delta F_{298} = -94,260 \text{ cal}$$

This value happens to correspond very closely to the heat of combustion, for $\Delta H^\circ_{298} = -94,052$ cal.   This is far from being the case in the combustion of graphite to carbon monoxide, for which we shall find the equation

$$C(s) + \tfrac{1}{2}O_2(g) = CO(g)$$
$$\Delta F_{298} = -32,778 \text{ cal} \qquad \Delta H^\circ_{298} = -26,416 \text{ cal}$$

It is evident here that, if graphite were burned to carbon monoxide and the heat of combustion were all converted into work, this work would still be less than the maximum possible work of the reaction.

Free Energies from Third-law Entropies and Heat Data.   In addition to the various methods which measure free energy directly, the third law of thermodynamics provides a very important indirect method through the equation given earlier.

$$\Delta F = \Delta H - T \, \Delta S \qquad (15\text{-}2)$$

We noted before that even by the beginning of the present century there were extensive data available on the ΔH of various chemical reactions. Until the development of the third law, however, there was no source of ΔS values which did not essentially require a measurement of ΔF.   The

---

[1] The value of ΔF will vary according to the kind of carbon chosen.   The figures given are for graphite in the 1-atm standard state.   The oxygen and carbon dioxide are in the perfect-gas standard state.

third law provided a means of obtaining from heat-capacity measurements at low temperatures the absolute entropy of each substance in a reaction. If such entropy values are available for each reactant and each product, then ΔS may be calculated and combined with a calorimetrically determined ΔH according to Eq. (15-2) to obtain ΔF.

It is sometimes difficult or impossible to find a catalyst which will cause a given reaction to take place with reasonable speed, whether spontaneously or in a galvanic cell, but it is always possible to cool each reactant and product to a very low temperature and measure each of their heat capacities up to the temperature of interest. Consequently, the third law opened a great new opportunity for the prediction of the free-energy changes of reactions which are too slow to be measured by direct methods. Furthermore it has been found that for many types of substances the entropy values show regular trends. This allows estimates of entropy values which are as accurate as many experimental heat-of-reaction data justify. Latimer made particularly effective use of these methods in his work on the free energies of inorganic substances.[1] Similar methods are equally useful for organic compounds. Some of the systems for estimating entropies are discussed in Chapter 32; others are mentioned at other points throughout the book. In addition to experimental entropy measurements, the third law also provides the basis for the calculation of entropies of gases when enough information is available about the individual quantum energy levels of their molecules. These methods are given in Chapter 27.

An interesting application of the third-law method concerns the reaction

$$C(\text{diamond}) = C(\text{graphite})$$

This reaction was known to take place in the direction diamond to graphite at high temperature and low pressure, but the first and, to the present date, the only precise information about ΔF comes from heat and entropy data. The heats of combustion of diamond and of graphite have been measured many times, and the difference is found to be $\Delta H^{\circ}_{298} = -450$ cal/mole. The calculation of the entropy of graphite from heat capacities was discussed in Chapter 11, and the result was 1.37 cal/deg mole at $298.15°K$. A similar calculation for diamond yields 0.58 cal/deg mole. Consequently

$$\Delta F^{\circ}_{298} = -450 - 298 \times 0.79 = -685 \text{ cal/mole}$$

From this result and the densities of diamond and graphite one may calculate by Eq. (10-16) that ΔF would be zero at approximately 22,000

[1] W. M. Latimer, "The Oxidation States of the Elements and Their Potentials in Aqueous Solutions," 2d ed., Prentice-Hall, Inc., Englewood Cliffs, N.J., 1952.

atm pressure at room temperature and that at higher pressures diamond becomes the stable form of carbon.

Another example is the reaction of isomerization of 1-butene to cyclobutane, which has not been observed directly as yet.   The heat of combustion of each hydrocarbon has been measured,[1,2] and subtraction of the results yields

$$\text{1-butene(g)} = \text{cyclobutane(g)} \qquad \Delta H_{298} = 6.40 \text{ kcal/mole}$$

The entropy of each substance[1,3] has also been obtained by the methods of Chapter 11 from low-temperature heat-capacity measurements on the solid and liquid together with heat-of-fusion and heat-of-vaporization measurements.   There results $\Delta S_{298} = -9.61$ cal/deg mole and

$$\Delta F_{298} = 6.40 - 298.15(-9.61) \times 10^{-3} = 9.96 \text{ kcal/mole}$$

Consequently we may predict that at $298°K$ cyclobutane would react almost completely to 1-butene.

The Addition of Free-energy Equations.   We have designated standard reference states for the elements to which heats of formation of compounds could be referred.   We have also designated standard states for all substances for tabulation of heat-capacity, heat-content, and entropy data.   Since F is defined in terms of H and S, these conventions will apply to F.   As the free energy of a gas does not approach a constant at low pressures, standard-state free energies of formation are given for the gases in the *hypothetical perfect-gas state at 1 atm*.   The expression for the standard entropy of a gas was given in Eq. (12-6),

$$\text{s}° = \lim_{P \to 0} (\text{s}_P + R \ln P) \qquad (12\text{-}6)$$

From this one may derive the corresponding equation for the standard free energy

$$\text{F}° = \lim_{P \to 0} (\text{F}_P - RT \ln P) \qquad (15\text{-}3)$$

The methods of calculation of the free energy of a gas in this hypothetical state are given in Chapter 16.   For the moment, we note that the difference between the free energy of a gas in the hypothetical 1-atm perfect-gas standard state and the free energy of a real gas at 1 atm is rather small for most gases.

---

[1] Values for 1-butene are summarized in F. D. Rossini et al., "Selected Values of Physical and Thermodynamic Properties of Hydrocarbons," Carnegie Press, Pittsburgh, 1953 edition and subsequent revisions.

[2] S. Kaarsemaker and J. Coops, *Rec. trav. chim.*, **71**, 261 (1952).

[3] G. W. Rathjens et al., *J. Am. Chem. Soc.*, **75**, 5629, 5634 (1953).

The same reasons which justified our adding and subtracting thermo-chemical equations also permit our combining free-energy equations in a similar manner. We have obtained two such equations, namely,

$$Pb(s) + Hg_2Cl_2(s) = PbCl_2(s) + 2Hg(l) \qquad \Delta F^\circ_{298} = -24{,}709 \text{ cal}$$
$$Hg(l) + \tfrac{1}{2}Cl_2(g) = \tfrac{1}{2}Hg_2Cl_2(s) \qquad \Delta F^\circ_{298} = -25{,}182 \text{ cal}$$

Multiplying the latter by 2 and adding to the former, we find

$$Pb(s) + Cl_2(g) = PbCl_2(s) \qquad \Delta F^\circ_{298} = -75{,}073 \text{ cal}$$

This is the reaction which occurs in a cell with one electrode of chlorine and the other of lead and lead chloride. For such a cell, therefore, $\mathcal{E}_{298} = 75{,}073/(2 \times 23{,}062.3) = 1.628$ volts. So likewise by combining the two equations for the combustion of carbon, which we have obtained above, we may write,

$$CO + \tfrac{1}{2}O_2 = CO_2 \qquad \Delta F^\circ_{298} = -61{,}452 \qquad \Delta H^\circ_{298} = -67{,}636 \text{ cal}$$

THE FREE-ENERGY CHANGE AS A FUNCTION OF THE TEMPERATURE

It will be understood that when we speak of a free-energy change, without further qualification, we refer to an isothermal process. Thus we have obtained the difference between the free energy of 1 mole of lead chloride at 25°C and the free energy of its elements at the same temperature.

Such a value of ΔF° will ordinarily be different when we carry out the isothermal process at some other temperature, and it is evidently of much importance to be able to calculate ΔF° at other temperatures when it is known at some one temperature.

Since we apply Eq. (10-15) both to the substances consumed and to the substances produced, we may write

$$\left(\frac{\partial \Delta F}{\partial T}\right)_P = -\Delta S = \frac{\Delta F - \Delta H}{T} \qquad (15\text{-}4)$$

The subscript P in the first term merely indicates that the pressure on each substance is to be the same whether we are working at one temperature or another. Thus, if we know ΔF, at one temperature, for the conversion of 1 mole of liquid ammonia at 1 atm into ammonia vapor at 10 atm, then Eq. (15-4) shows us how to calculate ΔF at some other temperature when likewise 1 mole of liquid ammonia at 1 atm is converted into ammonia vapor at 10 atm.

Indeed, since the pressures are always stated or implied in our chemical equations and quantities such as ΔF, ΔS, and ΔH pertain to the reaction as expressed in this chemical equation, we may henceforth without ambiguity omit the subscript P in Eq. (15-4).

The Gibbs-Helmholtz Equation.   We are now led immediately to a valuable formula for the temperature coefficient of the emf of a reversible galvanic cell.   Combining Eqs. (15-1) and (15-4), we have

$$-n\mathfrak{F}\frac{d\mathcal{E}}{dT} = -\Delta S = \frac{-n\mathfrak{F}\mathcal{E} - \Delta H}{T} \qquad (15\text{-}5)$$

or, in simpler form,

$$\mathcal{E} + \frac{\Delta H}{n\mathfrak{F}} = T\frac{d\mathcal{E}}{dT} \qquad (15\text{-}6)$$

Thus for the cell which we have described above, in which lead reacts with mercurous chloride, $\mathcal{E}_{298} = 0.5357$ volt, and this value increases by 0.000145 volt/deg.   Hence we may calculate the heat of this reaction, namely,

$$\Delta H^{\circ}_{298} = 2 \times 23{,}062.3(298.15 \times 0.000145 - 0.5359) = -22{,}730 \text{ cal}$$

and this is more accurate than any of the calorimetric values.

The Integration of the Free-energy Equation.   Equation (15-4) may be put into several alternative forms, two of which will be useful from time to time.   Thus, by the use of differential calculus, we find

$$\frac{d(\Delta F/T)}{dT} = -\frac{\Delta H}{T^2} \qquad (15\text{-}7)$$

$$\frac{d(\Delta F/T)}{d(1/T)} = \Delta H \qquad (15\text{-}8)$$

Either of these equations is suitable for direct integration if we know ΔH as a function of T.   Sometimes the variation of ΔH is of such character that the integration can best be performed by graphical methods.   Ordinarily, however, when we are dealing with reactions over a range of temperature, the values of C_P are conveniently represented by equations of the algebraic type, and we may express ΔC_P and ΔH as we have previously done in Eqs. (6-17)* and (6-18)*, namely,

$$\Delta C_P = \Delta a + \Delta bT + \Delta cT^{-2}$$
$$\Delta H = \Delta H_I + \Delta aT + \tfrac{1}{2}\Delta bT^2 - \Delta cT^{-1}$$

Substituting this last expression in Eq. (15-7), and integrating, we find,

$$\frac{\Delta F}{T} = \frac{\Delta H_I}{T} - \Delta a\ln T - \tfrac{1}{2}\Delta bT - \tfrac{1}{2}\Delta cT^{-2} + I \qquad (15\text{-}9)^{*}$$

or $\qquad \Delta F = \Delta H_I - \Delta a\,T\ln T - \tfrac{1}{2}\Delta bT^2 - \tfrac{1}{2}\Delta cT^{-1} + IT \qquad (15\text{-}10)^{*}$

The quantity I is the integration constant and may be evaluated when we know ΔF at some one temperature in the temperature range for which the ΔC_P equation is valid. As a simple illustration of the use of this equation, let us consider again the conversion of rhombic to monoclinic sulfur,

$$S(\text{rhombic}) = S(\text{monoclinic})$$

We have already seen that for this reaction $\Delta F^\circ_{368.5} = 0$. Eastman and McGavock[1] give the heat-capacity difference of monoclinic and rhombic sulfur. Their results have been corrected for contamination of their monoclinic sulfur by rhombic sulfur. Over the limited range 298 to 369°K, the resulting heat capacities can be represented by

$$\Delta c^\circ_P = 0.085 + 0.66 \times 10^{-3}T$$

and

$$\Delta H^\circ = \Delta H^\circ_I + 0.085T + 0.33 \times 10^{-3}T^2$$

Taking[2] $\Delta H^\circ_{368.5} = 96$ cal, we find $\Delta H^\circ_I = 20$ cal. Hence Eq. (15-10)* assumes the form

$$\Delta F^\circ = 20 - 0.085T \ln T - 0.33 \times 10^{-3}T^2 + IT \qquad (15\text{-}11)*$$

Now substituting[3] the value $\Delta F^\circ = 0$ at $T = 368.5$, we find $I = 0.571$, whence we may calculate ΔF° at other temperatures; thus $\Delta F^\circ_{298} = 16$ cal. This positive value corresponds to the fact that the rhombic form is more stable at this temperature.

When we have such expressions for ΔF as a function of the temperature for a number of different reactions, we may combine the chemical equations and at the same time combine the free-energy equations, term by term, and thus obtain new equations for new reactions. This procedure will be illustrated in many of our later calculations.

The Free-energy Function. With a sufficient number of terms, equations of the type of Eq. (15-10)* can suffice for the calculation of free-energy changes with temperature with as high an accuracy as might be desired. However, the evaluation of many terms can be tedious, and it is more convenient to use a tabular method. It is not normally convenient to tabulate ΔF° for a reaction, as ΔF° often changes rapidly with temperature, and interpolations are accurate only with tabulation at close temperature intervals. The ideal function would vary slowly enough with temperature to allow accurate interpolation; it would also be derived from the properties of a single substance.

The free-energy function $(F^\circ_T - H^\circ_0)/T$ meets these requirements. In Chapter 27, it is shown that this function can be directly calculated for gases when the energy levels are known from spectroscopic data. For

---

[1] E. D. Eastman and W. C. McGavock, *J. Am. Chem. Soc.*, **59**, 145–151 (1937).

[2] E. D. West, *J. Am. Chem. Soc.*, **81**, 29 (1959).

[3] In the arithmetical operations connected with the use of such a free-energy equation it is sometimes expedient to employ Eq. (15-9), thus solving first for $\Delta F/T$. This frequently gives a higher accuracy with less effort.

solids for which the third law of thermodynamics is applicable, the free-energy function may be calculated by graphical evaluation of either of two sets of double integrals,

$$\frac{F_T^\circ - H_0^\circ}{T} = -\frac{1}{T} \int_0^T dT' \int_0^{T'} \frac{c_P^\circ}{T''} dT'' \tag{15-12a}$$

$$\frac{F_T^\circ - H_0^\circ}{T} = -\int_0^T (T')^{-2} dT' \int_0^{T'} c_P^\circ dT'' \tag{15-12b}$$

Also one has the relationship

$$s_T^\circ = \frac{H_T^\circ - H_0^\circ}{T} - \frac{F_T^\circ - H_0^\circ}{T} \tag{15-13}$$

Since the first integration in Eq. (15-12a) gives the entropy, the enthalpy $H_T^\circ - H_0^\circ$ may be calculated from Eq. (15-13) without a separate integration. If one uses Eq. (15-12b), the first integration gives $H_T^\circ - H_0^\circ$ and s_T° can then be calculated from Eq. (15-13).

The use of the free-energy function will be illustrated for the reaction S(rhombic) = S(monoclinic), for which we have already evaluated ΔF°, using analytical expressions. In Table 15-1 are tabulated values of heat contents and entropies which have been evaluated from heat capacities determined calorimetrically and the derived free-energy functions at 240, 298.15, 350, and 368.5°K.

TABLE 15-1. FREE-ENERGY FUNCTIONS OF SULFUR, CAL/DEG MOLE

T, °K	Rhombic			Monoclinic			$\dfrac{\Delta F_T^\circ - \Delta H_0^\circ}{T}$
	$\dfrac{H_T^\circ - H_0^\circ}{T}$	s_T°	$\dfrac{F_T^\circ - H_0^\circ}{T}$	$\dfrac{H_T^\circ - H_0^\circ}{T}$	s_T°	$\dfrac{F_T^\circ - H_0^\circ}{T}$	
240	3.12	6.50	−3.38	3.21	6.62	−3.41	−0.03
298.15	3.53	7.62	−4.09	3.65	7.82	−4.17	−0.08
350	3.83	8.51	−4.68	3.97	8.74	−4.77	−0.09
368.5	3.93	8.81	−4.88	4.08	9.07	−4.99	−0.11

For the reaction S(rhombic) = S(monoclinic)

$$\frac{\Delta F_T^\circ - \Delta H_0^\circ}{T} = \left(\frac{F_T^\circ - H_0^\circ}{T}\right)_{mc} - \left(\frac{F_T^\circ - H_0^\circ}{T}\right)_{rh}$$

$$\frac{\Delta F_T^\circ}{T} = \left(\frac{\Delta F_T^\circ - \Delta H_0^\circ}{T}\right) + \frac{\Delta H_0^\circ}{T} \tag{15-14}$$

From the value of $(\Delta F_T^\circ - \Delta H_0^\circ)/T$ of -0.11 cal/deg at 368.5°K, where $\Delta F^\circ = 0$, one obtains from Eq. (15-14) $\Delta H_0^\circ = 40.5$ cal. To obtain ΔF_{298}°, $\Delta F_{298}^\circ/298.15 = -0.08 + {}^{40.5}\!/_{298.15} = 0.06$; $\Delta F_{298}^\circ = 18$ cal. With tabulated free-energy functions available, it is obviously much easier to calculate ΔF° at a variety of temperatures than it would be to use the analytical equation for ΔF°. When one is interested in a wide temperature range that requires a complex analytical function for c_P°, if accuracy is to be maintained, the advantage of the tabulated free-energy functions is even more evident.

In addition to free-energy functions based on the heat content at the absolute zero, free-energy functions based on 298.15°K are also tabulated. Thus

$$\frac{F_T^{\circ} - H_{298}^{\circ}}{T} = \frac{H_T^{\circ} - H_{298}^{\circ}}{T} - s_T^{\circ} \qquad (15\text{-}15)$$

This function can be related to the 0°K free-energy function by

$$\frac{F_T^{\circ} - H_{298}^{\circ}}{T} = \frac{F_T^{\circ} - H_0^{\circ}}{T} - \frac{H_{298}^{\circ} - H_0^{\circ}}{T} \qquad (15\text{-}16)$$

When the free-energy functions of reactants and products are combined, one obtains ΔF_T° from ΔH_{298}° by

$$\frac{\Delta F_T^{\circ}}{T} = \frac{\Delta F_T^{\circ} - \Delta H_{298}^{\circ}}{T} + \frac{\Delta H_{298}^{\circ}}{T} \qquad (15\text{-}17)$$

In general the use of free-energy functions based on either 0 or 298°K is similar. The 0°K free-energy functions of gases are obtained directly from statistical calculations as discussed in Chapter 27 and thus are more frequently found in journal papers. However, when low-temperature heat capacities are not available, one can still use Eqs. (15-15) and (15-17) if the entropy is known from equilibrium measurements, whereas one cannot use Eqs. (15-12) and (15-14). On the other hand, when data are available for using the 0°K free-energy functions, one always knows $H_{298}^{\circ} - H_0^{\circ}$ and can calculate the 298°K free-energy function through use of Eq. (15-16). Thus the latter function has a wide range of applicability. The main reason, however, for using the 298°K function is that most heats of formation are tabulated at 298.15°K and are not readily available at 0°K. Also the excellent summaries of heat content and entropy data that have been published by K. K. Kelley[1] are organized for easy calculation of the 298°K free-energy function. Thus from the three quantities s_{298}°, $H_T^{\circ} - H_{298}^{\circ}$, and $s_T^{\circ} - s_{298}^{\circ}$ which he tabulates, one obtains the free-energy function by

$$\frac{F_T^{\circ} - H_{298}^{\circ}}{T} = \frac{H_T^{\circ} - H_{298}^{\circ}}{T} - (s_T^{\circ} - s_{298}^{\circ}) - s_{298}^{\circ} \qquad (15\text{-}18)$$

In general, when one is accumulating data for a thermodynamic calcu-

[1] K. K. Kelley, *U.S. Bur. Mines Bulls.* 584, 1960, and 477, 1950, and K. K. Kelley and E. G. King, *U.S. Bur. Mines Bull.* 592, in press.

lation, one will find that the free-energy functions available will be based on 0°K for some of the compounds involved and on 298°K for others. It will be necessary to convert to a common basis. If low-temperature heat-capacity data are missing for any of the products or reactants, it will usually be necessary to convert to 298°K but occasionally sufficiently good estimates of the missing low-temperature heat contents can be made to allow use of the 0°K free-energy functions with introduction of little error. If sufficient data are available for use of either basis, one will convert to the basis which involves the fewest conversions considering the number of free-energy functions and heats of formation to be converted.

A few sample calculations will readily convince the reader that the use of free-energy functions, when they are available, greatly simplifies the calculation of standard free-energy changes at various temperatures. Interpolations are readily made and, in general, higher accuracy can be obtained than through use of analytical expressions with a reasonable number of terms such as Eq. (15-10)*.

Free-energy functions are tabulated for a variety of compounds in the chemical literature, most frequently in the temperature range 298 to 2000°K. A number of compilations exist which have brought together free-energy functions of related substances. A bibliography of such sources is given at the end of this chapter.

Equilibrium Calculations. Through use of Eq. (15-10)*, (15-14), or (15-17), one can calculate values of $\Delta F°$ for reactions of interest at desired temperatures. For example, we have calculated $\Delta F°$ for the reaction S(monoclinic) = S(rhombic) for a variety of temperatures, and these results indicate that monoclinic sulfur is unstable with respect to rhombic sulfur at 1 atm pressure below 368.5°K and stable above 368.5°K. However, the standard free-energy value gives the direction of the reaction only under the standard conditions of 1 atm. Which is the stable form at pressures other than 1 atm? We must calculate the free-energy change for

$$\text{S(monoclinic, } P) = \text{S(rhombic, } P) \qquad \Delta F_1 \qquad (15\text{-}19)$$

To obtain ΔF_1 from the free-energy change under standard conditions,

$$\text{S(monoclinic, 1 atm)} = \text{S(rhombic, 1 atm)} \qquad \Delta F°$$

one needs the free-energy changes with pressure from Eq. (10-16),

$$\text{S(monoclinic, 1 atm)} = \text{S(monoclinic, } P) \qquad \Delta F_2 = v_m(P-1)$$
$$\text{S(rhombic, 1 atm)} = \text{S(rhombic, } P) \qquad \Delta F_3 = v_r(P-1)$$

where v_m and v_r, the molal volumes of monoclinic and rhombic sulfur, are assumed constant. Thus, for reaction (15-19),

$$\Delta F_1 = \Delta F° - \Delta F_2 + \Delta F_3 = \Delta F° + (v_r - v_m)(P-1) \qquad (15\text{-}20)$$

Tammann[1] reports $v_m - v_r = 0.4$ cc at temperatures above 100°C and at high pressure. Thus the free energy of monoclinic sulfur increases more rapidly with increase in pressure than does the free energy of rhombic sulfur. Above 368.5°K, ΔF_1 for the conversion will be negative for small pressures, but it will change sign at sufficiently large values of P. At each temperature, there is one value of P for which the free energies of the two forms are the same. This is obtained from

$$\Delta F_1 = 0 = \Delta F^\circ + (v_r - v_m)(P - 1)$$

At 380°K or 11.5° above the transition at 1 atm,

$$\Delta F^\circ = 0.26 \times 11.5 = 3 \text{ cal}$$

as Δc_P can be neglected for such a temperature interval. Thus

$$P = 1 + 308$$

or roughly 300 atm at which rhombic and monoclinic sulfur are at equilibrium at 380°K.

The above steps illustrate a general type of procedure that will be used in carrying out equilibrium calculations. One first starts with the ΔH° for the reaction in question, usually given at 25°C. One then uses either the free-energy functions or the heat capacities and entropies of the reactants and products to calculate ΔF° at the temperature of interest. Finally, one calculates the free-energy change for each reactant and product for the change from its standard state to its state in the equilibrium system. The algebraic combination of the free-energy terms yields for the equilibrium change of state

$$\Delta F = \Delta F^\circ + \Sigma_p \Delta F_p - \Sigma_r \Delta F_r = 0 \qquad (15\text{-}21)$$

The first term, ΔF°, is a specific number for each temperature. The two sums over products and reactants, respectively, represent summations of terms which are functions of pressure or, as we shall see later, composition and other variables which influence the free energy and therefore the equilibrium conditions.

The same procedure can be illustrated for a vaporization process. If one is interested in the equilibrium Cu(l) = Cu(g) at 1457°K, one can readily interpolate between -30.71 and -30.46 cal/deg mole for the $(\Delta F^\circ - \Delta H_{298}^\circ)/T$ values at 1400 and 1500°K obtained from the tabulation of Stull and Sinke[2] to obtain -30.57 at 1457°K. This value then combined with $\Delta H_{298}^\circ = 81,100$ cal yields $\Delta F_{1457}^\circ/1457 = 25.09$ cal/deg mole or

---

[1] G. Tammann, "Kristallisieren und Schmelzen," p. 274, J. A. Barth, Leipzig, 1903.

[2] D. R. Stull and G. C. Sinke, Thermodynamic Properties of the Elements, *Advances in Chem. Ser.*, no. 18, 1956.

$\Delta F^\circ_{1457} = 36,560$ cal/mole. As this is a standard free-energy change, it corresponds to the change of state from copper liquid at 1 atm to copper vapor behaving as a perfect gas at 1 atm pressure. The large positive value indicates that copper vapor at 1 atm is very unstable relative to copper liquid at $1457°K$.

Let us consider the free-energy change for the vaporization under conditions other than the standard conditions.

$$Cu(l, P) = Cu(p.g., P) \qquad \Delta F_1 \qquad (15\text{-}22)$$

To obtain ΔF_1 from the free-energy change under standard conditions,

$$Cu(l, 1 \text{ atm}) = Cu(p.g., 1 \text{ atm}) \qquad \Delta F^\circ$$

one needs the free-energy changes with pressure from Eqs. (10-16) and (10-44)*,

$$Cu(l, 1 \text{ atm}) = Cu(l, P) \qquad \Delta F = v(P - 1)$$

where v is the molal volume of liquid copper and

$$Cu(p.g., 1 \text{ atm}) = Cu(p.g., P) \qquad \Delta F = RT \ln P$$

Thus for reaction (15-22),

$$\Delta F_1 = \Delta F^\circ - v(P - 1) + RT \ln P \qquad (15\text{-}23)*$$

From the value of 7.7 cc for the molal volume of liquid copper, it is apparent that the $v(P - 1)$ term is very small for pressures below 1 atm. The last term of Eq. (15-23)* indicates that the free energy of the vapor is decreased as the pressure is decreased, and one would expect the sign of ΔF_1 to change from positive to negative as P decreases. There is one value of P for which the free energies of the liquid and gas are the same. This is P_{sat}, and it is obtained from

$$0 = \Delta F_1 = \Delta F^\circ - v(P - 1) + RT \ln P$$

from which one obtains $P_{sat} = 10^{-5.48}$ atm.

When one is dealing with equilibria involving pure solids and liquids under moderate pressures such that their free energies do not differ materially from standard free energies and with gases at low enough pressures so that perfect-gas behavior can be assumed, then Eq. (15-21) takes a relatively simple form. All the ΔF_i terms are either zero or of the form $\Delta F_i = RT \ln P_i$, which represents the free-energy change when a gas is taken from its standard state to a low pressure where it behaves as a perfect gas. Thus for the simple equilibria such as that between liquid and gaseous water, Eq. (15-21) reduces to

$$0 = \Delta F^\circ + RT \ln P \qquad (15\text{-}24)*$$

where ΔF° is the standard free energy of vaporization of liquid water to gaseous water in the hypothetical 1-atm perfect-gas standard state and P is the equilibrium partial pressure of water.

If, on the other hand, the deviation from perfect-gas behavior is significant, the ΔF_i terms for the free-energy difference between the gas in the hypothetical perfect-gas state and the real gas at an equilibrium pressure p_i, corresponding to a fugacity f_i, are of the form

$$\Delta F_i = RT \ln f_i = RT \left(\ln \frac{f_i}{p_i} + \ln p_i \right) \qquad (15\text{-}25)$$

where f_i/p_i, the fugacity coefficient, is a measure of the deviation from perfect-gas behavior. Knowledge of f_i/p_i for a given p_i, which can be determined from the equation of state of the gas by the methods discussed in Chapter 16, corresponds to knowledge of the difference between the free energy of the hypothetical perfect-gas standard state and the free energy of the real gas at partial pressure p_i.

Let us now consider a general gaseous reaction in which l moles of the substance L and m moles of M, etc., give q moles of Q, etc., as represented by the equation

$$l\mathrm{L} + m\mathrm{M} + \cdots = q\mathrm{Q} + r\mathrm{R} + \cdots$$

Let ΔF be the free-energy change in this reaction when the substances are in any given states,

$$\Delta F = (qF_Q + rF_R + \cdots) - (lF_L + mF_M + \cdots) \qquad (15\text{-}26)$$

and let ΔF° be the free-energy change when each substance is in its standard state,

$$\Delta F^\circ = (qF_Q^\circ + rF_R^\circ + \cdots) - (lF_L^\circ + mF_M^\circ + \cdots) \qquad (15\text{-}27)$$

If f_L, f_M, etc., represent the fugacities in the nonstandard states, then we have

$$l(F_L - F_L^\circ) = lRT \ln f_L = RT \ln f_L^l \qquad (15\text{-}28)$$

and so on.

Combining the several equations we find

$$\Delta F - \Delta F^\circ = RT \ln \frac{f_Q^q f_R^r \cdots}{f_L^l f_M^m \cdots} \qquad (15\text{-}29)$$

The important quotient appearing in the last term we may call the fugacity quotient of the gaseous reaction.

If the gas pressures are low enough so that the difference between real-gas behavior and perfect-gas behavior is negligible, Eq. (15-29) reduces to

$$\Delta F - \Delta F^\circ = RT \ln \frac{p_Q^q p_R^r \cdots}{p_L^l p_M^m \cdots} \qquad (15\text{-}30)^*$$

When $\Delta F°$ is obtained for a number of reactions, it may immediately be found for other reactions by addition or subtraction. When a reaction is the formation of a substance from its elements in the standard reference states, as, for example, $H_2(g) + \frac{1}{2}O_2(g) = H_2O(l)$, then $\Delta F°$ is known as the $\Delta F°$ of formation, or, more loosely, as the free energy of formation of the compound.

THE EQUILIBRIUM CONSTANT

Whenever we meet with a case of chemical equilibrium, we immediately acquire important information regarding the free-energy change in the reaction concerned. For in such a reaction, by Eq. (13-4),

$$\Delta F = 0$$

Hence, for equilibrium, by Eq. (15-29)

$$\Delta F° = -RT \ln \frac{f_Q^q f_R^r \cdots}{f_L^l f_M^m \cdots} \tag{15-31}$$

At a given temperature $\Delta F°$ is a constant, and therefore the condition of equilibrium is that the fugacity quotient shall also be constant. The value of this quotient, when the system is in equilibrium, we shall call the equilibrium constant K and write

$$\Delta F° = -RT \ln K \tag{15-32}$$

THE CHANGE OF EQUILIBRIUM CONSTANT WITH THE TEMPERATURE

The standard free-energy change $\Delta F°$ varies with the temperature according to Eqs. (15-7) and (15-8),

$$d\left(\frac{\Delta F°}{T}\right) = -\frac{\Delta H°}{T^2}\,dT = \Delta H°\,d\left(\frac{1}{T}\right) \tag{15-33}$$

where $\Delta H°$ is the sum of the heat contents of the substances produced, less the sum of the heat contents of the substances consumed when each substance is in its standard state. Now, substituting for $\Delta F°$, by Eq. (15-32), we obtain the well-known equation of van't Hoff,

$$\frac{d \ln K}{dT} = \frac{\Delta H°}{RT^2} \tag{15-34}$$

or, in another form,

$$\frac{d(R \ln K)}{d(1/T)} = \frac{4.5758\,d \log K}{d(1/T)} = -\Delta H° \tag{15-35}$$

These equations enable us to calculate the rate at which the equilibrium constant is changing with the temperature, when $\Delta H°$ is known; and if we know $\Delta H°$ as a function of the temperature, they may be integrated. Thus from the value of the equilibrium at any one temperature we may calculate its value at any other temperature. We are going to have frequent opportunity to illustrate such calculations, and therefore we shall confine ourselves at this time to the description of an interesting method of solving the converse problem, which consists in calculating the heat of a reaction from the temperature variation of its equilibrium constant.

We see from Eq. (15-35) that if we should plot $-R \ln K$ against $1/T$ the slope of the curve at each point must give the value of $\Delta H°$. Such a method is especially useful in numerous cases where the only information which we possess regarding the heat of a reaction is derived from the equilibrium measurements themselves.

As an example, we may employ the measurements of Preuner and Schupp[1] and of Randall and Bichowsky[2] on the equilibrium between hydrogen, sulfur, and hydrogen sulfide, at temperatures above 1000°K. Now at these temperatures sulfur vapor is in the form of S_2, a species which exists to no measurable extent at ordinary temperatures and is therefore not subject to the usual kind of calorimetric investigation. The reaction is written

$$H_2(g) + \tfrac{1}{2}S_2(g) = H_2S(g) \qquad K = \frac{[H_2S]}{[H_2][S_2]^{\frac{1}{2}}}$$

where $[H_2S]$ represents f_{H_2S} or approximately p_{H_2S} at low to moderate pressures. The results are given in Table 15-2, and the data of the last two columns are plotted in Fig. 15-1. The individual points fall beau-

TABLE 15-2. EQUILIBRIUM BETWEEN HYDROGEN, SULFUR, AND HYDROGEN SULFIDE

T	$\log K$	$-R \ln K$	$1/T$
1023	2.025	−9.272	0.0009775
1103	1.710	−7.830	0.0009066
1218	1.305	−5.975	0.0008210
1338	0.964	−4.414	0.0007474
1362	0.902	−4.130	0.0007342
1405	0.793	−3.631	0.0007117
1473	0.643	−2.944	0.0006789
1537	0.490	−2.244	0.0006506
1667	0.257	−1.177	0.0005999

[1] G. Preuner and W. Schupp, *Z. physik. Chem.*, **68**, 157 (1909).
[2] M. Randall and F. R. Bichowsky, *J. Am. Chem. Soc.*, **40**, 368 (1918).

tifully upon a smooth curve which is nearly a straight line.[1]   The slope at each point is $\Delta H°$.   Thus, at $1025°K$, $\Delta H° = -21,000$; at $1675°K$, $\Delta H° = -21,500$ cal; and this trend agrees with that calculated from the specific heats of the gases.

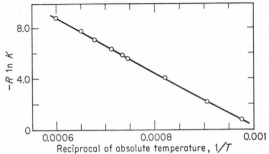

FIG. 15-1. Equilibrium between hydrogen, sulfur, and hydrogen sulfide.

If the heat-capacity equations are available for all reactants and products, an improved procedure is possible.   From Eq. (15-9)*, with $\Delta F°/T = -R \ln K$, one obtains

$$-R \ln K = \frac{\Delta H_I°}{T} - \Delta a \ln T - \tfrac{1}{2} \Delta b \, T - \tfrac{1}{2} \Delta c \, T^{-2} + I$$

$$\frac{\Delta H_I°}{T} + I = \sum = -R \ln K + \Delta a \ln T + \tfrac{1}{2} \Delta b \, T + \tfrac{1}{2} \Delta c \, T^{-2} \quad (15\text{-}36)*$$

At each temperature for which a value of K is known, Σ, the right-hand side of Eq. (15-36)* can be evaluated.   As the left-hand side of Eq. (15-36)* is a linear function of $1/T$, a plot of Σ against $1/T$ gives a straight line with a slope of $\Delta H_I°$.   From the value of $\Delta H_I°$, one can then use Eq. (6-18)* to obtain $\Delta H°$ at any temperature in the range 298 to $2000°K$ for which the heat-capacity equations are applicable.   From the intercept of the Σ versus $1/T$ plot, one obtains I.   Substitution of the values of the two integration constants $\Delta H_I°$ and I into Eq. (15-9)* or (15-10)* allows ΔF and therefore K to be calculated at any temperature in the range of applicability of the $C_P°$ equations.

The use of the Σ plot is a more satisfactory method of obtaining the heat of reaction than the plot of log K versus $1/T$ given in Fig. 15-1, as it is easier to get the slope of a straight line than of a curved line.   We may use H_2S again as an example.   From the heat-capacity equations given

---

[1] Because such curves usually approximate to straight lines, this method is also very advantageous for the interpolation or extrapolation of the equilibrium constant. It is frequently convenient to plot log K instead of $-R \ln K$, which serves equally well for interpolation, but in this case the slope must be multiplied by -4.5758 in order to obtain the value of ΔH.

in Table 6-4, one obtains, for the reaction $H_2(g) + \frac{1}{2}S_2(g) = H_2S(g)$,

$$\Delta c_P^o = -3.07 + 2.10 \times 10^{-3}T - 0.13 \times 10^5 T^{-2} \qquad \text{cal/deg}$$
$$\Delta H^o = \Delta H_I^o - 3.07T + 1.05 \times 10^{-3}T^2 + 0.13 \times 10^5 T^{-1} \qquad (15\text{-}37)*$$
$$\Sigma = -R \ln K - 3.07 \ln T + 1.05 \times 10^{-3}T - 0.065 \times 10^5 T^{-2}$$
$$= \frac{\Delta H_I^o}{T} + I \qquad (15\text{-}38)*$$

A comparison of the Σ-plot method with the method of Fig. 15-1 using the H_2S data is carried out in Problem 15-12.

If we had reversed the equation of this reaction, which is the same as taking the reciprocal of the former equilibrium constant, we would write

$$H_2S(g) = H_2(g) + \frac{1}{2}S_2(g) \qquad K = \frac{[H_2][S_2]^{\frac{1}{2}}}{[H_2S]}$$

In such a case we speak of the reaction as a dissociation, and this class of reaction presents some interesting features which are discussed further in the following section.

THE DISSOCIATION CONSTANT

Since the pioneer work of Sainte-Claire Deville,[1] many of the most important reversible reactions which have been studied at high temperatures are those which may be classified as dissociations.

In such valuable technical processes as the catalytic union of sulfur dioxide and oxygen to form sulfur trioxide or the Haber process for the synthesis of ammonia, the yield under given conditions of temperature and pressure is limited by the equilibrium, i.e., by the balance between the process of formation of sulfur trioxide or ammonia and the process of dissociation.

An interesting type of dissociation is one in which a molecule breaks up to form two like molecules. Let us consider the dissociation of iodine vapor at high temperatures, according to the equation

$$I_2(g) = 2I(g) \qquad K = \frac{[I]^2}{[I_2]} \qquad (15\text{-}39)$$

At a temperature of 1000°C or above, the dissociation is nearly complete,[2] but, at 600°C, K is about 0.0002, so that if I_2 is at atmospheric pressure, I will be at 0.014 atm (and the degree of dissociation will be 0.007).

If, therefore, we were working with this gas in the neighborhood of atmospheric pressure, we should find Boyle's law to be approximately

[1] Sainte-Claire Deville, *Ann. chem. pharm.*, **135,** 94 (1865).

[2] M. L. Perlman and G. K. Rollefson, *J. Chem. Phys.*, **9,** 362 (1941).

obeyed and would take the fugacity of I_2 as equal to the total pressure of the gas.

If now the pressure were lowered, we should pass through a range in which the gas would be found to deviate very markedly from the law of the perfect gas, until at low pressures, say one-millionth of an atmosphere, the dissociation would be practically complete and Boyle's law would be once more obeyed.

Strictly speaking, it is outside the province of thermodynamics to speculate concerning the cause of such anomalous behavior. We may choose any formula for the substance iodine, such as I, or I_2, or I_7, and then carry through all our thermodynamic calculations, ultimately arriving at the same conclusions, irrespective of our choice. On the other hand, it may be far more convenient to use one formula rather than another, and in such a case as we are now discussing there is nothing to be lost, and much to be gained, by accepting the view that we have in iodine vapor two distinct molecular species, I_2 and I, and by explaining the anomalous behavior of the gas as due to equilibrium between these two species. Thus in the range of low dissociation we may say approximately that the fugacity of I_2 is equal to the total pressure,[1] while from Eq. (15-39) the fugacity of I is proportional to the square root of that pressure.

Indeed if the association of monatomic iodine to form diatomic iodine and the dissociation of diatomic iodine to form monatomic iodine were both very slow reactions, it might be possible to isolate both I_2 and I as separate pure substances. Such separation has actually been made in the case of two forms of elementary sulfur, and we may recall in this connection our discussion of pure substances in Chapter 2.

When, from the dissociation constant of this reaction, we calculate its standard free-energy change, we may be dealing with a process which is not realizable in practice. We write

$$I_2(g) = 2I(g) \qquad \Delta F^\circ = -RT \ln K$$

where ΔF° is the increase in free energy which would attend an isothermal process which uses up 1 mole of pure I_2 and produces 2 moles of pure I, both at atmospheric pressure, or, more strictly, both at unit fugacity. In using such equations it will make no difference to us whether the various substances considered are in states which are experimentally realizable. Thus we can calculate from Eq. (15-11) the change in free energy in going from rhombic sulfur to monoclinic sulfur at 100° whether or not it is possible to obtain rhombic sulfur in the metastable superheated condition or not. Early in this chapter we have in fact calculated the free-energy difference between rhombic sulfur and monoclinic sulfur at 380°K at 1 atm, where rhombic sulfur is unstable, in order to be able to calculate the pressure at which rhombic sulfur would become stable at that temperature.

Use of Free-energy Functions in Treating Equilibrium Data.
If the function $\Delta(F^\circ - H_0^\circ)/T$ is available for the reaction under study, a particularly simple and accurate treatment can be given which yields the

[1] In the range of negligible dissociation, if the gas law is not obeyed, we cannot determine the fugacity by experiments at lower and lower pressures as our practice has been hitherto; but we may use the methods of Chapter 16 and by extrapolating to zero pressure find approximately what the behavior of the gas would be if there were no dissociation.

heat of the reaction as well as the free energy as a function of temperature. Each equilibrium measurement yields a value of $\Delta F_T^\circ / T$ from Eq. (15-32), and by use of Eq. (15-14) one has

$$\Delta H_0^\circ = T \left[\frac{\Delta F_T^\circ}{T} - \Delta \left(\frac{F_T^\circ - H_0^\circ}{T} \right) \right] \qquad (15\text{-}40)$$

Thus one obtains a series of values of ΔH_0° from a series of values of the equilibrium constant, and if the data are accurate, all these values of ΔH_0° will be the same. An alternate treatment yielding ΔH_{298}° follows in a corresponding manner from Eq. (15-17) if values of $\Delta (F_T^\circ - H_{298}^\circ)/T$ are available. Since the free-energy functions are almost invariably obtained by methods based upon the third law of thermodynamics, this treatment of equilibrium data is commonly called the *third-law* method.

Often temperature-dependent errors are difficult to eliminate from the equilibrium measurements, and while the resulting equilibrium constants or free energies of reaction are approximately correct, the temperature coefficient and the corresponding heat of reaction from the second law [Eq. (15-8)] may be very greatly in error. The third-law method will also yield the heat of reaction when ΔF° values have been determined over too small a temperature interval to determine the temperature coefficient accurately.

An excellent example of the advantage of the third-law method is provided by the dissociation reaction

$$\mathrm{F_2(g)} = 2\mathrm{F(g)}$$

The function $(F^\circ - H_0^\circ)/T$ may be calculated[1] for F_2 and F from spectroscopic data by the methods to be given in Chapter 27, and then values of ΔH_0° are obtained[2] from each of the direct dissociation-constant measurements of Doescher[3] and Wise.[4] These values are shown plotted with respect to temperature in Fig. 15-2. It is clear that some temperature-dependent errors have arisen in the measurement of Wise, which would affect the value of ΔH° if calculated by Eq. (15-8) or (15-35). But the third-law treatment not only points out this difficulty but also would allow a reasonably accurate ΔH_0° to be obtained from the mean of the individual values.

It is possible, of course, to obtain nonconstant ΔH_0° values from accurate equilibrium data because of erroneous $\Delta(F^\circ - H_0^\circ)/T$ functions, but in the case of F and F_2 there seems no reason to suspect these functions.

[1] L. G. Cole, M. Farber, and G. W. Elverum, Jr., *J. Chem. Phys.*, **20**, 586 (1952).

[2] J. G. Stamper and R. F. Barrow, *Trans. Faraday Soc.*, **54**, 1592 (1958).

[3] R. N. Doescher, *J. Chem. Phys.*, **20**, 330 (1952).

[4] H. Wise, *J. Phys. Chem.*, **58**, 389 (1954).

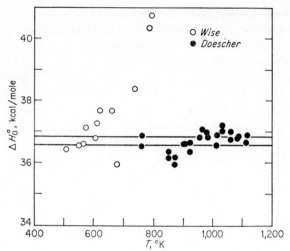

Fig. 15-2. Values of ΔH_0° of dissociation of F_2 from combination of individual equilibrium measurements with $(F_0^\circ - H_0^\circ)/T$ data. [*J. G. Stamper and R. F. Barrow, Trans. Faraday Soc.*, **54**, 1592 (1958), *fig.* 1.]

Also in this case the data of Doescher offer strong confirmation of the correctness of the free-energy functions.

The data shown in Fig. 15-2 yield 36.71 ± 0.13 kcal/mole for the ΔH_0° of dissociation of F_2.

BIBLIOGRAPHY OF FREE-ENERGY COMPILATIONS

In Appendix 7, tables of free-energy functions from 298 to 2000°K and values of heats of formation at 298°K are given for many substances, which allow calculation of ΔF° values over a range of temperature. These tables are restricted to rather common substances and generally to data of high accuracy. If data are desired for less common substances or if less accurate data can be used, there are many journal papers which can be located through the use of *Chemical Abstracts*, which gives tables of free-energy functions and heats of formation. In recent years, summaries of current papers containing thermodynamic data have been presented each year in *Annual Reviews of Physical Chemistry*.[1] In addition, the following list presents a number of comprehensive compilations of free-energy functions and heats of formation which will be useful sources of information. Some of these compilations give values at smaller temperature intervals than Appendix 7 and therefore may allow easier interpolation.

1. F. D. Rossini et al., "Selected Values of Physical and Thermodynamic Properties of Hydrocarbons and Related Compounds," Carnegie Press, Pittsburgh, 1953; also loose-leaf supplements. Free-energy func-

[1] H. Eyring (ed.), Annual Reviews, Inc., Palo Alto, Calif.

tions based on 0°K and other thermodynamic data for a large variety of hydrocarbons in the temperature range 0 to 1500°K have been compiled by *Research Project* 44 of the American Petroleum Institute. Data are also presented for elemental carbon, hydrogen, oxygen, and nitrogen and for CO, CO_2, NO, OH, and H_2O gases up to 5000°K.

2. D. D. Wagman and W. H. Evans, *Natl. Bur. Standards (U.S.) Ser.* III *Compilation,* 1947–1956. Free-energy functions based on 0°K for 100 inorganic substances and 30 organic compounds in the range of 0 to 5000°K are being distributed in loose-leaf form to most libraries. This is a portion of a larger compilation effort of the Chemistry Division of the National Bureau of Standards of which *Series* I and II have been published by F. Rossini, D. Wagman, W. Evans, S. Levine, and I. Jaffe, as *National Bureau of Standards Circular* 500, 1952. *Circular* 500 lists heats, entropies, and free energies of formation at 298°K for the elements and their compounds. Some ΔH values are also given at 0°K, and thermodynamic data are given for phase transitions. Supplements to all series have been published in the *Journal of Research of the National Bureau of Standards.*

3. D. R. Stull and G. C. Sinke, Thermodynamic Properties of the Elements, *Advances in Chem. Ser.* 18, November, 1956. This is the most complete compilation of thermodynamic data for the elements which presents free-energy functions based on 298°K for elements in both the condensed and gaseous states from 298 to 3000°K.

4. K. K. Kelley, Contributions to the Data on Theoretical Metallurgy: XIII, High Temperature Heat-content, Heat-capacity, and Entropy Data for Inorganic Compounds, *U.S. Bur. Mines Bull.* 584, 1960; XI, Entropies of Inorganic Substances, *U.S. Bur. Mines Bull.* 477, 1950; XIV, Entropies of Inorganic Substances, and K. K. Kelley and E. G. King, *U.S. Bur. Mines Bull.* 592, in press, present a critical summary of s°_{298}, $H^{\circ}_{T} - H^{\circ}_{298}$, and $s^{\circ}_{T} - s^{\circ}_{228}$ data at 100° intervals from which $(F^{\circ} - H^{\circ}_{298})/T$ values can be readily calculated. Theoretical Metallurgy: XII, Heats and Free Energies of Formation of Inorganic Oxides, *U.S. Bur. Mines Bull.* 542, 1954, by J. P. Coughlin, presents a critical summary of ΔH° and ΔF° data and estimated values up to 2000°K.

5. L. Brewer, L. A. Bromley, P. W. Gilles, and N. L. Lofgren, Papers 6 and 8, in L. L. Quill (ed.), "The Chemistry and Metallurgy of Miscellaneous Materials—Thermodynamics," McGraw-Hill Book Company, Inc., New York, 1950. Experimental and estimated free-energy functions based on 298°K have been tabulated for virtually all known halides in the range 298 to 1500°K.

6. R. Hultgren et al., Selected Values for the Thermodynamic Properties of Metals and Alloys, *Univ. Calif. (Berkeley) Inst. Eng. Research, Minerals Research Lab.;* issued in loose-leaf form at irregular intervals

since 1955. Free-energy functions based on 298°K are presented for many metals up to 2500°K along with other thermodynamic data including solution data.

7. W. M. Latimer, "The Oxidation States of the Elements and Their Potentials in Aqueous Solutions," 2d ed., Prentice-Hall, Inc., Englewood Cliffs, N.J., 1952, has critically reviewed all thermodynamic data for inorganic systems and some organic systems at room temperature with emphasis on aqueous solutions. Also many estimated values are given.

8. T. Hilsenrath, C. W. Beckett, W. S. Benedict, L. Fano, H. J. Hoge, J. F. Maci, R. L. Nuttall, Y. S. Touloukian, and H. W. Woolley, Thermal Properties of Gases, *Natl. Bur. Standards Circ.* 564, 1955. Free-energy functions based on 0°K are given for Ar, CO_2, H_2, N_2, O_2, and H_2O up to 5000°K. This publication is part of the program of the Thermodynamics Section of the Heat and Power Division of the U.S. Bureau of Standards. More recent publications include *Natl. Bur. Standards Monograph* 20, Ideal Gas Thermodynamic Functions and Isotope Exchange Functions for the Diatomic Hydrides, Deuterides, and Tritides, by L. Haar, A. S. Friedman, and C. W. Beckett (in press), and the paper Thermodynamic Properties at High Temperatures: Ideal Gas Thermal Functions to 25,000°K for Diatomic Molecules, Oxygen, Nitrogen, Nitric Oxide, and Their Molecule Ions, by C. W. Beckett and L. Haar, Institution of Mechanical Engineers, London, 1958.

9. M. G. Ribaud, *French Air Ministry Pub.* 266, 1952. Methods of calculating free-energy functions are reviewed, and free-energy functions based on 0°K are tabulated for saturated hydrocarbons up to 1500°K and for many other inorganic and organic substances up to 4000°K.

10. H. Zeise, "Thermodynamik auf den Grundlagen der Quantentheorie, Quantenstatistik und Spektroskopie," Bd. III, "Ergebnisse in Tabellarischer and Graphischer Form," S. Hirzel, Leipzig, 1954–1957.

11. O. Kubaschewski and E. L. Evans, "Metallurgical Thermochemistry," 3d ed., Pergamon Press, London, 1958. Tables of thermodynamic data of particular interest to those dealing with metallurgical problems are presented, along with discussions of experimental methods of determining data as well as methods of estimating unknown data.

12. J. L. Margrave, "Proceedings of the Symposium on High Temperature—A Tool for the Future," Berkeley, Calif., June, 1956; sponsored by Stanford Research Institute and the University of California and published by Stanford Research Institute, Palo Alto, Calif. Free-energy functions based on 298°K are compiled for gaseous atoms up to 5000°K. [Also published by T. J. Katz and J. L. Margrave, *J. Chem. Phys.*, **23**, 983 (1955).] Nomographs for calculating translational, rotational, and electronic contributions to the gaseous free-energy functions are also given.

13. J. L. Margrave, app. V, in J. O'M. Bockris, J. L. White, and J. D. Mackenzie (eds.), "Physico-chemical Measurements at High Temperatures," Academic Press, Inc., New York, 1959. Free-energy functions are given at 500° intervals up to 5000°K for gases and up to 2000°K for many condensed phases.

In addition to the sources listed above, there are now a number of preliminary tabulations of free-energy functions which are issued by United States government agencies and other organizations as mimeographed reports and which serve to fill gaps in the literature until permanent publications become available.

PROBLEMS

✓ **15-1.** From the densities of diamond and graphite, 3.51 and 2.55 g/cc, respectively, calculate the change of ΔF with pressure for the conversion reaction, and verify the statement in the text that the equilibrium pressure is approximately 22,000 atm at room temperature.

✓ **15-2.** A galvanic cell in which occurs the reaction $Ag(s) + \frac{1}{2}HgCl_2(s) = AgCl(s) + Hg(l)$ gives $\varepsilon^{\circ}_{298} = 0.0455$ volt.[1] For this reaction $\Delta H^{\circ}_{298} = 1276$ cal. Calculate ΔF°_{298}, ΔS°_{298}, and $d\varepsilon^{\circ}/dT$.   (Here is a case in which ΔH° and ΔF° have opposite signs.)

15-3. For the reaction $CO(g) + \frac{1}{2}O_2(g) = CO_2(g)$, find the complete free-energy equation from the data in this chapter and that obtained in Problem 6-6. Calculate ΔF°_{2000}.   What can be said regarding the stability of carbon dioxide at 2000°K?

✓ **15-4.** Use the free-energy functions from Appendix 7 to calculate

$$\frac{\Delta F^{\circ} - \Delta H^{\circ}_0}{T} \text{ and } \Delta F^{\circ}_{2000} \quad \text{for } CO(g) + \frac{1}{2}O_2(g) = CO_2(g)$$

✓ **15-5.** A cell, operating at constant temperature and pressure, undergoes a change of volume ΔV when n equivalents pass. Show that, if the pressure on the whole cell is changed, the emf changes according to the equation

$$\frac{d\varepsilon}{dP} = -\frac{\Delta V}{n\mathfrak{F}}$$

15-6. A galvanic cell with one electrode of hydrogen at a partial pressure of 0.9 atm and another electrode of mercury and solid mercurous chloride, with an electrolyte over which the partial pressure of hydrogen chloride gas is 0.01 atm, has an emf at 25°C of 0.0110 volt. The cell reaction is

$$\frac{1}{2}H_2(g, 0.9 \text{ atm}) + \frac{1}{2}Hg_2Cl_2(s) = Hg(l) + HCl(g, 0.01 \text{ atm})$$

Calculate ΔF and ΔF°.

15-7. We also find from emf measurements

$$\frac{1}{2}Cl_2(g) + Hg(l) = \frac{1}{2}Hg_2Cl_2(s) \qquad \Delta F^{\circ}_{298} = -25,182 \text{ cal}$$

Find ΔF°_{298} for the reaction

$$\frac{1}{2}H_2(g) + \frac{1}{2}Cl_2(g) = HCl(g)$$

✓ **15-8.** In the "water-gas reaction" $CO_2 + H_2 = CO + H_2O(g)$, equilibrium would

[1] R. H. Gerke, *J. Am. Chem. Soc.*, **44**, 1684–1704 (1922).

be reached at 1538°K when the four gases have, respectively, the following partial pressures: 0.10, 0.10, 0.10, and 0.24 atm.   Calculate K, ΔF°_{1538}, and ΔH°_0.

15-9. Another reaction for which ΔH° cannot be obtained by calorimetric methods is the one studied by M. L. Perlman and G. K. Rollefson.[1]

For the reaction $I_2(g) = 2I(g)$, calculate ΔH° at 900°K and at 1200°K from the data of the following table.   By extrapolation obtain K_{800} and ΔF°_{800}.

T, °K........	872	973	1073	1173	1274
K............	1.81×10^{-4}	1.80×10^{-3}	0.0108	0.0480	0.167

15-10. From the fact that $I(g)$ is monatomic, and from Table 6-4, derive the equation

$$I_2(g) = 2I(g) \qquad \Delta F^{\circ} = \Delta H_I - 1.00T \ln T + 0.07 \times 10^{-3}T^2 - 0.085 \times 10^5 T^{-1} + IT$$

$$\text{or} \quad \Sigma = -R \ln K + 1.00 \ln T - 0.07 \times 10^{-3}T + 0.085 \times 10^5 T^{-2} = \frac{\Delta H_I}{T} + I$$

Use the data of Problem 15-9 to calculate Σ at each temperature.   Plot Σ versus $1/T$ and obtain ΔH_I and I.   Calculate ΔH°_{298} and compare with the value from Appendix 7.

15-11. Calculate ΔF° for $I_2(g) = 2I(g)$ at 500°C.   Find the degree of dissociation of the gas when the total pressure is 1 atm; $\tfrac{1}{1000}$ atm; $\tfrac{1}{1,000,000}$ atm.

15-12. Equation (15-37)[*] gives ΔH as a function of temperature for the reaction $H_2(g) + \tfrac{1}{2}S_2(g) = H_2S(g)$.   Calculate the change in ΔH between 1675 and 1025°K, and show that the result agrees in sign and in order of magnitude with the difference obtained in the text from equilibrium measurements alone.   Use the data of Table 15-2 to calculate Σ from Eq. (15-38)[*] at each temperature and to obtain ΔH°_I and I from a plot of Σ versus $1/T$.   Calculate the degree of dissociation of H_2S gas at 2000°K at a total pressure of 1 atm. ΔF° for $HS(g)$ at 2000°K is 8930 cal.   Will consideration of HS influence the above calculations?

15-13. At 0°C and a pressure of 87 mm Hg, the density[2] of an equilibrium mixture of NO_2 and N_2O_4 is 0.84 times the density calculated for pure N_2O_4.   Assume that both gases are perfect, and calculate the degree of dissociation of N_2O_4, the dissociation constant K, and ΔF°_{273} for $N_2O_4(g) = 2NO_2(g)$.

15-14. Refer to Problems 15-3 and 15-4 to find the percentage of CO_2 which would be dissociated if the gas is heated to 2000°K at atmospheric pressure.   (As a first approximation assume the partial pressure of CO_2 to be 1 atm, and, thus obtaining the partial pressures of CO and O_2, subtract these from unity, and obtain the partial pressure of CO_2 which is to be used in the second approximation.)

15-15. Calculate the equilibrium partial pressure for all vapor species present at greater than 10^{-4} atm when $CCl_4(g)$ is heated to 1000°K in the presence of graphite. The total pressure is 1 atm.   Use data from Appendix 7.

15-16. (a) Use the data of Appendix 7 to calculate the equilibrium constant at 500°K for the reaction $Hg(g) + Cl_2(g) = HgCl_2(g)$.   (b) Calculate the per cent of $HgCl_2$ gas at 0.01 atm decomposed to the elements at 500°K.   (c) Of the halogen gas that results from the decomposition of the $HgCl_2$ at 0.01 atm, what fraction is monatomic chlorine?

[1] *Loc. cit.*

[2] E. Natanson and L. Natanson, *Ann. Physik*, (3)**24**, 454 (1885), **27**, 606 (1886); K. Shreber, *Z. physik. Chem.*, **24**, 651 (1897).

16

REAL GASES

We have indicated on a number of occasions that consideration would be given to the properties of real gases as they differ from those of perfect gases. This discussion was postponed until now because we wish to make extensive use of the fugacity, which was defined in the last part of Chapter 14.

The thermodynamic properties of a gas may be represented by an equation of state which gives the volume as a function of pressure and temperature or, conversely, the pressure as a function of volume and temperature. Such properties as the fugacity, enthalpy, and entropy are related to derivatives or integrals of the equation-of-state function. Thus we desire an equation of state which not only represents accurately the volume of the gas but has a mathematical form which can be differentiated and integrated readily to yield other thermodynamic functions. The actual volumetric behavior of gases at high pressures is so complex, however, that such an equation of state has not been obtainable. Consequently we shall have to discuss several methods of calculation which have advantages for particular purposes.

Virial Equations of State. In 1901 Onnes first used a power series to express the volume of a gas. This form is known as a virial equation of state.

$$PV = A + \frac{B'}{V} + \frac{C'}{V^2} + \frac{D'}{V^3} + \cdots \qquad (16\text{-}1)$$

Here A is the first virial coefficient, B' is the second virial coefficient, etc. Each coefficient is a function of temperature. An alternate form is a series of positive powers of the pressure, which is more convenient for most purposes.

$$PV = A + BP + CP^2 + DP^3 + \cdots \qquad (16\text{-}2)$$

It may be shown that A is the same quantity in Eq. (16-1) as in Eq. (16-2) and that $B' = AB$ provided that B and B' are evaluated in the low-pressure range. All the higher coefficients are different, and their interrelationship is more complex. We shall employ primarily the expansion in pressure.

It is clear that the first virial coefficient is just the perfect-gas-law term. Hence, if we deal now with the molal volume, we have

$$Pv = RT + \text{B}P + cP^2 + \text{D}P^3 + \cdots \qquad (16\text{-}3)$$

Fugacity; Free Energy. From Eq. (14-14) we have the differential relationship between fugacity and pressure,

$$\left(\frac{\partial \ln f}{\partial P}\right)_T = \frac{v}{RT} \qquad (14\text{-}14)$$

which we may now integrate. First, however, let us make a rearrangement. We take the ratio of fugacity to pressure and obtain

$$\frac{\partial \ln (f/P)}{\partial P} = \frac{\partial \ln f}{\partial P} - \frac{\partial \ln P}{\partial P} = \frac{v}{RT} - \frac{1}{P} \qquad (16\text{-}4)$$

This equation may be integrated from a low pressure P_1 to our final pressure P_2.

$$\ln \frac{f_2}{P_2} - \ln \frac{f_1}{P_1} = \int_{P_1}^{P_2} \left(\frac{v}{RT} - \frac{1}{P}\right) dP$$

But the fugacity has been defined to be equal to the pressure at zero pressure; hence we have

$$\ln \frac{f_2}{P_2} = \int_0^{P_2} \left(\frac{v}{RT} - \frac{1}{P}\right) dP \qquad (16\text{-}5)$$

If we now substitute Eq. (16-3) into Eq. (16-5), we have

$$\ln \frac{f_2}{P_2} = \int_0^{P_2} \left(\frac{\text{B}}{RT} + \frac{cP}{RT} + \frac{\text{D}P^2}{RT} + \cdots\right) dP$$

or

$$\ln \frac{f}{P} = \frac{\text{B}P}{RT} + \frac{cP^2}{2RT} + \frac{\text{D}P^3}{3RT} + \cdots \qquad (16\text{-}6)$$

The volume of nitrogen at 0°C and pressures up to 200 atm was expressed by Otto, Michels, and Wouters[1] with an equation which we have converted to a molal basis in units of cubic centimeters and atmospheres.

$$Pv = 22{,}414.6 - 10.281P + 0.065189P^2 + 5.1955 \times 10^{-7}P^4$$
$$- 1.3156 \times 10^{-11}P^6 + 1.009 \times 10^{-16}P^8 \qquad (16\text{-}7)^*$$

[1] J. Otto, A. Michels, and H. Wouters, *Physik. Z.*, **35,** 97 (1934); similar equations are given for other temperatures up to 150°C.

Thus we obtain for the fugacity of nitrogen at 0°C

$$\ln \frac{f}{P} = -4.5867 \times 10^{-4}P + 1.4542 \times 10^{-6}P^2 + 5.7948$$
$$\times 10^{-12}P^4 - 0.9782 \times 10^{-16}P^6 + 5.627 \times 10^{-22}P^8 \quad (16\text{-}8)^*$$

The relationship between the fugacity ratio and the free-energy difference for a pair of states A and B was given by Eq. (14-10) to be

$$\text{F}_B - \text{F}_A = RT \ln \frac{f_B}{f_A} \qquad (14\text{-}10)$$

The standard state for free energy was defined as a hypothetical ideal gas at 1 atm, which implies a fugacity of 1 atm.   Hence

$$\text{F} - \text{F}^\circ = RT \ln f \qquad (16\text{-}9)$$

which shows the simple relationship between the fugacity of a gas and the departure of its free energy from the standard free energy.   It is convenient to tabulate values of (or to give equations for) either f/P or $\ln (f/P)$ for real gases, since the fugacity is usually the preferred measure of escaping tendency, but calculation of free energy from such values is very simple.

Real Gases at Low Pressure; the Second Virial Coefficient. There are a number of valuable relationships which are limited in their application to gases at low pressures, where the deviation from the perfect-gas law is small.   In this region the higher virial coefficients may be ignored and our attention concentrated on the second virial coefficient, which gives the first-order deviation from the perfect-gas law.   For example,

$$P\text{v} = RT + \text{B}P \qquad (16\text{-}10)^*$$

and

$$\ln \frac{f}{P} = \frac{\text{B}P}{RT} \qquad (16\text{-}11)^*$$

or

$$\frac{f}{P} = e^{\text{B}P/RT}$$

but since $e^x = 1 + x$ when x is small and we have assumed $\text{B}P/RT$ to be small,

$$\frac{f}{P} = 1 + \frac{\text{B}P}{RT} = \frac{RT + \text{B}P}{RT}$$

Hence from Eq. (16-10)*

$$\frac{f}{P} = \frac{P\text{v}}{RT} \qquad (16\text{-}12)^*$$

and if we substitute P_i for RT/v, which is the pressure that an ideal gas would exert at the given volume, we obtain

$$\frac{f}{P} = \frac{P}{P_i} \qquad (16\text{-}13)^*$$

Thus the actual pressure lies between the fugacity and the pressure calculated from the perfect-gas law and is the geometrical mean of the two.   The values in Table 16-1 for nitrogen at 0°C show that this result is an excellent approximation up to 10 atm and is in error by only 1.5 per cent at 100 atm.

In 1907 D. Berthelot[1] proposed the equation of state

$$Pv = RT + \frac{9R}{128}\frac{PT_c}{P_c}\left(1 - 6\frac{T_c^2}{T^2}\right) \qquad (16\text{-}14)^*$$

where P_c and T_c are the critical pressure and temperature of the gas.   It is evident that this equation is in the virial form with just a second virial coefficient whose value is

$$\text{B} = \frac{9R}{128}\frac{T_c}{P_c}\left(1 - 6\frac{T_c^2}{T^2}\right) \qquad (16\text{-}15)^*$$

This equation has proved to be quite satisfactory for the estimation of small deviations from the ideal-gas law when data are not available for the particular gas in question.

We may rearrange Eq. (16-15)* to the form

$$\frac{\text{B}P_c}{RT_c} = \frac{9}{128}\left(1 - 6\frac{T_c^2}{T^2}\right) \qquad (16\text{-}16)^*$$

whereby we see that $\text{B}P_c/RT_c$ has the same functional dependence on T_c/T for all substances.   This is an example of a "reduced" equation of state; we shall discuss such equations further in a later section.   Figure 16-1 shows the data for the second virial coefficient for a number of substances in the reduced form developed above.   We see that the curve of the Berthelot equation represents the experimental data about as well as any single curve could.   It is also apparent that the points for various substances deviate significantly from a single curve.   We shall present an improved but more complex equation for the second virial coefficient in Appendix A-1.

If we now substitute Berthelot's expression into Eq. (16-11)*, we obtain the fugacity of a Berthelot gas.

$$\ln\frac{f}{P} = \frac{9T_cP}{128TP_c}\left(1 - 6\frac{T_c^2}{T^2}\right) \qquad (16\text{-}17)^*$$

[1] D. Berthelot, *Trav. mem. bur. intern. poids mesures*, no. 13 (1907).

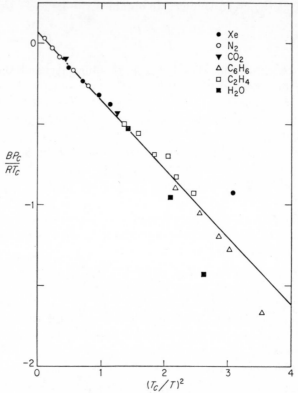

FIG. 16-1. The reduced second virial coefficients for several substances.   The Berthelot equation gives the straight line.

Also after expansion of the logarithm

$$\frac{f}{P} \cong 1 + \frac{9T_cP}{128TP_c}\left(1 - 6\,\frac{T_c{}^2}{T^2}\right)$$

$$f \cong P + \frac{9T_cP^2}{128TP_c}\left(1 - 6\,\frac{T_c{}^2}{T^2}\right) \qquad (16\text{-}18)^*$$

Calculation of Fugacity by Graphical Methods.   While Eq. (16-6) gives a convenient expression for the fugacity when the virial coefficients are known, it is also easy to integrate Eq. (16-5) graphically.   Figure 16-2 shows a graph in which $(v/RT) - (1/P)$ is plotted as the ordinate and P as the abscissa.   The data are for nitrogen at $0°C$.   The area between the curve and the axis from zero pressure out to the desired value of P yields the integral in Eq. (16-5), which in turn equals $\ln(f/P)$. We note that there is a negative area at low pressure but a positive area at high pressures.   Table 16-1 gives a summary of the calculation.

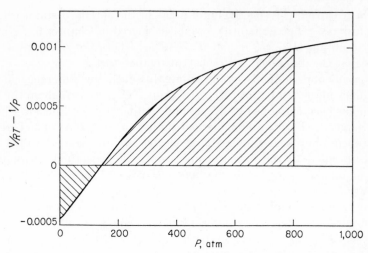

FIG. 16-2. The graphical integration of the fugacity of N_2 at 800 atm. The data are for 0°C.

TABLE 16-1. THE FUGACITY OF NITROGEN AT 0°C†

P, atm	$\dfrac{v}{RT} - \dfrac{1}{P}$	$\dfrac{f}{P}$	$\dfrac{Pv}{RT} = \dfrac{P}{P_i}$
1	−0.00045	0.99955	0.99955
10	−0.00043	0.9956	0.9957
50	−0.00030	0.9812	0.9850
100	−0.000146	0.9703	0.9854
150	+0.000020	0.9672	1.0030
200	0.000181	0.9721	1.0363
300	0.000451	1.0055	1.1353
400	0.000641	1.062	1.2566
600	0.000874	1.239	1.5242
800	−0.000995	1.495	1.7964
1000	0.001070	1.839	2.070

† J. Otto, A. Michels, and H. Wouters, *Physik. Z.*, **35,** 97 (1934); also B. H. Sage and W. N. Lacey, "Thermodynamic Properties of the Lighter Paraffin Hydrocarbons and Nitrogen," American Petroleum Institute, New York, 1950.

The last column in the table allows us to test Eq. (16-13)*. According to that equation the values in the last two columns should be equal at low pressures.

The Enthalpy and Heat Capacity of Real Gases. In earlier discussions we stated that the heat capacity or enthalpy of a real gas could be taken as the sum of the quantity for the perfect gas and a term giving the effect of gas imperfection. We shall write $H°$ and $C_p°$ for the enthalpy and heat capacity, respectively, of a gas in the standard state, which is the ideal gas. These quantities were considered in Chapter 6. Now we wish to obtain $(H - H°)_T$ and $(C_p - C_p°)_T$, where the subscript T indicates that the difference is at constant temperature.

Experimental methods of obtaining this enthalpy difference include the direct measurement of $(\partial H/\partial P)_T$ proposed by Buckingham[1] in 1903, which has been employed by Keyes and Collins[2] and by Eucken, Clusius, and Berger.[3] This is a flow method similar to that in a Joule-Thomson experiment, but heat is introduced electrically to make the final temperature equal to the initial temperature. This method is applicable only at temperatures below the inversion point, where the gas cools on expansion.

One may also employ the conventional free expansion of a gas without heating to obtain enthalpy data indirectly. We saw in Chapter 5 that, if we prevent heat transfer to the gas, this is a constant-enthalpy process. Thus, if our gas is X, we may write

$$X(g, P_1, T_1) = X(g, P = 0, T_2) \qquad \Delta H = 0$$

But if we have the heat capacity of the ideal gas $C_p°$, we may readily calculate the enthalpy difference from T_2 to T_1 at low pressure. Consequently

$$(H_{P_1} - H°)_{T_1} = H_{T_2}° - H_{T_1}° = \int_{T_1}^{T_2} C_p° \, dT \qquad (16\text{-}19)$$

For example, if oxygen is expanded from 200 atm and 298°K to a very low pressure, it is found that the final temperature is 248°, or 50° lower than the initial temperature. Since the heat capacity of oxygen is 6.95 cal/deg mole in this range, we find $(H - H°)_T$ at 200 atm and 298°K to be -347 cal/mole.

Enthalpy values may also be obtained from volumetric data on the gas. By rearrangement of Eq. (10-24) we obtain

$$\left(\frac{\partial H}{\partial P}\right)_T = V - T\left(\frac{\partial V}{\partial T}\right)_P \qquad (16\text{-}20)$$

Since any gas approaches the perfect-gas state at low pressure, we need only integrate Eq. (16-20) at constant temperature from zero to our final pressure to obtain $(H - H°)_T$.

$$(H - H°)_T = \int_0^{P'} \left(\frac{\partial H}{\partial P}\right)_T dP = \int_0^{P'} \left[V - T\left(\frac{\partial V}{\partial T}\right)_P\right] dP \qquad (16\text{-}21)$$

[1] E. Buckingham, *Phil. Mag.*, **6**, 518 (1903).
[2] F. G. Keyes and S. C. Collins, *Proc. Natl. Acad. Sci. U.S.*, **18**, 328 (1932).
[3] A. Eucken, K. Clusius, and W. Berger, *Z. tech. Physik*, **13**, 267 (1932).

Substitution of the virial equation into Eq. (16-21) yields

$$(\text{H} - \text{H}°)_T = \left(\text{B} - T\frac{d\text{B}}{dT}\right)P + \left(\text{C} - T\frac{d\text{C}}{dT}\right)\frac{P^2}{2} + \cdots \quad (16\text{-}22)$$

In order to obtain the heat capacity of a real gas we may differentiate Eq. (16-20) and (16-21) with respect to temperature at constant pressure.

$$\left(\frac{\partial C_P}{\partial P}\right)_T = -T\left(\frac{\partial^2 \text{V}}{\partial T^2}\right)_P \quad (16\text{-}23)$$

$$(C_{P'} - C_T°)_T = -T\int_0^{P'} \left(\frac{\partial^2 \text{V}}{\partial T^2}\right)_P dP \quad (16\text{-}24)$$

The heat capacity of a gas may be measured by the flow method discussed in Chapter 6 or calculated from velocity of sound or $(\partial T/\partial P)_S$ measurements as described in Chapter 10. Measured values of C_P at different pressures may be related to $(\partial^2 V/\partial T^2)_P$ according to Eqs. (16-23) and (16-24) and thereby provide a stringent test of an equation of state. For example, differentiation of Eq. (16-22) yields

$$(C_P - C_P°)_T = -T\left[\left(\frac{d^2\text{B}}{dT^2}\right)P + \left(\frac{d^2\text{C}}{dT^2}\right)\frac{P^2}{2} + \cdots\right] \quad (16\text{-}25)$$

At low pressures only the second virial coefficient is important. Waddington, Todd, and Huffman[1] measured the heat capacity of n-heptane at several pressures in a flow calorimeter and obtained the values of $(\partial C_P/\partial P)_T$ shown in Table 16-2. These values are compared with those calculated from Eq. (16-25) with various equations of state.

TABLE 16-2. PRESSURE VARIATION OF THE MOLAL HEAT CAPACITY OF GASEOUS n-HEPTANE

T	$(\partial C_P/\partial P)_T$, cal/deg atm			
	Experimental	Calculated from		
		Eq. (16-15)*	Eq. (16-37)*	Eq. (A1-6)*
357.10	1.45	0.65	1.28	1.32
373.15	1.15	0.57	1.06	1.03
400.40	0.71	0.46	0.79	0.72
434.35	0.45	0.36	0.57	0.50
466.10	0.27	0.29	0.42	0.37

The results of Table 16-2 show that the Berthelot equation (16-14)* gives only a rough estimate of the change of heat capacity with pressure. The more complex equations, however, which fit the volumetric data accurately, do yield useful estimates of $(\partial C_P/\partial P)_T$.

[1] G. Waddington, S. S. Todd, and H. M. Huffman, *J. Am. Chem. Soc.*, **69**, 22 (1947).

The Entropy of Real Gases.   The equation for the change in entropy with pressure was obtained in Chapter 10, where Eq. (10-17) is derived,

$$\left(\frac{\partial S}{\partial P}\right)_T = -\left(\frac{\partial V}{\partial T}\right)_P \tag{10-17}$$

Then one can immediately write

$$(S_P - S_0)_T = -\int_0^P \left(\frac{\partial V}{\partial T}\right)_P dP \tag{16-26}$$

But this equation is not directly useful because the entropy of a gas is infinite at zero pressure.   In order to avoid this difficulty, the standard state was defined as the hypothetical ideal gas at 1 atm pressure.   Since $(\partial \mathrm{v}/\partial T)_P$ for a perfect gas is R/P, we have

$$(\mathrm{s}_{P'} - \mathrm{s}^\circ)_T = -\int_0^{P'} \left(\frac{\partial \mathrm{v}}{\partial T}\right)_P dP + \int_0^1 \frac{R}{P} dP$$

$$= \int_0^{P'} \left[\frac{R}{P} - \left(\frac{\partial \mathrm{v}}{\partial T}\right)_P\right] dP - R \ln P' \tag{16-27}$$

Substitution of the virial equation and a change of sign yields

$$(\mathrm{s}^\circ - \mathrm{s})_T = R \ln P + \left(\frac{d\mathrm{B}}{dT}\right) P + \left(\frac{d\mathrm{c}}{dT}\right)\frac{P^2}{2} + \cdots \tag{16-28}$$

These equations yield finite values which may be employed to obtain the entropy of a real gas.

The integrals in Eqs. (16-21), (16-24), and (16-27) can be evaluated graphically in the same manner as was illustrated in Fig. 16-2 for the fugacity calculation, but the temperature derivatives of the volume are required instead of the volume itself.   Also, with the virial expansion one requires the temperature derivatives of the virial coefficients.

Volume as the Independent Variable.   It is equally possible to treat the properties of real gases by using volume instead of pressure as the independent variable. We shall merely indicate a few initial steps.   Consider the molal-work-content function of the real gas

$$d\mathrm{A} = -P \, d\mathrm{v} \tag{16-29}$$

and the hypothetical molal work content A^* of an ideal gas

$$d\mathrm{A}^* = -\frac{RT}{\mathrm{v}} d\mathrm{v} \tag{16-30}$$

The difference between the molal work content of real gas at pressure P' and volume v' and ideal gas in the hypothetical standard state at 1 atm is

$$(\mathrm{A} - \mathrm{A}^\circ)_T = \int_\infty^{\mathrm{v}'} \left(\frac{RT}{\mathrm{v}} - P\right) d\mathrm{v} + RT \ln \frac{RT}{\mathrm{v}'} \tag{16-31}$$

since the volume in the standard state is RT. We note that the integral is from the infinite volume, where the real gas approaches ideal behavior, to the volume of interest v'. If the integral is to be integrated graphically, a rearrangement is made to the form,

$$(A - A^\circ)_T = RT \ln \frac{RT}{v'} + \int_{v=\infty}^{v=v'} (Pv^2 - RTv)\, d\left(\frac{1}{v}\right) \qquad (16\text{-}32)$$

whereupon a plot of $v(Pv - RT)$ versus $(1/v)$, both of which remain finite, yields the desired result.

Appropriate temperature derivatives and other manipulations of Eqs. (16-31) and (16-32) yield other thermodynamic functions as integrals of v or $1/v$. We shall not pursue this further in general terms but rather shall turn to the virial expansion in $1/v$ in the form

$$\frac{Pv}{RT} = 1 + \frac{B}{v} + \frac{C''}{v^2} + \frac{D''}{v^3} + \cdots \qquad (16\text{-}33)$$

This form has the advantage over Eq. (16-1) in that the second virial coefficient B is now identical to B for the pressure expansion, Eq. (16-3). Substitution of Eq. (16-33) into (16-31) yields

$$(A - A^\circ)_T = RT\left(\ln \frac{RT}{v} + \frac{B}{v} + \frac{C''}{2v^2} + \cdots\right) \qquad (16\text{-}34)$$

The fugacity is obtained as follows:

$$\ln f = \frac{F - F^\circ}{RT} = \frac{A - A^\circ}{RT} + \frac{PV}{RT} - 1$$

$$= \ln \frac{RT}{v} + \frac{2B}{v} + \frac{3C''}{2v^2} + \cdots \qquad (16\text{-}35)$$

This may be rearranged to show its equivalence to Eq. (16-6) as far as the term in B First note that

$$\ln \frac{f}{p} = -\ln \frac{Pv}{RT} + \frac{2B}{v} + \frac{3C''}{2v^2} + \cdots$$

but expansion in series of $\ln (Pv/RT)$ yields

$$\ln \frac{f}{P} = \frac{B}{v} + \frac{C'' + B^2}{2v^2} + \cdots \qquad (16\text{-}36)$$

Other thermodynamic quantities may be obtained from these equations by appropriate transformations such as were used in previous sections.

Continuity of Liquid and Gas Phases; the Critical Point. The discovery by Andrews[1] of the critical phenomenon and its implications contributes very greatly to an understanding of the relation between a liquid and its vapor. He found that, while the ordinary condensation of a vapor or evaporation of a liquid involves a discontinuity between the two phases, it is nevertheless possible to pass from liquid to vapor, or from vapor to liquid, by a process in which the substance remains perfectly homogeneous. Thus, if a liquid under high pressure is heated above its critical point, the pressure then reduced isothermally to a small

---

[1] T. Andrews, *Phil. Mag.*, **4**, 39, 150 (1870).

value, and the temperature finally reduced to its original value, the substance will have passed by imperceptible gradations from liquid to vapor.

This idea that the liquid and vapor phases are essentially one state of matter is clearly brought out by an inspection of the isotherms on a *P-V* diagram. In Fig. 16-3 each continuous curve shows, at a single

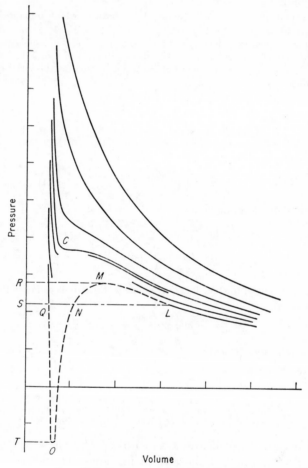

Fɪɢ. 16-3. Isotherms of carbon dioxide.

temperature, the measured molal volume of a pure phase of carbon dioxide at various pressures. At the highest temperature (represented by the curve farthest from the axes) the gas is nearly perfect, and the curve approaches the regular hyperbola. As the temperature is lowered, deviations from the gas law become more and more pronounced, until at $t = 31.35°C$ a curve is reached which at the point C has zero slope, that is, $dP/dV = 0$. This is the critical point. Below this point we find two phases capable of coexistence, and two separate curves at the same temperature are experimentally determined, one for the change in volume of the liquid with pressure, and one for the change in volume of the vapor with pressure.

However, it is attractive to assume that these two branches are really parts of a single curve and that a single mathematical equation can be found to fit both vapor and liquid regions. Thus the two branches would be joined by some such curve as the broken line QONML in the figure.

The famous equation of van der Waals satisfies these conditions qualitatively, but it yields only the roughest sort of agreement with the actual curves in the liquid region. The equation proposed by Benedict, Webb, and Rubin[1] yields reasonably good agreement, but it is so complex as to discourage its general use. Their equation has the form

$$P = RTd + \left(B_0RT - A_0 - \frac{C_0}{T^2}\right)d^2 + (bRT - a)d^3$$
$$+ a\alpha d^6 + \frac{cd^2}{T^2}(1 + \gamma d^2)e^{-\gamma d^2} \quad (16\text{-}37)^*$$

where d is the density, that is, $1/v$, and A_0, B_0, C_0, a, b, c, α, and γ are eight constants which are to be determined for each substance. These constants have been evaluated for several hydrocarbons and other substances.

Let us now return to the broken line in Fig. 16-3, which we assume to have been generated by an equation of the type under discussion. Indeed the left-hand portion QO is subject to experimental verification since liquids may be studied at pressures less than their vapor pressure—even at negative pressures.[2] While the central portion of the curve has never been and may never be realized, it is nevertheless instructive to follow the value of the molal free energy along the whole range of the broken curve. The horizontal line QNL represents the vapor pressures, and the molal free energy F is therefore the same at Q and at L. So also the total change in free energy in passing from L to Q by any path must be zero. Remembering that the free energy of a pure substance always increases with increasing pressure and diminishes with decreasing pressure, we see that, in proceeding from L to M, F increases by $\int v\, dP$, namely, by the area LMRS. From M to N it diminishes by the area MNSR, so that at N the free energy is greater than at the beginning by the difference between these areas, namely, the small area LMN. From N to O it continues to diminish by the area NOTS, and from O to Q it increases by OQST, so that from N to Q it decreases by the area NOQ. But the free

[1] M. Benedict, G. B. Webb, and L. C. Rubin, *J. Chem. Phys.*, **8,** 334 (1940), **10,** 747 (1942).

[2] L. J. Briggs, *J. Chem. Phys.*, **19,** 970 (1951).

energy is the same at L and Q, and therefore the area LMN is equal to the area NOQ.   This is a well-known theorem.

The calculation outlined in the previous paragraph may be carried out quantitatively for the Benedict equation of state and the vapor pressure determined.   Conversely, the observed vapor-pressure curve for a substance may be used to help determine the constants in the equation of state.

Corresponding States; Reduced Equations of State.   In view of the complexities which have arisen in the mathematical representation of equations of state, it behooves us to give further attention to graphical methods together with any simplifying general principles which may be applicable.   One such principle is the proposal of van der Waals that all substances would have the same equation of state when expressed in

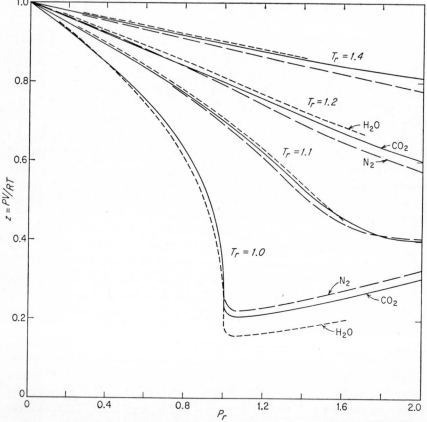

FIG. 16-4. A test of the hypothesis of corresponding states.   Isotherms of nitrogen (long-dashed lines), carbon dioxide (solid lines), and steam (short-dashed lines) as a function of reduced pressure.

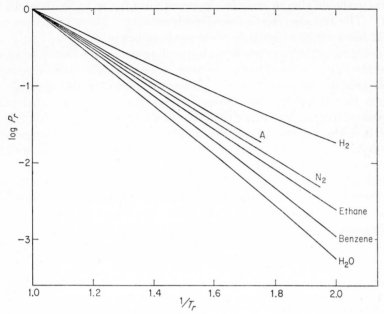

Fɪɢ. 16-5. The reduced-vapor-pressure curves for several liquids.

terms of the reduced variables, $T_r = T/T_c$, $P_r = P/P_c$, $V_r = V/V_c$, which are the ratios of the actual variables to the values of the critical temperature, pressure, and volume, respectively. In our discussion of the second virial coefficient we have already made some use of this type of equation. In the more general region comprising both gas and liquid phases it is best to study graphically the variation of the *compressibility factor*,

$$z = \frac{Pv}{RT} \tag{16-38}$$

which would, of course, be a constant of unity if the substance behaved as a perfect gas.

The general pattern of behavior of the compressibility factor as a function of reduced pressure is shown in Fig. 16-4, where curves are given for N_2, CO_2, and H_2O at several reduced temperatures. We see that on the reduced basis the volumetric behavior of various substances is similar but not exactly the same. Charts and tables of the compressibility factor as a function of reduced temperature and pressure have been presented by several authors.[1] These charts represent the average behavior of the common substances for which data are available. Hougen and Watson[1]

---

[1] See, for example, O. A. Hougen and K. M. Watson, "Chemical Process Principles," pt. 2, "Thermodynamics," chap. 12, John Wiley & Sons, Inc., New York, 1947.

estimate the maximum error in compressibility factor to be about 15 per cent. The average error is considerably smaller. These charts are very useful for work in which their accuracy is sufficient.

According to the principle of corresponding states all substances should have the same reduced-vapor-pressure equation. The actual reduced-vapor-pressure curves for several substances are shown in Fig. 16-5. It is clear that there are large differences—at $T_r = 0.5$ the reduced vapor pressure of hydrogen is higher than that of water by a factor of 30. However, it is also apparent from Fig. 16-5 that the reduced-vapor-pressure curves fall into a single family of curves, the members of which differ from one another in their slope. This slope of the reduced-vapor-pressure curve has been taken[1] as an additional characterizing parameter in a more precise system of correlation of imperfect-gas behavior. This system is presented in Appendix 1 together with the tables necessary for its application.

PROBLEMS

16-1. Complete the comparison of Eqs. (16-36) and (16-6) by substitution for v. What relationship between c and c″ is indicated by this comparison? Is this the same relationship between c and c″ obtained from Eqs. (16-3) and (16-33)?

16-2. Derive an expression for $(E - E°)_T$ in terms of the virial coefficients B, C, etc., of Eq. (16-3).

16-3. Calculate f/P for N_2 at 50 atm and 273°K from the Berthelot equation. Compare with the value in Table 16-1.

16-4. The van der Waals equation may be solved for P more easily than for V; hence the procedure outlined in fine print and Eq. (16-31) is appropriate. Obtain $\ln f$ for a van der Waals gas.

16-5. Derive the Berthelot equation for the change of heat capacity with pressure. Check the values given in Table 16-2 by substitution of the critical constants for *n*-heptane into your equation.

16-6. Derive the equation for $(\partial C_P/\partial P)_T$ from Eq. (A1-6), and check one of the entries in Table 16-2 ascribed to that equation.

16-7. By further partial differentiation of Eq. (10-23) obtain the following equation, and interpret its meaning:

$$\left(\frac{\partial C_V}{\partial V}\right)_T = T\left(\frac{\partial^2 P}{\partial T^2}\right)_V \tag{16-39}$$

16-8. From Eq. (16-23) and from Problem 5-11, show that for air at 60°C $(\partial^2 V/\partial T^2)_P = -0.000035$ cc/g deg². Since $\partial V/\partial T$ is about 3 cc/deg for 1 g of air, we see that, while this coefficient is not constant, it is changing by only one-thousandth of 1 per cent per degree.

16-9. The simple equation $Pv = RT(1 + 6.4 \times 10^{-4}P)$ with P in atmospheres fits the data for hydrogen at 25°C within about 0.5 per cent up to 1500 atm. Calculate the fugacity of hydrogen at 1000 atm.

[1] K. S. Pitzer, *J. Am. Chem. Soc.*, **77**, 3427 (1955); K. S. Pitzer, D. Z. Lippmann, R. F. Curl, Jr., C. M. Huggins, and D. E. Petersen, *J. Am. Chem. Soc.*, **77**, 3433 (1955).

17

SOLUTIONS

We have now completed the presentation of the general principles of thermodynamics together with examples of application to one-component systems. Next we turn to systems of variable composition, i.e., gaseous, liquid, or solid solutions. Most chemical processes involve solutions; hence their treatment is of the utmost importance.

The pure substances from which a solution may be prepared are called the components, or constituents, of the solution. There is always something arbitrary in the choice of these components; thus a given aqueous sulfuric acid could be prepared equally well from H_2SO_4 and H_2O or from SO_3 and H_2O. We are at liberty to choose either of these pairs as the components of the solution in question, or indeed we might even consider all three substances, SO_3, H_2SO_4, H_2O, as the components. However, it is always possible to state the *minimum* number of pure components from which the solution may be made, and this number plays an important role in thermodynamics. In the case we have just cited this number is two, and such a solution is called a binary solution, or sometimes, with less precision, a binary mixture.

Under ordinary circumstances the extensive properties of a pure substance are determined by pressure, temperature, and amount, and its intensive properties by pressure and temperature alone. Likewise we shall assume, unless the contrary is especially stated, that the extensive properties of a solution are determined by pressure, temperature, and the amount of each constituent and its intensive properties by pressure, temperature, and the relative amounts of the several constituents, or, in other words, by pressure, temperature, and *composition*.

MEASURES OF COMPOSITION

The composition of a solution is most advantageously expressed by the ratio of the number of moles of each component to the total number of moles. Thus, if a solution contains n_1 moles of substance 1, n_2 moles of

199

substance 2, and n_3 moles of substance 3, the *mole fraction* of the first substance, which we shall designate by x_1, is given by the formula,

$$x_1 = \frac{n_1}{n_1 + n_2 + n_3} \qquad (17\text{-}1)$$

and it is evident that

$$x_1 + x_2 + x_3 = 1$$

In the case of a binary solution we note that

$$x_1 + x_2 = 1 \qquad dx_1 = -dx_2 \qquad (17\text{-}2)$$

In this case it is convenient to use also the *mole ratio*, which for substance 1 is defined as $n_1/n_2 = x_1/x_2$.

When we have such an amount of a given solution that

$$n_1 + n_2 + \cdots = 1$$

or $n_1 = x_1$, etc., we say that we have 1 mole of the solution. Its volume will be called the molal volume of the solution and designated, as in the case of a pure substance, by v. In general,

$$\mathrm{v} = \frac{V}{n_1 + n_2 + \cdots} \qquad (17\text{-}3)$$

In dealing with dilute solutions it is convenient to speak of the component present in the largest amount as the *solvent*, while a substance present in small amount is called a *solute*. But it is really immaterial which constituent is called the solvent, and we may even apply this term at times to a component present in a relatively small amount.

In a very dilute solution the mole fraction of a solute is proportional to the number of moles of solute in a fixed amount of solvent and also proportional to the concentration, which is the number of moles per unit volume of solution.

It would be simpler, perhaps, to express all compositions in terms of mole fractions, but in the case of aqueous solutions it has become the almost universal custom to express numerical data either as the concentration (moles per liter of solution) or as the molality (moles per 1000 g of water). The latter method is the one which we shall usually adopt, and when we speak of a molal or-tenth molal aqueous solution, for example, HCl(0.1 M), we shall refer to the number of moles of solute in 1000 g (55.51 moles) of water, which we may denote by m. This method has certain advantages; for example, the molality of a given solution is independent of the temperature, while the concentration is not.

ESCAPING TENDENCY IN SOLUTIONS

Let us now consider with some care a simple process in which one component is transferred to or from a solution.   Suppose only component 1 of a liquid solution has appreciable volatility; then the gas phase above the solution will consist of pure gas.   Thus at room temperature H_2SO_4 has practically zero vapor pressure, and the gas in equilibrium with aqueous sulfuric acid is pure water vapor.   We may write

$$H_2O(g, P \text{ atm}) = H_2O(\text{soln}, x_1)$$

where P is the gas pressure and x_1 the mole fraction in the solution.

Now, if this system is at equilibrium, we have seen that $dF = 0$ for an infinitesimal transfer of water from gas to solution.   But the free energy of the entire system, gas and solution, is merely the sum of the free energies of the two phases.   Thus

$$F = F(g) + F(\text{soln}) \tag{17-4}$$

where $F(g)$ and $F(\text{soln})$ are the free energies of gas and solution, respectively.   Since the gas is pure water vapor, its free energy is just the molal free energy $F_1(g)$ multiplied by the number of moles of gas n_g, that is, $F(g) = n_g F_1(g)$.   However, the free energy of the solution depends on the composition as well as the total amount.   Hence we must regard $F(\text{soln})$ as a function of n_1 and n_2, the number of moles of water and sulfuric acid, respectively, as well as the temperature and pressure.   The total differential is

$$dF(\text{soln}) = \left(\frac{\partial F}{\partial T}\right)_{P,n_1,n_2} dT + \left(\frac{\partial F}{\partial P}\right)_{T,n_1,n_2} dP$$
$$+ \left(\frac{\partial F}{\partial n_1}\right)_{P,T,n_2} dn_1 + \left(\frac{\partial F}{\partial n_2}\right)_{P,T,n_1} dn_2 \tag{17-5}$$

and if we hold P, T, and n_2 constant, we have

$$dF(\text{soln}) = \left(\frac{\partial F}{\partial n_1}\right)_{P,T,n_2} dn_1$$

The change in total free energy on transfer of dn_1 moles of water from gas to solution is then

$$dF = dF(g) + dF(\text{soln})$$
$$= -F_1(g) \, dn_1 + \left(\frac{\partial F}{\partial n_1}\right)_{P,T,n_2} dn_1$$

and if this is at equilibrium with $dF = 0$, we have

$$F_1(g) = \left[\frac{\partial F(\text{soln})}{\partial n_1}\right]_{P,T,n_2} \tag{17-6}$$

In Chapter 14 we saw that the molal free energy was a measure of escaping tendency where there was but a single component. We now see that the corresponding measure of escaping tendency from a solution is the partial derivative of the free energy with respect to the number of moles of that component at constant pressure and temperature. We call this the *partial molal free energy* and adopt for it the symbol $\bar{F}$. Thus in a binary solution

$$\bar{F}_1 = \left(\frac{\partial F}{\partial n_1}\right)_{P,T,n_2} \qquad \bar{F}_2 = \left(\frac{\partial F}{\partial n_2}\right)_{P,T,n_1} \tag{17-7}$$

We may now discuss the equilibrium in terms of equality of chemical potential instead of the total free energy. In these terms we say simply that the escaping tendency or chemical potential of component 1, the water, must be the same in both gas and solution. We see that Eq. (17-6) and the definitions following it constitute just this condition.

The Chemical Potential. Let us now restate in somewhat more formal and general terms the results of the preceding section. Also we shall make use of the symbol μ_i for the *chemical potential* of the ith component. For solutions the partial molal free energy is the chemical potential

$$\mu_i = \bar{F}_i = \left(\frac{\partial F}{\partial n_i}\right)_{P,T,n_j} \tag{17-8}$$

where all n's are constant except n_i. The total differential of the free energy is

$$dF = \left(\frac{\partial F}{\partial T}\right) dT + \left(\frac{\partial F}{\partial P}\right) dP + \sum_i \left(\frac{\partial F}{\partial n_i}\right) dn_i$$

$$= -S\, dT + V\, dP + \sum_i \mu_i\, dn_i \tag{17-9}$$

where in the second line we have substituted the familiar symbols for the quantities represented by the various partial derivatives. Usually our chemical systems are subject to constant temperature and pressure, whence dT and dP are zero, and we see again that the partial molal free energies are the chemical potentials which govern the transfer of the chemical substances.

Occasionally we deal with a system under constant-volume and -temperature conditions. Then we wish to consider the work-content function A.

$$dA = dF - P\, dV - V\, dP$$

If we substitute Eq. (17-9), we have

$$dA = -S\, dT - P\, dV + \sum_i \mu_i\, dn_i \tag{17-10}$$

Since Eq. (17-10) is the total differential of A, μ_i must be $(\partial A/\partial n_i)_{T,V}$ as well as $(\partial F/\partial n_i)_{T,P}$. These two quantities may also be shown to be equal to one another by other methods.

The condition for equilibrium under constant-temperature and -volume conditions is that $dA = 0$ [Eq. (13-2)]. Thus we see that the same chemical potentials μ_i govern the transfer of chemical substances for constant-volume and for constant-pressure conditions.

By similar use of the basic relationships between E, H, A, and F, we may derive the equations

$$dH = T \, dS + V \, dP + \sum_i \mu_i \, dn_i \qquad (17\text{-}11)$$

$$dE = T \, dS - P \, dV + \sum_i \mu_i \, dn_i \qquad (17\text{-}12)$$

Indeed the last equation was chosen as a definition of the chemical potential by Gibbs. However, physical systems are never studied under conditions in which the amounts of material are changed while the total entropy is held constant. Thus we shall not use Eqs. (17-11) and (17-12).

Other Partial Molal Quantities. While the partial molal free energy is of primary interest because it governs the direction of transfer of chemical substances, other partial molal quantities are also useful and necessary in thermodynamic work with solutions.

If we take the derivative of the partial molal free energy with respect to pressure at constant temperature and composition, we have

$$\left(\frac{\partial \bar{F}_i}{\partial P}\right)_T = \frac{\partial}{\partial P}\left(\frac{\partial F}{\partial n_i}\right)_{P,T} = \frac{\partial}{\partial n_i}\left(\frac{\partial F}{\partial P}\right)_T = \left(\frac{\partial V}{\partial n_i}\right)_{P,T} = \bar{V}_i \qquad (17\text{-}13)$$

where we have used the fact that the order of differentiation is immaterial and have omitted subscripts indicating composition variables held constant. The partial molal volume is seen to be the measure of the rate of change of partial molal free energy with pressure.

By entirely analogous methods we find that all our equations relating total thermodynamic functions apply equally to the partial molal functions. We record just a few of the most important (in terms of μ_i as well as $\bar{F}_i$).

$$\left(\frac{\partial \mu_i}{\partial P}\right)_T = \left(\frac{\partial \bar{F}_i}{\partial P}\right)_T = \bar{V}_i \qquad (17\text{-}14)$$

$$\left(\frac{\partial \mu_i}{\partial T}\right)_P = \left(\frac{\partial \bar{F}_i}{\partial T}\right)_P = -\bar{S}_i \qquad (17\text{-}15)$$

$$\bar{H}_i = \bar{F}_i + T\bar{S}_i = \mu_i + T\bar{S}_i \qquad (17\text{-}16)$$

$$\left(\frac{\partial \bar{H}_i}{\partial T}\right)_P = \bar{C}_{P,i} \qquad (17\text{-}17)$$

In all cases the composition is held constant in these derivatives.

It is possible to define partial molal quantities wherein variables other than pressure and temperature are held constant, for example, $(\partial A/\partial n)_{T,V}$, which was mentioned above. However, we shall use such quantities so rarely that no abbreviated notation is needed. The bar notation and the unqualified term *partial molal*, such as $\bar{v}_i$ and *partial molal volume*, will always refer to partial derivatives at constant pressure and temperature.

Let us pause a moment to note the physical meaning of a partial molal quantity. We take the volume since it is easily visualized. The derivative $(\partial V/\partial n_1) = \bar{v}_1$ is the rate of change of the total volume of the solution with the amount of component 1. If we start with a very large amount of solution, then $\bar{v}_1$ is the increase in its volume when 1 mole of substance 1 is added. Unlike the molal volume of a single substance, the partial molal volume may be either positive or negative. We shall see presently that, with dilute solutions of magnesium sulfate, the addition of more solute actually diminishes the total volume of the solution.

The Fugacity of a Dissolved Substance. Just as in the case of a one-component system, the fugacity is a convenient measure of the escaping tendency for many purposes. As before we define the fugacity of a solution constituent by the equation

$$\left(\frac{\partial \ln f_1}{\partial \bar{F}_1}\right)_T = \frac{1}{RT} \tag{17-18}$$

or
$$\bar{F}_1 = RT \ln f_1 + B_1 \tag{17-19}$$

where B_1 is that constant for a given substance at a given temperature which makes the fugacity equal to the pressure in the ideal-gas state.

The statement that a system is in equilibrium when, and only when, the fugacity of every substance is constant throughout holds also for systems which contain solutions. The numerical value of the fugacity of any constituent of a solution may therefore be obtained, just as in the case of a pure solid or liquid, if we can ascertain the fugacity of the constituent in the vapor phase over the solution.

Also by methods identical with those used in deriving Eqs. (14-14) and (14-18), we find

$$\left(\frac{\partial \ln f_1}{\partial P}\right)_{T,x} = \frac{\bar{v}_1}{RT} \tag{17-20}$$

and
$$\left(\frac{\partial \ln f_1}{\partial T}\right)_{P,x} = \frac{\bar{H}_1^* - \bar{H}_1}{RT^2} \tag{17-21}$$

where $\bar{H}_1^* - \bar{H}_1$ is the heat absorbed per mole when a small quantity of constituent 1 evaporates from the solution into a vacuum.

METHODS OF DETERMINING PARTIAL MOLAL QUANTITIES

Several methods of calculating partial molal quantities for solutions will be presented below. We take the volume for these examples, but the methods are equally applicable to the entropy, enthalpy, heat capacity, or similar quantities. Before presenting these methods the definition of apparent molal quantities is needed.

Apparent Molal Volume. The apparent molal volume, denoted by the symbol[1] ϕV, is a quantity which is related to the partial molal volume and which is frequently to be found in the literature. If a solution contains n_1 moles of a substance which we may call the solvent and n_2 moles of a substance which we may call the solute, then $V - n_1 v_1^{\circ}$ is the difference between the volume of the solution and the volume of the pure solvent which it contains, and the apparent molal volume is defined as

$$\phi V = \frac{V - n_1 v_1^{\circ}}{n_2} \tag{17-22}$$

This quantity is related to the partial molal volume $\bar{v}_2$, as is a ratio of finite increments to the corresponding differential coefficient. Therefore these two quantities approach identity as n_2 approaches zero, i.e., as the solution approaches infinite dilution. The apparent molal quantities have little thermodynamic utility, but we shall occasionally use them as a step toward the determination of the partial molal quantities. Table 17-1 gives the apparent molal volume of sodium chloride at 25°C at several concentrations in aqueous solution. Here, taking n_1 fixed at 55.51 moles (1000 g), $n_2 = m$, the molality. The value at $m = 0$ is extrapolated and gives not only the apparent molal volume but also the partial molal volume of sodium chloride in an infinitely dilute aqueous solution.

TABLE 17-1. APPARENT MOLAL VOLUME, CC, OF SODIUM CHLORIDE IN AQUEOUS SOLUTION[†]

m	0	0.1	0.2	0.3	0.4	0.5	1.0
ϕV	16.63	17.20	17.44	17.63	17.80	17.94	18.52

† A. Kruis, *Z. physik. Chem.*, **26B**, 181 (1934); and H. E. Wirth, *J. Am. Chem. Soc.*, **62**, 1128 (1940).

Method I (Analytical). If the volume of a solution is known as a function of the composition, the partial molal volume of a constituent may be found by partial differentiation with respect to the amount of

---

[1] The superscript ϕ before a symbol will be used generally to indicate an apparent molal quantity, for example, ϕH, etc.

that constituent.   Thus the data of Table 17-1 may be adequately represented by the empirical equation[1]

$$\phi V = \frac{V - 55.51 v_1^{\circ}}{n_2} = 16.6253 + 1.7738 n_2^{1/2} + 0.1194 n_2$$

or $\qquad V = 55.51 v_1^{\circ} + 16.6253 n_2 + 1.7738 n_2^{3/2} + 0.1194 n_2^2$

and, differentiating with respect to n_2 (n_1 constant),

$$\bar{v}_2 = \frac{\partial V}{\partial n_2} = 16.6253 + 2.6607 n_2^{1/2} + 0.2388 n_2$$

Thus at 0.5 M, $\bar{v}_2 = 18.63$ cc, while $\phi V = 17.94$ cc.   This analytical
method is usually less expeditious than the graphical methods, which are
about to be discussed.

Method II (Graphical).   If V, the volume of solution containing a
fixed amount of solvent, is known for several values of n_2, we may plot V
against n_2 and the slope of the curve at any point is $\bar{v}_2$.   This is illustrated
in Fig. 17-1, which shows the volumes of aqueous solutions of magnesium

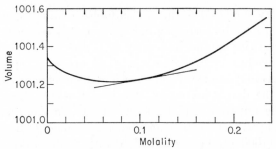

FIG. 17-1. Volumes of solutions containing magnesium sulfate and 1000 g of water
at 18°C.

sulfate,[2] at 18°C.   The volume at every point is the volume of a solution
containing 1000 g of water (thus n_1 is constant) and n_2 moles of magnesium sulfate.   The partial molal volume of the latter is given at each
point by the slope of the tangent, and it will be seen that in this case $\bar{v}_2$ is
negative in dilute solutions, passes through zero at about 0.07 M, and
becomes positive at higher concentrations.   This is the most obvious
graphical method of finding partial molal volume, but there are other
methods which, in point of speed and accuracy, are preferable.   This is
true of the following method.

Method III (Graphical).   If we have the apparent molal volumes
of the solution and plot them against some measure of composition, say
m or $m^{1/2}$, one obtains a graph such as Fig. 17-2.   From Eq. (17-22) with

[1] I. M. Klotz, "Chemical Thermodynamics," p. 194, Prentice-Hall, Inc., Englewood
Cliffs, N.J., 1950.
[2] T. Kohlrausch and W. Hallwachs, *Ann. Physik*, (3)**53,** 14 (1894).

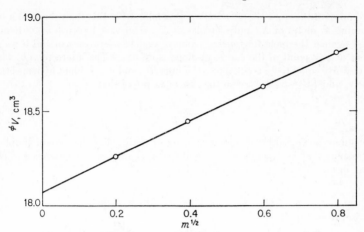

FIG. 17-2. The apparent molal volume of aqueous HCl.

$n_2 = m$

$$V = m(^\phi V) + \frac{1000}{M_1} v_1^\circ$$

and $\qquad \bar{v}_2 = \dfrac{\partial V}{\partial m} = {^\phi V} + m \dfrac{d(^\phi V)}{dm}$ $\qquad\qquad$ (17-23)

Thus the value of $\bar{v}_2$ is just the sum of $^\phi V$ and m times the slope $d(^\phi V)/dm$. For electrolytes it is sometimes preferable to plot versus $m^{1/2}$, and then one has

$$\bar{v}_2 = {^\phi V} + \tfrac{1}{2} m^{1/2} \frac{d(^\phi V)}{dm^{1/2}} \qquad\qquad (17\text{-}24)$$

Figure 17-2 shows the data of Wirth[1] for aqueous HCl plotted against $m^{1/2}$. Since this gives very nearly a straight line, it is easy to obtain accurate values of the slope and hence of $\bar{v}_2$.

Method IV (Graphical, Method of Intercepts[2]). Suppose that we have the molal volume v (of the solution as a whole) tabulated against the mole fraction.   Let us plot this molal volume against the mole fraction, as in Fig. 17-3.   Now, if we draw a tangent to the curve at any

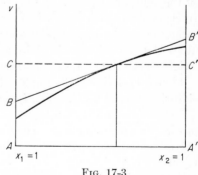

FIG. 17-3

point, the intercept of this tangent upon the ordinate $x_1 = 1$ is equal to $\bar{v}_1$; and the intercept corresponding to $x_2 = 1$ equals $\bar{v}_2$.

[1] H. E. Wirth, *J. Am. Chem. Soc.*, **62**, 1128 (1940).

[2] H. W. B. Roozeboom, "Die Heterogenen Gleichgewichte," II-1, p. 288, Friedrich Vieweg und Söhn, Braunschweig, Germany, 1904.

The proof of this very simple theorem is as follows: If V is the volume of a solution containing n_1 moles of X_1 and n_2 moles of X_2, then $\mathrm{v} = V/(n_1 + n_2)$. If now v is plotted against the mole fraction x_2, namely, against $n_2/(n_1 + n_2)$, and if we draw a tangent at any point of the curve, its slope is $d\mathrm{v}/dx_2$. The intercept AB, which we wish to prove equal to $\bar{\mathrm{v}}_1$, is equal to AC minus BC, and it is evident by inspection that $AC = \mathrm{v}$, and $BC = x_2 \, d\mathrm{v}/dx_2$, so that we must prove that

$$\mathrm{v} - x_2 \frac{d\mathrm{v}}{dx_2} = \bar{\mathrm{v}}_1 \tag{17-25}$$

Now x_2 may be varied by changing n_1 or n_2, or both. For the sake of simplicity we shall assume that n_2 is kept constant and that a change in x_2 is produced solely by a change in n_1. We have then from the definition of v, Eq. (17-3),

$$d\mathrm{v} = \frac{dV}{n_1 + n_2} - V \frac{dn_1}{(n_1 + n_2)^2} \tag{17-26}$$

Likewise, from Eq. (17-1),

$$dx_2 = \frac{-n_2 \, dn_1}{(n_1 + n_2)^2} = -x_2 \frac{dn_1}{n_1 + n_2} \tag{17-27}$$

whence we find

$$x_2 \frac{d\mathrm{v}}{dx_2} = -\frac{dV}{dn_1} + \frac{V}{n_1 + n_2} = -\bar{\mathrm{v}}_1 + \mathrm{v} \tag{17-28}$$

which is identical with Eq. (17-25). In the same way we prove that $A'B'$ is equal to $\bar{\mathrm{v}}_2$.

It is evident that, if we had chosen the gram instead of the mole as the unit amount of a substance, our equations would all have been of the same form. In place of the mole fraction we should have used the gram fraction (weight fraction), and in place of the partial molal volume we could define the partial specific volume of one component of a solution as the change in volume of the solution per gram of that component. Thus we may employ the method of intercepts by plotting the specific volume of the solution, which is the reciprocal of the density, against the weight fraction, and this enables us to use directly data as they are ordinarily tabulated in the literature. The intercepts of the tangent give immediately the partial specific volumes, and these multiplied by the molal weights give the corresponding partial molal quantities.

Figure 17-4 shows the application of this method to data on aqueous sulfuric acid, at 25°C. The tangent at 60 weight per cent, $x_2 = 0.216$, cuts the two limiting ordinates at 0.964 and 0.465, these being the two partial specific volumes. The former multiplied by 18 gives $\bar{\mathrm{v}}_1$, the partial molal volume of water in the solution, namely, 17.4 cc. The latter multiplied by 98 gives $\mathrm{v}_2 = 45.6$ cc, the partial molal volume of the sulfuric acid.

The variation of the partial molal volumes with the mole fraction, for two different types of solutions, is further illustrated in Figs. 17-5 and 17-6. Here for convenience we give as ordinates the partial molal volume of each constituent, less the molal volume of that constituent in the pure

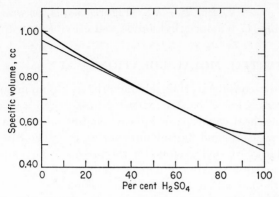

FIG. 17-4. Specific volume of aqueous sulfuric acid at 25°C.

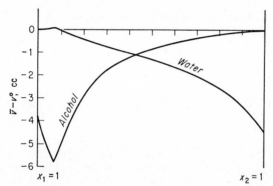

FIG. 17-5. Partial molal volumes of water and alcohol.

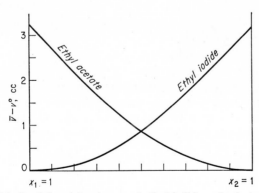

FIG. 17-6. Partial molal volumes of ethyl iodide and ethyl acetate.

state, namely, $\bar{v}_1 - v_1^0$ and $\bar{v}_2 - v_1^0$. Figure 17-5 is for water and ethyl alcohol, and Fig. 17-6 is for ethyl iodide and ethyl acetate.[1]

GENERAL PARTIAL MOLAL EQUATIONS

We may now consider certain characteristics common to all partial molal quantities. Let Y be any extensive property of a given solution, such as volume, heat capacity, or internal energy, which is a function of temperature, pressure, and the amounts of the several constituents. For the sake of clarity we shall assume for the present that temperature and pressure are constant, so that Y depends only upon $n_1, n_2, \ldots$

We define the partial molal values by the equations

$$\bar{Y}_1 = \left(\frac{\partial Y}{\partial n_1}\right)_{P,T,n_2,n_3,\ldots} \qquad \bar{Y}_2 = \left(\frac{\partial Y}{\partial n_2}\right)_{P,T,n_1,n_3,\ldots} \qquad (17\text{-}29)$$

Now, by the chief equation of partial differentiation,

$$dY = \left(\frac{\partial Y}{\partial n_1}\right)_{P,T,n_2,n_3,\ldots} dn_1 + \left(\frac{\partial Y}{\partial n_2}\right)_{P,T,n_1,n_3,\ldots} dn_2 + \cdots \quad (17\text{-}30)$$

or $\qquad dY = \bar{Y}_1 \, dn_1 + \bar{Y}_2 \, dn_2 + \cdots \qquad\qquad\qquad (17\text{-}31)$

It is evident that these partial molal quantities such as $\bar{Y}_1$ are not extensive but intensive properties of the solution. They depend, therefore, not upon the total amount of each constituent, but only upon the composition, i.e., upon the *relative* amounts of the several constituents.

If, therefore, to a given solution at constant temperature and pressure we add the several constituents simultaneously, keeping their ratios constant, these partial molal quantities will remain constant. We may therefore integrate Eq. (17-31), keeping $n_1, n_2, \ldots$ in constant proportions, and find

$$dY = (\bar{Y}_1 x_1 + \bar{Y}_2 x_2 + \cdots) \, dn$$
$$Y = (\bar{Y}_1 x_1 + \bar{Y}_2 x_2 + \cdots)n$$

and $\qquad Y = n_1 \bar{Y}_1 + n_2 \bar{Y}_2 + \cdots \qquad\qquad\qquad (17\text{-}32)$

In deriving Eq. (17-32) we did not limit ourselves to any special values of $n_1, n_2, \ldots$. Hence this equation, being entirely general, can be differentiated with respect to any change of composition, however this change is produced (whether by addition or subtraction of infinitesimal amounts of any or all of the components). This general differentiation gives

$$dY = n_1 \, d\bar{Y}_1 + \bar{Y}_1 \, dn_1 + n_2 \, d\bar{Y}_2 + \bar{Y}_2 \, dn_2 + \cdots \quad (17\text{-}33)$$

[1] Calculated by J. H. Hildebrand, from the experiments of J. C. Hubbard, *Phys. Rev.*, (1)**30**, 740 (1910); *Z. physik. Chem.*, **74**, 207 (1910).

and this equation combined with Eq. (17-31) gives

$$n_1 \, d\bar{Y}_1 + n_2 \, d\bar{Y}_2 + \cdots = 0 \qquad (17\text{-}34)$$

which shows, for any infinitesimal alteration in composition, at constant temperature and pressure, the relation between the change in any one $\bar{Y}_i$ and the change in all the others. Equations (17-32) and (17-34) may for brevity be called the *partial molal equations.*

These equations assume a number of special forms, which are frequently useful. Thus, if we are dealing with 1 mole of solution, Eqs. (17-32) and (17-34) become

$$Y = x_1 \bar{Y}_1 + x_2 \bar{Y}_2 + \cdots \qquad (17\text{-}35)$$
$$x_1 \, d\bar{Y}_1 + x_2 \, d\bar{Y}_2 + \cdots = 0 \qquad (17\text{-}36)$$

We may regard the number of moles of one constituent, say n_1, as the main variable. Then if, as a reminder, we indicate also the constancy of P and T, Eq. (17-34) takes the form

$$n_1 \left(\frac{\partial \bar{Y}_1}{\partial n_1} \right)_{P,T} + n_2 \left(\frac{\partial \bar{Y}_2}{\partial n_1} \right)_{P,T} + \cdots = 0 \qquad (17\text{-}37)$$

Similarly Eq. (17-36) becomes

$$x_1 \left(\frac{\partial \bar{Y}_1}{\partial x_1} \right)_{P,T} + x_2 \left(\frac{\partial \bar{Y}_2}{\partial x_1} \right)_{P,T} + \cdots = 0 \qquad (17\text{-}38)$$

While the derivation just given is quite sufficient, it may be helpful to some readers to verify Eq. (17-37) by another method. We recall that each partial molal quantity is already a partial derivative. Hence

$$\frac{\partial \bar{Y}_2}{\partial n_1} = \frac{\partial}{\partial n_1} \left(\frac{\partial Y}{\partial n_2} \right) = \frac{\partial}{\partial n_2} \left(\frac{\partial Y}{\partial n_1} \right) = \frac{\partial \bar{Y}_1}{\partial n_2} \qquad (17\text{-}39)$$

which follows because the order of differentiation is immaterial. Then Eq. (17-37) becomes

$$n_1 \left(\frac{\partial \bar{Y}_1}{\partial n_1} \right) + n_2 \left(\frac{\partial \bar{Y}_1}{\partial n_2} \right) + \cdots = 0 \qquad (17\text{-}40)$$

Each of these derivatives gives the change in $\bar{Y}_1$ per mole addition of the specified component. But if we add these components in the same proportion in which they were originally present (n_1, n_2, etc.), then the composition of the solution has not changed and the value of $\bar{Y}_1$ must remain unchanged. Thus we see that the expression in Eq. (17-40) should be equal to zero.

These equations which were derived in the present section are of fundamental importance. It is interesting to note that they do not depend on thermodynamics but follow from mathematics for any extensive quantity which is a continuous and single-valued function of the composition of the solution.

It will be noted that, while n_1 can vary without any change in n_2, n_3, etc., we cannot change x_1 without some change in x_2, x_3, etc. So, in a binary solution,

$$dx_1 = -dx_2$$

It is also to be noted that, in spite of the similarity of Eqs. (17-37) and (17-38), $d\bar{F}_1/dn_1$ has a very different meaning from $d\bar{F}_1/dx_1$. Thus in a binary mixture, when n_2 is constant,

$$dx_1 = \frac{n_2}{(n_1 + n_2)^2} dn_1 \tag{17-41}$$

Applications to Binary Solutions. If we apply Eq. (17-38) to a mixture of two components and for later convenience choose x_2 as the chief variable,

$$x_1 \frac{\partial \bar{F}_1}{\partial x_2} + x_2 \frac{\partial \bar{F}_2}{\partial x_2} = 0 \tag{17-42}$$

or

$$\frac{\partial \bar{F}_1/\partial x_2}{\partial \bar{F}_2/\partial x_2} = -\frac{x_2}{x_1} = -\frac{x_2}{1 - x_2} \tag{17-43}$$

If $\bar{F}_1$ and $\bar{F}_2$ are plotted against x_2, the two differential coefficients are the slopes of the two curves, and therefore the slope of one curve is entirely determined by the slope of the other curve and the composition. Thus at 50 mole per cent, where $x_1 = x_2$, the slopes of the two curves must be equal and opposite in sign. If at any composition one of the curves has a maximum, the other curve will have a minimum at the same composition. These statements are well illustrated by the curves of Figs. 17-5 and 17-6.

The Special Case of an Infinitely Dilute Solution. For finite values of x_1 and x_2, if $\partial \bar{F}_1/\partial x_2 = 0$, then $\partial \bar{F}_2/\partial x_2 = 0$. But the limiting case, where one of the mole fractions is zero, deserves comment. In an infinitely dilute solution of component 2 in component 1, where we may write $x_2/x_1 = 0$, it is evident that either $\partial \bar{F}_1/\partial x_2$ is zero, or $\partial \bar{F}_2/\partial x_2$ is infinite. In other

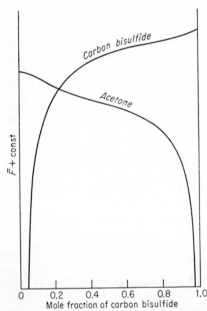

Fig. 17-7. Partial molal free energies in solutions of carbon bisulfide and acetone.

words, when x_2 is zero, either the curve of Υ_1 becomes horizontal, or the curve of Υ_2 becomes vertical. These two types are illustrated, respectively, in Figs. 17-6 and 17-7. There is no difficulty in working with functions of the first type. Partial molal volumes are seen in Figs. 17-5 and 17-6 to belong to the first type, and it is found that enthalpies and heat capacities also constitute examples of the first type. But it is inconvenient, to say the least, to deal with the infinite slope and the absence of a finite intercept in the second type. Since we shall find that the partial molal free energy follows the second type of behavior, it is often more convenient to use the related function, fugacity.

CHANGE OF ESCAPING TENDENCY WITH COMPOSITION

In treating the effect of pressure and temperature upon the escaping tendencies of the components of a solution, we have obtained equations which are identical in form with those found for pure substances. We turn now to a problem which has no analogy in a system of pure phases.

In order to obtain an exact expression for the effect of change of composition upon the escaping tendencies of the various constituents of a solution, we may employ the general equation for partial molal quantities derived earlier in this chapter [Eq. (17-38)]. Introducing free energy into this equation, we obtain a formula of fundamental importance, first obtained by Gibbs in 1875,

$$x_1 \left(\frac{\partial \bar{F}_1}{dx_1} \right)_{P,T} + x_2 \left(\frac{\partial \bar{F}_2}{dx_1} \right)_{P,T} + \cdots = 0 \qquad (17\text{-}44)$$

The equation does not permit us to determine in any case how the escaping tendency of each constituent changes with its mole fraction, nor is this possible from thermodynamics alone. But if we have experimentally solved this problem for one constituent of a binary mixture, then it is solved for the other constituent by Eq. (17-44).

The trend of the partial molal free energy of each constituent in a mixture of carbon bisulfide and acetone is shown[1] in Fig. 17-7. The mole fraction is plotted horizontally so that the left-hand axis of ordinates represents pure acetone and the right-hand axis of ordinates, pure carbon bisulfide.

In any plot such as Fig. 17-7, Eq. (17-44) tells us that, when the two mole fractions are equal, the slopes of the two curves are equal and opposite in sign; when $x_1 = \frac{1}{4}$ and $x_2 = \frac{3}{4}$, $\partial \bar{F}_2/\partial x_1 = -\frac{1}{3} \partial \bar{F}_1/\partial x_1$, and so on. In general, if one of the curves is known and we know a single

---

[1] Since we never attempt to obtain the numerical value of free energy, but only its change when we pass from one condition to another, each curve can be shifted vertically at will. The ordinates give therefore in each case $\bar{F}$ + constant.

point on the other curve, the slope of the second curve is determined at that point. Hence it is possible, by graphical or by analytical methods, to build up the second curve through the range of composition over which the first curve is given.

The curves in Fig. 17-7 suggest that $\bar{F}$ is one of the partial molal quantities, mentioned earlier, which numerically approach infinity as the corresponding mole fraction approaches zero. That this is the general characteristic of partial molal free energy is demonstrated by a consideration of the fugacity and its relation to vapor pressure.

Writing as before,

$$\bar{F}_1 = RT \ln f_1 + B_1 \qquad (17\text{-}45)$$

we see that, if, in any given condition, $\bar{F}_1$ and f_1 have finite values, B_1 also is finite. Now the vapor pressure of any constituent of a solution must approach zero as its mole fraction approaches zero. But as the vapor pressure approaches zero, it becomes equal to the fugacity. Therefore, when $x_1 = 0$, $f_1 = 0$. Hence $RT \ln f_1 = -\infty$, and $\bar{F}_1 = -\infty$.

We have not explained how we obtained the values of $\bar{F}$ plotted in Fig. 17-7. In fact, of the various methods which are employed in the determination of partial molal free energy, there are several which we are not yet in a position to discuss; but the simplest and one of the most useful of these methods consists in determining the vapor pressure, and thence the fugacity, of the constituents of a solution.

If we use fugacity in place of free energy, Eq. (17-44) becomes

$$x_1 \left(\frac{\partial \ln f_1}{\partial x_1} \right)_{P,T} + x_2 \left(\frac{\partial \ln f_2}{\partial x_1} \right)_{P,T} + \cdots = 0 \qquad (17\text{-}46)$$

In a binary solution, where $dx_1 = -dx_2$, this equation becomes

$$x_1 \left(\frac{\partial \ln f_1}{\partial x_1} \right)_{P,T} = x_2 \left(\frac{\partial \ln f_2}{\partial x_2} \right)_{P,T} \qquad (17\text{-}47)$$

When the vapors are nearly perfect gases, we may substitute partial pressures[1] for fugacities and obtain the approximate equation

$$x_1 \left(\frac{\partial \ln p_1}{\partial x_1} \right)_{P,T} = x_2 \left(\frac{\partial \ln p_2}{\partial x_2} \right)_{P,T} \qquad (17\text{-}48)^*$$

This approximate equation was found by Duhem and Margules[2] and it is

---

[1] The *partial pressure* is defined as the product of the total pressure of vapor times the mole fraction in the vapor, for example, $p_1 = x_1 p$. See also the last section of Chapter 21.

[2] P. Duhem, Dissolutions et Mélanges III, *Trav. Univ. Lille*, **III**(13), 79 (1894); M. Margules, *Sitzber. Akad. Wiss. Wien, Math.-naturw. Kl., Abt. IIa*, **104**, 1243 (1895).

in this form that these fundamental relations between escaping tendency and composition have received their most extensive experimental verification, especially in the work of Zawidzki,[1] who determined, over the whole range of concentration, the partial vapor pressures for numerous binary liquid mixtures. We shall reproduce in Figs. 17-8 to 17-11 his curves for the four pairs, propylene bromide–ethylene bromide, carbon

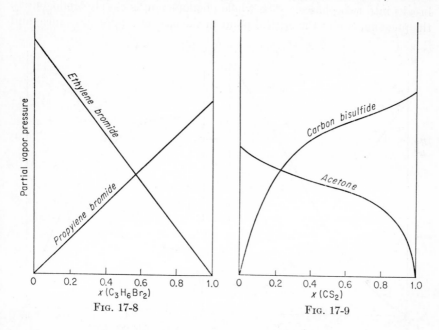

FIG. 17-8

FIG. 17-9

bisulfide–acetone, acetone-chloroform, and pyridine-water. No discernible difference would be made in these figures if we plotted the actual fugacity in place of the partial pressures.

Although a reference to Zawidzki's original paper will show better the excellence of the quantitative agreement, an inspection of the curves shows that their slopes conform to Eq. (17-48)*, which may also be put in the alternative form,

$$\frac{\partial p_1/\partial x_1}{\partial p_2/\partial x_2} = \frac{p_1/x_1}{p_2/x_2} \qquad (17\text{-}49)*$$

One of the simplest corollaries of this equation is that, if one of the curves is a straight line across the entire range of composition, the other is also a straight line across the entire range. For if the first curve is straight, $\partial p_1/\partial x_1 = p_1/x_1$, whence $\partial p_2/\partial x_2 = p_2/x_2$. Such a case is illustrated in Fig. 17-8.

[1] J. von Zawidzki, *Z. physik. Chem.,* **35,** 129 (1900).

The Critical Mixing Point. If we study, over a range of temperature, the fugacity–mole-fraction curve of one of the components of such a mixture as that represented in Fig. 17-11, we often find that as the temperature is lowered the curvature becomes more pronounced, as illustrated in Fig. 17-12. Finally a temperature T' is reached, where the curve is horizontal at a point C. If the solution at C is further cooled, it breaks into two phases. The whole phenomenon is closely analogous to the phenomenon of the critical point in the case of a pure substance, and

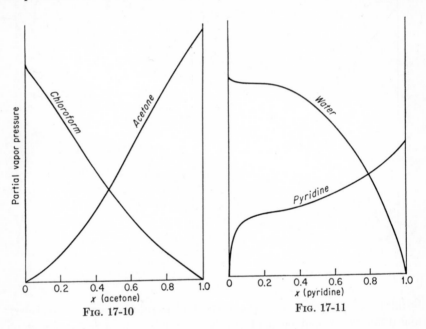

FIG. 17-10   FIG. 17-11

therefore the temperature at this point is known as the critical mixing temperature[1] and the composition as the critical composition.

At this critical point, since the curve is horizontal,

$$\frac{\partial f_1}{\partial x_1} = 0 \qquad \frac{\partial \bar{F}_1}{\partial x_1} = 0 \tag{17-50a}$$

Hence for the other component, by Eqs. (17-46) and (17-44),

$$\frac{\partial f_2}{\partial x_1} = 0 \qquad \frac{\partial \bar{F}_2}{\partial x_1} = 0 \tag{17-50b}$$

---

[1] In peculiar cases, we find two substances which are miscible in all proportions *below* a certain critical mixing temperature and which form two phases above that temperature.

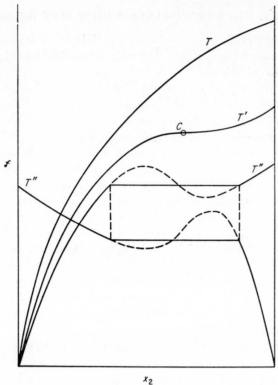

$$x_2$$

Fig. 17-12

We note also that the critical point in Fig. 17-12 is not a simple maximum or minimum but rather is a point of inflection. Thus the second derivatives are also zero.

$$\frac{\partial^2 f_1}{\partial x_1{}^2} = 0 \qquad \frac{\partial^2 \mathbb{F}_1}{\partial x_1{}^2} = 0 \qquad (17\text{-}51a)$$

$$\frac{\partial^2 f_2}{\partial x_1{}^2} = 0 \qquad \frac{\partial^2 \mathbb{F}_2}{\partial x_1{}^2} = 0 \qquad (17\text{-}51b)$$

If, below the critical mixing temperature, we could prevent the separation into two phases, we should expect to obtain for both constituents something like the dotted portions of the two curves T''', each having a maximum and a minimum, and a certain region in which the escaping tendency would diminish with an increase in mole fraction.   In practice, however, except for a small degree of possible supersaturation, separation into two phases occurs.   This separation must occur in such measure as to make the escaping tendency of either constituent the same in both phases.

EQUILIBRIUM BETWEEN PHASES WHICH MAY BE SOLUTIONS

The types of equilibrium in systems containing two or more substances are so numerous and so complex that we must content ourselves here with a few simple illustrations.

The properties of a given amount of a binary solution are determined by temperature, pressure, and composition, the latter being fixed when the mole fraction of either constituent is given. When two phases in equilibrium are present, for example, solid water and brine, or the two liquid phases obtained by mixing ether and water, or aqueous hydrochloric acid and its vapor, the system is restricted to two degrees of freedom and in this respect resembles a single phase of a pure substance.

Change of Eutectic Temperature with Pressure. With three phases in equilibrium the case is analogous to two phases of a pure substance, with one degree of freedom. Thus the whole system is determined when we choose the temperature, or the pressure, or the composition of one phase which contains both constituents. As an example of such a system, we may consider a solid salt, ice, and a saturated solution at the eutectic point. If the temperature is changed, the pressure, as well as the composition of the solution, must change by a fixed amount in order to maintain equilibrium.

In this very simple case the rate of change of the eutectic pressure with the temperature may be immediately ascertained by the methods of Chapter 10. But it will be instructive to go through the more complicated calculation in which we investigate the escaping tendency of each substance in the several phases. If we denote by F_1' the molal free energy of water in the solid state and by $\bar{F}_1$ its partial molal free energy in the solution, the first condition of sustained equilibrium is that $F_1' = \bar{F}_1$ and $dF_1' = d\bar{F}_1$. But we may write

$$d\bar{F}_1 = \frac{\partial \bar{F}_1}{\partial T} dT + \frac{\partial \bar{F}_1}{\partial P} dP + \frac{\partial \bar{F}_1}{\partial x_1} dx_1 \qquad (17\text{-}52)$$

and for the ice, since it is a pure phase,

$$dF_1' = \frac{\partial F_1'}{\partial T} dT + \frac{\partial F_1'}{\partial P} dP \qquad (17\text{-}53)$$

Equating these two expressions, and substituting the values of the temperature and pressure coefficients, by means of Eqs. (17-14) and (17-15) and the corresponding equations for the pure substances, we find

$$(s_1' - \bar{s}_1) \, dT + (\bar{v}_1 - v_1') \, dP + \frac{\partial \bar{F}_1}{\partial x_1} dx_1 = 0 \qquad (17\text{-}54)$$

and likewise for the second constituent,

$$(\mathbf{s}_2' - \mathbf{\bar{s}}_2)\, dT + (\mathbf{\bar{v}}_2 - \mathbf{v}_2')\, dP + \frac{\partial \mathbf{\bar{F}}_2}{\partial x_1}\, dx_1 = 0 \qquad (17\text{-}55)$$

Now the remaining differential coefficients are eliminated, in accordance with Eq. (17-44), if we multiply the first equation by x_1 and the second by x_2 and then add. Hence

$$(x_1 \mathbf{s}_1' - x_1 \mathbf{\bar{s}}_1 + x_2 \mathbf{s}_2' - x_2 \mathbf{\bar{s}}_2)\, dT = (x_1 \mathbf{v}_1' - x_1 \mathbf{\bar{v}}_1 + x_2 \mathbf{v}_2' - x_2 \mathbf{\bar{v}}_2)\, dP \qquad (17\text{-}56)$$

But we recall that, by Eq. (17-35), $x_1\mathbf{\bar{s}}_1$ and $x_2\mathbf{\bar{s}}_2$ together equal s, the entropy of 1 mole of the solution, while $x_1\mathbf{s}_1' + x_2\mathbf{s}_2'$ gives the entropy of the corresponding amounts of the two solids. The algebraic sum of all these entropy terms is therefore ΔS, the increase in entropy when 1 mole of the solution is formed from the two solids. Similarly the volume terms together give ΔV in the same process, and we obtain an equation which we might have taken directly from Eq. (10-14). It is identical with the equation for the change of pressure with temperature when only one component is present and only two phases, namely,

$$\frac{dP}{dT} = \frac{\Delta S}{\Delta V} = \frac{\Delta H}{T\, \Delta V} \qquad (17\text{-}57)$$

The Solubility Curve of a Dissociable Solute. Let us consider one further case in which we deal simultaneously with an equilibrium between phases and an equilibrium in a chemical reaction. We may select for such a study the solubility curve of a hydrated salt such as $CaCl_2 \cdot 6H_2O$.

If we are dealing with a system of two components and one of our variables is fixed, as when the pressure is kept constant at 1 atm, a system of two phases has one remaining degree of freedom and either temperature or composition may be arbitrarily varied, but not both. Thus, if the two components are $CaCl_2$ and H_2O and the two phases are $CaCl_2 \cdot 6H_2O$ and a solution, the relation between the temperature and the composition of the solution may be represented by a continuous curve, as in Fig. 17-13, where temperature is the ordinate, and the mole fraction x_2 of $CaCl_2$ in the solution is the abscissa.

At the maximum point of the curve, where the composition of the solution is the same as that of the solid, any finite addition of either H_2O or $CaCl_2$ will lower the equilibrium temperature. But if in the solution there is any appreciable dissociation of $CaCl_2 \cdot 6H_2O$ into its constituents, an infinitesimal addition of either component does not change the equilibrium temperature. In other words, the point in question is a true maximum of a continuous curve, and not a cusp, or a point of intersection of two curves.   This theorem, due to Lorentz and Stortenbeker,[1] may be demonstrated as follows:

Although this system is one of two *independent* constituents, we may, if we choose, consider three constituents present in the solution, namely, n_1 moles of H_2O, n_2 moles of $CaCl_2$, and n_3 moles of $CaCl_2 \cdot 6H_2O$, with partial molal free energies $\bar{F}_1$, $\bar{F}_2$, and $\bar{F}_3$.   At the melting point of the

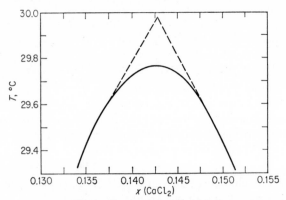

Fig. 17-13. Solubility of $CaCl_2 \cdot 6H_2O$.

pure hydrate, its composition is that of the solution, and therefore $x_1 = 6x_2$ (however far dissociation may occur according to the equation $CaCl_2 \cdot 6H_2O = CaCl_2 + 6H_2O$).

Let us now find the effect of adding dn_1 moles of water.   By Eq. (17-34), we have, at constant temperature and pressure,

$$n_1 \, d\bar{F}_1 + n_2 \, d\bar{F}_2 + n_3 \, d\bar{F}_3 = 0 \qquad (17\text{-}58)$$

or
$$n_2(6 \, d\bar{F}_1 + d\bar{F}_2) + n_3 \, d\bar{F}_3 = 0 \qquad (17\text{-}59)$$

Now $CaCl_2 \cdot 6H_2O$, $CaCl_2$, and H_2O are in equilibrium and must so remain. Therefore, by Eq. (13-4),

$$\Delta F = \bar{F}_3 - 6\bar{F}_1 - \bar{F}_2 = 0 \qquad d\bar{F}_3 = 6 \, d\bar{F}_1 + d\bar{F}_2$$

Combining this with Eq. (17-59), we find

$$n_2 \, d\bar{F}_3 + n_3 \, d\bar{F}_3 = 0 \qquad d\bar{F}_3 = 0 \qquad (17\text{-}60)$$

In other words, the escaping tendency of $CaCl_2 \cdot 6H_2O$ in the liquid phase is not changed by an infinitesimal addition of water, and it will therefore remain in equilibrium with the solid $CaCl_2 \cdot 6H_2O$ without change in the equilibrium temperature.

Of course thermodynamics is unable to predict how flat such a curve as that of Fig. 17-13 will be.   This must depend upon the extent to which

[1] W. Stortenbeker, *Z. physik. Chem.*, **10**, 194 (1892).

the compound dissociates, for if there were no dissociation at all in the solution, adding $CaCl_2$ or H_2O would be like adding some foreign constituent and the continuous curve would be replaced by two curves intersecting at the melting point, as illustrated by the dotted lines in the figure.

PROBLEMS

17-1. What is the mole fraction of the solute in a molal aqueous solution? What is its mole ratio?

17-2. What is the mole fraction of water in aqueous ethyl alcohol which is 50 per cent by weight?

17-3. At 15° the density of aqueous sulfuric acid containing 5 moles/liter is 1.2894. Calculate the molality of H_2SO_4 and its mole fraction.

17-4. What is the molal volume of the solution used in the preceding problem?

17-5. Show that the volume of a solution containing 1000 g of water is $V = (1000 + mM)/d$, where m is the molality, M is the molal weight of solute, and d is the density of the solution.

17-6. In a mixture of water and ethyl alcohol in which the mole fraction of the water is 0.4, the partial molal volume of the alcohol is 57.5 cc, and the density of the mixture is 0.8494. Calculate the partial molal volume of the water.

17-7. Table 17-2 gives Hubbard's values of the specific volumes (cubic centimeters per gram) at 50° of mixtures of ethyl iodide and ethyl acetate, at several compositions, expressed in fractions by weight of ethyl iodide. Plot as in method IV, and find from the curve the partial molal volume of ethyl acetate when $x_1 = x_2 = 0.50$ (not when weight fraction is 0.50).

TABLE 17-2

Wt fraction of ethyl iodide	Sp vol at 50°	Wt fraction of ethyl iodide	Sp vol at 50°
0.00000	1.15866	0.74566	0.70040
0.19082	1.04358	0.82792	0.64814
0.35007	0.94654	0.89093	0.60749
0.49517	0.85737	0.94970	0.56927
0.59741	0.79377	1.00000	0.53622
0.68529	0.73880		

17-8. For each of the first four data of Table 17-2, calculate the number of moles of ethyl iodide to 1 mole ethyl acetate and also the volume occupied by an amount of solution containing one mole ethyl acetate. Obtain the apparent molal volumes, and determine more accurately than in the preceding exercise some one value of $\bar{v}_2$.

17-9. In Table 17-3, m represents the molality and d the density at 25° of aqueous solutions of potassium chloride. Determine the partial molal volume of the latter at several different molalities.

TABLE 17-3

m	0	0.1668	0.2740	0.3885	0.6840	0.9472
d	0.99707	1.00490	1.00980	1.01271	1.02797	1.03927

17-10. In order to determine the partial molal V-T coefficient it is necessary to know the volume of a solution at different compositions and at different temperatures.　If the volume of a certain amount of some solution is given over a small range by the equation $V = 20(n_1 + n_2) + 0.006T(n_1 + n_2) + 0.20n_1$ cc, calculate $\partial \bar{v}_1 / \partial T = \partial^2 V / \partial T\, \partial n_1$ in this range.

17-11. The molal volume of sodium chloride in the solid state is greater than its partial molal volume in a saturated solution.　How does the solubility change with the pressure?

17-12. Let dn_2 moles of X_2 be added to a system composed of two or more phases; let the total free energy be taken as the sum of the free energies of the several phases of the system, $F = F' + F'' + \cdots$.　The partial molal free energies $\bar{F}_2'$, $\bar{F}_2''$, etc., are equal.　Let $dn_2 = dn_2' + dn_2'' + \cdots$, where dn_2' is the amount going into the first phase, and so on.　Show that here, as in the case of a single phase, $\partial F / \partial n_2 = \bar{F}_2' = \bar{F}_2'' = \cdots$.

17-13. Show that the partial molal entropy of a solution approaches an infinite positive value with increasing dilution.

17-14. Show that in the curves T'' of Fig. 17-12 the maximum in the curve for one constituent must come at the same composition as the minimum in the curve for the other.

17-15. Show diagrammatically the sort of fugacity-composition curves which would characterize a solution possessing two critical mixing temperatures between which separation into two phases occurs.

17-16. A constant-boiling solution is one in which evaporation causes no change in the composition of the liquid.　In other words, the composition of liquid and gaseous phases must be identical.　If the vapors may be assumed to be perfect gases, then the ratio of the two partial pressures is equal to the ratio of the mole fraction in the liquid.　Hence show that

$$\frac{dp_1}{dx_1} + \frac{dp_2}{dx_1} = 0$$

and that the total vapor pressure $P = p_1 + p_2$ is a maximum or minimum.　This is an important principle in the theory of distillation.

17-17. Show from the phase rule that, under constant atmospheric pressure, a sodium chloride solution in the presence of the solid salt has one degree of freedom. Let us calculate in this case the change in the fugacity of water in the saturated solution as the temperature varies.　Let f_1 be the fugacity of water in the solution, f_2' that of solid salt, and f_2 that of the dissolved salt.　We write $d \ln f_2 = d \ln f_2'$ and further

$$d \ln f_1 = \frac{\partial \ln f_1}{\partial T} dT + \frac{\partial \ln f_1}{\partial x_1} dx_1$$

$$d \ln f_2 = \frac{\partial \ln f_2}{\partial T} dT + \frac{\partial \ln f_2}{\partial x_1} dx_1$$

$$d \ln f_2' = \frac{\partial \ln f_2'}{\partial T} dT$$

From these equations show that

$$\frac{d \ln f_1}{dT} = \frac{\mathrm{H}_1^* - \bar{\mathrm{H}}_1}{RT^2} + \frac{x_2}{x_1} \frac{\mathrm{H}_2' - \bar{\mathrm{H}}_2}{RT^2}$$

where the numerators of the two last fractions are the negative of the differential heats of solution of attenuated water vapor and of solid salt.

17-18. Show, by combining two or more equations of the type of (17-20) or (17-21) by means of Eq. (17-35), that

$$x_1 \left(\frac{\partial \ln f_1}{\partial P} \right)_{T,x} + x_2 \left(\frac{\partial \ln f_2}{\partial P} \right)_{T,x} + \cdots = \frac{\mathrm{v}}{RT} \tag{17-61}$$

and
$$x_1 \left(\frac{\partial \ln f_1}{\partial T} \right)_{P,x} + x_2 \left(\frac{\partial \ln f_2}{\partial T} \right)_{P,x} + \cdots = \frac{\mathrm{H}^* - \mathrm{H}}{RT^2} \tag{17-62}$$

where v is the volume and $\mathrm{H}^* - \mathrm{H}$ is the ideal heat of vaporization of 1 mole of the solution.[1]

[1] Substituting the vapor pressure for the fugacity in Eqs. (17-61) and (17-62), we may obtain two approximate equations which are simplifications of an equation obtained by G. Kirchhoff, *Ann. Physik*, (2)**104,** 612 (1858), and of one obtained by W. Nernst, "Theoretische Chemie," Ferd. Enke Verlag, Stuttgart, 1909. See G. N. Lewis, *Proc. Am. Acad.*, **43,** 259 (1907); *Z. physik. Chem.*, **61,** 129 (1907).

18

THE IDEAL SOLUTION

When two liquids are mixed, it is often possible to predict with close approximation the properties of the solution from the properties of the pure constituents and the amounts of these constituents. Thus, if we are dealing with an extensive property like the volume, it may frequently be assumed that the volume of a solution is the sum of the volumes of the pure substances of which it is composed. But when water and alcohol are mixed, there is a large diminution in total volume. So also the internal energy is sometimes far from additive, as illustrated by the large evolution of heat when water and sulfuric acid are mixed.

Nevertheless the fact that these phenomena excite comment shows that we consider them as exceptions to the normal results of mixing. Indeed it is a fact of observation that two liquids which are similar in chemical composition and in physical properties, especially if they are of the less polar type, mix with little change in volume or energy.

Moreover, a pair of such liquids approximates to Raoult's law;[1] that is to say, it gives the linear type of curve shown in Fig. 17-8, where the vapor pressure is proportional to the mole fraction. Such a result might be predicted from the kinetic theory, for if the molecules of the two components are so closely identical that the forces between unlike molecules are the same as those between like molecules, it would result from the laws of chance that the number of molecules of either constituent escaping into the vapor phase would be proportional to the relative number of those molecules in the liquid.

Among organic isomers, pairs of substances may be found whose properties are so nearly identical that we may expect no measurable departure from linearity in the curves of vapor pressure against mole fraction. An even closer approach to identity is furnished by isotopes, which resemble one another so closely that the separation of a solution of two isotopes into its components presents a problem of the greatest difficulty.

[1] F. M. Raoult, *Compt. rend.*, **104,** 1430 (1887); *Z. physik. Chem.*, **2,** 353 (1888).

We are thus led to the conception of a perfect, or *ideal*, solution with properties which are not perhaps possessed by any actual solution, just as for the perfect gas an ideal substance was invented with properties not possessed by any real substance.

In order to focus our attention upon the solution itself and not upon another phase in equilibrium with it, let us consider the fugacities rather than vapor pressures and define the perfect solution as one in which the fugacity of each constituent is proportional to the mole fraction of that constituent.   Indeed we go still further and require that this proportionality exist at every pressure and at every temperature.

In accordance with our definition, we therefore write for the first component of a perfect solution,

$$f_1 = f_1^{\circ} x_1 \qquad (18\text{-}1)^*$$

where f_1° at a given temperature and pressure is a constant.   If we can proceed over the whole range of concentration to $x_1 = 1$, f_1° appears as the fugacity of the pure constituent.   In the ordinary case where the solution in question is a liquid, f_1° represents the fugacity of the pure component in the liquid state.   Similar relationships hold for other components.

In terms of the partial molal free energy, Eq. (17-18) shows that for a perfect solution

$$d\bar{\mathrm{F}}_1 = RT\, d\ln f_1 = RT\, d\ln x_1 \qquad (18\text{-}2)^*$$

or
$$\left(\frac{\partial \bar{\mathrm{F}}_1}{\partial x_1}\right)_{P,T} = \frac{RT}{x_1} \qquad (18\text{-}3)^*$$

Integrating between two compositions,

$$\bar{\mathrm{F}}_1'' - \bar{\mathrm{F}}_1' = RT \ln \frac{x_1''}{x_1'}$$

and if one of these compositions is the pure component with $x_1 = 1$ and free energy[1] $\bar{\mathrm{F}}_1^{\circ}$, we have

$$\bar{\mathrm{F}}_1 - \bar{\mathrm{F}}_1^{\circ} = RT \ln x_1 \qquad (18\text{-}4)^*$$

Benzene and toluene form nearly ideal solutions.   Suppose that we add 1 mole of pure benzene to a large volume of a solution of benzene in toluene in which the mole fraction of benzene is $\frac{1}{4}$.   The decrease in free energy in this spontaneous process is given by Eq. (18-4)*,

$$C_6H_6(l) = C_6H_6(\text{in toluene soln}, x = 0.25) \qquad \Delta \bar{\mathrm{F}} = -RT \ln 4$$

Our assumption that the ideal-solution law holds at all temperatures and pressures makes the law more comprehensive than may be evident at

---

[1] Strictly the free energy of pure liquid component 1 is the standard free energy $\bar{\mathrm{F}}_1^{\circ}$ only at 1 atm total pressure.   The differences in free energy of solids or liquids less than 1 atm are very small however, and will be ignored at this point.   Equation (18-4)* holds exactly for a perfect solution if $\bar{\mathrm{F}}_1$ relates to 1 atm total pressure also.

first sight.   If we differentiate Eq. (18-4)* with respect to pressure, we note that $RT \ln x_1$ is independent of pressure and have

$$\frac{\partial \bar{F}_1}{\partial P} - \frac{\partial F_1^o}{\partial P} = \bar{v}_1 - v_1^o = 0 \tag{18-5}*$$

Thus the partial molal volume is equal to the molal volume of the pure component v_1^o at all compositions.   Since the corresponding equations apply to all components, there can be no change in volume on mixing the pure components to form the solution.

In a similar fashion we may differentiate Eq. (18-4)* with respect to temperature,

$$\frac{\partial \bar{F}_1}{\partial T} - \frac{\partial F_1^o}{\partial T} = -\bar{s}_1 + s_1^o = R \ln x_1 \tag{18-6}*$$

and by combining Eq. (18-4)* with this result,

$$\bar{H}_1 - H_1^o = (\bar{F}_1 - F_1^o) + T(\bar{s}_1 - s_1^o) = 0 \tag{18-7}*$$

We see at once that the heat of mixing must be zero since the partial molal heat content of each component is the same as the molal heat content of the pure component.   The result for the entropy is also important.   The entropy of 1 mole of an ideal binary solution exceeds that of the pure components by

$$\Delta S = x_1(\bar{s}_1 - s_1^o) + x_2(\bar{s}_2 - s_2^o) = -R(x_1 \ln x_1 + x_2 \ln x_2) \tag{18-8}*$$

EQUILIBRIUM BETWEEN A PURE SOLID AND AN IDEAL LIQUID SOLUTION

Sometimes a binary solution, which might be expected to behave throughout as a perfect solution, cannot be studied over the whole range of composition owing to the appearance of a new phase of one of the pure constituents.   Thus, if benzene is gradually removed from a solution of benzene and naphthalene, it might be possible with care to proceed to pure liquid naphthalene in a supercooled condition; and the solutions over the whole range would undoubtedly obey very nearly the law of the perfect solution.

Ordinarily, however, before reaching the highest concentrations in naphthalene, that substance will separate out in the crystalline state. The resulting system, of pure solid naphthalene and a solution of naphthalene in benzene, furnishes an interesting study.   We shall assume that the solution is a perfect one and therefore that the partial molal volume and the partial molal heat content of the dissolved naphthalene are equal to the molal volume and the molal heat content of supercooled liquid naphthalene.   If f_2^o is the fugacity of the latter, then for any mole fraction of naphthalene $f_2 = f_2^o x_2$.

Let us now see how the solubility of naphthalene in benzene changes with the pressure and the temperature. We consider pressure first.

Change of Solubility with Pressure. If f_2 and f_2' represent the fugacity of the naphthalene in the solution and in the pure solid state, then if any change in condition occurs which leaves the solution saturated, these two quantities must remain equal, or we may write

$$d \ln f_2 = d \ln f_2'$$

Since the temperature is constant, $\ln f_2'$ depends only upon the pressure, while $\ln f_2$ depends upon pressure and composition. Therefore

$$\frac{\partial \ln f_2'}{\partial P} dP = \frac{\partial \ln f_2}{\partial P} dP + \frac{\partial \ln f_2}{\partial x_2} dx_2 \qquad (18\text{-}9)$$

Substituting for the first two differential coefficients from Eqs. (14-14) and (17-20), and writing Eq. (18-3)* in the form

$$\frac{\partial \ln f_2}{\partial x_2} = \frac{1}{x_2} \qquad (18\text{-}10)^*$$

we find

$$\frac{v_2'}{RT} dP = \frac{\bar{v}_2}{RT} dP + \frac{dx_2}{x_2} \qquad (18\text{-}11)^*$$

or, expressing in the equation the constancy of temperature,

$$\left(\frac{\partial \ln x_2}{\partial P} \right)_T = \frac{v_2' - \bar{v}_2}{RT} \qquad (18\text{-}12)^*$$

Since x_2 may be used as a measure of the solubility, this equation shows how we may calculate the change of solubility with the pressure in the case of a perfect solution.

For the system under consideration x_2 is about 0.50 at 323°K. The molal volume of solid naphthalene is about 115 cc, while $\bar{v}_2$, which is also the molal volume of pure supercooled liquid naphthalene, may be calculated to be 128 cc by extrapolating the values given for liquid naphthalene at higher temperatures. Since $R = 82.06$,

$$\left(\frac{\partial \ln x_2}{\partial P} \right)_T = -0.0004 \text{ atm}^{-1}$$

Thus the solubility diminishes with increasing pressure, x_2 changing by 0.04 per cent per atmosphere. In such a system, which involves only liquids and solids, the effect of pressure upon an equilibrium is so small as to be ordinarily neglected except when we deal with very large changes in pressure.

Change of Solubility with Temperature. A much more important equation is obtained when we consider the change of solubility with temperature, at constant pressure. Proceeding precisely as before, and using Eq. (18-10)* with Eqs. (14-18) and (17-21), we find,

$$\frac{\partial \ln x_2}{\partial T} = \frac{\bar{\mathrm{H}}_2 - \mathrm{H}_2'}{RT^2} = \frac{\Delta\bar{\mathrm{H}}}{RT^2} \qquad (18\text{-}13)*$$

Here $\Delta\bar{\mathrm{H}}$ is the differential heat of solution, but it is also the heat of fusion of naphthalene at the same temperature, since $\bar{\mathrm{H}}_2$, by the law of the perfect solution, is equal to the heat content of the supercooled liquid naphthalene. Over a small range of temperature $\Delta\bar{\mathrm{H}}$ may be taken as constant, in which case a simple integration gives

$$\ln \frac{x_2''}{x_2'} = \int_{T'}^{T''} \frac{\Delta\bar{\mathrm{H}}}{RT^2}\, dT = \frac{-\Delta\bar{\mathrm{H}}}{R}\left(\frac{1}{T''} - \frac{1}{T'}\right) = \frac{\Delta\bar{\mathrm{H}}}{R}\left(\frac{T'' - T'}{T''T'}\right) \quad (18\text{-}14)*$$

This equation was first obtained by Schröder,[1] who also made a thorough experimental investigation of solubilities of substances of this type. For example, he found that naphthalene dissolves in benzene at 61°C until $x_2 = 0.689$. Naphthalene melts at 79.9°C, and the heat of fusion is 4560 cal/mole. Substitution of these values in Eq. (18-13)* yields a calculated value of $x_2 = 0.692$, which agrees very satisfactorily with the measured value.

The $\Delta\bar{\mathrm{H}}$ of fusion was assumed to be a constant in Eq. (18-13)*, but the heat capacity of a liquid is frequently higher than that of the same material in the solid state. Thus the $\Delta\bar{\mathrm{H}}$ of fusion should be expressed as a function of temperature before integration of Eq. (18-13)* if high accuracy is desired over a wide temperature range. A next approximation is to assume $\Delta\bar{c}_P$ of fusion to be constant, which yields

$$\Delta\bar{\mathrm{H}} = \Delta\bar{\mathrm{H}}' + \Delta\bar{c}_P(T - T') \qquad (18\text{-}15)*$$

$$\begin{aligned}
\ln \frac{x_2''}{x_2'} &= \int_{T'}^{T''} \frac{(\Delta\bar{\mathrm{H}}' - T'\,\Delta\bar{c}_P)}{RT^2}\, dT + \int_{T'}^{T''} \frac{\Delta\bar{c}_P}{RT}\, dT \\
&= \frac{(\Delta\bar{\mathrm{H}}' - T'\,\Delta\bar{c}_P)(T'' - T')}{RT''T'} + \frac{\Delta\bar{c}_P}{R}\ln \frac{T''}{T'} \qquad (18\text{-}16)*
\end{aligned}$$

Here T' may be any fixed temperature, but we ordinarily take it to be the melting point. Then $\Delta\mathrm{H}'$ is the heat of fusion at the melting point, and $x_2' = 1$.

Equation (18-16)* was applied by Pitzer and Scott[2] to solutions of xylene isomers in one another. If we choose p-xylene and m-xylene as an example, the pertinent data are given in Table 18-1. The values for

$$\Delta\bar{c}_P = 3R = 5.96 \text{ cal/deg}$$

[1] I. Schröder, *Z. physik. Chem.*, **11**, 449 (1893).
[2] K. S. Pitzer and D. W. Scott, *J. Am. Chem. Soc.*, **65**, 803 (1943).

are obtained by extrapolation of the curves for solid and liquid to the melting point and are only approximate.

TABLE 18-1. SOLUBILITY OF m- AND p-XYLENES

m-Xylene: $T' = 225.27°K$, $\Delta \bar{h}' = 2765$ cal/mole, $\Delta \bar{c}_P \cong 3R$
p-Xylene: $T' = 286.39°K$, $\Delta \bar{h}' = 4090$ cal/mole, $\Delta \bar{c}_P \cong 3R$
Eutectic temperature: Calculated 220.3°K, observed 220.2°K
Eutectic composition: Calculated 13.0% para, observed 12.9% para

In this case the heat of fusion of each component is known, and we calculate the solubility of each in the liquid solution which we have assumed to be ideal. At the eutectic point both solids are in equilibrium with the same solution. This point is found from the pair of solubility equations; a graphical solution such as is shown in Fig. 18-1 is usually most convenient. The close agreement found in Table 18-1 for the xylene solution indicates close conformity to the ideal-solution law in that case. This result is reasonable in view of the close similarity of the two isomeric components.

It is interesting to note that the specific properties of the solvent (i.e., the other component) do not enter at all. Hence the solubility in *mole-fraction units* at a given temperature and pressure for one substance is the same in all solvents which yield ideal solutions. Some results of Schröder[1] illustrate this situation very well. He measured the solubility of p-dibromobenzene in several solvents. The results are shown in Fig. 18-2. The curve of solubility with temperature is almost identical for benzene, bromobenzene, and carbon bisulfide but is quite different for ethyl alcohol, which would not be expected to form an ideal solution.

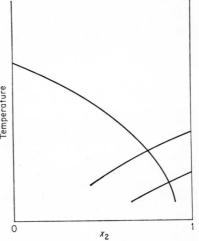

FIG. 18-1. Curves for solid–ideal-liquid-solution equilibria for the three xylenes. The intersections represent the eutectic points for the pairs; extensions of curves below each eutectic temperature represent solutions supersaturated with respect to one of the two components.

The points in Fig. 18-2 may be interpreted as expressing the solubility of p-dibromobenzene in the several solvents, or the lowering of the freezing point of p-dibromobenzene by the same substances.

In this chapter we have gone rather fully into the theory of the ideal solution, notwithstanding the fact that in the ordinary calculations of

[1] *Loc. cit.*

thermodynamic chemistry we do not often meet with solutions which can be regarded as perfect. We have, however, presented in such detail the properties of these solutions, partly because the ideal solution is in some sense the norm with respect to which we may classify solutions in general; partly because, in dealing with a solution whose properties are unknown, we may, as a first approximation, apply the simple law in anticipation of later experimental investigation of the solution in question; but especially

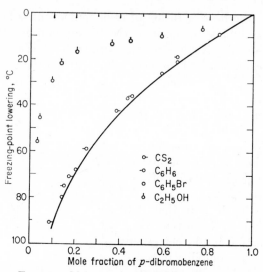

Fɪɢ. 18-2. Mole fraction of *p*-dibromobenzene.

because in certain respects every solution whatsoever approaches the perfect solution as a limit, as it becomes more and more dilute. This important theorem will be the subject of the next chapter.

PROBLEMS

18-1. Assuming a perfect solution, what is the ratio between the fugacity of pure ether and the fugacity of ether in a solution containing 10 g of ether, 10 g of benzene, and 10 g of carbon bisulfide?

18-2. Predict the freezing point of a 1 mole per cent solution of any ideal solute in benzene. The melting point of pure benzene is 5.53°C and the heat of fusion 2351 cal/mole.

18-3. Calculate the minimum work necessary to separate 1 mole of pure benzene from (*a*) a large volume of an equimolal solution of benzene and toluene; (*b*) a solution composed of exactly 1 mole of benzene and 1 mole of toluene.

18-4. Investigate the effect of the $\Delta \bar{c}_P$ term on the change of solubility of *p*-xylene with temperature. Compare the results calculated from Eq. (18-16)* with those from Eq. (18-14)* at 220.3°K. Note data in Table 18-1.

18-5. A solution containing 1 mole of A and 2 moles of B is mixed with a solution containing 3 moles of A and 4 moles of B in a reversible manner at 25°C. Assume all solutions to be perfect. What is the maximum available work?

19

THE LAWS OF THE DILUTE SOLUTION

In the development of physical chemistry the dilute solution has played a very important role. Consequently, both in theory and in experiment, the study of very dilute solutions has been given a prominence perhaps even greater than it intrinsically deserves. The various laws of the dilute solution are really laws of the infinitely dilute solution. It remains for experiment or other theory to show how far into the range of finite concentration these laws may be employed without material error.

There are several of these laws which, although discovered independently, may be shown to be thermodynamically deducible from one another. We may mention van't Hoff's law of osmotic pressure, the laws for the lowering of the vapor pressure and of the freezing point of a solvent obtained by Raoult and van't Hoff, and Henry's law for the vapor pressure of the solute. Because they are thermodynamically related, we shall see that if any one of these laws is assumed as an empirical fact the others necessarily follow. But they cannot all be obtained from thermodynamics alone without some empirical basis.

It is, however, astonishing to find what a small basis of experimental fact is necessary for the development of the whole theory of dilute solutions. Let us consider a solution in which the mole fraction of the solvent is x_1 and that of the solute is x_2. If x_2 is infinitesimal, we call the solution infinitely dilute. Plotting f_2, the fugacity of the solute, against x_2, as we did in Chapter 17, the slope of this curve at the point $x_2 = 0$ might be infinite, finite, or zero. We shall assume as an experimental fact that this slope is finite, namely, that when $x_2 = 0$, df_2/dx_2 is finite.[1] From this one assumption all the laws of the dilute solution may be deduced.

---

[1] When the solute dissociates, as when hydrochloric acid is dissolved in water, we have the only known type of exception to the above rule. In such case $df_2/dx_2 = 0$ when $x_2 = 0$. If, however, we take the products of dissociation as our solute, instead of the dissociable substance, then the rule as stated holds for these cases also. However, since we are going to devote a later chapter to this problem of dissociation in solution, we shall proceed now with the understanding that the laws of the infinitely dilute solution will not be applied, without further explanation, to cases in which the solute dissociates.

Henry's Law. In the first place, we may show that this assumption leads directly to the exact formulation of Henry's law. It is an elementary proposition in calculus that, in the immediate neighborhood of the point where $x = 0$ and $y = 0$, $dy/dx = y/x$. We have shown that $f_2 = 0$ when $x_2 = 0$, and therefore in the immediate neighborhood of this point, i.e., in the region of the infinitely dilute solution,

$$\frac{\partial f_2}{\partial x_2} = \frac{f_2}{x_2} = \text{constant} \qquad (19\text{-}1)^*$$

or
$$f_2 = kx_2 \qquad (19\text{-}2)^*$$

In other words, at constant pressure and temperature, the f_2-x_2 curve starts out as a straight line, and in the infinitely dilute solution the fugacity of the solute is proportional to its mole fraction. By the nature of the proof, it is evident that this equation holds whether the solvent is a pure substance or is itself a solution. Thus, in a mixture of any number of constituents, if one of these constituents is present in a very small amount, its fugacity is proportional to its mole fraction.

In this respect, therefore, any infinitely dilute solution has the character of a perfect solution. But the value of the constant k cannot be predicted. It depends upon the nature of both solute and solvent, and only in the special case of a solution which is perfect over the whole range of composition does the f_2-x_2 curve remain a straight line. It is therefore only in this case that k can be identified with the fugacity of a pure phase of the solute.

We have seen in Chapter 17 that when a solution becomes very dilute the mole fraction of the solute becomes proportional to the concentration c and to the molality m. Hence, with different constants,

$$f_2 = k_c c \qquad f_2 = k_m m \qquad (19\text{-}3)^*$$

Equations $(19\text{-}2)^*$ and $(19\text{-}3)^*$ express in exact form the idea of Henry's law and are entirely true at infinite dilution. Henry's law in its original form states that the vapor pressure of the solute is proportional to its concentration, or

$$p_2 = k_c c \qquad (19\text{-}4)^*$$

This is also true at infinite dilution, where f_2 and p_2 are both infinitesimal and therefore equal to one another. If then at a finite concentration we find deviations from Henry's law, it may be that the concentration is too high for Eq. $(19\text{-}3)^*$ to hold or the vapor pressure may be high enough to depart measurably from the fugacity.

Nernst's Law of Distribution between Two Solvents. If a very small amount of alcohol is added to a two-phase mixture, such as benzene and water, so that the alcohol solution in each phase may be regarded as infinitely dilute, the fugacity of the alcohol in each phase is proportional to its mole fraction in that phase. For equilibrium the fugacities in both phases must be the same, and therefore, as the amount of solute is varied, the mole fraction in one phase must remain proportional to the mole fraction in the other phase.[1] Also the concentration or molality in one phase will be proportional to the concentration or molality in the other.

In mathematical form, we have for the solute in the first phase $f_2 = kx_2$ and in the second $f_2' = k'x_2'$. Then, for equilibrium,

$$\frac{x_2'}{x_2} = \frac{k}{k'} = \text{constant} \qquad (19\text{-}5)^*$$

Raoult's Law. We have previously found for constant temperature and pressure the general equation, Eq. (17-47),

$$x_1 \frac{\partial \ln f_1}{\partial x_1} = x_2 \frac{\partial \ln f_2}{\partial x_2} = \left(\frac{x_2}{f_2}\right)\left(\frac{\partial f_2}{\partial x_2}\right)$$

where we have written the right-hand member again in an equivalent form. Now in the infinitely dilute solution we apply Eq. (19-1)* and find the right side of this equation to be unity. Therefore

$$d \ln f_1 = d \ln x_1 \qquad (19\text{-}6)^*$$

and by integration

$$f_1 = f_1^\circ x_1 = f_1^\circ(1 - x_2) \qquad (19\text{-}7)^*$$

where f_1° is the value of f_1 when $x_1 = 1$ and is therefore the fugacity of the pure solvent. Thus we see that in the infinitely dilute solution, regardless of the nature of the solute, the fugacity of the solvent is proportional to its mole fraction. Hence in this respect every infinitely dilute solution is a perfect solution. In so far as the vapor of the solvent behaves like a perfect gas, we may substitute vapor pressure for fugacity and write

$$p_1 = p_1^\circ x_1 = p_1^\circ(1 - x_2) \qquad (19\text{-}8)^*$$

This is Raoult's law.[2] It is well illustrated by Figs. 17-8 to 17-11, which show that the vapor pressure of the solvent is always lowered by a small addition of the solute just as though the mixture were going to behave as a perfect solution, although large deviations from the perfect solution may appear at higher concentration.

---

[1] W. Nernst, *Z. physik. Chem.*, **8**, 110 (1891).

[2] F. M. Raoult, *Compt. rend.*, **104**, 130 (1887); *Z. physik. Chem.*, **2**, 353–373 (1888), put this in a different form. Equation (19-8)* is easily rearranged to $(p_1^\circ - p_1)/p_1^\circ = x_2$, which shows that the fractional lowering of the vapor pressure of the solvent is equal to the number of moles of solute in 1 mole of solution.

We may put Eq. (19-6)* in other forms, for example,

$$\frac{df_1}{dx_1} = \frac{f_1}{x_1}$$

Since we are dealing with an infinitely dilute solution, x_2 is negligible compared with x_1 and therefore $x_1 \cong 1$.   Hence we may write this equation in the form

$$d \ln f_1 = dx_1 = -dx_2 \qquad (19\text{-}9)*$$

or, expressing in the equation itself the constancy of temperature and pressure,

$$\left(\frac{\partial \ln f_1}{\partial x_2} \right)_{P,T} = -1 \qquad (19\text{-}10)*$$

Van't Hoff's Law of Freezing-point Lowering.   If a mixture of ice and water is placed in a thermostat at 0°C and a certain amount of solute is added to the water, the escaping tendency of the water becomes less than that of the ice and the ice will disappear.   If the same thing is done, not in a thermostat, but in a thermally insulated vessel, a part of the ice will melt and thus the mixture will be cooled, until once more the escaping tendency of the ice will be the same as that of the water in the solution; for by Eq. (17-21) ice, having a smaller heat content than water, will suffer the greater diminution in escaping tendency as the temperature is lowered.

We may put this into quantitative form, using the fugacity as a measure of escaping tendency.   If f_1' is the fugacity of the ice and f_1 that of the water, and if we add solute, lowering temperature at the same time, so that f_1 and $\ln f_1$ remain equal to f_1' and $\ln f_1'$, then $\ln f_1'$ will change only because of the change of temperature, while $\ln f_1$ changes both because of the change of temperature and because of the addition of solute.   Thus

$$\frac{\partial \ln f_1'}{\partial T} dT = \frac{\partial \ln f_1}{\partial T} dT + \frac{\partial \ln f_1}{\partial x_2} dx_2 \qquad (19\text{-}11)$$

and substituting the values of the partial differential coefficients, by Eqs. (19-10),* (14-18), and (17-21), we find

$$\frac{dT}{dx_2} = -\frac{RT^2}{\bar{H}_1 - \bar{H}_1'} \qquad (19\text{-}12)*$$

and, in such dilute solution, $\bar{H}_1 - \bar{H}_1'$ is the heat of fusion of the pure solvent.

This is the familiar formula[1] for the lowering of the freezing point in a dilute solution.[2]   It shows that the freezing point depends only upon the

[1] J. H. van't Hoff, *Z. physik. Chem.,* **1,** 481 (1887).

[2] While x_2 is the number of moles of solute in 1 mole of solution, it may also be regarded here as the number of moles of solute in 1 mole of solvent.

molal composition, and not upon the nature of the solute. All solutes, whether they give perfect or imperfect solutions at higher concentrations, produce the same lowering of the freezing point per mole at high dilution.[1] This is well illustrated in a diagram which we have already given in Fig. 18-2 while discussing the lowering of the freezing point of p-dibromobenzene by benzene and alcohol. It will be noticed that the molal lowering of the freezing point caused by these two substances is very different when they are present in large amount but becomes identical for the two as their concentration diminishes.

The Freezing-point Lowering When a Solid Solution Appears.[2] Equation (19-12)* was derived on the assumption that the solid phase of the solvent was practically pure. This is the usual case, but there are instances, which indeed become very numerous at high temperatures, in which the solute proves to be appreciably soluble in the solid phase of the solvent. Let x_1 and x_2 be the mole fractions of solvent and solute in the liquid solution and x_1' and x_2' the corresponding mole fractions in the solid solution. Since the properties of the solid phase now depend upon the composition as well as upon the temperature, we may repeat the operation which gave us Eq. (19-12)*, merely adding one term, and find

$$\frac{\bar{H}_1 - \bar{H}_1'}{RT^2} dT = -(dx_2 - dx_2') \qquad (19\text{-}13)^*$$

assuming that both solid and liquid solutions are infinitely dilute.

Now, by Eq. (19-5)*, if k/k' is the distribution coefficient of the solute between the solid and liquid phases, $k/k' = dx_2'/dx_2$, which, with Eq. (19-13)*, gives a result which is the same as that of Eq. (19-12)*, except that a numerical coefficient is involved which becomes unity when the solute remains entirely in the liquid solution, namely,

$$\frac{dT}{dx_2} = \left(\frac{k}{k'} - 1\right) \frac{RT^2}{\bar{H}_1 - \bar{H}_1'} \qquad (19\text{-}14)^*$$

Here again $\bar{H}_1 - \bar{H}_1'$, for an infinitely dilute solution, is the same as the heat of fusion of the pure solvent.

Let us once more emphasize the fact that such equations are derived only for the limiting case of an infinitely dilute solution, and thermodynamics does not predict how far they will maintain their validity in the range of finite concentrations. It may well be that, in the solid solutions which we have been discussing, marked deviations from the laws of the dilute solution appear at much lower concentrations than they ordinarily do in liquid solutions.

---

[1] Always excepting the type of solute mentioned in a previous footnote, and to be discussed later, namely, the solute which dissociates.

[2] J. H. van't Hoff, *Z. physik. Chem.*, **5**, 322 (1890).

Osmotic Pressure.   Upon adding a solute to any solvent the escaping tendency of the solvent decreases.   We have seen how the vapor pressure of the solvent is decreased by addition of solute at constant temperature. There are several methods by which we can raise the relative escaping tendency of solvent from the solution and bring it into equilibrium again with pure solvent.   In the most common of these methods the temperature is decreased below the freezing point of the pure solvent.   We have just discussed the resulting equation for the freezing-point lowering.   In another method pressure is applied to the solution but not to the pure solvent.

Certain membranes, known as semipermeable membranes, permit the free passage of water but do not permit the passage of certain substances dissolved in water.   In Fig. 19-1, A represents such a semipermeable

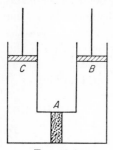

FIG. 19-1

membrane between the space AB, which contains an aqueous solution, and the space AC, which contains pure water.   If the pressures are the same on both sides, the escaping tendency of the water is less in the solution and the phenomenon of osmosis will occur; i.e., water will flow through the membrane from left to right.   But if the pressure upon the piston B is increased (or that upon the piston C is diminished) to the point where the escaping tendency of the water is the same on both sides, then equilibrium prevails, and water will pass through the membrane in neither direction.   If P° is the pressure upon the pure solvent and P is the pressure upon the solution when osmotic equilibrium is maintained, $P - P^\circ$ is the osmotic pressure.   The botanist Pfeffer[1] was the first to obtain reproducible osmotic pressure measurements in 1877.

The law of osmotic pressure for dilute solutions was obtained by van't Hoff.[2]   The osmotic pressure must restore the fugacity of the solvent f_1 in the solution to that of the pure solvent.   Hence we write

$$d \ln f_1 = 0 = \frac{\partial \ln f_1}{\partial P} dP + \frac{\partial \ln f_1}{\partial x_2} dx_2 \qquad (19\text{-}15)$$

and, substituting Eqs. (19-10)* and (17-20),

$$\frac{dP}{dx_2} = \frac{d(P - P^\circ)}{dx_2} = \frac{RT}{\bar{v}_1} \qquad (19\text{-}16)^*$$

Now $P - P^\circ = 0$ when $x_2 = 0$, and therefore, by the rule that we used in obtaining Eq. (19-1)*, $d(P - P^\circ)/dx_2 = (P - P^\circ)/x_2$.   Furthermore

[1] W. Pfeffer, "Osmotische Untersuchungen," W. Engelmann, Leipzig, 1877.
[2] J. H. van't Hoff, *Z. physik. Chem.*, **1**, 481 (1887).

in the infinitely dilute solution $\bar{v}_1$ is equal to v_1, the molal volume of the pure solvent; and x_2, the number of moles of solute in 1 mole of solution, is also the number of moles of solute in 1 mole of solvent. Thus we have the familiar equation of van't Hoff,

$$P - P^\circ = \frac{x_2 RT}{v_1} \qquad (19\text{-}17)^*$$

Van't Hoff himself used the equation

$$P - P^\circ = \frac{n_2 RT}{V} \qquad (19\text{-}18)^*$$

where n_2 is the number of moles of solute in the volume V of the solution. This is identical with Eq. (19-17)* in the infinitely dilute solution.

The formal similarity of Eq. (19-18)* to the perfect-gas equation has led to attempts to interpret osmotic pressure in terms analogous to the kinetic theory of gases. We do not believe that consideration of the collisions of solute molecules with the membrane constitutes a useful approach to an understanding of osmotic pressure. Rather one should note the tendency of any gas to fill its container and of any solute to diffuse throughout the solvent accessible to it. The driving force in each case is the entropy, i.e., the increase in probability which arises for distribution of particles throughout a large volume as compared with a small volume. We expect the probability relationship to be the same in each case. Also, from Eq. (7-9)* for the entropy of the gas we have

$$\frac{\partial s}{\partial v} = \frac{R}{v} \qquad (19\text{-}19)^*$$

In dilute solution one may combine Eq. (18-6)* with $x_1 \propto 1/V$ to show that

$$\frac{\partial \bar{s}}{\partial v} = \frac{R}{V} \qquad (19\text{-}20)^*$$

where V is the volume of the solvent (or the solution) containing 1 mole of solute. Thus, as expected, the entropy increase is the same for the dilution of a gas or the solute in a dilute solution. Now we may use Eq. (10-13):

$$P = -\left(\frac{\partial A}{\partial V}\right)_T = -\left(\frac{\partial E}{\partial V}\right)_T + T\left(\frac{\partial S}{\partial V}\right)_T$$

Since $\partial E/\partial v$ is zero for the perfect gas and $\partial \bar{E}/\partial v$ is zero for the solute in solution, we find

$$P = \frac{nRT}{V}$$

which is the familiar equation for gas pressure or osmotic pressure provided that we supply the appropriate definitions for n and V, respectively.

We see that both the freezing-point lowering and the osmotic pressure measure the effect of the solute on the fugacity of the solvent. The choice of the preferable method for a given investigation depends on several factors: (1) the availability of a truly semipermeable membrane, (2) the absence of solubility of the solute in the solid solvent, and (3) the precision of measurement of the pressure or temperature difference. With low-molecular-weight solutes in solvents such as water or benzene, which seldom form solid solutions, the freezing-point method is usually better.   The situation is reversed, however, for solutions of very high-molecular-weight solutes, where the molal concentration of solute is exceedingly low, because the osmotic pressure is the more sensitive measurement.   Also, it is easier to obtain a membrane permeable to the solvent but not to the solute if the latter comprises very large molecules.

Dilute Solution in a Mixed Solvent.   If, instead of a pure solvent, we consider a solvent of two (or more) constituents in fixed proportion, a dilute solution of another substance in this mixed solvent will have many of the characteristics of a dilute solution in a pure solvent.   Thus, if the two constituents of the solvent are called components 1 and 2 and the solute is component 3, the fugacity of the solute f_3 may be assumed, at infinite dilution, to be proportional to its mole fraction x_3.   The general equation for a ternary mixture [Eq. (17-46)] is

$$x_1 \left(\frac{\partial \ln f_1}{\partial x_3} \right)_{T,P} + x_2 \left(\frac{\partial \ln f_2}{\partial x_3} \right)_{T,P} + x_3 \left(\frac{\partial \ln f_3}{\partial x_3} \right)_{T,P} = 0 \qquad (19\text{-}21)$$

If component 3 is infinitely dilute,

$$f_3 = kx_3 \qquad d \ln f_3 = d \ln x_3 \qquad \frac{d \ln f_3}{dx_3} = \frac{1}{x_3} \qquad (19\text{-}22)*$$

and
$$x_1 \left(\frac{\partial \ln f_1}{\partial x_3} \right)_{T,P} + x_2 \left(\frac{\partial \ln f_2}{\partial x_3} \right)_{T,P} = -1 \qquad (19\text{-}23)*$$

or, more simply,

$$x_1 \, d \ln f_1 + x_2 \, d \ln f_2 = -dx_3 \qquad (19\text{-}24)*$$

As this equation shows, it is impossible from thermodynamics alone to calculate how much the fugacity of each constituent of the solvent is lowered by the addition of solute; in fact, one of the two fugacities might actually be increased.

This indeed happens, especially when the solute is very insoluble in one of the constituents of the solvent.   Thus in a mixture of nearly equal parts of water and alcohol the fugacity of the water is increased by the addition of a small amount of benzene.   It is an interesting experiment to bring such a mixture of alcohol and water in contact with ice in a well-insulated vessel and observe the rise of the freezing point upon the addition of benzene.

If in the experiment just mentioned the mixture had been contained, not in an insulated vessel, but in a thermostat, the increased escaping tendency of the water in the solution would have manifested itself by a separation of ice, until the change in composition thus produced led to a new state of equilibrium.

We may obtain an interesting general formula for such a case.   Let us consider the mixed solvent of components 1 and 2 in equilibrium with solid 2 in a thermostat. Since the temperature is constant, and since therefore f_2 is constant because of the equilibrium with the solid, the result of adding a solute is given by a simplified form of Eq. (19-24)*, namely,[1]

$$x_1 \, d \ln f_1 = -dx_3 \qquad (19\text{-}25)*$$

Since in this equation x_3/x_1 is the number of moles of solute to 1 mole of the first solvent, the expression is very similar to that for the lowering of the fugacity of a single solvent [Eq. (19-9)*]. The meaning of Eq. (19-25)* may be illustrated as follows: If a saturated solution of salt in 1000 g of water is in contact with solid salt in a thermostat and a small amount of sugar is added, the fugacity of the water is lowered by the same percentage as that resulting when the same amount of sugar is added to 1000 g of pure water.

Deviations from the Laws of Dilute Solutions. It is beyond the scope of thermodynamics to predict the deviation from the laws of dilute solutions on the basis of the molecular properties of the solvent and solute. Nevertheless, a very simple argument gives such useful information that its inclusion seems desirable.

In a dilute solution the solute molecules usually have only solvent molecules for near neighbors. Henry's law applies as long as solute molecules never approach one another closely enough to interact significantly, because the escaping tendency is then strictly proportional to the number of solute molecules in the fixed amount of solvent. The intermolecular interaction potential between nonpolar molecules or between rotating dipolar molecules[2] falls off as r^{-6}, where r is the intermolecular distance. Thus the interaction energy decreases very rapidly with increased separation.

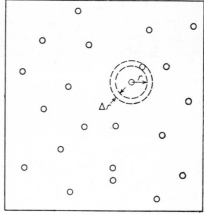

If we consider spherical shells in space round a given solute molecule (see Fig. 19-2), the probability of finding another solute molecule in the shell at radius r is proportional

Fig. 19-2

to the shell volume or to r^2. But the interaction energy decreases as r^{-6}; hence the net effect decreases as r^{-4}. Thus in an approximate treat-

---

[1] This equation was first obtained by G. N. Lewis, *Proc. Am. Acad.*, **43**, 259 (1907); *Z. physik. Chem.*, **61**, 129 (1907).

[2] See, for example, W. J. Moore, "Physical Chemistry," 2d ed., pp. 424–426, Prentice-Hall, Inc., Englewood Cliffs, N.J., 1955.

ment we may ignore distant portions of the solution and consider just the probability that a pair of solute molecules are next to one another. In the liquid state a given molecule has about 10 immediate neighbors. Thus at mole fraction 0.01 there is about a 10 per cent chance of finding another solute molecule next to a given solute molecule. At mole fraction 0.001 this probability drops to 1 per cent. For most purposes this probability of 1 per cent is small enough to have a very small effect on the thermodynamic properties, and we may regard the solution to be dilute.

Since the probability of finding another solute molecule next to a given solute molecule is directly proportional to the concentration of the solute, we may expect the initial deviation from Henry's law to be linear in the concentration.

$$\frac{f_2}{kx_2} = 1 + k'x_2 + \cdots \qquad (19\text{-}26)$$

This conclusion, that the initial deviation depends on x_2 to the first power, depends only on the fact that the intermolecular forces fall off faster than r^{-2}, that is, on their short range nature. It is true even for nonrotating dipoles where the interaction potential is proportional to r^{-3}.

There are two principal exceptions to the conclusions just reached. If the solute molecules are very large, such as high-polymer molecules, then the probability that a part of one solute molecule is close to a part of another solute molecule is much greater. In fact one may use a similar argument but replace the mole fraction by the volume fraction. The second exception arises for molecules with electrical charges, ions, because the intermolecular potential then decreases only as r^{-1}. The effect of distant portions of the solution is small only if there are precisely equal concentrations of positive and negative charge. In the case of electrolytes we must abandon the conclusion that the initial deviation from Henry's law will be linear in the concentration. We shall return to this problem in Chapters 22 and 23, where we shall find that terms in the half power of the concentration appear.

PROBLEMS

19-1. Derive equations for $\bar{v}_2$ and $\bar{H}_2$ in terms of the Henry's-law constant and its derivatives.

19-2. The change in volume or heat content on dilution with additional solvent is zero for an ideal solution. Derive these properties for a nonideal solution which is so dilute that it conforms to Henry's law.

19-3. Derive an expression for the change in solubility (mole fraction of solute) with the temperature for a case in which the solubility is small.

19-4. When an infinitesimal amount of solute is added, find the lowering of the freezing point of water (a) per mole of solute added to 1 mole of water; (b) per mole of solute added to 1000 g of water.   The heat of fusion of ice is 1436 cal/mole.

19-5. Show again the analogy between a solution at its eutectic point and a pure liquid at its freezing point by obtaining an equation for the lowering of the eutectic point of a binary solution through the addition of an infinitesimal amount of a third substance.   Thus may be obtained an expression identical in form with Eq. (19-12)*.[1] The method used will be analogous to that used in deriving Eq. (17-57).

19-6. Derive an exact equation for the rise in the boiling point of a solvent caused by an infinitesimal addition of a nonvolatile solute.

19-7. The boiling point of a constant-boiling mixture is changed by the addition of a nonvolatile solute according to the same law as that which applies in the case of a simple solvent.[1]   Demonstrate this theorem in the following manner.   Let f_1 and f_2 be the fugacities in a binary liquid solution and f_1' and f_2' in the gaseous phase.   Let x_1 and x_2, x_1' and x_2' be the mole fractions.   For equilibrium $d \ln f_1 = d \ln f_1'$, and $d \ln f_2 = d \ln f_2'$.   In a constant-boiling mixture $x_1 = x_1'$, and $x_2 = x_2'$.   Hence, multiplying and adding,

$$x_1 \, d \ln f_1 + x_2 \, d \ln f_2 = x_1' \, d \ln f_1' + x_2' \, d \ln f_2' \tag{19-27}$$

This is true for any sort of change.   If an infinitesimal amount of a nonvolatile solute component 3 is added at constant pressure, and if the temperature is changed also to maintain equilibrium, then the first member of Eq. (19-27) is due to changes in T and x_3, while the second member is due to changes in T alone.   By means of the fundamental equation of partial differentiation, substitute for the several terms in Eq. (19-27), and then simplify by using Eq. (19-23)* and the values of the several differential coefficients.

19-8. All experiments to be discussed in this problem begin with the same solution containing $\frac{1}{2}$ mole of water and $\frac{1}{2}$ mole of ethyl alcohol.   Very small amounts of benzene were added to this solution, and the partial pressure of ethyl alcohol was determined as a function of the amount of benzene added.   The results can be represented by the equation $P_{\text{EtOH}} = 29e^{-12n_b}$ mm Hg, where n_b represents moles of benzene added.   If the partial pressure of water in the original solution was 12 mm Hg, derive an equation showing the variation of partial pressure of water as a function of amount of benzene added, and calculate the partial pressure of water when 0.01 mole of benzene has been added.

[1] See G. N. Lewis, *Proc. Am. Acad.*, **43,** 259 (1907); *Z. physik. Chem.*, **61,** 129 (1907).

20

THE USEFUL FUNCTIONS, ACTIVITY,
ACTIVITY COEFFICIENT,
AND EQUILIBRIUM CONSTANT

Our task of presenting the basic ideas and methods of thermodynamics is nearly completed.   There are a few matters of fundamental importance which it has seemed expedient to postpone until after the classic methods of thermodynamics have been fully treated.   But in the main we shall henceforth direct our attention more particularly to the questions which arise in the application of the fundamental principles to specific numerical calculations.

Of all the applications of thermodynamics to chemistry, none has in the past presented greater difficulties, or been the subject of more misunderstanding, than the one involved in the calculation of what has rather loosely been called the free energy of dilution, namely, the difference in the chemical potential or partial molal free energy of a dissolved substance at two concentrations.   We shall therefore, in this and in subsequent chapters, give much attention to the various special methods whereby this important quantity may be simply and precisely determined.

If we consider one constituent of a solution at two different concentrations, but at the same temperature, and if by some method we determine for this constituent $\Delta F = \bar{F} - \bar{F}'$ between these concentrations, then we have also determined the ratio of the fugacities.   For, by Eq. (17-45),

$$\mu - \mu' = \bar{F} - \bar{F}' = RT \ln \frac{f}{f'} \tag{20-1}$$

Now we often have occasion to determine and to use such a ratio of fugacities when it is impossible or inexpedient to determine the numerical value of either fugacity; for example, we may be dealing with an almost involatile substance.   It has therefore proved advantageous to consider, at a given temperature, the ratio between the fugacity f of a substance in some given state and its fugacity $f°$ in some state which, for temporary convenience, is chosen as a *standard state*. This relative fugacity[1] is called the *activity* and is denoted by a. Thus in the standard state the activity is unity, $a^\circ = 1$, while in any other state it is given by the equations,

$$RT \ln a = \mu - \mu^\circ = \bar{F} - \bar{F}^\circ \qquad (20\text{-}2)$$

where, at a given temperature, the chemical potential in any state is μ and in the standard state $\mu^\circ = \bar{F}^\circ$.

Since in any state the activity at a given temperature is always proportional to the fugacity, we may write $d \ln a = d \ln f$. Thus for the change of activity with pressure we write from Eq. (14-14),

$$\left(\frac{\partial \ln a}{\partial P}\right)_T = \left(\frac{\partial \ln f}{\partial P}\right)_T = \frac{v}{RT} \qquad (20\text{-}3)$$

Now, if for the purpose of some calculation we choose any one state of a substance as the standard state, for example, liquid water at atmospheric pressure, we shall consider this the standard state at different temperatures. Thus, in the example chosen, if we should take the activity of liquid water as unity, the activity of water in any solution would be given by the ratio of its fugacity to that of liquid water, at each temperature.

It would evidently avoid confusion if once for all we should choose for a given substance its standard state. This consideration, however, is outweighed by the practical advantage of being able at any time to choose the standard state or states best adapted for a particular problem. For reasons which will become more apparent when we enter upon free-energy calculations, we shall choose as standard now one state and now another, as convenience dictates, although to avoid confusion this choice must in each case be clearly stated. For as we proceed, we shall find it desirable to choose different standard states for a substance, not only in different problems, but even in a single problem. However, we may in advance lessen the arbitrariness of our procedure by setting down certain rules for choosing the standard state to which we shall almost invariably adhere.

---

[1] Sometimes called also the relative activity; for G. N. Lewis first defined the absolute activity, as he defined the fugacity, by reference to the attenuated gaseous state, in which the activity was taken as equal to the concentration [*Proc. Am. Acad.*, **43,** 259 (1907); *Z. physik. Chem.*, **61,** 129 (1907)]. Furthermore, in 1939, R. H. Fowler and E. A. Guggenheim called the quantities $\lambda_i = \exp(\mu_i/RT)$ absolute activities ("Statistical Thermodynamics," p. 66, Cambridge University Press, New York, 1939). We shall not use either sort of absolute activity.

CHOICE OF A STANDARD STATE

1. For a Gas: $a = f$; **That Is,** $a/P = 1$ **When** $P = 0$. In Chapters 12 and 15 we have briefly referred to a 1-atm perfect-gas standard state for tabulation of entropies and free energies, and in Chapter 16 we have presented the procedure for calculating standard entropy and free-energy values from experimental data and the reverse procedure for calculating values for a real gas at any pressure from the standard values. At this time we can characterize this standard state more explicitly. Since the activity of a substance at a given temperature is always proportional to its fugacity, it will be convenient in the case of a gas to make the activity equal to the fugacity; in other words, we choose at each temperature as standard state the one in which the fugacity is unity. For a perfect gas this is the same as making the activity unity at unit pressure;[1] but in general the activity will not be exactly equal to the pressure. It is therefore to be borne in mind that, while the standard state of a liquid or solid is always taken at unit pressure, the standard state of a gas is one in which, not the pressure but the fugacity, is unity.

As a matter of fact we wish to go a little further than this. The standard state, as we shall use the term, implies not only that the fugacity is unity but that the heat capacity, heat content, etc., are those of the gas at infinite attenuation. Our standard state therefore is a hypothetical one and corresponds to no real state of the gas.

In a manner similar to the presentation of Eqs. (12-5)* and (12-6) in Chapter 12, we note that the difference between the free energy of a perfect gas at some low pressure P and at 1 atm is given by $\mathrm{F}_P - \mathrm{F}° = RT \ln P/1$. At low pressures, the real gas approaches the ideal-gas state, and one has the equation for the standard free energy

$$\mu° = \mathrm{F}° = \lim_{P \to 0} \left(\mathrm{F}_P - RT \ln \frac{P}{1} \right) \tag{15-3}$$

For the standard-state activity or fugacity, one has $RT \ln f° = \lim_{P \to 0} \{RT \ln [f/(P/1)]\}$ or $f° = \lim_{P \to 0} [f/(P/1)]$. The standard fugacity is equal to the limiting f/P ratio of unity at low pressures. The $P/1$ term arising from the free-energy change of a perfect gas from pressure P to a pressure of 1 atm is dimensionless, and thus $f°$ has the dimensions of pressure.

The characterization of the standard state for a gas is illustrated in Fig. 20-1, where the fugacity of a real gas is plotted against the pressure. At low pressures the fugacity vs. pressure curve approaches a limiting slope of unity, which is indicated by the dotted line and represents perfect-gas behavior. For purposes of tabulating heat capacities and

---

[1] In the case of a mixed gas we shall also make the activity equal to the fugacity for each constituent.

heat contents, it is sufficient to characterize the gaseous standard state as corresponding to a state of the gas at a low enough pressure such that the difference between the curve for the real gas and the perfect gas is less than experimental error.   To specify the entropy or the free energy, one must also give a specific pressure or fugacity.   This is done by fixing the gaseous standard state at a value of fugacity equal to unity as indicated

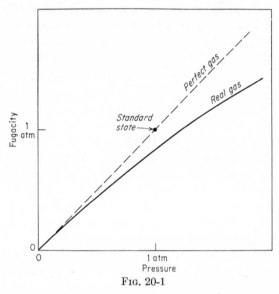

FIG. 20-1

by the point of the dotted curve of Fig. 20-1.   To relate this hypothetical state to any real state of the gas, one must move along the dotted line which represents perfect-gas behavior until the difference between it and the real curve is negligible.   Then one must move along the real curve by the methods given in Chapter 16 until one has reached the desired state.   It may seem to be an unnecessary complication to use such a hypothetical state as a standard state.   However, many of our calculations will be applied to systems where the difference between the perfect-gas behavior and the real-gas behavior is small and the difference is easily calculated as a second-order correction.   Under those circumstances, the use of the perfect-gas standard state is quite advantageous. When we deal with actual equilibrium calculations in detail, it will be obvious that even when deviations from ideality are large there are additional advantages in choosing the perfect-gas standard state for which the dependence of properties upon pressure and temperature is readily separable.

2. For a Liquid or Solid Which May Act as a Solvent: $a/x = 1$ **When** $x = 1$. We have noted previously that the pure liquid or pure solid at 1 atm pressure is chosen as the standard state for the tabulation of heat capacities, heat contents, entropies, free energies, and fugacities. This same standard state is applicable to solutions also. It is always used for the component in largest amount, e.g., the solvent, and is frequently used for all components. If a pure liquid or a pure solid at atmospheric pressure be chosen as the standard state, or state of unit activity, at each temperature, we may change the activity of such a liquid or solid by dissolving in it a very small amount of another substance. Then, by Raoult's law, the fugacity, and therefore the activity a, is proportional to the mole fraction x_1. But, by definition, $a = 1$ for the pure solvent when $x_1 = 1$. Thus, for the solution,

$$\frac{a_1}{x_1} = 1 \qquad\qquad (20\text{-}4)*$$

an equation that is valid for the infinitely dilute solution and as far into the range of finite concentrations as Raoult's law is obeyed. In a region where it is not obeyed a series of values of a_1/x_1 shows immediately the degree of departure from that law.

By our definition of a_1, the activity of the solvent is equal to f_1/f_1°. Now, if the solvent has a measurable vapor pressure, we may let p_1 be the measured vapor pressure of the solvent from the solution and p_1° be the vapor pressure of the pure solvent. Then, if we may assume that the vapor behaves as a perfect gas, $p_1 = f_1$ and $p_1^{\circ} = f_1^{\circ}$, so that

$$a_1 = \frac{p_1}{p_1^{\circ}} \qquad\qquad (20\text{-}5)*$$

and this equation, as we shall see, furnishes one of the most useful means of calculating activities. If the vapor is not a perfect gas, we may still determine $a_1 = f_1/f_1^{\circ}$ by the methods of Chapter 16.

When the vapor pressure of the solvent is not measurable, numerous other methods of determining the activity are available. Thus, if we should measure the emf of some galvanic cell with a silver electrode and then substitute for the silver a solid solution of gold in silver, the emf would be found to change, owing to the fact that the gold lowers the escaping tendency of the silver. A careful investigation of this sort would enable us, by methods which we shall illustrate later in this chapter, to determine the activity of the silver in a series of gold-silver alloys.

3. For a Solute:[1] **Either $a_2/x_2 = 1$ When $x = 0$ or $a_2/m = 1$ When $m = 0$.** In the typical case of a solution at infinite dilution, we have Henry's law, which states that the fugacity of the solute, and therefore its activity, is proportional to its mole fraction. It is therefore con-

---

[1] The case of electrolytes will be discussed in Chapter 22.

venient to choose the standard state of a solute A_2 so that at 1 atm pressure and at infinite dilution the ratio of the activity to the measure of concentration is unity.   Thus, if composition is being given as the mole fraction, one chooses $a_2/x_2 = 1$ when $x_2 = 0$.   It is probably unfortunate, but most data on aqueous solutions are reported in terms of the molality, i.e., the number of moles per 1000 g of water.   In the very dilute region molality becomes proportional to mole fraction; hence Henry's law requires the activity to become proportional to the molality, and one may define $a_2/m = 1$ at 1 atm pressure when $m = 0$.

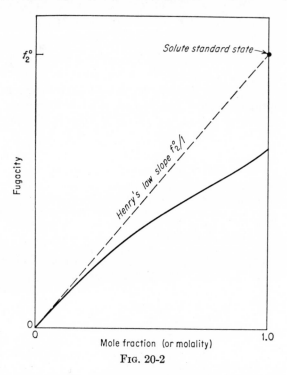

$$f_2^o$$

Solute standard state

Henry's law slope $f_2^o/1$

Fugacity

0

0   Mole fraction (or molality)   1.0

Fig. 20-2

These definitions yield hypothetical states as standard states and in this respect are exactly like the definition for a gas.   The standard state is not the real state of unit activity but rather a hypothetical ideal solution in which the partial molal enthalpy and heat capacity of the solute have the values of the infinitely dilute solution but the free energy and entropy correspond to unit activity.   Figure 20-2 illustrates this definition of the solute standard state.   At low mole fraction the plot approaches asymptotically the Henry's-law slope, which is indicated by the dotted line.   However, it is not sufficient just to refer to the infinitely dilute solution since there the partial molal free energy becomes minus infinity, as shown in Fig. 17-7.   Hence one specifies unit activity as the standard state.

In mathematical terms the equations for the very dilute solution from Chapter 19 yield

$$f^\circ = \lim_{x_2 \to 0} \frac{f}{x_2} \qquad \text{or} \qquad f^\circ = \lim_{m \to 0} \frac{f}{m/1}$$

and for the standard chemical potential

$$\mu^\circ = \bar{F}^\circ = \lim_{x_2 \to 0} \left(\bar{F}_{x_2} - RT \ln x_2 \right)$$

or

$$\mu^\circ = \bar{F}^\circ = \lim_{m \to 0} \left[\bar{F}_m - RT \ln \left(m/1 \right) \right]$$

where the respective forms for the mole fraction and molality systems are given.   In the case of molality $m/1$ is written to remind us that the unit molality state has the dimension moles per kilogram, whereas mole fraction is a pure number.

In dilute solution the deviation of a_2/m from unity measures once more the departure of the solution from the perfect solution, but in concentrated solutions, where the molality is not proportional to the mole fraction, this is no longer the case.   For nearly every purpose the mole fraction furnishes the most advantageous method of measuring composition, and the employment of this measure in aqueous as well as in nonaqueous solutions is to be encouraged.   In the meantime, however, many of the existing data in aqueous solutions are expressed in terms of molality (or, what is very nearly the same, the concentration), and, unless otherwise stated, our unity of activity in aqueous solutions will be the one based upon molality.

We offer no general rule for fixing the standard state of a solute in a mixed solvent.   If in any problem the constituents of the mixed solvent remain in the same relative amounts, the case can be treated just as though the solvent were pure.   In other cases it will usually be best to use as standard state one defined with respect to a solution in one of the pure solvents.

Let us make sure that there is no misunderstanding of the way in which we are to use the term activity.   We have seen in a previous chapter that in a state of equilibrium the fugacity of a given substance is the same in every phase, or in every part of a system.   Since the activity is defined as the relative fugacity, if we should choose, for the whole system, a single standard state of the given substance, its activity would be the same in every part of the system.   On the other hand, since we have decided to use for the substance in question different standard states in different phases, its activities in the several phases which are in equilibrium will not be equal but will nevertheless remain proportional to one another as long as equilibrium persists, the factor of proportionality depending upon the choice of standard states.

Thus, in the case of a solvent in equilibrium with its vapor, the activity in the vapor phase being defined in the section on the activity of a gas and that in the liquid phase being defined in the section on the activity of a liquid solvent, the ratio of these two activities remains constant no matter how the concentration of solute changes (and this ratio is equal to f_1°, the fugacity of the pure liquid).   We shall meet a similar case when we discuss the distribution of a solute between two solvents.

EFFECT OF PRESSURE ON ACTIVITY

Since small pressure changes have very little effect on the chemical potential of liquid or solid phases, these pressure effects are frequently ignored.   This is in contrast to the case for gases, where pressure effects are large and the fugacity of a component would never be regarded as independent of the pressure.   However, accurate work requires the consideration of pressure effects on condensed-phase chemical potentials. It is best to consider the mixing of components to form a solution at constant pressure and temperature; hence we consider the pressure effect on each component in its reference state.

The term *standard state* and the symbol $^\circ$ are reserved for the condition of 1 atm pressure (or the ideal gas), and the term *reference state* will be used for states reached from standard states by a change of pressure. The *activity in a reference state* at a pressure P_2, for which we shall adopt the symbol Γ, is given by the integral of Eq. (20-3),

$$\ln \Gamma = \frac{\mu' - \mu^\circ}{RT} = \int_1^{P_2} \frac{\mathrm{v}}{RT} \, dP \qquad (20\text{-}6)$$

where the lower limit of integration is 1 atm pressure.   If a solute reference state is being used, v is the partial molal volume $\bar{\mathrm{v}}$ at infinite dilution; otherwise v is the molal volume of the pure liquid or solid.

THE ACTIVITY COEFFICIENT

Starting with the fundamental thermodynamic properties, energy and entropy, we have found it useful to introduce a number of derived functions, such as H, A and F, f, and now a.   We have discussed in detail in Chapters 10 and 13 the value of introducing functions such as A and F. The free energy has one disadvantage in that it varies from minus infinity to plus infinity.   Such a large variation makes it difficult to estimate whether a particular value is reasonable or not and to relate it to directly determined equilibrium properties such as pressure and composition. One can readily eliminate the variation from minus infinity to plus infinity by taking the antilogarithm of the free energy.   This has been done to obtain quantities like fugacity and activity, which are always positive. Also by equating fugacity to pressure at very low pressures one has a quantity which is of the order of magnitude of the pressure. This gives one more of a physical feeling for the fugacity than one would have for a free-energy value.   Likewise in ideal solutions, close to 1 atm pressure, the activity is equal to the mole fraction (or molality) so that, when the deviations from ideality are not large, one has a close relationship between the activity and the experimentally observable composition.

When one comes to the problem of tabulating thermodynamic quantities, one should use a function which is slowly varying so that interpolations can be carried out accurately without tabulation at very close intervals.   For this reason free-energy functions were introduced for tabulation of free-energy values; we have already considered in detail the means by which one can calculate the change of free energy of the standard state of a pure substance with temperature, and we shall shortly see that the procedures for calculating the variations of partial molal free energy of a solute standard state are very similar.   However, to complete an equilibrium calculation, one must do more than determine the variation of free energy or fugacity of a standard state with temperature. One must evaluate the free-energy terms given in Eq. (15-21) corresponding to the free-energy change from the standard state to the actual equilibrium state.   The equilibrium state may differ from the standard state owing to the pressure not being 1 atm or because of the fact that the equilibrium state corresponds to a solution where the components are at reduced activities.   Thus it is of very great importance to know the free energies of dilution of the components of a solution in order to be able to carry out equilibrium calculations.   If one were to tabulate activities as a function of composition, one could readily calculate changes in chemical potential from Eq. (20-2).   However, activities vary too rapidly with composition, and it is useful to find a function which shows a smaller variation so that one can interpolate readily.   Thus it is useful to express the activity as the product of three factors.   First is the *activity of the reference state* Γ, which expresses the effect of pressure on activity and which may be calculated from knowledge of the molal volume; second is the mole fraction (or molality), which expresses the major effect of composition on activity; and third is the *activity coefficient* γ, which expresses the deviation from ideal-solution behavior.   Thus the activity coefficient will remain near unity unless the solution deviates widely from ideal behavior.   Where mole fraction is used as the composition measure, we have for the ith component

$$\gamma_i = \frac{a_i}{\Gamma_i x_i} \qquad \text{and} \qquad \mu_i - \mu_i^\circ = RT \ln x_i \gamma_i \Gamma_i \qquad (20\text{-}7a)$$

It is also useful to express γ_i in terms of fugacities,

$$\gamma_i = \frac{f_i}{f_i^\circ \Gamma_i x_i} = \frac{f_i}{f_i' x_i} \tag{20-8a}$$

where f_i° is the fugacity of the standard state at 1 atm and f_i' is the fugacity in the reference state at the actual pressure P' of the solution.

If molality is used as the measure of composition for a solute (component 2), the corresponding equations are

$$\gamma_2 = \frac{a_2}{\Gamma_2 m_2} \quad \text{and} \quad \mu_2 - \mu_2^\circ = RT \ln m_2 \gamma_2 \Gamma_2 \tag{20-7b}$$

and

$$\gamma_2 = \frac{f_2}{f_2^\circ \Gamma_2 m_2} = \frac{f_2}{f_2' m_2} \tag{20-8b}$$

In approximate calculations for low pressures one may frequently set $\Gamma_i \cong 1$ and ignore that factor in the equations. In precise work the Γ_i must be considered, but if the solutions are nearly ideal, the activity coefficients are nearly independent of pressure. The pressure derivative of Eq. (20-8) yields

$$\frac{\partial \ln \gamma_i}{\partial P} = \frac{\bar{v}_i - v'}{RT} \tag{20-9}$$

and $\bar{v} - v'$ is zero for an ideal solution and very much smaller than v' or $\bar{v}$ for nearly ideal solutions.

In most of our subsequent work we shall omit the factor Γ because the system is at 1 atm pressure or so near to 1 atm that the difference is insignificant.

When composition is expressed as mole fraction, a dimensionless quantity, both activity and activity coefficient are also dimensionless. When one uses molality or concentration, either the activity or the activity coefficient, but not both, can be made dimensionless. As the activity coefficient is the quantity that is invariably tabulated, it would seem to be preferable to make it the dimensionless quantity. Actually, the same procedure has been followed for gases, where the activity was taken equal to the fugacity with the dimensions of pressure. Thus the fugacity coefficient f/P is dimensionless. For dimensional consistency, this requires that Eq. (20-2) read

$$\frac{a}{1} = \frac{f}{f^\circ} \quad \mu - \mu^\circ = RT \ln \frac{a}{1}$$

where $a/1$ is the activity of a gas expressed in atmospheres divided by unit activity also expressed in atmospheres. For a solution whose composition is given in molality, $a/1$ is the activity of the solute expressed in molality divided by unit activity also expressed in molality. A corresponding expression applies to activities based on concentration. When this convention is followed, all activity coefficients are dimensionless.

The above convention makes no practical difference, as the numbers used are the same whether activity or activity coefficient is made dimensionless, and both activities and activity coefficients will in general be different for a given solution depending upon which composition units are used. Activities and activity coefficients based on concentration in moles per liter are sometimes used for limited purposes. As we have noted earlier, concentration units are inconvenient for thermodynamic purposes because the concentration of a given solution varies with temperature. Since concentration activity coefficients are close to molality activity coefficients, they can readily be confused with one another and the tabulation of the concentration activity coefficients should be discouraged.

THE NUMERICAL CALCULATION OF ACTIVITY AND ACTIVITY COEFFICIENTS

The Activity and Activity Coefficient of a Solvent from Its Vapor Pressure over a Solution. The vapor pressure of a component is a particularly convenient measure of its activity. Sometimes the vapor deviates significantly from the perfect-gas law, and the fugacity must be calculated by the methods of Chapters 16 and 21; in other cases at sufficiently low pressures the vapor may be assumed to be a perfect gas, and $f_i = p_i$. In any case, if x_1 is the mole fraction of a solvent, a_1 its activity, γ_1 its activity coefficient, f_1 the fugacity of its vapor over the solution, and f_1° the fugacity of the pure solvent, then, since the activity of the pure liquid solvent is taken as unity,

$$a_1 = \frac{f_1}{f_1^\circ} \quad \text{and} \quad \gamma_1 = \frac{f_1}{x_1 f_1^\circ}$$

Thallium amalgams were studied at 325°C by Hildebrand and Eastman,[1] and under these conditions the mercury vapor may be taken to be a perfect gas. Table 20-1 gives the values of $a_1 = p_1/p_1^\circ$ measured at several values of the mole fraction. The third column gives γ_1, whose difference from unity shows the degree of departure from the perfect solution.

TABLE 20-1. ACTIVITY OF MERCURY IN THALLIUM AMALGAMS AT 325°C

x_1	$a_1 = p_1/p_1^\circ$	γ_1
0.957	0.955	0.998
0.942	0.938	0.996
0.915	0.901	0.985
0.893	0.875	0.980
0.836	0.803	0.961
0.742	0.690	0.930
0.664	0.602	0.907
0.614	0.548	0.893
0.497	0.433	0.871
0.347	0.293	0.844
0.202	0.166	0.822

[1] J. H. Hildebrand and E. D. Eastman, *J. Am. Chem. Soc.*, **37**, 2452 (1915).

Activity and Activity Coefficient of a Solute from Its Vapor Pressure.   When a solution is in equilibrium with the vapor of the solute A_2, we may measure the vapor pressure of A_2 over a range of concentration and by knowing the fugacity of the vapor at each pressure we may obtain the activity of the solute in the solution.   When we may assume that the vapor is a perfect gas, a_2, the activity in the solution, may be taken as proportional to p_2, the vapor pressure of the solute. Hence, as we pass from some mole fraction x_2, to an infinitely dilute solution of mole fraction x_2^*, $a_2/p_2 = a_2^*/p_2^*$.   Now by Henry's law x_2/p_2 approaches at infinite dilution a constant value, which may be denoted by x_2^*/p_2^*.   But by definition of our standard state $a_2^* = x_2^*$, and therefore at any concentration,

$$a_2 = \frac{p_2}{p_2^*/x_2^*} \qquad \gamma_2 = \frac{a_2}{x_2} = \frac{p_2/x_2}{p_2^*/x_2^*} \qquad (20\text{-}10)^*$$

Lewis and Storch[1] measured the vapor pressure of bromine from solutions of bromine[2] (Br_2) in carbon tetrachloride (CCl_4) ranging from $x_2 = 0.004$ to $x_2 = 0.025$, and they found, over the whole range, a constant value of p_2/x_2, namely, 0.539.   Thus within the accuracy of their experiments the deviation between the Henry's-law curve and the actual curve was negligible, and the value of p_2^*/x_2^* or the value of the fugacity for the solute standard state is equal to 0.539 atm.   Also γ_{Br_2} relative to the solute standard state of bromine in CCl_4 is unity over this range of concentrations within experimental error.   However, if we should attempt to extrapolate to $x_2 = 1$, we should find for the vapor pressure of pure bromine 0.539 atm, while the actual vapor pressure of pure bromine is 0.280 atm.   Thus we see that the solute standard state which we have chosen for bromine, in its dilute solutions in carbon tetrachloride, differs materially from the standard state of pure liquid bromine.

This is illustrated in Fig. 20-3, where the line A shows the values that would be calculated from the vapor pressure of pure bromine, assuming a perfect solution.   The line B is an extrapolation of the straight line experimentally obtained in the dilute solutions, while the actual curve for the whole range would be of the form of C.   In other words, B represents

---

[1] G. N. Lewis and H. Storch, *J. Am. Chem. Soc.*, **39**, 2544 (1917).

[2] It must constantly be borne in mine that the mole fraction is ambiguous unless the chemical formulas are stated.   The mole fraction of bromine in a solution containing 160 g bromine and 154 g carbon tetrachloride is $\frac{1}{2}$ if we are considering the formula Br_2 but $\frac{2}{3}$ if we are considering the formula Br.   While theoretically thermodynamics is not concerned with the way in which the molecular species or the formula is defined, in practice it would produce awkward complications to use any formula except Br_2 for bromine solutions as Henry's law would not hold for bromine in carbon tetrachloride unless the formula Br_2 is used.   This is equivalent to saying that bromine is found empirically to have the molal weight 160 in solution.

the extrapolation given by Henry's law from dilute solutions of bromine in carbon tetrachloride, while A represents the extrapolation by Raoult's law from dilute solutions of carbon tetrachloride in bromine. The ratio of 0.539 : 0.280, which is also the ratio of the slopes of B and A, gives us at once the ratio of the fugacity of bromine in the two standard states, the first of which being the one we have chosen for bromine in dilute solution in carbon tetrachloride, and the second (pure bromine) being the one we would have chosen if we had taken bromine as a solvent. Also the specification of the activity coefficient of bromine will depend upon the standard state with respect to which the γ is referred and to the standard state from which partial molal free energies of dilution are calculated.

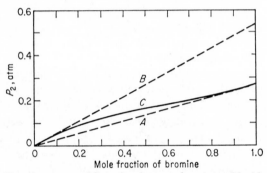

Fig. 20-3. Vapor pressure of bromine from carbon tetrachloride solutions.

Thus in the mole-fraction range from 0.004 to 0.025 the activity coefficient of bromine with respect to the solute standard state of bromine is unity. On the other hand, the activity coefficient of bromine relative to pure liquid bromine, which is given by $p_2/p_2^o x_2$, is $^{0.539}\!/_{0.280}$, or 1.92. With respect to the pure liquid standard state of bromine, γ increases from the unit value at high concentrations of bromine to a limiting value of 1.92 at very low concentrations of bromine, as is readily seen from Eqs. (20-7). For a given solution, the ratio of activity coefficients relative to the two standard states is just the inverse of the ratio of the fugacities of the two standard states.

If the vapor of the solute cannot be regarded as a perfect gas, our procedure is not seriously complicated. Let us consider the vapor pressure of CO_2 from aqueous solutions (or the solubility of CO_2 in water). At a given temperature let a_2 be the activity and m the molality of CO_2 in the aqueous phase. Let p_2 be the vapor pressure of CO_2 and f_2 the fugacity, which is the same in both phases. Then, since a_2 is always proportional to f_2,

$$\frac{a_2}{f_2} = \frac{a_2^*}{f_2^*}$$

The second member represents the values in a very dilute solution, where f_2^*, being small, is equal to p_2^* and a_2^* by definition of the solute standard state may be taken equal to m^*. Hence

$$\frac{a_2}{f_2} = \frac{m^*}{p_2^*} \tag{20-11}$$

Here the second member is the limit approached at infinite dilution by the "solubility coefficient" m/p_2. In the present case, at 0°C, $m^*/p_2^* = 0.076$, and $a_2 = 0.076 f_2$.

Thus, consider a solution which at 0°C gives a pressure of CO_2 amounting to 5 atm. From the equation of state of CO_2, the fugacity is 4.88 atm, and the activity of CO_2 in the solution is $0.076 \times 4.88 = 0.371$.

Determination of Activity from the Distribution of a Solute between Two Solvents.

Instead of finding the distribution of a substance between a solution and a gaseous phase, in which its activity is known, we may study the distribution of a solute between two immiscible solvents, so that if its activity at different concentrations is known in one solvent it may be calculated in the other.

Thus, when a very small amount of bromine (Br_2) is shaken up with carbon tetrachloride and water[1] (slightly acidified to prevent the hydrolysis of the bromine), it was found by Lewis and Storch[2] that the molality m in the aqueous phase, divided by the mole fraction x_2 of bromine in the carbon tetrachloride phase, is 0.371 at 25°C. In other words, at infinite dilution $m^*/x_2^* = 0.371$. If we denote by a_w the activity of bromine in the water phase and by a_t its activity in the other, these two activities must be proportional to one another, and since we define the standard states so that at infinite dilution $a_w^* = m^*$ and $a_t^* = x_2^*$,

$$\frac{m\gamma_w}{x_2\gamma_t} = \frac{a_w}{a_t} = \frac{a_w^*}{a_t^*} = \frac{m^*}{x_2^*} \tag{20-12}$$

Thus, $a_w = 0.371 a_t$. But we have seen, earlier in this chapter, that in the carbon tetrachloride phase, up to $x_2 = 0.025$, $a_t = x_2$, $\gamma_t = 1$, and therefore in this range $a_w = 0.371 x_2$. If then we wish to know the activity of bromine in any dilute aqueous solution, we find its mole fraction in

---

[1] Strictly speaking we measure the distribution, not between pure water and pure carbon tetrachloride, but between water saturated with carbon tetrachloride and carbon tetrachloride saturated with water. In the present case the difference is not important, but this is by no means always true. In the interesting case studied by G. N. Lewis and G. H. Burrows, *J. Am. Chem. Soc.*, **34**, 1515 (1912), where urea was shaken up with the two-phase system of ethyl acetate and water, the results would have been vastly different if the distribution between urea in pure water and urea in pure ethyl acetate could have been obtained, for the escaping tendency of urea from ethyl acetate solutions varies greatly with a small change in the water content of the ester. (Indeed the solubility of solid urea in ethyl acetate containing water was found to be nearly proportional to the water content.)

[2] *Loc. cit.*

a carbon tetrachloride solution which maintains equilibrium with the aqueous phase and multiply this mole fraction by 0.371.

If iodine (I_2) is shaken with glycerine and carbon tetrachloride, and if its mole fraction in the former phase is x_g and in the latter x_t, then x_g/x_t should be constant in very dilute solutions. At the moderate concentrations studied at 25°C by Landau[1] this constancy no longer holds. He found that, when x_g varies from 0.0001 to 0.002, x_g/x_t changes from 0.50 to 0.40. If we assume for iodine (as for bromine) that γ_t is unity and the activity in carbon tetrachloride is equal to x_t, we may find in the glycerine phase the ratio of activities at the two concentrations. Or if we consider the activity coefficient in glycerine, $\gamma_g = a_g/x_g$, this ratio evidently increases 20 per cent in going from the more dilute to the more concentrated solution.

Activity from Measurements of Electromotive Force. In most galvanic cells the emf depends upon the composition of the electrolyte. We shall defer consideration of such cells until we have more fully studied the thermodynamic properties of electrolytes, but there is still an important class of cells of which the emf depends solely upon the condition of the electrodes.

If two electrodes of the same metal in the same condition are placed in the same electrolyte, a small current results in the transfer of a certain amount of metal from one electrode to the other,[2] but this is a process which is attended by no change in free energy, and the reversible emf of the cell is zero, as might have been foretold also from the symmetry of its construction. If, however, the two electrodes differ from one another, e.g., if the electrodes are in different states of strain, or if one of the electrodes is of pure copper and the other is of copper containing some impurity such as silver in solid solution, the activity of the copper will not be the same in the two electrodes and the cell will have an emf.

Thus, in the second of these cases, the process which occurs in the cell and the change in free energy accompanying this process may be expressed by the equation

$$Cu(s) = Cu(\text{in the solid soln})   \Delta F = RT \ln \frac{a}{a'} = RT \ln a = RT \ln \gamma x$$

if a is the activity, x is the mole fraction, and γ is the activity coefficient in the solid solution and if a', the activity of pure copper, is taken as $a°$, or unity. Hence from (15-1),

$$-n\mathcal{E}\mathfrak{F} = RT \ln a = RT \ln \gamma x   \mathcal{E} = \frac{RT}{n\mathfrak{F}} \ln a = \frac{RT}{n\mathfrak{F}} \ln \gamma x   (20\text{-}13)$$

---

[1] M. Landau, *Z. physik. Chem.*, **73**, 200 (1910).

[2] At the same time a certain amount of electrolyte, determined by the transference number, is carried from one pole to the other, but here also, since the transfer is between two regions of equal concentration, the free-energy change is zero.

Ordinarily in a case of this kind the emf is entirely independent of the particular electrolyte which is employed.   But in the present instance it is interesting to note that the nature of the electrolyte may affect ε by determining the value of n; for if the electrolyte is a pure cupric salt, $n = 2$, but if it is a pure cuprous salt, $n = 1$.   With given electrodes ε is therefore twice as great in the latter case as in the former.

A more common type of cell is furnished when the electrode is a metallic solution in which the more electropositive metal is the solute. Let us consider a cell whose electrodes are two thallium amalgams of different composition, both immersed in the same solution of electrolyte. Now, when a current passes through such a cell, the only thing of thermodynamic significance which occurs in the cell is the carrying of a certain quantity of thallium out of one amalgam into the other, a process which may be expressed as follows:

$$\text{Tl(in Tl amalg; } x_2') = \text{Tl(in Tl amalg; } x_2)$$

If a_2 and a_2' and γ_2 and γ_2' are the activities and activity coefficients of the thallium, when the mole fractions of the thallium are x_2 and x_2',

$$\Delta F = RT \ln \frac{a_2}{a_2'} = RT \ln \frac{\gamma_2 x_2}{\gamma_2' x_2'} = -n\varepsilon\mathfrak{F} \qquad (20\text{-}14)$$

Introducing the numerical values of $\mathfrak{F}$ and R from Appendix 10, and using common logarithms,

$$\varepsilon = -\frac{0.00019841\,T}{n} \log \frac{\gamma_2 x_2}{\gamma_2' x_2'} \qquad (20\text{-}15)$$

If we fix $n = 1$ by using a pure thallous salt, it is evident that the emf is independent of the particular salt chosen and also independent of its concentration and of the solvent in which it is dissolved.   In fact, at a given temperature, it is determined completely by the properties of the two amalgams.

In very dilute amalgams, where the activity asymptotically approaches proportionality to the mole fraction and the activity coefficient approaches a limit, we may predict the emf of the concentration cell by means of the equation

$$\varepsilon = -\frac{0.00019841\,T}{n} \log \frac{x_2}{x_2'} \qquad (20\text{-}16)*$$

On the other hand, in the concentrated amalgams, we may use the measured values of ε to find how the activity and activity coefficient vary with the composition.

In order to illustrate this method, we may employ the data obtained for such cells by Richards and Daniels[1] at 20°C. At this temperature Eq. (20-15) becomes

$$\log a_2 = \frac{-\mathcal{E}}{0.05816} + \log a_2' \qquad (20\text{-}17)$$

Now we may let x_2 be the mole fraction and a_2 the activity of thallium in any amalgam and x_2' and a_2' the corresponding values in some particular amalgam chosen for reference. Then $\mathcal{E}$ is the emf of any concentration cell of which one electrode is the amalgam of fixed mole fraction x_2' and the other is an amalgam of any mole fraction x_2.

While this equation permits the immediate calculation of the ratio of activities between two amalgams which have been studied experimentally, it is frequently desirable to obtain this ratio by interpolation between two

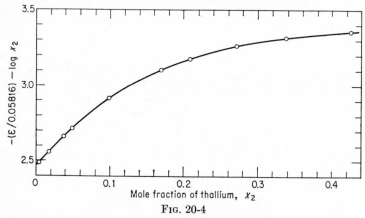

Fig. 20-4

compositions that have not been directly investigated. Especially it is desirable to obtain not merely such ratios of activities but also the value of each activity, which can be found if we are able to extrapolate our data to infinite dilution, where $a_2 = x_2$ if the solute standard state is chosen. For such purpose of extrapolation and interpolation we shall once more employ the expedient of plotting the difference between the experimental data and a function chosen with regard to simplification at infinite dilution.

Thus, subtracting $\log x_2$ from both sides of Eq. (20-17),

$$\log \gamma_2 = \log \frac{a_2}{x_2} = \left(\frac{-\mathcal{E}}{0.05816} - \log x_2 \right) + \log a_2' \qquad (20\text{-}18)$$

If now we plot the quantity in parentheses against x_2, as in Fig. 20-4, we note that when $x_2 = 0$ we have by definition $\gamma_2 = 1$, or $\log \gamma_2 = 0$. The

[1] T. W. Richards and F. Daniels, *J. Am. Chem. Soc.*, **41**, 1732 (1919). In discussing, here and elsewhere, the thermodynamic properties of thallium amalgams, we follow the paper of G. N. Lewis and M. Randall, *J. Am. Chem. Soc.*, **43**, 233 (1921).

value of the ordinate where the curve cuts the vertical axis is therefore equal to $- \log a_2'$. Thus we determine the activity in the reference amalgam. The value of the limiting ordinate, subtracted from the ordinate at any other value of x_2, gives at once the corresponding $\log \gamma_2$. The curve is drawn from the experimental data given in Table 20-2.

TABLE 20-2. ACTIVITY AND ACTIVITY COEFFICIENT OF THALLIUM
IN AMALGAMS AT 20°C

x_2	$-\mathcal{E}$	$\dfrac{-\mathcal{E}}{0.05816} - \log x_2$	γ_2	a_2
0	$-\infty$	2.4689	1	0
0.003259	0	2.4869	1.042	0.003396
0.01675	0.04555	2.5592	1.231	0.02062
0.03723	0.07194	2.6660	1.574	0.05860
0.04856	0.08170	2.7184	1.776	0.08624
0.0986	0.11118	2.9177	2.811	0.2772
0.1680	0.13552	3.1045	4.321	0.7259
0.2074	0.14510	3.1780	5.118	1.061
0.2701	0.15667	3.2610	6.196	1.674
0.3361	0.16535	3.3159	7.031	2.363
0.4240	0.17352	3.3558	7.707	3.268
0.428 (sat)†	0.17387	3.3580	7.75	3.316
Tl(supercooled)			8.3	8.3

† The next to the last value of a_2 is for saturated amalgams containing an excess of solid thallium. It is to be noted that G. N. Lewis and C. L. von Ende, *J. Am. Chem. Soc.*, **32**, 732 (1910), assumed the potential of solid thallium to be the same as that of a saturated thallium amalgam, since at ordinary temperatures thallium and mercury had been shown by Kurnakov and Pushin to form no compound, and since Sucheni had shown that mercury does not dissolve appreciably in solid thallium. Richards and Daniels find by preliminary experiment that the saturated amalgam has a lower potential than pure thallium by 0.0025 volt. Hence the statement of Sucheni must be incorrect, and solid thallium must dissolve several per cent of its own weight of mercury. It will be observed that in the higher concentrations the values of γ_2 are rapidly approaching a limit, about 8.3. This value will give γ_2 for pure supercooled liquid thallium, a quantity which we would have taken as unity if we had chosen to regard liquid thallium as a solvent. Thus between the two standard states, one of which makes $\gamma_2 = 1$ when $x_2 = 1$, and the other of which makes $\gamma_2 = 1$ when $x_2 = 0$, the ratio of fugacities is 8.3.

We have chosen $x_2' = 0.00326$, so that $-\mathcal{E}$ is equal to the value of the emf given by Richards and Daniels. The first column gives the mole fraction of thallium; the second gives $-\mathcal{E}$ (the emf between an amalgam of the mole fraction given and one in which $x_2' = 0.00326$). In the next column is given $-\mathcal{E}/0.05816 - \log x_2$, plotted as ordinate in Fig. 20-4; the fourth gives the values of γ_2 obtained from the plot by the method just described.

Namely, $\log \gamma_2$ is obtained by subtracting the intercept on the axis of ordinates ($- \log a_2' = 2.4689$) from the values in column 3. The last column gives the values of a_2.

CALCULATION OF THE ACTIVITY AND ACTIVITY COEFFICIENT OF ONE COMPONENT OF A SOLUTION WHEN THE ACTIVITY OF THE OTHER IS KNOWN

When we know any partial molal quantity for one constituent of a binary mixture over a range of compositions, we have seen in previous chapters how it is possible to find the change in the corresponding partial molal quantity for the other constituent over the same range. Thus for changes in partial molal free energy, due to isothermal changes in composition, we have by Eq. (17-44)

$$d\bar{F}_1 = - \frac{x_2}{x_1} d\bar{F}_2 \tag{20-19}$$

or by integration

$$\bar{F}_1'' - \bar{F}_1' = - \int_{x_2'}^{x_2''} \frac{x_2}{x_1} d\bar{F}_2 \tag{20-20}$$

The only difference between the use of this equation and of similar equations for V, C_P, H, etc., is that F of a solute, instead of approaching a finite value at infinite dilution, approaches $- \infty$. For this reason it is a great convenience to replace the free energy by the activity, according to the differential form of Eq. (20-2),

$$d\bar{F} = RT \, d \ln a \tag{20-21}$$

Then Eq. (20-19) becomes

$$d \ln a_1 = - \frac{x_2}{x_1} d \ln a_2 \tag{20-22}$$

or

$$\log \frac{a_1''}{a_1'} = - \int_{x_2'}^{x_2''} \frac{x_2}{x_1} d \log a_2 \tag{20-23}$$

Now by plotting x_2/x_1 against $\log a_2$ we could integrate this equation graphically, merely by determining the area under the curve between two limits. We have, however, observed before that, of two exact equations which are mathematically identical, one may give far more accurate results than another when employed in practice in an arithmetical or graphical computation. The method we have just suggested of integrating Eq. (20-22) does not in fact, without the expenditure of an undue amount of labor, permit a very accurate calculation of the change in $\log a_1$ between two finite limits of concentration, and still less does it lend itself to an extrapolation to infinite dilution.

Graphical Computation. We may, however, by a simple expedient, turn this equation into another which is equally simple and which is extremely well suited for use in our graphical methods.

Noting that by Eq. (17-2)

$$dx_1 = -dx_2 \qquad x_1 \, d\ln x_1 = -x_2 \, d\ln x_2 \qquad d\ln x_1 = -\frac{x_2}{x_1} \, d\ln x_2$$

$$(20\text{-}24)$$

we may subtract the last equation from Eq. (20-22) and find[1]

$$d\ln \gamma_1 = -\frac{x_2}{x_1} \, d\ln \gamma_2 \qquad (20\text{-}25)$$

or, by using common logarithms and integrating,

$$\log \gamma_1'' - \log \gamma_1' = -\int_{x_2'}^{x_2''} \frac{x_2}{x_1} \, d\log \gamma_2 \qquad (20\text{-}26)$$

If now we integrate graphically, by plotting x_2/x_1 as ordinates against $\log \gamma_2$, the area under the curve between the points corresponding to the

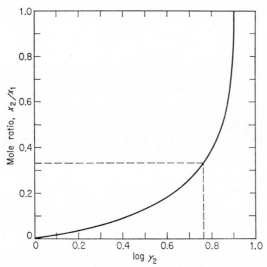

Fig. 20-5. Activity of thallium in thallium amalgams at 20°C.

two compositions shows the difference in the two values of $\log \gamma_1$.

If one of the two compositions be taken at infinite dilution, i.e., if $x_2' = 0$, then by definition $\gamma_1' = a_1'/x_1' = 1$ and its logarithm is 0. Hence

$$\log \gamma_1'' = -\int_0^{x_2''} \frac{x_2}{x_1} \, d\log \gamma_2 \qquad (20\text{-}27)$$

[1] The activity of the reference state is a function of pressure but not of solution composition; hence Γ does not appear in Eq. (20-25) even if the pressure differs significantly from 1 atm.

where the integral now is the whole area under the curve from the origin (where $\log \gamma_2 = 0$) to the given composition.

We shall now illustrate the use of this equation by determining the activity of mercury in thallium amalgams, by means of the activities of thallium which were obtained in Table 20-2 from measurements of emf. Figure 20-5 shows x_2/x_1 plotted against $\log \gamma_2$. By measuring the areas to ordinates which correspond to round mole fractions (illustrated in the figure by the ordinate corresponding to $x_2 = 0.25$, $x_2/x_1 = 0.333$) we obtain the values of γ_1 given in the last column of Table 20-3. The approximate interpolated values of γ_2 are also given for later reference.

TABLE 20-3. ACTIVITY COEFFICIENT OF MERCURY (AND OF THALLIUM) IN AMALGAMS AT 20°C

x_2	x_2/x_1	γ_2	γ_1
0	0	1	1
0.005	0.00502	1.06	0.9998
0.01	0.0101	1.15	0.999
0.05	0.0526	1.80	0.986
0.1	0.111	2.84	0.950
0.2	0.250	4.98	0.866
0.3	0.428	6.60	0.790
0.4	0.667	7.57	0.734
0.5	1.000	7.98	0.704

Let us now use the information on the activity of mercury in thallium amalgams at 325°C, presented in Table 20-1, to calculate the activity of thallium. Interchange of the subscripts in Eq. (20-26) yields

$$\log \gamma_2'' - \log \gamma_2' = - \int_{x_1'}^{x_1''} \frac{x_1}{x_2} \, d \log \gamma_1 \qquad (20\text{-}26a)$$

The corresponding graph is shown in Fig. 20-6. We have, however, to decide upon the reference state for thallium, which is liquid at this temperature. We may choose a solvent type of definition and make pure liquid thallium the reference state. In this case the integral of Eq. (20-26a) extends from $x_1' = 0$, where $\gamma_2' = 1$, to the composition of interest, and we require an area such as A in Fig. 20-6.

On the other hand we may decide to use the solute reference state for thallium. This would facilitate comparison of the data at 325°C with that at temperatures below the melting point of thallium, where the pure liquid no longer exists. The solute-reference-state definition requires that $\gamma_2 = 1$ at $x_1' = 1$ or $x_1'/x_2' = \infty$. Consequently, the integral in Eq. (20-26a) corresponds to the area B in Fig. 20-6. It is evident at once that we are now in difficulty because the ordinate becomes infinite

within the range of integration. The graph indicated in Fig. 20-6 is no longer suitable, and we must either rearrange our equations into a more suitable form for graphical integration or use an analytical integration of an empirical equation in the trouble-
some region. Before illustrating these procedures, however, we wish to emphasize that this difficulty arises from the use of the infinitely dilute solution for reference purposes and that use of the pure liquid reference state avoids the difficulty. Thus the solvent type of reference state should be employed wherever feasible.

A Special Graphical Method.
In the last section we encountered difficulties in the graphical calculation of activities based on the solute standard state from solvent activity data. These can be overcome by a mathematical rearrangement together with use of theoretical principles of the limiting behavior of dilute solutions. Consider the special function called the practical osmotic coefficient, which was first defined by Bjerrum,[1]

$$\phi = -\frac{\ln a_1}{r} \quad \text{with } r = \frac{x_2}{x_1} \quad (20\text{-}28)$$

$$d\phi = -r^{-1} d\ln a_1 + r^{-2} \ln a_1 \, dr \quad (20\text{-}29)$$

Solution for $d \ln a_1$ and substitution into Eq. (20-22) yields, after rearrangement,

$$d \ln a_2 = d\phi + \phi \, d \ln r$$

Next we subtract $d \ln r$ from each side and integrate from $r = 0$ to the final composition r'.

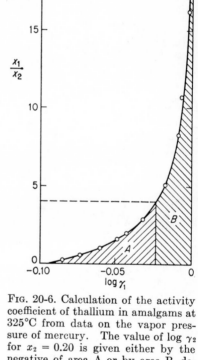

FIG. 20-6. Calculation of the activity coefficient of thallium in amalgams at 325°C from data on the vapor pressure of mercury. The value of log γ_2 for $x_2 = 0.20$ is given either by the negative of area A or by area B, depending on the choice of reference state.

$$\ln \frac{a_2'}{r'} = \phi(r') - \phi(0) + \int_0^{r'} \frac{\phi - 1}{r} \, dr \quad (20\text{-}30)$$

Let us consider now the behavior of the osmotic coefficient in the region

[1] N. Bjerrum, *Z. Elektrochem.*, **24**, 259 (1907).

of small r.   The discussion of the last section of Chapter 19 led to Eq. (19-26) for the limiting behavior of dilute nonelectrolyte solutions.   If we convert this result to an activity coefficient of the solute we find

$$\gamma_2 = 1 + k'x_2 + \cdots = 1 + k'r + \cdots$$
$$\ln \gamma_2 = k'r + \cdots \tag{20-31}$$

Substitution into Eq. (20-25) yields

$$d \ln \gamma_2 = k' \, dr \cdots = -r^{-1} d \ln \gamma_1$$

and integration from $r = 0$ to some small r

$$\ln \gamma_1 = -\tfrac{1}{2}k'r^2 + \cdots \tag{20-32}$$

and also $\ln x_1 = -\ln (1 + r) = -r + \dfrac{r^2}{2} \cdots$

Hence $\phi = -\dfrac{\ln \gamma_1 + \ln x_1}{r} = 1 + \dfrac{k' - 1}{2} r + \cdots \tag{20-33}$

and we see that $\phi(0) = 1$; also that $(\phi - 1)/r$ approaches $(k' - 1)/2$ as $r \rightarrow 0$.   The general theory (last section of Chapter 19) does not yield a numerical value of this limit $(k' - 1)/2$, only the fact that it is finite for nonelectrolytes.   This is sufficient, however, to make the integral in

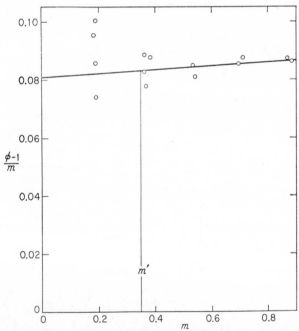

FIG. 20-7. The calculation of the activity of sucrose by use of Eq. (20-34) and iso-piestic data giving the activity of the solvent, water.

Eq. (20-30) satisfactory for graphical calculation provided that the experimental values of ϕ are sufficiently accurate in the dilute region.

This method of calculation is readily adapted to use with data reported in molality. For a solvent of molecular weight M_1, $r = mM_1/1000$, and $\phi = -1000(M_1m)^{-1} \ln a_1$. Then, at $m = m'$,

$$\ln \gamma_2 = \ln \frac{a'_2}{m'} = \phi' - 1 + \int_0^{m'} \frac{\phi - 1}{m} \, dm \qquad (20\text{-}34)$$

Figure 20-7 illustrates the use of Eq. (20-34) with the isopiestic data of Scatchard, Hamer, and Wood[1] on sucrose. The integral in Eq. (20-34) is given by the area under the curve from $m = 0$ out to the final molality m'.

Since the denominator of the integrand $(\phi - 1)/m$ or $(\phi - 1)/r$ is proportional to the concentration, the effect of experimental errors is greatly increased in the dilute region. The measurements on sucrose are very precise; yet the error becomes very noticeable at molalities below 0.5.

We saw in Fig. 20-6 that the mercury vapor-pressure data at 325°C sufficed to give reasonably accurate activity coefficients for thallium in the amalgams provided that the solvent standard state was selected for thallium as well as for mercury. If we were to use the solute standard state instead for thallium, the resulting plot for the integration of Eq. (20-30) is shown in Fig. 20-8. It is apparent that the data in the region of r less than 0.2 are not accurate enough to yield more than a very crude estimate of the required area.

An Analytical Method. If the activity of one component of a solution is expressed with satisfactory accuracy by an empirical equation, then Eq. (20-26) or (20-26a) may be integrated analytically to yield the activity coefficient of the other component. We continue to use the thallium amalgams at 325°C as an example and take the empirical equation selected by Hildebrand and Eastman to represent their measurements of the vapor pressure of mercury. With the assump-

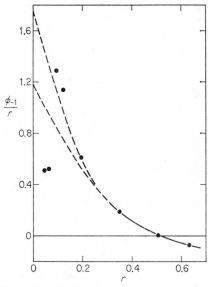

Fig. 20-8. A plot for the integral of Eq. (20-30) of the data on the vapor pressure of mercury in thallium amalgams at 325°C. The difficulty of extrapolating to the solute standard state is apparent.

[1] G. Scatchard, W. J. Hamer, and S. E. Wood, *J. Am. Chem. Soc.*, **60**, 3061 (1938).

tion that the fugacity of mercury is equal to the pressure, which is justified by equation-of-state data for mercury, one obtains

$$\log \gamma_1 = \log \frac{p_1}{p_1^\circ} - \log x_1 = - \frac{0.0960}{(1 + 0.263r)^2} \qquad (20\text{-}35)$$

where $r = x_1/x_2$ (note that in the preceding section r was defined with the subscripts interchanged). Substitution of Eq. (20-35) in Eq. (20-26a) and integration[1] yields

$$\log \gamma_2'' - \log \gamma_2' = \frac{0.960}{0.263} \left[\frac{2}{1 + 0.263r''} - \frac{1}{(1 + 0.263r'')^2} \right.$$
$$\left. - \frac{2}{1 + 0.263r'} + \frac{1}{(1 + 0.263r')^2} \right] \qquad (20\text{-}36)$$

Again we have the choice of our reference state with $r' = \infty$ for the solute definition or $r' = 0$ for the solvent definition (and in either case $\log \gamma_2' = 0$). The results are given in Table 20-4.

TABLE 20-4. ACTIVITY COEFFICIENTS OF MERCURY AND THALLIUM IN AMALGAMS
AT 325°C FROM THE EMPIRICAL EQUATION

x_2	γ_1	Reference state	
		$x_2 = 0$	$x_2 = 1$
		γ_2	γ_2
0.00	(1.00)	(1.00)	0.43
0.10	0.98	1.53	0.660
0.20	0.95	1.86	0.803
0.30	0.92	2.05	0.886
0.40	0.89	2.17	0.935
0.50	0.87	2.23	0.964
0.60	0.85	2.28	0.981
0.70	0.83	2.30	0.991
0.80	0.82	2.31	0.996
0.90	0.81	2.32	0.999
1.00	0.80	2.32	(1.000)

The analytical method gives no insight as to the accuracy of the results, and if we had only used this method, we would have no reason to favor one set of values of γ_2 in Table 20-4 over the other. From our use of graphical

---

[1] The fundamental equation for the integration, as given in various tables of integrals, is

$$\int \frac{x\, dx}{(1 + cx)^3} = \frac{1}{c^2} \left[\frac{-1}{1 + cx} + \frac{1}{2(1 + cx)^2} \right]$$

methods, however, we know that the γ_2 values based upon the solute reference state are subject to a much greater uncertainty.   This uncertainty is associated with the integral in the very dilute region where r approaches infinity; hence the error would change all γ_2 values based on the solute reference state by the same factor.   Thus the ratio γ_2''/γ_2' for a pair of solutions of finite composition is unaffected by the choice of reference state.

THE EQUILIBRIUM CONSTANT

When the free-energy change in a chemical reaction is measured, the substances involved are often in states which are chosen by the experimenter either arbitrarily or to meet the demands of the experimental conditions.   One substance may be in the gaseous state at high pressure, or another may be a constituent of some concentrated solution.   The change of free energy as measured would be of little service in further calculations if we could not obtain from it the free-energy change of the same reaction when each substance taking part in it is in some simple state or, better, in its standard state.

When Lewis and Burrows[1] determined the first free energy of formation of an organic substance, urea, it proved to be a simple matter to obtain the change in free energy attending the conversion of ammonium carbonate into urea and water, in a very concentrated aqueous solution of ammonium carbonate.   The great difficulties of the investigation came in determining for each substance the difference between the free energy in that concentrated solution and in some standard state.

If $\bar{\text{F}}$ is the partial molal free energy of a substance in any state, and F° that in the standard state, the difference, according to Eq. (20-2), is given as

$$\bar{\text{F}} - \text{F}^\circ = RT \ln a$$

The problem of converting free energies in various states into free energies in standard states is therefore the problem of determining the activities of the various substances concerned.

Let us consider a reaction such as

$$3Ag_2O(s) + 2NH_3(aq, 2.0 \text{ M})$$
$$= 6Ag(s) + 3H_2O(l, \text{ with } NH_3 \text{ 2.0 M}) + N_2(g, 100 \text{ atm})$$

and let us assume that ΔF has been determined for the reaction as written, i.e., for the reaction in which 3 moles of solid silver oxide disappear, together with 2 moles of ammonia from a 2 M aqueous solution, while

[1] G. N. Lewis and G. H. Burrows, *J. Am. Chem. Soc.*, **34**, 1515 (1912).

6 moles of solid silver, 3 moles of water (in the same ammonia solution), and 1 mole of nitrogen gas at 100 atm are formed.[1]

Knowing the value of ΔF, we could also find the change in free energy in the same reaction, with every substance in its standard state, if we knew $\bar{\mathrm{F}} - \mathrm{F}^{\circ}$ for each substance. The silver oxide and the silver are already in their standard states, and $\bar{\mathrm{F}} - \mathrm{F}^{\circ} = 0$. By investigating the vapor pressure of ammonia solutions, or by one of the other methods described in previous chapters, we could determine the activity of the ammonia and write $\mathrm{F} - \mathrm{F}^{\circ} = RT \ln a$. So also the activity of the water is not that of pure water but is several per cent lower owing to the presence of the ammonia, a difference which can readily be determined. The activity of the nitrogen could be calculated roughly by assuming it to be a perfect gas or more accurately by the methods of Chapter 16. We are then in a position to determine ΔF°, that is, the change of free energy for the reaction when each substance is in its standard state. The reaction is then written

$$3Ag_2O(s) + 2NH_3(aq) = 6Ag(s) + 3H_2O(l) + N_2(g)$$

which indicates, in the absence of any further specification, that each substance is at unit activity.

In Chapter 15, the equilibrium constant was introduced in Eqs. (15-31) and (15-32) in terms of the fugacities of the reactants and products of a reaction. We can now express the equilibrium constant in terms of activities and activity coefficients, following the same steps as given in Eqs. (15-26) to (15-29) for a general reaction of the type

$$lL + mM + \cdots = qQ + rR + \cdots$$

Let ΔF be the free-energy change in this reaction when the substances are in any given states,

$$\Delta F = (q\mathrm{F}_Q + r\mathrm{F}_R + \cdots) - (l\mathrm{F}_L + m\mathrm{F}_M + \cdots)$$

and let ΔF° be the free energy change when each substance is in its standard state,

$$\Delta F^{\circ} = (q\mathrm{F}_Q^{\circ} + r\mathrm{F}_R^{\circ} + \cdots) - (l\mathrm{F}_L^{\circ} + m\mathrm{F}_M^{\circ} + \cdots)$$

If a_L, a_M, etc., represent the activities in the nonstandard states, then we have

$$l(\mathrm{F}_L - \mathrm{F}_L^{\circ}) = RT \ln a_L^l \tag{20-37}$$

and so on.

---

[1] All the substances except nitrogen are taken at atmospheric pressure. If we had taken them all at 100 atm, the difference in free energy thus produced in these liquid and solid phases could ordinarily be neglected or, if the precision of the work should warrant, the difference could be calculated by Eq. (20-3).

Combining the several equations, we find

$$\Delta F - \Delta F^\circ = RT \ln \frac{a_Q^q a_R^r \,\cdots}{a_L^l a_M^m \,\cdots} \qquad (20\text{-}38)$$

The important quotient appearing in the last term we may call the activity quotient of the reaction. By determining this quotient we may calculate ΔF° when ΔF is known, and conversely.

Whenever we meet with a case of chemical equilibrium, we immediately acquire important information regarding the free-energy change in the reaction concerned. For in such a reaction, by Eq. (13-4),

$$\Delta F = 0$$

Hence, for equilibrium, by Eq. (20-38)

$$\Delta F^\circ = -RT \ln \frac{a_Q^q a_R^r \,\cdots}{a_L^l a_M^m \,\cdots} \qquad (20\text{-}39)$$

At a given temperature ΔF° is a constant, and therefore the condition of equilibrium is that the activity quotient shall also be constant. The value of this quotient, when the system is in equilibrium, we shall call the equilibrium constant K and write

$$\Delta F^\circ = -RT \ln K \qquad (20\text{-}40)$$

Some of the substances taking part in the reaction, such as pure liquids or solids at approximately 1 atm pressure, have unit activity at a given temperature. It is frequently the custom to retain in the equilibrium constant only the activities of gases and of substances in solution, where the activity varies markedly with the pressure or the composition.

Let us consider a case in which the substances of variable activity are constituents of a perfect solution. Thus in a homogeneous mixture of methyl bromide, ethyl chloride, ethyl bromide, and methyl chloride we might, from our knowledge of these substances, expect the activity of each substance to be approximately equal to its mole fraction. Hence for the reaction

$$CH_3Cl + C_2H_5Br = CH_3Br + C_2H_5Cl$$

we should expect an equilibrium to be attained which would satisfy the condition

$$K = \frac{x_{CH_3Br} x_{C_2H_5Cl}}{x_{CH_3Cl} x_{C_2H_5Br}} \qquad (20\text{-}41)^*$$

The great majority of equilibrium measurements have been made in systems involving gases at moderate pressures and dilute aqueous solutions. In such cases it is frequently possible, with adequate accuracy, to replace the activities of the gases by their pressures and of the solutes by their molalities.   We then have the *mass law* in the form which we owe to Guldberg and Waage[1] and to van't Hoff,[2] a generalization which will always be esteemed as one of the milestones in the progress of chemistry toward an exact science.

After writing the equation for a given reaction it will be our convention to write the equilibrium constant with the activities of the substances produced in the numerator and of the substances consumed in the denominator.

It is sometimes convenient to denote the activity of a gas by its formula in brackets and the activity of any constituent of an aqueous solution by its formula in parentheses.   To illustrate these conventions, we write

$$\tfrac{1}{2}N_2(g) + \tfrac{3}{2}H_2(g) = NH_3(g) \qquad\qquad K = \frac{[NH_3]}{[N_2]^{\frac{1}{2}}[H_2]^{\frac{3}{2}}}$$

$$N_2(g) + 3H_2(g) = 2NH_3(g) \qquad\qquad K = \frac{[NH_3]^2}{[N_2][H_2]^3}$$

$$NH_3(g) = \tfrac{1}{2}N_2(g) + \tfrac{3}{2}H_2(g) \qquad K = \frac{[N_2]^{\frac{1}{2}}[H_2]^{\frac{3}{2}}}{[NH_3]}$$

$$NH_3(aq) = NH_3(g) \qquad\qquad K = \frac{[NH_3]}{(NH_3)}$$

If in these equations we substitute for the activities the partial pressures or molalities, the quotients are not strictly constant at finite pressures and concentrations but approach the true equilibrium constant (the quotient of activities) as the gases approach zero pressure, and the concentrations of the solutes approach zero.

It is frequently desirable to express the activity of components of solutions in terms of activity coefficients.   According to Eq. (20-7)

$$a_i = x_i\gamma_i\Gamma_i \qquad \text{or} \qquad a_i = m_i\gamma_i\Gamma_i$$

The factor Γ_i is also appropriate for the activity of a pure solid or liquid at a pressure other than 1 atm.   Since the effect of pressure on gases is very large, it is usually best to consider their fugacities directly.   If desired, one may use a fugacity coefficient f_i/p_i to express gas imperfection and then $f_i = p_i(f_i/p_i)$.   We may illustrate these methods with the reaction

$$2H_2(g) + CH_3COOH(x_1) = C_2H_5OH(x_2) + H_2O(x_3)$$

where the acetic acid, ethanol, and water form a liquid solution of the composition indicated.   Combination of the values of free energy of

[1] C. M. Guldberg and P. Waage, "Etudes sur les affinités chimiques," Brögger and Christie, Christiania, 1867.

[2] J. H. van't Hoff, *Z. physik. Chem.*, **1**, 481 (1887).

formation[1] yields $\Delta F° = -4.7$ kcal at $298.15°$K, and

$$K = 2.8 \times 10^3 = \frac{a_{C_2H_5OH}a_{H_2O}}{a_{H_2}^2 a_{CH_3COOH}}$$

But $a_{CH_3COOH} = \Gamma_1\gamma_1 x_1$, and similar expressions hold for ethanol with subscript 2 and water with subscript 3; hence

$$K = \frac{\Gamma_2\gamma_2 x_2\Gamma_3\gamma_3 x_3}{(f/p)_{H_2}^2 p_{H_2}^2 \Gamma_1\gamma_1 x_1}$$

This expression may be rearranged to

$$K = \frac{x_2 x_3}{p_{H_2}^2 x_1} \frac{\gamma_2\gamma_3}{(f/p)_{H_2}^2\gamma_1} \frac{\Gamma_2\Gamma_3}{\Gamma_1}$$

where the first factor on the right-hand side is the usual approximate equilibrium quotient for the ideal-gas and ideal-solution case and the second quotient contains the correcting factors to the real state at low pressure, i.e., the activity coefficients and f/p for the gas. It is customary to ignore the third factor $\Gamma_2\Gamma_3/\Gamma_1$ unless the pressure is high; let us now check the magnitude of this factor. If we assume that the liquid volumes remain constant, we obtain

$$\ln \frac{\Gamma_2\Gamma_3}{\Gamma_1} = \frac{(v_2° + v_3° - v_1°)(P - 1)}{RT}$$

and substitution of numerical values yields

$$\ln \frac{\Gamma_2\Gamma_3}{\Gamma_1} = 0.79 \times 10^{-3}(P - 1)$$

if P is in atmospheres. It is evident that a pressure of a few atmospheres has little effect, but the Γ factor becomes very significant for pressures in the 1000-atm range.

CHOICE OF STANDARD STATE WHEN ASSOCIATION OR DISSOCIATION OCCURS

The previous examples have been ones where the choice of components has been rather obvious. In connection with the discussion of the distribution of bromine between carbon tetrachloride and water, we have noted that the choice of the molecular weight of bromine has a degree of arbitrariness but that there is a great advantage to choosing molecular weights corresponding to the major constituents in the solution to ensure that Henry's law and Raoult's law will be approached in the limit of low

[1] F. D. Rossini et al., *Selected Values of Chemical Thermodynamic Properties*, *Natl. Bur. Standards (U.S.) Cir.* 500, 1952.

concentration of solute.   In some instances, there may be a number of major constituents corresponding to a given component, and the most suitable choice of components may not be immediately obvious.   A number of examples will be discussed to illustrate the procedure for choosing components and corresponding standard states.

The Activity of a Solute Which Forms Compounds with the Solvent.   We frequently have to deal with solutions in which, for one reason or another, it is assumed that the solute forms compounds with the solvent.   These are known in general as *solvates*.   Thus many substances dissolved in water are assumed to form hydrates, with one or more molecules of water combined with each molecule of solute.   This assumption is sometimes mere hypothesis, but often it rests upon very substantial evidence.   Until the present time, however, it has been difficult to determine with any degree of certainty the relative amounts of unhydrated substance and of the various possible hydrates.   How, then, are we to treat such compounds in our thermodynamic work?

The simplest method of disposing of this question would be to ignore the existence of such hydrates, and this would be entirely justifiable, since thermodynamics is not compelled to take cognizance of the various molecular species which may exist in a system, particularly when the existence of such species cannot be absolutely demonstrated.   Nevertheless, it will be frequently more convenient, as well as more consistent with chemical usage, to include these hydrates in our consideration, especially since by a simple device we may do this without really complicating our procedure.   We may illustrate this device by a concrete case.

When ammonia dissolves in water, it is supposed to form, although in unknown amount, at least one hydrate, the monohydrate, which may be written $NH_3 \cdot H_2O$ or NH_4OH.   In other words, we assume the reaction

$$NH_3(aq) + H_2O(l) = NH_4OH(aq)$$

Now, if (NH_3), (H_2O) and (NH_4OH) represent the several activities,

$$\frac{(NH_4OH)}{(NH_3)(H_2O)} = K$$

At high dilution the activity of the water is constant and equal to unity; therefore the ratio of (NH_4OH) to (NH_3) is constant, and the two molalities are also approximately proportional to one another.   But we do not know any of these quantities separately; we only know the gross or stoichiometric molality m, as determined, for example, by the number of moles of gaseous ammonia which have been dissolved in 1 kg of water.   In very dilute solution this is the same as the number of moles of NH_4OH which would be dissolved in 1 kg of water to produce the same concentration.

In ignorance of the individual concentrations, we may arbitrarily take the standard state of each substance in such manner that at infinite dilution its activity is equal to the gross molality m. Or, in other words we assume such standard states as to make $K = 1$. The activities of the two substances then remain equal as long as $(H_2O) = 1$ (but in concentrated solutions their ratio is the activity of the water).

Such a definition introduces an equal simplicity into the free-energy equation, since it makes $\Delta F° = 0$. The same method may be employed in all similar cases, although it is possible that this method might be abandoned if we ever should succeed in determining quantitatively the actual concentrations of the individual species. If there is no reason to know the amounts of the individual species and as long as equilibrium is maintained among these species, the above procedure is the simplest for treating these solutions.

Dissociating Solutes. One often encounters solutes that dissociate to simple species to an extent which depends upon the temperature, the concentration, and the nature of the solvent. If either the parent mole-

TABLE 20-5. DISSOCIATION AT 0°C OF N_2O_4, DISSOLVED IN CHLOROFORM

x_1	x_2''	x_2'	$\dfrac{(x_2')^2}{x_2''} \times 10^3$
0.00	1.00	0.00094	88
0.27	0.73	0.00080	87
0.46	0.54	0.00067	83
0.70	0.30	0.00045	67
0.875	0.125	0.00029	66
0.934	0.066	0.00019	52
0.950	0.050	0.00015	43
0.963	0.037	0.00012	35
0.982	0.018	0.00010	49

cule or one of the products of dissociation is colored and the other colorless, one can readily determine the degree of dissociation colorimetrically. Even when both are colored, one can use a spectrophotometer and observe each species at the wavelength of its characteristic absorption band. By a simple colorimetric method, Cundall[1] made an exhaustive investigation of the dissociation of N_2O_4 in various solvents, and especially in chloroform. Since chloroform, NO_2, and N_2O_4 might be expected to form nearly perfect solutions, it will be interesting to see how well Cundall's results can be represented by Eq. (20-39) with $a_i = x_i$.

We have calculated from his data at 0°C the various figures needed for this comparison, and in Table 20-5 we give in the several columns: x_1, the

[1] J. T. Cundall, *J. Chem. Soc.*, **59**, 1076 (1891); **67**, 794 (1895).

mole fraction of chloroform; x_2'', the mole fraction of N_2O_4; x_2', the mole fraction of NO_2; and the ratio $(x_2')^2/x_2''$, which should be constant according to Eq. (20-39). We see that it changes only twofold in passing from the liquid containing no chloroform to the dilute solution in chloroform; and in the concentrated solution, when the mole fraction of chloroform rises from 0 to 0.5, the ratio changes by only 6 per cent. The small variation from constancy which exists is evidently due to the fact that, over this wide range of composition, the activities of NO_2 and N_2O_4 are not quite proportional to their mole fractions.

We have chosen this illustration partly also for the sake of discussing another point which without explanation might prove troublesome. In the table it will be noted that the amounts of NO_2 are so small that x_1 and x_2'' are together practically equal to unity. At higher temperatures this is no longer the case. We might, for example, find a mixture containing 80 moles of $CHCl_3$, 10 moles of NO_2 and 10 moles of N_2O_4, and we might write: (1) $x_1 = 0.80$, $x_2' = 0.10$, and $x_2'' = 0.10$. It must be emphasized, however, that this procedure is in some ways arbitrary and due to our foreknowledge of the dissociation of N_2O_4. We would have obtained a mixture of the same composition if we had taken 15 moles of pure N_2O_4 and 80 moles of $CHCl_3$, in which case, if we followed the normal usage and paid no attention to the possible dissociation, we would have written: (2) $x_1 = 80/(80 + 15)$ and $x_2 = 15/(80 + 15)$. Again, we might have reached the same composition had we used 30 moles of NO_2 and 80 moles of $CHCl_3$. In this case, paying no attention to the possible association of the NO_2, we would have written: (3) for $CHCl_3$, $x_1 = 80/(80 + 30)$ and, for NO_2, $x_2 = 30/(80 + 30)$.

Suppose that we were ignorant of this phenomenon of dissociation and were to study the activities of the solvent and the solute in solutions of N_2O_4 in chloroform at $0°C$, by any of the methods employed in the preceding chapters. We should of course reckon the mole fractions by method (2) suggested above; and plotting a_2, the activity of N_2O_4, against its mole fraction as in Fig. 20-9, we should undoubtedly find in moderately dilute solutions that a_2 would be roughly proportional to x_2 and might easily be confused with the dashed line "no dissociation." But careful measurements in the very dilute region would yield a curve, such as the solid line in Fig. 20-9, which becomes proportional to x_2^2 at extreme dilution. Thus, at $x_2 = 0$, $da_2/dx_2 = 0$. Likewise a_1 measured at moderate concentration would be found to be near the value calculated from Raoult's law for no dissociation and the difference might be ascribed to experimental error. But at extreme dilution the curve for a_1 approaches the dashed line of slope double the Raoult's-law slope, which is marked "complete dissociation" in Fig. 20-9.

This seems to be, and is in fact, a direct contradiction of one of the fundamental laws of the infinitely dilute solution (obtained in Chapter 19), but it will be recalled that the whole development of that chapter was based on the assumption that df_2/dx_2 is finite on approaching infinite dilution where $x_2 = 0$. This also means that da_2/dx_2 is finite.[1] However, as we pointed out at that time, there is one exception to this rule, namely, in the case where the solute dissociates.

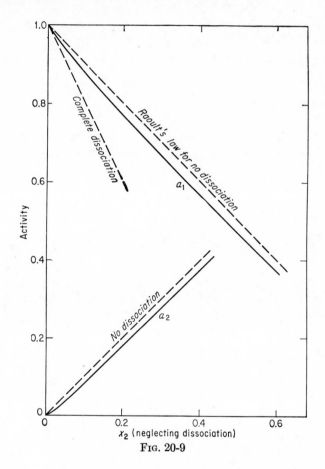

Fig. 20-9

Fortunately, in such cases we may eliminate these exceptions to the laws of the dilute solution by considering as solute, not the undissociated substance, but the products of dissociation. Thus, if in place of Fig. 20-9 we had plotted the activity of $CHCl_3$ and of NO_2 against the mole fraction of NO_2, as in the above method (3), we should obtain in extremely

---

[1] At least, if we do not make so absurd a choice of a standard state as to make the ratio of a to f zero or infinite.

dilute solutions such a plot as that of Fig. 20-10, where it is evident that both Henry's law and Raoult's law are obeyed at first, although even in rather dilute solutions marked deviations from these laws would appear, which now would be attributed to association of NO_2.

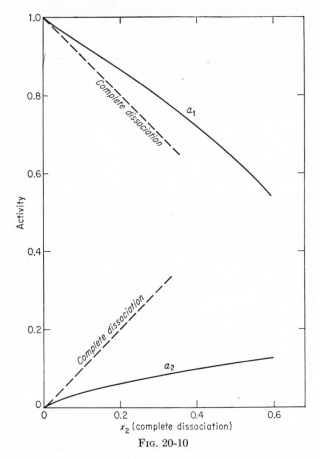

F<small>IG</small>. 20-10

We have entered at some length into this discussion of the thermodynamic properties of a solution in which dissociation occurs, partly because of its intrinsic importance in thermodynamic theory, and partly also because this discussion furnishes a simple introduction to the study of solutions of electrolytes.

The above discussion of the NO_2-N_2O_4 solution in chloroform solution is directly applicable to gases which dissociate or polymerize, and, in fact, both NO_2 and N_2O_4 molecules can be observed in gaseous NO_2, and the previous discussion of variation of fugacities in terms of mole fractions can be restated for gaseous NO_2 in terms of the variation of fugacity with pressure or concentration. One need not consider dissociation of N_2O_4 to $2NO_2$ but if one does not, then the P-V-T behavior of the gas is most abnormal. Thus it is usually preferable to deal with a dissociating gas as a mixture of two separate molecular species.

Before leaving the problem of dissociation, however, we wish to emphasize that there is an inherent arbitrariness in the dissociation treatment. Thus one cannot distinguish between gas imperfection of NO_2 and association to N_2O_4—either can explain a decrease of volume below that calculated from the perfect-gas law for NO_2. Only by assuming a hypothetical behavior for unassociated NO_2 is it possible to calculate the amount of N_2O_4. If the gas density is low, it is reasonable to assume the perfect-gas law and an association treatment is must useful. However, at high densities one would expect substantial gas imperfection even without association. Under these conditions an association treatment becomes quite arbitrary, and it is frequently more useful to apply the virial-coefficient treatment [Eq. (16-1) or (16-2)], with the realization that abnormal values will be found for the virial coefficients.

It is beyond the scope of our discussion to consider fully the statistical methods of dealing with a dissociating gas, but we wish to remark that the same ambiguity arises with respect to the calculation of properties of the dimeric species and the gas imperfection of the monomer. The properties may be calculated statistically by the virial-coefficient method if the potential energy of molecular interaction is known. It makes no difference, in principle, whether the interaction is only that of van der Waals forces or is that of a chemical bond. It is also possible to use the dissociation treatment if consistent, but arbitrary, definitions are adopted for the precise dividing boundary between a dimeric molecule and two monomers.

TREATMENT OF INFINITE IONIC OR ATOMIC CONDENSED PHASES WITH NO CHARACTERISTIC MOLECULAR UNITS

In many systems such as the bromine–carbon tetrachloride system one can from available chemical information immediately write down the most logical formulas to be used as components. In systems such as the ferrous oxide system, the titanium-oxygen system, the copper-sulfur system, and other such systems where there are no discrete molecular units larger than the individual atoms or ions, the choice of components is not particularly obvious. For example, in treating the ferrous oxide system, which extends in composition from $FeO_{1.05}$ to $FeO_{1.19}$, one could use as the components iron and oxygen, or FeO and oxygen, or iron and Fe_3O_4, or quite a variety of components. The component FeO would seem to be one of the first choices, but it is at a disadvantage since the composition FeO itself is not thermodynamically stable, and thus one cannot obtain data for it. Also the variation of partial molal free energy of iron and oxygen is so complex that it is difficult to extrapolate to a

FeO composition.   Under such circumstances, the choice of components and standard states can be quite arbitrary.   We have noted in Chapter 2 that Darken and Gurry[1] chose $FeO_{1.0477}$ as a component when dealing with solid ferrous oxide phases.   For the liquid phase, they chose an arbitrary composition $FeO_{1.07}$ for the standard state, and their liquid system was treated as a solution of components $FeO_{1.07}$ and oxygen.   Under such circumstances, one is at a disadvantage since neither Henry's law nor Raoult's law will hold, and one cannot use any of the solute standard states that we have set up.   However, it is quite possible to carry out accurate thermodynamic calculations under such circumstances by designating any arbitrary standard state and referring partial molal free energies, activities, and activity coefficients to this arbitrary standard state, which would normally be taken at some composition in the stable-solution range.   Thus one can readily obtain the free-energy difference in going from the standard state to any other state.

PROBLEMS

20-1. Calculate the activity of liquid water at 25°C and at 100 atm pressure, assuming it to be incompressible.

20-2. What is the activity of water at 100°C in a solution from which the vapor pressure of water is 700 mm Hg?

20-3. For the process $Hg(l) = Hg(in Tl amalg, x_2 = 0.60)$, calculate ΔF_{598}.

20-4. We may show that the activity of bromine in aqueous solution is nearly proportional to the molality, up to the saturated solution.   First show, from data given in this chapter, that, for the vapor pressure of bromine from a dilute aqueous solution at 25°C, $p = 1.45m$.   If this proportionality continued up to the point where $p = 0.280$, which is the vapor pressure of pure bromine, we could calculate the molality of a solution which would be in equilibrium with liquid bromine, or, in other words, the solubility of bromine.   Compare with the value thus obtained the solubility at 25°C, measured by Jakowkin,[2] namely, 34.0 g/1000 g of water.   The data of Lewis and Storch were obtained for slightly acidified aqueous solutions, whereas the measurements of Jakowkin were made with pure water.

20-5. The distribution of $HgCl_2$ between benzene and water at 25°C has been obtained by Linhart.[3]   He expresses concentration (moles per liter) in the two phases by c_B and c_W.   These are given in Table 20-6, together with the ratio of c_W to c_B at the various concentrations.   In these dilute solutions $c_W = m$, the molality.   Assuming the activity of $HgCl_2$ in benzene to be proportional to c_B, and taking the activity in the water equal to m in the most dilute solution, find the activity in the most concentrated of the aqueous solutions.

20-6. In Fig. 20-4, $(-\varepsilon/0.05816 - \log x_2)$ might equally well have been plotted against any function of x_2, instead of against x_2 itself.   If instead of x_2 we had used x_2/x_1, show that with such a plot we could have dispensed with Fig. 20-5 and could have obtained $\log \gamma_1$ without obtaining $\log \gamma_2$.

20-7. At 325°C, find, from Table 20-4, the emf of a concentration cell with two thallium amalgam electrodes, $x_2 = 0.40$, and $x_2 = 0.80$.

[1] L. S. Darken and R. W. Gurry, *J. Am. Chem. Soc.*, **67**, 1398 (1945); **68**, 798 (1946).

[2] *Z. physik. Chem.*, **20**, 19 (1896).

[3] *J. Am. Chem. Soc.*, **37**, 258 (1915).

TABLE 20-6. DISTRIBUTION OF MERCURIC CHLORIDE BETWEEN WATER AND BENZENE

c_B	c_W	m/c_B
0.0210	0.2866	13.65
0.0174	0.2326	13.38
0.01222	0.1578	12.91
0.00880	0.1112	12.64
0.00524	0.0648	12.35
0.000618	0.00738	11.95
0.000310	0.00369	11.90
0.000155	0.001845	11.90

20-8. From mixtures of acetone, mole fraction $= x_1$, and chloroform, mole fraction $= x_2$, J. von Zawidzki[1] obtained at 35.2°C the following vapor pressures p_1 of acetone (in millimeters of mercury):

x_1	1.000	0.9405	0.8783	0.8165	0.7103	0.5750	0.3378	0.1978	0.0823
p_1	344.5	322.9	299.7	275.8	230.7	173.7	79.1	38.0	13.4

Calculate γ_1, and by plotting as in Fig. 20-6 calculate, for the chloroform, γ_2 when $x_2 = 0.20$, 0.40, and 0.60. Use the solvent type of standard state for chloroform as well as for acetone. Try also to establish γ_2 on the basis of the solute standard state, and estimate the accuracy attainable with these data.

20-9. The activity of component 1 in a solution is given by the equation

$$R \ln a_1 = R \ln x_1 + A x_2{}^2 + B x_2{}^3$$

where x_1 and x_2 are the respective mole fractions and A and B are constants. The equation is valid over the full composition range from pure liquid 1 to pure liquid 2. Derive an expression for the activity of component 2.

20-10. At 60°C aniline and water mixtures form two liquids of composition 4.4 and 93.4 per cent aniline by weight. Assume that Raoult's law holds for the abundant component and Henry's law holds for the dilute component in each phase. Calculate for each phase the activity coefficient of the dilute component on the basis of a solvent standard state. What is the difference in free energy between the solute and solvent standard state for each component?

20-11. From values in Appendix 7 calculate the equilibrium constant for the reaction n-pentane = 2-methyl butane (isopentane) in the gas phase at 298°K. Calculate also the equilibrium constant for the liquid phase. The vapor pressures of the two pentanes, in millimeters of mercury, are given by the equations

n-Pentane: $\qquad\qquad \log_{10} p = 6.8522 - \dfrac{1064.6}{T - 41.1}$

Isopentane: $\qquad\qquad \log_{10} p = 6.7897 - \dfrac{1020.0}{T - 40.0}$

What is the composition of the equilibrium liquid at 1 atm and 298°K if an ideal solution is assumed? The liquid densities are 0.621 for n-pentane and 0.615 for isopentane. Calculate the values of Γ for each component at 100 atm and the equilibrium composition at 100 atm and 298°K.

[1] *Z. physik. Chem.*, **35**, 129 (1900).

21

NONELECTROLYTE SOLUTIONS

We have now presented the purely thermodynamic principles which relate to solutions and have given examples of their application. Beyond these principles the real basis of knowledge of solution properties is direct experimental observation, and our presentation will be based principally upon generalization from direct experimental studies. However, one can predict solution properties from statistics for assumed simple models. Such simple models seldom constitute accurate representations of real solutions, but in favorable cases the approximation is useful. Also these model calculations generate forms of equations which are useful even though the predicted numerical constants may be in error.

In this chapter we shall consider all solutions except electrolytes. The special long-range nature of electrostatic forces yields special problems which are best considered separately in Chapters 22 and 23. A complete treatment of even nonelectrolyte solutions would require a book larger than this one. We present here only a few of the results which seem well established and most useful.

Primary consideration will be given to liquid solutions since these are most important. In most cases the same equations need only an appropriate interpretation for use in the gas or solid phase.

STATISTICAL TREATMENT OF IDEAL SOLUTIONS

In Chapter 18 the properties of ideal solutions were considered on the basis of the simple postulate of Raoult's law. We may now give a statistical calculation based on a molecular model which gives further insight into the nature of an ideal solution. Assume that there are N_1 and N_2 molecules of the two types, respectively, and that they are completely interchangeable without affecting the internal energy. Thus we may take the positions of all these molecules at some instant as constituting an array of $N_1 + N_2$ sites. We calculate the multiplicity of a random distribution. The first molecule can be placed on any of the $N_1 + N_2$ sites, the second on any of the $N_1 + N_2 - 1$ empty sites, the third on any of the $N_1 + N_2 - 2$ empty sites, and so on. The total number of possibilities is $(N_1 + N_2)(N_1 + N_2 - 1) \cdots = (N_1 + N_2)!$. However, the molecules of the same kind are not distinguishable from one another; consequently we must divide by $N_1!$, which is the number of possible interchanges of molecules of the first kind, and by $N_2!$ for the interchanges of molecules of the second kind. Thus the number of distinguishable arrangements of the $N_1 + N_2$ molecules is

$$W_{\text{mixed}} = \frac{(N_1 + N_2)!}{N_1!N_2!} \tag{21-1}*$$

For the pure components before mixing one has the trivial values

$$W_1 = \frac{N_1!}{N_1!} = 1 \qquad W_2 = \frac{N_2!}{N_2!} = 1$$

The entropy of mixing is given by the multiplicity ratio

$$\Delta S_{\text{mix}} = k \ln \frac{W_{\text{mix}}}{W_1 W_2} = k \ln \frac{(N_1 + N_2)!}{N_1!N_2!} \tag{21-2}*$$

Since these numbers are large, we may use Stirling's approximation[1] for the factorials

$$\ln N! = N \ln N - N \tag{21-3}*$$

Then we find

$$\begin{aligned} \Delta S_{\text{mix}} &= k[(N_1 + N_2) \ln (N_1 + N_2) - N_1 \ln N_1 - N_2 \ln N_2] \\ &= -k \left(N_1 \ln \frac{N_1}{N_1 + N_2} + N_2 \ln \frac{N_2}{N_1 + N_2} \right) \\ &= -k(N_1 \ln x_1 + N_2 \ln x_2) \end{aligned} \tag{21-4}*$$

where the mole fractions x_1 and x_2 are introduced in the final step. This is the ideal, or Raoult's-law, entropy of mixing. In terms of the numbers of moles n_1 and n_2, the result is

$$\Delta S_{\text{mix}} = -R(n_1 \ln x_1 + n_2 \ln x_2) \tag{21-5}*$$

and the partial molal entropies are

$$\begin{aligned} \bar{s}_1 - s_1^\circ &= -R \ln x_1 \\ \bar{s}_2 - s_2^\circ &= -R \ln x_2 \end{aligned} \tag{21-6}*$$

These results were obtained in Eqs. (18-6)* and (18-8)* in our initial discussion of ideal solutions.

This derivation is equally appropriate to liquid or solid solutions. It is also easy to derive Raoult's law for mixtures of perfect gases or, with somewhat greater complexity, to make the derivation for mixtures of real

---

[1] In case the reader is unfamiliar with Stirling's formula, he may note that $\ln N! \cong \int_1^N \ln y \, dy = [y \ln y - y]_1^N = N \ln N - N + 1$. If N is large, the 1 may be neglected.

gases with appropriately equal imperfection properties. In all cases except the perfect gas the assumptions imply that the molecules have the same size and that the intermolecular forces between pairs of like molecules of each type, as well as between unlike molecules, are all the same. These are very stringent conditions. Molecules which differ only by isotopic substitution[1] provide the examples which fully conform to this model, and indeed they form ideal solutions.

Actually, if we allow even 1 or 2 per cent deviation, we find that a considerably broader range of molecular properties still yields ideal behavior. Examples include[2] (1) benzene and toluene, (2) 1,2-dibromoethane and 1,2-dibromopropane and (3) methyl iodide and chloroform.

From the point of view of statistical theory, it is not clear at present just why these systems with greater differences in molecular properties still yield ideal solutions. The toluene molecule is substantially larger than that of benzene; consequently they cannot be regarded as interchangeable. It seems likely that small differences in molecular size and small differences in intermolecular force yield opposite deviations from Raoult's law and that there is a cancellation of opposing effects in some cases.

The nature of these nearly ideal solutions will be considered further after equations have been presented for nonideal systems.

REGULAR SOLUTIONS

In 1929 Hildebrand called attention to the very great similarity in the behavior of a class of nonideal solutions which he named *regular solutions*. Such solutions are characterized by the absence of any specific interaction between molecules such as hydrogen bonding, acid-base association, etc. The pure components usually show the properties we associate with normal liquids or fluids (see Appendix 1), although other classes of substance such as metals may yield regular solutions also. These criteria are necessarily somewhat vague because there are no sharp boundaries to this category of solutions. Nevertheless, the classification has proved to be very useful.

Regular solutions differ from ideal solutions in that the intermolecular forces are no longer equal. Also the molecules may be more unequal in size. However, these differences are sufficiently moderate so that thermal energy still yields very nearly random mixing. Consequently the entropy of mixing has approximately the Raoult's-law value, although

[1] The quantum-mechanical effects of change in mass of light isotopes may even exclude some examples of this type.

[2] J. H. Hildebrand and R. L. Scott, "The Solubility of Nonelectrolytes," 3d ed., p. 210, Reinhold Publishing Corporation, New York, 1950.

it is influenced by the volume change on mixing, which is not ordinarily zero.

Equal-size Molecules. Let us consider first the subcategory of regular solutions where the components have molecules of exactly the same size. There have been many statistical studies[1] of this system, all of which yield as leading terms the equation (for two components),

$$\Delta F_{\text{mix}} = RT(n_1 \ln x_1 + n_2 \ln x_2) + (n_1 + n_2)x_1 x_2 w \qquad (21\text{-}7)*$$

Here w is the characteristic parameter which measures the deviation from Raoult's law. The excess free energy of mixing is defined as the deviation from Raoult's law and is consequently just the second term in Eq. (21-7)*,

$$\Delta F_{\text{mix}}^{E} = (n_1 + n_2)x_1 x_2 w \qquad (21\text{-}8)*$$

In the molecular models w is related to the difference in the energy of attraction of unlike molecules as compared with the mean of the energies for pairs of like molecules. As an illustration, let us assume the molecules to be located on a lattice of $N_a + N_b$ sites which has a coordination number of Z. If the components are segregated, then there are $\frac{1}{2}ZN_a$ nearest-neighbor interactions of molecules of component a and $\frac{1}{2}ZN_b$ interactions for component b. Random mixing yields a probability equal to the mole fraction $x_a = N_a/(N_a + N_b)$ that any site contains a molecule of a and a probability x_b for a molecule b. The probabilities of nearest-neighbor interactions of the type a-a, a-b, and b-b are therefore x_a^2, $2x_a x_b$, and x_b^2, respectively, where the factor 2 arises because we are counting both a-b and b-a interactions.

Now let us assume that the interaction energy is ϵ_{aa} when two a molecules are adjacent and similarly ϵ_{ab} for an a-b pair and ϵ_{bb} for two b molecules. Then the energy of the random mixtures is

$$
\begin{aligned}
E(\text{mixed}) &= \tfrac{1}{2}Z(N_a + N_b)(\epsilon_{aa}x_a^2 + 2\epsilon_{ab}x_a x_b + \epsilon_{bb}x_b^2) \\
&= \tfrac{1}{2}Z(N_a + N_b)^{-1}(\epsilon_{aa}N_a^2 + 2\epsilon_{ab}N_a N_b + \epsilon_{bb}N_b^2) \qquad (21\text{-}9)*
\end{aligned}
$$

The energy of the unmixed components is

$$E_a + E_b = \tfrac{1}{2}Z(N_a \epsilon_{aa} + N_b \epsilon_{bb}) \qquad (21\text{-}10)*$$

and the change in energy of mixing is

$$
\begin{aligned}
\Delta E_{\text{mix}} &= \tfrac{1}{2}Z(N_a + N_b)^{-1}N_a N_b(2\epsilon_{ab} - \epsilon_{aa} - \epsilon_{bb}) \\
&= (n_a + n_b)x_a x_b w \qquad\qquad\qquad\qquad (21\text{-}11)*
\end{aligned}
$$

where

$$w = \tfrac{1}{2}ZN_0(2\epsilon_{ab} - \epsilon_{aa} - \epsilon_{bb}) \qquad (21\text{-}12)*$$

[1] K. F. Herzfeld and W. Heitler, *Z. Elektrochem.*, **31**, 536 (1925); W. Heitler, *Ann. Physik*, (4)**80**, 630 (1926); see also E. A. Guggenheim, "Mixtures," chap. IV, Oxford University Press, New York, 1952. The equations, which were derived even earlier by van Laar and will be discussed later in this chapter, reduce to Eq. (21-7)* for equal molecular volumes of the two components.

If the mixing is random as assumed above, the entropy of mixing is ideal. If in addition there is no volume change, the ΔE of Eq. (21-11)* may be identified with the excess free energy of mixing in Eq. (21-8)*. Actually Eq. (21-8)* is probably valid to a higher approximation than is established in the derivation given above.

The interaction energies are dominated by the forces of attraction; consequently all the ϵ's are negative. It is found that the normal interactions which arise in regular solutions yield a value of ϵ_{ab} which is less negative than the mean value $\frac{1}{2}(\epsilon_{aa} + \epsilon_{bb})$. Consequently we expect w to be positive.

Thermodynamic Properties.   We turn now to the thermodynamic consequences of the regular solution equation for equal-sized molecules and to comparisons with actual systems.   The partial molal free energies, or chemical potentials, are

$$\begin{aligned} \mu_1 - \mu_1^\circ &= RT \ln x_1 + wx_2{}^2 \\ \mu_2 - \mu_2^\circ &= RT \ln x_2 + wx_1{}^2 \end{aligned} \tag{21-13}*$$

Note the appearance of x_1 in the first term and x_2 in the second for μ_1 and vice versa for μ_2 and also that the pure components are the standard reference states.   The activities are

$$\begin{aligned} \ln a_1 &= \frac{\mu_1 - \mu_1^\circ}{RT} = \ln x_1 + x_2{}^2 \frac{w}{RT} \\ \ln a_2 &= \frac{\mu_2 - \mu_1^\circ}{RT} = \ln x_2 + x_1{}^2 \frac{w}{RT} \end{aligned} \tag{21-14}*$$

and the activity coefficients ($\gamma_i = a_i/x_i$) are simply

$$\begin{aligned} \ln \gamma_1 &= x_2{}^2 \frac{w}{RT} \\ \ln \gamma_2 &= x_1{}^2 \frac{w}{RT} \end{aligned} \tag{21-15}*$$

Figure 21-1 shows two typical examples of regular solutions with curves indicating the accuracy of agreement with the equations.   Reasonable agreement is obtained even if the component molecules are somewhat different in size, as benzene and cyclohexane.   Table 21-1 lists the values of w for several systems together with other properties.   We note that w has a small but significant temperature dependence.   This appears in the entropy of mixing, which is

$$\begin{aligned} \Delta S_{\text{mix}} &= -\frac{\partial \Delta F_{\text{mix}}}{\partial T} \\ &= -R(n_1 \ln x_1 + n_2 \ln x_2) - (n_1 + n_2)x_1 x_2 \frac{\partial w}{\partial T} \tag{21-16}* \end{aligned}$$

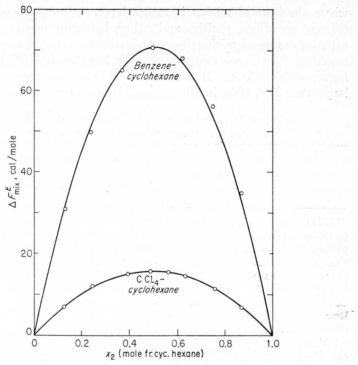

Fɪɢ. 21-1. The excess free energy of mixing vs. mole fraction cyclohexane.   The points are experimental; the curves are theoretical for regular solutions [Eq. (21-8)*].

The first term is the Raoult's-law expression for an ideal solution, and the second term represents the difference.   It is sometimes called the excess entropy of mixing and is presented separately.

$$\Delta S_{\text{mix}}^{E} = -(n_1 + n_2)x_1 x_2 \frac{\partial w}{\partial T} \qquad (21\text{-}17)^*$$

The heat of mixing follows readily from Eqs. (21-7)* and (21-16)*,

$$\Delta H_{\text{mix}} = (n_1 + n_2)x_1 x_2 \left(w - T\frac{\partial w}{\partial T} \right) \qquad (21\text{-}18)^*$$

Since the heat of mixing is zero in an ideal solution, this entire quantity is a deviation or excess quantity.   The change of heat capacity on mixing is given by the temperature derivative of ΔH.

$$\Delta C_{P,\text{mix}} = -(n_1 + n_2)x_1 x_2 T \frac{\partial^2 w}{\partial T^2} \qquad (21\text{-}19)^*$$

Examination of the various data in Table 21-1 shows little systematic behavior beyond the fact that all the quantities are small for some systems and that the excess entropy remains less than 0.5 cal/deg even for cases where the heat of mixing is rather large. Also the heat and excess entropy are always positive and $R \ln \gamma$ is usually positive. The heat-capacity change on mixing, however, is sometimes positive and sometimes negative. The excess entropy is usually less than $(\Delta H/T)$, as is required for a positive $R \ln \gamma$. But sometimes $R \ln \gamma$ is as much as three times larger than ΔS^E, while in other cases it is smaller or even essentially zero.

TABLE 21-1. PROPERTIES OF SOME REGULAR SOLUTIONS
$n_1 = n_2 = x_1 = x_2 = 0.50$, cal/mole deg or cal/mole

	T	$R \ln \gamma$ $= \dfrac{w}{4T}$	$\Delta H_{\text{mix}} =$ $\dfrac{1}{4}\left(w - T\dfrac{\partial w}{\partial T}\right)$	$\Delta S^E_{\text{mix}} =$ $-\dfrac{1}{4}\dfrac{\partial w}{\partial T}$	$\Delta C_{P,\text{mix}} =$ $-\dfrac{T}{4}\dfrac{\partial^2 w}{\partial T^2}$
CCl$_4$-benzene[a]	298	0.065	26.1	0.022	0.37
CCl$_4$-benzene	343	0.052	42.9	0.073	
CCl$_4$-cyclohexane[b]	313	0.051	34	0.056	
Benzene-cyclohexane[b]	293	0.26	200	0.42	−0.8
Benzene-cyclohexane	343	0.18	162	0.29	
Benzene-1,2-C$_2$H$_4$Cl$_2$[a]	298	0.018	15	0.03	0.0
Benzene-1,2-C$_2$H$_4$Cl$_2$[a]	343	0.011	14	0.03	
Benzene-toluene[c]	353	−0.007	11	0.037	−0.1
CS$_2$-acetone[a]	308	0.81	349	0.32	0.8
n-C$_5$F$_{12}$-n-C$_5$H$_{12}$[d]	277	1.06	370	0.27	
cy-C$_6$F$_{12}$-1,3,5-(CH$_3$)$_3$cy-C$_6$H$_9$[e]	338	0.99	475	0.40	

[a] L. A. K. Staveley, W. I. Tupman, and K. R. Hart, *Trans. Faraday Soc.*, **51**, 323 (1955).

[b] G. Scatchard, S. E. Wood, and J. M. Mochel, *J. Phys. Chem.*, **43**, 119 (1939); *J. Am. Chem. Soc.*, **61**, 3206 (1939), **62**, 712 (1940).

[c] C. H. Cheesman and W. R. Ladner, *Proc. Roy. Soc.* (*London*), **A229**, 387 (1955); A. P. Rollet, G. Elkaim, P. Toledano, and M. Senez, *Compt. rend.*, **242**, 2560 (1956).

[d] J. H. Simons and R. D. Dunlap, *J. Chem. Phys.*, **18**, 335 (1950).

[e] J. S. Rowlinson, private communication to R. L. Scott, *J. Phys. Chem.*, **62**, 136 (1958).

It is interesting, also, to note the conditions for the critical temperature for phase separation. The criteria were discussed in Chapter 17.

$$\frac{\partial \ln a_1}{\partial x_2} = -\frac{1}{1-x_2} + 2x_2\frac{w}{RT_c} = 0$$

$$\frac{\partial^2 \ln a_1}{\partial x_2^2} = -\frac{1}{(1-x_2)^2} + 2\frac{w}{RT_c} = 0$$

$$\frac{\partial \ln a_2}{\partial x_2} = \frac{1}{x_2} - 2(1-x_2)\frac{w}{RT_c} = 0$$

$$\frac{\partial^2 \ln a_2}{\partial x_2^2} = -\frac{1}{x_2^2} + 2\frac{w}{RT_c} = 0$$

(21-20)*

These equations are satisfied by the conditions $x_1 = x_2 = 0.5$ and $w/RT_c = 2$, or $T_c = w/2R$. At this point the activity coefficient of either component is given by $\ln \gamma = 0.5$ or $\gamma = 1.65$ and $a = 0.82$. One can then predict that the total vapor pressure of a solution at the critical point for phase separation will be $P = 0.82(p_1^\circ + p_2^\circ)$, that is, 82 per cent of the total vapor pressure for the two pure components. This result may be compared with the two extreme cases. An equimolal ideal solution would have $P = 0.50(p_1^\circ + p_2^\circ)$, whereas two liquids completely insoluble in one another would have a total vapor pressure $P = p_1^\circ + p_2^\circ$.

Unsymmetrical Systems, Unequal Molal Volumes. The equations derived above for solutions with components having equal molal volumes are found to apply in good approximation to systems where the volumes differ only moderately. Thus the molal volume of cyclohexane is 109.1 cc, whereas that of benzene is 89.8 cc at 30°C. In Fig. 21-1 the points for this system fall just slightly to the right of the theoretical curve of Eq. (21-8)* for equal volumes. Where the difference in molal volume of the components is greater, more unsymmetrical curves are obtained.

In 1906 van Laar[1] published a treatment of binary liquid solutions based upon the van der Waals equation. The excess free energy of mixing[2] is given by the equation

$$\Delta F_{\text{mix}}^E = \mathbb{Q}_{12} \frac{b_1 b_2 n_1 n_2}{n_1 b_1 + n_2 b_2} \qquad (21\text{-}21)*$$

Here b_1 and b_2 are the molecular-volume parameters of the van der Waals equation for components 1 and 2, and $\mathbb{Q}_{12}$ is a parameter expressing the deviation from ideal-solution behavior, which will be discussed further. From Eq. (21-21)* one may readily derive the expressions for the activity coefficients of the two components,

$$\ln \gamma_1 = \frac{\mathbb{Q}_{12} b_1}{RT} z_2^{\,2} \qquad (21\text{-}22a)*$$

$$\ln \gamma_2 = \frac{\mathbb{Q}_{12} b_2}{RT} z_1^{\,2} \qquad (21\text{-}22b)*$$

where z_1 and z_2 are measures of composition defined by the equations

$$z_1 = \frac{n_1 b_1}{n_1 b_1 + n_2 b_2} \qquad z_2 = \frac{n_2 b_2}{n_1 b_1 + n_2 b_2} \qquad (21\text{-}23)$$

The limitation of van Laar's treatment to the inexact van der Waals equation was unfortunate and unnecessary. In 1931 Scatchard[3] removed

[1] J. J. van Laar, "Sechs Vortrage über das Thermodynamische Potential," Vieweg-Verlag, Brunswick, Germany, 1906; *Z. physik. Chem.*, **72**, 723 (1910).

[2] Actually van Laar obtained this expression first for the heat of mixing and subsequently derived equations for vapor pressures which are equivalent to Eqs. (21-21)* and (21-22)*.

[3] G. Scatchard, *Chem. Rev.*, **8**, 321 (1931).

this limitation and derived equations which may be obtained from Eqs. (21-21)* and (21-22)* by substitution of the molal volumes v_1^o and v_2^o for b_1 and b_2. Scatchard's arguments are approximate but are quite plausible for the case of regular solutions. The composition variables, z_1 and z_2, now become the volume fractions

$$\phi_1 = \frac{n_1 v_1^o}{n_1 v_1^o + n_2 v_2^o} \qquad \phi_2 = \frac{n_2 v_2^o}{n_1 v_1^o + n_2 v_2^o} \qquad (21\text{-}24)$$

The excess free energy of mixing to form 1 mole of solution takes the very simple form

$$\Delta F_{\text{mix}}^E = \alpha_{12} v \phi_1 \phi_2 \qquad (21\text{-}25)^*$$

where v is the molal volume of the solution, $x_1 v_1^o + x_2 v_2^o$. The Scatchard equations fit quite well the experimental data for many binary systems composed of normal liquids.

There are other systems, particularly solutions of metals, whose properties fit the van Laar type of equations provided that the ratio of b_1 to b_2 is adjusted arbitrarily in each case to a value which has no relation to either the van der Waals constants or the molal volumes. In some of these last cases α_{12} may even be negative, whereas in regular solutions α_{12} is always positive.

A few examples are given in Table 21-2 of systems which may be fitted by Eq. (21-22)*.

TABLE 21-2. SYSTEMS REPRESENTED BY THE ASYMMETRIC EQUATIONS
(21-22a)*, (21-22b)*

Components 1-2	T, °K	b_2/b_1	$\alpha_{12} b_1 / RT$
Mercury-tin†	596	3.7	0.51
Mercury-cadmium†	595	0.53	−3.2
Carbon disulfide–isopentane‡	298	1.94	0.58
n-Heptane-benzene§	343	1.50	0.33

† J. H. Hildebrand, A. H. Foster, and C. W. Beebe, *J. Am. Chem. Soc.*, **42**, 545 (1920).

‡ J. Hirshberg, *Bull. soc. chim. Belg.*, **41**, 163 (1932).

§ I. Brown and A. H. Ewald, *Australian J. Sci. Research*, **A4**, 198 (1951).

The carbon disulfide–isopentane system is well fitted with a b_2/b_1 value selected to be the ratio of the molal volumes. But for the metallic systems b_2/b_1 has no resemblance to the volume ratio, which is not surprising, since the orbitals of the valence electrons are grossly modified in forming metallic solutions.

If in these equations b_2/b_1 is independent of temperature, then one may express the heat of mixing and excess entropy of mixing readily in terms of α_{12} and $\partial\alpha_{12}/\partial T$. Where b_2/b_1 is equal to the ratio of molal volumes, it is likely that this situation will arise. However, in other cases there is every reason to assume that b_2/b_1 may be a function of temperature, and in that case the equation for the heat of mixing becomes much more complicated. There are too few experimental data to justify a general discussion at this time of the temperature dependence of b_2/b_1.

Prediction of Deviation from Ideality.[1] There is no theory at present which will predict accurately the various results in Tables 21-1 and 21-2. Nevertheless, it is of interest in some cases to have even a crude estimate of the deviation from Raoult's law for a system which has not been studied experimentally. Thus we wish an estimate of α_{12} or, in the case of systems with components of approximately equal molal volume, of w/v which is then equal to α_{12}. The van Laar expression in terms of the van der Waals equation parameters is

$$\alpha_{12} = \left(\frac{a_1^{1/2}}{b_1} - \frac{a_2^{1/2}}{b_2}\right)^2 \qquad (21\text{-}26)*$$

but Scatchard in his 1931 derivation, which was described in the preceding section, obtained

$$\alpha_{12} = \left[\left(\frac{\Delta E_1}{v_1}\right)^{1/2} - \left(\frac{\Delta E_2}{v_2}\right)^{1/2}\right]^2 \qquad (21\text{-}27)*$$

where the ΔE's are molal energies of vaporization (approximately $\Delta H_{vap} - RT$) and the v's are molal volumes of the pure liquid components. Subsequently Hildebrand and Wood[2] derived Eq. (21-27)* by approximate statistical calculations based upon molecular-distribution functions. Equation (21-27)* is to be preferred in practically all cases in comparison to Eq. (21-26)*.

Equation (21-27)* may be simplified for components with equal molal volumes. However, if the volumes differ even slightly, it is best to use the full expression of Eq. (21-27)* for α_{12}. The average of v_1 and v_2 may then be taken in the calculation of $w = \alpha_{12}v$ for the simple symmetrical equations (21-13) to (21-20).

The comparison of calculated and observed values of α_{12} in Table 21-3 shows that this method yields agreement within a few tenths in $\alpha_{12}v_1/RT$. This is useful for estimates when experimental data are inaccessible. Relatively simple experiments, however, will yield a more reliable value.

---

[1] Additional methods of prediction are given by I. Prigogine "Molecular Theory of Solutions," North-Holland Publishing Company, Amsterdam, 1957.

[2] J. H. Hildebrand and S. E. Wood, *J. Chem. Phys.*, **1**, 817 (1933).

TABLE 21-3. OBSERVED AND ESTIMATED CONSTANTS FOR DEVIATION
FROM RAOULT'S LAW

System	T	$\left(\dfrac{\alpha_{12}v_1}{RT}\right)_{obs}$	$\left(\dfrac{\alpha_{12}v_1}{RT}\right)_{Eq.\,(21-27)}$*
Benzene-cyclohexane........	343	0.36	0.11
Benzene-CS_2..............	298	0.47	0.11
SnI_4-$SiCl_4$.................	298	3.8	4.4
SnI_4-benzene..............	298	1.9	1.7

POLYMER SOLUTIONS

Let us turn now to a consideration of the properties of solutions of very large nonpolar polymer molecules in ordinary solvents. Rubber in benzene and polystyrene in acetone are examples. The heat of mixing in such cases is approximated by the more general equations of the two preceding sections, but the entropy of mixing deviates grossly from the Raoult's-law value even when the heat of mixing is zero.

The mole fraction is not a satisfactory measure of composition for high-polymer solutions. Frequently the molecular weight of the polymer is not accurately known, but in any case it is so high as to make mole fractions inconvenient. The weight or mass fraction is good for fundamental work, since it is independent of pressure or temperature for a given solution. Theoretical discussions usually yield results in terms of the volume fraction, e.g., Eq. (21-25)*. The volume fraction was defined in Eq. (21-24).

The system rubber-benzene has been studied very carefully and extensively by Gee and collaborators.[1] They measured the osmotic pressure of dilute solutions and the benzene vapor pressure over concentrated solutions. Measurements were made at different temperatures in order to separate heat and entropy effects; also direct calorimetric measurements were made of the heat of mixing. Their results are shown as the solid curves in Figs. 21-2 and 21-3. The molecular weight of the rubber is very large; consequently the slope of the activity and entropy curves should be practically zero at the zero-concentration axis. This is observed, but beyond this one point both the entropy and activity curves deviate greatly from those predicted for an ideal solution.

The large discrepancy in concentrated solutions between the observed entropy and the ideal entropy can be explained on the basis that the individual segments of the polymer molecule have considerable freedom of

[1] G. Gee and L. R. G. Treloar, *Trans. Faraday Soc.*, **38**, 147 (1942); G. Gee and W. J. C. Orr, *Trans. Faraday Soc.*, **42**, 507 (1946).

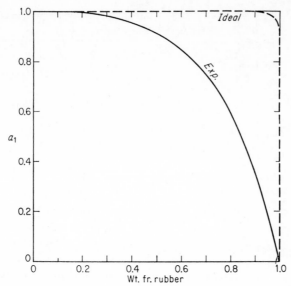

FIG. 21-2. The activity of benzene in rubber-benzene solutions compared with the prediction for an ideal solution.

motion. Flory[1] and Huggins[2] assumed a lattice model with a polymer molecule occupying a sequence of solvent-sized sites. They showed that, in the first approximation, the entropy of mixing was given by the replacement of mole fraction by volume fraction in the Raoult's-law expression.

$$\Delta S_{\text{mix}} = -R(n_1 \ln \phi_1 + n_2 \ln \phi_2) \tag{21-28}*$$

The partial molal entropy of the solvent is found by differentiation to be

$$\bar{s}_1 - s_1^\circ = -R\left[\ln(1 - \phi_2) + \phi_2\left(1 - \frac{v_1^\circ}{v_2^\circ}\right)\right] \tag{21-29}*$$

We may show that this approaches Raoult's law in very dilute solutions by expanding both expressions in powers of x_1 or ϕ_1. Raoult's law becomes

$$\bar{s}_1 - s_1^\circ = R\left(x_2 + \frac{x_2^2}{2} + \frac{x_2^3}{3} + \cdots\right) \tag{21-30}*$$

while Eq. (21-29)* becomes

$$\bar{s}_1 - s_1^\circ = R\left(\phi_2 \frac{v_1^\circ}{v_2^\circ} + \frac{\phi_2^2}{2} + \frac{\phi_2^3}{3} + \cdots\right) \tag{21-31}*$$

But both $\phi_2 v_1^\circ / v_2^\circ$ and x_2 approach n_2/n_1 as the concentration approaches zero.

In the more concentrated solutions Eq. (21-29)* may be simplified in a

[1] P. J. Flory, *J. Chem. Phys.*, **9**, 660 (1941), **10**, 5 (1942).
[2] M. L. Huggins, *Ann. N.Y. Acad. Sci.*, **43**, 1 (1942).

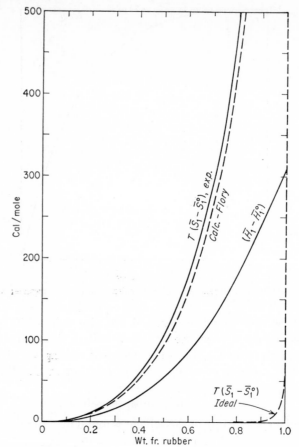

FIG. 21-3. The partial molal entropy and enthalpy of benzene in rubber-benzene solutions. The predictions for the entropy by ideal solution and Flory-Huggins theories are shown.

different fashion for very high polymers. Then v_1°/v_2° is negligible as compared with unity, and

$$(\bar{\mathsf{s}}_1 - \mathsf{s}_1^\circ) = -R[\ln{(1 - \phi_2)} + \phi_2] \qquad (21\text{-}32)^*$$

The curve calculated from this equation is shown in Fig. 21-3 as the dashed line labeled "Flory." The close similarity with the experimental curve is apparent.

Partial molal entropy data for the solvent in several other polymer-solution systems are shown in Fig. 21-4. Here it is seen that the agreement with the Flory-Huggins equation for the rubber-benzene system was apparently accidental and that the true situation is more complex. However, the large deviations from the Flory-Huggins equation are in

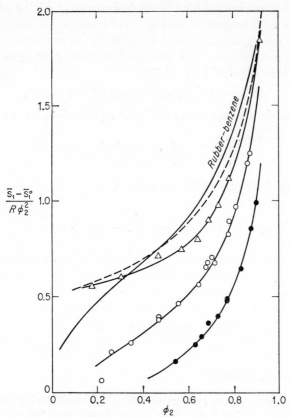

F IG. 21-4. Comparison of observed partial molal entropies of the solvent (points and solid lines) with the Flory-Huggins curve [Eq. (21-28)*] (broken line). (*Paul J. Flory, "Principles of Polymer Chemistry," p. 518, fig. 113, Cornell University Press, Ithaca, N.Y., 1953.*) [*The rubber-benzene curve is from G. Gee and L. R. G. Treloar, Trans. Faraday Soc.,* **38,** *147 (1942); G. Gee and W. J. C. Orr, Trans. Faraday Soc.,* **42,** *507 (1946). Others are polydimethylsiloxane in benzene, Δ, from M. J. Newing, Trans. Faraday Soc.,* **46,** *613 (1950), and polystyrene in methyl ethyl ketone,* ●, *and in toluene,* ○, *from C. E. H. Bawn, R. F. J. Freeman, and A. R. Kamaliddin, Trans. Faraday Soc.,* **46,** *677 (1950).*]

the direction of the ideal-solution equation. Thus the Flory-Huggins and the Raoult equations for the entropy of mixing appear to give the limits between which will lie the entropy of mixing of solutions of this type.

The excess free energy of mixing of polymer solutions arising from heat-of-mixing effects is reasonably well represented by Eq. (21-25)*. If we consider a solution with n_1 moles of solvent and write w_1 for the product $\alpha_{12}v_1^{\circ}$, Eq. (21-25)* becomes

$$\Delta F^E_{\text{mix}} = w_1 n_1 \phi_2 \qquad (21\text{-}33)^*$$

This excess free energy is added to the free energy of mixing which would arise if there were no heat of mixing. For a polymer solution, however, this is given not by Raoult's law but rather by an expression such as the Flory-Huggins equation. For zero heat of mixing $\Delta F_{mix} = - T \Delta S_{mix}$; hence from Eqs. (21-28)* and (21-33)*

$$\Delta F_{mix} = RT(n_1 \ln \phi_1 + n_2 \ln \phi_2) + w_1 n_1 \phi_2 \qquad (21\text{-}34)^*$$

The appropriate temperature derivative gives the corresponding heat of mixing,

$$\Delta H_{mix} = \left(w_1 - T \frac{dw_1}{dT} \right) n_1 \phi_2 \qquad (21\text{-}35)^*$$

Also one may derive the activity of the solvent, which is frequently the measurable quantity,

$$\ln a_1 = \ln (1 - \phi_2) + \phi_2 \left(1 - \frac{v_1^o}{v_2^o} \right) + \phi_2^2 \left(\frac{w_1}{RT} \right) \qquad (21\text{-}36)^*$$

While Eq. (21-36)* reproduces the major deviations of polymer-solution properties from ideal behavior, it does not ordinarily yield quantitative agreement with experimental data. Further refinements are discussed in works specially devoted to polymer-solution thermodynamics. This is currently an active research field, and many new developments may be anticipated.

GASEOUS SOLUTIONS

At low pressures, where pure gases follow the ideal-gas law, mixed gases follow the ideal-solution law. This simple behavior is implied in the discussion of Chapter 15 of chemical equilibria between gaseous substances. But at high pressure the properties of gaseous solutions may be nonideal, and we now turn to a consideration of this problem. We assume that the properties of each pure gaseous component are known in the pressure range of interest.

Two very simple postulates for mixed gases were made long ago. The first, by Dalton, is that the pressure of a gaseous solution is the sum of the pressures which each gas would exert if it alone occupied the entire volume at the same temperature. While this fits a few systems quite well, it is quite unsatisfactory in most cases (outside the ideal-gas region). A second and more generally useful postulate is that of Amagat,[1] who proposed that the volume of a gaseous solution was the sum of volumes of the components each at the temperature and total pressure of the solution. It may be shown that, if Amagat's law of additive volumes holds at all pressures up to the pressure of interest, then the ideal-solution law

[1] E. H. Amagat, *Ann. Chem. Phys.*, (5)**19**, 384 (1880); *Compt. rend.*, **127**, 88 (1898).

holds for the gaseous solution; i.e., the fugacity of the ith component is

$$f_i = x_i f_i^\circ \qquad (18\text{-}1)^*$$

where x_i is the mole fraction and f_i° is the fugacity of the pure gas at the temperature and total pressure of the solution.

Lewis and Randall[1] wrote:

It seems reasonable to suppose that the solution of a given pair of substances will be more nearly perfect when the density of the solution is less, or, in other words, when the average distance between the molecules is greater. We shall therefore expect gaseous solutions to be much more nearly perfect than corresponding liquid solutions, and since we find even among liquids numerous cases in which there is a close approach to the perfect solution, it is likely that almost any gaseous solution may be regarded as nearly perfect.

Experimental results have shown that this very plausible postulate is not generally correct, although many particular systems of mixed gases do follow the ideal-solution law at high pressures.

It is found that many gaseous-solution systems show more nearly ideal behavior at liquidlike densities (and high pressures) than at moderate densities comparable with the critical density. Figure 21-5 shows an example of this sort. The molal volume of carbon dioxide–n-butane solution at 411°K and 680 atm is very nearly a linear function of mole fraction, but at 68 atm the volume deviates widely from linear behavior, i.e., from Amagat's law.

An adequate theory for the general prediction of the nonideality of gaseous solutions is not yet available, but the subject is now under active study, and improvements may be expected. A useful postulate is

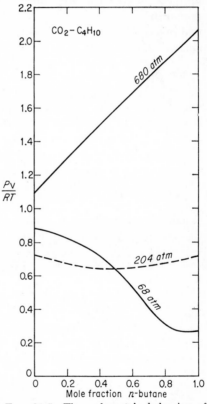

FIG. 21-5. The volumetric behavior of gaseous solutions of carbon dioxide and n-butane at 411°K and the pressures indicated. [*Data of R. H. Olds, H. H. Reamer, B. H. Sage, and W. N. Lacey, Ind. Eng. Chem.*, **41**, 475 (1949).]

[1] G. N. Lewis and M. Randall, "Thermodynamics and the Free Energy of Chemical Substances," 1st ed., p. 226, McGraw-Hill Book Company, Inc., New York, 1923.

that of Kay[1] that a gaseous solution follows the behavior of a hypothetical pure substance with pseudocritical constants computed as molal averages of the critical constants of the components. This implies the assumption of corresponding-states behavior for pure substances. Pitzer and Hultgren[2] recently proposed a generalization of the pseudocritical postulate applicable to the acentric-factor theory for pure substances (Appendix 2) and examined the deviation of several systems therefrom.

If there exist sufficiently extensive volumetric data for the solution system, then the fugacities of the components may be calculated by application of thermodynamics. The partial molal volume of each component must be determined (see Chapter 17), and then Eq. (16-4) may be employed, with the partial molal volume replacing the molal volume of a single component.

$$\frac{\partial \ln (f_i/P)}{\partial P} = \frac{\bar{v}_i}{RT} - \frac{1}{P} \tag{21-37}$$

The fugacity of a component approaches the limit $f_i = x_i P$ at low pressure, where the ideal-gas and ideal-solution laws apply. Since x_i is not a function of P, Eq. (21-37) may be integrated to a final total pressure P' to yield

$$\ln \frac{f_i'}{x_i P'} = \int_0^{P'} \left(\frac{\bar{v}_i}{RT} - \frac{1}{P} \right) dP \tag{21-38}$$

Whenever the partial molal volume is available as a function of pressure, this equation yields fugacity over that pressure range.

PROBLEMS

21-1. Show from Eq. (21-37) that, if Amagat's law of additive volumes hold for all pressures up to P', then the ideal-solution law [Eq. (18-1)*] holds for the gaseous mixture.

21-2. Predict the total vapor pressure of a solution made from 50 g each of benzene and $C_2H_4Cl_2$ at 323°K. Note data in Table 21-1.

21-3. Calculate the solubility of solid SnI_4 in liquid $SiCl_4$ at 298°K. Use the observed value of $\alpha_{12} v_1 / RT$ given in Table 21-3. The melting point and heat of fusion of SnI_4 are 417.7°K and 4.48 kcal/mole. Assume Δc_P° of fusion to be negligible.

21-4. The melting point of white phosphorus (P_4) is 317.4°K, and the heat of fusion is 600 cal/mole of P_4. The solubilities of white phosphorus in carbon disulfide at various temperatures were determined by Cohen and Inouye[3] to be:

t,°C	−10	−7.5	−5	−3.5	−3.2	−2.5	0	5	10
x_{P_4}	0.219	0.255	0.307	0.345	0.607	0.647	0.727	0.794	0.844

[1] W. B. Kay, *Ind. Eng. Chem.*, **28**, 1014 (1936).
[2] K. S. Pitzer and G. O. Hultgren, *J. Am. Chem. Soc.*, **80**, 4795 (1958).
[3] E. Cohen and K. Inouye, *Z. physik. Chem.*, **72**, 411 (1910).

If Δc_P° of fusion is neglected, $\Delta F^\circ = 600 - 1.89T$ for $P_4(s) = P_4(l)$. If Eq. (21-13)* is valid, a plot of $\Delta F^\circ + RT \ln x_{P_4}$ versus $x_{CS_2}^2$ should yield a straight line. Make such a plot, and obtain the value of w in Eq. (21-13)*. Note that experimental errors in the data become greatly magnified at large x_{P_4}.

21-5. Liquid silver is immiscible with a number of liquid metals such as chromium, manganese, nickel, and uranium. To apply Eqs. (21-22a)* and (21-22b)* to the estimation of the activity coefficients, one can calculate $\alpha_{12}b/RT$ from Eq. (21-27)* and the molal volumes and energies of vaporization of the metals. In Appendix 7, the heats of sublimation of the metals are given at 298°K. The enthalpies required to heat the gas from 298 to 2200°K and to heat the solid at 298°K to the liquid at 2200°K, as given by Stull and Sinke,[1] were used to calculate the heats of vaporization given in Table 21-4. Densities of liquid metals are not generally available, and the molal volumes given in Table 21-4 are those of the solid metals at 298°K.

TABLE 21-4. MOLAL HEATS OF VAPORIZATION AT 2200°K AND MOLAL VOLUMES AT 298°K

	ΔH_{2200}°, cal	V_{298}, cc
Ag............	61,600	10.3
Cr............	89,500	7.3
Ni............	93,900	6.6
U............	113,800	12.7

Neglecting the expansion upon heating from 298 to 2200°K, calculate $\alpha_{12}b$ for each component of the binary silver solutions, and calculate the mutual solubilities of silver liquid and each of the other liquids for comparison with experimental values as given in Table 21-5. As a first approximation one can assume that the mutual

TABLE 21-5. MUTUAL SOLUBILITIES OF METAL LIQUIDS

System	Temperature, °K	Saturated compositions, atomic %
Ag-Cr............	1718	3.5% and 85% Ag[†]
Ag-Ni............	1708	2% and 98% Ag[‡]
Ag-U............	1405	0.5% and 97.4% Ag[§]

[†] A. T. Gugoriev, E. N. Sololovakaia, and M. I. Kruglova, *Moscow Univ. Vestnik Ser. Fiz. Mat. Estest. Nauk*, **9**, 77–81 (1954).

[‡] Average of values quoted by M. Hansen, "Constitution of Binary Alloys," McGraw-Hill Book Company, Inc., New York, 1958.

[§] R. W. Buzzard, F. P. Fickle, and J. J. Park, *J. Research Natl. Bur. Standards*, **52**, 149 (1954).

solubilities are so small that the activities of the metals are not appreciably reduced from unity and that the solubilities are simply given by the reciprocal of the activity coefficients. The b_1/b_2 ratio may be obtained from the experimental values at one temperature and then used to calculate mutual solubilities at other temperatures with better accuracy than through use of the molal volumes of the pure metals.

[1] D. R. Stull and G. C. Sinke, Thermodynamic Properties of the Elements, *Advances in Chem. Ser.* 18, 1956.

22

ELECTROLYTE SOLUTIONS

Because of the prime importance of water in biological systems and the widespread use of water in many industrial processes, the thermodynamic properties of aqueous solutions are of considerable value. The study of aqueous salt solutions poses some special problems. Because the salts are quite nonvolatile compared with water and usually have extremely small solubilities in solvents which are not miscible with water, it is not possible to apply some of the methods discussed in Chapter 20 for the determination of partial molal free energies of dilution or activity coefficients. However, it is possible to construct galvanic cells whose cell reactions correspond to the removal or addition of salt to the solution and therefore offer a means of determining the partial molal free energy of a salt in solution. Although cells can yield quite accurate activity coefficients for solutions up to concentrations of several molal, some types of cells do not usually yield sufficiently accurate results for solutions less concentrated than 0.05 M. Because of this limitation, it is sometimes difficult to establish directly the solute standard state from cell data alone.

Since it is not usually possible to measure directly the vapor pressure of the salt in aqueous solutions, whereas there are ways of measuring the fugacity or activity of the water, a number of methods have been developed that depend upon the determination of the activity of water in a salt solution and then the subsequent calculation of the activity coefficient of the salt through the use of the Gibbs-Duhem equation. The vapor pressure of water in equilibrium with aqueous electrolyte solutions has been determined by a number of vapor-pressure methods, both static and dynamic. However, once the vapor pressure of water has been established accurately as a function of concentration for one solution, it is not necessary to make additional absolute determinations of the water vapor pressure for other solutions. It is possible to use the isopiestic method, which depends upon the equilibration of one solution with another until the vapor pressure of the water in equilibrium with both solutions is the same. This method is a particularly convenient and accurate one and has been very extensively used to obtain data for many salt solutions.   However, normally the results are not reliable below about 0.1 M.   Since the variation of activity coefficients of electrolyte solutions is quite appreciable at concentrations below the limit of the isopiestic method, it is necessary to have another method for the determination of the activity in very dilute solutions in order to establish the solute standard state for the salt.   Some galvanic cells are satisfactory; also the measurement of the depression of the freezing point is capable of yielding values of the activity of water to a very high degree of accuracy. These are the principal methods of obtaining activity coefficients of aqueous salt solutions at concentrations below 0.1 M.   The details of the freezing-point method of determining activity coefficients will be discussed in Chapter 26.

Since most salts are solid and have a limited solubility in water, it is not convenient to use a pure liquid standard state for the salt.   Also most salt solutions of interest are relatively dilute, and a solute standard state would be of more convenience than a standard state corresponding more closely to highly concentrated solutions.   In addition, there are some weak electrolytes which change their character quite completely from dilute to concentrated solutions, and the establishment of a solute standard state is then important.

We have noted several times earlier and particularly in the discussion of NO_2-chloroform ($CHCl_3$) solutions in Chapter 20 that the proper choice of the molecular weight for a solute may considerably simplify the nature of variation of partial molal free energies or activity coefficients with composition.   In particular, it is desirable to choose the molecular weight to correspond to the main solute constituent in a dilute solution to ensure an approach to Henry's law at low concentration; for if the solution behavior does not approach Henry's law at low concentrations, it is not possible to establish a solute standard state.

One common way to determine empirically the formula of a solute in solution is by determination of the freezing-point lowering as solute is added.   In Chapter 18, Eq. (18-13)* for the variation of solubility with temperature, assuming a perfect solution, can be used for the variation of freezing point with composition of the solution.   If we convert the mole fraction to molality, using the relationship $d \ln x_w = -dm/(55.5 + m)$, where x_w is the mole fraction of water in the salt solution in equilibrium with ice and m is the molality of the solute, then Eq. (18-13)* becomes

$$\frac{dm}{dT} = -\frac{55.5 \Delta H}{RT^2} \qquad (22\text{-}1)^*$$

where ΔH is the heat of fusion of ice and m has been neglected with respect to 55.5, as this equation is only to be used for very dilute solutions.

Taking $T = 273.15°K$ and using the heat of fusion of ice, one obtains $dT/dm = -1.860$ deg kg/mole.   Equation (22-1)* offers us a convenient method for determining the proper molecular weight to be used for a solute to ensure that Henry's law will be approached in dilute solutions of the solute.   In the derivation of Eq. (18-13)*, from which Eq. (22-1)* has been derived, Raoult's law was assumed for the water.   This would be true if Henry's law holds for the solute as we have shown in Chapter 19 in Eqs. (19-6)* and (19-8)*.

In Fig. 22-1 is shown a smoothed curve of the freezing-point lowering $\Delta T/m$ versus m for acetic acid solutions[1] compared with the limiting

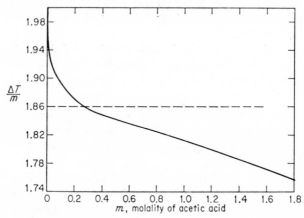

Fig. 22-1. Freezing-point depression of aqueous acetic acid solutions.

value required by Eq. (22-1)*.   At high molalities of acetic acid, there is an appreciable difference between the experimental and the limiting values.   However, Eq. (22-1)* is limited to very low concentrations of m, and the deviations at high molality are to be anticipated.   It will be noted that, at molalities of about 0.5, the experimental values appeared to be rapidly approaching the theoretical limiting value of 1.860.   On this basis, one would feel justified in extrapolating to infinite dilution in the usual manner and then determining the activity of acetic acid in the various concentration solutions by the procedures used in Chapter 20. However, the experimental values for very dilute solutions do not continue to approach the limiting value of 1.860 but soon exceed this value and continue to increase with dilution.   This is shown by the measurements of Hausrath,[2] given in the second column of Table 22-1.   This behavior is not peculiar to acetic acid but is characteristic of many acids, for example, dichloroacetic acid exhibits the same phenomena at even

[1] E. R. Jones and C. R. Bury, *Phil. Mag.*, (7)**3**, 1032 (1927).

[2] H. Hausrath, *Ann. Physik*, (4)**9**, 548 (1902).

higher concentrations.   We learn from measurements of the same author that the $\Delta T/m$ is 3.47 at 0.018 M, 3.60 at 0.008 M, and 3.71 at 0.003 M. Here it is evident that the molal lowering is approaching 3.72 deg kg/ mole, or twice the value required by Eq. (22-1)* as a limit.

This case shows so near an analogy to N_2O_4 dissolved in chloroform, which we studied in Chapter 20, that it is natural to assume here also some kind of dissociation whereby a molecule of acetic acid yields two new molecules.   This dissociation is negligible at the higher concentrations, becomes noticeable in the more dilute solutions, and approaches completion at infinite dilution.

Later we are going to discuss more fully the nature of this dissociation, and to show how the dissociation constant may be calculated from measurements of a quite different sort.   Such measurements give us the dissociation constant of acetic acid,[1] 1.657×10^{-5} at 0°C.   From this dissociation constant we find the degree of dissociation α at each of the three concentrations of Table 22-1.   Since m is the number of moles of acetic acid dissolved in 1 kg of water, the total molality would be $2m$ in the case of complete dissociation or, for incomplete dissociation, $(1 + \alpha)m$.   This coefficient $1 + \alpha$ is also known as the van't Hoff factor i.

TABLE 22-1. MOLAL-FREEZING-POINT LOWERING IN AQUEOUS ACETIC ACID

m	$\Delta T/m$, observed	α	$i = 1 + \alpha$	$\Delta T/m$, calculated
0.001	2.05	0.12	1.12	2.08
0.003	2.01	0.07	1.07	1.99
0.010	1.96	0.04	1.04	1.93
0.035	1.93	0.02	1.02	1.90

Assuming now that each species obeys the law of the dilute solution, we may calculate $\Delta T/m$ as equal to $1.860(1 + \alpha)$.   The values so calculated are given in the last column of Table 22-1 and agree, well within the limits of experimental error, with the figures of Hausrath.

This dissociation of acetic acid is of a type first explained by Arrhenius's brilliant theory of electrolytic dissociation, which was the subject of bitter contention for many years after its inception but which is now universally accepted.

It is assumed that a molecule of acetic acid dissociates to form a positively charged and a negatively charged molecule, the former belonging to the same molecular species (hydrogen ion) as exists in all aqueous acid solutions and the latter (acetate ion) common to all aqueous solutions of acetates.   These ions, or charged molecules, are assumed to be responsible for the phenomenon of electrolytic conduction, and it is by a study of the

[1] Data of H. S. Harned and R. W. Ehlers, *J. Am. Chem. Soc.*, **55**, 652 (1933).

conductivity of electrolytes that much of our information regarding these substances has been obtained.

While our discussion of electrolytes will concern aqueous solutions almost exclusively, solutions in other ionizing solvents such as methanol and ammonia show similar properties and may be treated by similar methods.

THE CONDUCTIVITY AND THE DISSOCIATION OF WEAK ELECTROLYTES

It is supposed that the conductivity due to any ion is the product of two factors, (1) the concentration of that ion, and (2) its mobility. If the mobility is assumed to be independent of the concentration (Kohlrausch's law, which is now known to be only approximate), the ratio of conductivity to ion concentration must be constant. If this be assumed for both ions of a substance like acetic acid, then the conductivity of this electrolyte divided by the ion concentration is constant but the conductivity divided by the gross concentration, which is usually expressed for this purpose in equivalents per liter, will vary according to the degree of dissociation. This latter quotient, called the equivalent conductivity and denoted by Λ, must approach a limiting value $\Lambda°$ as we approach infinite dilution and complete dissociation.

On the assumption of constant ionic mobility, $\Lambda/\Lambda°$ gives the degree of dissociation at each concentration, and $m\Lambda/\Lambda°$ gives the molality of each ion, while $m(1 - \Lambda/\Lambda°)$ gives the molality of the undissociated substance. Let us consider the quotient K_Λ defined by the equation

$$\frac{m(\Lambda/\Lambda°)^2}{1 - \Lambda/\Lambda°} = K_\Lambda \qquad (22\text{-}2)$$

If we now assume further that the activity of each substance is equal to its molality, the quotient K_Λ is equal to K, the equilibrium constant, for such a reaction as $XY = X^+ + Y^-$. We thus have the well-known dilution law of Ostwald.

If any one of the above assumptions is false we cannot expect K_Λ to be a constant nor is it in practice found to be constant except for a certain class of electrolytes, and then only in dilute solutions. However, it seems reasonable to conclude that the several assumptions approach complete validity at infinite dilution and therefore that the limiting value approached by K_Λ at infinite dilution is the true dissociation constant K.

To illustrate a case in which the several assumptions appear to be valid over a considerable range of concentrations, we may cite the measurements by MacInnes and Shedlovsky[1] of the conductivity of acetic acid

[1] D. A. MacInnes and T. Shedlovsky, *J. Am. Chem. Soc.*, **54**, 1429 (1932).

solution at 25°C.   Table 22-2 gives in the first column the concentration, and in the second the values of K_A calculated from Eq. (22-2).   MacInnes and Shedlovsky corrected for the variation of ionic mobility with concentration by comparison with the change Λ with concentration for strong electrolytes like hydrochloric acid, sodium chloride, and sodium acetate. They also corrected for interionic attraction, using the Debye-Hückel theory that will be discussed in Chapter 23.   The two corrections oppose each other to some extent and yield the results in the last column of Table 22-2.

TABLE 22-2. DISSOCIATION CONSTANT OF ACETIC ACID AT 25°C

c	$K_A \times 10^5$	$K \times 10^5$
0.00002801	1.759	1.753
0.00011135	1.769	1.754
0.0002184	1.770	1.752
0.0010283	1.780	1.751
0.002414	1.791	1.752
0.005912	1.802	1.750
0.02	1.806	1.740
0.05	1.801	1.726
0.1	1.793	1.700
0.2	1.746	1.653

In addition to error due to the assumption of constant ionic mobility and proportionality between the activity and concentration of the ions for which the K_A values have not been corrected, the values in both columns 2 and 3 are in error at higher concentrations owing to the fact apparent from Fig. 22-1 that the activity of the undissociated acid is less than the molality.   The ratio attains 0.94 in 1 M solution.[1]   Below 0.02 M, however, the constancy of K leaves nothing to be desired.   Dividing by the density of water to correct from concentration to molality, we may conclude that $K = 1.758 \times 10^{-5}$ and therefore, for the free energy of the reaction,

$$CH_3COOH = H^+ + CH_3COO^- \qquad \Delta F^\circ_{298} = -RT \ln K = 6487 \text{ cal}$$

Like acetic acid, a large number of acids and bases, and also a few salts, obey the Ostwald dilution law in aqueous solution over a measurable

---

[1] It must be constantly borne in mind that, by thermodynamic necessity, at a given temperature K is a constant at all concentrations.   If therefore, in addition to determining the activity of the undissociated acid in molal solution, we should also determine the activities of the ions, the product of the latter divided by the former would be 1.758×10^{-5}.   How such ion activities are measured, and how interpreted, will be fully discussed in this and following chapters.

range of concentrations. These are generally classified as weak electrolytes, a term which we may also use to include substances, about to be discussed, that are themselves ions but are capable of further ionization.

The Two Dissociation Constants of Carbonic Acid. Let us consider the case of carbonic acid, which, when dissolved in pure water, shows no measurable ionization except in accordance with the reaction

$$H_2CO_3(aq) = H^+ + HCO_3^-$$

The conductivity of water containing varying amounts of carbon dioxide has been measured by Shedlovsky and MacInnes.[1] The measurements

TABLE 22-3. DISSOCIATION CONSTANT OF H_2CO_3 AT 25°C

m	$K \times 10^7$
0 001878	4.310
0.001924	4.304
0.00512	4.324
0.01086	4.322
0.02165	4.326
0.03274	4.324

at 25° lead to the values of K_A at the several concentrations given in Table 22-3, from which we find $K = 4.31 \times 10^{-7}$, and therefore for the reaction

$$H_2CO_3(aq) = H^+ + HCO_3^- \qquad \Delta F^\circ_{298} = 8683 \text{ cal}$$

By the convention of Chapter 20 concerning hydrates we have

$$CO_2(aq) + H_2O(l) = H_2CO_3(aq) \qquad \Delta F^\circ = 0$$

Hence we might equally well write

$$CO_2(aq) + H_2O(l) = H^+ + HCO_3^- \qquad \Delta F^\circ_{298} = 8683 \text{ cal}$$

and likewise, for this reaction,[2]

$$K = \frac{(H^+)(HCO_3^-)}{(CO_2)(H_2O)} = 4.31 \times 10^{-7}$$

It is not experimentally possible to continue such measurements as those of Table 22-3 into the range of extreme dilution, where the ion HCO_3^- would itself largely dissociate according to the equation

$$HCO_3^- = H^+ + CO_3^=$$

[1] T. Shedlovsky and D. A. MacInnes, *J. Am. Chem. Soc.*, **57**, 1705 (1935).

[2] The rate of conversion of unhydrated to hydrated carbon dioxide is slow enough to allow the determination of the amount of H_2CO_3 actually present. K. F. Wissbrun, D. M. French, and A. Patterson, *J. Phys. Chem.*, **58**, 693 (1954), thus obtain 1.72×10^{-2} for the first ionization of hydrated carbon dioxide. See D. M. Kern, *J. Chem. Educ.*, **37**, 14 (1960), for a review of H_2CO_3 data.

But by indirect methods, which we need not discuss at this point, it is possible to calculate the constant of this dissociation.   Thus Harned and Scholes[1] have measured the second dissociation constant of carbonic acid,

$$K_{298} = \frac{(H^+)(CO_3^=)}{(HCO_3^-)} = 4.69 \times 10^{-11} \qquad \Delta F^\circ_{298} = 14{,}091 \text{ cal}$$

Now we may combine this result with that obtained for the first dissociation, either by adding the values of ΔF° or by multiplying the two values of K, and we thus find for the reaction

$$H_2CO_3(\text{aq}) = 2H^+ + CO_3^= \qquad K_{298} = \frac{(H^+)^2(CO_3^=)}{(H_2CO_3)} = 2.02 \times 10^{-17}$$
$$\Delta F^\circ_{298} = 22{,}776 \text{ cal}$$

The Effect on Temperature on Ionization.   The ionization constant of a weak acid or base may be determined at various temperatures, and when such results are examined, a surprising fact emerges.   For most equilibria a plot of log K versus $1/T$ yields a very nearly straight line, but such a plot of log K for acetic acid yields the very pronounced curve of Fig. 22-2.

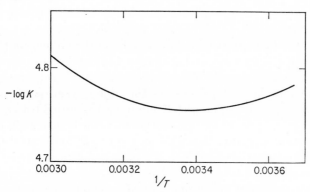

FIG. 22-2. The ionization constant of acetic acid as a function of $1/T$.

This curvature arises from a large negative ΔC_P° for the ionization reaction.   Pitzer[2] examined the data for a number of weak acids and found ΔC_P of ionization to range from -34 to -50 cal/deg mole with a mean near -40.   He further noted that the entropy of ionization is also nearly the same for various weak acids of a given type.   Thus for a first ionization ΔS° is approximately -20 cal/deg mole, while for a second ionization (such as that of HCO_3^-) the values fall near -30 cal/deg mole. These generalizations are useful for estimating the change of ionization

[1] H. S. Harned and S. R. Scholes, *J. Am. Chem. Soc.*, **63**, 1706 (1941).

[2] K. S. Pitzer, *J. Am. Chem. Soc.*, **59**, 2365 (1937).

constant with temperature when direct experimental data are unavailable. In particular, one should never assume ΔC_P to be zero for an ionization reaction; however, a value near -40 cal/deg mole is usually a good estimate for aqueous solutions near room temperature. These matters are considered further in Chapters 25 and 32.

STRONG ELECTROLYTES

In addition to weak electrolytes which show dissociation properties such as those of acetic acid, there are many aqueous electrolytes which show nearly complete dissociation. The latter substances are known as *strong electrolytes*. Furthermore the values of K_Λ calculated from conductances for typical strong electrolytes are not even approximately constant but rather decrease by a factor of 10 or more as the concentration is decreased from 0.1 M to 0.0001 M as shown in Table 22-4 for

TABLE 22-4

m	K_Λ
0.0005	0.036
0.0001	0.051
0.01	0.164
0.1	0.615

potassium chloride. In the first two decades of the present century there was an extended discussion, and at times considerable controversy, about the proper interpretation of these confusing properties of strong electrolytes.[1] We now know that most of the difficulty arose because ionic solutes were expected to conform closely to ideal-solution behavior. But the long-range nature of electrostatic forces is such that ionized solutes deviate substantially from Henry's law at very low concentrations. The theoretical work, particularly that of Debye and Hückel, which demonstrated this point will be given in Chapter 23. As soon as this abnormal behavior was accepted, it became apparent that typical strong electrolytes were either completely dissociated in dilute solution or so nearly fully dissociated that the concentration of undissociated molecules was of no consequence.

The thermodynamic methods for the treatment of strong electrolytes on the basis of 100 per cent dissociation were developed and applied, largely by Lewis[2] and his associates, many years before the theory of the interionic forces and the resulting nonideal-solution properties was published. Before proceeding to the presentation of these thermodynamic

[1] An interesting account of some of the early discussions of strong electrolytes is given in the first edition of the present book on pp. 315–323.

[2] G. N. Lewis, *Proc. Am. Acad.*, **43**, 259 (1907); *Z. physik. Chem.*, **61**, 129 (1907); **70**, 212 (1909); *J. Am. Chem. Soc.*, **34**, 1631 (1912).

methods, however, some further consideration of the difficulties of definition of dissociation seems desirable.

What Do We Mean by Degree of Dissociation? While we have this question before us and before we return to our purely thermodynamic treatment, it may be of interest to view for a moment the logical implications of such a term as degree of dissociation.

Let us consider the equilibrium in the vapor phase, between diatomic and monatomic iodine, and at such a temperature that on the average each molecule of I_2, after it has been formed by combination of two atoms, remains in the diatomic condition 1 min before it redissociates. During this minute such a molecule will traverse several miles in a zigzag path, and after its dissociation each of its constituents will traverse a similar path before it once more combines with another atom. If we imagine an instantaneous photograph of such a gaseous mixture, with such enormous magnifying power as to show us the molecules as they actually exist at any instant, then by counting the single and double molecules we should doubtless find the same degree of dissociation which is actually determined by physicochemical methods.

On the other hand, if we should choose a condition in which the dissociation and reassociation occurs 10^{13} or 10^{14} times as frequently, the atoms of the dissociated molecules would hardly emerge from one another's sphere of influence before they would once more combine with each other or with new atoms. In such a case the time required in the process of dissociation would be comparable with the total time during which the atoms would remain free, and even our imaginary instantaneous photograph would not suffice to tell us the degree of dissociation. For, first, it would be necessary to know how far apart the constituent atoms of a molecule must be to warrant our calling the molecule dissociated. But such a decision would be arbitrary; and, according to our choice of this limiting distance, we should find one or another degree of dissociation.

Until a problem has been logically defined, it cannot be experimentally solved; and it seems evident in such a case as we are now considering that, just as we should obtain different degrees of dissociation by different choices of the limiting distance, so we should expect to find different degrees of dissociation when we come to interpret different experimental methods.

Now it is generally agreed that ionic reactions are among the most rapid of chemical processes, and it is in just such reactions that we should expect to find difficulty in determining, either logically or experimentally, a really significant value of the degree of dissociation.

On the whole, we must conclude that the degree of dissociation and the concentration of the ions are quantities which cannot be defined without some degree of arbitrariness.   If consistency is maintained, no inaccuracy is introduced into a purely thermodynamic treatment.   The problem is quite analogous to the one discussed in Chapter 20 with respect to treatment of NO_2, where one can arbitrarily choose to treat the vapor as a mixture of monomer and dimer or to take only a single component and to treat the interactions between monomers by a virial equation.   Likewise in solutions of electrolytes, we can arbitrarily treat the system as a mixture of undissociated molecules and of ions, or we can treat the electrolyte as being completely dissociated, but taking into account the interaction between ions.

The practical decision of whether or not to recognize undissociated molecules in a strong electrolyte solution may be influenced by the existence of nonthermodynamic properties which distinguish clearly between dissociated and undissociated species.   Thus Raman spectra show distinct features characteristic of HNO_3 molecules and NO_3^- ions, and it is possible to use the intensities of these spectral bands to calculate the degree of dissociation of nitric acid in concentrated solutions.[1]   In dilute aqueous solution the HNO_3 band disappears, and we know that HNO_3 behaves as a typical strong electrolyte.   It should be realized, however, that the selection of the Raman spectrum as the criterion of dissociation is arbitrary, that other experimental phenomena[2] might give different values of the degree of dissociation, and that it is possible to give an exact thermodynamic treatment on the assumption of complete dissociation.

ACTIVITY COEFFICIENTS OF STRONG ELECTROLYTES

For purposes of establishing the solute standard state, the treatment of strong electrolytes is quite unambiguous.   The evidence that we shall review in more detail clearly indicates that Henry's law will be approached in the limit of zero molality only if complete ionization is assumed.   For weak electrolytes like acetic acid, it will be necessary to make two types of extrapolations to infinite dilution in order to establish the solute standard state for the ions as well as the undissociated molecule.   The procedure will be discussed in more detail after we have established a procedure for treating strong electrolytes.

Since the above discussion has shown that it is not profitable, at least for thermodynamic purposes, to attempt to determine separately the concentration of the undissociated substance and of the ions of a *strong* electrolyte, we may therefore employ the same expedient that we have used in the similar case of solvated solutes in studying the activities.

---

[1] O. Redlich and G. C. Hood, *Discussions Faraday Soc.*, **24,** 87 (1957), and references there cited.

[2] M. Eigen, *Discussions Faraday Soc.*, **24, 25** (1957).

Symmetrical Electrolytes. It is our custom to denote the activity of a solute by a_2. If this solute is a substance like sodium chloride, we may denote by a_+ and a_-, respectively, the activities of cation and anion, while a_2 is called the activity of undissociated NaCl, or, more simply, of NaCl. In the case of a binary electrolyte like this, the thermodynamic equation of chemical equilibrium takes the form,

$$\frac{a_+ a_-}{a_2} = K \qquad (22\text{-}3)$$

where at any given temperature K is an exact constant.

At infinite dilution, we make the activity of each ion of sodium chloride equal to its molality, which, assuming complete dissociation, is equal to the stoichiometrical molality of NaCl. Now in the complete absence of any reliable information as to the concentration of the undissociated salt, we shall find it extremely convenient to choose our standard state of that substance so that the K in Eq. (22-3) becomes unity. We thus define the activity of NaCl as the product of the activities of its two ions,

$$a_+ a_- = a_2 \qquad (22\text{-}4)$$

In the same manner we may treat possible intermediate ions of a strong electrolyte. Thus if we assume that barium chloride gives the intermediate ion $BaCl^+$, its activity multiplied by the activity of Cl^- will be written equal to the activity of $BaCl_2$ and likewise the activity of $BaCl^+$ will be written as equal to the product of the activities of Ba^{++} and Cl^-. This convention gives $\Delta F^\circ = 0$ for each of the reactions

$$BaCl_2 = Ba^{++} + 2Cl^-$$
$$BaCl_2 = BaCl^+ + Cl^-$$
$$BaCl^+ = Ba^{++} + Cl^-$$

An interesting case arises when a strong electrolyte dissociates to give an ion which itself dissociates as a weak electrolyte. Thus NaH_2PO_4 is a strong electrolyte, and we write

$$NaH_2PO_4 = Na^+ + H_2PO_4^- \qquad \Delta F^\circ = 0 \qquad K = 1$$

but $H_2PO_4^-$ is a weaker electrolyte than acetic acid, and the next ion $HPO_4^=$ is much weaker still. Pitzer[1] has reviewed the phosphoric acid ionization data and gives

$$H_2PO_4^- = H^+ + HPO_4^= \qquad K_{298} = 6.2 \times 10^{-8} \qquad \Delta F^\circ_{298} = 9830 \text{ cal}$$
$$HPO_4^= = H^+ + PO_4^{3-} \qquad K_{298} = 10^{-12} \qquad \Delta F^\circ_{298} = 16{,}370 \text{ cal}$$

Our methods in this chapter are designed primarily for strong electrolytes. In the treatment of a mixed type, such as monosodium phosphate, we shall ordinarily combine these methods with those for weak electrolytes. Nevertheless, this is a matter of choice, and it is perfectly rigorous to apply the methods of the present chapter to any weak electrolyte if we are unable, or if we do not choose, to fix separately the standard states of the ions and of the undissociated substance.

[1] K. S. Pitzer, *J. Am. Chem. Soc.*, **59**, 2365 (1937).

At infinite dilution, the molality of anion and cation being the same,

$$a_+ = a_- = a_2^{1/2} \tag{22-5}$$

How far this equality of the two ion activities extends into the range of finite concentration is a question which experiment alone can decide. However, if we consider the geometrical mean of the two ion activities, denoted by $a_\pm$, we see that at all concentrations

$$a_\pm = (a_+ a_-)^{1/2} = a_2^{1/2} \tag{22-6}$$

and this mean activity will play an important part in our calculations.

The mean activity of the ions, $a_\pm$, divided by the molality of the electrolyte, gives the activity coefficient, which is denoted by $\gamma_\pm$.

$$\gamma_\pm = \frac{a_\pm}{m}$$

Single-ion Activities.   In order to determine the activity of a single ion in a solution, it is necessary to measure experimentally some process which transfers that single charged species into or out of the solution. The enormous magnitude of space-charge energies prevents such measurements by conventional methods which require transfer of macroscopic amounts of material.   In principle there is the possibility of measurements in which only a few ions are transferred under conditions in which the net electrical charge of the solution is also measured or controlled. The authors are not aware that any macroscopic measurements of this sort have been successfully carried out; hence all existing experimental information is limited to mean activity coefficients and partial molal properties of neutral combinations of ions.   Nevertheless, single-ion properties are potentially measurable.

Several attempts have been made to infer single-ion properties from theoretical arguments.   While some of these arguments are quite persuasive, there is usually some point of real uncertainty and we shall not attempt to discuss such matters here.

Unsymmetrical Electrolytes.   So far we have been considering the case of a binary electrolyte such as KCl or $CuSO_4$.   When we treat the more complicated types, such as K_2SO_4, $K_4Fe(CN)_6$, and $La_2(SO_4)_3$, our equations become a little more complicated.   If an electrolyte dissociates into $\nu \, (= \nu_+ + \nu_-)$ ions according to the equation $X = \nu_+ X^+ + \nu_- X^-$, we write for equilibrium

$$(a_+^{\nu_+})(a_-^{\nu_-}) = a_2 \qquad a_\pm = (a_2)^{1/\nu} \tag{22-7}$$

When one is dealing with an electrolyte more complicated than a 1-1 electrolyte, it is convenient to consider separately the concentrations m_+ and m_- of the ions and also to consider the individual activity coefficients of the ions, for example, $\gamma_- = a_-/m_-$. Inasmuch as it is not usually feasible to carry out measurements corresponding to the removal of ions of a single charge from solution, individual activity coefficients may not have any operational significance. However, it is quite proper and of some utility to use individual activity coefficients in our discussion as long as they are applied as products or ratios which are operationally significant.

If we apply Eq. (20-39) to the reaction

$$X = \nu_+ X^{+z+} + \nu_- X^{-z-} \quad \text{with } \nu_+ z_+ = \nu_- z_-$$

$$K = \frac{m_+^{\nu_+} m_-^{\nu_-} \gamma_+^{\nu_+} \gamma_-^{\nu_-}}{a_2} \tag{22-8}$$

Since we do not choose to determine separately the concentration of undissociated and ionized molecules, we follow our usual convention of choosing our standard state such that there would be no difference in the standard free energies of dissociated and undissociated forms and therefore that K is unity in Eq. (22-8). Thus we obtain

$$a_\pm = a_2^{1/\nu} = (m_+^{\nu_+} m_-^{\nu_-} \gamma_+^{\nu_+} \gamma_-^{\nu_-})^{1/\nu} \tag{22-9}$$

Likewise we shall take a geometric mean of the ion activity coefficients

$$\gamma_\pm = (\gamma_+^{\nu_+} \gamma_-^{\nu_-})^{1/\nu} \tag{22-10}$$

Substituting into Eq. (22-9) one obtains

$$\gamma_\pm = \frac{a_\pm}{(m_+^{\nu_+} m_-^{\nu_-})^{1/\nu}} = \frac{a_\pm}{m_\pm} \tag{22-11}$$

This equation is applicable to any solution, whether the positive and negative ions are added together as a single salt or as a mixture of salts. If we are adding a single salt to the solution, then $m_+ = \nu_+ m$ and $m_- = \nu_- m$, where m is the molality of the salt added. For such a solution of a pure salt, one then obtains

$$\gamma_\pm = \frac{a_\pm}{m(\nu_+^{\nu_+} \nu_-^{\nu_-})^{1/\nu}} \tag{22-12}$$

Since our activity coefficients refer to the solute standard state, we wish to ensure that the activity coefficient will become equal to unity at infinite dilution. In a solution of barium chloride, which is dilute enough, $a_+ = m$, $a_- = 2m$, and $a_\pm = [(m)(2m)^2]^{1/3} = 2^{2/3}m$. By defining the activity coefficient by Eq. (22-12), $\gamma = a_\pm/(2^{2/3}m)$, one confirms that it does become equal to unity at infinite dilution. In the case of $La_2(SO_4)_3$, which gives two positive and three negative ions, the corresponding factor is $(2^2 3^3)^{1/5}$.

Having established these conventions, we are now ready to undertake the arduous but very interesting and important task of determining the activity of electrolytes dissolved in water. We may study the activity a_2 of the undissociated electrolyte or the geometrical mean of the activities of the ions, namely, $a_\pm = a_2^{1/\nu}$. The establishment of the solute standard state for electrolytes and the application of the above conventions will be illustrated by examining the vapor pressures of the hydrogen halides in equilibrium with their aqueous solutions. Few electrolytes are sufficiently volatile to permit the determination of their activities from their own vapor pressures. However, when this is possible, it furnishes a method of great simplicity and will be particularly useful for presentation of a concrete example of the use of the conventions for electrolytes.

ACTIVITY FROM THE VAPOR PRESSURE OF THE SOLUTE

Bates and Kirschman[1] have made a careful study at 25°C of the partial pressures of hydrogen chloride, bromide, and iodide over their aqueous solutions. Their results are given in Table 22-5, in which the first

TABLE 22-5. ACTIVITY COEFFICIENTS OF HYDROGEN HALIDES AT 25°C

m	HCl		HBr		HI	
	$p \times 10^4$, atm	$k_1\gamma$	$p \times 10^4$, atm	$k_2\gamma$	$p \times 10^4$, atm	$k_3\gamma$
4	0.2395	0.001222				
5	0.6974	0.001669				
6	1.842	0.002263	0.01987	0.0002351	0.00750	0.0001444
7	4.579	0.003058	0.04868	0.0003152	0.02395	0.0002213
8	11.10	0.004171	0.1171	0.0004280	0.08555	0.0003664
9	25.39	0.005586	0.2974	0.0006058	0.3882	0.0006928
10	55.26	0.007436	0.7763	0.0008815	1.737	0.001317
11	...	...	1.987	0.001280		

column gives the molality; the second, fourth, and sixth, the partial pressures in atmospheres of the halides; and the third, fifth, and seventh, a quantity which is proportional to the activity coefficient. If a_2 is the activity of one of the undissociated halides and $a_\pm$ is the mean activity of its ions, we have defined the activity coefficient as $a_\pm/m = a_2^{1/2}/m$. Hence, if we consider a_2 as proportional to the vapor pressure p, then $p^{1/2}/m$ is a quantity proportional to the activity coefficient and may be written as $k\gamma_\pm$.

If these measurements could be carried out to high dilutions, k could be determined at the limit approached by $k\gamma$ at infinite dilution. This,

[1] S. J. Bates and H. D. Kirschman, *J. Am. Chem. Soc.*, **41**, 1991 (1919).

however, is experimentally impossible in the present case, but the activity coefficients of the hydrogen halide can also be determined by the freezing-point method, and by galvanic-cell methods, which we are to review in more detail shortly, to very low concentrations. By combination of data from these three sources one can construct a complete vapor-pressure vs. molality curve for hydrogen chloride as given in Fig. 22-3. If HCl were a nonelectrolyte, this curve would be expected to take the form of Fig. 20-2 and approach a limiting finite slope of pressure vs. molality

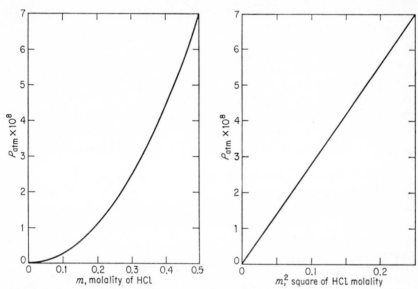

FIG. 22-3. Partial pressure of HCl in water at 25°C.

FIG. 22-4. Partial pressure of HCl in water at 25°C.

corresponding to Henry's law. This limiting slope would become the standard-state fugacity by the usual convention for establishment of the solute standard state. However, for the hydrochloric acid solution the limiting slope is seen to be zero. From our previous discussion it is clear that this is due to the fact that hydrochloric acid is dissociated into ions, and Henry's law will not be approached if the ionization is not recognized. For Henry's law to be approached, the fugacity must approach proportionality to m^2. In Fig. 22-4, the derived partial pressure curve of HCl is plotted against m^2. At low pressures of HCl this would be equivalent to a plot of fugacity of HCl against m^2. It is seen that a finite limiting slope is approached and the standard fugacity of the solute standard state is taken as the limiting slope of fugacity versus m^2. This limit is also equal to the fugacity of the hypothetical solution obtained by extrapolating the limiting slope to unit molality. Thus the above equations for $a_\pm$ and $\gamma_\pm$

correspond to a standard state with a fugacity given by $f° = \lim\limits_{m\to 0} f/m^2$, where both the fugacity and molality are for the solute.   This is equivalent to establishing a solute standard state for each ion independently by the general procedure that has been outlined in Chapter 20.   Since we do not determine individual ion fugacities independently, the above procedure corresponds to establishing the product of the individual-ion standard-state fugacities.   The plot of f_{HCl} versus m^2 in Fig. 22-4 can be converted to a plot of a_{HCl} versus m^2 by dividing the ordinate by the standard-state fugacity, thus obtaining a limiting slope of unity.   This corresponds to the equations

$$a = \frac{f}{f°} = a_{\pm}{}^2 = m^2\gamma_{\pm}{}^2 \tag{22-13}$$

and $\gamma_{\pm}$ must approach unity as molality approaches zero.   The corresponding equations for the standard-state fugacity for salts more complicated than the 1-1 salts can readily be derived from Eq. (22-11).

ACTIVITY COEFFICIENTS FROM ELECTROMOTIVE FORCE

The Activity of Hydrochloric Acid.   If in a cell filled with aqueous hydrochloric acid of given composition we have a hydrogen electrode (at a partial pressure of hydrogen of 1 atm) and an electrode of mercury and solid mercurous chloride, then for 1 equiv of electricity passing through the cell the following reaction occurs:

$$\tfrac{1}{2}H_2(g, 1 \text{ atm}) + \tfrac{1}{2}Hg_2Cl_2(s) = Hg(l) + HCl(aq) = Hg(l) + H^+ + Cl^-$$

Since $n = 1$, the change in free energy is

$$\Delta F = -n\mathfrak{F}\mathcal{E} = -\mathfrak{F}\mathcal{E}$$

If the temperature is fixed and the partial pressure of the hydrogen is maintained at 1 atm, the emf will depend solely upon the molality of the acid.

If we measure two such cells, the combination is called a concentration cell without liquid junction, or, less accurately, a concentration cell without transference.   If in the first cell the molality is m and in the second m', the difference between the two values of the emf, $\mathcal{E}$ and $\mathcal{E}'$, measures the change in free energy per mole in the transfer of acid from m' to m, $HCl(aq, m') = HCl(aq, m)$,

$$\Delta F = \mathbf{F}_2 - \mathbf{F}_2' = RT \ln \frac{a_2}{a_2'} = -\mathfrak{F}(\mathcal{E} - \mathcal{E}') \tag{22-14}$$

If in the one cell we have HCl at unit activity, so that we may write

$$\mathbf{F}_2' = \mathbf{F}_2° \qquad \mathcal{E}' = \mathcal{E}° \qquad a_2' = 1 \tag{22-15}$$

then at any other concentration the partial molal free energy and the activity of the solute are given by the equation

$$\bar{F}_2 - \bar{F}_2^0 = RT \ln a_2 = -\mathfrak{F}(\mathcal{E} - \mathcal{E}^\circ) \qquad (22\text{-}16)$$

Or if we wish to deal with the mean activity of the ions, we note that $a_{\pm}^2 = a_2$, or $2 \ln a_{\pm} = \ln a_2$, and

$$2RT \ln a_{\pm} = 2RT \ln m\gamma_{\pm} = -\mathfrak{F}(\mathcal{E} - \mathcal{E}^\circ) \qquad (22\text{-}17)$$

Such cells have been measured over a wide range of concentration and over a considerable range of temperature by numerous investigators and will be illustrated by the recent precise measurements of Hills and Ives.[1] Similar studies involving the cell using silver chloride instead of mercurous chloride have been carried out between 0 and 90° and over the composition range 0.003 to 16 M.[2-4] Measurements have been made on hydrobromic acid solutions by using a silver bromide electrode.[5,6] We can easily obtain the ratio of the activity of hydrochloric acid between any two concentrations from Eq. (22-14), but the problem of determining $\mathcal{E}^\circ$, and thence the absolute value of the activity at any one concentration, is a far more difficult one, since it involves extrapolation to infinite dilution. Any such extrapolation, depending upon emf measurements alone, must give large weight to measurements at high dilution, where the difficulties in securing accurate values for the emf are very great.

The extrapolation to infinite dilution can best be made by employing an expedient entirely similar to that which we used in Eq. (20-18). Changing Eq. (22-17) to common logarithms, and introducing the several numerical factors, we find

$$0.1183 \log a_{\pm} = \mathcal{E}^\circ - \mathcal{E} \qquad \text{volts} \qquad (22\text{-}18)$$

and to obtain the activity coefficient we subtract $0.1183 \log m$ from each member, thus:

$$0.1183 \log \frac{a_{\pm}}{m} = 0.1183 \log \gamma_{\pm} = \mathcal{E}^\circ - (\mathcal{E} + 0.1183 \log m) \quad (22\text{-}19)$$

At infinite dilution the first member disappears. If we plot the quantity in parentheses from the data of Hills and Ives as ordinate, against any function of m as abscissa, we can see that the limit approached by the ordinate at infinite dilution is equal to $\mathcal{E}^\circ$, which we take as 0.2680 volt.

[1] G. J. Hills and D. J. G. Ives, *J. Chem. Soc.*, **1951**, 318.

[2] H. S. Harned and R. W. Ehlers, *J. Am. Chem. Soc.*, **54**, 1350 (1932); **55**, 2179 (1933).

[3] R. G. Bates and V. E. Bower, *J. Research Natl. Bur. Standards*, **53**, 283 (1954).

[4] G. Åkerlof and J. W. Tare, *J. Am. Chem. Soc.*, **59**, 1855 (1937).

[5] H. S. Harned, A. S. Keston, and J. G. Donelson, *J. Am. Chem. Soc.*, **58**, 2989 (1936).

[6] B. B. Owen and L. Foering, *J. Am. Chem. Soc.*, **58**, 1575 (1936).

At any concentration this value, subtracted from the ordinate, gives the value of $-0.1183 \log \gamma_{\pm}$.

This plot which we have just described is given in Fig. 22-5, with $m^{1/2}$ as abscissa, not only since this serves to compress the wide range of concentration into reasonable compass, but also since it should give a curve

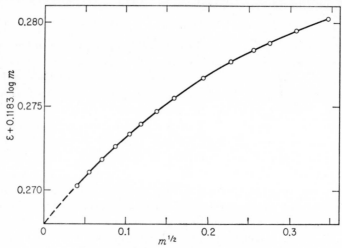

FIG. 22-5. Galvanic-cell data for HCl.

approximating to linearity at high dilution (for reasons which will be discussed later).

Table 22-6 gives the detailed data for HCl at 25°C from all the sources listed above.   The several columns show the molality of the hydrochloric acid, its mean activity coefficient, the mean activity of the ions, the activity of HCl, and finally the partial molal free energy of HCl less that in the standard state, $\bar{F}_2 - \bar{F}_2^{\circ} = RT \ln a_2 = 2RT \ln a_{\pm} = 2RT \ln m\gamma_{\pm}$.

The activity coefficient of any electrolyte relative to the solute standard starts at unity at zero molality and very frequently reaches a minimum. In the present case γ passes through the minimum even below $\frac{1}{2}$ M; from there it rises rapidly, passing unity between 1 M and 2 M, and reaching 42.4 at 16 M.   While in dilute solutions some workers have considered the activity coefficient as a thermodynamic measure of the degree of dissociation, such an idea loses all value in these concentrated solutions, where it is misleading to speak of a "degree of dissociation" of over 4000 per cent.

The enormous difference between the electrolyte solutions which we are now considering and the solutions that we have discussed in Chapter 21 is best seen in the great variation of a_2, the activity of HCl as such. For example, a_2 increases half a million times between 1 M and 16 M. Since this quantity, a_2, may be regarded as proportional to the vapor

Table 22-6. Activities in Aqueous Hydrochloric Acid Solutions at 25°C

m	$\gamma_\pm$	$a_\pm$	a_2	$\bar{F}_2 - \bar{F}_2^\circ$, cal
0.0005	0.975	0.000488	0.00000024	−9030
0.001	0.965	0.000965	0.00000093	−8230
0.002	0.952	0.001904	0.00000363	−7420
0.005	0.928	0.00464	0.00002154	−6370
0.01	0.904	0.00904	0.0000817	−5580
0.02	0.875	0.0175	0.0003060	−4794
0.05	0.830	0.0415	0.001721	−3771
0.1	0.796	0.0796	0.00634	−2999
0.2	0.767	0.1534	0.02352	−2222
0.3	0.756	0.2268	0.0514	−1759
0.4	0.755	0.3020	0.0912	−1432
0.5	0.757	0.3785	0.1433	−1151
0.6	0.763	0.4578	0.2096	− 926
0.7	0.772	0.540	0.2919	− 730
0.8	0.783	0.634	0.3928	− 554
0.9	0.795	0.716	0.512	− 397
1	0.809	0.809	0.655	− 251
2	1.009	2.018	4.072	+ 832
3	1.316	3.948	15.59	1627
4	1.762	7.048	49.68	2314
5	2.38	11.19	141.6	2935
6	3.22	19.32	373.3	3509
7	4.37	30.59	935.7	4053
8	5.90	47.20	2,228	4567
9	7.94	71.46	5,106	5059
10	10.44	104.4	10,900	5508
12	17.25	207	42,850	6319
14	27.3	382	146,100	7046
16	42.4	678	460,000	7726

pressure of HCl, we may compare the values of Table 22-5 and Table 22-6. Assuming from the latter table that $\gamma_\pm = 2.38$ at 5 M, k_1 of the former table becomes 0.000701. The several values of $k_1\gamma_\pm$ divided by this constant give the following values of $\gamma_\pm$ which, except for the last concentration, are in excellent agreement with Table 22-6.

Table 22-7. Activity Coefficients of HCl

m	4	5	6	7	8	9	10
$\gamma_\pm$	1.74	(2.38)	3.23	4.36	5.95	7.97	10.61

The extrapolation of Fig. 22-5, corresponding to extrapolating Eq. (22-19) to infinite dilution and unit value of the activity coefficient, is equivalent to the previous procedure discussed for establishing the solute standard state. It is clear that the data at low molality are very critical for fixing the extrapolation of zero molality. As will be discussed in Chapter 23, both Bronsted[1] and Lewis and Randall[2] found empirically that log $\gamma_{\pm}$ varied with the square root of the molality at low molality, and the interionic theory of Debye and Hückel[3] not only confirmed that log $\gamma_{\pm}$ approached proportionality to the square root of the molality at low molality, but their theory also yielded the constant of proportionality. The substitution of this limiting law into Eq. (22-19) shows that the quantity on the right-hand side which is plotted on the ordinate of Fig. 22-5 should approach asymptotically to the dashed line, which represents the limiting proportionality to the square root of the molality, as the molality approaches zero. This is the reason for making the plot of Fig. 22-5 a plot against the square root of the molality. However, appreciable deviations from the limiting Debye-Hückel law occur at very low molalities. A more satisfactory method of carrying out the extrapolation to infinite dilution to establish the solute standard state is to use an equation[4] such as (23-35). For a 1-1 electrolyte

$$\log \gamma_{\pm} = -A_{\gamma} \frac{m^{\frac{1}{2}}}{1 + m^{\frac{1}{2}}} + Bm \qquad (22\text{-}20)*$$

A_{γ} is a numerical coefficient arising from the Debye-Hückel theory; at 25°C, $A_{\gamma} = 0.511$. Substitution of this equation for log $\gamma_{\pm}$ in Eq. (22-19) and rearranging of terms yields

$$\mathcal{E}^{o\prime} = \mathcal{E}^{o} - 0.1183Bm = \mathcal{E} + 0.1183 \left(\log m - 0.511 \frac{m^{\frac{1}{2}}}{1 + m^{\frac{1}{2}}} \right) \qquad (22\text{-}21)*$$

Now, if $\mathcal{E}^{o\prime}$, as given by the quantity on the right-hand side, is plotted against m, a linear plot is obtained as shown in Fig. 22-6, which, with reliable data for the low molality range, can be extrapolated fairly unambiguously to a limiting $\mathcal{E}^{o}$ value. This extrapolation procedure thus makes use of the limiting behavior of the activity coefficient as established by the Debye-Hückel theory, as well as the functional dependence upon molality of the next higher term as established empirically.[4,5] In principle one should add terms in higher powers of m to the right side of Eq. (22-20)* or else regard B as a function of m. However, the linearity

[1] J. N. Bronsted, *J. Am. Chem. Soc.*, **44**, 938 (1922).

[2] G. N. Lewis and M. Randall, *J. Am. Chem. Soc.*, **43**, 1112 (1921).

[3] P. Debye and E. Hückel, *Physik. Z.*, **24**, 185, 384 (1923), **25**, 97 (1924).

[4] E. A. Guggenheim, *Phil. Mag.*, **19**, 588 (1935). See also E. Hückel, *Physik Z.*, **26**, 95 (1925); G. Scatchard, *Chem. Rev.*, **19**, 309 (1936).

[5] E. A. Guggenheim and J. C. Turgeon, *Trans. Faraday Soc.*, **51**, 747 (1955).

of the plot in Fig. 22-6 indicates that Eq. (22-20)* is a good approximation in dilute solutions with B constant. The additional information implied by Eq. (22-20)* is of great aid in making reliable and consistent extrapolations from experimental data at finite concentrations to the solute standard state at infinite dilution.

In Chapter 24 galvanic cells will be discussed in more detail, and a number of other types of cells which can also yield activity coefficients of

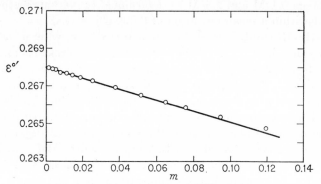

FIG. 22-6. Extrapolation of HCl cell data.

electrolytes will be considered. The extrapolation to infinite dilution to establish the solute standard state will be similar to that illustrated here for hydrochloric acid solutions.

ACTIVITY FROM THE VAPOR PRESSURE OF THE SOLVENT

The method of calculating a_2, the activity of the solute, from a_1, the activity of the solvent, we have already fully discussed in Chapter 20. It is applicable without modification to electrolytic solutions. In dilute solutions vapor-pressure measurements are far less accurate than emf measurements or freezing-point measurements, which we shall discuss in Chapter 26, but in concentrated solutions they furnish a very satisfactory means of determining the ratio of the activities of the solute between two concentrations.

The vapor pressure of water in equilibrium with an aqueous solution can be determined directly by either static or dynamic methods. The direct static method is perhaps best illustrated by the apparatus of Gibson and Adams[1] as used by Shankman and Gordon.[2] Measurements of the activity of water could be reproduced within 1 part in 2000.

The transpiration method of determining activities of water involves the saturation of an inert gas by the aqueous solution. Measurements

[1] R. E. Gibson and L. H. Adams, *J. Am. Chem. Soc.*, **55**, 2679 (1933).

[2] S. Shankman and A. R. Gordon, *J. Am. Chem. Soc.*, **61**, 2370 (1939).

by Becktold and Newton[1] indicate that measurements on calcium chloride and barium chloride yield water activities with a probable error of 1 part in 10,000. Brown and Delaney[2] have recently revived a method depending upon a temperature difference between solvent and solution to equalize the vapor-pressure difference, using a very sensitive differential manometer to observe the null point. Robinson[3] has analyzed their data and concludes that the activity of water in potassium chloride solutions between 0.1 M and 2.3 M as determined by various methods is in agreement within a mean deviation of 0.002 per cent.

Isopiestic Method.   Once the vapor pressure of water as a function of molality has been determined accurately for one solution, it is not necessary to repeat absolute measurements of the activity of water for other solutions. The isopiestic method as introduced by Bousfield[4] and refined by Sinclair[5] is a simple, convenient method which yields accurate results. The experimental arrangement consists of a glass desiccator containing a copper block in good thermal contact with several dishes. The desiccator is maintained in a thermostat and rocked to agitate the solutions. Depending upon the concentration, at least 1 to 4 days is necessary for the approach to equilibrium within the accuracy of the analysis, which is about 0.1 per cent. When equilibrium has been reached, all solutions have the same vapor pressure of water. From analyses of the solutions including a reference solution for which the vapor pressure of water is accurately known as a function of molality, one then has the partial pressure of water in equilibrium with the solution being studied. Using the Gibbs-Duhem equation as expressed in Eq. (20-25) or (20-26), one can obtain the activity coefficient of the solute from the determination of the vapor pressure and therefore the activity and activity coefficient of the water as a function of molality. As isopiestic methods are usually not sufficiently accurate for molalities less than 0.1, such measurements alone do not allow an accurate extrapolation to the solute standard state and these data must be supplemented by other measurements.

In order to use the vapor pressure of water to determine the activity coefficients of the solutes, it is necessary, especially in dilute solutions, to determine the activity of water to a very high degree of accuracy, for the activity coefficient of the solute may differ from unity by a considerable

[1] M. F. Becktold and R. F. Newton, *J. Am. Chem. Soc.*, **62**, 1390 (1940).

[2] O. L. I. Brown and C. M. Delaney, *J. Phys. Chem.*, **58**, 255 (1954).

[3] R. A. Robinson, *J. Phys. Chem.*, **60**, 501 (1956).

[4] W. R. Bousfield, *Trans. Faraday Soc.*, **13**, 401 (1918).

[5] D. A. Sinclair, *J. Phys. Chem.*, **37**, 495 (1933). See also R. A. Robinson and D. A. Sinclair, *J. Am. Chem. Soc.*, **56**, 1830 (1934); and G. Scatchard, W. J. Hamer, and S. E. Wood, *J. Am. Chem. Soc.*, **60**, 3061 (1938).

amount, and yet the activity of the solvent differs from unity by only a very small amount.   To express better the deviation from ideal behavior of the solvents, Bjerrum[1] introduced another function, termed the osmotic coefficient of the solvent.[2]   The practical osmotic coefficient ϕ, which we have already used in Chapter 20, is defined by

$$\ln a_1 = -\frac{\nu m M_1}{1000} \phi \qquad (22\text{-}24)$$

For an ideal dilute solution under 1 atm pressure, ϕ is unity.[3]   The osmotic coefficient is particularly useful for treating isopiestic data.   The equilibration of two solutions until their water vapor pressures become equal and determination of their molalities yields the isopiestic ratio

$$R = \frac{\nu_2 m_2}{\nu_3 m_3} \qquad (22\text{-}25)$$

where ν_2 is the number of ions produced by 1 mole of the reference solute and ν_3 is the number of ions produced by 1 mole of the solute under study. Knowing the osmotic coefficient of the reference solute, one then obtains directly the osmotic coefficient of the other solute from Eq. (22-24), which defines the osmotic coefficients.

$$\phi_3 = R\phi_2 \qquad (22\text{-}26)$$

From knowledge of the osmotic coefficient, one can apply the Gibbs-Duhem equation to obtain the activity coefficient.

Thus expressing Eq. (20-22) in molality with $x_2/x_1 = m/(1000/M_1)$ we obtain

$$\frac{1000}{M_1} d \ln a_1 = -m\, d \ln a_2 = -\nu m\, d \ln a_\pm = -\nu m\, d \ln (\gamma_\pm m_\pm) \quad (22\text{-}27)$$

---

[1] N. Bjerrum, *Z. Elektrochem.*, **24**, 259 (1907); *Proc. 7th Intern. Congr. Appl. Chem.*, London, 1909.

[2] In terms of mole fraction, the rational osmotic coefficient g is defined by

$$\ln a_1 = g \ln x_1 \qquad (22\text{-}22)$$

For an ideal solution, $g = 1$.   One can express the mole fraction of the solvent in terms of molality as $\ln x_1 = -\ln (1 + \nu m M_1/1000)$.   By expanding $\ln x_1$ in a series, one obtains

$$\ln a_1 = -g \left[\frac{\nu m_1 M_1}{1000} - \frac{\frac{1}{2}(\nu m_1 M_1)^2}{1000^2} + \cdots \right] \qquad (22\text{-}23)$$

[3] When the pressure of the solution differs appreciably from 1 atm, it is convenient to modify Eq. (22-24) to

$$\ln \frac{a_1}{\Gamma_1} = \frac{-\nu m M_1}{1000} \phi \qquad (22\text{-}24a)$$

where Γ_1 is the activity of the pure-water reference state under the pressure of the solution.   This ϕ approaches unity as molality approaches zero even in a solution under such a pressure that the activity does not approach unity.

From Eq. (22-12), one sees that $d \ln m_{\pm} = d \ln m$ and with Eq. (22-24)

$$\frac{1000}{M_1} d \ln a_1 = -\nu m \, d \ln (\gamma_{\pm} m) = -\nu \, d(\phi m) \qquad (22\text{-}28)$$

$$d \ln \gamma_{\pm} + d \ln m = \frac{1}{m} \phi \, dm + d\phi$$

$$d \ln \gamma_{\pm} = d\phi + (\phi - 1) \, d \ln m$$

$$\ln \gamma_{\pm} = \phi - 1 + \int_0^{m'} (\phi - 1) \, d \ln m \qquad (22\text{-}29)$$

Since both ϕ and $\gamma_{\pm}$ are known for the reference solution, it is more useful to relate $\gamma_3 = \gamma_{\pm}$ of solute 3 to $\gamma_2 = \gamma_{\pm}$ of solute 2. Equating Eq. (22-28) for both solutions with equal water activities, we obtain

$$-\nu_2 m_2 \, d \ln (\gamma_2 m_2) = -\nu_3 m_3 \, d \ln (\gamma_3 m_3)$$

$$d \ln \gamma_3 = R \, d \ln (\gamma_2 m_2) - d \ln m_3$$

$$= d \ln \gamma_2 + (R - 1) \, d \ln (\gamma_2 m_2) + d \ln \frac{m_2}{m_3}$$

$$\ln \gamma_3 = \ln \gamma_2 + \ln R + \int_0^{m'} (R - 1) \, d \ln (\gamma_2 m_2) \qquad (22\text{-}30)$$

since γ_3, γ_2, and R all approach unity as m_2 approaches zero. A plot of $(R - 1)/\gamma_2 m_2$ versus $\gamma_2 m_2$ yields the integral of Eq. (22-30) by graphical integration.

Normally it is not possible to obtain accurate values of R and thus accurate values of ϕ_3 below 0.1 M, and the evaluation of the integral of Eq. (22-30) to allow proper extrapolation to the solute standard state is not possible from isopiestic measurements alone. However, it is often possible to determine activity coefficients or osmotic coefficients from freezing-point measurements or cell measurements below 0.1 M and thus evaluate the integral between zero and 0.1 M.

It is of interest to evaluate the integral of Eq. (22-30) at low molalities by using Eq. (22-20)*, which is usually valid up to 0.1 M, and to determine the limiting variation of R with molality. By integration of Eq. (22-28) or simplification of Eq. (23-39)*, it is seen that Eq. (22-20)* can be expressed in terms of the osmotic coefficient for a 1-1 electrolyte as

$$1 - \phi = \frac{2.303}{3} A_{\gamma} m^{1/2} \sigma(m^{1/2}) - \frac{2.303}{2} Bm \qquad (22\text{-}31)\text{*}$$

where B from Eq. (22-20)* is $(2/2.303)\beta$ and σ is a function which approaches unity at low m and which is defined in Eq. (23-40). From Eq. (22-26), $R - 1 = (\phi_3 - \phi_2)/\phi_2$, and at low molalities, where ϕ_2 approaches unity and m_1 approaches m_2, one can see from Eq. (22-31)* that the quantity $(R - 1)/m$ should approach $\beta_3 - \beta_2$. This conclusion can be confirmed more rigorously by combining the various equations and expanding in appropriate series. The process is cumbersome, however, and will not be given in detail. The result

$$\frac{R-1}{m_2} = \beta_3 - \beta_2 = \frac{2.303}{2} (B_3 - B_2) \qquad (22\text{-}32)^*$$

is valid in the limit as $m \to 0$ and may be used to aid in the evaluation of the integral in Eq. (22-30) when values of B_2 and B_3 are available from freezing-point data or cell measurements.

Generally Eq. (22-30) is most simply applied in the form

$$\ln \gamma_3'' = \ln \gamma_2'' + \ln R'' + \int_{m_2'}^{m_2''} \frac{R-1}{\gamma_2 m_2} d\gamma_2 m_2 + \ln \frac{\gamma_3'}{\gamma_2' R'} \qquad (22\text{-}33)$$

where the single primes refer to the lowest molality at which isopiestic data are reliable and at which γ_2' and γ_3' are available from freezing-point depression or cell data. γ_2' and γ_3' relate not to salt solutions of the same molality but to molalities m_2' and $m_3' = \nu_2 m_2'/\nu_3 R$, which are found to be in isopiestic equilibrium. The same is true of γ_3'' and γ_2''. In a similar manner, Eq. (22-29) can be evaluated in the form

$$\ln \frac{\gamma_\pm''}{\gamma_\pm'} = \phi'' - \phi' + \int_{m'}^{m''} (\phi - 1) \, d \ln m \qquad (22\text{-}34)$$

where $\gamma_\pm'$ and ϕ' for m', the lowest molality at which isopiestic data are reliable, are calculated from cell or freezing-point data by the use of Eqs. (22-20)* and (22-31)* or their equivalent. Although Eqs. (22-20)* and (22-31)* have been presented for a 1-1 salt to illustrate their use, the more general forms for any type of salt, including mixtures, are given in Eqs. (23-35)*, (23-38)*, and (23-39)*. It will be noted that Eq. (22-29) and (22-30) and the resulting equations (22-33) and (22-34) were derived in a general form applicable to complex salts as well as 1-1 salts. The careful evaluation of the terms in Eqs. (22-33) and (22-34) corresponding to the portions of the integrals of Eqs. (22-29) and (22-30) at low molalities is very important to ensure that extrapolation to the solute standard state has been properly carried out.

Osmotic-pressure Method. Of the methods for determination of the activity coefficient of electrolytes which depend upon the measurement of the activity of the solvent, the freezing-point-depression method, which will be discussed in Chapter 26, and the isopiestic method, discussed in this chapter, are the most commonly used methods. However, the osmotic-pressure method has some important applications, particularly for solutes with large molecular weights. The concept of osmotic pressure has been discussed in Chapter 19, and the limiting equation for dilute solutions has been given in Eq. (19-17)*. Using the method of Eq. (19-15) with molality in place of mole fraction and activity in place of

fugacity, one obtains

$$\frac{\partial(P - P°)}{\partial m} = - \frac{RT}{\bar{v}_1} \frac{\partial \ln a_1}{\partial m} \tag{22-35}$$

If the pressure and molality remain small enough so that $\bar{v}_1$ remains essentially constant, Eq. (22-35) integrates to

$$P - P° = - \frac{RT}{\bar{v}_1} \ln a_1 \tag{22-36}*$$

and substitution of Eq. (22-24) yields

$$P - P° = \frac{RT}{\bar{v}_1} \frac{\nu M_1}{1000} m\phi \tag{22-37}*$$

Thus measurements of the osmotic pressure yield values of the osmotic coefficient which can be treated by Eq. (22-29) to yield activity coefficients.

As an indication of the magnitude of the osmotic-pressure effect, one can note that a 0.01 M solution of solute in water will result in a lowering of the activity of water by 0.018 per cent for a perfect nonelectrolyte or a lowering in the vapor pressure of water at room temperature of 0.0045 mm Hg. In contrast to these very small effects, the osmotic pressure for such a solution would amount to 182 mm Hg. It is clear that, potentially, the osmotic-pressure method has possibilities of high accuracy for very dilute solutions. A number of experimental difficulties have handicapped the method, as one must have a membrane which is permeable to solvent molecules but impermeable to ions and one must achieve extreme temperature uniformity. Frazer and Patrick[1] obtained a perfect membrane for nonvolatile solutes by equilibrating solution and solvent through the vapor phase with a porous disk supporting the pressure difference, as illustrated schematically in Fig. 22-7. Instead of restoring equilibrium between the solvent and the solution by applying pressure to the solution, the solvent is subjected to a negative pressure, an effect that we have briefly discussed in Chapter 16. A disk with pores between 1 and 10 μ holds the column of liquid. Lack of equilibrium is indicated by distillation of solvent. Williamson[2] has refined the method and has been able to achieve temperature uniformity within millionths of a degree. The method has been of considerable use for high-polymer

FIG. 22-7. Porous-disk osmometer.

[1] I. Frazer and W. Patrick, *Z. physik. Chem.*, **130**, 691 (1927).
[2] A. G. Williamson, *Proc. Roy. Soc. (London)*, **195**, 97 (1948).

solutions and polyelectrolytes. As these solutions are normally studied at such dilution that the partial molal volume of water in Eq. (22-35) may be considered constant and equal to the molal volume of water, the osmotic-pressure measurements are directly converted into osmotic-coefficient values by Eq. (22-37)*. The use of negative pressure measurements on the solvent simplifies Eqs. (22-36)* and (22-37)* to the extent that one would replace the partial molal volume of water in the solution by the molal volume of pure water as a rigorous expression even at high molalities. Also the osmotic coefficient and the activity coefficients determined from these measurements would apply to a solution at 1 atm and would not have to be corrected for pressure.

If the osmotic-pressure measurements are made by applying pressure to the solution, the resulting activity coefficients relate to the pressure of measurement. We have already given in Eq. (20-9) the relationship of activity coefficient to pressure, which could be used to reduce such results to the standard pressure of 1 atm.

ACTIVITY COEFFICIENTS OF ELECTROLYTES FROM DIFFUSION EXPERIMENTS

Activity coefficients can be obtained from measurement of diffusion since the driving force for diffusion is an activity gradient. Onsager and Fuoss[1] have derived an equation for the diffusion coefficients of a dilute electrolyte solution. Their equation reduces to the limiting law at zero molality derived previously by Nernst.[2] The limitations of the Debye-Hückel derivation discussed in Chapter 23 are also applicable here, and one would expect the theoretical equations to be limited to very dilute solutions. However, the conductometric method of measuring diffusion coefficients can be applied with very high accuracy to very dilute solutions of electrolytes. Harned and Nutall[3] report an accuracy of 0.2 per cent even down to 0.0005 M. Also Harned[4,5] has shown that the deviations from the limiting equation for diffusion are due primarily to deviations of the activity coefficients from unity, and Harned[5] tabulates activity coefficients for a number of salts determined by this method for concentrations of 0.01 M to as low as 0.0001 M in many instances. For salts more complex than the 1-1 salts, the theory is limited to even lower concentrations, perhaps to below 0.002 M.

[1] L. Onsager and R. M. Fuoss, *J. Phys. Chem.*, **36**, 2689 (1932).

[2] W. Nernst, *Z. physik. Chem.*, **2**, 613 (1888).

[3] H. S. Harned and R. L. Nuttall, *J. Am. Chem. Soc.*, **69**, 736 (1947).

[4] H. S. Harned, *Discussions Faraday Soc.*, **24**, 7 (1957).

[5] H. S. Harned, in W. J. Hamer (ed.), "The Structure of Electrolytic Solutions," pp. 152–159, John Wiley & Sons, Inc., New York, 1959.

TABULATION OF ACTIVITY COEFFICIENTS

To be able to carry out equilibrium calculations for aqueous solutions, it is necessary to know activity coefficients or free energies of dilution as a function of concentration. The activity coefficient of an electrolyte changes quite rapidly with concentration. The activity coefficients are often known within several tenths of a per cent, and very extensive tables would be required with tabulations at quite close concentration intervals in order to allow accurate interpolation consistent with the accuracy of the data. Therefore it would be desirable to tabulate a more slowly varying function to permit tabulation at wider concentration intervals.

There are a number of deviation functions that can be used for tabulating activity coefficients. The most direct procedure would be to start with the limiting Debye-Hückel term as given in Eq. (23-20)*, which is the only term rigorously given by theory, and to treat the remainder of the interaction by a power series of virial coefficients as for gases. For most salts, it would be necessary to use not only a second but a third and higher virial coefficients, even at rather low concentrations. Another alternative is to take a fixed set of higher terms corresponding to the average of many salts and then use only the term in m to the first power, i.e., the second virial coefficient, as an adjustable parameter. This in effect is the procedure followed by Guggenheim in using Eq. (22-20)* or (23-35)*, which may be readily seen if one expands $m^{1/2}(1 + m^{1/2})^{-1}$. The second term of the expansion gives a term linear in molality and therefore corresponds to an addition to the B coefficient. However, the following terms of the expansion correspond to a fixed set of higher virial coefficients. If these higher terms derived from $(1 + m^{1/2})^{-1}$ do not correspond to the actual behavior for a given salt, B will show a variation with molality.

Scatchard[1] has used B as a deviation function, which we shall write as $B^{\cdot}$ to distinguish it from the limit at zero concentration. One obtains $B^{\cdot}$ for a 1-1 salt by subtracting the first term of Eq. (22-20)* from $\log \gamma$ and dividing by the molality.

$$B^{\cdot} = \frac{\log \gamma_{\pm} + A_{\gamma}m^{1/2}/(1 + m^{1/2})}{m} \tag{22-38}$$

The limiting variation of $B^{\cdot}$ with m at low molalities will not in general approach zero unless the $1 + m^{1/2}$ term has accurately represented the higher terms. Unfortunately, there are very few data with sufficient accuracy to allow the determination of the variation of $B^{\cdot}$ with molality at molalities below 0.1. Guggenheim and Turgeon[2] have reviewed the

[1] G. Scatchard, *Chem. Rev.*, **19**, 309 (1936).
[2] E. A. Guggenheim and J. C. Turgeon, *Trans. Faraday Soc.*, **51**, 747 (1955).

accurate freezing-point measurements and cell measurements which are available for low molalities and have shown that the $1 + m^{1/2}$ term of Eq. (22-20)* does adequately account for the higher virial coefficients of many 1-1 electrolytes at low molalities, although noticeable deviations are apparent around 0.1 M. Scatchard[1] has suggested that a better average of the higher virial coefficients would be provided by a $1 + 1.5m^{1/2}$ term. Our review of the data indicates little basis for preference since some substances are slightly better fitted by one form and others by the other. We shall adopt the $1 + m^{1/2}$ form because of its greater simplicity and because more extensive tables of B values have been reported on that basis.

In Fig. 22-6, the cell data for HCl are plotted so as to obtain the value of $B^{\cdot}$ from the slope. The HCl data show remarkable constancy of $B^{\cdot}$ almost up to 0.1 M. The effect of increasing contribution of higher virial coefficients as one goes to salts with larger $B^{\cdot}$ values is seen in Fig. 22-8,

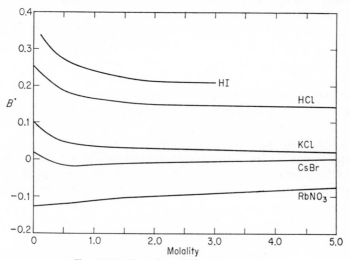

FIG. 22-8. Variation of $B^{\cdot}$ with molality.

where $B^{\cdot}$ is plotted vs. molality for $RbNO_3$, CsBr, KCl, HCl, and HI. The larger the value of B, the more rapid the variation of $B^{\cdot}$ at low molalities. However, a change of 0.01 in $B^{\cdot}$ corresponds at 0.01 M to a change of only 0.02 per cent in γ. It is thus clear that the variation of $B^{\cdot}$ at low molalities will be of concern only with data of very high accuracy and then primarily for salts with very large or very small $B^{\cdot}$ values.

The fact that $B^{\cdot}$ varies with molality in the same direction for almost all salts points the way to an even more accurate and simpler method of

---

[1] G. Scatchard, in W. J. Hamer (ed.), "The Structure of Electrolytic Solutions," p. 9, John Wiley & Sons, Inc., New York, 1959.

tabulating activity coefficient data. Åkerlöff and Thomas[1] have suggested that the difference in the $B^\cdot$ values of any two salts might be independent of molality. Although the difference in the $B^\cdot$ values does not prove to be quite a constant at all molalities, it varies much more slowly than individual $B^\cdot$ values and thus serves as a very useful function.

KCl is a convenient reference salt since the activity coefficients of KCl are known quite accurately over a wide molality range. Also the activity coefficients of KCl vary only slowly at high concentrations; hence it is easy to interpolate values at intermediate concentrations. $\Delta B^\cdot$, the $B^\cdot$ value of any salt minus the $B^\cdot$ value of KCl at the same molality, is simply calculated by subtracting log γ_{KCl} from log γ_{MX} and dividing the result by m.

$$\log \frac{\gamma_{MX}}{\gamma_{KCl}} = m \, \Delta B^\cdot \qquad (22\text{-}39)$$

In Appendix 4, values of $-\log \gamma_{\pm}$ of KCl are tabulated at close molality intervals, and values of the quantity $\Delta B^\cdot = B_{MX}^\cdot - B_{KCl}^\cdot$ are tabulated versus m. Because of the slowly varying nature of this quantity with m, interpolations may be made very accurately, and one can obtain $\Delta B^\cdot$ for any salt at any desired molality with high accuracy. $\Delta B^\cdot$ multiplied by m and then added to log γ_{KCl} at the interpolated molality yields log γ_{MX}. This procedure provides activity-coefficient data as accurately as is justified by the original data with a minimum of computations. Since isopiestic data, which comprise the greatest bulk of the activity coefficient data, are related to the activity coefficients of KCl, the values of $\Delta B^\cdot$ tabulated in Appendix 4 are independent of any subsequent changes in γ_{KCl} due to newer data or to any revisions of the value of the limiting Debye-Hückel slope. If any changes in the activity coefficient values of KCl should result from newer data, the tabulated values of γ_{KCl} can be revised to agree with the new results without any need to alter the $\Delta B^\cdot$ values unless new data should become available for a given salt.

As the accuracy of the activity coefficients calculated from the tabulations in Appendix 4 depends upon the accuracy of the KCl data, a careful evaluation was made of the data for KCl below 0.1 M. The data have been evaluated by use of Eq. (22-38) and the resulting $B^\cdot$ values plotted in Fig. 22-9. By plotting the data in the form of the deviation function $B^\cdot$, minor discrepancies can be readily uncovered. The data for KCl are in excellent agreement and fix a value at 0.1 M of $B^\cdot = 0.087$ kg/mole. Most of the analytical expressions which have been used to represent the data have been chosen to fit well at higher molalities, thus causing slight distortions of the data at lower molalities, which produce large changes in $B^\cdot$. Also, the equations incorporated values of the limiting Debye-

[1] G. Åkerlöff and H. C. Thomas, *J. Am. Chem. Soc.*, **56**, 593 (1934).

Hückel slope different from the most recent value, which cause large changes in $B\cdot$ at low molalities. However, use of the original data, rather than equations which have smoothed the data, yield excellent agreement at low molalities with a value of B of 0.10.

The constant value of B given in Tables A4-2 and A4-7 for KCl and CaCl$_2$ at low molalities is not to be taken as implying that the $1 + I^{1/2}$ term of Eq. (23-35) does precisely represent all possible higher virial coefficients so that B truly does become a constant at low molality. The

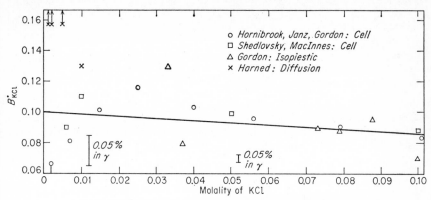

FIG. 22-9. $\dfrac{1}{m}$ [log $\gamma_{\text{KCl}} + A_\gamma m^{1/2}(1 + m^{1/2})^{-1}$] below 0.1 M.

use of a constant value of B is merely indicative of the fact that it is very difficult to determine the variation of $B\cdot$ with molality at very low molalities and, conversely, since the calculation of activity coefficients at very low molalities is quite insensitive to errors in the value of B chosen, there is no practical advantage to attempting to represent the variation of $B\cdot$ with molality. We have previously shown that the single parameter B is quite adequate to represent even the most accurately available data at molalities below 0.03 to 0.05 M, and for data of typical accuracy the single parameter is adequate to molalities as high as 0.1 M.

In obtaining the $\Delta B\cdot$ values, it is also important not to use equations which have smoothed the data unless the different salts have been treated in essentially identical manners. Thus the $\Delta B\cdot$ values for NaCl relative to KCl such as are given in Fig. 22-10 would show very great discrepancies if tabulated smooth values were taken from the literature, while the original experimental data lead to quite concordant results with a $\Delta B\cdot$ value of 0.049 and with no variation below 0.1 M within experimental error.

It is important to note that the simple relationship between log γ and ϕ expressed in Eqs. (22-20)* and (22-31)* with the same value of B in both equations is applicable only at low molalities, where the variation of $B\cdot$

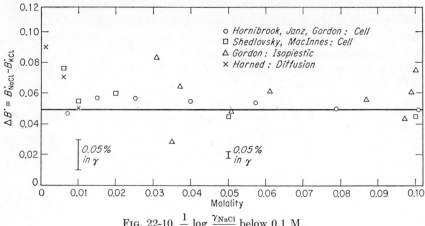

FIG. 22-10. $\dfrac{1}{m} \log \dfrac{\gamma_{\mathrm{NaCl}}}{\gamma_{\mathrm{KCl}}}$ below 0.1 M.

with molality can be neglected.   This restriction also applies to the equations used to obtain $\Delta B\cdot$ values at molalities of 0.1 M and below.

In order to relate these $\Delta B\cdot$ values to the osmotic coefficient, we may expand $\Delta B\cdot$ in a power series,

$$\Delta B\cdot = \Delta B \left(1 + m \frac{\Delta C}{\Delta B} + \cdots\right) \tag{22-40}$$

Then we may show that

$$\phi_{\mathrm{MX}} - \phi_{\mathrm{KCl}} = \frac{2.303m}{2}\left(\Delta B + \frac{4m\,\Delta C}{3} + \cdots\right) \tag{22-41}$$

$$\phi_{\mathrm{MX}} - \phi_{\mathrm{KCl}} = \frac{2.303m\,\Delta B\cdot}{2}\left(1 + \frac{m\,\Delta C}{3\,\Delta B} + \cdots\right) \tag{22-42}$$

The number of terms needed in the power series quickly becomes very large for representing the behavior at molalities much above 0.1 M, and ϕ is most easily related to $\log \gamma$ at higher molalities by Eq. (22-34), which is evaluated graphically.   However, for molalities close to 0.1 M, Eq. (22-42) can be useful for relating ϕ to γ.

PROBLEMS

22-1. Use data in Appendix 4 to calculate $\gamma_\pm$ and $a_\pm$ for NaCl in 4.3 M and 4.1 M solution at 25°C.   What is the ratio of the fugacities of NaCl for the two solutions?

22-2. The solubility of $KBrO_3$ is 0.48 M at 25°C.   Use the data of Appendix 4 to calculate ΔF°_{298} for $KBrO_3(s) = KBrO_3(aq)$.

22-3. For $Ba(NO_3)_2 = Ba(NO_3)_2(aq)$, $\Delta F^\circ_{298} = 3200$ cal.   Calculate the molality of saturated $Ba(NO_3)_2$ solution, using the data of Table A4-8.

22-4. For molalities sufficiently dilute such that $\gamma_\pm$ may be taken equal to unity, calculate $a_\pm$ and a_2 for solutions of $PbCl_2$, $PbFCl$, $AlCl_3$, and $Al_2(SO_4)_3$ at molality m. What is $a_\pm$ of $Al_2(SO_4)_3$ in a mixture containing $m_{Na_2SO_4}$ and m_{AlCl_3}?

22-5. For 0.05 M and 0.5 M KCl at 25°C, compare $\gamma_\pm$ calculated from Eq. (22-20)*, using $B = 0$ with $\gamma_\pm$ calculated with $B^\cdot$ obtained from Table A4-2b.

22-6. From Table 22-6, calculate ε at 25°C for a HCl concentration cell without liquid junction with $m_1 = 0.05$ and $m_2 = 0.5$ M. What is the ratio of HCl pressures for these two solutions?

22-7. Use the data of Table A4-7a and Eq. (22-37)* to calculate the osmotic pressure for 0.1 M $CaCl_2$ at 25°C.

22-8. Harned[1] reports the following $\gamma_\pm$ values for KCl obtained from diffusion methods at 25°C:

m	0.0005	0.001	0.002	0.005	0.01
$\gamma_\pm$	0.975	0.9651	0.9517	0.9272	0.9016

Calculate B_{KCl}, using Eq. (22-20)*. Owing to the uncertainty of an electrophoretic correction, these $\gamma_\pm$ values are uncertain by 0.2 per cent at 0.0005 M and 0.1 per cent at 0.01 M. What uncertainty does this introduce into the values of B obtained from the data?

22-9. Derive the equation

$$\frac{F - F^\circ}{3RT} = -m\phi + 2.303m \log 2^{2/3}m\gamma_\pm$$

for aqueous $CaCl_2$, where $F - F^\circ$ is the total free energy relative to pure water and $CaCl_2$ in the solute standard state of a solution containing 1 kg water and m moles of $CaCl_2$.

22-10. Smith and Robinson[2] report the following solutions to be in isopiestic equilibrium at 25°C:

m_{KCl}	0.4810	0.6292	0.6872	0.8530	0.9950	1.065	1.110	1.123
m_{NaOAc}	0.4580	0.5886	0.6378	0.7850	0.9010	0.9610	1.008	1.013

Use the ΔB_{NaOAc} value from Appendix 4 to guide the extrapolation of a plot of $(R - 1)/\gamma_{KCl}m_{KCl}$ versus $\gamma_{KCl}m_{KCl}$ to zero molality. Calculate values of γ_{NaOAc} at 0.5 and 1 M.

[1] H. S. Harned, in W. J. Hamer (ed.), "The Structure of Electrolytic Solutions," paper 10, p. 152, John Wiley & Sons, Inc., New York, 1959.

[2] E. R. B. Smith and R. A. Robinson, *Trans. Faraday Soc.*, **38**, 70 (1942).

23

THEORIES OF ELECTROLYTE SOLUTIONS

The theory of electrolyte solutions begins, of course, with the brilliant discovery of Arrhenius that certain solutes dissociate into electrically charged species, ions. The next major advance, to which many authors[1-6] contributed during the first two decades of the twentieth century, was the realization that the deviations from ideality in dilute strong electrolytes are fundamentally different from the deviations in nonelectrolyte solutions. In electrolytes the long-range nature of the electrostatic force between ions yields effects which have no counterpart in solutions of neutral molecules. Milner[5] first attempted a theoretical treatment of this phenomenon and obtained an accurate but complicated expression which he was able to evaluate only approximately. Debye and Hückel[7] effected a great simplification by approximations in setting up the problem and obtained simple results of great utility. Their approximations have been the subject of countless further papers, and we shall return presently to a consideration of the Debye-Hückel equations and their region of validity.

Before turning to the quantitative theory, however, let us consider certain general ideas. If the particles of a solute are distributed with reasonable uniformity, then the number of solute molecules or ions within a spherical shell of radius r and thickness dr about a given particle will be approximately proportional to the volume of the shell, $4\pi r^2\, dr$, and to the concentration of the solute. The effect of the solute within this shell upon the energy of the central solute particle is given by the energy of interaction times the probable number of interacting particles. If the

[1] W. Sutherland, *Phil. Mag.*, (6)**3**, 161 (1902), **12**, 1(1906).

[2] A. A. Noyes, address before the International Congress of Arts and Sciences, St. Louis, *Technol. Quart.*, **17**, 293 (1904).

[3] N. Bjerrum, *Proc. 7th Intern. Congr. Appl. Chem.*, London, 1909; *Z. Elektrochem.*, **24**, 321 (1918); *Medd. Vetenskapsakad. Nobelinst.*, vol. 5, no. 16 (1919).

[4] G. N. Lewis, *Z. physik. Chem.*, **70**, 212 (1909).

[5] S. R. Milner, *Phil. Mag.*, (6)**23**, 551 (1912), (6)**25**, 742 (1913).

[6] J. C. Ghosh, *J. Chem. Soc.*, **113**, 149, 267, 790 (1918).

[7] P. Debye and E. Hückel, *Physik. Z.*, **24**, 185, 334 (1923), **25**, 97 (1924).

energy of interaction falls off with distance faster than r^{-2}, then it is clear that the dominant effects will come from solute molecules near the central one.   This is the situation for all types of nonelectrolyte solutions.   But the energy of ionic interaction depends on r^{-1}; hence the net interaction of distant solute ions on a central ion may be very important.   Indeed the total interaction energy in an electrolyte converges to a finite value only because the ratio of positive to negative charge in a shell rapidly approaches unity as the radius increases.   It is this long-range nature of the electrostatic interaction which makes it qualitatively different.

The existence of long-range forces between ions by no means excludes the simultaneous existence of short-range forces of the various sorts which operate between neutral molecules.   If there is a strong enough short-range attraction, then ions associate just as neutral molecules do. Typical weak electrolytes, such as acetic acid, exemplify this behavior. Such solutions are treated by combining the usual equation for a chemical dissociation equilibrium with that for the special interionic effects of the dissociated ions.   It was Bjerrum[1] who emphasized that electrostatic forces alone could cause association.   In water, this effect is important only for multiply charged ions, but in solvents of low dielectric constant it is always important.   After presenting the general interionic-attraction theory we shall return to a consideration of these ion-association effects.

Interionic-attraction Theory.   It is first assumed that the chemical potential of an electrolyte in dilute solution may be separated into a term $\mu_2{}^{el}$ for the electrostatic interaction of the ions and a term $\mu_2^{\circ} + \nu RT \ln m$ for the ideal (Henry's-law) behavior of a dilute solute which dissociated into ν particles per molecule.   From the definition of the mean activity coefficient one has

$$\nu RT \ln \gamma_{\pm} = \mu_2 - \mu_2^{\circ} - \nu RT \ln m = \mu_2{}^{el}$$

$$\ln \gamma_{\pm} = \frac{\mu_2{}^{el}}{\nu RT} \tag{23-1}$$

Similarly, the chemical potential of the solvent may be separated into an electrostatic-interaction term $\mu_1{}^{el}$ and a term $\mu_1^{\circ} - \nu m RT (M_1/1000)$ for the dilute ideal solution.   Here M_1 is the molecular weight of the solvent.   Then the practical osmotic coefficient ϕ, defined in Eq. (22-24), is given by

$$(1 - \phi)\nu m RT \frac{M_1}{1000} = \mu_1{}^{el}$$

$$1 - \phi = \frac{\mu_1{}^{el}}{\nu m RT} \frac{1000}{M_1} \tag{23-2}$$

The electrostatic chemical potentials of solvent and solute are related

[1] N. Bjerrum, *Kgl. Danske Videnskab. Selskab, Mat.-fys. Medd.*, vol. 7, no. 9 (1926).

to one another, of course, by the Gibbs-Duhem equation and to the total electrostatic free energy by the equation

$$F^{el} = n_1\mu_1{}^{el} + n_2\mu_2{}^{el} \tag{23-3}$$

It is interesting to note next the variables which might affect F^{el} and to see how much can be concluded from dimensional reasoning alone.   For simplicity let us consider a binary electrolyte with ionic charges $\pm ze$ in a solvent of dielectric constant D.   Other factors which might enter are the thermal energy kT and the ion concentration, which may be written N/V, where N is the number of pairs of ions.   While the size, shape, and other properties of the ions as well as other properties of the solvent may affect F^{el} at finite concentrations, we expect that these variables will disappear in the limit at infinite dilution.   The only dimensionless combination of the other factors is $(N/V)^{1/3}(z^2e^2/DkT)$; consequently F^{el}/RT, the activity coefficient, and the osmotic coefficient should be functions of this quantity.

Milner[1] made the first serious effort to apply statistical thermodynamics to the ionic interaction problem.   He noted the extreme difficulties of an exact solution and attempted an approximate but direct evaluation of F^{el}.   Milner presented his primary results in the form of a table of values of a quantity very similar to the osmotic coefficient and showed by graphs that his function fitted the freezing-point data for several aqueous strong electrolytes.   In a footnote he gave a closed expression, also approximate, for the very dilute solution, which may be rearranged into

$$1 - \phi = \frac{\pi}{3}\left[\left(\frac{N}{V}\right)^{1/3}\frac{z^2e^2}{DkT}\right]^{3/2} \tag{23-4}*$$

The important feature of this result is the functional dependence on the three-halves power of the bracketed quantity, which we selected from dimensional reasoning.   This corresponds to a proportionality of both $1 - \phi$ and $\ln \gamma_\pm$ to the half power of the concentration.   Had this functional relationship been fully recognized and accepted in 1913, even without the precisely correct coefficient, progress in electrolyte theory would have been more rapid, because the Debye-Hückel result differs from Milner's only by the factor $(2/\pi)^{1/2}$.

In the decade between Milner's work and that of Debye and Hückel, both Bronsted[2] and Lewis and Randall[3] came to adopt on an empirical basis the functional form

$$1 - \phi = \alpha m^{1/2}$$

[1] S. R. Milner, *Phil. Mag.*, (6)**23**, 551 (1912); (6)**25**, 742 (1913).

[2] J. N. Bronsted, *J. Am. Chem. Soc.*, **44**, 983 (1922).

[3] G. N. Lewis and M. Randall, *J. Am. Chem. Soc.*, **43**, 1112 (1921); see also "Thermodynamics and the Free Energy of Chemical Substances," 1st ed., pp. 342–346, McGraw-Hill Book Company, Inc., New York, 1923.

Lewis and Randall found α to be in the range 0.3 to 0.45, while Bronsted assigned $\alpha = 0.32$ for 1-1 electrolytes in water at 0°C. These values are close to the Debye-Hückel result, which gives $\alpha = 0.374$.

Debye-Hückel Theory. In 1923 Debye and Hückel presented a simple theory of interionic-attraction effects which nevertheless retains all essential features in the limit of infinite dilution. Their result has been of enormous value in the practical treatment of electrolyte solutions by providing a limiting law for the extrapolation to zero concentration of not only activities but also heat contents, heat capacities, volumes, etc.

For the calculation of the electrical contribution to the chemical potential μ^{el} the solvent is assumed to constitute only a continuous dielectric in which the ions interact according to Coulomb's law with dielectric constant D. Furthermore it is assumed that the average effect on a given ion of all other ions may be obtained from a continuous charge distribution, or "ionic atmosphere," calculated in the following manner.

If the electrical potential at a given point is ψ, the energy of an ion of charge $z_j e$ is $z_j e \psi$. We may expect the concentration of such ions N_j' (particles per cubic centimeter) to be related to the concentration at zero potential N_j by Boltzmann's law [see Eq. (27-4) and also Eq. (A3-10)],

$$N_j' = N_j \exp\left(-\frac{z_j e \psi}{kT}\right) \tag{23-5}$$

The local charge density is obtained by summing the ionic charge concentrations over all species of ions,

$$\rho = \sum_j z_j e N_j \exp\left(-\frac{z_j e \psi}{kT}\right) \tag{23-6}$$

At this point we introduce the basic equation[1] of Poisson, relating the electrostatic potential to the charge distribution

$$\nabla^2 \psi = -\frac{4\pi}{D}\rho \tag{23-7}$$

where ∇^2 is a differential operator, which is $\partial^2/\partial x^2 + \partial^2/\partial y^2 + \partial^2/\partial z^2$ in cartesian coordinates. The average charge distribution around a single ion will have spherical symmetry. Consequently, it is convenient to use spherical polar coordinates, whereupon the angles may be omitted and the Poisson equation becomes

$$\frac{1}{r^2}\frac{d}{dr}\left(r^2 \frac{d\psi}{dr}\right) = -\frac{4\pi}{D}\rho \tag{23-8}$$

Equation (23-6) for the charge density may now be substituted and a

[1] See Eq. (A8-6) and associated footnote.

solution sought.[1]   The exponential form of Eq. (23-6) is inconvenient, to say the least, for the solution of the differential equation.   Since the other approximations can be expected to be valid only in very dilute solutions, where the ions are usually far apart, it is reasonable to expand the exponentials in series,

$$\rho = \sum_j z_j e N_j - \sum_j \frac{N_j z_j^2 e^2 \psi}{kT} + \sum_j \frac{N_j z_j^3 e^3 \psi^2}{2k^2 T^2} + \cdots \qquad (23\text{-}9)$$

Since the ions come from a neutral solute compound, the first sum must be zero.   Also, if the electrolyte is a simple binary type such as NaCl, $CaSO_4$, etc., the third sum is zero.   In any case all terms are dropped except the second, whereupon Eq. (23-8) becomes

$$\frac{1}{r^2} \frac{d}{dr} \left(r^2 \frac{d\psi}{dr} \right) = \kappa^2 \psi \qquad (23\text{-}10)^*$$

with

$$\kappa^2 = \frac{4\pi e^2}{DkT} \sum_j N_j z_j^2 \qquad (23\text{-}11)$$

Equation (23-10)* is a familiar type of differential equation with the general solution

$$\psi = A \frac{e^{\kappa r}}{r} + B \frac{e^{-\kappa r}}{r} \qquad (23\text{-}12a)^*$$

where A and B are constants to be determined from the physical conditions of the problem.   Since the potential must remain finite at large values of r, A must be zero.   It is assumed that there is some distance a of closest approach of other ions to the central ion.   For $r < a$ there are no other ions present, and the potential near an ion of charge z_j is given by the familiar electrostatic formula

$$\psi_j = \frac{z_j e}{Dr} + C_j \qquad (23\text{-}12b)^*$$

where C_j is a constant which arises because of the presence of other ions outside the region $r < a$.   At $r = a$ both the potential ψ and the electric field $-\partial \psi / \partial r$ must be continuous.   These two conditions permit the evaluation of both B and C, which are found to be

$$B_j = \frac{z_j e}{D(1 + \kappa a)} e^{\kappa a} \qquad (23\text{-}13a)^*$$

$$C_j = - \frac{z_j e \kappa}{D(1 + \kappa a)} \qquad (23\text{-}13b)^*$$

One self-consistency check on this treatment is the calculation of the total charge density associated with the ionic atmosphere of a central ion

[1] E. A. Guggenheim, *Trans. Faraday Soc.*, **55**, 1714 (1959); **56**, 1152 (1960).

of charge $z_i e$.   Substitution of (23-13a)* into (23-9) as simplified yields

$$\rho_i = - \frac{\kappa^2 z_i e}{4\pi r_i} \frac{e^{\kappa a} e^{-\kappa r_i}}{1 + \kappa a}$$

and integration over all space with $r > a$ yields

$$\int_a^\infty \rho_i 4\pi r_i^2 \, dr_i = - \frac{z_i e}{1 + \kappa a} e^{\kappa a} \int_a^\infty e^{-\kappa r_i} \kappa^2 r_i \, dr_i = -z_i e$$

Thus the charge of the central ion is exactly compensated by an equal but opposite charge distribution in the solution.

We are now prepared to complete the calculation of $\mu_i{}^{el}$ for the ion i. This is the electrical free energy associated with the introduction of one ion into the solution multiplied by Avogadro's number.   One assumes that the solution contains the full concentration N_j of each type of ion which establishes the value of κ.   The one additional particle is first introduced in a hypothetical uncharged state and in a second step is imagined to be gradually charged to $z_i e$.   The first step requires negligible electrical energy.   The energy of adding each increment of charge dq is just that increment times the environmental potential arising from all the other ions.   This potential is just the quantity C given in Eq. (23-13b)*; hence

$$\frac{\mu_i{}^{el}}{N_0} = \int_0^{z_i e} - \frac{\kappa q \, dq}{D(1 + \kappa a)}$$

$$= - \frac{z_i{}^2 e^2 \kappa}{2D(1 + \kappa a)} \qquad (23\text{-}14)*$$

$$\ln \gamma_i = - \frac{z_i{}^2 e^2 \kappa}{2DkT(1 + \kappa a)} \qquad (23\text{-}15)*$$

This charging process, which was first proposed by Guntleberg,[1] is simpler than that used by Debye and Hückel, wherein all ions in the solution are charged simultaneously and therefore κ varies during the charging process. Both methods yield the same result.

Our primary interest in the Debye-Hückel theory is to obtain a limiting law valid in the region of very low concentration for the activity coefficient and other properties.   In this region κ becomes very small, and the product κa becomes negligible as compared with unity.   Consequently the term $1 + \kappa a$ in the denominator is omitted for limiting-law expressions, although it will be introduced again for some expressions used at higher concentrations.   The reason for the introduction of the distance of closest approach a in this initial derivation is to avoid the use of Eq. (23-9) in the region close to an ion where the approximation of expanding the exponential in Eq. (23-6) is clearly subject to doubt.   Since such a finite distance of closest approach is physically reasonable, there is no

---

[1] N. Bjerrum, *Z. physik. Chem.*, **119**, 145 (1926); see p. 155.

objection to its introduction, but, as we have seen, the exact value of a drops out of the limiting expression for low concentration.

The mean activity coefficient, which is the thermodynamically defined and measured quantity, is given as follows for an electrolyte dissociating into ν_+ ions of charge z_+e and ν_- ions of charge z_-e:

$$(\nu_+ + \nu_-) \ln \gamma_\pm = \nu_+ \ln \gamma_+ + \nu_- \ln \gamma_-$$

and also
$$\nu_- z_- = -\nu_+ z_+$$

Substitution of Eq. (23-15) with the omission of the $1 + \kappa a$ factor yields, after simplification,

$$\ln \gamma_\pm = -\frac{e^2 \kappa |z_+ z_-|}{2DkT} \qquad (23\text{-}16)^*$$

Before substituting κ from Eq. (23-11) let us note the definition of ionic strength as proposed originally by Lewis and Randall,[1]

$$I = \frac{1}{2} \sum_i m_i z_i^2 \qquad (23\text{-}17)$$

which is evidently just the molality of a 1-1 electrolyte. The next step is the conversion from the molality to concentration in ions per cubic centimeter. This shift to a volumetric basis should be emphasized because it leads to additional terms in the derived functions for enthalpy, heat capacity, etc.

$$\sum_j N_j z_j^2 = \frac{2N_0 d_1}{1000} I = \frac{2N_0 M_1}{1000 v_1} I \qquad (23\text{-}18)$$

Here d_1 is the density of the solvent, M_1 its molecular weight, v_1 its molal volume.

Now we may convert κ to conventional macroscopic units,

$$\kappa^2 = \frac{8\pi e^2 N_0 d_1}{1000 DkT} I \qquad (23\text{-}19)$$

and the final equation for the activity coefficient is

$$\log \gamma_\pm = -A_\gamma |z_+ z_-| I^{1/2} \qquad (23\text{-}20)^*$$

$$A_\gamma^2 = \frac{2\pi N_0 d_1}{1000 \times 2.303^2} \left(\frac{e^2}{DkT}\right)^3 \qquad (23\text{-}21)$$

For water[2] at 25°C, $A_\gamma = 0.511$. It varies only slowly with tempera-

[1] G. N. Lewis and M. Randall, *J. Am. Chem. Soc.*, **43**, 1112 (1921).

[2] Note that the coefficient A_γ as defined here for use with I in molality differs by the factor $d_1^{1/2}$ from the quantity called A by Robinson and Stokes and $S_{(f)}$ by Harned and Owen. Guggenheim's α is $2.303 A_\gamma$ since he uses base e logarithms but I in molality.

ture. In water at 0°C, $A_\gamma = 0.492$, while at 100°C, $A_\gamma = 0.596$. A table of values is given in Appendix 4 for various temperatures.

One can readily show from Eqs. (20-34) and (22-28) that the osmotic coefficient of the solvent is related to the activity coefficient of the solute by the equation

$$1 - \phi = -\frac{1}{m} \int m \, d \ln \gamma_\pm \tag{23-22}$$

Substitution of Eq. (23-20)* yields, after several steps,

$$1 - \phi = (2.303/3) A_\gamma |z_+ z_-| I^{1/2} \tag{23-23}*$$

Since we now have the partial molal free energy associated with the long-range electrostatic interactions, it is straightforward thermodynamics to obtain the corresponding terms for enthalpy, heat capacity, and volume. The difference in partial molal heat content from that of the standard state is called the *relative molal heat content* with the symbol L.

$$\mathrm{L}_2 = \bar{H}_2 - \bar{H}_2^\circ = -\nu R T^2 \left(\frac{\partial \ln \gamma_\pm}{\partial T} \right)$$

$$= \frac{\nu}{2} |z_+ z_-| I^{1/2} \left[(2 \times 2.303) R T^2 \left(\frac{\partial A_\gamma}{\partial T} \right) \right] \tag{23-24}*$$

Examination of Eq. (23-21) for A_γ shows that there are two temperature-dependent quantities d_1 and D in addition to the explicit appearance of T. After the derivative is taken, all the factors in brackets above may be collected as a quantity A_H; hence

$$\mathrm{L}_2 = A_H \left(\frac{\nu}{2} \right) |z_+ z_-| I^{1/2} \tag{23-25}*$$

$$A_H = -2.303 R T^2 A_\gamma 3 \left[\frac{1}{T} + \frac{\partial \ln D}{\partial T} + \frac{\alpha}{3} \right] \tag{23-26}$$

where α is the coefficient of volumetric expansion $V^{-1}(\partial V/\partial T)$. The value of A_H for water at 25°C is 688 cal/mole.

For the relative molal heat capacity at constant pressure, which is given the symbol J, one has

$$\mathrm{J}_2 = \bar{C}_2 - \bar{C}_2^\circ = \frac{\partial \mathrm{L}_2}{\partial T}$$

$$= A_J \left(\frac{\nu}{2} \right) |z_+ z_-| I^{1/2} \tag{23-27}*$$

$$A_J = 2.303 R A_\gamma \left(\frac{3}{2} \right) \left[1 + 2 \frac{T}{D} \frac{\partial D}{\partial T} + 5 \left(\frac{T}{D} \frac{\partial D}{\partial T} \right)^2 + \frac{2T^2}{DV} \frac{\partial D}{\partial T} \frac{\partial V}{\partial T} \right.$$

$$\left. + \frac{2}{3} \frac{T}{V} \frac{\partial V}{\partial T} + \left(\frac{T}{V} \frac{\partial V}{\partial T} \right)^2 - \frac{2T^2}{D} \frac{\partial^2 D}{\partial T^2} - \frac{2}{3} \frac{T^2}{V} \frac{\partial^2 V}{\partial T^2} \right] \tag{23-28}$$

In this expression α has been replaced by the equivalent function of V, which may be taken uniformly as the specific volume or the molal volume of the solvent. The appearance of the second derivative of D with respect to temperature makes it difficult to obtain an accurate value of A_J. Present data yield the value 10.4 cal/deg mole for water at 25°C, but this is still subject to possibly 10 per cent uncertainty.

In order to obtain the interionic-attraction effect on molal volume, the pressure derivatives are required.

$$\bar{V}_2 - \bar{V}_2^0 = \nu RT \frac{\partial \ln \gamma_{\pm}}{\partial P} = A_V \left(\frac{\nu}{2}\right) |z_+ z_-| I^{\frac{1}{2}} \qquad (23\text{-}29)*$$

$$A_V = 2.303 RT A_\gamma \left[3 \frac{\partial \ln D}{\partial P} - \frac{\partial \ln V}{\partial P} \right] \qquad (23\text{-}30)$$

Range of Validity of Debye-Hückel Theory. It is evident, even without close examination, that many of the assumptions of the Debye-Hückel theory can be valid only in the case of very dilute solutions. As we remarked before, many papers have been written examining these assumptions, and in every case it is agreed that the result is valid in the limit of zero concentration. A recent and rigorous analysis is that by Kirkwood and Poirier.[1] We shall accept this result and turn our attention to the question of the range of validity of the Debye-Hückel equation. If it were valid only below 10^{-6} M, it would be of little practical importance and in some circumstances this is nearly the situation. Fortunately in most cases the range of validity is broader.

Some ideas will be clarified by a simple calculation. Let us calculate the probable number of positive ions in a spherical shell of radius r around a central positive ion, i.e., the radial-distribution function of other positive ions. We make the same calculation for negative ions of the same absolute charge z. In each case we shall ignore, initially, the mutual interaction of the other ions; hence

$$N_+ r^2 \, dr = N r^2 \, dr \exp\left(-\frac{z^2 e^2}{DrkT}\right) \qquad (23\text{-}31a)*$$

$$N_- r^2 \, dr = N r^2 \, dr \exp\left(+\frac{z^2 e^2}{DrkT}\right) \qquad (23\text{-}31b)*$$

Figure 23-1 shows $N_+ r^2$ and $N_- r^2$ plotted as functions of a reduced radius, $r/z^2 r_0$, in which $r_0 = e^2/DkT$, which is 7.13Å for water at room temperature. The long-dashed line is $N r^2$, which would be the distribution in the absence of interaction. Actually the effect of ions other than the central ion on one another causes the distributions to take a course like the short-dashed lines and to approach the long-dashed line at large r.

[1] J. G. Kirkwood and J. C. Poirier, *J. Phys. Chem.*, **58**, 591 (1954).

Two features of Fig. 23-1 are important to us.   First we note that the distribution of ions of like sign, r^2N_+, has become very small by $r \cong 0.7z^2r_0$ or $5\mathring{A}$ for a univalent electrolyte in water.   Since the crystallographic radii of even the largest simple ions are less than half this amount, the short-range forces between like ions will not ordinarily be important.

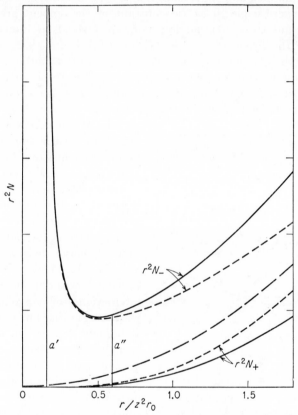

FIG. 23-1. Radial distribution of ions of like charge r^2N_+ and of unlike charge r^2N_- around a central ion.   In the absence of interionic forces both distributions would follow the long-dashed line.

For higher z values or lower dielectric constants this conclusion is even stronger.   This result arises, of course, from the electrical repulsion of like charges and is a welcome simplification of our problem.

The second feature of Fig. 23-1 is the sharp rise of the distribution of unlike ions r^2N_- at small values of r.   This is the purely electrostatic-association phenomenon emphasized by Bjerrum and will always occur unless a short-range repulsive force between the unlike ions prevents it. It is the common but crude approximation to assume a fixed distance a of closest approach of unlike ions.   Evidently if this distance is large, such as a'' in Fig. 23-1, then the electrostatic association will be of no importance.   For univalent electrolytes in water this is usually the case.   But for polyvalent electrolytes the opposite is frequently the case, for example, a' in Fig. 23-1.   In this case it seems to us that the use of a fixed distance of closest approach and the assumption at this distance of Coulomb's law with the macroscopic dielectric constant of the solvent are nonsense. The important facts are that ion pairs are formed, as indicated qualitatively by the rise of the curve r^2N_- for small r, and that they may be treated by the conventional association equilibrium method.   Then the details of interactions at short distances are all absorbed in the equilibrium constant.   There is some ambiguity in the apportionment of the region near the minimum in the r^2N_- curve (at $r = \frac{1}{2}z^2r_0$) between associated ion pairs and separated ions, but it turns out that as long as one is consistent no difficulty arises in practical calculations.

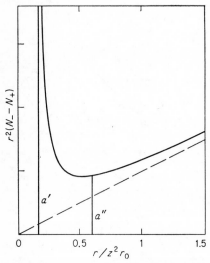

Figure 23-2 shows the difference between the distributions of unlike and like ions again as a function of r/z^2r_0.   This is the radial distribution of charge density of the ion atmosphere.   The solid curve is calculated without consideration of interactions of surrounding ions with one another but uses the exact Boltzmann expression, whereas the straight dashed line is the approximation used in the Debye-Hückel theory.   Consideration of environmental ion interactions at finite concentrations would cause both curves to fall off eventually, but this effect may be ignored in discussing phenomena at small values

FIG. 23-2. The radial distribution of charge of the ion atmosphere. The difference between the solid and dashed lines is the error of the Debye-Hückel approximation.

of r.   The difference between the solid line, terminated at a' or a'', and the dashed line represents the error in the simple Debye-Hückel result as we have presented it.   We see that this error can have either sign.   If the distance of closest approach is large, a'', for example, then the effect of the excess probability implied by the dashed line to the left of a'' exceeds that of the deficiency of the dashed line to the right of a''.   The result is an activity coefficient larger than that of Eq. (23-20)*.   It is clear, however, that for a sufficiently small a, such as a', the error can easily be in the opposite direction.   The measured activity-coefficient curves for NaCl and ZnSO$_4$ in Fig. 23-3 show these two types of deviation. ZnSO$_4$ is an example of ion association, which, however, is moderate enough so that it may be ignored in calculation of $\gamma_\pm$. In each case the results approach the limiting law at infinite dilution.

In the initial presentation of Debye-Hückel theory a distance a was introduced for the closest approach of another ion to the central ion. In Eq. (23-15)* we saw that the effect was to introduce a factor $1 + \kappa a$ in the denominator of the expression for the logarithm of the activity coefficient. The result of the introduction of a reasonable value for a is a curve such as that for NaCl in Fig. 23-3. Mayer[1] has derived a more

rigorous ionic solution theory, which exhibits the Debye-Hückel limiting law and which can, at least in principle, be employed to calculate thermodynamic properties well up into the experimental concentration range. The Mayer theory must be used with some physical model for the interactions of ions in order to get numerical results. Poirier[2] has carried out numerical calculations based on the Debye-Hückel model and obtains good agreement with experiment for a variety of valence types below $I = 0.1$ M by adjustment of the ion-size parameters. The a values reported in the literature for the

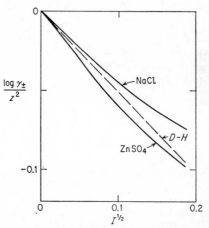

FIG. 23-3. The activity coefficients of NaCl and ZnSO$_4$ compared with the Debye-Hückel limiting law.

Debye-Hückel or the Mayer-Poirier formulas do not correlate in any straightforward manner with crystallographic radii. In view of the physical model this situation is not surprising, but it does lead us to omit further discussion of these a values and of the theories from which they arise.

The essential conclusion of this section is that the Debye-Hückel limiting law is valid and that it may always be used to guide extrapolations to zero concentration. Failure of experimental values for a given electrolyte to approach the limiting-law slope at a concentration such as 0.001 M indicates that ion association is still important.

Mixed Electrolytes, Less Dilute Solutions. It is convenient to consider in the same section the effects of mixing different strong electro-

[1] J. E. Mayer, *J. Chem. Phys.*, **18**, 1426 (1950).

[2] J. C. Poirier, *J. Chem. Phys.*, **21**, 965, 972 (1953); for an excellent review see H. S. Frank and M. S. Tsao, *Ann. Rev. Phys. Chem.*, **5**, 43 (1954).

lytes and the treatment of less dilute solutions, whether single electrolytes or mixed. Two principles emerged from the previous sections which are applicable here. First, there will be specific short-range interactions between ions of unlike charge but not between ions of like charge. This principle of *specific ion interaction* was stated by Bronsted[1] in 1922. Second, in mixed electrolytes the long-range electrostatic interaction between ions of like or unlike sign requires that all ions be considered in calculating the charge density throughout the solution. This is accomplished by including all ions present in the sum

$$I = \frac{1}{2} \sum_i m_i z_i^2 \qquad (23\text{-}17)$$

which completes the definition of the *ionic strength*. Thus in a mixed electrolyte such as HCl and $CaCl_2$ the ionic strength is calculated by summing $z_i^2 m_i$ for H^+, Ca^{++}, and Cl^-. Then the mean activity coefficients of HCl and $CaCl_2$ are given to the limiting-law approximation by

$$\log \gamma_{\pm}(\text{HCl}) = -A_\gamma I^{1/2}$$
$$\log \gamma_{\pm}(\text{CaCl}_2) = -A_\gamma |2| I^{1/2}$$

The solvent activity may be calculated from this law for a mixed solute. The deviation from the ideal osmotic pressure, freezing-point lowering, or related quantity is given by the relationship[2]

$$1 - \phi = \frac{2.303}{3} A_\gamma \frac{\left(\sum_i z_i^2 m_i\right)^{3/2}}{2^{1/2} \sum_i m_i}$$

$$= \frac{2.303}{3} A_\gamma \frac{2 I^{3/2}}{\sum_i m_i} \qquad (23\text{-}32)$$

In preparation for discussion of more concentrated solutions let us first return to Eq. (23-15)* with the term in the denominator for the distance of closest approach. With the introduction of the definitions of Eqs. (23-17), (23-19), and (23-21) one has

$$\log \gamma_i = -A_\gamma z_i^2 \frac{I^{1/2}}{1 + B_\gamma a_i I^{1/2}} \qquad (23\text{-}15a)^*$$

[1] J. N. Bronsted, *J. Am. Chem. Soc.*, **44**, 877 (1922). The neglect of short-range interactions between ions of like sign is, of course, an approximation which becomes less satisfactory at high concentration [see H. L. Friedman, *J. Chem. Phys.*, **32**, 1134, 1351 (1960)]. Also, in unsymmetrical (2-1, 3-1, etc.) electrolytes, theory yields a term, constant $I \ln I$, which is like the limiting law in depending only on the ion charges and solvent properties, but this term appears to be negligible in practical calculations.

[2] See Eq. (34-33) for the definition of ϕ in a mixed electrolyte solution.

where the coefficient B_γ involves numerical factors, the dielectric constant, and the temperature. Its value for water at room temperature is 0.33 if a_i is in angstroms. While the inclusion of the term in a_i improves agreement with observed activity coefficients, Eq. (23-15a)* becomes inadequate at relatively low concentration even if a_i is adjusted for each substance. Additional terms are required, but before introducing them another aspect should be considered.

It is desirable to adopt a form of equation for more concentrated solutions which may be used for mixed electrolytes as well as for single electrolytes. In a multicomponent system one has the condition

$$\frac{\partial \bar{F}_i}{\partial n_j} = \frac{\partial^2 F}{\partial n_j \, \partial n_i} = \frac{\partial \bar{F}_j}{\partial n_i}$$

and substitution of the relationship of activity coefficient to partial molal free energy yields, for the ion species i and j in a mixed electrolyte the following:

$$\frac{\partial \ln \gamma_i}{\partial m_j} = \frac{\partial \ln \gamma_j}{\partial m_i} \tag{23-33}$$

One can readily show that Eq. (23-33) is satisfied for expressions of the type of Eq. (23-15a)* for two different species only if a_i is the same for each species. Furthermore, although different ions certainly have different radii, we know that the term involving a_i in Eq. (23-15a)* does not constitute an accurate representation of the effect of ion size. Hence it seems best to account for effects of different ionic radius by additional terms and to adopt a standard a value for all substances in the term $B_\gamma a_i I^{1/2}$ in Eq. (23-15a)*.

The first-order effects of specific ion interactions may be combined with the limiting law in an equation also suggested by Bronsted[1] in 1922. For a simple M-X electrolyte this takes the form

$$1 - \phi = \alpha z^2 I^{1/2} - \beta_{MX} m \tag{23-34a}$$
$$2.303 \log \gamma_\pm = -3\alpha z^2 I^{1/2} + 2\beta_{MX} m \tag{23-34b}$$

Bronsted showed that these equations fitted experimental data quite well up to about $I = 0.1$, but he fitted α empirically at a value somewhat smaller than that given subsequently by Debye-Hückel theory. If one uses the theoretical value of α, Eq. (23-34)* corresponds to the use of $a = 0$ in Eq. (23-15a)*. In 1935 Guggenheim[2] suggested that the use of a standard value of a in the vicinity of 3 A would improve the empirical success of the Bronsted type of equation without introducing any serious

[1] J. N. Bronsted, *J. Am. Chem. Soc.*, **44**, 938 (1922).
[2] E. A. Guggenheim, *Phil. Mag.*, **19**, 588 (1935).

complications.   This yields (still for a simple M-X electrolyte)

$$\log \gamma_{\pm} = -A_{\gamma}|z_+z_-| \frac{I^{\frac{1}{2}}}{1 + I^{\frac{1}{2}}} + B_{MX}m \qquad (23\text{-}35)*$$

This equation fits very well most experimental data up to $I = 0.1$ with the theoretical Debye-Hückel coefficient A_{γ} and a single adjustable constant B_{MX} for each electrolyte, and its use in empirical representation of data has been discussed in Chapter 22.

The Bronsted-type equation has the important advantage that the specific interaction terms are consistent with Eq. (23-33) for a mixed electrolyte.   Thus, if we take the Guggenheim form for the first term, the equation for a particular cation M′ of charge $z_{M'}$ is

$$\log \gamma_{M'} = -A_{\gamma} \frac{z_{M'}^2 I^{\frac{1}{2}}}{1 + I^{\frac{1}{2}}} + \sum_{X} B_{M'X}m_X \qquad (23\text{-}36)*$$

where the sum includes all anions X present with each specific ion interaction constant for interaction with X multiplied by the molality of that anion.   No terms are present for other cations.   The corresponding equation for the activity coefficient of a particular anion X′ of charge $z_{X'}$ is

$$\log \gamma_{X'} = -A_{\gamma} \frac{z_{X'}^2 I^{\frac{1}{2}}}{1 + I^{\frac{1}{2}}} + \sum_{M} B_{MX'}m_M \qquad (23\text{-}37)*$$

One may readily verify that Eq. (23-33) is satisfied by Eqs. (23-36)* and (23-37)*, which may be combined to yield the experimentally measurable mean activity coefficient of M′X′.

$$\log \gamma_{\pm} = -A_{\gamma}|z_+z_-| \frac{I^{\frac{1}{2}}}{1 + I^{\frac{1}{2}}} + \frac{\nu_+}{\nu_+ + \nu_-} \sum_{X} B_{M'X}m_X$$

$$+ \frac{\nu_-}{\nu_+ + \nu_-} \sum_{M} B_{MX'}m_M \qquad (23\text{-}38)*$$

This equation is readily simplified for a pure electrolyte of any type.   Friedman[1] discusses the accuracy of Eq. (23-38)* at low molalities.   Applications of this equation are given in Chapter 34.

Guggenheim frequently uses base e logarithms and the quantity $\beta = 2.303B/2$.   Guggenheim and Turgeon[2] recently revised earlier tables of β or B values.   Some of their results are given in Table 23-1.   The short-range interaction effects vary with temperature and pressure; values are given in Table 23-1 for both 0 and 25°C where available.

[1] H. L. Friedman, *J. Chem. Phys.*, **32**, 1134, 1351 (1960).

[2] E. A. Guggenheim and J. C. Turgeon, *Trans. Faraday Soc.*, **51**, 747 (1955).

TABLE 23-1. SPECIFIC ION-INTERACTION CONSTANTS
FROM GUGGENHEIM AND TURGEON
$\beta = 1.151B$

Electrolyte	β at 0°C	β at 25°C	B at 25°C
HCl..............	0.25	0.27	0.23
HBr..............	......	0.33	0.29
HI...............	......	0.36	0.31
HClO$_4$.............	......	0.30	0.26
HNO$_3$.............	0.16		
LiCl..............	0.20	0.22	0.19
NaF..............	......	0.07	0.06
NaCl..............	0.11	0.15	0.13
NaI...............	......	0.21	0.18
NaClO$_3$.............	0.0	0.10	0.09
NaClO$_4$.............	0.05	0.13	0.11
NaBrO$_3$.............	......	0.01	0.01
NaIO$_3$.............	−0.41		
NaNO$_3$.............	−0.04	0.04	0.03
RbCl..............	......	0.06	0.05
CsCl..............	......	0.0	0.0
TlNO$_3$.............	......	−0.36	−0.31

Equation (23-38)* can, of course, be transformed by the Gibbs-Duhem equation into a form suitable for solvent equilibrium effects. The B values of Eq. (23-38)* are independent of m [in contrast to the B· values of Eq. (22-38)]; hence the osmotic coefficient takes the form

$$1 - \phi = \frac{2.303}{3} A_\gamma \frac{2I^{3/2}\sigma(I^{1/2})}{\sum_M m_M + \sum_X m_X} - \frac{2 \sum_M \sum_X \beta_{MX} m_M m_X}{\sum_M m_M + \sum_X m_X} \qquad (23\text{-}39)^*$$

where the sums in M cover all cations and the sums in X cover all anions. Now β is more convenient than B. Equation (23-39)* reduces to Eq. (22-31)* for a single 1-1 electrolyte. The factor $\sigma(I^{1/2})$ is a function of ionic strength defined by

$$\sigma(y) = \frac{3}{y^3}\left[1 + y - \frac{1}{1+y} - 2\ln(1+y)\right]$$
$$= 1 - 3(\tfrac{3}{4}y - \tfrac{3}{5}y^2 + \tfrac{4}{6}y^3 - \tfrac{5}{7}y^4 \cdots) \qquad (23\text{-}40)$$

Finally we must emphasize that the equations of this section are not derived rigorously from any detailed molecular model but rather are based on general arguments such as were given in the preceding section. Let us review these briefly. The first term in Eq. (23-38)* is an approximation to the theoretically predicted behavior for all interionic effects at distances longer than a fixed value such as a'' in Fig. 23-2.   The $1 + I^{\frac{1}{2}}$ form of the denominator corresponds to the approximate Debye-Hückel derivation for a distance of closest approach of about 3 A.   This distance is not to be taken too seriously, for neither is the derivation accurate nor is the model of a continuous solvent of constant dielectric constant realistic at this interionic distance.   However, this first term does give the correct limiting behavior as $m \to 0$ and provides a satisfactory and uniform general term to which specific interionic effects are to be added. These specific-effect terms account for all short-range interactions.   Thus they include the influence of the molecular nature of the solvent, the direct short-range forces between the ions, etc.   Ion association leads to negative B_{MX} values.   The Bm terms give, of course, only the first-order effect of the various molecular phenomena just mentioned.   Higher-order terms also occur but are negligible below 0.1 M for 1-1 electrolytes with the present accuracy of measurement.   At higher concentrations one may add further terms or may regard B as a function of concentration. In the case of strong ion association it is best to use the full chemical-equilibrium treatment of a weak electrolyte.

PROBLEMS

23-1. Verify the derivation of Eq. (23-16)* from (23-15)* for the limit of low concentration with the general electrolyte $A_{\nu_+}B_{\nu_-}$.

23-2. What is the increase in partial molal entropy of NaCl in aqueous solution for dilution from 0.01 to 0.001 M?   Assume first the limiting law; then consider terms involving B and dB/dT.

23-3. Make a table of the particular forms of Eqs. (23-38)* and (23-39)* for 1-1, 1-2, 1-3, and 2-2 pure electrolytes in terms of the molality of the complete salt.   Check the derivation of Eq. (23-39)* from Eqs. (23-38)* and (22-38) for the case of the 1-2 salt.

23-4. Show that the Bronsted specific-interaction terms in Eqs. (23-36)* and (23-37)* are consistent with Eq. (23-33).

23-5. Estimate the activity coefficient of very dilute TlCl in an aqueous solution at 25°C as a function of the molality of added NaNO₃.   How would the solubility of TlCl in aqueous NaNO₃ differ from that in pure water?   Note data in Table 23-1.

24

GALVANIC CELLS AND ELECTRODE POTENTIALS

We have already considered the relationship between the potential of a galvanic cell and the free-energy change of the chemical reaction which accompanies passage of current through the cell. In this chapter a number of additional topics related to cells will be discussed. These include a detailed analysis of the various chemical potentials in a cell, the definition of half-cell potentials, and a number of examples of the use of cells to determine important thermodynamic quantities.

Half-cell potentials are useful in the discussion of oxidation-reduction chemistry since they are proportional to free-energy changes per equivalent and therefore indicate immediately the possible direction of a given redox reaction. Thus many authors appropriately include in tables of half-cell potentials many values obtained by indirect methods. Since our interest is primarily the use of cells to obtain quantitative free-energy values, we merely recognize these indirect methods here but shall discuss them elsewhere as methods of obtaining free energies, entropies, etc.

To be useful for accurate thermodynamic purposes, a galvanic cell must be reversible, and the chemical change which occurs on passage of current must be known and precisely defined. Not all cells meet these criteria even though they may yield a measurable voltage. Other cells, which would seem to be satisfactory in theory, are unsatisfactory because one or both electrodes polarize, i.e., fail to react reversibly when a minute current is passing. Electrode polarization is a complex subject which falls into the general domain of chemical kinetics.[1] For thermodynamic purposes we confine our attention to electrodes which are reversible at least with the small current necessary for precise potentiometric measurement.

One more possible difficulty which must be mentioned is the physical state of the electrode materials. Solids and in particular solid metals can differ in physical state because of strains or crystal imperfections to a degree which appreciably affects the measured voltage of the cell. The

---

[1] For a recent review of electrode kinetics see J. O'M. Bockris, "Modern Aspects of Electrochemistry," pp. 180–276, Academic Press Inc., New York, 1954.

standard state of the metal is the most stable state; hence these effects can be in only one direction.   We cannot hope to give a full discussion of the means by which unstrained solids may be produced, but in general spongy metal produced by electrolysis is less likely to be strained than rolled or polished bar or sheet material.[1]

CHEMICAL POTENTIAL IN A GALVANIC CELL

It is assumed that the reader is familiar with the general nature of galvanic cells.   Nevertheless the exact relationship of the various thermodynamic quantities in a simple example will be carefully considered because there are novel aspects which have not arisen in other systems.

The measured emf of a galvanic cell is the electrical potential difference between two wires of the same metal attached to the two electrodes. This will be written $\mathcal{U}_2 - \mathcal{U}_1$.   Our task is to relate this potential difference to the properties of other components of the cell.   This is best done by considering the escaping tendency of electrons, which may be measured by μ_{e^-}, their partial molal free energy, or chemical potential.

When electrical connection is made between two pieces of different metals, electrons flow instantly until the escaping tendency of electrons is the same in each metal.   While this requires that initially neutral metals become charged, the magnitude of the electron excess or deficiency is extremely small.   Likewise, when a zinc electrode is placed in an electrolyte solution, zinc ions dissolve from or deposit onto the electrode until the escaping tendency of the zinc ions is the same in the electrode and the solution.   Again the charge transferred is so small that ratio of valence electrons to zinc ions in the zinc metal is still almost exactly 2.   Consequently one may write

$$\mu_{Zn^{++}} + 2\mu_{e^-} = \mu_{Zn}$$

where μ_{Zn} is a constant at a given temperature and pressure.   Then, in the solution

$$\mu_{Zn^{++}}(\text{in soln}) = \mu_{Zn^{++}}(\text{in metal})$$
$$\mu_{e^-} = \tfrac{1}{2}[\mu_{Zn} - \mu_{Zn^{++}}(\text{in soln})]$$

This result shows the relationship between the escaping tendency of the zinc ion in solution and that of electrons in the electrode or a wire attached to the electrode.

---

[1] The difficulty of reproducing the physical state of solid Ag and AgCl is such that Bates et al., *J. Chem. Phys.*, **25,** 361 (1956), suggest the use of γ_{HCl} as a means of standardization of individual electrodes.   They recommend 0.904 for γ_{HCl} at 25°C.

Consider next a complete cell such as is indicated by the following succession of symbols for the various parts:

$$\text{Pt, } H_2(g) | HCl(aq, \text{ sat with } H_2) | HCl(aq, \text{ sat with AgCl}) | \text{AgCl, Ag}$$

The concentration of HCl in the two aqueous phases is the same. Ordinarily the solubilities of both AgCl and H_2 in the aqueous HCl are assumed to be so small that the two aqueous solutions may be regarded as identical. Thus the cell may be described as "without liquid junction." Nevertheless, it is important to realize that the solution near the silver chloride electrode is saturated with AgCl and substantially free from H_2, whereas the solution near the hydrogen electrode is saturated with hydrogen and free from AgCl. If the same portion of the solution contained both H_2 and AgCl, these might spontaneously react, although the rate would probably be too slow to affect the electrical-potential measurements in this case.

At the silver chloride electrode the following equations apply:

$$\mu_{Cl^-}(\text{in AgCl}) = \mu_{Cl^-}(\text{in aq HCl})$$
$$\mu_{Ag^+}(\text{in Ag metal}) = \mu_{Ag^+}(\text{in AgCl})$$
$$\mu_{Ag^+} + \mu_{Cl} = \mu_{AgCl}$$
$$\mu_{Ag^+} + \mu_{e^-} = \mu_{Ag}$$

These yield

$$\mu_{e^-}(\text{in Ag}) = \mu_{Ag} - \mu_{AgCl} + \mu_{Cl}$$

At the hydrogen electrode one has

$$\mu_{H^+}(\text{in aq HCl}) + \mu_{e^-}(\text{in Pt}) = \tfrac{1}{2}\mu_{H_2}$$

These may be combined to yield

$$\mu_{e^-}(\text{in Ag}) - \mu_{e^-}(\text{in Pt}) = \mu_{Ag} + (\mu_{H^+} + \mu_{Cl^-}) - \mu_{AgCl} - \tfrac{1}{2}\mu_{H_2} \quad (24\text{-}1)$$

The possibility of electrical charging of substances, provided that the charges are small, does not affect the escaping tendency of neutral species; hence μ_{H_2}, μ_{AgCl}, and μ_{Ag} may be identified with the ordinary molal free energies of these substances. The escaping tendency of a charged species is affected by the net electrical charge of the phase in which it is located.[1] However, whatever the effect of such a net charge in the aqueous solution may be upon the escaping tendency of the H^+ ion, it will have exactly the opposite effect on the escaping tendency of the Cl^- ion. Thus the sum $\mu_{H^+} + \mu_{Cl^-}$ is not affected by such a net charge and may be identified with the ordinary chemical potential, or partial molal free energy, of HCl for the aqueous solution present in the cell.

[1] In order to emphasize this additional effect upon the escaping tendency of ions, Guggenheim uses the term *electrochemical potential* [E. A. Guggenheim, *J. Phys. Chem.*, **33,** 842 (1929)].

There is a second complication in the interpretation of the escaping tendency of H^+ and Cl^-. The solution near the Ag, AgCl electrode is saturated with AgCl but presumably free from H_2, whereas the solution near the hydrogen electrode is saturated with H_2 but free from Ag^+. Thus there will be diffusion of Ag^+ toward the hydrogen electrode (where at equilibrium it will be reduced) and a diffusion of H_2 in the opposite direction. Also the dissolving of AgCl increases the concentration of Cl^- above that of H^+ in the solution near the Ag, AgCl electrode. All these effects approach zero as the solubilities of AgCl and H_2 approach zero and are probably negligible in this particular cell. In other cells of similar type, but in which the substances are more soluble, the errors from these complications may become significant.

Let us now complete the analysis of the cell indicated by the diagram, now abbreviated in a conventional manner to Pt, H_2, HCl(aq), AgCl, Ag. If positive electricity passes from left to right (or negative electricity from right to left) through the cell in the amount of 1 equiv, i.e., 1 mole of electrons, the net reaction is

$$\tfrac{1}{2}H_2(g) + AgCl(s) = Ag(s) + H^+(aq) + Cl^-(aq)$$

The free-energy change for this chemical reaction is clearly given by the right side of Eq. (24-1) and is thus measured by the difference in escaping tendency of electrons between the two electrodes. But the measured emf of the cell ε is the difference in electrical potential between wires attached to the two electrodes and is conventionally taken as

$$\varepsilon = \mathcal{V}_2 - \mathcal{V}_1 \qquad (24\text{-}2)$$

where 2 refers to the right electrode and 1 to the left electrode. Also the electrical potential is defined for a unit positive test charge, whereas electrons are negative and have a charge $\mathcal{F}$ (the Faraday constant) per mole. Hence

$$\varepsilon = \mathcal{V}_2 - \mathcal{V}_1 = -\frac{\mu_e\text{-(in Ag)} - \mu_e\text{-(in Pt)}}{\mathcal{F}} \qquad (24\text{-}3)$$

or in general

$$\varepsilon = \mathcal{V}_2 - \mathcal{V}_1 = -\frac{\mu_e\text{-}(2) - \mu_e\text{-}(1)}{\mathcal{F}} = -\frac{\Delta F}{n\mathcal{F}} \qquad (24\text{-}4)$$

The last equation introduces the free-energy change for the cell reaction ΔF divided by the number of equivalents of electricity n flowing through the external circuit when that amount of reaction occurs in the cell.

The final result of Eq. (24-4) conforms to the basic thermodynamic principles, since $n\mathcal{F}\varepsilon$ is the electrical work done in the external circuit when the cell changes free-energy content by $-\Delta F$.

CONVENTIONS REGARDING THE SIGN OF ELECTROMOTIVE FORCE

Let us now direct our attention specifically to the name and sign to be ascribed to the measured emf and to the related definitions and conventions.[1] We continue to consider as an example the cell A:

$$\text{Pt, H}_2, \text{HCl(aq), AgCl, Ag}$$

together with the definitions associated with Eq. (24-2). For cell A in particular this yields

$$\mathcal{E}_A = \mathcal{U}_{\text{Ag,AgCl}} - \mathcal{U}_{\text{H}_2} \qquad (24\text{-}2')$$

If all substances involved have their standard partial molal free energies, then these potentials are *standard* potentials and are so indicated.

$$\mathcal{E}_A^\circ = \mathcal{U}_{\text{Ag,AgCl}}^\circ - \mathcal{U}_{\text{H}_2}^\circ = 0.222 \text{ volt} \qquad (24\text{-}2'')$$

The chemical reaction of a cell is, by convention, written in the direction such that it corresponds to positive electricity flowing through the cell from left to right (or negative electricity flowing through the cell from right to left). Thus the reaction for cell diagram A is

$$\tfrac{1}{2}\text{H}_2(\text{g}) + \text{AgCl(s)} = \text{H}^+(\text{aq}) + \text{Cl}^-(\text{aq}) + \text{Ag(s)}$$

If the cell emf $\mathcal{E}$ is positive (as it will be for this cell at 1 atm of hydrogen pressure and 1 M hydrochloric acid), then positive electricity spontaneously tends to flow from the right electrode to the left electrode in the external circuit and from left to right within the cell. Consequently a positive $\mathcal{E}$ corresponds to a chemical reaction which tends to take place spontaneously and to a negative ΔF for the reaction. As noted in the preceding section, electrical work received by the cell at constant temperature and pressure is ΔF, and

$$\Delta F = -n\mathfrak{F}\mathcal{E} \qquad (24\text{-}4')$$

where $\mathfrak{F}$ is the Faraday constant and n is the number of equivalents of electricity flowing per mole of reaction as written.

The conventions and equations given above for the *cell* emf $\mathcal{E}$ are entirely consistent with those of the first edition of this book. The *electrical potentials* $\mathcal{U}$ were first introduced by Gibbs[2] in 1877 with the words "the electrical potentials in pieces of the same kind of metal connected

---

[1] These definitions are essentially those recommended by the International Union of Pure and Applied Chemistry in 1953 (*Compt. rend. 17th conf. union intern. chim. pure et appl.*, 1953, pp. 82–85).

[2] "Collected Works of J. Willard Gibbs," pp. 332–349, Yale University Press, New Haven, Conn., reprinted 1948; also p. 429.

with the two electrodes." Since each quantity has certain advantages in further applications, we shall continue to use both and wish to emphasize certain differences at this point.

If we write cell A in the reverse direction, securing cell B,

$$\text{Ag, AgCl, HCl(aq), } H_2, \text{ Pt}$$

the emf $\mathcal{E}$ has the opposite sign, that is, $\mathcal{E}_B = -\mathcal{E}_A$, but the expression $\mathcal{U}_{Ag,AgCl} - \mathcal{U}_{H_2}$ has a definite and unchanged value regardless of the diagram used or any other adequate description of the cell. However, while the relationship of $\mathcal{E}$ to ΔF is fully specified in Eq. (24-4) and the associated definitions and remains unchanged when the cell is reversed, the relationship of $\mathcal{U}_{Ag,AgCl} - \mathcal{U}_{H_2}$ to ΔF becomes definite only after the chemical reaction has been written in one direction or the other. If one accepts the cell diagram and the convention that the chemical reaction corresponds to positive electricity flowing through the cell from left to right, then one may write

$$\Delta F = n\mathcal{F}(\mathcal{U}_{left} - \mathcal{U}_{right}) \tag{24-5}$$

which follows at once from the rearrangement of Eq. (24-4). Thus we see that the Gibbs electrode potentials have certain advantages in dealing directly with galvanic cells but that the definitions associated with the cell emf are necessary in relating cell information unambiguously to chemical reactions.

HALF CELLS AND ELECTRODE POTENTIALS

Galvanic cells can always be divided into two parts, one associated with each electrode, and these respective parts or half cells may then be recombined in different pairs to yield additional cells. There are, of course, problems associated with the joining of half cells. Sometimes their electrolytes are incompatible, and junction is possible, if at all, only by interposing a third electrolyte. We shall lay aside for the present these problems and those of the electrical potentials associated with complex junctions and assume that the half cells to be considered can be combined without appreciable junction potential into complete cells.

It is obvious that there is a great simplification if a table of half-cell potentials will suffice to allow the calculation of the emf of all cells. It is well known that this is possible provided that a particular half cell is selected as a reference point. The hydrogen electrode is generally accepted as this reference electrode whose potential is assigned the value zero when the fugacity of hydrogen gas is 1 atm and the activity of hydrogen ion is 1 M. Actually one does not determine the activity of a single ion; when a whole cell is considered, we have seen that its emf depends on a combination of activities corresponding to electroneutrality.

Thus we may write

$$\text{Pt, H}_2\text{, H}^+\text{:} \qquad \mathcal{E}^\circ = \mathcal{U}^\circ = \Delta F^\circ = 0.000$$

with the chemical half reaction

$$\tfrac{1}{2}\text{H}_2 = \text{H}^+ + e^-$$

The electron indicated by e^- is, of course, delivered through the connecting wire into the external circuit. Since the electrons always cancel when half reactions are combined into whole cell reactions, it is necessary only that their state be the same. However, in view of the zero value of ΔF° for the hydrogen half reaction, it is possible in a formal way to say that the partial molal free energy of the electron in its standard state is that which exists in equilibrium in any wire connected to a standard hydrogen half cell.

It is with respect to potentials of half cells that the present literature contains a confusing variety of conventions. We shall present three systems, which comprise the various alternates consistent with the conventions for the whole cell presented above.

If we represent the half cell by the diagram: electrode, electrolyte [for example, Ag, AgCl, Cl$^-$(aq)], then we assign to this half cell the $\mathcal{E}$ value of the complete cell constructed by adding the half cell H$^+$(aq), H$_2$, Pt. This corresponds to cell B of the preceding section and to the chemical reaction

$$\text{Ag(s)} + \text{H}^+\text{(aq)} + \text{Cl}^-\text{(aq)} = \text{AgCl(s)} + \tfrac{1}{2}\text{H}_2\text{(g)}$$

with the standard values at 298°K: $\mathcal{E}^\circ = -0.222$ volt, $\Delta F^\circ = 5120$ cal/mole. Since the half-cell chemical reaction

$$\text{Ag(s)} + \text{Cl}^-\text{(aq)} = \text{AgCl(s)} + e^-$$

is written in the oxidation direction, these half cell emf values are frequently called *oxidation potentials*. The particularly extensive compilation by Latimer[1] uses this convention and is entitled "Oxidation Potentials."

If we present the half cell by the reversed diagram: electrolyte, electrode, then the corresponding complete cell is reversed and in the emf convention the sign of $\mathcal{E}$ is reversed. Thus one has the following:

$$\text{Cl}^-\text{(aq), AgCl, Ag}$$
$$\text{AgCl(s)} + e^- = \text{Ag(s)} + \text{Cl}^-\text{(aq)}$$
$$\mathcal{E}^\circ = 0.222 \text{ volt} \qquad \Delta F^\circ = -5120 \text{ cal/mole}$$

---

[1] W. M. Latimer, "Oxidation Potentials," 2d ed., Prentice-Hall, Inc., Englewood Cliffs, N.J., 1952.

While half-cell electromotive force values in this system have been given various names, the term *reduction potentials* would seem most appropriate.

The third alternative[1,2] is the use of the Gibbs electrical potentials. If we define the standard hydrogen half-cell potential to be zero ($\mho_{H_2}^{\circ} = 0$), then there is no ambiguity in the sign of the remaining half-cell potentials. Thus from Eq. (24-2'') $\mho_{Ag,AgCl}^{c} = +0.222$ volt. Since this potential value is associated with the electrode rather than the direction of the assumed chemical half reaction, the term *electrode potential* seems appropriate for these values. In this case the relationship to the free energy depends on the direction of the assumed half reaction. Thus for

Oxidation at an *anode*[3]

$$\Delta F_e = +n\mathfrak{F}\mho_e \qquad (24\text{-}6a)$$

Reduction at a *cathode*

$$\Delta F_e = -n\mathfrak{F}\mho_e \qquad (24\text{-}6b)$$

where it has been assumed throughout that for the standard hydrogen electrode $\Delta F_{H_2}^{\circ} = 0$, $\mho_{H_2}^{\circ} = 0$.

While any one of these systems is self-consistent and can be used to obtain unambiguous results, there is a natural tendency to associate the sign and magnitude of the numerical value pertaining to a given electrode with that terminal of the actual cell rather than with the tendency of a half reaction to proceed one way or the other. Thus the Gibbs electrode potentials are possibly preferable for half-cell tabulations.[4] However, the unambiguous relationship of the whole-cell emf $\mathcal{E}$ with the free energy change ΔF makes it desirable to standardize upon the relationship

[1] A. J. deBethune, *J. Electrochem. Soc.*, **102**, 288C (1955). Gibbs's symbol was V, and deBethune suggests its adoption. We feel that the script $\mho$ does no violence to Gibbs and avoids confusion with volume V.

[2] A fourth system was proposed by J. B. Ramsey, *J. Electrochem. Soc.*, **104**, 255 1957, which uses the partial molal free energy of electrons in the electrode. There result values equal in magnitude but opposite in sign to the Gibbs electrode potentials. This paper gives a very clear discussion of the thermodynamics of galvanic cells.

[3] An anode is an electrode where oxidation is occurring or is assumed to occur by the chemical reaction written, and a cathode an electrode where reduction is occurring or is assumed to occur, irrespective of whether the reaction is spontaneous or is forced by an external emf.

[4] In their desire to associate the signs of tabulated half-cell potential with observed cell voltages, many writers have criticized the oxidation-potential system without realizing the limitations of the possible alternatives. W. M. Latimer recognized the desirability of a uniform sign convention and indicated willingness to adopt another system provided that it was free from logical objections such as he mentioned in his last communications on this subject [*J. Am. Chem. Soc.*, **76**, 1200 (1954); *Proc. 7th Meeting Intern. Comm. Electrochem. Thermodynam. and Kinet.*, 1955, p. 176]. Unfortunately Latimer had not been able to consider the deBethune proposal at the time of his death.

$\mathcal{E} = \mathcal{V}_{right} - \mathcal{V}_{left} = \mathcal{V}_{cathode} - \mathcal{V}_{anode}$ and then to use the quantity $\mathcal{E}$ in the thermodynamic treatment of whole cells. Let us illustrate this procedure with an example. Table 24-3 near the end of this chapter contains the entries

$$\frac{1}{2}Cl_2 + e^- = Cl^- \qquad \mathcal{V}° = 1.359 \text{ volts}$$
$$AgCl + e^- = Ag + Cl^- \qquad \mathcal{V}° = 0.222$$
$$H^+ + e^- = \frac{1}{2}H_2 \qquad \mathcal{V}° = 0.0000$$

Consider a cell made up of a chlorine electrode, that is, Cl_2, at 1 atm on an inert electrode M, an aqueous chloride solution (e.g., sodium chloride), and a silver–silver chloride electrode. The diagram for the cell is

$$M, Cl_2, Cl^-(aq), AgCl, Ag$$

and the chemical equation which corresponds is

$$AgCl(s) = Ag(s) + \frac{1}{2}Cl_2(g)$$

The chlorine electrode may be seen to be the positive electrode of the cell directly from the tabulated $\mathcal{V}°$ values since the value 1.359 is more positive than 0.222. But the cell emf is

$$\mathcal{E}° = \mathcal{V}°_{right} - \mathcal{V}°_{left} = 0.222 - 1.359$$
$$= -1.137 \text{ volts}$$

where the negative value indicates at once that the chemical reaction tends to go in the reverse direction; i.e., silver tends to react spontaneously with chlorine to yield silver chloride.

We note further that the convention of positive electricity flowing through the cell from left to right implies that the left electrode is the anode and the right electrode the cathode. Thus one could instead have described the cell as a chlorine anode and a silver–silver chloride cathode with an aqueous chloride electrolyte. Then the same calculation follows,

$$\mathcal{E}° = \mathcal{V}°_{cathode} - \mathcal{V}°_{anode} = 0.222 - 1.359$$
$$= -1.137 \text{ volts}$$

and one concludes from the negative result that spontaneous reaction in this direction is impossible but that electrolysis is possible if the applied potential exceeds 1.137 volts.

EFFECT OF TEMPERATURE AND PRESSURE ON THE ELECTROMOTIVE FORCE OF A CELL

The well-known temperature derivative of the free energy [Eq. (15-4)] when applied to the emf of a cell [Eq. (24-4) or (24-4′)] yields an expression for the entropy change of the cell reaction,

$$\Delta S = n\mathfrak{F} \left(\frac{\partial \mathcal{E}}{\partial T} \right)_P \tag{24-7}$$

and by further use of familiar equations

$$\Delta H = -n\mathfrak{F} \left[\mathcal{E} - T \left(\frac{\partial \mathcal{E}}{\partial T} \right)_P \right] \tag{24-8}$$

Cell potentials are frequently measurable to sufficiently high precision to make the temperature derivative of useful accuracy also. Indeed, a further derivative of Eq. (24-8) yields heat-capacity values which sometimes have useful accuracy.

$$\Delta C_P = n\mathfrak{F}T \left(\frac{\partial^2 \mathcal{E}}{\partial T^2} \right)_P \tag{24-9}$$

Again the cell Pt, H_2, HCl(aq), AgCl, Ag is taken as an example, with the chemical reaction

$$\tfrac{1}{2}H_2(g) + AgCl(s) = Ag(s) + H^+(aq) + Cl^-(aq)$$

Bates and Bower[1] measured these cells with particularly great care at 5° intervals over the range 0 to 90°C, and their results at 0.1 M HCl may be expressed by the equation

$$\mathcal{E} = 0.35510 - 0.3422 \times 10^{-4}t - 3.2347 \times 10^{-6}t^2$$
$$+ 6.314 \times 10^{-9}t^3 \qquad \text{volts}$$

where t is the temperature in degrees centigrade. From Eq. (24-8) one obtains

$$\Delta H = -8405 - 40.75t + 0.0447t^2 + 2.91 \times 10^{-4}t^3 \qquad \text{cal/mole}$$
$$\Delta H = -9391 \text{ cal/mole at 25°C}$$

This result is probably about as accurate as that obtained by indirect combination of calorimetric measurements. Application of Eq. (24-9) yields

$$\Delta C_P = -40.75 + 0.0894t + 8.73 \times 10^{-4}t^2 \qquad \text{cal/deg mole}$$
$$\Delta C_P = -37.97 \text{ cal/deg mole at 25°C}$$

The partial molal heat capacity of aqueous hydrochloric acid at 25°C has been measured calorimetrically by Gucker and Schminke,[2] and the value for 0.1 M solution is -29 cal/deg mole. The heat capacities of hydrogen, silver, and silver chloride have been measured accurately,[3] and, combining

[1] R. G. Bates and V. E. Bower, *J. Research Natl. Bur. Standards*, **53**, 283 (1954).

[2] F. T. Gucker and K. H. Schminke, *J. Am. Chem. Soc.*, **54**, 1358 (1932).

[3] See F. D. Rossini et al., Selected Values of Chemical Thermodynamic Properties, *Natl. Bur. Standards (U.S.) Circ.* 500, 1952.

all these values, one obtains, at 25°C, $\Delta C_P = -38.5$ cal/deg.   In this case the calorimetric value is probably more accurate.   But it is impressive that a value obtained from the second temperature derivative of a cell potential comes as close as this.

It should be remembered that the example just cited represents unusually precise results.   Only a few cell measurements yield comparably reliable second derivatives and heat capacities.

The pressure derivative of the free energy yields the volume change for the reaction.   In terms of the cell emf this becomes

$$\Delta V = -n\mathfrak{F}\left(\frac{\partial \mathcal{E}}{\partial P}\right)_T \tag{24-10}$$

Large volume changes are always associated with the consumption or evolution of gas, and these effects are normally treated by the method of the next section.   Nevertheless Eq. (24-10) remains available for the calculation of ΔV whenever desired.

ACTIVITY AND ELECTROMOTIVE FORCE

The definitions and discussion of activity in Chapters 20 and 22 may be applied directly to cell emf through the free-energy relationships. Let us consider a cell such that, when n equivalents of electricity pass in the defined direction, the reaction is

$$b\mathrm{B} + c\mathrm{C} + \cdots = q\mathrm{Q} + r\mathrm{R} + \cdots$$

When each of these substances has unit activity, we write the change of free energy as $\Delta F°$ and the electromotive force as $\mathcal{E}°$.   In the more general case, where the activities are not unity, the corresponding values of ΔF and of $\mathcal{E}$ are given immediately by Eq. (24-11), which reads

$$\Delta F = \Delta F° + RT \ln \frac{a_Q^q a_R^r \cdots}{a_B^b a_C^c \cdots} \tag{24-11}$$

and since $\Delta F = -n\mathfrak{F}\mathcal{E}$,

$$\mathcal{E} = \mathcal{E}° - \frac{RT}{n\mathfrak{F}} \ln \frac{a_Q^q a_R^r \cdots}{a_B^b a_C^c \cdots} \tag{24-12}$$

If we avoid those irregularities in the surface conditions of metal electrodes which we have discussed in a preceding section, the activities of the various cell constituents will ordinarily depend, at given temperature, only upon pressure and concentration.   The pressure effect is usually negligible except when some one of the substances concerned is a gas. In such a case the activity, which is equal to the fugacity, may be found by the methods of Chapter 16.   It frequently suffices to assume a perfect gas and to set the activity equal to the partial pressure.

When the hydrogen electrode came into common use, it was important to show that this electrode gives a reversible emf, and one which is in accord with the assumed cell reaction. The simplest test was to study the effect of pressure upon the emf of a cell with hydrogen and calomel electrodes in hydrochloric acid, the cell reaction being

$$H_2 + Hg_2Cl_2 = 2Hg + 2HCl(aq)$$

If pure hydrogen passes through such a cell against a given external pressure, the partial pressure of the hydrogen is equal to the difference between the external pressure and the vapor pressure of water from the solution. By varying the external pressure, the partial pressure of the hydrogen, p, is also changed. If this is the only variable, we may write by Eq. (24-12)

$$\mathcal{E} = \text{constant} + \frac{RT}{2\mathfrak{F}} \ln p$$

This equation was first tested by Lewis,[1] who allowed the hydrogen in such a cell to escape against a variable excess pressure. The results (at 25°C) are given in Table 24-1. The excess pressure is given in centimeters of water. (Between the fourth and fifth experiments the excess pressure was raised to 100 cm.) The very satisfactory agreement between the observed and the calculated values, together with the lack of appreciable polarization in the electrode, makes it certain that we are dealing here with a well-defined and reversible reaction.

TABLE 24-1. EFFECT OF PRESSURE UPON THE HYDROGEN ELECTRODE

Δp	0	37	63	84	84	63	37
$\mathcal{E}$, observed	0.40089	0.40134	0.40163	0.40190	0.40189	0.40164	0.40138
$\mathcal{E}$, calculated	0.40089	0.40134	0.40165	0.40189	0.40189	0.40165	0.40134

In order to illustrate more generally the relation between the emf of a cell and the activity of the cell constituents, we may take a case which was fully studied experimentally and thermodynamically by Dolezalek,[2]

$$H_2, H_2SO_4(aq), PbSO_4, PbO_2, Pb$$

with the cell reaction

$$H_2(g) + PbO_2(s) + H_2SO_4(m) = PbSO_4(s) + 2H_2O \text{ (with } H_2SO_4 \text{ at } m)$$

By expressing as in Chapters 15 and 20 the activity of a gaseous substance by the formula in brackets and that of a substance in solution by

---

[1] G. N. Lewis and M. Randall, *J. Am. Chem. Soc.*, **36**, 1969 (1914).

[2] F. Dolezalek, *Z. Elektrochem.*, **5**, 533 (1899); a very recent paper on this cell is by P. Rüetschi, R. T. Angstadt, and B. O. Cahan, *J. Electrochem. Soc.*, **106**, 547 (1959).

its formula in parentheses, Eq. (24-12) becomes

$$\varepsilon = \varepsilon^\circ - \frac{RT}{2\mathfrak{F}} \ln \frac{(H_2O)^2}{(H_2SO_4)[H_2]} \qquad (24\text{-}13)$$

the solid substances being omitted, since their activity may be taken as unity.   Thus the emf varies with the pressure of hydrogen and with the composition of the solution (which affects the activity of both H_2SO_4 and H_2O).   By measuring the emf of some one cell, and by placing the activity of the hydrogen equal to its partial pressure, while the activity of the water and that of the sulfuric acid ($a_2 = a_\pm{}^3$) are determined by the methods which we have illustrated in the preceding chapters, we obtain the value of ε°.   This is the emf of the cell when each substance is in its standard state, namely, for the reaction

$$H_2(1 \text{ atm}) + PbO_2(s) + H_2SO_4(a_2 = 1) = PbSO_4(s) + 2H_2O(l)$$

In a case like the preceding it makes no difference whether we deal with the activity of the electrolyte as a whole or with the activities of the individual ions.   Thus we might have written the reaction

$$H_2 + PbO_2 + 2H^+ + SO_4^= = PbSO_4 + 2H_2O$$

in which case we should have in place of Eq. (24-13)

$$\varepsilon = \varepsilon^\circ - \frac{RT}{2\mathfrak{F}} \ln \frac{(H_2O)^2}{(H^+)^2(SO_4^=)[H_2]} \qquad (24\text{-}14)$$

But the two equations are identical, since our conventions regarding activities of electrolytes[1] make $(H_2SO_4) = (H^+)^2(SO_4^=)$.

ELECTROMOTIVE FORCE AND EQUILIBRIUM CONSTANT

Since the standard free-energy change of a reaction may be calculated at once from an equilibrium constant by Eq. (15-32) and since, moreover, it is related to the standard emf of a galvanic cell by the equation $\Delta F^\circ = -n\mathfrak{F}\varepsilon^\circ$, we may determine ε° for a cell from the equilibrium constant of the reaction occurring within the cell.   Thus

$$-RT \ln K = \Delta F^\circ = -n\mathfrak{F}\varepsilon^\circ \qquad (24\text{-}15)$$

and $$\varepsilon^\circ = \frac{RT}{n\mathfrak{F}} \ln K \qquad (24\text{-}16)$$

or at $298.15°K = 25°C$

$$\varepsilon^\circ_{298} = \frac{0.05915}{n} \log K_{298} \qquad (24\text{-}17)$$

---

[1] Here we treat H_2SO_4 as a strong electrolyte even with respect to the second ionization.

If we construct a cell Ag, Ag^+, Fe^{++}, Fe^{3+}, Pt, where Pt stands for any inert electrode, the cell reaction may be written

$$Ag + Fe^{3+} = Ag^+ + Fe^{++}$$

Now the equilibrium in this reaction has been studied by Noyes and Brann,[1] who allowed silver and ferric nitrate to form ferrous nitrate and silver nitrate until equilibrium was established. They worked at various concentrations and by extrapolation to infinite dilution were able to obtain the true equilibrium constant in the above reaction, namely, $K_{298} = (Fe^{++})(Ag^+)/(Fe^{3+}) = 0.128$. Hence from Eq. (24-17), where $n = 1$, $\mathcal{E}^{\circ}_{298} = -0.0528$. This value agrees within the limits of error with one obtained from emf measurements.

CONCENTRATION CELLS WITH LIQUID JUNCTION

In general, cells which involve a junction between two different solutions give an emf which varies with the kind and with the physical nature of the liquid junction. We shall not discuss the various approximate methods which have been devised to deal more or less reliably with such cells.[2] In case the two solutions are of different concentrations of the same electrolyte, however, an exact theoretical treatment is possible and excellent experimental results have been obtained.

Brown and MacInnes[3] and Janz and Gordon[4] measured very carefully the cell

$$\text{Ag, AgCl, NaCl}(m_A), \text{NaCl}(m_B), \text{AgCl, Ag}$$

and obtained accurately reproducible results. The reactions at each electrode are straightforward, but the region of variable concentration between the two solutions must be considered carefully. In Fig. 24-1 at

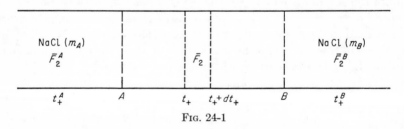

FIG. 24-1

[1] A. A. Noyes and B. F. Brann, *J. Am. Chem. Soc.*, **34**, 1016 (1912).

[2] See H. S. Harned and B. B. Owen, "The Physical Chemistry of Electrolytic Solutions," 3d ed., Reinhold Publishing Corporation, New York, 1958, for a discussion of some of these, including particularly methods of pH determination (p. 442).

[3] A. S. Brown and D. A. MacInnes, *J. Am. Chem. Soc.*, **57**, 1356 (1935).

[4] G. J. Janz and A. R. Gordon, *J. Am. Chem. Soc.*, **65**, 218 (1943).

A and to the left of A the molality of the salt is constant and equal to m_A; the transference number of the cation is t_+^A, and the partial molal free energy of the NaCl is $\bar{F}_2^A$. Somewhere between A and B the concentration varies in some continuous but unspecified manner, until at B, and to the right, the molality is constant at m_B, the transference number at t_+^B, and the partial molal free energy at $\bar{F}_2^B$.

Now when the cell operates and a small negative current passes through the cell from right to left, the total change in free energy, per equivalent of electricity, is determined if we know the amount of salt being added or removed at each point, and also the value of $\bar{F}_2$ at each point.

The amount leaving the region to the left of A is t_+^A, the amount entering the region to the right of B is t_+^B, and the difference is taken from (or given to) the intervening region where the gradient occurs.   In the region to the right of B the free energy increases by $t_+^B\bar{F}_2^B$; in the region to the left of A it increases by $-t_+^A\bar{F}_2^A$; and in the intervening region there is a summation of effect which may be analyzed as follows: Consider some infinitesimal region (indicated by the space between the dashed lines) in which the transference number varies from t_+ on the one side to $t_+ + dt_+$ on the other, and in which the partial molal free energy is $\bar{F}_2$. The amount of salt leaving this region is dt_+; hence the increase in free energy in the total intermediate region is $- \int_A^B \bar{F}_2 \, dt_+$.

Therefore for the whole cell

$$\Delta F = t_+^B\bar{F}_2^B - t_+^A\bar{F}_2^A - \int_A^B \bar{F}_2 \, dt_+$$

and upon integration by parts this simplifies to

$$\Delta F = \int_A^B t_+ \, d\bar{F}_2 = 2RT \int_A^B t_+ \, d\ln\left(m\gamma_\pm\right) \qquad (24\text{-}18)$$

Since this ΔF was for 1 equiv of electricity, the emf of the cell is

$$\varepsilon = - \frac{2RT}{\mathcal{F}} \int_A^B t_+ \, d\ln\left(m\gamma_\pm\right) \qquad (24\text{-}19)$$

If one carries through the same analysis for the more complex case of the electrolyte $(M^{z+})_{\nu+}(X^{z-})_{\nu-}$ the result is

$$\varepsilon = - \frac{RT}{\mathcal{F}} \left(\frac{1}{z_+} + \frac{1}{z_-}\right) \int_A^B t_+ \, d\ln\left(m\gamma_\pm\right) \qquad (24\text{-}20)$$

provided that the electrode reaction produces or consumes the anion. We may therefore calculate the emf of any such concentration cell if we know, over a range of concentration, the values of the transference number and of the partial molal free energy (which may, for example, be obtained from the cells without liquid junctions).   It is necessary only to plot at several concentrations the values of t_+ against the values of $\log m\gamma_\pm$, and the area under the curve, gives immediately the desired result.

Conversely, if potentials of such cells are known as well as the transference number of the solution as a function of concentration, then the activity coefficient may be calculated. There is no simple inversion of Eq. (24-19) for the calculation of the activity coefficient, but successive-approximation methods are available which yield any accuracy justified by the data. Janz and Gordon followed such methods with a series of cells, where m_A was held near 0.05 M and m_B varied from 0.01 to 0.10 M. A typical portion of their results are listed in Table 24-2, where the values of $t_+{}^B$ are also given, as well as the values of ε calculated from the equation they selected for the activity coefficient of NaCl at 25°C.

$$\log \gamma_\pm = -0.5049 \frac{m^{\frac{1}{2}}}{1 + 1.350m^{\frac{1}{2}}} + 0.031m$$

The new values of physical constants require an increase of the Debye-Hückel slope from 0.5049 to 0.511 with an appropriate readjustment of the other constants, but this will not affect the excellent agreement between observed and calculated cell potentials in Table 24-2.

TABLE 24-2. CONCENTRATION CELLS WITH TRANSFERENCE;
NaCl AT 25°C; DATA OF JANZ AND GORDON†

m_A	m_B	$t_+{}^B$	E_{obs}, mv	E_{calc}, mv
0.049903	0.009963	0.3918	30.391	30.400
0.049741	0.010021	0.3918	30.220	30.215
0.049819	0.019955	0.3903	17.112	17.110
0.049794	0.029960	0.3893	9.440	9.435
0.049741	0.079333	0.3863	−8.581	−8.570
0.049846	0.099872	0.3853	−12.720	−12.720

† G. J. Janz and A. R. Gordon, *J. Am. Chem. Soc.*, **65**, 218 (1943).

CELLS WITHOUT LIQUID JUNCTION

A typical cell without liquid junction was discussed in detail earlier in this chapter. Data from such a cell were used in Chapter 22 to determine the activity coefficient of aqueous HCl. At this point we wish to consider several other cells selected to determine dissociation constants of weak electrolytes.

Dissociation Constant of a Weak Acid. Harned and Ehlers[1] measured the cell

$$\text{Pt, H}_2 |\ \text{HOAc}(m_1), \text{NaOAc}(m_2), \text{NaCl}(m_3)\ |\text{AgCl, Ag}$$

[1] H. S. Harned and R. W. Ehlers, *J. Am. Chem. Soc.*, **54**, 1350 (1932).

where OAc^- is the acetate ion. The ratios of the molalities $m_1:m_2:m_3$ were held constant in a series of solutions. The potential of this cell, like that of the cell discussed at the beginning of this chapter, depends on the activity product $a_{H^+}a_{Cl^-}$. Substitution of the dissociation constant of acetic acid,

$$K_a = \frac{a_{H^+}a_{OAc^-}}{a_{HOAc}} = \frac{m_{H^+}m_{OAc^-}}{m_{HOAc}} \frac{\gamma_{H^+}\gamma_{OAc^-}}{\gamma_{HOAc}} \qquad (24\text{-}21)$$

into the usual equation for the cell emf yields

$$\mathcal{E} = \mathcal{E}° - \frac{RT}{\mathcal{F}} \ln \frac{m_{Cl^-}m_{HOAc}}{m_{OAc}} - \frac{RT}{\mathcal{F}} \ln \frac{\gamma_{Cl^-}\gamma_{HOAc}}{\gamma_{OAc^-}} - \frac{RT}{\mathcal{F}} \ln K_a \qquad (24\text{-}22)$$

where $\mathcal{E}°$ is the standard potential of the cell

$$\text{Pt, } H_2| \text{ HCl(aq) } |AgCl, \text{ Ag}$$

considered earlier.

The function of molalities in the second term in Eq. (24-22) may be written $m_3(m_1 - m_{H^+})/(m_2 + m_{H^+})$. The molality of hydrogen ion constitutes a very small correction. It suffices in this case to take

$$m_{H^+} \cong K_a \frac{m_{HOAc}}{m_{OAc}} \cong K_a \frac{m_1}{m_2} \qquad (24\text{-}23)*$$

with an approximate value of K_a.

The third term in Eq. (24-22) involves the ratio of γ_{Cl^-} to γ_{OAc^-}; consequently the D-H (Debye-Hückel) limiting-law terms cancel. Likewise, if one uses the Guggenheim equation (23-37)*, the term in $I^{1/2}/(1 + I^{1/2})$ cancels. The remaining effects of the activity coefficients of the ions, together with the activity coefficient of the acetic acid, yield an initial dependence on molality to the first power. Hence we define

$$\mathcal{E}' = \mathcal{E} + \frac{RT}{\mathcal{F}} \ln \frac{m_3(m_1 - m_{H^+})}{m_2 + m_{H^+}}$$

and plot $\mathcal{E}'$ versus m_1 or any other of the molalities. The result is a linear plot in the dilute region whose intercept at $m = 0$ is $\mathcal{E}° - (RT/\mathcal{F}) \ln K$. Substitution of the value of $\mathcal{E}°$ then gives K_a, which is found to be 1.754×10^{-5} at 25°C.

The slope of the plot $\mathcal{E}'$ versus molality is a function of the difference of ion-interaction parameters B for sodium acetate and sodium chloride together with the similar coefficient giving the dependence of $\ln \gamma_{HOAc}$ on concentration of these solutions. The comparison of this slope with other data is given in Chapter 34.

The method just explained for acetic acid has been applied very successfully to various weak acids with dissociation constants of about 10^{-4} or less. When applied to more highly dissociated acids, however, the correction for m_{H^+} is no longer trivial, and an approximation such as Eq. (24-23)* does not suffice. Various successive-approximation schemes have been used to obtain m_{H^+} in such cases, and apparent convergence was obtained. It has been found, however, that these solutions are not unique[1] and that many values published for acids with K greater than 10^{-4} are unreliable by amounts increasing with K.

In Chapter 23 we noted that a low Bronsted-Guggenheim B parameter indicated incompleteness of dissociation. In a weak electrolyte MX the use of both a dissociation constant and a B_{MX} parameter introduces a redundancy, which makes each quantity ambiguous unless the other is defined. This is part of the difficulty mentioned in the preceding paragraph. These problems will be discussed further in Chapter 34.

Dissociation Constant of Water. Measurements of cells of the type

$$\text{Pt, H}_2|\ \text{MOH}(m_1),\ \text{MCl}(m_2)\ |\text{AgCl, Ag}$$

with M an alkali metal, may be used in connection with $\mathcal{E}^\circ$ for the cell with the same electrodes and HCl electrolyte to obtain the ionic dissociation constant of water[2]

$$K_w = a_{H^+}a_{OH^-} = m_{H^+}m_{OH^-}\gamma_{H^+}\gamma_{OH^-} \qquad (24\text{-}24)$$

In this case the emf is

$$\mathcal{E} = \mathcal{E}^\circ - \frac{RT}{\mathcal{F}}\ln\frac{m_2}{m_1} - \frac{RT}{\mathcal{F}}\ln\frac{\gamma_{Cl^-}a_{H_2O}}{\gamma_{OH^-}} - \frac{RT}{\mathcal{F}}\ln K_w \qquad (24\text{-}25)$$

and the effect of water ionization on m_{OH^-} is negligible. Again the D-H limiting-law terms cancel in the activity-coefficient factor.

We define K' by

$$-\log K' = \frac{\mathcal{F}(\mathcal{E} - \mathcal{E}^\circ)}{2.303RT} + \log\frac{m_2}{m_1} \qquad (24\text{-}26)$$

Hence

$$-\log K' = -\log K_w - \log\frac{\gamma_{Cl^-}a_{H_2O}}{\gamma_{OH^-}} \qquad (24\text{-}27)$$

We use the Bronsted-Guggenheim approximation [Eq. (23-37)*] to the ion activity coefficients and the Raoult's-law approximation for the water activity.

$$\log a_{H_2O} \cong \log x_1 \cong -\frac{2(m_1 + m_2)}{2.303 \times 55.5} = -0.016(m_1 + m_2)$$

The second term in Eq. (24-27) is then

$$-\log\frac{\gamma_{Cl^-}a_{H_2O}}{\gamma_{OH^-}} = (B_{MOH} - B_{MCl} + 0.016)(m_1 + m_2) \qquad (24\text{-}28)$$

[1] W. J. Hamer, "The Structure of Electrolyte Solutions," chap. 15, John Wiley & Sons, Inc., New York, 1959.

[2] $a_{H_2O} = 1$ is implied; strictly, $K_w = a_{H^+}a_{OH^-}/a_{H_2O}$.

and a plot of $-\log K'$ versus $m_1 + m_2$ should have the slope $B_{MOH} - B_{MCl} + 0.016$.

Guggenheim and Turgeon[1] treated the data of Harned and his collaborators in this manner with the result shown in Fig. 24-2. The result-

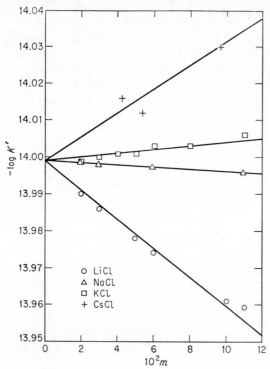

FIG. 24-2 [*E. A. Guggenheim and J. C. Turgeon, Trans. Faraday Soc.*, **51**, 757 (1955), *fig.* 6.]

ing K_w is 1.002×10^{-14} at 25°C. The slopes of the various curves agree, within experimental error, with the available knowledge of the B's for the various chlorides and hydroxides.

Cells with Solid Electrolyte. Some crystalline solids show ionic conductance at high temperatures, and such materials may serve as electrolytes in galvanic cells.[2] A good example is silver iodide, which becomes an electrolytic conductor above 146°C. The cell

$$Ag(s)|\ AgI(s)\ |Ag_2S(s),\ S(l),\ graphite$$

has been found[2] to give reversible potentials in the range 150 to 425°C. The cell reaction is

$$2Ag(s) + S(l) = Ag_2S(s)$$

[1] E. A. Guggenheim and J. C. Turgeon, *Trans. Faraday Soc.*, **51**, 747 (1955).
[2] K. Kiukkola and C. Wagner, *J. Electrochem. Soc.*, **104**, 308, 379 (1957).

Hence the cell potential gives immediately the free energy of formation of Ag_2S. Such cells have also been employed[1] to study solid compounds and solutions such as Ag in Ag-Sb alloy and Ag_2S in Ag_2S-Sb_2S_3. In the former case the cell is Ag| AgI |Ag-Sb alloy and in the latter Pt, S(l), Ag_2S| AgI |Ag_2S-Sb_2S_3, S(l), Pt. In either case 1 mole of Ag^+ is transferred per equivalent through the AgI from the left to the right electrode. The cells with the Ag_2S-Sb_2S_3 electrode gave a constant potential of 43 mv in the composition range Ag_3SbS_3 to $AgSbS_2$ and 98 mv in the range $AgSbS_2$ to Sb_2S_3 at 275°C. These results show that no significant solid solubility arises in this system and give the free energies of formation of the compounds. For example, the reaction

$$Ag_2S + Sb_2S_3 = 2AgSbS_2$$

involves two equivalents of Ag^+ transferred through the cell, and

$$\Delta F = -2\mathcal{F}\mathcal{E} = -4.5 \text{ kcal at } 548°K$$

Cells with solid electrolyte frequently give more satisfactory results at high temperature than those with liquid electrolyte. All diffusion processes become more rapid at high temperature, and the unwanted diffusion of material from one electrode to the other may become troublesome. This diffusion is, of course, greatly reduced in a solid electrolyte.

Electronic conductance is commonly more important than ionic conductance in solids, and electronic conductance effectively short-circuits the cell; consequently only a limited number of solids are suitable for solid electrolytes. If the electronic conductance is a small but not negligible fraction of the ionic conductance, then a correction may be made. But the electronic-transference number is usually a sensitive function of impurity concentration or deviation from simple salt composition; consequently the corrections for electronic conductance are complex and beyond the scope of the present discussion.[2]

STANDARD ELECTRODE POTENTIALS

Many types of cells have been employed to obtain the standard electrode potentials of the elements and the potentials of other important electrodes such as Hg_2Cl_2, Hg, etc. Liquid junctions are to be avoided if possible; otherwise some limiting extrapolation is made which should eliminate the junction potential. It is impractical to survey this field, but a few examples may suggest the types of cells employed. The treatment of the results should be apparent in most cases from the discussions already presented.

[1] A. G. Verduch and C. Wagner, *J. Phys. Chem.*, **61**, 558 (1957).

[2] For a discussion of these problems, see K. Kiukkola and C. Wagner, *J. Electrochem. Soc.*, **104**, 308 (1957).

Electrode Potentials of Metals. If the metal is stable in contact with the aqueous solution and the metal chloride is soluble, then a cell of the type

$$M| \; MCl_z(aq) \; |Hg_2Cl_2, \; Hg$$

yields $\mathcal{E}^\circ = \mathcal{V}^\circ_{Hg_2Cl_2,Hg} - \mathcal{V}^\circ_M$ and $\mathcal{V}^\circ_M$ can be readily calculated. Also from the concentration dependence of the emf the activity coefficients of MCl_z can be obtained.

Sometimes it is preferable to use a saturated amalgam of M in place of the solid metal. If mercury is insoluble in the solid metal, the result is unchanged; otherwise a correction must be made for the reduction of activity of M by solution of Hg.

These methods are successful for such metals as Tl, Cd, Zn, Fe, etc. In case the metal chloride is insoluble but the sulfate is soluble, an analogous method may be used with the $PbSO_4$, Pb electrode. Thus one may measure $\mathcal{V}^\circ$ for the half cell Cu, Cu^{++} in the cell

$$Cu| \; CuSO_4(aq) \; |PbSO_4, \; Pb$$

Extrapolation to the solute standard state is more difficult with multiply charged ions present; hence the chloride system is to be preferred wherever it is applicable.

Some metals react spontaneously with water. But in some such cases a dilute amalgam of the metal reacts reversibly with the aqueous metal ion without any appreciable spontaneous reaction with water. This is the case for sodium. Hence one may measure the cell

$$Na \; in \; Hg(x_2)| \; NaCl(aq) \; |Hg_2Cl_2, \; Hg$$

and obtain thereby the potential of the sodium amalgam electrode. To relate this to sodium metal one must obtain the activity of Na in the amalgam. If another solvent[1] is available, which does not react with the metal but which dissolves a salt of the metal with some ionization, then one may measure a second cell which relates the metal to the amalgam. In the case of sodium a suitable solvent is ethylamine with NaI the salt, and the cell is

$$Na| \; NaI \; in \; ethylamine \; |Na \; in \; Hg(x_2)$$

In this case the potential is independent of the NaI concentration.

By this two-stage method the standard potentials have been measured for Na, K, and some other active metals. There is always a risk that there is still some spontaneous reaction with water, which causes error in this type of cell; hence some caution is appropriate in accepting the results.

---

[1] There is an extensive literature on cells with nonaqueous electrolyte solutions; for a summary see H. Strehlow, *Z. Elektrochem.*, **56**, 827 (1952).

Electrode Potentials of the Halogens.   An inert electrode, such as platinum, in the presence of chlorine and chloride ion yields a reversible emf.   Possible examples of complete cells are Pt, $H_2|$ HCl(aq) $|Cl_2$, Pt and Ag, AgCl$|$ HCl(aq) $|Cl_2$, Pt.   In the former the $\mathcal{E}°$ gives $\mathcal{V}°_{Cl_2,Cl^-}$ directly, but the actual cell potentials at finite concentrations of HCl must be extrapolated to the solute standard state at infinite dilution.   The second cell potential is independent of the activity of Cl^- and yields $\mathcal{V}°_{Cl_2,Cl^-} - \mathcal{V}°_{AgCl,Ag}$.   By the latter method Randall and Young[1] found $\mathcal{V}°_{Cl_2,Cl^-} = 1.358$ volts.

The principal complication in measurements of the chlorine electrode arise from the very substantial solubility of chlorine in water and the subsequent reactions $Cl_2 + Cl^- = Cl_3^-$ and

$$Cl_2 + H_2O = HOCl + H^+ + Cl^-$$

It is possible to avoid these difficulties by keeping the partial pressure of Cl_2 very low and then correcting the potential to the standard value for 1 atm fugacity of Cl_2.

The potentials of the bromine and iodine electrodes may be measured by methods analogous to those for the chlorine system but modified in view of the liquid and solid states, respectively.

Electrode Potentials by Indirect Methods.   Many active metals as well as the halogen fluorine react spontaneously with water, and their potentials in aqueous-solution systems have not been measured.   Indirect methods may be used, however, to obtain the free energy of the equivalent reaction.   In the case of fluorine this is

$$\tfrac{1}{2}H_2(g) + \tfrac{1}{2}F_2(g) = H^+(aq) + F^-(aq)$$

and
$$\Delta F° = -\mathfrak{F}\mathcal{E}°_{reaction} = \mathfrak{F}\mathcal{E}°_{F^-,F_2} = -\mathfrak{F}\mathcal{V}°_{F_2,F^-}$$

where $\mathcal{E}°_{F^-,F_2}$ is the potential of the half cell written for oxidation, i.e., the reverse of the direction in the equation above.   For a metal yielding the ion M^{+z} the reaction is

$$M + zH^+ = M^{+z} + \frac{z}{2} H_2$$

with
$$\Delta F° = -\mathfrak{F}\mathcal{E}° = \mathfrak{F}\mathcal{V}°_{M,M^{+z}}$$

In these cases the heat of the reaction may be obtained from thermal measurements and the entropies of all substances from the third law by methods described in Chapters 11, 12, 25, and 27.

Summary of Standard Electrode Potentials.   The most extensive summary of information on electrode potentials is given by Latimer[2] in

[1] M. Randall and L. E. Young, *J. Am. Chem. Soc.*, **50**, 989 (1928).

[2] W. M. Latimer, "Oxidation Potentials," 2d ed., Prentice-Hall, Inc., Englewood Cliffs, N.J., 1952; see also A. J. deBethune, T. S. Licht, and N. Swendeman, *J. Electrochem. Soc.*, **106**, 616 (1959).

the book "Oxidation Potentials." Latimer includes many values calculated indirectly from other thermodynamic sources or inferred from the direction in which oxidation-reduction reactions are known to take place. Tables of free energies were listed at the end of Chapter 15, and these free-energy data are, of course, equivalent to electrode potentials.

Table 24-3 lists the standard potentials of some of the more common electrodes.

TABLE 24-3. POTENTIALS OF HALF CELLS
$\mathcal{E}^\circ$ = oxidation potential, $\mathcal{U}^\circ$ = electrode potential, volts at 25°C

	$\mathcal{E}^\circ$	$\mathcal{U}^\circ$
$K = K^+ + e^-$	2.925	-2.925
$Ca = Ca^{++} + 2e^-$	2.87	-2.87
$Na = Na^+ + e^-$	2.714	-2.714
$La = La^{3+} + 3e^-$	2.52	-2.52
$Mg = Mg^{++} + 2e^-$	2.37	-2.37
$Th = Th^{4+} + 4e^-$	1.90	-1.90
$U = U^{3+} + 3e^-$	1.80	-1.80
$Al = Al^{3+} + 3e^-$	1.66	-1.66
$Zn = Zn^{++} + 2e^-$	0.763	-0.763
$U^{3+} = U^{4+} + e^-$	0.61	-0.61
$Fe = Fe^{++} + 2e^-$	0.440	-0.440
$Eu^{++} = Eu^{3+} + e^-$	0.43	-0.43
$Pb + SO_4^= = PbSO_4 + 2e^-$	0.356	-0.356
$Tl = Tl^+ + e^-$	0.3363	-0.3363
$V^{++} = V^{3+} + e^-$	0.255	-0.255
$Pb = Pb^{++} + 2e^-$	0.126	-0.126
$H_2 = 2H^+ + 2e^-$	0.000	0.000
$Ag + Br^- = AgBr + e^-$	-0.071	0.071
$Ag + Cl^- = AgCl + e^-$	-0.222	0.222
$2Hg + 2Cl^- = Hg_2Cl_2 + 2e^-$	-0.2676	0.2676
$U^{4+} + 2H_2O = UO_2^{++} + 4H^+ + 2e^-$	-0.334	0.334
$Cu = Cu^{++} + 2e^-$	-0.337	0.337
$V^{3+} + H_2O = VO^{++} + 2H^+ + e^-$	-0.361	0.361
$2I^- = I_2(s) + 2e^-$	-0.5355	0.5355
$Fe^{++} = Fe^{3+} + e^-$	-0.771	0.771
$2Hg = Hg_2^{++} + 2e^-$	-0.789	0.789
$Ag = Ag^+ + e^-$	-0.7991	0.7991
$Hg_2^{++} = 2Hg^{++} + 2e^-$	-0.920	0.920
$Au + 4Cl^- = AuCl_4^- + 3e^-$	-1.00	1.00
$VO_2^{++} + 3H_2O = V(OH)_4^+ + 2H^+ + e^-$	-1.00	1.00
$2Br^- = Br_2(l) + 2e^-$	-1.0652	1.0652
$Tl^+ = Tl^{3+} + 2e^-$	-1.25	1.25
$2Cl^- = Cl_2(g) + 2e^-$	-1.3595	1.3595
$PbSO_4 + 2H_2O = PbO_2 + SO_4^= + 4H^+ + 2e^-$	-1.685	1.685
$Ag^+ = Ag^{++} + e^-$	-1.98	1.98
$O_2 + H_2O = O_3 + 2H^+ + 2e^-$	-2.07	2.07
$2F^- = F_2(g) + 2e^-$	-2.87	2.87

PROBLEMS

24-1. What is the potential at 25°C of the cell H_2, Pt, HCl(aq), AgCl, Ag if the solution is saturated at $P_{H_2} = \frac{1}{2}$ atm and $P_{HCl} = \frac{1}{2}$ atm?

24-2. Calculate from the data in Table 24-3 the equilibrium constants for the reactions $O_3 + 2Ag^+ + 2H^+ = O_2 + H_2O + 2Ag^{++}$ and $3U^{4+} + 2H_2O = 2U^{3+} + UO_2^{++} + 4H^+$.

24-3. Consider the four cells (A) H_2, Pt, HCl(aq)m_1, HCl(aq)m_2, Pt, H_2; (B) Ag, AgCl, HCl(aq)m_1, HCl(aq)m_2, AgCl, Ag; (C) H_2, Pt, HCl(aq)m_1, AgCl, Ag; (D) H_2, Pt, HCl(aq)m_2, AgCl, Ag. (*a*) Derive the equation for the potential of cell A. (*b*) What exact relationships, if any, are there between the various cell potentials?

24-4. Derive the equation which is analogous to Eq. (24-20) but which applies if the electrode reaction produces or consumes the positive ion.

24-5. The quantity pH may be defined by pH $= - \log a_{H^+}$, but its precise measurement is beset by the difficulties of determination of single-ion activities. Consider the cell H_2, Pt|aq soln HA(m_1), NaA(m_2), NaCl(m_3)|AgCl, Ag, where HA is any weak acid. Relate the pH as defined above to the potential of this cell and any other quantities required. Discuss the possible assumptions with respect to any of these additional quantities which are not directly measurable.

24-6. The cell[1] H_2, Pt, solution of C_2H_5OH saturated with CH_3NH_2 at P_2 and with solid CH_3NH_3Cl, AgCl, Ag shows the potential $\varepsilon = 0.697$ volt at 25°C when the pressure of H_2 is 0.893 atm and the solution has a partial pressure of methylamine of 4.15×10^{-3} atm. Write the cell reaction, and find the equilibrium. Combine this result with other appropriate data to determine the equilibrium constant for the reaction $CH_3NH_3Cl(s) = CH_3NH_2(g) + HCl(g)$.

24-7. What is the reaction for the cell Pt, S(l), Ag_2S| AgI |(Ag_3SbS_3 + $AgSbS_2$), S(l), Pt? Given the potential $\varepsilon = 43$ mv of this cell at 548°K and the data in the text for related cells, calculate ΔF for the reaction

$$3Ag_2S + Sb_2S_3 = 2Ag_3SbS_3$$

[1] J. G. Aston and F. L. Gittler, *J. Am. Chem. Soc.*, **77**, 3175 (1955).

25

THERMAL PROPERTIES OF SOLUTIONS

In our discussion of partial molal quantities in Chapter 17, we have defined partial molal entropy, partial molal enthalpy (or heat content) and partial molal heat capacity in Eqs. (17-15) to (17-17). In Chapters 21 and 23, we have presented theoretical equations for these quantities in certain nonelectrolyte and aqueous-electrolyte solutions. At this time we would like to review some of the experimental data and the methods of treating these data to obtain values of the partial molal quantities.

Up to this point the main emphasis has been on the partial molal free energy, which is of prime importance in equilibrium calculations. However, to get the variation of partial molal free energy with temperature or pressure, one must know partial molal heat contents, partial molal volumes, and other similar properties.

In Chapter 17 we also discussed the means of calculating partial molal quantities from experimental measurements, with volume taken as the example. The same methods are directly applicable to partial molal enthalpies and heat capacities.

PARTIAL MOLAL HEAT CAPACITIES

Richards and Daniels[1] give the heat capacity per gram (specific heat) of liquid thallium amalgams at various weight per cents. We may plot their results in Fig. 25-1 and obtain the values of $\bar{c}_P$ by the method of intercepts described in Chapter 17. Thus, by laying a ruler tangent to the curve at any composition, we read off the two intercepts at 0 and 100 per cent thallium. At 25 g per cent, which is 25.3 mole per cent, the intercepts are 0.144 and 0.1767. These results are in joules/deg g, so that, if we divide them by 4.184 and multiply by the atomic weights, respectively, of mercury and thallium, we obtain for mercury $\bar{c}_{P1} = 6.93$ and for thallium $\bar{c}_{P2} = 8.62$ cal/deg. In the same way we have read off the values at several round mole fractions, as shown in Table 25-1.

[1] T. W. Richards and F. Daniels, *J. Am. Chem. Soc.*, **41**, 1732 (1919).

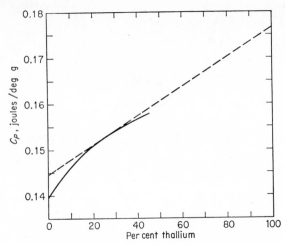

FIG. 25-1. Heat capacity of thallium amalgams.

TABLE 25-1. PARTIAL MOLAL HEAT CAPACITIES IN THALLIUM AMALGAMS
AT ABOUT 30°C, CAL/DEG

x_2	$\bar{c}_{P1}$	$\bar{c}_{P2}$
0.00	6.70	10.20
0.05	6.72	9.81
0.10	6.75	9.54
0.15	6.80	9.15
0.20	6.85	8.82
0.25	6.93	8.62
0.30	6.97	8.50
0.35	7.02	8.40
0.40	7.05	8.34
1.00 (extrapolated)	....	8.2

As an example from the domain of aqueous salt solutions, we give, in Table 25-2, the results of Randall and Rossini[1] on sodium chloride solutions. The several columns show (1) the molality m, (2) the apparent molal heat capacity, ϕC_P, (3) the heat capacity of the solution containing 1000 g of water and $n_2 = m$ moles of salt, (4) the heat capacity per gram (specific heat), (5) the partial molal heat capacity of the water, and (6) that of the salt. The partial molal heat capacities are given at even molalities along with smoothed values of the other data.

[1] M. Randall and F. D. Rossini, *J. Am. Chem. Soc.*, **51**, 323 (1929); see also F. D. Rossini, *Bur. Standards J. Research*, **7**, 47 (1931).

TABLE 25-2. HEAT CAPACITIES OF AQUEOUS SODIUM CHLORIDE SOLUTIONS
AT 25°C, CAL/DEG

m	ϕC_P	C_P (total)	Sp ht	$\bar{c}_{P1}$	$\bar{c}_{P2}$
0.00	-22.1	998.90	0.9989	17.9956	-22.1
0.01	-21.3	998.69	0.9981	17.9955	-20.9
0.02	-21.0	998.48	0.9973	17.9953	-20.3
0.04	-20.4	998.08	0.9957	17.9949	-19.4
0.05	-20.2	997.89	0.9950	17.9944	-19.0
0.10	-19.15	996.98	0.9912	17.9916	-17.00
0.20	-17.35	995.43	0.9839	17.984	-14.20
0.35	-15.30	993.54	0.9736	17.969	-11.15
0.50	-13.65	992.07	0.9639	17.951	-8.65
0.75	-11.35	990.39	0.9488	17.913	-5.25
1.00	-9.50	989.40	0.9347	17.866	-2.30
1.25	-7.80	989.15	0.9218	17.814	0.25
1.50	-6.28	989.48	0.9097	17.756	2.60
2.00	-3.50	991.90	0.8881	17.621	6.90
2.50	-1.03	996.32	0.8692	17.469	10.65

Randall and Rossini used a twin calorimeter which gave the difference in heat capacity between pure water and the solution and thus yielded ϕC_P directly. Their ϕC_P values were multiplied by 1.0003 to convert to defined calories. The heat capacity of the total solution is obtained by

$$C_{soln} = \phi C_P m + 998.9$$

In Chapter 17 we obtained Eq. (17-24) relating apparent and partial molal volumes. Transformation to heat capacities yields

$$\bar{c}_{P2} = \phi C_P + \tfrac{1}{2} m^{1/2} \frac{d\,\phi C_P}{dm^{1/2}} \tag{25-1}$$

One can use the graphical method III for determining partial molal quantities described earlier in Chapter 17 with a plot of ϕC_P versus $m^{1/2}$ as shown in Fig. 25-2.

One might anticipate a plot similar to that of log $\gamma_{\pm}$ versus $m^{1/2}$ as in Fig. 22-3, where a limiting proportionality of $\log_{\pm}$ to $m^{1/2}$ was approached asymptotically at very low molality. In contrast, Fig. 25-2 shows a linear relationship at high molality. Rossini[1] has made similar plots for a number of salts and has found similar behavior. However, the slopes differ from salt to salt and do not correspond to the limiting Debye-Hückel slope, and no theories have been developed to explain this

[1] F. D. Rossini, *J. Research Natl. Bur. Standards*, **7**, 47 (1931).

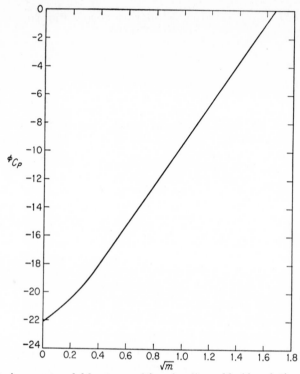

Fig. 25-2. Apparent molal heat capacities of sodium chloride solutions at 25°C.

behavior. In view of the success of the Debye-Hückel theory in predicting the behavior of partial molal free energies in very dilute salt solutions, one has every confidence in Eqs. (23-25)* and (23-27)* as limiting equations for enthalpy and heat capacity. Also Young and Machin[1] have clearly demonstrated that a change in slope must occur at low molalities for heat capacities. The Debye-Hückel equation does not predict the limiting $\bar{c}_P^o$ value; it merely gives the limiting slope. Thus there is a question of how to join the available data to the limiting slope. In addition, the limiting slope given by Eq. (23-28) involves the first and second derivatives of the dielectric constant of water, which impart an uncertainty of perhaps as much as 10 per cent to A_J. Guggenheim and Prue[2] have used the equivalent of the following equation for a 1-1 electrolyte,

$$^\phi C_P - \bar{c}_P^o = A_J m^{1/2} \left[(1 + m^{1/2})^{-1} - \frac{\sigma(m^{1/2})}{3} \right]$$
$$- 2.303 RT^2 \left(\frac{2}{T} \frac{dB}{dT} + \frac{d^2B}{dT^2} \right) m \quad (25\text{-}2)^*$$

[1] T. F. Young and J. S. Machin, *J. Am. Chem. Soc.*, **58**, 2254 (1936).
[2] E. A. Guggenheim and J. E. Prue, *Trans. Faraday Soc.*, **50**, 710 (1954).

which is derived from Eq. (22-20)*, to calculate the variation of $^\phi C_P$ below 0.05 M.   The derivatives of B with respect to T can be obtained in Appendix 4, and their evaluation from heat of dilution data is to be described later in this chapter.   For NaCl solutions at 25°C,

$$\frac{dB}{dT} = 0.82 \times 10^{-3} \quad \text{and} \quad \frac{d^2B}{dT^2} = -4 \times 10^{-5}$$

which yield $-12.2m$ cal/deg mole for the last term of Eq. (25-2)*.   Thus for 0.05 M NaCl, one calculates

$$^\phi C_P - \bar{c}_P^\circ = 1.9 \text{ cal/deg}$$

Although the two derivatives of B as well as A_J have high absolute uncertainties, there is considerable cancellation of error and the 1.9 value is uncertain by not more than 0.2 cal/deg mole.   The combination of this value with an experimental value of $^\phi C_P$ at 0.05 M yields c_P°.   This calculation is not quite straightforward in that even the remarkably accurate heat-capacity measurements of Randall and Rossini begin to show a considerable range of scatter as the molality approaches 0.05 M. On the other hand, if one tries to apply Eq. (25-2)* up to 0.1 M, there is significant deviation from the experimental data unless one regards the B derivatives as functions of molality [see Eq. (A4-7) with additional terms obtainable from differentiation of Eq. (A4-4)].   By examining the variation of these derivatives with molality it is possible to extrapolate to a reasonable $\bar{c}_P^\circ$ value as indicated in Fig. 25-2, where $\bar{c}_P^\circ$ for NaCl is found to be -22.1 ± 0.5 cal/deg.   The almost exact agreement of this value with the value of -22.2 given by Guggenheim and Prue is somewhat accidental.[1]

With the value of $\bar{c}_P^\circ$ fixed for NaCl, the corresponding values for other salts may be fixed in a much simpler way.   From Eq. (25-2)*

$$^\phi C_{MX} - {}^\phi C_{KCl} = (\bar{c}_{MX}^\circ - \bar{c}_{KCl}^\circ) - 2.303RT^2 \left(\frac{2}{T} \frac{d\Delta B}{dT} + \frac{d^2\Delta B}{dT^2} \right) m \quad (25\text{-}3)^*$$

Thus a plot of $^\phi C_{MX} - {}^\phi C_{KCl}$ against m yields $\bar{c}_{MX}^\circ - \bar{c}_{KCl}^\circ$ as the intercept at zero molality, and the slope yields values of the term involving the derivatives of B.   In obtaining the values given in Appendix 4, plots of this type were first actually made for pairs of salts of similar behavior such as NaCl and NaBr, NaI and KI, etc., as these plots have very small slopes and one can obtain quite accurate intercepts.   Then all the salts were related to KCl, also applying the principle of additivity of partial

---

[1] In addition to the changes introduced by more recent values of A_H and A_J, there is apparently an error in the value of the $\frac{2}{3}(T^2/V)\, d^2V/dT^2$ term used by Guggenheim and Prue.

molal heat capacities of ions at infinite dilution.   For example, the differences in partial molal heat capacities at infinite dilution for the pairs
NaI-KI, NaCl-KCl, NaNO$_3$-KNO$_3$, and NaBr-KBr were initially found
to be 5.8, 5.5, 5.4, and 5.7 cal/deg, respectively.   The final curves were
adjusted to agree with the value 5.6.   This smoothing was carried out
first to obtain the best NaCl curve as given in Fig. 25-2, from which the
absolute value of $\bar{c}_P^\circ$ was fixed for NaCl and therefore for all the salts
through use of Eqs. (25-2)* and (25-3)*.   The final results are tabulated
in Tables A4-5 and A4-6.

From the heat-capacity data together with heat-of-dilution data,
values of dB/dT and d^2B/dT^2 are obtained from which values of B can be
determined and therefore values of activity coefficients over a wide range
of temperature.   To extend this range as far as possible, one should also
know values of d^3B/dT^3.   Eigen and Wicke,[1] Rutskov,[2] Gucker and
Christens,[3] Ackermann,[4] and Cobble and Chris[5] have measured heat
capacities of aqueous solutions over a wide range of temperature which
show that ϕC_P goes through a maximum around 50 to 80°C.   The largest
part of the variation arises from $\bar{c}_P^\circ$, although J_2 also shows variation due
to increase of A_J with temperature and to a positive d^3B/dT^3 value.

This example shows that the partial molal heat capacity, like the
partial molal volume, may be negative.   In other words, the heat capacity of a dilute solution of sodium chloride is actually diminished by the
addition of a further amount of the salt.   Indeed this phenomenon, which
has not been elsewhere observed, proves to be of frequent occurrence in
aqueous solutions of electrolytes.

Thermodynamics exhibits no curiosity; certain things are poured into its hopper,
certain others emerge according to the laws of the machine, no cognizance being taken
of the mechanism of the process or of the nature and character of the various molecular
species concerned.   In thermodynamics a pure substance obeys the same rigorous
laws, whether its molecules are all of one sort, as we imagine them to be in hexane, or
extremely diversified, as we assume them to be in water.   In a thermodynamic
formula we may use the value of a partial molal quantity equally well whether it is
positive or negative, and if it proves to be the latter no explanation need be given.
It may therefore be out of place in a thermodynamic treatise to consider why it is
that partial molal volumes and heat capacities are so frequently small or even negative.   However, the fact is so characteristic of aqueous solutions, and the explanation
is so simple, that we may pause briefly for its consideration.

Various properties of water, especially the phenomenon of maximum density at 4°C,
can most readily be interpreted by assuming: that, in the neighborhood of its freezing
point, water contains a large amount of a molecular species formed by the aggregation

[1] M. Eigen and E. Wicke, *Z. Elektrochem.*, **55,** 354 (1951).

[2] A. P. Rutskov, *Zhur. Priklad. Khim.*, **21,** 820 (1948).

[3] F. T. Gucker and J. M. Christens, *Proc. Indiana Acad. Sci.*, **64,** 97 (1954).

[4] Th. Ackermann, *Z. Elektrochem.*, **62,** 411, 1143 (1958).

[5] J. W. Cobble and C. M. Chris, in press.

of simple molecules (possibly similar to ice); that this species occupies a larger volume than the simpler molecules of which it is constituted; and that it breaks up into these simpler molecules (with absorption of heat) as the temperature is raised.

It appears that any electrolyte dissolved in water in some way causes these aggregates to break up to a degree which increases with the concentration. Thus, when a small additional amount of the solute is added to an aqueous solution of an electrolyte, the total volume is increased by the volume occupied by the solute molecules but is diminished by the dissociation of the larger aggregates. The latter effect may predominate over the former, in which case the partial volume of the solute is negative. Again, the large heat capacity of water can be partly ascribed to the absorption of heat necessary to the dissociation of these aggregates as temperature increases. In so far as the addition of an electrolyte diminishes the number of these aggregates remaining to be dissociated, it lowers the heat capacity of the water, and if this effect predominates over the heat capacity which may be ascribed to the solute molecules themselves, the partial heat capacity of the solute is negative.

In cases of another type, where solvent and solute belong to the class of substances which we call normal liquids, the various partial molal quantities in a liquid solution do not differ greatly from the corresponding molal quantities for the pure liquid constituents.

As an extreme example of the abnormalities which sometimes are met in aqueous solutions, we give in Table 25-3 and in Fig. 25-3 the partial

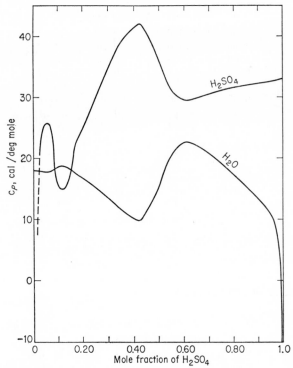

FIG. 25-3. Partial molal heat capacities of H_2O and H_2SO_4 in aqueous sulfuric acid.

TABLE 25-3. PARTIAL MOLAL HEAT CAPACITIES OF THE CONSTITUENTS OF
AQUEOUS SULFURIC ACID AT 25°C, CAL/DEG

n_1/n_2	x_2	$\bar{c}_{P1}$	$\bar{c}_{P2}$
∞	0.00	17.996	
55.506	0.0177	17.896	21.03
50	0.0196	17.875	22.01
25	0.0385	17.783	25.45
20	0.0476	17.771	25.72
17.5	0.0541	17.769	25.77
15	0.0625	17.806	25.18
12	0.0769	18.248	19.33
10	0.0909	18.567	15.80
8	0.111	18.660	14.92
6.25	0.138	18.348	17.02
5.75	0.148	17.995	19.13
5.0	0.167	17.352	22.58
4.0	0.200	16.657	25.72
3.0	0.250	15.329	30.24
2.0	0.333	12.111	38.12
1.50	0.400	10.20	41.47
1.35	0.426	9.76	42.10
1.00	0.500	15.06	36.29
0.85	0.541	20.00	31.70
0.65	0.606	22.80	29.52
0.40	0.714	20.28	30.74
0.25	0.800	17.22	31.72
0.10	0.909	12.84	32.47
0.04	0.962	9.8	32.67
0.01	0.990	− 5.4	32.96
0.001	0.999	−59	33.15
0	1.00	....	33.20

molal heat capacities of the constituents of aqueous sulfuric acid from
Giauque, Hornung, Kunzler, and Rubin.[1]

PARTIAL MOLAL ENTHALPY

In the same way that we have defined other partial quantities, we
define the partial molal enthalpy or heat content.   If the total enthalpy

[1] W. F. Giauque, E. W. Hornung, J. E. Kunzler, and T. R. Rubin, *J. Am. Chem. Soc.*, **82**, 62 (1960).

of any solution is H, we write for the first constituent,

$$\bar{H}_1 = \frac{\partial H}{\partial n_1} \tag{25-4}$$

We have obtained in Chapter 17

$$\left(\frac{\partial \bar{H}_1}{\partial T}\right)_P = (\bar{c}_{P1}) \tag{17-17}$$

In such algebraic equations the molal enthalpy may be freely used regardless of the possibility of its numerical evaluation. When we come to arithmetical calculations, we find a slight complication owing to our inability to state the absolute values of the enthalpy. We measure directly the heat capacity or the volume of a system; thus we state that the molal volume of ice at 0°C is 20 cc. But concerning the molal enthalpy of ice at 0°C we can only say that it is 1436 cal less than that of liquid water at the same temperature, or 69,908 cal less than the enthalpy of 1 mole of hydrogen gas and ½ mole of oxygen gas at the same temperature. For this reason, and also because of the immense confusion existing in the literature regarding various sorts of heat of solution, heat of dilution, and the like, we shall enter with some minuteness into the problem of the partial molal enthalpy.

RELATIVE ENTHALPIES AND HEAT CAPACITIES

While we cannot give the absolute magnitude of $\bar{H}$ for a substance in solution, we may ascertain how much greater or less this is than the enthalpy of the same substance in some chosen state. Thus, at any temperature, if we are dealing with the partial molal enthalpy of water in some solution in which water is the *solvent*, we may choose pure liquid water as the reference state and denote its molal heat content by H_1°. (This is identical with $\bar{H}_1^\circ$, which will denote the partial molal enthalpy of water in any infinitely dilute aqueous solution.)

This definition of a reference state for a solvent is consistent with that chosen in Chapter 20 for the activity of the solvent.[1] For a second component in a solution we also wish to choose a reference state which may be either the pure liquid, i.e., a second solvent, or the infinitely dilute solute reference state. In most of the examples to follow, the second component does not exist as a pure liquid at the temperature of the solution; hence we are forced to use the solute standard state. The enthalpy of the second component in its state is H_2°; if the infinitely dilute solute standard state is chosen, we write $\bar{H}_2^\circ$ to emphasize that it is a partial molal enthalpy.

---

[1] Strictly, the standard state is at 1 atm, while one may deal with a reference state at some other pressure. We shall ignore the effect of such pressure differences on heats of solution, since they are usually negligible.

The difference between the partial molal enthalpy of any constituent and the molal enthalpy of its reference state may be called the relative enthalpy and designated by L_1. Thus, for solvent and solute, respectively,

$$L_1 = \bar{H}_1 - \bar{H}_1^\circ \qquad L_2 = \bar{H}_2 - \bar{H}_2^\circ \tag{25-5}$$

When we are using the solute standard state for the second component, we have, at infinite dilution, $L_1 = 0$ and $L_2 = 0$. This is illustrated in Fig. 25-4.

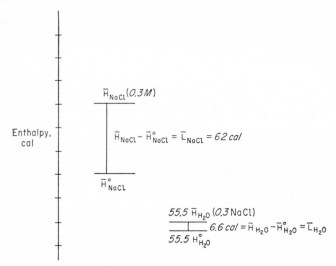

Enthalpy, cal

$\bar{H}_{NaCl}(0.3M)$

$\bar{H}_{NaCl} - \bar{H}_{NaCl}^\circ = \bar{L}_{NaCl} = 62\,cal$

$\bar{H}_{NaCl}^\circ$

$55.5\,\bar{H}_{H_2O}\,(0.3\,NaCl)$

$6.6\,cal = \bar{H}_{H_2O} - \bar{H}_{H_2O}^\circ = \bar{L}_{H_2O}$

$55.5\,H_{H_2O}^\circ$

FIG. 25-4. Relative partial molal enthalpy.

A similar convention is used for heat capacity at constant pressure, with the relative partial molal heat capacity given by

$$J_1 = \bar{C}_1 - C_1^\circ \qquad J_2 = \bar{C}_2 - \bar{C}_2^\circ \tag{25-6}$$

Using the same convention as for L, $J_1 = 0$ and $J_2 = 0$ at infinite dilution.

It is better to define the reference state and the relative enthalpy as we have done here than to state that we take the enthalpy in the reference state as zero. This could be done arbitrarily at some one temperature, but if any $\bar{H}^\circ$ is taken as zero at one temperature, it cannot be at another temperature, for it changes in accordance with Eq. (4-15).

Partial, or Differential, Heat of Solution. In discussing heats of solution in concentrated solutions, there are two quantities which must be carefully distinguished. When 1 mole of sodium chloride is dissolved in enough water to form a given solution, the heat absorbed is called the *total*, or *integral*, heat of solution. On the other hand, starting with the given solution of sodium chloride, if we add a further small quantity of the solute, the heat absorbed per mole is called the *partial*, or *differential*, heat of solution.

The total and the partial heats of solution become identical at infinite dilution, and therefore in dealing with very dilute solutions it is customary to disregard the distinction between them. However, in concentrated solutions the two quantities may differ widely. Thus, in the case of sulfuric acid, the total and partial heats of solution differ by 350 cal at 0.5 M and by 2000 cal at 5.0 M.

In our thermodynamic calculations it is usually the partial molal heat of solution that we need. Thus, if we are considering the thermodynamics of solid salt in contact with its saturated solution, we shall be interested, not in the heat absorbed when the solution is prepared from pure water and salt, but rather in the heat absorbed when an infinitesimal amount of salt dissolves in the almost saturated solution.

To a solution containing n_1 moles of water and n_2 moles of salt let us further add dn_2 moles of salt at

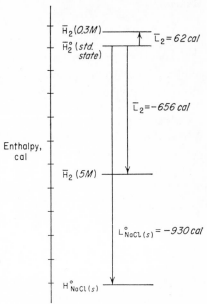

FIG. 25-5. Relative enthalpies.

constant temperature and pressure. If δq is the heat absorbed, then $\delta q/dn_2$ is the partial heat of solution of the salt, per mole. We note that δq is equal to the total increase in enthalpy. Representing the molal enthalpy of the solid salt by $\bar{H}_2(s)$, the enthalpy of the salt that is used is $\bar{H}_2(s)\, dn_2$; the increase in the enthalpy of the solution is, by definition, $\bar{H}_2\, dn_2$, so that

$$\frac{\delta q}{dn_2} = \bar{H}_2 - H_2(s) = L_2 - L_2(s) \tag{25-7}$$

where L_2 and $L_2(s)$ are the corresponding relative enthalpies.

In order to determine the value of $L_2(s)$, it is necessary only to determine the heat of solution of the solute in a large amount of pure solvent. In such case $L_2 = 0$, by our convention, and $\delta q/dn_2 = -L_2(s)$ as illustrated in Fig. 25-5. Thus at 25°C the heat of solution of 1 mole of solid sodium chloride in a very large amount of water is[1] 930 cal. Hence $L_2(s) = -930$ cal.

Although it may sound a little unusual, we shall speak also of the heat of solution of water in the aqueous solution of sodium chloride, for there is no essential difference between the introduction of the one or the other

[1] M. Randall and C. S. Bisson, *J. Am. Chem. Soc.*, **42,** 347 (1920).

constituent of the solution.   The equation for the partial heat of solution of the water is even simpler than that for the salt; for, pure water being taken as reference state, the relative heat content of the solvent is zero. Hence,

$$\frac{\delta q}{dn_1} = \text{L}_1 \tag{25-8}$$

If very small quantities of water are added to large quantities of salt solutions of various concentrations in a calorimeter, values of L_1 are directly obtained.   It was in this way that Randall and Bisson obtained the results of Table 25-5, which we shall shortly discuss.

Calculation of a Partial Molal Quantity for One Component, When That for the Other Component Is Known.   By our basic partial molal equation (17-32) we write for the total enthalpy of a solution,

$$H = n_1\bar{\text{H}}_1 + n_2\bar{\text{H}}_2 \tag{25-9}$$

Now if we define the total relative enthalpy as $L = H - n_1\text{H}_1^\circ - n_2\text{H}_2^\circ$, it is evident that

$$L = n_1\text{L}_1 + n_2\text{L}_2 \tag{25-10}$$

and at constant temperature and pressure, by Eq. (17-36),

$$x_1\,d\text{L}_1 + x_2\,d\text{L}_2 = 0 \tag{25-11}$$

This equation can be used in the same manner as was illustrated in Chapter 20 for the corresponding equation for partial molal free energy [Eq. (20-19)].   When applying Eq. (25-11) or the corresponding equations for heat capacity or volume, one does not encounter the difficulty of infinite values that made treatment of partial molal free energies awkward in dilute solutions.   The procedures described in Chapters 20 and 22 for evaluation of Eq. (20-25) are directly applicable to Eq. (25-11).

As an illustration of the application of Eq. (25-11), let us consider the investigation of Richards and Daniels[1] on the properties of thallium amalgams; their accurate determinations of the thermal changes accompanying changes of concentrations give almost immediately[2] the relative partial heat contents of thallium, L_2.   These values are given in the second column of Table 25-4, while the figures of the last column give the corresponding values for mercury, obtained graphically from the plot of Fig. 25-6 by the graphical integration corresponding to

$$\int d\text{L}_1 = - \int \frac{x_2}{x_1}\,d\text{L}_2 \tag{25-12}$$

---

[1] T. W. Richards and F. Daniels, *J. Am. Chem. Soc.*, **41**, 1732 (1919).

[2] See the full discussion of these data by G. N. Lewis and M. Randall, *J. Am. Chem. Soc.*, **43**, 233 (1921).

TABLE 25-4. RELATIVE PARTIAL MOLAL ENTHALPY IN THALLIUM AMALGAMS
AT 30°C, CAL

x_2	$\bar{L}_2$	$\bar{L}_1$
0.000	0	0.0
0.0250	212	− 3.1
0.0500	413	− 10.2
0.0863	671	− 28.1
0.1000	750	− 37.2
0.1070	800	− 41.8
0.1500	1013	− 75.8
0.2000	1195	−112
0.2500	1324	−144
0.3000	1415	−176
0.3500	1478	−207
0.4000	1520	−232
1.0000 (extrapolated)	1640	
Tl(s, sat with Hg)	805	

When $\bar{L}_2$ values are plotted as abscissas against x_2/x_1 as ordinates, as in Fig. 25-6, the negative value of the area under this curve between two limits is the change in $\bar{L}_1$ between these limits. If we take one of the limits at infinite dilution, the area under the curve from a certain point to $x_2/x_1 = 0$ gives immediately the value of $-\bar{L}_1$ at that point.

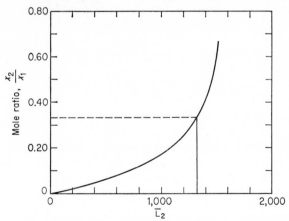

FIG. 25-6. Relative partial molal enthalpy of thallium in thallium amalgams.

For extrapolation to either the solvent standard state or the solute standard state, it is of great value to have knowledge of the limiting behavior. This has already been illustrated by consideration of heat-capacity data for aqueous sodium chloride solutions, which were given in Table 25-2 and Fig. 25-2, and we shall now treat the enthalpy data for sodium chloride solutions that are given in Table 25-5. The L_1 data were

TABLE 25-5. RELATIVE PARTIAL MOLAL ENTHALPIES OF THE CONSTITUENTS OF AQUEOUS SODIUM CHLORIDE AT 25°C

m	x_1/x_2	$\bar{L}_1$	$\bar{L}_2$
0	∞	0	0
0.278	200	0.2	69
0.370	150	0.3	42
0.555	100	0.9	−27
0.793	70	2.4	−153
1.110	50	4.0	−248
1.632	34	7.3	−368
2.13	26	11.0	−472
2.78	20	15.9	−594
3.47	16	20.0	−668
4.63	12	21.4	−687
5.55	10	15.6	−626
6.12 (sat)	9.04	11.5	−587

directly obtained by Randall and Bisson by additions of small amounts of water to large amounts of salt solution in a calorimeter. On the other hand, Gulbransen and Robinson[1] have added large amounts of water and thus measured integral heats of dilution of sodium chloride, which can be done more accurately. The treatment of both of these data will be illustrated. One can apply the Debye-Hückel theory from Chapter 23 to obtain the limiting behavior of the partial heat content of water at low molality. From Eqs. (23-25)* and (25-12) one obtains for aqueous solutions of 1-1 electrolytes

$$\bar{L}_1 = \frac{-A_H m^{3/2}}{166.53} \tag{25-13}*$$

In general the denominator would be $3000/M_1$. Comparison with the data of Randall and Bisson[2] shows that, at their lowest molalities, their data are far from the limiting behavior. Even the sign of $\bar{L}_1$ is opposite to the limiting behavior. Thus we must use the temperature derivative of Eq. (23-35)*, which is applicable to higher molalities than Eq. (23-25)*, and obtain for a 1-1 salt

$$\bar{L}_1 = \frac{-A_H m^{3/2} \sigma(m^{1/2})}{166.53} + \frac{2.303 RT^2 m^2}{55.51} \frac{dB}{dT} \tag{25-14}*$$

[1] E. A. Gulbransen and A. L. Robinson, *J. Am. Chem. Soc.*, **56**, 2637 (1934).
[2] *Loc. cit.*

The data in Appendix 4 show that dB/dT is either positive or at the most only slightly negative for all electrolytes.   Thus the sign of L_1 changes from negative at low molalities to a positive value at higher molalities for practically all aqueous electrolyte solutions at ordinary temperatures.

In Figs. 25-7a and b we have plotted the values of L_1 in aqueous sodium chloride, given in Table 25-5.   The value of L_2 at any point A is the total area to the left of AA' in Fig. 25-7a.   The value of L_1 reaches a maximum

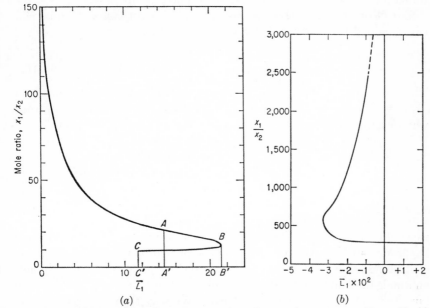

(a) (b)

FIG. 25-7. (a) Relative partial molal enthalpy of water in aqueous sodium chloride above 0.37 M; (b) relative molal enthalpy of water in aqueous sodium chloride below 0.2 M.

at B.   Here again L_2 is the total area under the upper curve, to the left of BB'.   At any point such as C, which represents the saturated solution, L_2 is equal to the area which we have just mentioned, *less* the area BCC'B'.   Likewise, the negative values of L_1 at low molalities (Fig. 25-7b) yield a negative area to the left of the $L_1 = 0$ line.[1]

In any case of this kind where we attempt to determine the area under a curve which runs to infinity, there is involved some uncertainty, but

---

[1] In the treatment of these data, Randall and Bisson in 1920 and Lewis and Randall in the first edition of this book in 1923 did not have available the predictions of the Debye-Hückel theory.   From their observation that L_1 was approaching zero at 0.2 M NaCl, they concluded that there would be no appreciable heat effects upon further dilution.   They extrapolated the plot of Fig. 25-7a with L_1 positive throughout the dilute region and did not include the negative area to the left of the $L_1 = 0$ line. Thus all their values of $-L_2$ were too large by 115 cal, the value of the area that was omitted.

one which is less, the more rapidly the quantity plotted approaches its limiting value. The limiting behavior given by Eq. (25-13)* is given by the dashed line and is used to guide the extrapolation. A more satisfactory procedure is to use the plot of Fig. 25-7 only to evaluate the change of L_2 above the lowest molality of Randall and Bisson. This was done to obtain the L_2 values of Table 25-5 with L_2 for the lowest molality being calculated from Eq. (25-14)* by using a dB/dT value obtained from integral heat-of-dilution data that we are now to discuss.

Total, or Integral, Heat of Solution. When 1 mole of sodium chloride is added to 1000 g of water, the process may be written

$NaCl(s) + 55.51H_2O(l)$
$$= soln(containing\ 1\ mole\ NaCl\ and\ 55.51\ moles\ H_2O)$$

The integral heat of solution is the value of $\Delta H = \Delta L$ for this process.

For 1 mole of the solid sodium chloride $L_2(s) = -930$ cal; L_1 for liquid water is taken as zero. The relative heat content of the solution, by Eq. (25-10), is $55.51L_1 + L_2$. By interpolation of Table 25-5 we find, at 1.0 M, that $L_1 = 3.4$ and $L_2 = -215$, whence $\Delta H = -27 + 930 = 903$. This is the integral heat of solution, as compared with 930 when the salt is dissolved in an infinite amount of water, and $-215 + 930 = 715$ when it is dissolved in an infinite amount of a molal solution.

Knowing the values of L_1 and L_2 and some one integral heat of solution, we could, conversely, calculate $L_2(s)$. In their most accurate experiment, Richards and Daniels added 0.0305 mole of thallium to amalgam already containing 1.199 moles of mercury and 0.1132 mole of thallium, producing an amalgam containing 1.199 moles of mercury and 0.1437 mole of thallium. The heat absorbed was 0.2 cal. We may therefore write,

0.0305 mole $Tl(s) + amalg(n_1 = 1.199, n_2 = 0.1132)$
$$= amalg(n_1 = 1.199, n_2 = 0.1437) \qquad \Delta H = \Delta L = 0.2\ cal$$

Let us now determine the total enthalpies of these two amalgams, for which the mole fractions are, respectively, $x_2 = 0.0863$ and $x_2 = 0.1070$. For the first amalgam, we find, from Table 25-4, $L_1 = -28.1$ and $L_2 = 671$, and $L = n_1L_1 + n_2L_2 = 42.3$. Likewise, for the second amalgam, $L = 64.8$ cal. Hence,

$$64.8 - 0.0305L_2(s) - 42.3 = 0.2 \qquad L_2(s) = 730\ cal$$

This is the heat which would be *evolved* if 1 mole of thallium were dissolved in an infinite amount of mercury. It is to be noted that Table 25-4 gives 805 as the heat content of solid thallium in contact with the saturated amalgam. We have noted in Chapter 20 in connection with Table 20-2 that such thallium contains mercury in solid solution, and therefore 805 is the relative partial molal enthalpy of thallium in this solid solution.

The methods of calculation are evidently similar when more solvent is added to a given solution. If to a certain amount of solution we add a very large amount of solvent, it will readily be seen that the total heat absorbed is equal to the value of $-L$ for the original solution. Randall and Bisson[1] found that 606 cal was absorbed when an amount of saturated solution of sodium chloride (6.12 M at 25°C) containing 1 mole of salt (and therefore $^{55.51}\!/_{6.12}$ moles of water) was diluted to 0.176 M. From the data of Gulbransen and Robinson[2] one calculates that 115 cal is evolved on additional dilution to infinite dilution; thus a total of 491 cal is absorbed upon addition of the saturated solution to a very large amount of water. This agrees, within the limits of error of experiment, with the values of $-L$ calculated from Table 25-5, namely,

$$587 - (^{55.51}\!/_{6.12})11.5 = 483 \text{ cal}$$

It would be possible, in this manner, to determine the values of L at various concentrations and then to determine the values of L_1 and L_2 by any of the methods illustrated by use of volume data in Chapter 17. Indeed it is not possible to give more than a few of the numerous methods of utilizing the thermal data as they are found in the literature. An excellent summary of the many varieties of graphical methods of obtaining partial molal properties is given by Young and Vogel.[3]

Heats of Dilution. Partial molal enthalpies can be obtained from the temperature variation of activity coefficients or by adding small amounts of solvent, but integral heats of dilution are generally the most accurate source of data. Calorimetric measurements of heats of dilution of solutions generally employ the twin-calorimeter method developed by Joule[4] and Pfaundler.[5] Richards and Gucker,[6] Lange and Robinson,[7] and Gucker, Ayres, and Rubin[8] present details of the method and reviews of work which has been done by using this technique. Lange[9] has recently reviewed the data on heats of dilution of electrolytes.

[1] *Loc. cit.*

[2] *Loc. cit.*

[3] T. F. Young and O. G. Vogel, *J. Am. Chem. Soc.*, **54**, 3025 (1932).

[4] J. P. Joule, *Mem. Manchester Lit. Phil. Soc.*, **2**, 559 (1849).

[5] L. Pfaundler, *Sitzber. Akad. Wiss. Wien*, **59**, 2145 (1849).

[6] T. W. Richards and F. T. Gucker, *J. Am. Chem. Soc.*, **47**, 1876 (1925).

[7] E. Lange and A. L. Robinson, *Chem. Rev.*, **9**, 89 (1931).

[8] F. T. Gucker, Jr., F. D. Ayres, and T. R. Rubin, *J. Am. Chem. Soc.*, **58**, 2118 (1936).

[9] E. Lange, in W. J. Hamer (ed.), "The Structure of Electrolytic Solutions," pp. 135–150, John Wiley & Sons, Inc., New York, 1959.

The method will be illustrated by the accurate integral heats of dilution of sodium chloride solutions measured by Gulbransen and Robinson[1] at four temperatures. For their most dilute solution of 0.00625 M, which was diluted to 0.00318 M, their observed heat effect was 24.6 cal/mole compared with 30 cal calculated from Eq. (23-25)*, the limiting Debye-Hückel equation. Thus it should not be difficult to extrapolate to the solute standard state by using the next term beyond the limiting term. In the derivation of Eqs. (25-13)* and (25-14)*, we have already considered the derivation of equations for L_1 and L_2 from activity-coefficient equations. For the integral enthalpy of an aqueous solution containing m moles of a 1-1 salt relative to infinite dilution

$$H - H° = L = A_H m^{3/2} \left[(1 + m^{1/2})^{-1} - \frac{\sigma(m^{1/2})}{3} \right] - 2.303 R T^2 \frac{dB}{dT} m^2$$

$$(25\text{-}15)*$$

Guggenheim and Prue[2] have used this equation to evaluate the data of Gulbransen and Robinson. The observed heat of dilution per mole of

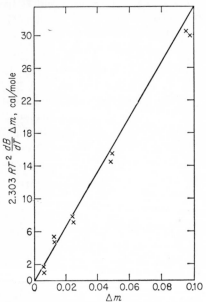

FIG. 25-8. Heat of dilution of NaCl solutions at 25°C.

solute between molality m_1 and m_2 corresponds to the difference $\Delta(L/m)$, since the amount of solute is fixed and only water is added. They plotted the difference between the observed heats of solution $\Delta(L/m)$ and that calculated from the first term of Eq. (25-15)*, $\Delta(L/m)^{St}$, against $m_1 - m_2$ and obtained values of $-2.303 R T^2 \, dB/dT$ from the slope. They also obtained dB/dT from the activity coefficients of sodium chloride at various temperatures as determined by Janz and Gordon[3] from cell measurements and have obtained agreement between the values of dB/dT obtained by the two methods.

Their calculations have been repeated using the more recent values of A_γ and A_H. Table 25-6 shows the data of Gulbransen and Robinson at 25°C and Fig. 25-8 shows the plot of the data. The data up to 0.05 M can be fitted by a straight line, thus substantiating Eq. (25-15)*,

[1] *Loc. cit.*

[2] E. A. Guggenheim and J. E. Prue, *Trans. Faraday Soc.*, **50**, 710 (1954).

[3] G. J. Janz and A. R. Gordon, *J. Am. Chem. Soc.*, **65**, 218 (1943).

but deviation from this straight line is noticeable at 0.1 M.   Young and Groenier[1] have applied the chord-area method[2] to the data of Gulbransen and Robinson to obtain limiting slopes within 20 to 30 per cent of the theoretical A_H values given in Table A4-1a.   The difference illustrates the difficulty of extrapolation to infinite dilution without guidance from theory.

TABLE 25-6.   HEAT OF DILUTION OF SODIUM CHLORIDE SOLUTIONS

m_1	m_2	Δm	$\Delta \left(\dfrac{L}{m}\right)$, cal/mole	$2.303RT^2 \dfrac{dB}{dT} \Delta m$, cal/mole
0.1	0.00257	0.0974	65.1	29.9
0.1	0.00507	0.0949	55.9	30.5
0.05	0.001285	0.0487	56.2	15.7
0.05	0.002535	0.0475	51.1	14.5
0.025	0.000642	0.0244	46.4	7.1
0.025	0.001280	0.0237	41.1	7.8
0.0125	0.000322	0.0122	34.6	4.7
0.0125	0.000634	0.0119	30.8	5.3
0.00625	0.000161	0.0061	27.6	1.0
0.00625	0.000318	0.0059	24.6	1.7

In Chapter 22 we have seen the value of using a reference salt for treatment of osmotic-coefficient data or activity-coefficient data in such a manner that the Debye-Hückel term cancels.   The use of this method for treating heat capacities of aqueous solutions has been described earlier in this chapter.   In like manner, from Eq. (25-15)*, one obtains

$$\frac{L_{MX} - L_{KCl}}{m^2} = \frac{{}^\phi L_{MX} - {}^\phi L_{KCl}}{m} = -2.303RT^2 \frac{d\Delta B}{dT} \qquad (25\text{-}16)^*$$

$$\frac{L_{MX} - L_{KCl}}{m} = -2 \times 2.303RT^2 \frac{d\Delta B}{dT} \qquad (25\text{-}17)^*$$

Integral heat-of-dilution data for solutions below 0.1 M have been evaluated by using Eq. (25-16)*, and the best values of the difference between dB/dT for MX and for KCl have been tabulated for many salts in Appendix 4.   For data of ordinary accuracy $d\Delta B/dT$ in Eqs. (25-16)* and (25-17)* may be considered constant for 1-1 salts up to about 0.05 M but for higher molalities or for data of exceptional accuracy, $d\Delta B^{\cdot}/dT$ must be considered a function of molality.

$$\frac{L_{MX} - L_{KCl}}{m} = -2 \times 2.303RT^2 \frac{d\Delta B^{\cdot}}{dT} \qquad (25\text{-}18)$$

[1] T. F. Young and W. L. Groenier, *J. Am. Chem. Soc.*, **58**, 187 (1936).
[2] T. F. Young and O. G. Vogel, *J. Am. Chem. Soc.*, **54**, 3030 (1932).

Values of $(L_{MX} - L_{KCl})/m$ and of $d\Delta B\cdot/dT$ are tabulated in Appendix 4 for several salts up to 5 M. Such a tabulation has the same advantage as that pointed out for activity coefficients, namely, that the values of $\Delta B\cdot$ or $d\Delta B\cdot/dT$ for salt MX compared with KCl do not depend upon the choice of the Debye-Hückel slopes and are thus not subject to revision when better values of the D-H slopes become available. The only result of the revision of the D-H slope would be a change in the values tabulated for KCl. Values of $d\Delta B\cdot/dT$ combined with values of $(J_{MX} - J_{KCl})/m$ from specific heat measurements also yield values of $d^2\Delta B\cdot/dT^2$ which will be found in Appendix 4.

CHANGE OF ACTIVITY COEFFICIENTS WITH TEMPERATURE

The main value of enthalpy and heat-capacity data to the chemist is their use in calculating the change of free energy with temperature. We shall particularly wish to calculate the variation of free energies of dilution or of activity coefficients with temperature. Consider the isothermal transfer of some constituent from a solution of one concentration to a solution at another concentration; then

$$\Delta F = \bar{F} - \bar{F}' = RT \ln \frac{a}{a'} = \frac{\gamma x}{\gamma' x'} \tag{25-19}$$

Now, if ΔH is the change in enthalpy accompanying this transfer, we may write, by Eq. (25-5),

$$\Delta H = \bar{H} - \bar{H}' = L - L' \tag{25-20}$$

where $L = \bar{H} - H°$ is the enthalpy referred to the chosen standard state. Hence by Eq. (15-7)

$$\frac{d(\Delta F/T)}{dT} = \frac{d(\bar{F} - \bar{F}')/T}{dT} = -\frac{\Delta H}{T^2} = -\frac{L - L'}{T^2} \tag{25-21}$$

where we are considering ΔF between two fixed concentrations, as a function of T. Combining with Eq. (25-19), we find

$$\frac{d \ln (a/a')}{dT} = -\frac{L - L'}{RT^2} \tag{25-22}$$

If the transfer in question is from a state which we have taken as standard, we may replace, in these equations, $\bar{F}'$ by $F°$, a' by $a° = 1$, and L' by $L° = 0$, and we have the fundamental equation for the change of activity or activity coefficient with temperature, at constant composition,

$$\frac{d \ln a}{dT} = \frac{d \ln \gamma}{dT} = -\frac{L}{RT^2} \tag{25-23}$$

This equation is applicable to any constituent of a solution.

If in any calculation L is assumed to be constant, the integration (always at constant composition) gives

$$R \ln a = \frac{\text{L}}{T} + \text{constant} \qquad (25\text{-}24)^*$$

or if we wish to consider the ratio of activity to mole fraction (or molality) and if we use common logarithms, with L in calories,

$$4.5758 \log \gamma = \frac{\text{L}}{T} + \text{constant} \qquad (25\text{-}25)^*$$

The integration constants may be found by determining a or γ at some one temperature.

On the other hand, if the partial molal heat capacities are known, L may be expressed as a function of the temperature by Eqs. (17-17) and (25-6).

$$\frac{d\text{L}}{dT} = \bar{c}_P - c_P^{\circ} = \text{J}$$

where J is the relative partial molal heat capacity in the given state. Except when the heat capacities are changing rapidly with the temperature or we are dealing with a very wide interval of temperature, we may then express L as a linear function of the temperature,

$$\text{L} = L_I + \text{J}T \qquad (25\text{-}26)^*$$

The integration of Eq. (25-23) then yields

$$4.5758 \log a = \frac{L_I}{T} - 2.303\text{J} \log T + \text{constant} \qquad (25\text{-}27)^*$$

and also

$$4.5758 \log \gamma = \frac{L_I}{T} - 2.303\text{J} \log T + \text{constant} \qquad (25\text{-}28)^*$$

Both these equations are for a solution of fixed composition.

We may illustrate these equations by showing how the tables for the activity of thallium and mercury in amalgams, which we gave in Chapter 20, may be compared with one another. For example, from the measurements of the emf of concentration cells at 20°C, we may calculate the vapor pressure of mercury over an amalgam containing 40 mole per cent of thallium at 325°C.

From Table 20-3, γ_1, for mercury at 20°, when $x_2 = 0.40$, is 0.734. For the same amalgam we find, by Table 25-1,

$$\text{J}_1 = 7.05 - 6.70 = 0.35 \text{ cal/deg}$$

and, by Table 25-4, $L_1 = -232$ cal at 30°C.   First let us assume L_1 constant and substitute our values in Eq. (25-25)*.   We first solve for the integration constant, and then, at $T = 598°K$, we find $\gamma_1 = 0.899$.

Now if we repeat this calculation, using Eq. (25-28)* and the values of J, we find, at 598°K, $\gamma_1 = 0.871$.   We may be sure that J must change very considerably.   If we assume that the solution approaches ideality as the temperature increases, then both J and L should decrease toward zero and the true γ at the higher temperature must lie between those obtained from the two calculations we have just made.   In fact, from the measurements of Hildebrand and Eastman (Table 20-4), $\gamma_1 = 0.89$.   Instead of calculating γ_1 we might calculate P_1/P_1°, which would be 0.60 from Raoult's law and which is found to be 0.534 by Hildebrand and Eastman, while we calculate that the value must lie between 0.539 and 0.522.

By precisely the same method we may calculate, in the same amalgam, the activity coefficient γ_2 of the thallium at 598°K.   Here, on account of the large values of L_2 and of J_2 (Tables 25-1 and 25-4), the uncertainty of extrapolation is greatly increased.   Thus we find for γ_2, at 598°, by the first method 2.00, by the second method 2.38, whereas the value obtained from the measurements at the high temperature (Table 20-4) is 2.17.

CORRELATION OF ACTIVITY COEFFICIENTS AND THERMAL QUANTITIES

In aqueous solutions, the extrapolation of activity coefficients over a wide temperature range is difficult because of the generally large values of relative partial molal enthalpies and heat capacities and their rapid change with molality.   This requires a more complete knowledge of the necessary thermal data and their temperature coefficients than is customarily available for most solutions.   In spite of the large amount of work that has been done on aqueous solutions, one often finds insufficient data for accurate extrapolation to temperatures far removed from room temperature.   It is thus desirable to be able to estimate some of the thermal quantities that might be lacking.   In Chapter 23, equations were presented for the calculation of the limiting Debye-Hückel terms for activity coefficient, enthalpy, and heat capacity, and in Appendix 4 values of these constants are tabulated over a wide temperature range. The value of B in Eqs. (22-20)* and (23-35)* is known for a very large number of aqueous salt solutions at either 25 or 0°.   Thus the extrapolation of activity coefficients or thermal data to other temperatures involves the knowledge of the temperature derivatives of B.

The terms of Eq. (22-20)* may be ascribed to three types of interaction. The limiting Debye-Hückel term represents the coulombic interaction between the ions and their ion atmospheres.   The $1 + I^{1/2}$ term in the denominator represents a repulsive term corresponding to the distance of nearest approach of the ions.   In Eq. (22-20)* this term does not contribute to the heat or heat capacity since it does not vary with temperature.   Any temperature contribution from the repulsive term is incorporated in the B coefficient.   As pointed out in Chapter 22, the division of interaction between these two terms is arbitrary.   The B coefficient is thus essentially a second virial coefficient for specific interaction between ions of unlike charge, including both attractive interactions and repulsive interactions such as those which are not completely accounted for by the $1 + I^{1/2}$ term of the denominator of the preceding term of Eq. (22-20)*. From this point of view the B coefficient corresponds to the equilibrium constant for interaction between a pair of unlike ions, excluding the coulombic type of interaction given by the Debye-Hückel limiting term.

One would expect that the partial molal entropy change resulting from the interaction of a pair of ions should be a simple and smoothly varying function of the degree of interaction.   If it is true that the entropy of interaction is simply related to the free energy of interaction, then one would expect that the heat of interaction would be simply related to the free energy of interaction.   To test this, one can plot dB/dT against B to determine whether or not there is any correlation.   In Fig. 25-9 are plotted values of the quantity $(L_{MX} - L_{KCl})/m$ against $\Delta B = B_{MX} - B_{KCl}$. The first quantity is proportional to $d\Delta B/dT$.   This plot is equivalent to plotting dB/dT against B, since the conversion involves addition of constant amounts to the ordinate and to the abscissa.   The figure shows that a correlation does exist, although some of the quantities deviate by a fair amount.   However, the uncertainties of many of the thermal values are large, and all the points which deviate considerably are points with large uncertainty.   It is reassuring that, if one takes only those points corresponding to accurate activity-coefficient and heat-of-dilution data, which are also consistent with freezing-point measurements, these points cluster very closely around the line which is drawn.   Thus it appears that this correlation should prove useful for estimating the temperature coefficient of B or relative partial molal heats of dilution when such data are lacking.

The question of a correlation between the first and second derivatives of B is more difficult to answer in that there are not many salts where both heats of dilution and partial molal heat capacities are known accurately. Figure 25-10 shows the result of a plot of available data of d^2B/dT^2 plotted versus ΔB.   There seems to be a reasonable correlation which should be of value for estimating the second derivative of B with respect to temperature.   The values for these plots have been obtained from Appendix 4 and are given for molalities of 0.1 M or lower.

These correlations allow one to predict activity coefficients, relative partial molal enthalpies, and relative partial molal heat capacities over a

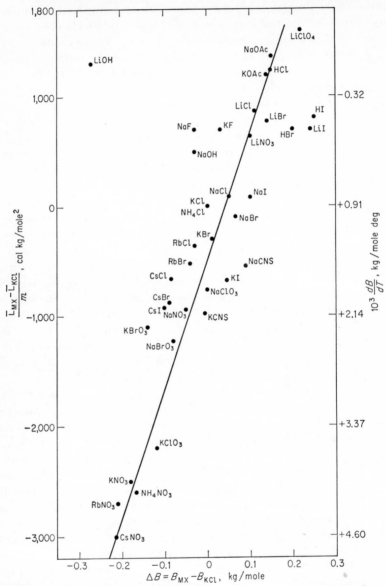

FIG. 25-9. Correlation of enthalpy of dilution with free energy of dilution.

range of temperatures.   In addition to these quantities, one often needs to know the limiting values at low molality of the partial molal heat capacity and partial molal entropy in order to predict variation of equilibrium constants with temperature, as has been noted in Chapter 22. We have noted that the limiting partial molal heat capacities of ions show

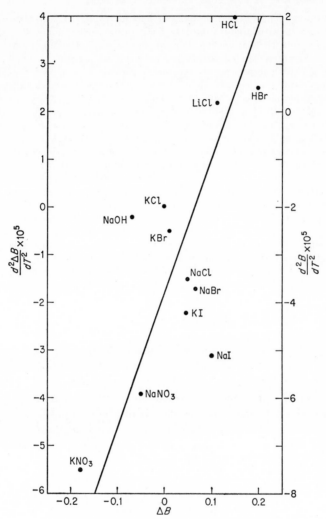

Fɪɢ. 25-10. Correlation of $d^2\Delta B/dT^2$ with ΔB.

a rapid change with temperature, with a maximum somewhere in the vicinity of 60 to 80°C. However, not enough data are available to indicate any correlations that would allow one to make reasonable predictions for this quantity. A number of correlations have been presented for variation of the standard partial molal entropy of ions as a function of radius, and these will be discussed in more detail in Chapter 32.

It is of interest to examine the correlation between B and its temperature coefficients. One would normally attribute a large positive value of B to lack of a specific attractive interaction between the pairs of ions involved.   It is interesting to note that for those salts with large B values the temperature coefficients of B are very small, corresponding to little heat effect.   For substances with ΔB values of 0.1 or greater or B values of greater than 0.2, dB/dT is zero or slightly negative, whereas the second coefficient of B with respect to temperature is zero or slightly positive.   Thus over a wide temperature range one would expect very little change of B with temperature, and the main thermal effect is due to the limiting Debye-Hückel term.   For salts of smaller B values, dB/dT is positive and becomes increasingly positive as B becomes smaller.   On the other hand, the second derivative is negative for such salts and becomes increasingly negative as B decreases, while the third derivative is positive. Thus one encounters opposing effects for the first, second, and third derivatives of B for almost all salts.   At temperatures above 50°C, it appears that for all salts dB/dT and d^3B/dT^3 are positive and d^2B/dT^2 is negative.   If these terms continue to high temperatures, one might expect the aqueous salt solutions to approach normal liquid-solution behavior.   It is the large value of these temperature coefficients and their opposing effect which makes it essential to consider all three temperature derivatives for extrapolation over any appreciable temperature range.   This is also true for the consideration of the variation of equilibrium constants taken with respect to the aqueous solute standard states in solution, as has been pointed out in Chapter 22, where it is shown that both the standard entropy and standard heat capacities must be considered for extrapolations over even relatively small temperature ranges.

PARTIAL MOLAL ENTROPIES

In most experimental studies of solutions the partial molal free energy is obtained from measurements such as the partial vapor pressures and the partial molal enthalpy by calorimetric measurements.   These two quantities may be combined to yield the partial molal entropy, which, although difficult to measure directly, is of great value for the insight it gives into the nature of the solution.   Also, as we shall note particularly in Chapter 32, it is more feasible to estimate the entropy than the free energy of an unknown substance, and this adds importance to the study of known entropies of solutions.   For nonelectrolytes the calculation of the ΔS of mixing of the pure components from heat-of-mixing and activity-coefficient or free-energy data is straightforward and requires no further comment.   Any of the methods described at the beginning of this chapter may then be used to obtain partial molal entropies.   Alternatively, if partial molal enthalpies are available,

$$\bar{s}_i - s_i^\circ = \frac{\bar{L}_i}{T} - R \ln a_i \tag{25-29}$$

In accordance with the selection of the standard state for other functions s_i° is the molal entropy of the pure liquid (or solid) if the solvent standard state is used, or in the case of the solute standard state it is the partial molal entropy in the hypothetical ideal solution of unit molality (or mole fraction).   In a manner analogous to that used several times before one may write for the solute standard state in molality units

$$\bar{s}^{\circ} = \lim_{m \to 0} (\bar{s}_m + R \ln m) \tag{25-30}$$

where $\bar{s}_m$ is the partial molal entropy at concentration m.

For consideration of the nature of solutions and comparison with theories such as those of Chapters 21 and 23, entropies of mixing or values of $\bar{s}_i - s_i^{\circ}$ are pertinent. However, chemical reactions in solution are determined by the ΔF of reaction, and this may be related to ΔH and ΔS of reaction. In order to obtain ΔS for a chemical reaction, one needs the absolute entropies of the reactants and products, including the absolute entropies of the constituents of the solution. We have considered the absolute entropies of pure substances in Chapters 11 and 12. Where the solvent standard state is used, s_i° is the entropy of the pure substance and the calculation of $\bar{s}_i$ from $\bar{s}_i - s_i^{\circ}$ and s_i° is obvious. For the solute standard state the situation is slightly more complex and will be illustrated by an example.

Entropies of Ions in Solution. The standard entropies of ions are obtained by adding to the entropy of the pure salt the standard entropy of solution, which is usually calculated from the familiar equation

$$\Delta S^{\circ} = \frac{\Delta H^{\circ} - \Delta F^{\circ}}{T}$$

For example, at 298.15°K the $\Delta \bar{H}^{\circ}$ of solution of KCl is 4.11 kcal/mole. The $\Delta \bar{F}^{\circ}$ of solution is obtained from the solubility 4.82 M and the activity coefficient of the saturated solution, $\gamma_{\pm} = 0.588$.

$$\Delta \bar{F}^{\circ} = -2RT \ln (\gamma m)_{\text{sat}} = -1.26 \text{ kcal/mole}$$

Combining these results, $\Delta \bar{s}^{\circ} = 18.0$ cal/deg mole, and since the entropy of solid KCl is 19.76,

$$\bar{s}_{K^+}^{\circ} + \bar{s}_{Cl^-}^{\circ} = 19.76 + 18.0 = 37.8 \text{ cal/deg}$$

It is not possible with present experimental methods to determine the entropies of K^+ and Cl^- separately. However, one may arbitrarily assign an entropy value to some one ion and then tabulate the corresponding values for various single ions. It is conventional to choose $\bar{s}_{H^+}^{\circ} = 0$. From the sum $\bar{s}_{H^+}^{\circ} + \bar{s}_{Cl^-}^{\circ} = 13.16$ one then obtains $\bar{s}_{Cl^-}^{\circ} = 13.16$ and $\bar{s}_{K^+}^{\circ} = 24.6$ cal/deg.

To illustrate values of thermodynamic properties of aqueous ions, Table 25-7 presents values of $\bar{c}_P^{\circ}$, $\bar{s}^{\circ}$, $\Delta \bar{H}_f^{\circ}$, and $\Delta \bar{F}_f^{\circ}$ at 298.15°K for a number of aqueous ions. The data are based on values given by the National Bureau of Standards[1] and Latimer[2] except for $\bar{c}_P^{\circ}$ values from

[1] F. D. Rossini et al., *Natl. Bur. Standards (U.S.) Circ.* 500, 1952.

[2] W. M. Latimer, "The Oxidation States of the Elements and Their Potentials in Aqueous Solutions," 2d ed., Prentice-Hall, Inc., Englewood Cliffs, N.J., 1952.

TABLE 25-7. THERMODYNAMIC DATA FOR AQUEOUS IONS AT 298.15°K,
CAL/DEG MOLE OR KCAL/MOLE

Ion	$\bar{C}_P^\circ$	$\bar{S}^\circ$	$\Delta\bar{H}_f^\circ$	$\Delta\bar{F}_f^\circ$
H^+	0	0	0	0
OH^-	−32.0	− 2.52	− 54.96	− 37.59
F^-	−29.5	− 2.3	− 78.66	− 66.08
Cl^-	−30.0	13.2	− 40.02	− 31.35
ClO_2^-	.....	24.1	− 17.18	2.74
ClO_3^-	−18	39	− 23.5	− 0.6
ClO_4^-	.....	43.2	− 31.41	− 2.47
Br^-	−31	19.29	− 28.90	− 24.57
I^-	−31	26.14	− 13.37	− 12.35
I_3^-	.....	57.1	− 12.4	− 12.31
$S^=$	.....	− 4	7.8	20.6
HS^-	.....	15.0	− 4.10	3.00
$SO_4^=$	−66	4.1	−216.90	−177.34
HSO_4^-	.....	30.52	−211.70	−179.94
$SeO_3^=$	.....	3.9	−122.39	− 89.33
$SeO_4^=$	.....	5.7	−145.3	−105.42
$HSeO_4^-$	.....	22.0	−143.1	−108.2
NH_4^+	16.9	26.97	− 31.74	− 19.00
$N_2H_6^{++}$	.....	19	− 4	22.5
$N_2H_5^+$	.....	31	− 1.7	21.0
$NH_2OH_2^+$	.....	37	− 30.7	− 13.54
NO_2^-	.....	29.9	− 25.4	− 8.25
NO_3^-	−18	35.0	− 49.37	− 26.43
PO_4^{3-}	.....	−52	−306.9	−245.1
$HPO_4^=$	.....	− 8.6	−310.4	−261.5
$H_2PO_4^-$	.....	21.3	−311.3	−271.3
$HAsO_4^=$	.....	0.9	−214.8	−169
$H_2AsO_4^-$	.....	28	−216.2	−178.9
$HCOO^-$	.....	21.9	− 98.0	− 80.0
HCO_3^-	.....	22.7	−165.18	−140.31
$CO_3^=$	.....	−12.7	−161.63	−126.22
CH_3COO^-	.....	20.8	−116.84	− 89.02
$C_2O_4^=$	.....	10.6	−195.7	−159.4
CN^-	.....	28.2	36.1	39.6
CNO^-	.....	31.1	− 33.5	− 23.6
Sn^{++}	.....	− 5.9	− 2.39	− 6.27
Pb^{++}	.....	5.1	0.39	− 5.81
Tl^+	.....	30.4	1.38	− 7.75
Zn^{++}	.....	−25.45	− 36.43	− 35.18
Cd^{++}	.....	−14.6	− 17.30	− 18.58
Hg^{++}	.....	− 5.4	41.59	39.38
Cu^{++}	.....	−23.6	15.39	15.53
Ag^+	.....	17.67	25.31	18.43
$Ag(NH_3)_2^+$	.....	57.8	− 26.72	− 4.16
$Ag(CN)_2^-$	.....	49.0	64.5	72.05

TABLE 25-7. THERMODYNAMIC DATA FOR AQUEOUS IONS AT 298.15°K,
CAL/DEG MOLE OR KCAL/MOLE (*Continued*)

Ion	$\bar{c}_P^\circ$	$\bar{s}^\circ$	$\Delta \bar{H}_f^\circ$	$\Delta \bar{F}_f^\circ$
$PtCl_4^=$		42	−123.4	− 91.9
$PtCl_6^=$		52.6	−167.4	−123.1
Fe^{++}		−27.1	− 21.0	− 20.30
Fe^{3+}		−70.1	− 11.4	− 2.53
$Fe(OH)^{++}$		−23.2	− 67.4	− 55.91
$FeNO^{++}$		−10.6	− 9.7	1.5
Mn^{++}		−20	− 53.3	− 54.4
MnO_4^-		45.4	−129.7	−107.4
$H_2BO_3^-$		7.3	−251.8	−217.6
BF_4^-		40	−365	−343
Al^{3+}		−74.9	−125.4	−115
Gd^{3+}		−43	−168.8	−165.8
Mg^{++}	4	−28.2	−110.41	−108.99
Ca^{++}	− 9	−13.2	−129.77	−132.18
Sr^{++}		− 9.4	−130.38	−133.2
Ba^{++}	−11	3	−128.67	−134.0
Li^+	14.2	3.4	− 66.55	− 70.22
Na^+	7.9	14.4	− 57.28	− 62.59
K^+	2.3	24.5	− 60.04	− 67.46
Rb^+	− 8.7	28.7	− 59.4	− 67.65
Cs^+	−18.7	31.8	− 62.6	− 70.8
UO_2^{++}		−17	−250.4	−236.4

Appendix 4, a few $\bar{c}_P^\circ$ values from Eigen and Wicke,[1] and values for $S^=$ and HS^- from Kury, Zielen, and Latimer.[2]   Single-ion values are based on assignment of zero values for $H^+(aq)$.

PROBLEMS

25-1. In 50 mole per cent sulfuric acid find, from Table 25-3, the change in $\bar{L}_1$ per degree rise in temperature.

25-2. Find the integral heat of solution of 0.1132 mole of thallium in 1.199 moles of mercury, and compare with the less accurate result of direct measurement, which gave approximately $q = -48$ cal.

25-3. What is the total relative heat content at 30°C of an amalgam containing 0.3 mole of thallium and 1.70 moles of mercury?

25-4. In Fig. 25-6 an ordinate is drawn at $x_2/x_1 = 0.333$ or $x_2 = 0.25$. Show that the area between the curve and the dashed line of that figure gives the total relative heat content of an amalgam containing 1 mole of mercury and 0.333 mole of thallium.

25-5. The measurements of Thomsen give q the heat absorbed when n_2 moles of gaseous HCl is dissolved in 1000 g of water. If $L_2(g)$ represents the relative heat

[1] M. Eigen and E. Wicke, *Z. Elektrochem.*, **55**, 354 (1951).

[2] J. W. Kury, A. J. Zielen, and W. M. Latimer, *J. Electrochem. Soc.*, **100**, 468 (1953).

content of the gas, show that $L_2(g)$ is about 17,300 cal/mole. Show further that $dq/dn_2 = \bar{L}_2 - L_2(g)$. By plotting q against n_2, find $\bar{L}_2 - L_2(g)$ at 1.0 M.

n_2	q	n_2	q
55.51	−298,000	2.775	−46,500
27.75	−315,300	1.110	−19,000
18.50	−247,200	0.555	− 9,550
11.10	−166,100	0.185	− 3,200
5.55	− 89,800		

25-6. Find the value of ΔH for each of the following changes at 25°C, using Appendixes 4 and 7 and Table 25-7.

$$KCl(s) = KCl(aq)$$
$$KCl(s) = KCl(sat, 4.81 \text{ M})$$
$$KCl(s) + 55.5H_2O = soln \text{ of } KCl + 55.5H_2O$$
$$KCl(s) = KCl(1 \text{ M})$$

25-7. The solubility of NaCl at 25°C is 6.14 M, and $\Delta\bar{H}° = 0.930$ kcal for NaCl(s) = NaCl(aq). (*a*) Use the activity-coefficient data of Appendix 4 and the value of $S_{298}° = 17.33$ cal/deg for NaCl(s) from Appendix 7 to calculate $\bar{S}°$ of NaCl(aq). (*b*) Also use the enthalpy data of Appendix 7 to calculate the temperature coefficient of solubility of NaCl.

25-8. Harned and Nims[1] measured the following type of cell:

$$Ag|AgCl|NaCl(m)|Na \text{ in } Hg|NaCl(0.1 \text{ M})|AgCl|Ag$$

The |Na in Hg| is a sodium amalgam which is sufficiently dilute so that the spontaneous reaction with water is negligible; thus it serves to transfer the Na^+ ion from one aqueous solution to the other. The cell reaction is NaCl(aq, m) = NaCl(aq, 0.1 M). Take the cell potentials listed below, extrapolate as required to $m = 0$, and calculate $\bar{L}_2$ and $\gamma_\pm$ at 0.1 and 0.5 M and 25°C. In the calculation of $\bar{L}_2$, obtain the appropriate heat-content equation from $\Delta\mathcal{E}/\Delta T = (\partial\mathcal{E}/\partial T)_P$ at each concentration, and then extrapolate to obtain $\bar{L}_2$.

m	$\mathcal{E}$ volts at		
	20°C	25°C	30°C
0.05	−0.03196	−0.03250	−0.03302
0.10	(0.0)	(0.0)	(0.0)
0.20	+0.03195	+0.03251	+0.03305
0.50	0.07438	0.07571	0.07700
1.00	0.10753	0.10955	0.11151

25-9. How much faster does the vapor pressure of water increase with temperature for a 2 M $CaCl_2$ solution compared with a 2 M KCl solution at 25°C? Use the data of Tables A4-2 and A4-7.

25-10. Enthalpy data are lacking for NH_4I. Freezing point depressions[2] yield $\gamma_\pm = 0.760$ at 0.1 M. Use the correlation of Figs. 25-9 and 25-10 to estimate $\bar{L}_2$ and $\bar{J}_2$ values for NH_4I. Calculate $\gamma_\pm$ at 25°C.

[1] H. S. Harned and L. F. Nims, *J. Am. Chem. Soc.,* **54,** 423 (1932).

[2] G. Scatchard and S. S. Prentiss, *J. Am. Chem. Soc.,* **54,** 2696 (1932).

25-11. Ackermann[1] reports a linear variation of ϕ_{C_P} with $m^{1/2}$ for NaCl between 10 and 120°C and gives the following values in cal/deg mole:

m	10°C	20°C	40°C	60°C	80°C	100°C	120°C
0	−28.0	−24.5	−18.6	−16.8	−18.1	−20.6	−26.8
0.5	−16.4	−14.0	−11.0	−10.1	−11.3	−13.5	−17.5
1	−11.2	− 9.4	− 7.8	− 7.4	− 8.6	−10.5	−13.4
2	− 4.7	− 3.0	− 2.8	− 3.4	− 4.5	− 6.2	− 7.5

(a) Convert these values to $\bar{J}_2$ values.

(b) Use Tables A4-1a, A4-4c, and A4-5b with Eq. (A4-8) and the assumption that $d^3B^{\cdot}/dT^3 = 0$ to calculate $\bar{J}_2$ at 10, 20, and 80°C and at 0.5 and 2 M. Compare with results from (a) and discuss the probable cause of discrepancies.

25-12. From the data and results of Problems 25-11, together with the data in Appendix 4, predict $\gamma_{\pm}$ of NaCl at 100°C, and compare with $\gamma_{\pm} = 0.644$ and 0.641 for 0.5 and 2 M at 100°C as determined from boiling-point elevations.[2]

[1] Th. Ackermann, *Z. Elektrochem.*, **62,** 411 (1958).
[2] R. P. Smith and D. S. Hirtle, *J. Am. Chem. Soc.*, **61,** 1123 (1939).

26

ACTIVITY COEFFICIENTS FROM
SOLID-LIQUID EQUILIBRIA

One of the most important methods of determining free energies of dilution or activity coefficients of solutions is the method based on the lowering of the freezing point of a solution by addition of solute. At the freezing point of the solution, $T = T_f - \theta$, where θ is the freezing-point depression in degrees, and for the solvent A,

$$A(s) = A(soln, x_A) \qquad \Delta \bar{F}_T = 0 \qquad (26\text{-}1)$$

To obtain partial molal free energies of the solute through use of the Gibbs-Duhem equation [Eq. (17-44) or (20-19)], one must know the isothermal variation of the partial molal free energy of the solvent with composition. Therefore the free-energy data obtained from freezing-point measurements over a range of temperatures must be converted to a single temperature. The freezing point of the pure solvent, T_f, is the most logical choice of reference temperature. Thus one must calculate $\Delta \bar{F}_f$ for $A(s) = A(l, x_A)$ at T_f from the knowledge of $\Delta \bar{F}_\theta = 0$ at $T = T_f - \theta$.

From Eq. (15-7) and using θ as a variable,

$$\frac{\partial(\Delta \bar{F}/T)}{\partial T} = -\frac{\partial(\Delta \bar{F}/T)}{\partial \theta} = -\frac{\Delta \bar{H}}{T^2} = -\frac{\Delta \bar{H}}{(T_f - \theta)^2} \qquad (26\text{-}2)$$

$$\frac{\partial \Delta \bar{H}}{\partial T} = \Delta \bar{c}_P = \Delta a + \Delta b T \qquad (26\text{-}3)$$

where $\Delta \bar{c}_P$, the difference between the partial molal heat capacity of the solvent in the solution and the molal heat capacity of the solid solvent, is taken to vary linearly with temperature. It is very unlikely that one would ever obtain data that would require consideration of higher terms, although higher terms are easily accommodated by the treatment to be used. It is to be noted that both Δa and Δb are functions of the composition of the solution.

$$\Delta \bar{H}_\theta - \Delta \bar{H}_f = \Delta a(T - T_f) + \tfrac{1}{2} \Delta b(T^2 - T_f{}^2)$$
or
$$\Delta \bar{H}_\theta - \Delta \bar{H}_f = -\theta[\Delta a + \tfrac{1}{2} \Delta b(2T_f - \theta)] \qquad (26\text{-}4)$$

where the subscript f indicates the quantity at the freezing point of the pure solvent.

$$\frac{\Delta \bar{F}_{\theta'}}{T'} - \frac{\Delta \bar{F}_f}{T_f} = \int_0^{\theta'} \frac{\Delta \bar{H}_f - (\Delta a + T_f \Delta b)\theta + \frac{1}{2} \Delta b \, \theta^2}{T_f^2 (1 - \theta/T_f)^2} \, d\theta \qquad (26\text{-}5)$$

where T' corresponds to the specific θ' and the integration with respect to temperature is at constant composition. If we expand the denominator in powers of θ/T_f and note that $\Delta a + T_f \Delta b$ is $\Delta \bar{c}_P$ at T_f, which we denote by $\Delta \bar{c}_f$, then one obtains

$$\frac{\Delta \bar{F}_{\theta'}}{T'} - \frac{\Delta \bar{F}_f}{T_f} = \int_0^{\theta'} \left[\frac{\Delta \bar{H}_f}{T_f^2} + \theta \left(\frac{2 \Delta \bar{H}_f}{T_f^3} - \frac{\Delta \bar{c}_f}{T_f^2} \right) \right.$$
$$\left. + \theta^2 \left(\frac{3 \Delta \bar{H}_f}{T_f^4} - \frac{2 \Delta \bar{c}_f}{T_f^3} + \frac{\Delta b}{2 T_f^2} \right) + \cdots \right] d\theta$$

After integration and omission of the prime marks on θ' and T',

$$\frac{\Delta \bar{F}_\theta}{T} - \frac{\Delta \bar{F}_f}{T_f} = \frac{\Delta \bar{H}_f}{T_f^2} \theta + \left(\frac{\Delta \bar{H}_f}{T_f^3} - \frac{\Delta \bar{c}_f}{2 T_f^2} \right) \theta^2 + \left(\frac{\Delta \bar{H}_f}{T_f^4} - \frac{2 \Delta \bar{c}_f}{3 T_f^3} + \frac{\Delta b}{6 T_f^2} \right) \theta^3$$
$$+ \cdots \qquad (26\text{-}6)$$

It is also possible to integrate Eq. (26-5) in closed form to

$$\frac{\Delta \bar{F}_\theta}{T} - \frac{\Delta \bar{F}_f}{T_f} = \frac{\Delta \bar{H}_f}{TT_f} \theta - \Delta a \left[\frac{\theta}{T} + \ln \left(1 - \frac{\theta}{T_f} \right) \right] - \frac{\theta^2}{2T} \Delta b \qquad (26\text{-}7)$$

This expression is convenient for large values of θ, but at smaller values the term θ/T almost exactly cancels $\ln (1 - \theta/T_f)$. Thus for moderate values of θ Eq. (26-6) is more convenient, and in most cases many of the terms in that expression are negligible.

At the freezing temperature of the solution, solid solvent and solution are at equilibrium, and $\Delta \bar{F}_\theta = 0$. But the pure solid and pure liquid solvent have the same free energy at T_f; hence $\Delta \bar{F}_f$ also applies to the change

$$A(l) = A(\text{soln}, x_A) \qquad \text{at } T_f$$

and the activity of the solvent in the solution at T_f is given by

$$RT_f \ln a_1 = \Delta \bar{F}_f$$

It is also convenient to note that

$$\Delta \bar{H}_f = \Delta H_f^\circ + \mathsf{L}_1$$

where ΔH_f° is the heat of fusion of the pure solvent and L_1 is the relative partial molal enthalpy of the solvent in the solution at T_f. Similarly one may write

$$\Delta \bar{c}_f = \Delta c_f^\circ + \mathsf{J}_1$$

where Δc_f° is the change of Δc_P on fusion of the pure solvent and J_1 is the relative partial molal heat capacity of the solvent.

The introduction of the several relationships just discussed into Eq. (26-6) yields

$$R \ln a_1 = -\left(\frac{\Delta \bar{\mathrm{H}}_f^\circ + \mathrm{L}_1}{T_f{}^2}\right)\theta - \left(\frac{\Delta \mathrm{H}_f^\circ + \mathrm{L}_1}{T_f{}^3} - \frac{\Delta c_f^\circ + \mathrm{J}_1}{2 T_f{}^2}\right)\theta^2$$
$$-\left(\frac{\Delta \mathrm{H}_f^\circ + \mathrm{L}_1}{T_f{}^4} - \frac{2}{3}\frac{\Delta c_f^\circ + \mathrm{J}_1}{T_f{}^3} + \frac{\Delta b}{6 T_f{}^2}\right)\theta^3 + \cdots \quad (26\text{-}8)$$

The same substitutions into Eq. (26-7) give

$$R \ln a_1 = -\left(\frac{\Delta \mathrm{H}_f^\circ + \mathrm{L}_1}{T T_f}\right)\theta + (\Delta c_f^\circ + \mathrm{J}_1)\left[\frac{\theta}{T} + \ln\left(1 - \frac{\theta}{T_f}\right)\right]$$
$$+ \Delta b\left[\frac{\theta^2}{2T} - \frac{T_f \theta}{T} + T_f \ln\left(1 - \frac{\theta}{T_f}\right)\right] \quad (26\text{-}9)$$

We note that Δb appears only in the coefficient of θ^3 in Eq. (26-8); hence the temperature coefficient of the heat capacity has no influence unless this last term is significant. Expansion of the coefficient of Δb in Eq. (26-9) likewise shows that all terms with θ to a lower power than 3 cancel exactly.

The determination of the freezing-point depression for solutions over a range of composition thus yields, through Eq. (26-8) or (26-9), values of the activity of the solvent as a function of composition, and one can readily obtain activity coefficients for the solute through the use of Eq. (20-27) and the procedures described in Chapter 20.

Osmotic Coefficients from Freezing-point Depressions. The freezing-point method has been particularly important for aqueous solutions, and applications to such solutions will now be illustrated in detail. Compositions of such solutions are commonly expressed in molality, and it is convenient to use the osmotic coefficient as defined in Eqs. (20-28) and (22-24).

$$R \ln a_1 = -\frac{\nu M_1 R}{1000} m\phi$$

The molal freezing-point-lowering constant is defined as

$$\lambda = \frac{M_1 R T_f{}^2}{1000\,\Delta \mathrm{H}_f^\circ} \quad (26\text{-}10)$$

Introduction of these last two expressions into Eq. (26-8) yields

$$\nu \lambda m\phi = \left(1 + \frac{\mathrm{L}_1}{\Delta \mathrm{H}_f^\circ}\right)\theta + \left(\frac{1}{T_f} - \frac{\Delta c_f^\circ}{2\,\Delta \mathrm{H}_f^\circ} + \frac{\mathrm{L}_1}{T_f\,\Delta \mathrm{H}_f^\circ} - \frac{\mathrm{J}_1}{2\,\Delta \mathrm{H}_f^\circ}\right)\theta^2$$
$$+ \left(\frac{1}{T_f{}^2} - \frac{2}{3}\frac{\Delta c_f^\circ}{T_f\,\Delta \mathrm{H}_f^\circ} + \frac{\Delta b}{6\,\Delta \mathrm{H}_f^\circ} + \frac{\mathrm{L}_1}{T_f{}^2\,\Delta \mathrm{H}_f^\circ} - \frac{2\mathrm{J}_1}{3T_f\,\Delta \mathrm{H}_f^\circ}\right)\theta^3 + \cdots \quad (26\text{-}11)$$

In Eq. (26-11) several quantities are divided into terms depending on properties of pure solid and liquid water and properties of the solution. This division is not essential—indeed the heat capacity of pure water cancels if the terms are combined. Usually, however, the division in Eq. (26-11) is more convenient, and the pertinent values of the properties of pure water[1] are: $\Delta H_f^\circ = 1436 \pm 1$ cal, $\Delta c_f^\circ = 9.1 \pm 0.05$ cal/deg. These values yield

$$\lambda = 1.860 \pm 0.001 \text{ deg kg/mole}$$

$$\frac{1}{T_f} - \frac{\Delta c_f^\circ}{2 \, \Delta H_f^\circ} = (4.9 \pm 0.2) \times 10^{-4}/\text{deg}$$

$$\frac{1}{T_f^2} - \frac{2 \, \Delta c_f^\circ}{3 T_f \, \Delta H_f^\circ} = -(2.1 \pm 0.1) \times 10^{-6}/\text{deg}^2$$

and Eq. (26-11) becomes

$$1.860\nu m\phi = \left(1 + \frac{L_1}{1436}\right)\theta + \left(4.9 + \frac{L_1}{39.2} - \frac{J_1}{0.287}\right) \times 10^{-4}\theta^2$$
$$+ \left(-2.1 + \frac{\Delta b}{0.86} + \frac{L_1}{107} - \frac{J_1}{0.59}\right) \times 10^{-6}\theta^3 + \cdots \quad (26\text{-}12)$$

One does not usually find tabulations of values of L_1 and J_1 for aqueous solutions, but they are easily calculated from values of the integral heats and heat capacities and values of L_2 and J_2 through use of the general partial molal equation [Eq. (17-32)]. When Eq. (22-31)* is applicable, e.g., below about 0.1 M for 1-1 electrolytes, L_1 may be readily obtained from

$$L_1 = -\left(\frac{A_H}{3 \times 55.5}\right) m^{3/2}\sigma(m^{1/2}) + \frac{2.303 R T^2}{55.5} \frac{dB}{dT} m^2 \quad (26\text{-}13)^*$$

with values of dB/dT given in the 0.1 M column of Tables A4-2b and A4-4b. The first term of Eq. (26-13)* is approximately $4m^{3/2}$ cal/mole for 1-1 electrolytes; hence L_1 is evidently very small for molalities of 0.1 or smaller. Indeed, it is found that in dilute solutions, where θ is less than $1°$, $L_1/1436$ and all the subsequent terms on the right side of Eq. (26-12) are usually smaller than the uncertainty in λ. Hence for such solutions

---

[1] $c_P^\circ(H_2O, l) = 18.16$ cal at 273.15°K from N. S. Osborne, H. F. Stimson, and D. C. Ginnings, *J. Research Natl. Bur. Standards*, **23**, 197 (1939). The heat of fusion has been recalculated by N. S. Osborne, *J. Research Natl. Bur. Standards*, **23**, 643 (1939), from measurements of H. C. Dickinson and N. S. Osborne, *Bull. Natl. Bur. Standards*, **12**, 49 (1915–1916). Osborne also gives $c_P^\circ(H_2O, s) = 9.11$ cal/deg at 273.15°K, while W. F. Giauque and J. W. Stout, *J. Am. Chem. Soc.*, **58**, 1144 (1936), obtain results about 0.6 per cent lower than Osborne. For dc_P°/dT of ice between -40 and $0°C$, Osborne gives 0.0336 cal/deg² mole, while the data of Giauque and Stout yield 0.031.

the very simple equation

$$\phi = \frac{\theta}{1.860\nu m} \qquad (26\text{-}14)*$$

is adequate.

To indicate the magnitude of the various terms in Eq. (26-12) at higher concentrations, the measurement of Scatchard and Prentiss[1] for 1.2455 M KCl will be considered. The result is given as

$$j = 1 - \frac{\theta}{\nu\lambda m} = 0.1254$$

and corresponds to $\theta = 4.04789°$ with the value of $\lambda = 1.858$ used by Scatchard and Prentiss.

For 1.2455 M KCl at 25°C, Randall and Rossini[2] give $\bar{c}_{P,1} = 17.84$; hence $J_1 = -0.14$ cal/deg. Eigen and Wicke[3] and Ackermann[4] present heat-capacity data that indicate that dJ_1/dT is proportional to $m^{3/2}$ with the constant of proportionality ranging from -0.001 to $+0.001$ for most salts between room temperature and the freezing point but with an extreme value of 0.01 for NaOH. From Table A4-2, the relative apparent and partial molal enthalpies of KCl at 25°C are, respectively, -57 and -245 cal at 1.2455 M. Thus $L_1 = 4.21$ at 25°C. With

$$J_1 = -0.14 \text{ cal/deg}$$

$L_1 = 7.7$ cal at 0°C. No reliable value can be given for Δb. From the trend of the heat capacities of pure ice and water near 0°C,

$$\frac{d\,\Delta c_f}{dT} = -0.047 \text{ cal/deg}^2 \text{ mole}$$

and we may assume this value to approximate Δb for the KCl solution. Substitution of these various values into Eq. (26-12) yields

$$
\begin{aligned}
1.860 \times 2.491\phi = \ &(1 + 0.0054) \times 4.048 \\
&+ (4.9 + 0.2 + 0.5) \times 10^{-4} \times 4.048^2 \\
+ (-2.1 - 0.05 &+ 0.07 + 0.24) \times 10^{-6} \times 4.048^3 + \cdots
\end{aligned} \qquad (26\text{-}15)
$$

where the terms are in the same order as before. In view of the uncertainty of nearly 0.1 per cent in λ, it is clear that the entire term in θ^3 is negligible for this solution. Also the terms in L_1 and J_1 in the coefficient of θ^2 contribute less than the uncertainty in λ. Consequently, an appropriate approximation to Eq. (26-12) for aqueous solutions with θ values from 1 to 5° is

$$1.860\nu m\phi = \left(1 + \frac{L_1}{1436}\right)\theta + 4.9 \times 10^{-4}\theta^2 \qquad (26\text{-}16)*$$

---

[1] G. Scatchard and S. S. Prentiss, *J. Am. Chem. Soc.*, **55**, 4355 (1933).

[2] M. Randall and F. D. Rossini, *J. Am. Chem. Soc.*, **51**, 323 (1929).

[3] M. Eigen and E. Wicke, *Z. Elektrochem.*, **55**, 354 (1951).

[4] Th. Ackermann, *Z. Elektrochem.*, **62**, 411, 1143 (1958).

The resulting ϕ for the 1.2455 M KCl solution is 0.8804 ± 0.0006 by Eq. (26-12) or 0.8801 ± 0.0006 by Eq. (26-16)*.

Many of the freezing-point data in the literature are expressed in terms of the function $j = 1 - \theta/\nu\lambda m$, which was introduced by Lewis and Randall in the first edition of this book.   It will be noted that the ϕ values from Eq. (26-14)* are exactly $1 - j$ values.   For larger molalities and therefore larger θ values, ϕ will differ from $1 - j$ owing to the additional terms in Eq. (26-12), but the difference between $1 - j$ and ϕ remains small.   Inasmuch as ϕ is now so widely used in connection with isopiestic data and other data which yield the activity of the solvent, it is desirable to use ϕ in place of $1 - j$.

In 1912, Lewis calculated a value of 1.858 for λ and this value was used in the first edition of this book and in evaluation of most aqueous-freezing-point data.   As we have noted earlier in this chapter, the best value available now is 1.860.   Thus the $1 - j$ values in the literature which are based upon $\lambda = 1.858$ should be multiplied by $^{1.858}\!/_{1.860}$, or 0.9989.   Such corrected j values may be taken as $1 - \phi$ values when Eq. (26-14)* is adequate or may be used to obtain the leading term of Eq. (26-16)* expressed as

$$\phi = \frac{\theta}{\nu m\lambda} + \frac{\bar{L}_1\theta}{1436\nu m\lambda} + \frac{4.9 \times 10^{-4}}{\nu m\lambda}\theta^2$$

$$1 - \phi = j + (1 - j)\left(\frac{0.002}{1.860} - \frac{\bar{L}_1}{1436} - 4.9 \times 10^{-4}\theta\right) \qquad (26\text{-}17)*$$

where j is a value from the literature based on $\lambda = 1.858$.

Evaluation of Accuracy of Freezing-point Data.   We are now in the position to evaluate the accuracy of the freezing-point method for obtaining osmotic coefficients of aqueous solutions.   There is the primary limitation of 0.07 per cent uncertainty in ϕ due to the uncertainty of 1 cal in the heat of fusion of ice.   At low molalities, error in the temperature determination is the most important limitation.   It appears to be possible to reduce errors in θ to $\pm 2 \times 10^{-5}$ deg.[1]   To achieve such accuracy requires great experimental skill.   The solubility of air in water is troublesome; hence nitrogen is used because of its smaller solubility.   Care in the thermocouple calibration is very important.   In actual practice the experimental uncertainty is usually at least ten times 2×10^{-5} deg, and thus error in temperature measurement becomes more important than the uncertainty in the heat of fusion when θ is less than 0.3°C.   An uncertainty of 1.0 cal in $\bar{L}_1$ will also produce an error of 0.07 per cent in ϕ.   However, $\bar{L}_1$ is less than 1.5 cal for most 1-1 salt solutions at 0.5 M, and, for many salt solutions, a 100 per cent error in $\bar{L}_1$ will not cause more than a 0.07 per cent error in ϕ even up to 1 M.   Thus $\bar{L}_1$ is of importance in correcting to 0°C only for concentrated salt solutions.   An error of 0.07 per cent in ϕ is caused by an error of $2/\theta$ cal/deg in $\bar{J}_1$.   Thus this is rarely a serious limitation except for very concentrated solutions.   In the example of 1.2455 M KCl, the entire magnitude of $\bar{J}_1$ is less than the allowable error.   The Δb term or the temperature coefficient of $\bar{c}_P$ does not have any influence on ϕ as long as one does not have to use terms above the θ^2 term of Eq. (26-12), and even in the θ^3 term its effect is smaller than that of the uncertainty in Δc_f°.

Thus careful work should yield ϕ values with an uncertainty of less than 0.1 per cent for molalities near 0.1 M or above.   From Eq. (20-34) or (22-29), one sees that γ_2 should be accurate to about 0.1 per cent at 0.1 M or above, provided that the

---

[1] C. Robertson and V. K. LaMer, *J. Phys. Chem.*, **35,** 1953 (1939); G. Scatchard, P. T. Jones, and S. S. Prentiss, *J. Am. Chem. Soc.*, **54,** 2676 (1932).

integration through the more dilute region has not introduced substantial error. However, at 10^{-3} M temperature errors of 2×10^{-5} to 2×10^{-4} deg cause errors of 0.5 to 5 per cent in ϕ and would cause errors of the same magnitude in γ. Consequently, it is possible to obtain accurate activity coefficients at higher concentrations only if theory allows one to extrapolate accurately through this region of very low concentration. The Debye-Hückel theory as extended by the Bronsted ion-interaction concept in Eq. (23-39)* for electrolytes or the simpler theory of Eq. (19-26) for nonelectrolytes does suffice for the required extrapolation; hence freezing-point-depression measurements do provide a source of highly accurate activity coefficients.

Establishment of Solute Standard State. As we noted in the treatment of isopiestic data and galvanic-cell data, it is best to evaluate the ratio of activity coefficients at higher molalities to the activity coefficients at some lower molality, for example, 0.1 M, and then to consider separately the extrapolation to the solute standard state. For molalities below 0.1 M the freezing-point depression is small enough and the heat of dilution small enough so that Eq. (26-14)* is quite adequate for calculating ϕ. Also, Eq. (22-31)* is applicable below 0.1 M. By substitution of Eq. (26-14)* into Eq. (22-31)* one obtains for a 1-1 salt

$$\theta - 2\lambda m \left[1 - \frac{2.303}{3} A_\gamma m^{\frac{1}{2}} \sigma(m^{\frac{1}{2}}) \right] = 2.303 B \lambda m^2 \qquad (26\text{-}18)^*$$

A plot of the quantity on the left-hand side of Eq. (26-18)* against m^2 will yield a line from whose slope one can obtain B. Figure 26-1 shows such a plot for the measurements of KCl by Adams,[1] Scatchard and Prentiss,[2] and Brown and Prue.[3] A straight line is obtained which verifies the validity of Eq. (22-31)* for molalities up to 0.06 M. Using $A_\gamma = 0.492$ from Appendix 4, $B = 0.059$ was obtained.

When it is desired to obtain the difference in B values of two salts from freezing-point measurements at low molalities, one may use the equation

$$\theta_1 - \theta_2 = 2.303 \lambda m^2 (B_1 - B_2) \qquad (26\text{-}19)^*$$

This procedure has the advantage of not using the limiting Debye-Hückel slope explicitly and provides a particularly sensitive deviation function. If j values are tabulated in place of θ, still for a 1-1 salt,

$$j_1 - j_2 = \frac{2.303}{2} m (B_2 - B_1) \qquad (26\text{-}20)^*$$

Freezing-point measurements at low molalities, where Eq. (22-20)* is applicable, yield values of B from which activity coefficients can be calculated. At higher molalities, Eq. (22-29) may be used to calculate values of the activity coefficients at 0°C from the osmotic coefficient.

[1] L. H. Adams, *J. Am. Chem. Soc.*, **37**, 481 (1915).
[2] *Loc. cit.*
[3] P. G. M. Brown and J. E. Prue, *Proc. Roy. Soc. (London)*, **A232**, 320 (1955).

Once the activity coefficients have been evaluated at 0°C, one can use partial molal heat-content and heat-capacity data to calculate the activity coefficients at other temperatures. Historically, the correction of activity coefficients from freezing-point data to 25°C has been an important process, as other methods of determining activity coefficients of

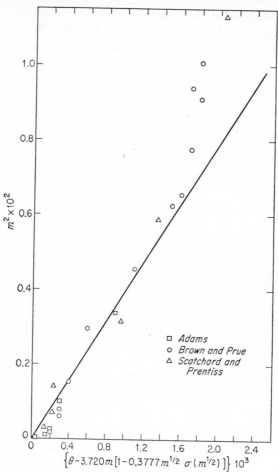

FIG. 26-1. Freezing-point data for aqueous KCl.

equal accuracy were not available. The isopiestic method now provides osmotic-coefficient and activity-coefficient data for many salts at room temperature.

Since the isopiestic method is of value only above 0.1 M, freezing-point data are still of value in fixing the 0.1 M activity coefficients and in aiding the extrapolation of the isopiestic data to infinite dilution. Also the high accuracy of freezing-point data well below 0.1 M allows their use to check theories of solution behavior in the dilute range.   In particular, freezing-point data constitute an important check on the reliability of Eqs. (22-20)* and (22-31)* with a single parameter at low molalities.   In addition, freezing-point data in conjunction with room-temperature data provide a check on the reliability of enthalpy and heat-capacity data.

Boiling-point-elevation data can be treated in the same manner as freezing-point-depression data except that the effect of variation of pressure must be considered, as discussed in Chapter 20.   Swietoslawski[1] has reviewed the development of the technique.   It does not compare in accuracy with the freezing-point method but is valuable in yielding data well above room temperature.   Smith and coworkers[2] have obtained valuable data on a few salt solutions in the range 60 to 100°C.

Nonelectrolyte Solutions.   The freezing points of aqueous solutions of nonelectrolytes have rarely been obtained with the highest possible accuracy and over a wide range of concentration.   As an example of the use of Eq. (26-14), we may use the freezing-point data for n-propanol aqueous solutions as determined by Webb and Lindsley[3] which are presented in Table 26-1.   Figure 26-2 is a plot of $(1 - \phi)/m$ versus m as required for the calculation of the activity coefficient of the solute by Eq. (20-34).   Evidently the drawing of the line through the experimental points is somewhat arbitrary.   Fortunately, the exaggeration produced by use of this method is so great that a considerable variation in the

TABLE 26-1. OSMOTIC COEFFICIENT OF AQUEOUS n-PROPANOL SOLUTIONS AT 0°C

m	θ	θ/m	$1 - \phi$	$(1 - \phi)/m$
0.00919	0.01702	1.8520	0.0043	0.47
0.01468	0.02721	1.8535	0.0035	0.24
0.01967	0.03618	1.8393	0.0111	0.56
0.03742	0.06884	1.8396	0.0110	0.29
0.04542	0.08342	1.8366	0.0125	0.27
0.05978	0.10977	1.8362	0.0127	0.21
0.07832	0.14274	1.8225	0.0201	0.26
0.10090	0.18383	1.8219	0.0204	0.20
0.11020	0.20131	1.8268	0.0177	0.16
0.13146	0.23929	1.8202	0.0213	0.16
0.14850	0.26983	1.8170	0.0230	0.15
0.17407	0.31684	1.8202	0.0213	0.12

[1] W. Swietoslawski, "Ebulliometry," Chemical Publishing Company, Inc., New York, 1937.

[2] R. P. Smith, *J. Am. Chem. Soc.*, **61**, 497 (1939); R. P. Smith and D. S. Hirtle, *J. Am. Chem. Soc.*, **61**, 1123 (1939); G. C. Johnson and R. P. Smith, *J. Am. Chem. Soc.*, **63**, 1351 (1941).

[3] T. J. Webb and C. H. Lindsley, *J. Am. Chem. Soc.*, **56**, 874 (1934).

curve causes very little difference in the final results. It is evident that $(1 - \phi)/m$ does approach a finite value as the molality approaches zero, as indicated by the general theory discussed at the end of Chapter 19 and in connection with Eq. (20-30).

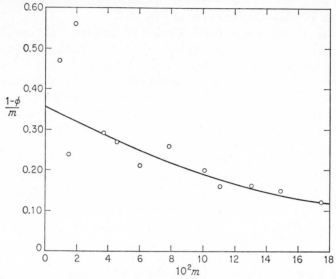

Fig. 26-2. Osmotic coefficients of n-propanol aqueous solutions at 0°C.

In Table 26-2 are tabulated values of $\phi - 1$ from the smoothed curve of Fig. 26-2 and values of the integral $\int_0^m [(1 - \phi)/m]\, dm$ obtained from

TABLE 26-2. Activity Coefficient of n-Propanol in Aqueous Solutions
at 0°C

m	$1 - \phi$	$\int \dfrac{1 - \phi}{m}\, dm$	$-\ln \gamma_2$
0.01	0.0034	0.0035	0.0069
0.02	0.0064	0.0068	0.0132
0.04	0.0112	0.0128	0.0240
0.05	0.0135	0.0156	0.0291
0.10	0.0190	0.0270	0.0460
0.15	0.0210	0.0355	0.0565
0.18	0.0216	0.0394	0.0610

the area under the curve of Fig. 26-2. Using Eq. (20-34), one can obtain immediately values of $-\ln \gamma_2$ for n-propanol by adding together the values of $\phi - 1$ in column 2 and values of the integral in column 3. Thus n-propanol in water shows a negative deviation from ideality with an activity coefficient of 0.941 with respect to the solute standard state at

$m = 0.18$.   Two freezing-point depressions of 1.953 and 9.698° at 1.09 and 5.74 M by Abegg[1] allow one to estimate the area of the extension of the curve of Fig. 26-2 to 5.74 M.   This yields an estimate of $\gamma = 0.8$ at 5.74 M by using Eq. (26-16)* without L_1.   For such large depressions, Eq. (26-9) or (26-12) would be used if more points were available to fix better the extension of the curve of Fig. 26-2.

Activity Coefficients from Phase Diagrams.   The emphasis in this chapter has been on the determination of activity coefficients from the variation with temperature of the solubility of the solid solvent.   In most systems, there are extensive liquidus curves for both components, and, in the same manner in which the activities of water were determined from the freezing-point depressions when ice was the saturating phase, activities of the other component can be determined from the variation of its solubility with temperature.   The eutectic point which corresponds to saturation of the liquid solution by both solids is a particularly valuable point, since activities of both components are determined simultaneously in the same solution, and one can check on the extension of equations for the variation of activity coefficients with composition which have been determined at the two extreme ends of the composition range.

Although the freezing-point method has been applied most extensively for the determination of activity coefficients or free energy of dilution of aqueous solutions, the method is widely applicable and will play an important role in the establishment of solution behavior in many types of solutions.   A large body of information has accumulated for oxide, metal, and halide systems, which has been expressed in the form of phase diagrams.   Only a very small portion of these data has been converted into thermodynamic form.   As the available phase diagrams cover a large range of types of solutions, it would be very valuable to obtain activity coefficients or free energies of dilution as well as entropies and enthalpies of dilution from these data.   Unless one has data under conditions where the solubility of the solid in the liquid is very small, the data normally correspond to simultaneous changes in composition and temperature. It is difficult to separate the effects of these two variables from phase-diagram data alone unless one has heat-of-dilution data which allow the use of Eq. (26-11) or (26-17).   Thus, it is often important to be able to apply theory to separate the effects of temperature and composition. For example, the correlation which has been pointed out in Chapter 25 between the heats and free energies of dilution is a type of correlation which is very valuable for evaluating phase-diagram data for which both temperature and composition vary simultaneously.

In Chapter 21 we have reviewed various assumptions which have been made in treating solutions, such as that made by the regular solu-

[1] R. Abegg, *Z. physik. Chem.*, **15**, 219 (1894).

tion theory, which allow one to separate the effects of temperature and composition. The regular solution type of equation for nonelectrolytes has been extended to fused salt solutions by using the entropy equations suggested by Temkin[1] and tested by the data of Flood,[2] Forland,[3] and others. A simple relationship is found between the free energy or enthalpy of mixing and the composition. Just as the use of the Debye-Hückel theory reduces the number of parameters that are required to represent the thermodynamic data of aqueous electrolyte solutions, theoretical equations which fix the functional form of entropy or enthalpy or free energy of dilution are extremely valuable in allowing evaluation of the vast number of data corresponding to solid-liquid equilibria. It is beyond the scope of this book to consider the many developments in the theories of solutions which are being extended to metals, fused oxides, and many other types of solutions.

In multicomponent systems, isothermal variations of solubility as a function of composition can be used to obtain activity coefficients as a function of composition. Such solubility measurements in aqueous salt solutions, particularly by Noyes and collaborators,[4] played an important role in the development of the concept of ionic strength by Lewis and Randall[5] and Bronsted[6] and in the development of Bronsted's theory of specific interaction. Multicomponent systems, including isothermal solubility measurements, will be considered in more detail in Chapter 34.

A common limitation to the evaluation of solubility data is lack of knowledge of the necessary heats of fusion. If one has sufficient information about the general form of the dependence of activity coefficient upon concentration, one may frequently obtain both the heat of fusion and activity coefficients from solubility measurements.

For example, if the solution is of a type for which application of the regular solution equations [Eqs. (21-15)* to (21-18)*] would be reasonable, and if dw/dT of Eqs. (21-16)* to (21-18)* as well as the second temperature derivative be neglected to allow use of Δc_f° of the pure phases and to reduce the number of parameters, Eq. (26-8) can be expressed as

$$RT_f \ln x_1 + x_2{}^2 w = -\left(\frac{\Delta_H f^\circ + x_2{}^2 w}{T_f}\right)\theta - \left(\frac{\Delta_H f^\circ + x_2{}^2 w - \frac{1}{2}T_f\,\Delta c_f}{T_f{}^2}\right)\theta^2 \cdots$$

$$(26\text{-}21)^*$$

[1] M. Temkin, *Acta Physiochim. U.R.S.S.*, **20**, 411–420 (1945).

[2] H. Flood, *Svensk Kem. Tidskr.*, **68**, 509 (1956); H. Flood, O. Fykse, and S. Urnes, *Z. Elektrochem.*, **59**, 364 (1955); H. Flood, T. Forland, and K. Grjotheim, *Z. anorg. Chem.*, **376**, 289 (1954).

[3] T. Forland, *J. Phys. Chem.*, **59**, 152 (1955).

[4] A. A. Noyes, W. Bray, W. D. Harkins, et al., *J. Am. Chem. Soc.*, **33**, 1643–1686, 1807–1873 (1911).

[5] G. N. Lewis, *J. Am. Chem. Soc.*, **34**, 1631 (1912); G. N. Lewis and M. Randall, *J. Am. Chem. Soc.*, **43**, 1112 (1921), and first edition of this book.

[6] J. N. Bronsted, *J. Am. Chem. Soc.*, **42**, 761 (1920), **44**, 877 (1922), **45**, 2898 (1923).

where x_1 and x_2 are mole fractions of solvent and solute, respectively, and where w and ΔH_f° are parameters to be determined from the solubility data alone if no other data are available. When both parameters are to be determined, it is important to have solubility data at both small x_1 and large x_1. The data at large x_1 are most important in fixing ΔH_f°, the heat of fusion of the solvent, and the data at small x_1 fix the value of w. As data are commonly available for the eutectic point of phase diagrams, such a point plus a solubility measurement at a large value of x_1 will fix both the heat of fusion and the value of w. A check on the assumptions made is provided by any additional data, as they will not yield consistent values of the two parameters if the regular solution approximation is not applicable to the particular solution under study.

Treatment of Systems with Extensive Solid Solubility.   The previous treatment of this chapter has been restricted to equilibria between solid and liquid phases where the solid phase has remained of fixed composition. The solute will always distribute itself between the solid and liquid phases, but, in many solutions such as the aqueous systems, the solid phase is sufficiently pure so that the variation of the free energy of the solid phase with composition can be neglected. In many other systems, neglect of solid solubility introduces a very large error. In Chapter 19, the freezing-point lowering when a solid solution appears has been considered for ideal solutions. The treatment of real systems does not require any new principles. Equations (26-1) to (26-9) must be modified to take into account the equation

$$A(s) = A(s, x_A) \tag{26-22}$$

Equations (26-3) to (26-7) for the change of ΔF of a reaction with temperature apply equally well to Eqs. (26-22) and (26-1) for pure solid going to either a liquid solution or a solid solution. For a solid solution in equilibrium with a liquid solution, Eq. (26-8) is directly applicable with addition of L_1 terms for the solid solution. Equation (26-9), likewise, is directly applicable, but it applies to the change of state

$$A(\text{solid soln}, x_s) = A(\text{liq soln}, x_l)$$

and
$$\Delta \bar{F}_{T_f} = RT_f \ln \frac{a_l}{a_s}$$

Thus one obtains a ratio of activities in the two phases in contrast to the activity in the liquid phase when solid solubility can be neglected. If the freezing-point depressions and the heats of dilution are large enough, partial molal heats of dilution and partial molal heat capacities corresponding to quantities needed for evaluation of Eq. (26-8) will be needed for both the liquid and solid solutions and Eq. (26-8) would read

$$\Delta \bar{F}_{T_f} = -\frac{\Delta H_f^\circ + L_l - L_s}{T_f}\theta - \left(\frac{\Delta H_f^\circ + L_l - L_s}{T_f^2} - \frac{\bar{C}_l - \bar{C}_s}{2T_f}\right)\theta^2$$
$$+ \cdots = RT_f \ln \frac{a_l}{a_s} \tag{26-23}$$

where subscripts l and s refer to liquid and solid phases, respectively. If data are available at small enough θ values so that the ratio of activities can be extrapolated to a low enough concentration to be equated to the ratio of mole fractions in the two phases, the solute standard states may be established in the two phases.   To obtain individual activities, it is necessary to have an independent determination of the activity in one of the phases, although it is sometimes possible to assume ideality in the liquid phase and thus obtain activity coefficients in the solid phase from the ratio of activities.

The problem of the distribution of solute between solid and liquid phases has become of very great practical importance in the production of semiconductors, and the behavior of such systems and the application of regular solution theory to such systems have been given by Thurmond and Struthers.[1]   The Fe-O system is one of the most carefully studied systems with extensive solid solubility, and the work of Darken and Gurry[2] is an excellent example of the treatment of such systems.   Many metallic phase diagrams involving both liquid and solid solutions have been treated thermodynamically and reported by Hultgren[3] in "Selected Values for the Thermodynamic Properties of Metals and Alloys."

PROBLEMS

26-1. The freezing-point depressions of aqueous NaOAc are reported[4] in terms of j values based on $\lambda = 1.858$.   Calculate ϕ values, and use Eq. (22-31)* to obtain B values.   How does B vary over the range of molality?

m_{NaOAc} .	0.00581	0.00766	0.01563	0.02348	0.03766	0.04977	0.05405	0.08087	0.11146
j	0.0256	0.0251	0.0335	0.0417	0.0447	0.0472	0.0467	0.0509	0.0522

26-2. Additional j values at higher m_{NaOAc} are:

m_{NaOAc} . .	0.1356	0.18475	0.23725	0.3203	0.4364	0.5681	0.6521
j	0.0534	0.0537	0.0517	0.0486	0.0406	0.0312	0.0251

Calculate $\gamma_\pm$ for NaOAc solutions at $m = 0.1$, 0.3 and 0.6.

26-3. Determine γ_{HOAc} in 1 M HOAc.   Use freezing-point depressions read from Fig. 22-1 after correcting for the effect of dissociation of HOAc by using $K_a = 1.66 \times 10^{-5}$ and the leading term of Eq. (23-35)* for the activity coefficients of the ions.

[1] C. D. Thurmond, *J. Phys. Chem.*, **57**, 827 (1953); C. D. Thurmond and J. D. Struthers, *J. Phys. Chem.*, **57**, 831 (1953).   See also C. D. Thurmond, F. A. Trumbore, and M. Kowalchik, *J. Chem. Phys.*, **25**, 799 (1956).

[2] L. S. Darken and R. W. Gurry, *J. Am. Chem. Soc.*, **67**, 1398 (1945), **68**, 798 (1946).

[3] R. Hultgren, Selected Values for the Thermodynamic Properties of Metals and Alloys, *Univ. Calif. (Berkeley) Inst. Eng. Research Minerals Research Lab.*, issued in looseleaf form at irregular intervals since 1955.

[4] G. Scatchard, S. S. Prentiss, and P. T. Jones, *J. Am. Chem. Soc.*, **56**, 807 (1934).

26-4. The freezing-point depression for NaCl is 15.140°C at 4 M. $\phi L = -395$ cal at 25°C, and $J_1 = -1.06$ cal/deg. Calculate the osmotic coefficient at 0°C for this solution, using additional data from Appendix 4.

26-5. Assume that Raoult's law holds for a KCl-KI liquid solution and that solid solubility can be neglected, and thereby calculate the temperature of first appearance of solid KI upon cooling a mixture of mole fraction 0.2KCl and 0.8KI. Pure KI melts at 955°K with a heat of fusion of 5740 cal/mole. The molal heat capacity of KI(s) is given by the equation $c_P = 12.1 + 1.95 \times 10^{-3}T$ cal/deg and, for KI(l), $c_P = 11.4 + 1 \times 10^{-3}$ cal/deg. Both salts are completely ionized, and, with Raoult's law holding, $\bar{L}_1 = 0$ and $\bar{L}_2 = 0$ for all concentrations.

Repeat the above calculations for a solution of mole fraction 0.8KI and mole fraction 0.2NaBr. Make the same assumptions, but note that both positive and negative ions differ in this case.

26-6. Use thermal data from Appendix 4 to correct the γ_{NaOAc} value for 0.1 M from Problem 26-2 to 25°C.

26-7. Obtain $\bar{L}_2$ and $\bar{J}_2$ and $\gamma_{\pm}$ values at 25°C for 0.5 M NaCl from Appendix 4, calculate $\gamma_{\pm}$ at 0°C, and compare with the value $\gamma_{\pm} = 0.673$ determined from freezing-point measurements.[1]

26-8. Guggenheim and Turgeon[2] have reviewed freezing-point data to obtain β values at 0°C, some of which are given in Table 23-1. Additional β values are:

Salt.....	LiBr	LiClO$_3$	LiClO$_4$	LiNO$_3$	LiOAc	NaBr	NaOAc
β........	0.30	0.25	0.35	0.23	0.19	0.20	0.26

Salt.....	KF	KCl	KBr	KClO$_3$	KIO$_3$	KNO$_3$	KOAc
β........	0.05	0.04	0.06	-0.19	-0.43	-0.26	0.30

Convert the β values to ΔB values based on KCl, and use the enthalpy and heat-capacity data summarized in Appendix 4 to convert to ΔB values at 25°C for comparison with the ΔB values in Appendix 4, which are largely based on isopiestic data. For which salts would you recommend a reevaluation of the isopiestic data or, on the basis of the correlation of Fig. 25-9, a reevaluation of the enthalpy data?

26-9. The Bi-Cu diagram as summarized by Hansen[3] consists of a simple eutectic diagram with no intermediate phases and very small solid solubility. The smoothed Cu liquidus curve is as follows:

T, °K........	1303	1260	1193	1098	1023	853	543
x_{Cu}..........	0.95	0.9	0.8	0.5	0.3	0.1	0.005 (eutectic)

The melting point of Cu is 1356°K. Derive an equation for large freezing-point depressions from Eq. (26-9) based upon the same assumptions used to obtain Eq. (26-21)*. Use the new equation or Eq. (26-21)*, as appropriate, to evaluate ΔH_f° and w. Does the constancy of w justify the approximations of Eq. (26-21)*?

[1] G. Scatchard and S. S. Prentiss, *J. Am. Chem. Soc.*, **55**, 4355 (1933).

[2] E. A. Guggenheim and J. C. Turgeon, *Trans. Faraday Soc.*, **51**, 747 (1955).

[3] M. Hansen, "Constitution of Binary Alloys," 2d ed., p. 308, McGraw-Hill Book Company, Inc., New York, 1958.

27

THERMODYNAMIC PROPERTIES OF IDEAL GASES CALCULATED FROM SPECTROSCOPIC AND OTHER MOLECULAR DATA

In Chapter 8 we discussed the statistical nature of the second law of thermodynamics and considered a few simple problems. Also in Chapter 21 certain solution properties were derived statistically. We shall now apply the principles of probability to molecular behavior in greater detail. Our aim is not a full treatment of statistical mechanics, which would require a volume as large as this one, but rather a limited number of the formulas which have proved very useful for the calculation of thermodynamic properties of ideal gases. In keeping with this purpose we shall make no attempt to give the most general or fundamental formulation of statistical mechanics but rather shall proceed as directly as possible to the desired results. The reader should consult one or more of the excellent treatises on statistical mechanics if he wishes to see the full scope of that important science which is so closely allied to thermodynamics.

We now know that an ideal gas is an idealized state of matter in which the individual gas molecules have no interaction upon one another. Thus, if we can calculate the properties of individual molecules, we should be able to calculate the properties of the ideal gas comprising these molecules.

It proves convenient for a number of reasons to consider the translational motion of entire molecules separately from all other motions which are internal within a given molecule. The quantized energy levels of the latter motions can be obtained by appropriate spectroscopic studies. Indeed there is a very extensive body of spectroscopic data on molecular energy levels.

Translation. In contrast to the internal motions, where spectroscopic data are available, the translational motion of a molecule must be treated by other methods.[1] The heat capacity of translation, $\frac{3}{2}R$ per

---

[1] In principle, the translation of a molecule in a fixed volume is quantized, but the energy levels have not yet been measured by spectroscopy.

mole, is calculated by elementary kinetic theory.   Likewise the entropy change on expansion was calculated in Eq. (7-9)*.

$$s_2 - s_1 = R \ln \frac{v_2}{v_1}$$

These two terms may be combined to yield an equation for the entropy of translation,

$$s = R(\ln v + \tfrac{3}{2} \ln T + \text{constant}) \qquad (27\text{-}1)^1$$

This equation gives everything needed for the application of the first and second laws of thermodynamics.   For use with the third law, however, we need the absolute value of s.

Sackur[2] showed that the constant in Eq. (27-1) for a number of substances could be expressed as the sum of a universal constant and a term $\tfrac{3}{2}R \ln M$, where M is the molecular weight of the molecule.   The subsequent development of quantum mechanics has given full theoretical confirmation of Sackur's empirical relationship and in addition yields the value of the universal constant, which was first found by Tetrode[3] (see Appendix 3).   The complete equation for the entropy of translation is

$$s = R(\ln v + \tfrac{3}{2} \ln T + \tfrac{3}{2} \ln M) + C$$
$$C = R\left(\frac{5}{2} + \frac{3}{2} \ln \frac{2\pi k}{h^2} - \frac{5}{2} \ln N_0 \right) \qquad (27\text{-}2)$$
$$= -11.074 \text{ cal/deg}$$

where M is the usual molecular weight and v is in cubic centimeters per mole.

Equation (27-2) may be tested by comparison with the entropies calculated from the third law of thermodynamics by using low-temperature heat-capacity measurements for the solid and entropies of vaporization for monatomic gases which have no internal energy levels within the thermal range.   These results are given in Table 27-1.

The other thermodynamic functions are readily obtained from Eq. (27-2), and the value $\tfrac{3}{2}RT$ for the internal energy $E^\circ - E_0^\circ$ of translation derived by elementary kinetic theory [and Eq. (A3-12)].   Also it is convenient to convert to the standard state of ideal gas at 1 atm.   The equations for the translational contributions are as follows:

---

[1] All equations in this chapter might be designated by * since they pertain to the ideal gas state.   The * is omitted throughout this chapter and is used hereafter only in circumstances where we wish to draw specific attention to the limitations on a given equation.

[2] O. Sackur, *Ann. Physik*, (4)**36**, 598 (1911), **40**, 67 (1913).

[3] H. Tetrode, *Ann. Physik*, (4)**38**, 434 (1912).

$$S_{tr}^{\circ} = R(\tfrac{3}{2} \ln M + \tfrac{5}{2} \ln T) - 2.315 \text{ cal/deg}$$

$$-\left(\frac{F^{\circ} - H_0^{\circ}}{T}\right)_{tr} = R(\tfrac{3}{2} \ln M + \tfrac{5}{2} \ln T) - 7.283 \text{ cal/deg} \qquad (27\text{-}3)$$

$$(H^{\circ} - H_0^{\circ})_{tr} = \tfrac{5}{2}RT$$

$$(C_P^{\circ})_{tr} = \tfrac{5}{2}R$$

Internal Energy Levels.[1] We now assume that we have the complete list of internal energy levels $\epsilon_0, \epsilon_1, \epsilon_2, \ldots, \epsilon_i, \ldots$ for the molecule (or atom). These values usually come from spectroscopic measurements. Quantum mechanics is of aid in interpreting the spectral data, and in some cases it is also possible to calculate the energy levels from theory alone.

TABLE 27-1. ENTROPIES OF MONATOMIC GASES AT 298.15°K[†]

	$\int C_P \, d \ln T$	Eq. (27-2)
Ne	35.01 ± 0.10	34.95 ± 0.01
Ar	36.95 ± 0.2	36.99 ± 0.01
Kr	39.17 ± 0.1	39.20 ± 0.01
Xe	$40.7 \ \pm 0.3$	40.54 ± 0.01

† K. K. Kelly, *U.S. Bur. Mines Bull.* 477, 1950, reviews critically the available data and gives original references.

We wish to calculate the properties of an ideal gas composed of these molecules in their equilibrium distribution among these various energy levels at a given temperature. For the moment let us consider a microscopic state to be defined by one of these internal energy levels. Each such microscopic state corresponds to certain definite quantum specifications not possessed in every particular by any other state. We assume that each microscopic state has the same a priori probability. By this we mean that, given equal opportunity to possess the energies necessary for their existence, all states are equally probable.

We may employ familiar thermodynamic equations to obtain the Boltzmann law for the distribution of molecules among these states. For the equilibrium between any two states we have

$$\Delta F^{\circ} = -RT \ln \frac{n''}{n'}$$

where n'' and n' are the numbers of molecules[2] in the respective states at equilibrium. Since the states have equal probability in the statistical

---

[1] This treatment follows closely that of W. F. Giauque, *J. Am. Chem. Soc.*, **52**, 4808 (1930).

[2] In this section we use lower-case n for number of molecules instead of number of moles.

sense, $\Delta s° = 0$ and

$$\Delta F° = \Delta E° = (\epsilon'' - \epsilon')N_0$$

Note that $E°$ is energy per mole, while ϵ is energy per molecule. Then we find

$$\frac{n''}{n'} = e^{-\Delta E°/RT} = e^{-(\epsilon''-\epsilon')/kT}$$

which is the familiar Boltzmann law.

It is found in many practical cases that there are several states of substantially the same energy. While we could carry through the calculations to follow with full accuracy by merely including the appropriate number of separate but equal terms for these several states of the same energy, it is more convenient to use multiple a priori probabilities. We designate by g_i the integral number of microscopic states which have the same energy within the accuracy needed in the calculation.

For convenience let us also assume that the lowest energy level has zero energy ($\epsilon_0 = 0$), which we may always arrange by an arbitrary shift in the energy scale. Also let the number of molecules in a single lowest state be n_0. Then the number in any other state of multiplicity g_i and an energy ϵ_i above zero will be

$$n_i = n_0 g_i e^{-\epsilon_i/kT} \tag{27-4}$$

and the total number of molecules is

$$N = n_0 g_0 + n_0 g_1 e^{-\epsilon_1/kT} + n_0 g_2 e^{-\epsilon_2/kT} + \cdots$$
$$= n_0 \sum_i g_i e^{-\epsilon_i/kT} \tag{27-5}$$

where the sum covers all the microscopic states.

The sum in Eq. (27-5) is given the name *partition function* (or sum over states) and the symbol Q, thus:

$$Q = \sum_i g_i e^{-\epsilon_i/kT} \tag{27-6}$$

We may now calculate the internal energy of all the molecules since we have the number of molecules in each state and the energy of each state.

$$(E - E_0)_{int} = \Sigma n_i \epsilon_i$$
$$= n_0 \Sigma \epsilon_i g_i e^{-\epsilon_i/kT}$$
$$= n_0 k T^2 \frac{dQ}{dT}$$

Since we usually do not know n_0, we eliminate this factor by Eqs. (27-5) and (27-6) and obtain for the internal energy per mole

$$(\mathrm{E} - \mathrm{E}_0)_{int} = \frac{N_0 k T^2}{Q} \frac{dQ}{dT}$$

$$= R T^2 \frac{d \ln Q}{dT} \tag{27-7}$$

The heat-capacity contribution follows directly by differentiation,

$$c_{int} = R \left(T^2 \frac{d^2 \ln Q}{dT^2} + 2T \frac{d \ln Q}{dT} \right) \tag{27-8}$$

Finally one may obtain the entropy contribution by integration.

$$
\begin{aligned}
(\mathrm{s}_T - \mathrm{s}_0)_{int} &= \int_0^T \frac{c}{T} \, dT \\
&= R \int_0^T \left(T \frac{d^2 \ln Q}{dT^2} + 2 \frac{d \ln Q}{dT} \right) dT \\
&= R \left[\ln Q + T \frac{d \ln Q}{dT} \right]_0^T \\
&= R \left(\ln Q_T - \ln g_0 + T \frac{d \ln Q_T}{dT} \right)
\end{aligned}
\tag{27-9}
$$

At $0°\mathrm{K}$ the partition function becomes just g_0, which has been substituted for Q_0 in the last step in the derivation above. In ordinary cases it is found that $g_0 = 1$, even if other g's have multiple values. In that case $\ln g_0 = 0$. Even in a case where g_0 is not unity we would take $S_0 = R \ln g_0$ in accordance with the general equation for the entropy in terms of probability. Thus in either event we obtain for our final result for the entropy contribution from internal energy levels the expression

$$\mathrm{s}_{int} = R \left(\ln Q + T \frac{d \ln Q}{dT} \right) \tag{27-10}$$

For practical computations it is frequently necessary to sum the partition function by direct numerical methods. Then the derivatives are not readily obtained, and it is best to work instead with the following functions:

$$Q' = \sum_i g_i \left(\frac{\epsilon_i}{kT} \right) e^{-\epsilon_i/kT} \tag{27-11}$$

$$Q'' = \sum_i g_i \left(\frac{\epsilon_i}{kT} \right)^2 e^{-\epsilon_i/kT} \tag{27-12}$$

These functions are readily computed at the same time that Q itself is calculated. It is easily shown that in these terms the thermodynamic functions are given by the following equations:

$$c_{int} = R\left[\frac{Q''}{Q} - \left(\frac{Q'}{Q}\right)^2\right] \tag{27-13}$$

$$(E - E_0)_{int} = RT\left(\frac{Q'}{Q}\right) \tag{27-14}$$

$$s_{int} = R\left(\ln Q + \frac{Q'}{Q}\right) \tag{27-15}$$

In Chapter 15 the great utility and convenience of the function $(F_T^\circ - H_0^\circ)/T$ was noted. For a perfect gas $H_0^\circ = E_0^\circ$, and for the internal energy contributions $F_{int} = A_{int} = E_{int} - TS_{int}$ since the volume affects only the translational energy contribution. Thus we obtain by combination of Eqs. (27-14) and (27-15) the very simple result

$$-\left(\frac{F^\circ - H_0^\circ}{T}\right)_{int} = R \ln Q \tag{27-16}$$

These equations, when combined with those for the translational contributions, suffice for the calculation of the thermodynamic properties of any ideal gas when the molecular energy levels are known. In some cases further rearrangements of the equations are convenient, and these will be mentioned together with the important results obtained in the next few pages.

Monatomic Gases at High Temperatures. The equations of the preceding section are appropriate for the calculation of the thermodynamic properties of monatomic gases at high temperatures, where excited electronic states become important. The energy levels are obtained by atomic spectroscopy, and the partition function is obtained by direct numerical summation. Table 27-2 gives a sample calculation for monatomic silicon gas at 5000°K. The energy-level values are taken from Moore.[1]

Rotational Contributions for Molecules. Even for relatively simple molecules the complete array of energy levels is so extensive that direct summation of the partition function becomes impractical. Fortunately the energy levels follow such a systematic pattern that new methods become feasible. It is found that each molecule has a characteristic pattern of relatively closely spaced energy levels, which are attributed to molecular rotation. This pattern is repeated at various higher energies. The larger energy differences are usually those of molecular vibrations, although excited electronic states may also arise. Let us now consider just the contribution of the rotational levels.

---

[1] C. E. Moore, Atomic Energy Levels, *Natl. Bur. Standards (U.S.) Circ.* 467, 1949.

TABLE 27-2. CALCULATION OF THERMODYNAMIC PROPERTIES OF Si GAS
AT 5000°K

State	ϵ_i, cm^{-1}	g_i	$\dfrac{\epsilon_i}{kT}$	$e^{-\epsilon_i/kT}$	$\dfrac{\epsilon_i}{kT}e^{-\epsilon_i/kT}$	$\left(\dfrac{\epsilon_i}{kT}\right)^2 e^{-\epsilon_i/kT}$
$^3P_0$	0	1	0.0	1.000	0.0	0.0
$^3P_1$	77.15	3	0.022	0.978	0.022	0.000
$^3P_2$	223.31	5	0.064	0.938	0.060	0.004
$^1D_2$	6,298.81	5	1.812	0.163	0.295	0.535
$^1S_0$	15,394.24	1	4.430	0.0119	0.053	0.234
$^3P_0^\circ$	39,683.10	1	11.419	0.00001	0.0001	0.001

$$Q = 9.451 \qquad Q' = 1.894 \qquad Q'' = 2.930$$

$$-\left(\frac{\text{F}^\circ - \text{H}_0^\circ}{T}\right)_{int} = R \ln 9.451 = 4.464 \text{ cal/deg mole}$$

$$(\text{H}^\circ - \text{H}_0^\circ)_{int} = RT(^{1.894}\!/_{9.451}) = 1991 \text{ cal/mole}$$

$$C_{int} = R[^{2.93}\!/_{9.451} - (^{1.894}\!/_{9.451})^2] = 0.536 \text{ cal/deg mole}$$

In the case of linear molecules, the rotational levels are given in good approximation by the simple formula

$$\epsilon_J = hcBJ(J + 1) \tag{27-17}$$

where B is a constant characteristic of the molecule and J is a quantum number taking integral values. B is in the spectroscopist's unit, the reciprocal centimeter, and must be multiplied by hc to yield energy. It is found that for symmetrical molecules such as N—N, O—C—O, etc., J may have only the alternate values. In some cases[1] the even values 0, 2, 4, . . . are allowed and in other cases the odd values 1, 3, 5, . . . , but for most of our purposes this difference will be of no consequence. For unsymmetrical molecules all integral values of J are permitted. All levels with J greater than zero are multiple with $g_J = 2J + 1$. This multiplicity may be verified by the application of electric or magnetic fields, which separate the energy levels and hence split the lines in the spectrum.

Quantum theory yields results in complete accord with spectroscopy and in addition gives for the constant B the formula

$$B = \frac{h}{8\pi^2 c I} \tag{27-18}$$

where I is the moment of inertia of the molecule about an axis through the center of gravity and perpendicular to the axis on which the atoms

---

[1] The orientation of the nuclear spin of the symmetrically placed nuclei determines whether even or odd values are allowed; see K. S. Pitzer, "Quantum Chemistry," pp. 208–210, Prentice-Hall, Inc., Englewood Cliffs, N.J., 1953.

lie. Thus, if the molecular dimensions are known, B may be calculated even though it has not been observed spectroscopically.

Spectroscopists usually report B in units of reciprocal centimeters, but in recent years values are also being given in megacycles per second (mc). In that case the factor hc in Eqs. (27-17) and (27-19) is replaced by $10^6 h$.

Let us now define the dimensionless quantity y as follows,

$$y = \frac{hcB}{kT} = \frac{h^2}{8\pi^2 I k T} \tag{27-19}$$

whereupon the partition function for rotation becomes

$$Q_r = \sum_J (2J + 1)e^{-J(J+1)y} \tag{27-20}$$

with the sum covering either all positive integral values of J or only alternate values as indicated above. For most substances the value of y is so small that the sum in Eq. (27-20) may be replaced by an integral.

$$Q_r = \frac{1}{\sigma} \int_0^\infty e^{-J(J+1)y}(2J + 1)\, dJ$$

$$= \frac{1}{\sigma y} \tag{27-21}$$

Here σ, the symmetry number, is 1 for the unsymmetrical molecules, where all values of J are allowed, and is 2 for the symmetrical molecules, where only alternate values are permitted.

The simple result of Eq. (27-21) is adequate for all cases with $y \leq 0.01$. At 300°K, y is 0.0012 for Cl_2, 0.0069 for O_2, 0.0095 for N_2, 0.050 for HCl, and 0.283 for H_2. Thus our approximation is adequate for many molecules and nearly adequate for all others except for the various isotopes of hydrogen. Hydrogen is best handled by direct numerical summation, and the results are available. By more elaborate mathematical methods one may show that the partition function[1] is

$$Q_r = \frac{1}{\sigma y}\left(1 + \frac{y}{3} + \frac{y^2}{15} + \cdots\right) \tag{27-22}$$

provided that $y \leq 0.3$. This result will be adequate for all practical cases except for the hydrogen isotopes.

We now introduce the result just obtained into Eqs. (27-7) to (27-10). We also note that $\ln (1 + x) = x - (x^2/2) + \cdots$ when x is small and that $dy/dT = -y/T$. The final results for linear molecules are then as follows:

[1] J. E. Mayer and M. G. Mayer, "Statistical Mechanics," p. 153, John Wiley & Sons, Inc., New York, 1940.

$$\ln Q_r = -\ln y - \ln \sigma + \frac{y}{3} + \frac{y^2}{90} + \cdots$$

$$-\frac{(\text{F} - \text{H}_0)_r}{T} = R\left(-\ln y - \ln \sigma + \frac{y}{3} + \frac{y^2}{90} + \cdots\right) \qquad (27\text{-}23)$$

$$(\text{H} - \text{H}_0)_r = RT\left(1 - \frac{y}{3} - \frac{y^2}{45} - \cdots\right) \qquad (27\text{-}24)$$

$$\text{s}_r = R\left(1 - \ln y - \ln \sigma - \frac{y^2}{90} - \cdots\right) \qquad (27\text{-}25)$$

$$\text{c}_r = R\left(1 + \frac{y^2}{45} + \cdots\right) \qquad (27\text{-}26)$$

It is interesting to note that the term $y/3$ drops out of the expressions for the entropy and heat capacity.

For nonlinear molecules the rotational-energy-level pattern is usually much more complicated than that for linear molecules. Consequently the necessary calculations to obtain the partition function are much longer even though they involve only the same principles as were employed for linear molecules. The result is best expressed in terms of the principal moments of inertia.

If cartesian axes are taken with their origin at the center of mass, the moments and products of inertia are defined as follows,

$$
\begin{aligned}
I_x &= \Sigma m_i(y_i^2 + z_i^2) & I_{xy} &= \Sigma m_i x_i y_i \\
I_y &= \Sigma m_i(x_i^2 + z_i^2) & I_{xz} &= \Sigma m_i x_i z_i \\
I_z &= \Sigma m_i(x_i^2 + y_i^2) & I_{yz} &= \Sigma m_i y_i z_i
\end{aligned}
\qquad (27\text{-}27)
$$

where m_i is the mass of ith atom whose coordinates are x_i, y_i, z_i. The sum covers all atoms in the molecule. For the partition function we need the value of the determinant

$$
D = \begin{vmatrix} I_x & -I_{xy} & -I_{xz} \\ -I_{xy} & I_y & -I_{yz} \\ -I_{xz} & -I_{yz} & I_z \end{vmatrix} \qquad (27\text{-}28)
$$
$$
= I_x I_y I_z - 2I_{xy} I_{yz} I_{xz} - I_x I_{yz}^2 - I_y I_{xz}^2 - I_z I_{xy}^2
$$

Parenthetically we may remark that it is always possible by rotation of axes to make all the products of inertia (that is, I_{xy}, I_{xz}, I_{yz}) zero. The new axes are known as the principal axes of inertia, and the determinant then reduces to the simple product of the three principal moments of inertia. If the molecule has elements of symmetry, then it is easy initially to select axes that make most or all of these products zero. One chooses symmetry axes and axes perpendicular to planes of symmetry. However, for our present purpose there is no need to find the principal axes since the determinant D is the quantity desired.

The high-temperature approximation for the partition function[1] is

$$Q_r = \frac{(\pi D)^{1/2}}{\sigma}\left(\frac{8\pi^2 kT}{h^2}\right)^{3/2} \qquad (27\text{-}29)$$

where σ is again the symmetry number. In this general case it is the number of rotational orientations which differ only in the exchange of identical particles. Thus a symmetrical, bent triatomic molecule such as H_2O has $\sigma = 2$; a pyramidal molecule such as NH_3 has $\sigma = 3$; but the planar molecule SO_3 has $\sigma = 6$ because it may be turned over by rotation about an S—O axis in addition to three rotational orientations about the threefold axis through the S. A tetrahedral molecule such as CH_4 has $\sigma = 12$, while an octahedral molecule such as SF_6 has $\sigma = 24$.

On insertion of the partition function [Eq. (27-29)] into Eqs. (27-7) to (27-10) we obtain the final equations for the rotation of nonlinear molecules.

$$-\frac{F - H_0}{T} = R\left(\frac{1}{2}\ln D + \frac{3}{2}\ln T - \ln\sigma + \frac{1}{2}\ln\frac{512\pi^7 k^3}{h^6}\right)$$

$$= R\left[\frac{1}{2}\ln(D \times 10^{117}) + \frac{3}{2}\ln T - \ln\sigma\right] - 3.014 \qquad (27\text{-}30)$$

$$(H - H_0)_r = \tfrac{3}{2}RT \qquad (27\text{-}31)$$

$$s_r = R\left[\frac{1}{2}\ln(D \times 10^{117}) + \frac{3}{2}\ln T - \ln\sigma\right] - 0.033 \qquad (27\text{-}32)$$

$$c_r = \tfrac{3}{2}R \qquad (27\text{-}33)$$

These equations are valid at sufficiently high temperatures for all nonlinear molecules. Because of the great variety of types of polyatomic molecules, it is difficult to give a precise statement of the low-temperature limit of validity of these formulas. There is certainly no significant error for any polyatomic molecule at room temperature (300°K), and calculations for various individual cases[2] indicate that these equations are usually reliable down to 100°K or lower.

Vibrational Contributions. The energy levels for a vibrating system of atoms, whether in a molecule or a crystal, are given by the equation

$$\epsilon = hc\omega(v + \tfrac{1}{2}) \qquad (27\text{-}34)$$

where ω is the frequency of the vibration in units of reciprocal centimeters and v is a quantum number which takes the integral values 0, 1, 2, ...

---

[1] See, for example, J. E. Mayer and M. G. Mayer, "Statistical Mechanics," p. 191, John Wiley & Sons, Inc., New York, 1940, or K. S. Pitzer, "Quantum Chemistry," pp. 211, 443, Prentice-Hall, Inc., Englewood Cliffs, N.J., 1953.

[2] K. S. Pitzer, *J. Chem. Phys.*, **7**, 251 (1939); C. C. Stephenson and H. O. McMahon, *J. Chem. Phys.*, **7**, 614 (1939); D. P. MacDougall, *Phys. Rev.*, **38**, 2296 (1931).

The true frequency is $\nu = c\omega$.   This formula applies to harmonic vibration, i.e., where the restoring force is strictly proportional to the displacement.   Atomic systems usually deviate slightly from the harmonic condition; consequently corrections for *anharmonicity* must be considered. It is found, however, that the harmonic formula always yields a good first approximation to the thermodynamic properties.

If the molecule has more than two atoms, it will have more than one vibrational motion.   A complete discussion of this situation would involve the classical theory of normal modes of vibration and the quantum theory of such systems.   The result, however, is quite simple.   For each mode of vibration there is a frequency ω_i, and the vibrational energy is just the sum of terms for the separate modes of vibration

$$\epsilon = hc \sum_i \omega_i \left(v_i + \frac{1}{2} \right) \tag{27-35}$$

The frequencies observed in the infrared or Raman spectra are, of course, the frequencies that appear in this equation.

Since in the approximation of Eq. (27-35) there are no cross terms between different vibrations, we shall find that their contributions separate conveniently.   We define $u_i = hc\omega_i/kT = 1.4387\omega_i/T$.   Then

$$Q_{vib} = \sum_{v_1} \sum_{v_2} \cdots \sum_{v_n} e^{-(u_1 v_1 + u_2 v_2 + \cdots + u_n v_n)}$$

where each of the v's independently has the values $0, 1, 2, \ldots$ Since $e^{-(u_1 v_1 + u_2 v_2 + \cdots + u_n v_n)} = e^{-u_1 v_1} e^{-u_2 v_2} \cdots e^{-u_n v_n}$, the multiple sum may be factored.

$$Q_{vib} = \left(\sum_{v_1} e^{-u_1 v_1} \right) \left(\sum_{v_2} e^{-u_2 v_2} \right) \cdots \left(\sum_{v_n} e^{-u_n v_n} \right)$$

$$= \prod_i \left(\sum_{v_i} e^{-u_i v_i} \right) \tag{27-36}$$

The thermodynamic functions all depend on $\ln Q$, and the product in Q becomes a sum.

$$\ln Q_{vib} = \sum_i \ln \left(\sum_{v_i} e^{-u_i v_i} \right) \tag{27-37}$$

Parenthetically we may remark that this separability of thermodynamic contributions into a sum is generally valid whenever the energy is a sum of independent terms.   By an analogous argument one may prove the separability of rotational, translational, vibrational, and other contributions to thermodynamic functions if the corresponding energy terms are independent in the energy-level formula.

We return now to the calculation of the partition function for a single vibration.

$$Q_{vib} = \sum_v e^{-uv}$$

$$= 1 + e^{-u} + (e^{-u})^2 + (e^{-u})^3 + \cdots$$

$$= \frac{1}{1 - e^{-u}} \tag{27-38}$$

It is very fortunate that this infinite sum can be written in a closed, simple form. The thermodynamic functions are as follows:

$$-\left(\frac{F - H_0}{T}\right)_{vib} = -R \ln (1 - e^{-u}) \tag{27-39}$$

$$(H - H_0)_{vib} = RT \frac{u}{e^u - 1} = RT \frac{ue^{-u}}{1 - e^{-u}} \tag{27-40}$$

$$S_{vib} = R \left[\frac{u}{e^u - 1} - \ln (1 - e^{-u})\right] \tag{27-41}$$

$$C_{vib} = R \frac{u^2 e^u}{(e^u - 1)^2} \tag{27-42}$$

where
$$u = \frac{h\nu}{kT} = \frac{hc\omega}{kT} = 1.4387 \frac{\omega}{T}$$

These are the formulas first obtained by Einstein in his treatment of a quantized vibration. The resulting functions are tabulated for convenience in Table 27-3.

Diatomic Molecules: Higher Approximations. The first approximation to the thermodynamic functions for a diatomic molecule is given by the rotational and vibrational terms developed in the preceding sections provided that the molecule has but a single electronic state within the thermal-energy range. This is commonly called the rigid-rotator–harmonic-oscillator approximation. This approximation is excellent at low temperatures, but at higher energies the spectroscopic energy levels deviate from the pattern assumed in the earlier treatment in three significant respects. Consequently, higher approximations are needed for calculations at high temperatures. An improved expression for the energy levels, which includes these three additional terms, is usually an adequate approximation. The symbols for the vibrational and rotational constants, ω and B, are now given the subscript e to show that they pertain to the equilibrium interatomic distance of minimum potential energy.

$$\frac{\epsilon}{hc} = \omega_e \left(v + \frac{1}{2}\right) - x_e \omega_e \left(v + \frac{1}{2}\right)^2 + B_e J(J + 1)$$

$$- DJ^2(J + 1)^2 - \alpha(v + \tfrac{1}{2})J(J + 1) \tag{27-43}$$

anharmonicity correction

Rotation Stretching.

TABLE 27-3. HARMONIC-OSCILLATOR FUNCTIONS†

u	$-(F - H_0)/RT$ $-\ln(1 - e^{-u})$	$(H - H_0)/RT$ $u/(e^u - 1)$	c/R $u^2 e^u/(e^u - 1)^2$
0.0	∞	1.0000	1.0000
0.1	2.3522	0.9508	0.9992
0.2	1.7078	0.9033	0.9967
0.3	1.3502	0.8575	0.9925
0.4	1.1096	0.8133	0.9868
0.5	0.9328	0.7708	0.9794
0.6	0.7959	0.7298	0.9705
0.7	0.6863	0.6905	0.9602
0.8	0.5966	0.6528	0.9483
0.9	0.5218	0.6166	0.9352
1.0	0.4587	0.5820	0.9207
1.1	0.4048	0.5489	0.9050
1.2	0.3584	0.5172	0.8882
1.3	0.3182	0.4870	0.8703
1.4	0.2832	0.4582	0.8515
1.5	0.2525	0.4308	0.8318
1.6	0.2255	0.4048	0.8114
1.7	0.2017	0.3800	0.7904
1.8	0.1807	0.3565	0.7687
1.9	0.1620	0.3342	0.7466
2.0	0.1454	0.3130	0.7241
2.2	0.1174	0.2741	0.6783
2.4	0.0951	0.2394	0.6320
2.6	0.0772	0.2086	0.5859
2.8	0.0627	0.1813	0.5405
3.0	0.0511	0.1572	0.4963
3.2	0.0416	0.1360	0.4536
3.4	0.0340	0.1174	0.4129
3.6	0.0277	0.1011	0.3743
3.8	0.0226	0.0870	0.3380
4.0	0.0185	0.0746	0.3041
4.2	0.0151	0.0639	0.2726
4.4	0.0124	0.0547	0.2436
4.6	0.0101	0.0467	0.2170
4.8	0.0083	0.0398	0.1928
5.0	0.0068	0.0339	0.1707
5.2	0.0055	0.0288	0.1508
5.4	0.0045	0.0245	0.1329
5.6	0.0037	0.0208	0.1168
5.8	0.0030	0.0176	0.1025

TABLE 27-3. HARMONIC-OSCILLATOR FUNCTIONS† (*Continued*)

u	$-(\text{F} - \text{H}_0)/RT$ $-\ln(1 - e^{-u})$	$(\text{H} - \text{H}_0)/RT$ $u/(e^u - 1)$	c/R $u^2 e^u/(e^u - 1)^2$
6.0	0.0025	0.0149	0.0897
6.5	0.0015	0.0098	0.0637
7.0	0.0009	0.0064	0.0448
7.5	0.0006	0.0042	0.0312
8.0	0.0003	0.0027	0.0215
8.5	0.0002	0.0017	0.0147
9.0	0.0001	0.0011	0.0100
9.5	0.0001	0.0007	0.0068
10.0	0.0000	0.0004	0.0045

† A table of these functions with closer spacing of u values is given by K. S. Pitzer, "Quantum Chemistry," app. 13, Prentice-Hall, Inc., Englewood Cliffs, N.J., 1953.

Here the first and third terms are the familiar vibration and rotation terms, which are strictly correct only for harmonic forces in vibration and for a rigid body in rotation. The first correction term, which involves x_e, arises from deviation of the actual forces in the molecule from the harmonic-force law. The fourth term, $DJ^2(J + 1)^2$, arises because the centrifugal force of rotation stretches the molecule and increases its moment of inertia. From Eq. (27-18) we see that an increase in moment of inertia has the effect of decreasing B. Since the stretching may be related to the centrifugal force and to the force holding the atoms together, one may relate D to B_e and ω_e. The relation obtained is

$$D = 4 \frac{B_e^3}{\omega_e^2} \qquad (27\text{-}44)$$

If the interatomic force does not follow the harmonic law, then the mean distance between atoms may vary with the vibrational energy. This effect leads to the last term.

Table 27-4 lists the various constants for a few diatomic molecules. We note that the three correction terms are relatively small as compared with their respective principal terms; consequently we may employ appropriate expansions in series to calculate their effect on the partition function.

Before proceeding, let us rearrange the energy-level expression into a form more convenient for our purpose.

$$\frac{\epsilon - \epsilon_0}{hc} = \omega_0 v - x\omega_0 v(v - 1) + B_0 J(J + 1)$$

$$- 4 \frac{B_e^3}{\omega_e^2} J^2(J + 1)^2 - \alpha v J(J + 1) \quad (27\text{-}45)$$

where
$$\omega_0 = \omega_e - 2x_e\omega_e$$
$$B_0 = B_e - \tfrac{1}{2}\alpha$$

$$x = \frac{x_e}{1 - 2x_e} \cong x_e$$

This form has the practical advantage that ω_0 gives the actual energy of the first excited vibrational state above the ground state and similarly

TABLE 27-4. CONSTANTS FOR A FEW DIATOMIC MOLECULES†
In reciprocal centimeters

	ω_e	$x_e\omega_e$	B_e	α	ν_{00}‡	Electronic state
H₂........	4395.2	118.0	60.809	2.993	0	$^1\Sigma_g^+$
HF.......	4138.5	90.07	20.939	0.770	0	$^1\Sigma^+$
N₂........	2359.6	14.46	2.010	0.0187	0	$^1\Sigma_g^+$
NO.......	1903.7	13.97	1.7046	0.0178	120.9	$^2\Pi_{3\!/\!2}$
	1904.0	13.97	1.7046	0.0178	0	$^2\Pi_{1\!/\!2}$
O₂........	1580.4	12.073	1.44566	0.01579	0	$^3\Sigma_g^-$
	1509.3	12.9	1.4264	0.0171	7,882.4	$^1\Delta_g$
	1432.7	13.95	1.40042	0.01817	13,120.9	$^1\Sigma_g^+$
Cl₂......	564.9	4.0	0.2438	0.0017	0	$^1\Sigma_g^+$
I₂........	214.57	0.613	0.03736	0.00012	0	$^1\Sigma_g^+$

† G. Herzberg, "Molecular Spectra and Molecular Structure," 2d ed., I. Spectra of Diatomic Molecules, D. Van Nostrand Company, Inc., Princeton, N.J., 1950.

‡ ν_{00} gives the energy of this electronic state above the lowest electronic state as measured between the ground vibration-rotation levels.

B_0 gives the actual rotational-level spacing in the ground vibrational state. Also we have included the theoretical value for D.

Now let us introduce the temperature and define additional quantities as follows:

$$\frac{\epsilon - \epsilon_0}{kT} = uv - xuv(v - 1) + yJ(J + 1) - 4\gamma^2 yJ^2(J + 1)^2$$
$$- \delta yvJ(J + 1) \quad (27\text{-}46)$$

where
$$u = \frac{hc\omega_0}{kT}$$

$$y = \frac{hcB_0}{kT}$$

$$\gamma = \frac{B_e}{\omega_e} = \frac{1}{2}\left(\frac{D}{B_e}\right)^{1\!/\!2}$$

$$\delta = \frac{\alpha}{B_0}$$

The partition function is
$$Q = \sum_v \sum_J (2J + 1)e^{-(\epsilon - \epsilon_0)/kT}$$

where the exponent is given in Eq. (27-46).   We deal first with the sum over J.   Let us examine further the exponential factor, which we must sum,

$$
\begin{aligned}
\exp\left[-yJ(J+1) + 4\gamma^2 yJ^2(J+1)^2 + \delta yvJ(J+1)\right] \\
= \exp\left[-yJ(J+1)\right]\exp\left[4\gamma^2 yJ^2(J+1)^2 + \delta yvJ(J+1)\right]
\end{aligned}
$$

The exponent in the second exponential factor is always small at values of J where the first factor is substantial.   Consequently, we may expand the second exponential in series and retain only the first two terms. Also, since y is small, we may replace the sum over J by an integral.   We further substitute $z = J(J+1)$, $dz = (2J+1)\,dJ$.

$$
Q = \sum_v e^{-uv+xuv(v-1)} \int_0^\infty e^{-yz}(1 + 4\gamma^2 yz^2 + \delta yvz)\,dz
$$

The integration is straightforward.   We may also add the small terms given in Eq. (27-22) which correct for the difference between the true sum over J and the integral provided that $y \leq 0.3$ and introduce the symmetry number.   We then have

$$
Q = \sum_{v=0}^\infty e^{-uv+xuv(v-1)} \frac{1}{\sigma y}\left(1 + \frac{8\gamma^2}{y} + \delta v + \frac{y}{3} + \frac{y^2}{15}\right) \tag{27-47}
$$

Again we expand the exponential of the small term which involves x and neglect all cross terms involving products of small quantities.

$$
Q = \sum_v e^{-uv} \frac{1}{y\sigma}\left[1 + xuv(v-1) + \delta v + \frac{8\gamma^2}{y} + \frac{y}{3} + \frac{y^2}{15}\right]
$$

Since u is not necessarily a small quantity, the sum over v must be taken exactly.   In order to do so we note that

$$
\sum_{v=0}^\infty e^{-uv} = (1 - e^{-u})^{-1}
$$

$$
-\frac{d}{du}\sum_v e^{-uv} = \sum_v v e^{-uv} = e^{-u}(1 - e^{-u})^{-2}
$$

$$
\frac{d^2}{du^2}\sum_v e^{-uv} = \sum_v v^2 e^{-uv} = e^{-u}(1 - e^{-u})^{-2} + 2e^{-2u}(1 - e^{-u})^{-3}
$$

By use of these expressions, we obtain the exact sum over the vibrational quantum number v,

$$Q = \frac{1}{y\sigma(1 - e^{-u})}\left[1 + \frac{2xu}{(e^u - 1)^2} + \frac{\delta}{e^u - 1} + \frac{8\gamma^2}{y} + \frac{y}{3} + \frac{y^2}{15}\right] \quad (27\text{-}48)$$

Finally, let us obtain the logarithm of Q, which we shall separate into three terms: the rigid-rotator term, the harmonic-oscillator term, and a term including all the higher terms introduced in this section. We recall that

$$\ln(1 + x) \cong x - \frac{x^2}{2} + \cdots \qquad \text{for } x \ll 1$$

$$\ln Q = \left(-\ln y - \ln \sigma + \frac{y}{3} + \frac{y^2}{90}\right) - \ln(1 - e^{-u})$$

$$+ \left[\frac{2xu}{(e^u - 1)^2} + \frac{\delta}{e^u - 1} + \frac{8\gamma^2}{y}\right] \quad (27\text{-}49)$$

$$= \ln Q_r + \ln Q_{vib} + \ln Q_{cor}$$

$$\ln Q_{cor} = \frac{2xu}{(e^u - 1)^2} + \frac{\delta}{e^u - 1} + \frac{8\gamma^2}{y} \quad (27\text{-}50)$$

We may simplify the expression for the correction terms further by the substitution $y = \gamma u$, which is exact except for the small differences between B_e and B_0, ω_e and ω_0. Then we proceed to obtain the corrections to the various thermodynamic functions.

$$\ln Q_{cor} = \frac{2xu}{(e^u - 1)^2} + \frac{\delta}{e^u - 1} + \frac{8\gamma}{u} \quad (27\text{-}51)$$

$$-\left(\frac{F - H_0}{T}\right)_{cor} = \left(\frac{R}{u}\right)\left\{2x\left[\frac{u^2}{(e^u - 1)^2}\right] + \delta\left[\frac{u}{e^u - 1}\right] + 8\gamma\right\} \quad (27\text{-}52)$$

$$(H - H_0)_{cor} = \left(\frac{RT}{u}\right)\left\{2x\left[\frac{u^2(2ue^u - e^u + 1)}{(e^u - 1)^3}\right]\right.$$

$$\left. + \delta\left[\frac{u^2 e^u}{(e^u - 1)^2}\right] + 8\gamma\right\} \quad (27\text{-}53)$$

$$S_{cor} = \frac{(H - H_0)_{cor}}{T} - \left(\frac{F - H_0}{T}\right)_{cor}$$

$$C_{cor} = \left(\frac{R}{u}\right)\left\{4x\left[\frac{u^3 e^u(2ue^u - 2e^u + u + 2)}{(e^u - 1)^4}\right]\right.$$

$$\left. + \delta\left[\frac{u^3 e^u(e^u + 1)}{(e^u - 1)^3}\right] + 16\gamma\right\} \quad (27\text{-}54)$$

The functions in brackets are tabulated in Tables 27-3 and 27-5. Thus correction terms may be obtained very easily once the values of x, δ, and γ are available.

TABLE 27-5. ANHARMONIC-OSCILLATOR FUNCTIONS

u	$\dfrac{u^2}{(e^u-1)^2}$	$\dfrac{u^2(2ue^u-e^u+1)}{(e^u-1)^3}$	$\dfrac{u^3e^u(e^u+1)}{(e^u-1)^3}$	$\dfrac{u^3e^u(2ue^u-2e^u+u+2)}{(e^u-1)^4}$
0.0	1.0000	1.0000	2.0000	1.0000
0.1	0.9041	0.9960	2.0000	0.9999
0.2	0.8160	0.9846	2.0000	0.9994
0.3	0.7355	0.9669	1.9999	0.9979
0.4	0.6615	0.9438	1.9998	0.9952
0.5	0.5940	0.9157	1.9995	0.9906
0.6	0.5327	0.8841	1.9990	0.9843
0.7	0.4768	0.8492	1.9981	0.9756
0.8	0.4261	0.8120	1.9967	0.9647
0.9	0.3802	0.7731	1.9948	0.9513
1.0	0.3387	0.7330	1.9923	0.9355
1.1	0.3012	0.6922	1.9889	0.9172
1.2	0.2675	0.6512	1.9846	0.8965
1.3	0.2372	0.6105	1.9791	0.8736
1.4	0.2100	0.5704	1.9725	0.8485
1.5	0.1856	0.5312	1.9646	0.8216
1.6	0.1638	0.4930	1.9552	0.7929
1.7	0.1444	0.4563	1.9443	0.7628
1.8	0.1271	0.4209	1.9316	0.7314
1.9	0.1116	0.3872	1.9175	0.6991
2.0	0.0980	0.3553	1.9014	0.6662
2.2	0.075	0.297	1.864	0.599
2.4	0.057	0.245	1.819	0.532
2.6	0.044	0.201	1.768	0.468
2.8	0.033	0.163	1.709	0.406
3.0	0.025	0.131	1.645	0.349
3.2	0.019	0.105	1.575	0.297
3.4	0.014	0.083	1.501	0.250
3.6	0.010	0.066	1.423	0.208
3.8	0.008	0.051	1.343	0.172
4.0	0.006	0.040	1.262	0.141
4.2	0.004	0.031	1.180	0.115
4.4	0.003	0.024	1.099	0.093
4.6	0.002	0.018	1.018	0.074
4.8	0.002	0.014	0.941	0.059
5.0	0.001	0.010	0.866	0.047
5.2	0.001	0.008	0.793	0.037
5.4	0.001	0.006	0.724	0.029
5.6	0.000	0.004	0.659	0.023
5.8	......	0.003	0.598	0.017

TABLE 27-5. ANHARMONIC-OSCILLATOR FUNCTIONS (*Continued*)

u	$\dfrac{u^2}{(e^u - 1)^2}$	$\dfrac{u^2(2ue^u - e^u + 1)}{(e^u - 1)^3}$	$\dfrac{u^3e^u(e^u + 1)}{(e^u - 1)^3}$	$\dfrac{u^3e^u(2ue^u - 2e^u + u + 2)}{(e^u - 1)^4}$
6.0	......	0.002	0.541	0.014
6.5	......	0.001	0.415	0.007
7.0	......	0.001	0.314	0.004
7.5	......	0.000	0.234	0.002
8.0	......	......	0.172	0.001
8.5	......	......	0.125	0.000
9.0	......	......	0.090	
9.5	......	......	0.065	
10.0	......	......	0.045	

Table 27-6 gives an illustration of the complete calculation for N_2 at $2000°K$.

TABLE 27-6. SAMPLE CALCULATION OF THE THERMODYNAMIC PROPERTIES OF A DIATOMIC MOLECULE INCLUDING ANHARMONICITY EFFECTS N_2 at $2000°K$. See Table 27-4 for molecular data

	$\dfrac{-(F° - H_0°)}{T}$	$c_P°$
Translation, $M = 28.016$.................	40.413	4.968
Rotation (rigid), $y = 0.001439$.............	11.627	1.987
Vibration (harmonic), $u = 1.6766$.........	0.411	1.580
Corrections:		
Vibration anharmonicity, $x = $ ███.*0.006.13*	0.002	0.023
Vibration-rotation, $\delta = 0.00935$..........	0.004	0.021
Rotation stretching, $\gamma = 0.000852$........	0.008	0.016
Total...........................	52.465	8.595

(handwritten annotations in margin: 8.535; $JanaF\ 8.601$; $.006$)

Polyatomic Molecules: Higher Approximations.

The energy levels of a polyatomic molecule deviate from the pattern calculated on the rigid-rotator–harmonic-oscillator model for exactly the same reasons that were mentioned for diatomic molecules in the preceding section. However, the number and variety of terms are very great for the general case. Consequently, it is preferable to develop the appropriate formulas for various classes of molecules. Also there is much less complete spectroscopic information with respect to these higher terms for polyatomic than for diatomic molecules. Pennington and Kobe[1] and Woolley[2] have presented formulas and tables for the thermodynamic functions which

[1] R. E. Pennington and K. A. Kobe, *J. Chem. Phys.*, **22**, 1442 (1954).
[2] H. W. Woolley, *J. Research Natl. Bur. Standards*, **56**, 105 (1956).

correspond to a sufficiently complete energy-level formula for molecules such as carbon dioxide, nitrous oxide, acetylene, or water. These higher approximations have been included for the molecules just mentioned and a few others, but the thermodynamic data for most polyatomic molecules are limited to the rigid-rotator–harmonic-oscillator approximation. There is one additional type of motion, however, which arises very commonly in moderately large molecules and therefore deserves attention at this point.

Internal Rotation. If a molecule contains groups of atoms connected by single electron-pair bonds, there arise internal rotational motions. Simple examples are H_3C—CH_3 and H_2N—OH, where the rotations are CH_3 versus CH_3 and NH_2 versus OH, respectively. Since the rotating groups are relatively close to one another, it is reasonable that there should be a significant change of potential energy (and therefore a force) associated with the internal rotational coordinate. We may divide internal-rotation problems into three categories depending on the magnitude of this potential.

If the potential is very large in comparison with kT, then we have a torsional vibration which may be treated like any other vibration. Twisting about a double bond, as in $H_2C\!\!=\!\!CH_2$, normally falls in this category.

If the potential change with angle of rotation is very small as compared with kT, then it may be ignored and we have a free rotation. The energy levels, from quantum mechanics, are

$$\epsilon = \frac{h^2K^2}{8\pi^2 I_r} \tag{27-55}$$

where K is a quantum number that can have all integral values positive and negative including zero and I_r is the reduced moment of inertia for the rotation, which will be defined further a little later. If one or both of the rotating groups has symmetrically placed atoms, then only certain values of K are allowed. It is always found that $1/n$th of the K values is allowed if there are n equivalent orientations. Thus the symmetry number for internal rotation n plays an entirely equivalent role to that of the symmetry number for over-all rotation. The partition function for free rotation is then

$$Q_f = \frac{1}{n} \sum_{K=-\infty}^{+\infty} e^{-K^2 h^2 / 8\pi^2 I_r kT}$$

We again find that it is an appropriate approximation to replace the sum by an integral.

$$Q_f = \frac{1}{n} \int_{-\infty}^{+\infty} e^{-K^2 h^2 / 8\pi^2 I_r kT} \, dK$$

$$= \frac{1}{n} \left(\frac{8\pi^3 I_r kT}{h^2} \right)^{\frac{1}{2}}$$

$$= \frac{2.7935}{n} \, (10^{38} I_r T)^{\frac{1}{2}} \tag{27-56}$$

The thermodynamic functions can be readily obtained from this partition function. As expected, we find $c = \frac{1}{2}R$. One of the few examples where free internal rotation has been observed is dimethyl cadmium, H_3C—Cd—CH_3. Here the rotating groups are separated by two bond lengths instead of one. Table 27-7 gives the numerical results for this substance.

TABLE 27-7. ENTROPY OF CADMIUM DIMETHYL, AN EXAMPLE OF FREE INTERNAL ROTATION,[†] CAL/DEG MOLE

Translation and over-all rotation	60.66
Vibration	8.76
Free internal rotation	2.93
Total, calculated	72.35
Experimental, from third law	72.40 ± 0.20

[†] J. C. M. Li, *J. Am. Chem. Soc.*, **78**, 1081 (1956).

The more common and more complex case of internal rotation involves a potential barrier which is neither very large nor very small as compared with kT. In this case one must seek the energy levels for an appropriate molecular model. Since we have a rotational coordinate, the potential energy must be a periodic function with period $2\pi/n$, where n is the number of times per revolution that the molecule returns to an equivalent position. Thus for ethane $n = 3$. We can expand the potential in a Fourier series, and it is found[1] that the first two terms suffice to give an excellent approximation.

$$V = c_0 + c_1 \cos n\phi + \cdots$$

$$= \frac{1}{2} V_0 (1 - \cos n\phi) \tag{27-57}$$

In the second line the expression is arranged so that the potential is zero when the angle ϕ is zero and the height of the potential peaks is V_0.

The quantum-mechanical problem of a rotation subject to this potential barrier has been solved, but no simple expression can be given for the energy levels. Rather the levels must be obtained from infinite but convergent continued fractions or some equivalent expressions. Also there

---

[1] K. S. Pitzer, *Discussions Faraday Soc.*, **1951**, no. 10, p. 66; K. S. Pitzer and J. L. Hollenberg, *J. Am. Chem. Soc.*, **75**, 2219 (1953); D. R. Herschbach, *J. Chem. Phys.*, **27**, 1420 (1957).

are complexities arising from the relationships between the rotation of the entire molecule and the internal rotation. It is beyond the scope of this presentation to discuss all these problems.

Fortunately, it is found[1] that the final results are relatively simple and that the contribution of an internal rotation to the thermodynamic properties can be presented in general tables as the functions of two variables. There are alternate choices for the two variables;[2] we shall take the ratio of the potential barrier to thermal energy, V_0/RT, and the reciprocal of the partition function which would apply if the rotation were free, $1/Q_f$.

Also we must further define the reduced moment of inertia. If the molecule consists of a pair of symmetrical coaxial tops, such as H_3C—CH_3 or H_3C—CCl_3, then

$$I_r = \frac{AB}{A + B} \tag{27-58}$$

where A and B are the moments of inertia of the respective tops about the axis of internal rotation. If there is one such symmetrical top attached to a less symmetrical structure, as in the case of H_3C—CH=CH_2, one may write

$$I_r = A \left(1 - \sum_{i=1}^{3} \frac{\alpha_i^2 A}{I_i} \right) \tag{27-59}$$

where A is the moment of inertia of the symmetrical top and the sum covers the three principal axes for over-all rotation. Then α_i is the direction cosine between the axis of internal rotation and the ith principal axis of the entire molecule, and I_i is the moment of inertia for over-all rotation about the ith axis.

The thermodynamic functions are given in Tables 27-8 to 27-13. It is more convenient to give a portion of the entropy and free-energy tables in terms of the difference between the function for restricted rotation and for free rotation. The formulas for the free-rotation values are given in the table headings. These tables are also applicable to even less symmetrical cases than have been mentioned so far. However, a variety of factors must be considered in each case; consequently the reader is referred to the appropriate papers in the original literature.[3]

[1] K. S. Pitzer and W. D. Gwinn, *J. Chem. Phys.*, **10**, 428 (1942); K. S. Pitzer, *J. Chem. Phys.*, **5**, 469 (1937); J. C. M. Li and K. S. Pitzer, *J. Phys. Chem.*, **60**, 466 (1956).

[2] K. S. Pitzer, *J. Chem. Phys.*, **5**, 469 (1937).

[3] K. S. Pitzer, *J. Chem. Phys.*, **14**, 239 (1946); J. E. Kilpatrick and K. S. Pitzer, *J. Chem. Phys.*, **17**, 1064 (1949).

TABLE 27-8. HEAT CAPACITY c, CAL/DEG

$1/Q_f$

V/RT	0.0	0.05	0.10	0.15	0.20	0.25	0.30	0.35	0.40	0.45	0.50	0.55	0.60	0.65	0.70	0.75	0.80	0.85	0.90	0.95
0.0	0.994	0.994	0.994	0.994	0.994	0.994	0.994	0.994	0.994	0.994	0.994	0.994	0.994	0.994	0.994	0.994	0.994	0.994	0.994	0.994
0.2	1.0035	1.003	1.003	1.002	1.001	1.000	0.999	0.998	0.998	0.998	1.000	1.000	1.000	1.000	1.000	1.000	1.000	0.999	0.999	0.999
0.4	1.0328	1.033	1.032	1.030	1.028	1.025	1.024	1.021	1.019	1.017	1.018	1.017	1.015	1.013	1.012	1.010	1.008	1.007	1.007	1.004
0.6	1.0801	1.080	1.079	1.076	1.073	1.068	1.065	1.060	1.056	1.051	1.049	1.046	1.041	1.036	1.031	1.026	1.021	1.017	1.014	1.011
0.8	1.1435	1.143	1.141	1.138	1.133	1.128	1.121	1.114	1.106	1.099	1.092	1.084	1.075	1.067	1.058	1.049	1.040	1.031	1.025	1.020
1.0	1.2203	1.219	1.217	1.212	1.206	1.199	1.190	1.180	1.169	1.157	1.144	1.131	1.118	1.105	1.091	1.078	1.065	1.052	1.040	1.031
1.5	1.4508	1.449	1.444	1.435	1.423	1.408	1.391	1.370	1.348	1.324	1.299	1.273	1.247	1.218	1.192	1.165	1.141	1.115	1.090	1.070
2.0	1.6778	1.695	1.687	1.673	1.655	1.632	1.606	1.574	1.541	1.505	1.465	1.424	1.382	1.341	1.300	1.258	1.218	1.180	1.146	1.113
2.5	1.9213	1.917	1.908	1.888	1.866	1.840	1.801	1.756	1.717	1.670	1.619	1.562	1.504	1.448	1.393	1.341	1.289	1.238	1.190	1.146
3.0	2.0989	2.095	2.082	2.062	2.033	1.996	1.952	1.900	1.846	1.794	1.732	1.663	1.597	1.532	1.466	1.401	1.337	1.276	1.217	1.164
3.5	2.2226	2.218	2.204	2.180	2.146	2.106	2.054	1.995	1.934	1.869	1.803	1.727	1.654	1.580	1.506	1.432	1.361	1.293	1.226	1.165
4.0	2.2989	2.294	2.276	2.249	2.213	2.168	2.110	2.048	1.980	1.907	1.834	1.754	1.674	1.593	1.513	1.435	1.359	1.286	1.215	1.148
4.5	2.3358	2.330	2.312	2.280	2.238	2.190	2.129	2.062	1.990	1.911	1.832	1.749	1.664	1.578	1.496	1.413	1.333	1.259	1.185	1.115
5.0	2.3447	2.338	2.318	2.285	2.241	2.186	2.120	2.056	1.972	1.890	1.808	1.718	1.631	1.543	1.457	1.373	1.292	1.214	1.140	1.068
6.0	2.3158	2.307	2.283	2.245	2.192	2.130	2.059	1.979	1.893	1.803	1.711	1.614	1.520	1.429	1.342	1.255	1.173	1.096	1.022	0.954
7.0	2.2650	2.256	2.228	2.185	2.126	2.055	1.973	1.883	1.787	1.688	1.588	1.487	1.390	1.296	1.207	1.120	1.040	0.962	0.890	0.826
8.0	2.2160	2.205	2.174	2.125	2.058	1.979	1.888	1.788	1.684	1.576	1.468	1.366	1.262	1.164	1.074	0.988	0.908	0.834	0.765	0.704
9.0	2.1762	2.164	2.130	2.074	1.999	1.909	1.808	1.699	1.587	1.474	1.362	1.250	1.144	1.048	0.956	0.869	0.789	0.717	0.652	0.593
10.0	2.1457	2.133	2.094	2.033	1.951	1.854	1.745	1.630	1.507	1.382	1.262	1.151	1.045	0.943	0.850	0.765	0.688	0.618	0.556	0.499
12.0	2.1053	2.089	2.043	1.972	1.877	1.763	1.636	1.502	1.365	1.233	1.107	0.989	0.877	0.774	0.682	0.600	0.528	0.463	0.407	0.358
14.0	2.0813	2.063	2.009	1.923	1.814	1.686	1.546	1.400	1.254	1.112	0.978	0.855	0.744	0.644	0.554	0.479	0.411	0.352	0.303	0.262
16.0	2.0657	2.044	1.983	1.887	1.764	1.622	1.468	1.311	1.156	1.009	0.873	0.749	0.639	0.542	0.457	0.387	0.324	0.272	0.229	0.194
18.0	2.0547	2.031	1.961	1.853	1.717	1.562	1.397	1.232	1.070	0.919	0.780	0.657	0.549	0.456	0.378	0.312	0.259	0.215	0.175	0.144
20.0	2.0465	2.020	1.944	1.827	1.678	1.510	1.333	1.158	0.991	0.837	0.701	0.580	0.477	0.389	0.316	0.256	0.208	0.168	0.135	0.109

441

TABLE 27-9. HEAT CONTENT $(H_T - H_0)/T$, CAL/DEG

$1/Q_f$

V/RT	0.0	0.05	0.10	0.15	0.20	0.25	0.30	0.35	0.40	0.45	0.50	0.55	0.60	0.65	0.70	0.75	0.80	0.85	0.90	0.95
0.0	0.994	0.994	0.994	0.994	0.994	0.994	0.994	0.994	0.994	0.994	0.994	0.994	0.994	0.994	0.994	0.994	0.994	0.994	0.994	0.994
0.2	1.1824	1.142	1.106	1.074	1.050	1.032	1.022	1.015	1.008	1.004	1.000	0.996	0.994	0.994	0.994	0.992	0.992	0.991	0.990	0.989
0.4	1.3515	1.300	1.249	1.200	1.151	1.106	1.073	1.051	1.036	1.025	1.015	1.006	0.999	0.994	0.992	0.990	0.990	0.988	0.986	0.985
0.6	1.5013	1.437	1.374	1.311	1.251	1.190	1.138	1.099	1.072	1.049	1.030	1.014	1.004	0.995	0.990	0.987	0.984	0.982	0.980	0.979
0.8	1.6326	1.556	1.482	1.411	1.340	1.272	1.211	1.157	1.114	1.077	1.048	1.026	1.009	0.996	0.984	0.980	0.976	0.974	0.972	0.971
1.0	1.7463	1.660	1.576	1.495	1.418	1.344	1.275	1.211	1.155	1.106	1.065	1.038	1.014	0.996	0.982	0.972	0.965	0.962	0.960	0.959
1.5	1.9610	1.856	1.753	1.654	1.561	1.472	1.385	1.306	1.230	1.164	1.103	1.059	1.019	0.987	0.962	0.945	0.932	0.922	0.916	0.915
2.0	2.0937	1.971	1.854	1.742	1.636	1.536	1.440	1.350	1.265	1.190	1.120	1.057	1.005	0.962	0.928	0.904	0.886	0.873	0.864	0.860
2.5	2.1660	2.031	1.900	1.779	1.662	1.550	1.448	1.351	1.260	1.179	1.104	1.032	0.972	0.922	0.882	0.850	0.827	0.811	0.801	0.796
3.0	2.1974	2.049	1.909	1.777	1.651	1.535	1.426	1.321	1.224	1.140	1.060	0.988	0.924	0.870	0.828	0.791	0.763	0.744	0.732	0.728
3.5	2.2033	2.043	1.893	1.753	1.621	1.497	1.382	1.275	1.176	1.088	1.006	0.933	0.868	0.811	0.765	0.727	0.697	0.676	0.663	0.659
4.0	2.1947	2.024	1.864	1.715	1.577	1.448	1.329	1.221	1.121	1.030	0.947	0.872	0.806	0.749	0.701	0.661	0.630	0.609	0.595	0.590
4.5	2.1791	1.998	1.829	1.673	1.529	1.394	1.273	1.162	1.061	0.968	0.884	0.810	0.744	0.687	0.638	0.599	0.567	0.545	0.531	0.526
5.0	2.1610	1.971	1.794	1.631	1.481	1.344	1.218	1.104	1.002	0.909	0.824	0.750	0.685	0.628	0.580	0.540	0.508	0.485	0.470	0.465
6.0	2.1264	1.918	1.727	1.552	1.392	1.247	1.115	0.999	0.893	0.799	0.714	0.644	0.580	0.523	0.476	0.437	0.406	0.383	0.368	0.361
7.0	2.0987	1.875	1.670	1.484	1.315	1.164	1.029	0.908	0.802	0.708	0.624	0.554	0.491	0.437	0.392	0.354	0.324	0.302	0.286	0.279
8.0	2.0784	1.840	1.623	1.427	1.251	1.095	0.955	0.833	0.725	0.631	0.549	0.480	0.420	0.368	0.326	0.290	0.261	0.239	0.223	0.215
9.0	2.0637	1.811	1.583	1.379	1.196	1.035	0.892	0.768	0.661	0.569	0.488	0.421	0.363	0.312	0.273	0.240	0.211	0.191	0.176	0.168
10.0	2.0529	1.787	1.548	1.335	1.147	0.982	0.838	0.715	0.608	0.515	0.437	0.370	0.314	0.269	0.231	0.200	0.174	0.154	0.140	0.132
12.0	2.0385	1.749	1.492	1.264	1.067	0.896	0.745	0.624	0.519	0.431	0.356	0.296	0.244	0.202	0.170	0.143	0.121	0.104	0.091	0.084
14.0	2.0295	1.717	1.441	1.202	0.997	0.823	0.672	0.551	0.450	0.365	0.297	0.240	0.195	0.158	0.127	0.103	0.084	0.072	0.062	0.056
16.0	2.0232	1.690	1.401	1.150	0.937	0.760	0.613	0.493	0.394	0.314	0.249	0.198	0.157	0.127	0.098	0.076	0.061	0.051	0.044	0.038
18.0	2.0185	1.663	1.363	1.102	0.886	0.707	0.561	0.443	0.347	0.271	0.211	0.164	0.128	0.099	0.077	0.060	0.047	0.036	0.029	0.026
20.0	2.0150	1.646	1.329	1.061	0.841	0.660	0.515	0.399	0.307	0.236	0.181	0.138	0.105	0.080	0.061	0.047	0.036	0.028	0.022	0.018

TABLE 27-10. FREE ENERGY $-f/T$, CAL/DEG

V/RT	$1/Q_f$														
	0.25	0.30	0.35	0.40	0.45	0.50	0.55	0.60	0.65	0.70	0.75	0.80	0.85	0.90	0.95
0.0	2.754	2.392	2.086	1.821	1.587	1.377	1.190	1.014	0.856	0.710	0.575	0.443	0.323	0.208	0.102
0.2	2.710	2.359	2.061	1.803	1.574	1.368	1.182	1.009	0.852	0.707	0.570	0.441	0.321	0.207	0.101
0.4	2.623	2.296	2.014	1.765	1.543	1.342	1.164	0.997	0.842	0.699	0.565	0.438	0.318	0.206	0.099
0.6	2.518	2.208	1.944	1.708	1.498	1.309	1.136	0.974	0.826	0.687	0.555	0.431	0.315	0.204	0.097
0.8	2.406	2.106	1.856	1.636	1.442	1.266	1.099	0.947	0.804	0.670	0.543	0.424	0.310	0.200	0.096
1.0	2.296	2.004	1.764	1.559	1.379	1.214	1.056	0.912	0.777	0.647	0.526	0.411	0.302	0.195	0.094
1.5	2.040	1.770	1.548	1.370	1.210	1.069	0.937	0.815	0.700	0.588	0.481	0.379	0.277	0.178	0.084
2.0	1.819	1.563	1.360	1.193	1.052	0.927	0.817	0.713	0.615	0.521	0.428	0.338	0.249	0.160	0.074
2.5	1.630	1.389	1.197	1.043	0.912	0.802	0.705	0.616	0.534	0.454	0.375	0.298	0.219	0.141	0.063
3.0	1.473	1.240	1.059	0.914	0.793	0.695	0.608	0.530	0.458	0.390	0.324	0.258	0.191	0.122	0.053
3.5	1.340	1.117	0.943	0.802	0.694	0.603	0.525	0.457	0.395	0.336	0.278	0.222	0.165	0.105	0.042
4.0	1.225	1.013	0.847	0.713	0.613	0.527	0.455	0.393	0.339	0.288	0.239	0.190	0.140	0.088	0.034
4.5	1.133	0.925	0.764	0.637	0.543	0.463	0.398	0.340	0.290	0.247	0.205	0.162	0.117	0.074	0.027
5.0	1.053	0.849	0.696	0.577	0.483	0.408	0.347	0.297	0.253	0.214	0.177	0.139	0.102	0.063	0.020
6.0	0.919	0.728	0.586	0.477	0.393	0.325	0.273	0.230	0.193	0.161	0.131	0.103	0.074	0.045	0.012
7.0	0.819	0.636	0.503	0.402	0.325	0.267	0.218	0.181	0.149	0.123	0.100	0.078	0.056	0.032	0.008
8.0	0.735	0.564	0.440	0.346	0.275	0.221	0.179	0.145	0.118	0.096	0.078	0.060	0.042	0.024	0.005
9.0	0.667	0.504	0.388	0.300	0.235	0.186	0.149	0.120	0.095	0.078	0.062	0.047	0.032	0.019	0.004
10.0	0.610	0.456	0.345	0.264	0.203	0.159	0.124	0.100	0.079	0.063	0.049	0.037	0.026	0.015	0.002
12.0	0.521	0.380	0.280	0.209	0.157	0.120	0.092	0.071	0.054	0.042	0.033	0.025	0.018	0.010	0.001
14.0	0.452	0.321	0.232	0.169	0.124	0.092	0.069	0.052	0.038	0.030	0.023	0.016	0.012	0.007	0.000
16.0	0.396	0.276	0.195	0.139	0.100	0.072	0.053	0.039	0.028	0.021	0.016	0.012	0.008	0.004	0.000
18.0	0.351	0.240	0.166	0.117	0.082	0.058	0.042	0.030	0.022	0.016	0.012	0.008	0.006	0.003	0.000
20.0	0.315	0.211	0.144	0.098	0.068	0.047	0.033	0.023	0.017	0.012	0.009	0.006	0.004	0.002	0.000

Table 27-14 gives an example of the calculations for a restricted internal rotation. The moments of inertia are calculated for rotation of the entire rigid molecule, i.e., as if the internal rotation were frozen. Other parameters are obtained as described in this and previous sections. In this example,[1] CH_3CCl_3, the product of the principal moments of inertia for the rigid molecule is $D = 6.14 \times 10^{-113}$ g cm² while the reduced moment for internal rotation is 5.25×10^{-40} g cm². The potential barrier to internal rotation was found from the infrared spectrum to be well fitted by Eq. (27-57) with $V_0 = 2967$ cal/mole. It is easier to interpolate in Table 27-13 than in Table 27-12, although either may be used. The results are given in Table 27-14.

Electronic States for Molecules. A large majority of molecules of importance at room temperature have electronic states in which the electrons are paired in a fashion to yield zero electron spin and zero orbital angular momentum. Thus the electronic state is a single quan-

[1] K. S. Pitzer and J. L. Hollenberg, *J. Am. Chem. Soc.*, **75**, 2219 (1953).

TABLE 27-11. FREE-ENERGY INCREASE FROM FREE ROTATION, $(F - F_f)/T$, CAL/DEG

$$F_f/T = -R \ln Q_f$$

V/RT	1/Q_f											
	0	0.05	0.10	0.15	0.20	0.25	0.30	0.35	0.40	0.45	0.50	0.55
0.0	0.0000	0.000	0.000	0.000	0.000	0.000	0.000	0.000	0.000	0.000	0.000	0.000
0.2	0.1937	0.154	0.117	0.085	0.061	0.044	0.033	0.025	0.018	0.013	0.009	0.005
0.4	0.3776	0.326	0.274	0.225	0.176	0.131	0.096	0.072	0.056	0.044	0.035	0.026
0.6	0.5516	0.489	0.424	0.361	0.298	0.236	0.184	0.142	0.113	0.089	0.068	0.054
0.8	0.7161	0.640	0.566	0.493	0.420	0.348	0.286	0.230	0.185	0.145	0.111	0.088
1.0	0.8711	0.784	0.699	0.617	0.537	0.461	0.389	0.322	0.262	0.208	0.163	0.129
1.5	1.2200	1.114	1.010	0.909	0.809	0.714	0.622	0.538	0.451	0.375	0.308	0.250
2.0	1.5182	1.395	1.276	1.159	1.045	0.935	0.829	0.726	0.628	0.535	0.450	0.371
2.5	1.7724	1.635	1.501	1.371	1.246	1.124	1.004	0.889	0.778	0.675	0.575	0.479
3.0	1.9893	1.839	1.693	1.552	1.415	1.282	1.152	1.027	0.907	0.794	0.682	0.576
3.5	2.1756	2.013	1.856	1.704	1.557	1.414	1.275	1.143	1.019	0.893	0.774	0.660
4.0	2.3366	2.163	1.996	1.833	1.676	1.525	1.379	1.239	1.108	0.974	0.850	0.732
4.5	2.4772	2.293	2.117	1.945	1.780	1.621	1.467	1.322	1.184	1.044	0.914	0.791
5.0	2.6012	2.408	2.221	2.042	1.868	1.703	1.543	1.392	1.244	1.104	0.969	0.841
6.0	2.8108	2.599	2.396	2.202	2.015	1.836	1.664	1.500	1.344	1.194	1.052	0.916
7.0	2.9833	2.755	2.537	2.328	2.129	1.936	1.757	1.583	1.418	1.262	1.111	0.971
8.0	3.1294	2.886	2.653	2.432	2.220	2.020	1.828	1.646	1.474	1.312	1.157	1.011
9.0	3.2563	2.998	2.753	2.520	2.298	2.087	1.888	1.698	1.520	1.351	1.192	1.039
10.0	3.3686	3.097	2.839	2.594	2.362	2.144	1.936	1.741	1.557	1.383	1.219	1.063
12.0	3.5602	3.263	2.982	2.718	2.468	2.233	2.013	1.806	1.612	1.429	1.258	1.096
14.0	3.7205	3.400	3.099	2.816	2.551	2.303	2.071	1.854	1.651	1.462	1.285	1.119
16.0	3.8584	3.517	3.197	2.897	2.618	2.358	2.116	1.891	1.682	1.486	1.305	1.135
18.0	3.9793	3.618	3.280	2.965	2.674	2.403	2.152	1.920	1.704	1.505	1.319	1.146
20.0	4.0872	3.707	3.353	3.024	2.720	2.440	2.181	1.942	1.722	1.519	1.331	1.155

tum state, and there is no electronic contribution to the thermodynamic properties at moderate temperatures. Exceptions among common substances are O_2, NO, and NO_2. Each of these three presents a slightly different situation.

Oxygen in its lowest energy form is said to have a $^3\Sigma_g^-$ electronic state. For the complete meaning of this symbol the reader should consult a text on molecular spectra. For our purpose it is sufficient to know that the symbol Σ means that there is no orbital angular momentum, while the preceding superscript 3 means that it is a triplet state with respect to electron spin; i.e., there are two spins coupled parallel, yielding a spin vector $S = 1$ and multiplicity $2S + 1 = 3$. While the electron spin

TABLE 27-12. ENTROPY s, CAL/DEG

V/RT	$1/Q_f$														
	0.25	0.30	0.35	0.40	0.45	0.50	0.55	0.60	0.65	0.70	0.75	0.80	0.85	0.90	0.95
0.0	3.748	3.386	3.079	2.814	2.580	2.371	2.182	2.009	1.850	1.703	1.567	1.438	1.316	1.203	1.097
0.2	3.743	3.382	3.076	2.811	2.578	2.369	2.180	2.003	1.848	1.701	1.563	1.433	1.312	1.196	1.091
0.4	3.730	3.370	3.065	2.801	2.568	2.359	2.170	1.996	1.837	1.691	1.555	1.428	1.307	1.193	1.085
0.6	3.709	3.347	3.043	2.780	2.547	2.340	2.151	1.980	1.823	1.677	1.541	1.415	1.295	1.184	1.076
0.8	3.679	3.318	3.013	2.750	2.519	2.315	2.125	1.957	1.800	1.654	1.523	1.399	1.284	1.171	1.068
1.0	3.638	3.279	2.974	2.714	2.485	2.279	2.094	1.928	1.774	1.629	1.499	1.377	1.262	1.153	1.052
1.5	3.512	3.156	2.854	2.600	2.376	2.173	1.997	1.833	1.685	1.552	1.428	1.310	1.201	1.094	1.000
2.0	3.355	3.004	2.709	2.458	2.241	2.048	1.874	1.718	1.578	1.450	1.332	1.224	1.122	1.024	0.936
2.5	3.180	2.836	2.548	2.303	2.091	1.907	1.739	1.589	1.456	1.335	1.224	1.126	1.031	0.942	0.860
3.0	3.008	2.667	2.380	2.138	1.933	1.756	1.576	1.456	1.330	1.217	1.114	1.021	0.936	0.855	0.779
3.5	2.838	2.500	2.218	1.978	1.782	1.610	1.458	1.323	1.206	1.100	1.004	0.919	0.841	0.769	0.703
4.0	2.678	2.343	2.069	1.834	1.643	1.475	1.328	1.199	1.087	0.988	0.901	0.821	0.748	0.683	0.623
4.5	2.528	2.199	1.926	1.698	1.511	1.348	1.209	1.086	0.978	0.884	0.804	0.730	0.662	0.607	0.551
5.0	2.396	2.068	1.798	1.579	1.392	1.233	1.097	0.982	0.881	0.794	0.716	0.648	0.588	0.535	0.486
6.0	2.166	1.844	1.585	1.370	1.192	1.040	0.915	0.808	0.715	0.637	0.568	0.509	0.457	0.412	0.372
7.0	1.983	1.665	1.411	1.204	1.033	0.891	0.774	0.672	0.588	0.516	0.453	0.401	0.357	0.319	0.285
8.0	1.830	1.519	1.272	1.071	0.906	0.770	0.660	0.566	0.486	0.422	0.366	0.320	0.281	0.248	0.220
9.0	1.703	1.397	1.156	0.962	0.804	0.674	0.570	0.483	0.407	0.350	0.300	0.258	0.223	0.195	0.171
10.0	1.593	1.295	1.060	0.872	0.719	0.596	0.496	0.414	0.348	0.293	0.248	0.211	0.180	0.154	0.134
12.0	1.417	1.125	0.904	0.728	0.588	0.476	0.388	0.315	0.255	0.213	0.176	0.146	0.122	0.101	0.084
14.0	1.275	0.994	0.783	0.620	0.492	0.388	0.309	0.247	0.196	0.157	0.126	0.100	0.084	0.069	0.056
16.0	1.157	0.890	0.688	0.533	0.414	0.322	0.251	0.196	0.155	0.119	0.092	0.075	0.059	0.048	0.038
18.0	1.058	0.801	0.609	0.464	0.353	0.270	0.205	0.158	0.121	0.093	0.072	0.056	0.042	0.034	0.026
20.0	0.975	0.727	0.542	0.405	0.303	0.228	0.170	0.129	0.097	0.073	0.056	0.042	0.032	0.024	0.018

introduces many complexities into the detailed spectrum, the net effect is to yield just an additional multiplicity factor of 3 for all rotational energy levels except the very lowest one, which remains a single level. Thus at all but the very lowest temperatures the effect of the electron spin is to add $R \ln 3$ to the entropy but to make no change in either enthalpy or heat capacity. These results apply in the absence of a magnetic field; the net electron spin has a major effect on the magnetic susceptibility.

At very high temperatures one must consider the excited electronic states of oxygen which are designated as $^1\Delta_g$ and $^1\Sigma_g^+$ and which lie 7882 and 13,121 cm^{-1}, respectively, above the ground state. The symbol $^1\Sigma_g^+$ implies the normal type of single state, but the symbol $^1\Delta_g$ indicates 2 units of orbital angular momentum ($\Lambda = 2$) and no net spin angular momentum. Again there are various detailed effects, but in the absence of a magnetic field the net result is just a multiplicity factor of 2 for all of the vibration-rotation states of the $^1\Delta_g$ electronic state.

TABLE 27-13. ENTROPY DECREASE FROM FREE ROTATION, $s_f - s$, CAL/DEG

$$s_f = R(\tfrac{1}{2} + \ln Q_f)$$

V/RT	1/Qf											
	0.0	0.05	0.10	0.15	0.20	0.25	0.30	0.35	0.40	0.45	0.50	0.55
0.0	0.0000	0.000	0.000	0.000	0.000	0.000	0.000	0.000	0.000	0.000	0.000	0.000
0.2	0.0049	0.005	0.004	0.004	0.004	0.004	0.004	0.003	0.003	0.002	0.002	0.002
0.4	0.0198	0.020	0.018	0.018	0.018	0.018	0.016	0.014	0.013	0.012	0.012	0.010
0.6	0.0440	0.044	0.043	0.043	0.040	0.039	0.039	0.036	0.034	0.033	0.031	0.028
0.8	0.0771	0.077	0.077	0.075	0.072	0.069	0.068	0.066	0.064	0.061	0.056	0.053
1.0	0.1185	0.118	0.117	0.115	0.112	0.110	0.107	0.105	0.100	0.095	0.092	0.086
1.5	0.2527	0.252	0.250	0.248	0.242	0.236	0.230	0.225	0.214	0.204	0.198	0.189
2.0	0.4182	0.417	0.415	0.410	0.402	0.393	0.382	0.370	0.356	0.339	0.323	0.308
2.5	0.6001	0.599	0.594	0.585	0.577	0.568	0.550	0.531	0.511	0.489	0.464	0.440
3.0	0.7856	0.783	0.777	0.768	0.757	0.740	0.719	0.699	0.676	0.647	0.615	0.581
3.5	0.9660	0.964	0.957	0.944	0.929	0.910	0.886	0.861	0.836	0.798	0.761	0.722
4.0	1.1356	1.133	1.126	1.111	1.094	1.070	1.043	1.011	0.980	0.937	0.896	0.855
4.5	1.2918	1.289	1.280	1.265	1.244	1.220	1.187	1.153	1.116	1.069	1.023	0.977
5.0	1.4339	1.431	1.421	1.404	1.380	1.352	1.318	1.281	1.235	1.188	1.138	1.086
6.0	1.6781	1.674	1.662	1.643	1.616	1.582	1.542	1.494	1.444	1.388	1.331	1.268
7.0	1.8783	1.874	1.860	1.837	1.807	1.765	1.721	1.668	1.610	1.547	1.480	1.411
8.0	2.0447	2.040	2.024	1.998	1.962	1.918	1.867	1.807	1.743	1.674	1.601	1.525
9.0	2.1864	2.180	2.163	2.134	2.095	2.045	1.989	1.923	1.852	1.776	1.697	1.612
10.0	2.3095	2.303	2.284	2.252	2.208	2.155	2.091	2.019	1.942	1.861	1.775	1.686
12.0	2.5155	2.508	2.485	2.447	2.394	2.331	2.261	2.175	2.086	1.992	1.895	1.793
14.0	2.6847	2.676	2.650	2.607	2.547	2.473	2.392	2.296	2.194	2.088	1.983	1.872
16.0	2.8289	2.819	2.788	2.740	2.674	2.591	2.496	2.391	2.281	2.166	2.049	1.930
18.0	2.9545	2.943	2.910	2.855	2.781	2.690	2.585	2.470	2.350	2.227	2.101	1.976
20.0	3.0659	3.054	3.017	2.956	2.872	2.773	2.659	2.537	2.409	2.277	2.143	2.011

The distribution of the oxygen among the three electronic states is governed by the thermodynamic properties of the three states. If the vibrational and rotational properties were the same, one could use the multiplicities of 3, 2, and 1 together with the energy differences to calculate the distribution. This electronic partition function would be

$$Q_{el} = 3 + 2e^{-7882hc/kT} + e^{-13,121hc/kT}$$

In the case of oxygen, however, this calculation would not be very accurate, for we see from Table 27-4 that the molecular constants are appreciably different in the three states. An accurate calculation would require the treatment of each electronic state as a distinct species, with a

TABLE 27-14. THERMODYNAMIC FUNCTIONS FOR METHYL CHLOROFORM, AN
EXAMPLE OF RESTRICTED INTERNAL ROTATION
$T = 286.53°K$, cal/deg mole

$S°$

Translation + rotation............	65.98
Vibration......................	8.07
Internal rotation................	2.16
Total, calculated..............	76.21
Experimental†..................	76.22 ± 0.16

† T. R. Rubin, B. H. Levedahl, and D. M. Yost, *J. Am. Chem. Soc.*, **66,** 279 (1944).

sum of the complete partition functions for each of the states yielding the total population.

The case of NO_2 is much simpler. Here we have just double multiplicity because of the one unpaired electron spin. This adds $R \ln 2$ to the entropy and $-(F° - H_0°)/T$ functions but has no effect on heat capacity or enthalpy.

The NO molecule represents a very interesting case. It has both orbital and spin angular momentum—1 unit of the former and ½ unit of the latter. This yields the pair of states $^2\Pi_{1/2}$ and $^2\Pi_{3/2}$, which are separated by only 120.9 cm^{-1} in energy.

The complete theory for this case is somewhat more complex, but the basic ideas are just those already illustrated.

One must always base a calculation of this type on the complete internal partition function for the substance. The differences arise in the energy-level patterns for various molecules and the possibilities of factoring the partition function. The equations given in this chapter suffice for most gases at temperatures up to about 2000°K, but there are further complexities which require extension of these methods in certain cases.

Free radicals and molecules of importance at high temperatures normally have much more loosely bound electrons, and low-lying electronic levels are much more common than for molecules stable at room temperature. Important electronic contributions to the thermodynamic properties can be expected for such molecules.

PROBLEMS

27-1. Calculate the free-energy function $-(F° - H_0°)/T$ for CO at 2000°K for the harmonic-oscillator–rigid-rotator approximation. The rotational constant $B = 1.931$ cm^{-1}, and the vibration frequency is $\omega = 2167$ cm^{-1}. Compare this result with the more accurate value in Appendix 7, and note the magnitude of the effects of anharmonicity, stretching, etc., at this temperature.

27-2. Calculate all the energy levels of HF within 2000 cm^{-1} of the lowest quantum state. Use molecular data from Table 27-4. Indicate the multiplicity of each state (ignore the factor of 4 for the nuclear spins which applies to all states). Obtain Q and Q' for 200°K by direct summation of Eqs. (27-6) and (27-11). Compare the directly summed value of Q with those obtainable from Eq. (27-21) or (27-22). Calculate $S°$ at 200°K, and note the difference between results from Eqs. (27-15) and (27-25).

27-3. Estimate the importance of the $^1\Delta$ state of O_2 for the function $-(F° - H°)/T$ at 3000°K. Do not make a detailed calculation; accuracy to a factor of 2 in the contribution of the $^1\Delta$ state is sufficient.

27-4. Estimate the entropy of the radical CH_2Cl at 350°K. The odd electron will contribute $R \ln 2$ of electronic entropy. Assume 120° angles at carbon with the distances 1.1 A for C—H and 1.7 A for C—Cl and that the C—Cl stretching vibration frequency is 730 cm^{-1} as in CH_3Cl. How many other vibrations has this radical, and what is the possible range in entropy if their frequencies are all above 1000 cm^{-1}?

27-5. Sodium vapor at the boiling point at 1163°K is a mixture of atoms and diatomic molecules. The molecules have single electronic states, and the molecular constants $\omega = 159$, $\omega x = 0.726$, $B = 0.1547$, $\alpha = 0.00079$, $D = 0.584 \times 10^{-6}$, all in reciprocal centimeters. The dissociation energy is $\Delta H_0° = 17{,}500$ cal/mole. The lowest atomic energy level is $^2S_{1/2}$; the next in importance are $^2P_{1/2}$ and $^2P_{3/2}$ at 16,956 and 16,973 cm^{-1}, respectively. Calculate the free-energy function for both Na and Na_2 at 1163°K, and find the proportion of Na_2 in the saturated vapor.

27-6. The Sackur-Tetrode formula [Eq. (27-3)] may be used to calculate the entropy and other properties of a gas of electrons provided that the density is low enough and the temperature high enough[1] and provided that the space charge is canceled by appropriate positive charges. Calculate the value of $s°$ for electrons at 5000°K.

27-7. From the spectrum of Cs gas, $\Delta E_0°$ for $Cs(g, {}^2S_{1/2}) = Cs^+(g, {}^1S_0) + e^-(g)$ is 31,406.71 cm^{-1} or 3.893 ev.† In Problem 27-6 we note that the entropy and heat capacity of electrons at low pressures can be calculated in the same manner as for other monatomic gases. Thus, for the ionization of Cs(g), $\Delta c_P = \frac{5}{2}R$ at temperatures below that where the first excited electronic level of Cs(g) has an appreciable population. From Eq. (27-3), calculate $\Delta s°$ and $\Delta F°$ for the ionization of Cs at 2000°K and the per cent ionization of Cs vapor at 2000°K and 10^{-4} atm. Note the electronic multiplicities.

---

[1] The quantity $(V/n)(2\pi mkT/h^2)^{3/2}$ must be significantly greater than unity. See K. S. Pitzer, "Quantum Chemistry," p. 115, Prentice-Hall, Inc., Englewood Cliffs, N.J., 1953.

† C. E. Moore, Atomic Energy Levels, *Natl. Bur. Standards (U.S.) Circ.* 467, vol. III, 1958.

28

IRREVERSIBLE PROCESSES NEAR EQUILIBRIUM; NONISOTHERMAL SYSTEMS; STEADY STATES

Thermodynamics generally deals with systems that are at equilibrium with respect to some processes but not with respect to others. Ideally the forbidden processes have zero rates, but actually it suffices that their rates be very small compared with the rates of allowed processes. Then the observed properties can still be compared with the theoretical result for the ideal case.

The velocity of physical or chemical processes is not treated by thermodynamics, although the statistical theories of rate processes draw much background information from thermodynamics. Also near equilibrium the various rate coefficients must satisfy certain restricting relationships in order that the thermodynamic equations will be fulfilled at the equilibrium state. These equations between rate coefficients then constitute a thermodynamics of near-equilibrium irreversible processes. A steady state is obtained when the properties of the system itself do not change with time but there is an irreversible flow of heat, electricity, or some substance through the system. In general, steady states are treated in terms of the theories of the rates of the various processes involved. However, when this irreversible flow causes only a small deviation from true equilibrium, thermodynamic methods are still useful.

Nonisothermal systems present many examples of steady states which are subject to thermodynamic treatment. Perfect heat insulators are not available, and hence heat flows through a system at a finite rate; but by making the temperature gradient small, one can always obtain a steady state which is only slightly shifted from the isothermal equilibrium state.

The classical theory of thermodynamics as developed by Carnot, Clausius, Kelvin, and others in the middle of the nineteenth century was clearly intended to apply to nonisothermal systems such as heat engines. The theory applies to the reversible changes which occur to the working fluid; simultaneous dissipative processes such as friction and nonisothermal heat conduction are regarded as independent and treated separately. In his classic paper "On the Dynamical Theory of Heat," Kelvin[1] assumes this separation of processes to be obvious in ordinary heat engines where he simply specifies "perfect engine," but he discusses the separation at length for thermoelectric phenomena as a separate postulate. Thermodynamics does not predict the absolute rate of the flow process; rather it yields relations between measurable quantities such as the thermoelectric potential and the Peltier heat effect, which occurs when electricity flows from one metal to another.

If the two or more processes can be clearly separated into noninteracting mechanisms of change and the steady irreversible flow affects only one, then a straightforward equilibrium treatment can be given for the others. For example, a slight leakage of the solute through a semipermeable membrane will not affect an osmotic-pressure experiment provided that the solution composition on each side is maintained constant by external means. The thermodynamic treatment of such isothermal systems is now so familiar that we shall devote no further attention to such examples. But in nonisothermal cases certain new features arise, and an example which illustrates these is the subject of the next section. There are, in addition, methods which avoid the identification of noninteracting processes, which will be presented later.

Thermomolecular Pressure, Thermoosmosis.[2] Take a vessel divided into two compartments by a barrier resistant to heat flow but permeable to the fluid substance, gas or liquid, which is contained in the vessel. The barrier may be a capillary, a porous plate, or a material dissolving the fluid. A temperature difference is established by steadily removing heat from the cooler compartment of temperature T and adding heat at an equal rate to the warmer compartment at $T + dT$. It is assumed that the heat conductance of the barrier, including the fluid in it, is so small that each compartment remains at constant temperature and the temperature gradient is within the barrier. The ordinary heat leakage through the barrier by the thermal conductance of the materials will have no direct interaction with the passage of fluid through the barrier. In general this system will develop a steady-state pressure difference dP which will depend not only upon the fluid but also upon the nature of the barrier. In gaseous systems this is usually termed the thermomolecular pressure difference, whereas with liquids the term thermoosmosis is more common.

[1] W. Thomson (Lord Kelvin), "Mathematical and Physical Papers," vol. I, art. XLVIII, University Press, Cambridge, 1882; reprints from articles in *Trans. Roy. Soc. Edinburgh*, 1851–1854.

[2] E. D. Eastman, *J. Am. Chem. Soc.*, **48**, 1482 (1926).

Consider the transfer of a small amount of fluid from the cooler to the hotter compartment.   In addition to the heat content of the substance itself, which increases by $d\text{H}$ per mole, there is usually a heat effect when the fluid enters the surface of the barrier and passes into the interior. This may be thought of as primarily a heat of solution, and, of course, the reverse heat effect occurs when the fluid comes out of the barrier into the other compartment.   Thus heat is transferred with the substance; and it should be emphasized that thermodynamics is not concerned with the detailed source, heat of solution, etc., but only with the total heat transferred per mole of substance.   This quantity is called the *heat of transfer*[1] and written as Q^* or Q^* per mole.

The criterion for the steady-state pressure difference is just that for a reversible equilibrium: zero change in entropy for the transfer of the fluid.   Per mole of fluid transferred the change in entropy of the fluid itself may be written

$$ds_1 = \left(\frac{\partial \text{s}}{\partial T}\right)_P dT + \left(\frac{\partial \text{s}}{\partial P}\right)_T dP$$

while the heat effects defined in the previous paragraph yield the additional entropy changes

$$ds_2 = -\frac{d\text{H}}{T} - \frac{\text{Q}^*}{T} + \frac{\text{Q}^*}{T + dT}$$

$$= -\left(\frac{\partial \text{H}}{\partial T}\right)_P \frac{dT}{T} - \left(\frac{\partial \text{H}}{\partial P}\right)_T \frac{dP}{T} - \frac{\text{Q}^* \, dT}{T^2}$$

Upon addition of $dS_1 + dS_2$ we note that the first terms cancel one another and the next two combine into a derivative of the free energy. The result is

$$dS = -\left(\frac{\partial \text{F}}{\partial P}\right)_T \frac{dP}{T} - \frac{\text{Q}^* \, dT}{T^2} = 0 \qquad (28\text{-}1)$$

Substitution of v for $\partial \text{F}/\partial P$ and rearrangement yields

$$\frac{dP}{dT} = -\frac{\text{Q}^*}{\text{v}T} \qquad (28\text{-}2)$$

Thermodynamics thus yields an equation relating the thermal pressure difference to the heat of transfer but, of course, yields no prediction about the magnitude of the effect.   While the pressure difference in thermo-osmosis, etc., has been observed by various workers, there are as yet no sufficiently accurate measurements of the heat of transfer to constitute any check on Eq. (28-2).   The calculated heats of transfer for thermo-

---

[1] Some authors use the term *reduced heat of transfer* to distinguish Q^* from the total enthalpy which moves with the substance, $\text{Q}^* + \text{H}$.   We shall always write Q^* and H separately.

osmosis of gases through a rubber membrane do, however, approximate the heats of solution of those gases in rubber.[1]

The kinetic theory of ideal gases predicts[2] for capillary passages which are small compared with the mean free path of the gas that the steady-state pressure is proportional to the square root of the temperature, or

$$\frac{d \ln P}{d \ln T} = \frac{1}{2} \qquad (28\text{-}3)$$

This requires $Q^* = -\frac{1}{2}RT$. As the size of the tube increases, this effect decreases, and for an open tube of diameter large compared with the mean free path $Q^* = 0$ and no thermal pressure differences occur.

GENERAL THEORY OF NEAR-EQUILIBRIUM PROCESSES

In the preceding section we considered an example of a near-equilibrium system in which one irreversible steady-state process of heat flow was readily distinguished from another process of matter flow at equilibrium. The latter process was affected, however, by the temperature difference between the two portions of the system. Ordinarily steady-state systems of this sort may be treated by similar methods. In nonisothermal systems there is always some pure heat conductance in addition to whatever other processes exist which transfer some component of matter (or electricity) along with its heat of transfer. Nevertheless, in some cases the selection of the independent processes may be ambiguous or inconvenient, and a more general theory is useful.

It is also necessary to consider whether or not the usual thermodynamic variables and properties are still valid when irreversible processes are proceeding at finite rates. One may picture the system as being subdivided into various portions and that suddenly all transfer of heat, matter, etc., between subdivisions is stopped. Each portion of the system will now come to equilibrium, and its temperature, pressure, entropy, etc., will be unambiguous. If it is possible to make these portions large enough to contain many molecules (and hence to have macroscopic properties) and yet small enough so that the original gradients within a given portion were small, then the finally measured properties may be assigned to the original system. There may also be regions of high gradient but which contain a negligible portion of the total matter, energy, etc.

Consideration of actual methods of measurement leads to similar conditions. For example, a thermometer must have thermal contact with

[1] R. J. Bearman, *J. Phys. Chem.*, **61**, 708 (1957).

[2] E. H. Kennard, "Kinetic Theory of Gases," p. 66, McGraw-Hill Book Company, Inc., New York, 1938.

a macroscopic amount of matter within the system.   The resulting temperature measurement will be unambiguous only if there are no significant temperature differences within the matter which is in thermal contact with the thermometer.

Prigogine[1] has considered these questions from the viewpoint of kinetic theory and statistical mechanics.   He concludes that the domain of validity of thermodynamic variables when transport processes are occurring is the range of linear-rate laws.   Chemical reactions must be sufficiently slow so that the Maxwellian distribution of molecular velocities is not significantly disturbed.   These conditions are clearly consistent with those developed from the other points of view.

Entropy Production in Heat Flow.   In earlier discussions we have considered the possibility of entropy production in irreversible processes. Let us now write down exact expressions for the entropy production in a few common situations.   Take a system comprising two regions at different temperatures T_1 and T_2 but each isothermal.   The entropy increase in region 1 is $dS_1 = \delta q_1 / T_1$ and for region 2 is $dS_2 = \delta q_2 / T_2$.   If heat flows from region 1 to region 2, $\delta q_1 = -\delta q_2$ and

$$dS_{\text{irr}} = \delta q_2 \left(\frac{1}{T_2} - \frac{1}{T_1} \right) = \delta q_2 \frac{T_1 - T_2}{T_1 T_2} \qquad (28\text{-}4)$$

The irreversible entropy increase[2] dS_{irr} must, of course, be positive; hence $T_2 < T_1$ in agreement with the well-known direction of spontaneous heat flow.   If the temperature difference is infinitesimal, Eq. (28-4) becomes

$$dS_{\text{irr}} = \delta q \, d \left(\frac{1}{T} \right) \qquad (28\text{-}5)$$

Entropy Production in Matter Flow.   Consider next the flow of δn moles of some component of matter (or of electricity) from one region of temperature T_1, where the chemical potential is μ_1, to a second region, where the temperature is T_2 and the chemical potential μ_2.   The partial molal entropy $\bar{s} = (\bar{H}/T) - (\mu/T)$, but we must also consider the heat of transfer Q^* defined earlier in this chapter.   The first law of thermodynamics requires that the sum $\bar{H} + Q^*$ be the same whether considered from region 1 or region 2.   The entropy increase for transfer of δn moles is then

$$dS_{\text{irr}} = \delta n \left[\left(\frac{\mu_1}{T_1} - \frac{\mu_2}{T_2} \right) + (\bar{H} + Q^*) \left(\frac{1}{T_2} - \frac{1}{T_1} \right) \right] \qquad (28\text{-}6a)$$

[1] I. Prigogine, *Physica*, **15**, 272 (1949); *J. Phys. Chem.*, **55**, 765 (1951).

[2] We shall sometimes omit the subscript irr when it seems obvious that only entropy increase and no entropy transfer is involved.

or for infinitesimal differences

$$dS_{irr} = \delta n \left[-d\left(\frac{\mu}{T}\right) + (\text{H} + \text{Q}^*)\, d\left(\frac{1}{T}\right) \right] \qquad (28\text{-}6b)$$

We see at once that the second term drops out if $T_2 = T_1$. This explains why the heat of transfer never needs to be considered in isothermal problems.

If there is but a single component and pressure is the only external force, the change of the ratio μ/T may be expressed as

$$d\left(\frac{\mu}{T}\right) = \frac{\text{V}}{T}\, dP + \text{H}\, d\left(\frac{1}{T}\right)$$

and Eq. (28-6) becomes

$$dS_{irr} = \delta n \left[-\frac{\text{V}}{T}\, dP + \text{Q}^*\, d\left(\frac{1}{T}\right) \right] \qquad (28\text{-}7)$$

which is equivalent to the result derived as Eq. (28-1).[1]

Entropy Production from Electric Current. An electric current may always be considered as a flow of one or more kinds of charged matter with the term $z_i \mathfrak{F}\, d\upsilon$ included in the change of chemical potential $d\mu_i$. Here z_i is the charge in protonic units, $\mathfrak{F}$ the Faraday constant, and υ the electrical potential. Electrons and other charged particles commonly have a heat of transfer; also other factors than electrical potential may affect the chemical potential of the charged species.

In the special case of an electric current in an isothermal system, one may write

$$dS_{irr} = \frac{\delta n \mathfrak{F}}{T}\, d\upsilon = \frac{I\, \delta t}{T}\, d\upsilon \qquad (28\text{-}8)$$

where I is the current and δt the time of flow.

Rate of Entropy Production. We now wish to discuss the rate of flow of heat, of some component of matter, or of electricity. It is customary to adopt the symbol J_i for the ith flux, which may be calories of heat per second, moles per second of matter, etc. In each of the examples of entropy production the resulting expression was the product of an extensive quantity δq or δn times some sort of potential difference. The latter may be regarded as the driving force for the flow in question. The symbol X_i is customary for the potential difference associated with the ith flux. The total rate of entropy production becomes

$$\frac{dS}{dt} = \dot{S} = \sum_i J_i X_i \qquad (28\text{-}9)$$

with X_i defined as the coefficient of J_i in the expression for $\dot{S}$.

[1] The material must remain in the same state—gas, liquid, or solid; otherwise the finite difference $\bar{\text{H}}_2 - \bar{\text{H}}_1$ enters the result.

It is frequently much easier to write out Eq. (28-9) for a given system in terms of simple flow quantities, heat, components of matter, etc., than it is to analyze the actual flow processes. Thus in the case of two regions with flow of heat and of a single component of matter we may write at once

$$\dot{S} = J_h \Delta \left(\frac{1}{T}\right) + J_m \left(-\frac{\mathrm{v}}{T} \Delta P\right) \qquad (28\text{-}10)$$

where J_m is the flow of matter and J_h is the total flow of heat whether associated with the matter as a heat of transfer or not. In view of our analysis above, the total heat flow J_h may be divided into that arising from heat of transfer $J_m Q^*$ and a remaining pure heat flow J_h'. In these terms Eq. (28-10) may be rewritten

$$\dot{S} = J_h' \Delta \left(\frac{1}{T}\right) + J_m \left[Q^* \Delta \left(\frac{1}{T}\right) - \frac{\mathrm{v}}{T} \Delta P\right] \qquad (28\text{-}11)$$

where the first term is clearly derivable from Eq. (28-5) and the second from Eq. (28-7).

Rate Equations. Let us now postulate a reasonable form for the rate equations governing the fluxes J_i in the region near equilibrium. The rates must become zero when all potential differences X_i are zero, and the fluxes certainly will reverse direction if all the potential differences change sign. Thus it is reasonable to write for the J's power-series expansions in the X's and to retain only the first-power terms for the near-equilibrium case. If we have identified truly independent flow processes, each flux should be governed by its own potential difference and be independent of other potentials. Thus

$$J_i' = L_i' X_i' \qquad (28\text{-}12)$$

where L_i' is a rate constant. If, however, we have written down fluxes and their associated potentials without regard to actual processes, as in Eq. (28-10), we may expect cross terms to arise. Since Eq. (28-9) for the entropy production and the accompanying definition of the potential differences must be valid for any set of fluxes and associated potentials, a relationship may be derived between off-diagonal rate constants. Let the fluxes and potentials J_i' and X_i' pertain to the independent processes with rate equations (28-12), while some other definition of fluxes J_i is given by the transformation

$$J_i = \sum_j \alpha_{ij} J_j' \qquad (28\text{-}13)$$

Then

$$\dot{S} = \sum_i \sum_j \alpha_{ij} J_j' X_i \qquad (28\text{-}14)$$

and X'_j, the coefficient of J', is

$$X'_j = \sum_i \alpha_{ij} X_i \qquad (28\text{-}15)$$

Substitution of these results and the rate equations for the independent processes [Eq. (28-12)] into Eq. (28-13) yields

$$J_i = \sum_j \alpha_{ij} L'_j X'_j$$

$$= \sum_j \sum_k \alpha_{ij} L'_j \alpha_{kj} X_k$$

If new rate constants are defined by

$$L_{ik} = \sum_j \alpha_{ij} L'_j \alpha_{kj} \qquad (28\text{-}16)$$

then

$$J_i = \sum_k L_{ik} X_k \qquad (28\text{-}17)$$

and it is apparent from Eq. (28-16) that

$$L_{ik} = L_{ki} \qquad (28\text{-}18)$$

These last equations are commonly called the Onsager relations.[1]   Onsager[2] derived them from the postulate of microscopic reversibility, using statistical arguments which we shall not repeat here.   Other derivations have been given.[3,4]   It is a matter of personal opinion which postulates constitute the most plausible basis for Eq. (28-18).   In some cases the independence of the several separate processes is clear-cut, and the derivation used here seems most straightforward.   In other cases, such as simultaneous fluxes of heat and electricity in metals, the separation is less evident, and other postulates may seem preferable.   In any case the Onsager relations are believed to be generally true, and the final decision will come from the agreement with experiment of results derived from these equations.

The second law of thermodynamics requires that the entropy production $\dot{S}$ be positive (or zero).   This places a limit on the magnitude of the off-diagonal rate constants L_{ij} $(i \neq j)$.   For example, with two fluxes

$$\dot{S} = L_{11} X_1{}^2 + (L_{12} + L_{21}) X_1 X_2 + L_{22} X_2{}^2 \geq 0$$

[1] In the presence of a magnetic field $\mathbf{H}$ the statistical mechanical theory yields a modification of Eq. (28-18):

$$L_{ik}(\mathbf{H}) = L_{ki}(-\mathbf{H}) \qquad (28\text{-}18a)$$

[2] L. Onsager, *Phys. Rev.*, **37**, 405 (1931), **38**, 2265 (1931).

[3] R. B. Parlin, R. J. Marcus, and H. Eyring, *Proc. Natl. Acad. Sci. U.S.*, **41**, 900 (1955).

[4] J. C. M. Li, *J. Chem. Phys.*, **29**, 747 (1958).

regardless of the sign and magnitude of either X_1 or X_2. This requires $L_{11} > 0$, $L_{22} > 0$, and $(L_{12} + L_{21})^2 < 4L_{11}L_{22}$. In general the quadratic form must be positive definite. While this restriction is unquestionably valid, it does not ordinarily yield any useful information in practical cases.

Before proceeding to other systems, let us see how the Onsager equations may be used to obtain Eq. (28-2) for thermoosmosis. The problem is formulated in terms of a flux of matter J_m and a flux of heat (other than the heat content of the matter) J_h. The entropy production is given in these terms by Eq. (28-10), which serves also to define the potentials

$$X_m = -\frac{v}{T}\Delta P \tag{28-19a}$$

$$X_h = \Delta\left(\frac{1}{T}\right) \tag{28-19b}$$

The rate equations are

$$J_m = L_{mm}X_m + L_{mh}X_h \tag{28-20a}$$
$$J_h = L_{hm}X_m + L_{hh}X_h \tag{28-20b}$$

The steady state in thermoosmosis is given by $J_m = 0$, whereby only a heat flow remains, as is required to maintain the temperature difference. Therefore

$$X_m = -X_h\frac{L_{mh}}{L_{mm}} = -X_h\frac{L_{hm}}{L_{mm}} \tag{28-21}$$

where the second equality follows because $L_{mh} = L_{hm}$ by Eq. (28-18).

In order to identify the ratio L_{hm}/L_{mm} with some physical quantity, consider the processes caused by a pressure difference in an isothermal system ($X_h = 0$). The heat transferred per mole of matter flowing is the heat of transfer

$$Q^* = \frac{J_h}{J_m} = \frac{L_{hm}}{L_{mm}} \tag{28-22}$$

Substitution of this result in Eq. (28-21) yields

$$X_m = -Q^*X_h$$

or

$$-\frac{v}{T}\Delta P = -Q^*\Delta\left(\frac{1}{T}\right) \tag{28-23}$$

This result is readily seen to be equivalent to Eq. (28-2) obtained before. Also this result could have been obtained at once by setting dS_{irr} to zero in Eq. (28-7), in other words, by requiring no irreversible entropy production for infinitesimal transfer of matter.

ELECTROKINETIC EFFECTS

The general theory of near-equilibrium processes forms a convenient basis for deriving the equations relating various electrokinetic phenomena.

These systems are isothermal, with two compartments separated by a porous membrane. There are two possible irreversible fluxes: the volumetric flow of liquid J and electric current I. The entropy production is readily obtained from Eq. (28-7) with $d(1/T) = 0$ and Eq. (28-8).

$$\dot{S} = J\frac{\Delta P}{T} + \frac{I\,\Delta \mathrm{U}}{T} \qquad (28\text{-}24)$$

and the potential gradients are self-evident. The linear-rate equations are

$$J = L_{11}\frac{\Delta P}{T} + L_{12}\frac{\Delta \mathrm{U}}{T} \qquad (28\text{-}25a)$$

$$I = L_{21}\frac{\Delta P}{T} + L_{22}\frac{\Delta \mathrm{U}}{T} \qquad (28\text{-}25b)$$

with the Onsager relationship $L_{12} = L_{21}$.

The four electrokinetic phenomena are given in terms of these rate constants. The *streaming potential* is the potential difference per unit pressure difference with zero electrical current.

$$\left(\frac{\Delta \mathrm{U}}{\Delta P}\right)_{I=0} = -\frac{L_{21}}{L_{22}} \qquad (28\text{-}26)$$

The flow of liquid at zero pressure difference caused by unit electrical current is called *electroosmosis* and is given by

$$\left(\frac{J}{I}\right)_{\Delta P=0} = \frac{L_{12}}{L_{22}} \qquad (28\text{-}27)$$

The *electroosmotic pressure* is the pressure difference per unit potential difference at zero flow of matter.

$$\left(\frac{\Delta P}{\Delta \mathrm{U}}\right)_{J=0} = -\frac{L_{12}}{L_{11}} \qquad (28\text{-}28)$$

The *streaming current* is the electrical current per unit volumetric flow at zero electrical potential difference.

$$\left(\frac{I}{J}\right)_{\Delta \mathrm{U}=0} = \frac{L_{21}}{L_{11}} \qquad (28\text{-}29)$$

From the Onsager relationship we see that the first quantity is just the negative of the second and the third is the negative of the fourth.

$$\left(\frac{\Delta \mathrm{U}}{\Delta P}\right)_{I=0} = -\left(\frac{J}{I}\right)_{\Delta P=0} \qquad (28\text{-}30)$$

$$\left(\frac{\Delta P}{\Delta \mathrm{U}}\right)_{J=0} = -\left(\frac{I}{J}\right)_{\Delta \mathrm{U}=0} \qquad (28\text{-}31)$$

The first of these, known as Saxen's relation, was first derived from kinetic considerations.

This derivation shows the convenience and elegance of the general theory for relating rate phenomena within the linear region to the ratios of potential gradients under steady-state conditions.

DIFFUSION AND THERMAL DIFFUSION

Isothermal diffusion in a two-component system is described by a single diffusion constant. This is the rate constant for the single process occurring. No near-equilibrium steady state is possible; likewise no Onsager relationship appears for a single process.

For diffusion in a system of three or more components, however, one can have near-equilibrium steady-state situations, and the methods of this chapter may be applied to relate certain of the diffusion coefficients to one another. In a three-component system four diffusion coefficients occur, but there is one thermodynamic relationship among them. Diffusion coefficients are usually defined and measured in terms of concentration, whereas the thermodynamic relationship involves activity or free energy. Only very recently have the necessary data been obtained to test the thermodynamic equation for three-component diffusion. The necessary manipulations to reduce the thermodynamic equations for comparison with multicomponent diffusion coefficients are quite lengthy and will not be presented here. Miller[1] has found in several three-component systems that agreement is obtained for the one Onsager relationship.

A two-component system in a temperature gradient yields a situation much like that of thermoosmosis. A concentration gradient develops in the steady state. This thermal-diffusion phenomenon is also called the Soret effect when the system is a liquid system. If no pressure gradient is present, the solvent must move as required to compensate for the volume change of solute motion and one has effectively a single-component material flow. The entropy production for transfer of 1 mole of solute from Eq. (28-6*b*) is

$$ds_{\text{irr}} = -d\left(\frac{\mu_2}{T}\right) + (\bar{H}_2 + Q_2^*)\, d\left(\frac{1}{T}\right)$$

and expansion of the first term with P, T^{-1}, and composition as independent variables yields

$$ds_{\text{irr}} = -\frac{\bar{V}_2}{T}\, dP - R\frac{\partial \ln a_2}{\partial \ln x_2}\, d\ln x_2 + Q_2^*\, d\left(\frac{1}{T}\right) \qquad (28\text{-}32)$$

[1] D. G. Miller, *J. Phys. Chem.*, **63**, 570 (1959).

In most cases the pressure gradient is negligible and the thermal-diffusion steady state $(dS_{irr} = 0)$ is given by

$$\frac{d \ln x_2}{dT} = - \frac{Q_2^*}{RT^2} \left(\frac{\partial \ln a_2}{\partial \ln x_2} \right)^{-1} \tag{28-33}$$

where the final factor is, of course, unity for an ideal solution.

Frequently molality is used as the composition variable, and one is interested in electrolyte solutions. Also the Soret coefficient σ is defined (sometimes with opposite sign), and one finds from Eq. (28-32)

$$\sigma = \frac{d \ln m}{dT} = - \frac{Q_2^*}{RT^2} \left\{ (\nu_+ + \nu_-) \left[1 + \frac{\partial \ln \gamma_\pm}{\partial \ln m} \right] \right\}^{-1} \tag{28-34}$$

where ν_+ and ν_- are the numbers of positive and negative ions, respectively, per mole of electrolyte and $\gamma_\pm$ is the usual mean activity coefficient. The total heat of transfer of the solute Q_2^* may be written as the sum over the component ions,

$$Q_2^* = \nu_+ Q_+^* + \nu_- Q_-^* \tag{28-35}$$

Heat of Transfer in Electrolytes. The Soret coefficient has been measured for many systems, and the corresponding heat-of-transfer values have been calculated from Eq. (28-33) or (28-34). Since electrolytes will be of interest again in connection with thermocells, we present in Table 28-1 the values of Q_2^* for several alkali halides as summarized by Agar[1] for 0.01 M aqueous solution at 25°C. The differences between the various fluoride-chloride pairs, etc., or between the lithium-sodium pairs, etc., are practically constant. Consequently one may conclude that at 0.01 M the heat of transfer is essentially an independent ionic property.

The values[1] of the heat of transfer for some other 1-1 electrolytes are considerably larger than any given in Table 28-1. For example, Q^* for both HCl and $N(C_2H_5)_4Cl$ is close to 3100 cal/mole at 25°C and 0.01 M.

TABLE 28-1. HEATS OF TRANSFER OF ALKALI HALIDES
In cal/mole at 0.01 M and 25°C

	Li^+		Na^+		K^+
		$(Na^+ - Li^+)$		$(K^+ - Na^+)$	
F^-	850	(802)	1652	(-279)	1373
$(F^- - Cl^-)$	(870)		(876)		(864)
Cl^-	-20	(796)	776	(-267)	509
$(Cl^- - Br^-)$	(-34)		(-38)		(-48)
Br^-	14	(800)	814	(-257)	557
$(Br^- - I^-)$	(597)		(581)		(584)
I^-	-583	(816)	233	(-260)	-27

[1] W. J. Hamer (ed.), "The Structure of Electrolytic Solutions," chap. 13 by J. N. Agar, John Wiley & Sons, Inc., New York, 1959.

The available data also show that for 1-1 electrolytes Q^* increases with temperature by about 40 cal/deg mole and decreases with increasing concentration approximately linearly in $m^{1/2}$. The slope $(dQ^*/dm^{1/2})$ is approximately -1900 cal $kg^{1/2}/mole^{3/2}$.

There is no successful interpretation as yet of the absolute magnitude of these heat-of-transfer values for strong electrolytes. Indeed the peculiar reversal of trend between Cl^- and Br^- in the halide series indicates that no simple interpretation will be possible. However, Agar has shown that the trend with concentration in very dilute solutions is consistent with a simple picture of the effect of a charge on a surrounding dielectric. In the region some distance from the ion, the electrical polarization of the dielectric reduces its entropy and thus releases heat. Consequently the motion of the ion releases heat ahead and absorbs heat behind, which contributes a positive heat of transfer. The volume of dielectric thus affected by a given ion decreases with concentration, which yields the decrease in the heat of transfer.

THERMOELECTRIC EFFECTS

The simplest thermoelectric device is a thermocouple comprising two different metals arranged as shown in Fig. 28-1. We shall also consider

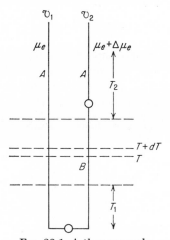

FIG. 28-1. A thermocouple.

somewhat more complex systems (Fig. 28-2) where an electrolytic solution replaces one of the metals. The two electrode-electrolyte junctions now comprise half cells, which are identical except for their temperature and possibly the electrolyte concentrations. Let us first consider the thermodynamics of such units from an over-all viewpoint. Later we shall trace the escaping tendency of electrons through a thermocouple and see that it yields the same result.

In each case we shall take $\mathcal{E}$ (or $d\mathcal{E}$), the emf of the cells, as the electrical potential of the wire from the hotter electrode or junction less that of the wire from the colder electrode or junction, that is, $\mathcal{E} = \mathcal{V}_2 - \mathcal{V}_1$ in Figs. 28-1 and 28-2.   Assume that 1 equiv of positive electricity passes through the system from terminal 1 to 2, that is, through the cold junction first and then the hot junction.   Assume also that the rate of transfer, i.e., the current, is so small that resistance heating is negligible.   If the system is at equilibrium, except for spontaneous heat conduction, then no net entropy change occurs, but electrical work $\mathfrak{F}\mathcal{E}$ is transferred to the surroundings.

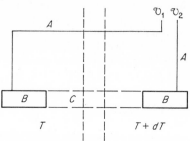

FIG. 28-2. A thermocell.   C is the electrolyte solution connecting the electrodes B, which are in turn connected to wires A.

The entropy change ΔS is absorbed from the surroundings as heat at the higher temperature, $T + dT$, and an equal entropy is emitted to the surroundings at T.   In the thermocell system some chemical change may occur at one electrode, but the reverse change occurs at the other electrode, and the net entropy change is an infinitesimal which may be neglected.   Now from the first law we see that

$$\mathfrak{F}\,d\mathcal{E} = (T + dT)\,\Delta S - T\,\Delta S = \Delta S\,dT$$
$$\mathfrak{F}\frac{d\mathcal{E}}{dT} = \Delta S = \Delta S_j + \Delta S_t \tag{28-36}$$

where we have divided ΔS into two terms to emphasize that it arises from two rather distinct sources. ΔS_j is the entropy change in the junction or at the electrode itself and is an ordinary entropy of chemical reaction involving the molal or partial molal entropies of the species formed or consumed.   On the other hand ΔS_t is the entropy absorbed at the higher temperature because of the heat of transfer of the species moving through the temperature gradient.

This division of ΔS, although convenient in our opinion, is not necessary.   It is equally possible to calculate ΔS by summation of the entropies of substances produced and remaining in the hot region together with the total entropy transported, $\bar{s}_i + q_i^*/T$, by substances leaving the hot region less the corresponding terms for substances consumed in or transported into the hot region.

Metallic Thermocouple.   In the thermocouple the heat absorbed on passage of an equivalent of positive electricity from metal B to A (see Fig. 28-1) is called the Peltier heat and is given the symbol $\Pi_{BA} = -\Pi_{AB}$.

Thus $T \Delta S = \Pi_{BA}$, and

$$\frac{d\mathcal{E}}{dT} = \frac{\Pi_{BA}}{\mathcal{F}T} \tag{28-37}$$

It is helpful to calculate the two terms of Eq. (28-36) separately. The electrode or junction effect is the reaction e^- (in A) $= e^-$ (in B) and in addition

$$\Delta S_j = \bar{S}_B - \bar{S}_A \tag{28-38}$$

where $\bar{S}_A$ and $\bar{S}_B$ are the partial molal entropies of electrons in the metals indicated. The electron flow in wire B carries the heat of transfer Q_B^* from the warmer to the colder region, while the electron flow in wire A carries Q_A^* in the opposite direction. Hence the net entropy absorbed at the hot junction from transfer phenomena ΔS_t is

$$\Delta S_t = \frac{Q_B^* - Q_A^*}{T} \tag{28-39}$$

The sum of these two effects yields ΔS, or the potential of the thermo-couple

$$\mathcal{F}\frac{d\mathcal{E}}{dT} = \left(\bar{S} + \frac{Q^*}{T}\right)_B - \left(\bar{S} + \frac{Q^*}{T}\right)_A \tag{28-40}$$

It is interesting to analyze the thermocouple by microscopic considera-tion of the chemical potential of electrons instead of the thermal effects of the whole system. The measured electrical potential difference times the Faraday constant is just the negative of the difference of chemical potential of electrons in the two wires of A at the top, that is, $\mathcal{F}\mathcal{E} = -\Delta\mu_e$.

Consider first a small section of one of the metals with an infinitesimal temperature difference from T to $T + dT$. For the transfer of 1 mole of electrons from T to $T + dT$ in metal B the entropy increase from Eq. (28-6) is

$$dS_B = -d\left(\frac{\mu_e}{T}\right)_B + (\bar{H}_B + Q_B^*) \, d\left(\frac{1}{T}\right) \tag{28-41}$$

where $\bar{H}_B + Q_B^*$ is the sum of the partial molal heat content of electrons and the heat of transfer of electrons in metal B. We assume here, as before, that the electron flow is so slow that the resistance heating is negligible.

The steady-state situation in a thermocouple with heat flow but no electric current is entirely analogous to that of thermoosmosis. Follow-ing any of the methods already described leads to the result that $dS_B = 0$ for electron flow and

$$d\left(\frac{\mu_e}{T}\right)_B = (\bar{H}_B + Q_B^*) \, d\left(\frac{1}{T}\right) \tag{28-42}$$

or

$$(d\mu_e)_B = -\left(\bar{S}_B + \frac{Q_B^*}{T}\right) dT \tag{28-43}$$

At the junctions μ_e is the same in each metal; hence the thermoelectric potential difference arises purely from the difference in $d\mu_e$ between the two metals in the temperature gradient.

$$\mathfrak{F}\frac{d\mathcal{E}}{dT} = -\frac{d\,\Delta\mu_e}{dT} = \Delta\mathfrak{s} + \frac{\Delta Q^*}{T} \tag{28-44}$$

or

$$\mathfrak{F}\mathcal{E} = -\Delta\mu_e = \int_{T_1}^{T_2}\left(\Delta\mathfrak{s} + \frac{\Delta Q^*}{T}\right)dT \tag{28-45}$$

where the Δ refers to the difference B minus A in each case, and we note that Eqs. (28-44) and (28-40) are equivalent.

This analysis has yielded exactly the same equation as before, but it emphasized the conclusion that the potential difference arises from the wire in the temperature gradient. Frequently the thermocouple wire is nonuniform in composition or other properties, and it is valuable to know that the portion of the wire in the temperature gradient determines the potential and that variations in the wire in isothermal regions have no effect.

Let us now consider more carefully a small segment of wire B in the temperature gradient. If 1 mole of electrons flows in the direction from temperature T to $T + dT$, the entropy drawn in at T is $\mathfrak{s}_B(T) + T^{-1}Q_B^*(T)$, while that passing out at $T + dT$ is $\mathfrak{s}_B(T + dT) + (T + dT)^{-1}Q_B^*(T + dT)$. The difference is

$$\left(\frac{-\sigma_B}{T}\right)dT = \left(\frac{\bar{c}_B}{T} + \frac{1}{T}\frac{dQ_B^*}{dT} - \frac{Q_B^*}{T^2}\right)dT \tag{28-46}$$

where $\bar{c}_B$ is the partial molal heat capacity of electrons in metal B. Since no entropy was created or destroyed in this reversible electron flow, this additional entropy $(-\sigma_B/T)\,dT$ must have flowed into the wire as heat from the surroundings. This heat absorbed, $-\sigma_B$, is the Thomson heat, which was first predicted to occur by thermodynamic argument by Thomson (Lord Kelvin) and later found experimentally. The minus sign before σ_B occurs because it was originally defined for a flow of positive electricity.

Differentiation of Eq. (28-44) with respect to temperature yields

$$\mathfrak{F}\frac{d^2\mathcal{E}}{dT^2} = \frac{\Delta\bar{c}}{T} + \frac{1}{T}\frac{d\,\Delta Q^*}{dT} - \frac{\Delta Q^*}{T^2}$$

which is readily combined with Eq. (28-46) to obtain

$$\frac{d^2\mathcal{E}}{dT^2} = \frac{-\Delta\sigma}{\mathfrak{F}T} \tag{28-47}$$

The basic phenomenological equations of thermoelectricity are Eqs. (28-37) and (28-47), which relate the electrical potential difference to the Peltier and Thomson heats. These relationships could have been derived without separate reference to the thermal properties of the electrons in metals A and B and their heats of transfer. We have presented the derivation in this more complex form to emphasize the relationships to the corresponding quantities in thermoosmosis and in electrochemical thermocells, to which we return in the next section. Before proceeding, however, we note that a thermoelectric potential difference would arise if the heats of transfer were zero provided that the partial molal heat capacities and entropies of electrons were different in the two metals. But likewise the effect might arise purely from differences in heat of transfer. Apportionment of the experimental Peltier and Thomson heats between the differences in equilibrium thermal properties and the differences in heats of transfer is possible only if one or the other can be determined separately. While the partial molal entropies, etc., of electrons are measurable in principle, the enormous magnitude of space-charge energies prevents practical measurement of such properties of charged species. Thus at present quantum-statistical theory of electron motion in metals provides the only source of information about the division between the equilibrium partial molal quantities and the heat of transfer.

The quantity $\bar{s}_i + Q_i^*/T$ for electrons in a given metal is measurable, however, and we shall see that it is also measurable for individual ions in solution. Because of the importance of this quantity it is frequently given a symbol and name, and the literature contains some inconsistent or confusing selections. We shall follow Temkin and Khoroshin[1] and Agar and Breck[2] in the symbol $\bar{\bar{s}}$ with the double bar on top and the latter authors for English terminology with *transported entropy*. Thus for the ith species

$$\bar{\bar{s}}_i = \bar{s}_i + \frac{Q_i^*}{T}$$

It is to be noted that the *transported entropy* $\bar{\bar{s}}_i$ is the sum of the partial molal entropy $\bar{s}_i$ and the entropy of transfer $s_i^* = Q_i^*/T$. Since there is still risk of confusion, the sum will be written explicitly in many cases.

From Eq. (28-46) one can readily deduce that for electrons in a metal

$$\bar{\bar{s}}_{T'} = \left(\bar{s} + \frac{Q^*}{T} \right)_{\text{at } T'} = \int_0^{T'} \left(-\frac{\sigma}{T} \right) dT \qquad (28\text{-}48)$$

since $\bar{s}$ and Q^*/T are zero at $0°K$ by the third law of thermodynamics. If the total entropy of a substance is zero and negative entropies are impossible, then the partial molal entropies of components must be zero

---

[1] M. I. Temkin and A. V. Khoroshin, *Zhur. Fiz. Khim.*, **26,** 500 (1952).
[2] J. N. Agar and W. G. Breck, *Trans. Faraday Soc.*, **53,** 167 (1957).

and motion of such components cannot transfer entropy. Data on metals at low temperatures are consistent with this interpretation.[1]

The individual Thomson coefficients are obtained from measurements of the change in temperature gradient in wires with the amount of current flowing. Temkin and Khoroshin[2] have integrated Eq. (28-48) for electrons in copper and platinum and find that at 298°K

Pt:
$$\bar{\bar{s}} = \left(\bar{s} + \frac{Q^*}{T} \right) = 0.104 \text{ cal/deg mole}$$

Cu:
$$\bar{\bar{s}} = \left(\bar{s} + \frac{Q^*}{T} \right) = -0.045 \text{ cal/deg mole}$$

Thermocells. An electrochemical cell with a single uniform electrolyte and two identical electrodes yields zero emf, of course, if it is isothermal. But if one half cell is at a different temperature from the other, a potential difference may develop, just as it does in a metallic thermocouple. Such systems are known as thermocells and were first studied by Gockel[3] in 1885.

The electrolyte connecting the two half cells is subject to thermal diffusion (Soret effect), and if this is allowed to take place, the two half cells will differ in electrolyte composition as well as in temperature. It is possible, however, to make the electrical measurements before significant thermal diffusion occurs, and this yields an initial emf $\mathcal{E}_{in}$. After the thermal-diffusion steady state is established, a different emf $\mathcal{E}_{st}$ may be measured.

To consider the thermodynamics of a thermocell, we return to our first analysis of thermoelectric systems and note Eq. (28-36),

$$\mathfrak{F} \frac{d\mathcal{E}}{dT} = \Delta S_j + \Delta S_t \tag{28-36}$$

where ΔS_j is the entropy change of the electrode reaction of a thermocell and ΔS_t is the entropy absorbed from the hot region because of heat-of-transfer effects. The former quantity is the usual ΔS of the half-cell reaction for the passage of 1 equiv in the direction of positive electricity from electrolyte to electrode.

The heat-of-transfer effects become more complex in an electrolyte solution than in a metallic wire. If we assume uniform composition, i.e., neglect thermal diffusion, then various ions of charge z_i and molal heat of

---

[1] It is interesting to note that σ and $\bar{s}$ are zero even at finite temperatures for electrons in superconductors. See D. Shoenberg, "Superconductivity," p. 86, Cambridge University Press, New York, 1952.

[2] *Loc. cit.*

[3] A. Gockel, *Wied. Ann.*, **24,** 618 (1885).

transfer Q_i^* carry portions of the current indicated by their transference numbers t_j. The net heat of transfer for 1 equiv of positive electricity is

$$Q_{in}^* = \sum_i \frac{t_i Q_i^*}{z_i} \qquad (28\text{-}49)$$

where the subscript *in* indicates that this is the initial state before thermal diffusion.

If instead we assume that thermal diffusion is rapid compared with the electrical flow in the cell, then the only ion transferred through the temperature gradient in the electrolyte is the ion which enters the electrode reaction. Hence

$$Q_{st}^* = \frac{Q_j^*}{z_j} \qquad (28\text{-}50)$$

where *st* refers to the stationary state.

In either the initial state or the stationary state the total entropy absorbed at the hot electrode by transfer effects includes also the electrons in the wire and is

$$\Delta S_t = - \frac{1}{T} [Q_{e^-}^* + Q_{ion}^*] \qquad (28\text{-}51)$$

where Q_{ion}^* is either Q_{in}^* or Q_{st}^* as appropriate and $Q_{e^-}^*$ is the heat of transfer of the electron in the metal wire passing through the temperature gradient.

This completes our general analysis of a thermocell. Let us turn to examples with Ag-AgBr electrodes and M^+Br^- aqueous electrolytes as studied by Eastman and his collaborators.[1] The electrode reaction in the direction defined is

$$AgBr + e^-(\text{in wire}) = Ag + Br^-(aq)$$
$$\Delta S_j = s_{Ag} + s_{Br^-} - s_{AgBr} - s_{e^-} \qquad (28\text{-}52)$$

The entropy from heat of transfer in the initial state of uniform composition is

$$\Delta S_t = \frac{t_- Q_{Br^-}^* - t_+ Q_{M^+}^* - Q_{e^-}^*}{T} \qquad (28\text{-}53)$$

and the initial-cell emf is

$$\mathfrak{F} \left(\frac{d\mathcal{E}}{dT} \right)_{in} = s_{Ag} - s_{AgBr} + s_{Br} + \frac{t_- Q_{Br^-}^*}{T} - \frac{t_+ Q_{M^+}^*}{T} - s_{e^-} - \frac{Q_{e^-}^*}{T} \qquad (28\text{-}54)$$

In the final stationary state the result is somewhat less complex,

$$\mathfrak{F} \left(\frac{d\mathcal{E}}{dT} \right)_{st} = s_{Ag} - s_{AgBr} + \left(s_{Br^-} + \frac{Q_{Br^-}^*}{T} \right) - \left(s_{e^-} + \frac{Q_{e^-}^*}{T} \right) \qquad (28\text{-}55)$$

[1] J. C. Goodrich et al., *J. Am. Chem. Soc.*, **72**, 4411 (1950).

Note that the terms in parentheses in Eq. (28-55) for Br^- and e^- comprise the transported entropy $\bar{s}_i$ in each case.   Thus the stationary-state result is more nearly analogous to that for the simple thermocouple than is the case for the initial state.   Also, since all other quantities are measurable independently, $\bar{s}_{Br^-}$ may be determined from Eq. (28-55).

The difference between the initial and stationary-state emf values yields the heat of transfer of the electrolyte.   For the Ag-AgBr cell the result is

$$\mathcal{F}\frac{d}{dT}\left(\mathcal{E}_{st} - \mathcal{E}_{in}\right) = \frac{t_+}{T}\left(Q_{M^+}^* + Q_{Br^-}^*\right) \tag{28-56}$$

One can readily see from Eqs. (28-49) and (28-50) that this is a general result.

Agar and Breck[1] measured both $\mathcal{E}_{in}$ and $\mathcal{E}_{st}$ for cells with thallium amalgam (5 per cent) electrodes and various thallous salts (Tl^+X^-) as electrolytes.   The equations for the emf are

$$\mathcal{F}\left(\frac{d\mathcal{E}}{dT}\right)_{in} = s_{Tl} - s_{Tl^+} - \frac{t_+ Q_{Tl^+}^*}{T} + \frac{t_- Q_{x^-}^*}{T} - \bar{s}_{e^-} \tag{28-57}$$

$$\mathcal{F}\left(\frac{d\mathcal{E}}{dT}\right)_{st} = s_{Tl} - \left(s_{Tl^+} + \frac{Q_{Tl^+}^*}{T}\right) - \bar{s}_{e^-} \tag{28-58}$$

The second equation shows that $\mathcal{E}_{st}$ should be independent of the anion X^- so long as $\bar{s}_{Tl^+}$ for the thallous ion is unaffected.   The comparison was made at 0.10 M, which is hardly dilute enough to make interionic effects negligible, but the values of $\bar{s}_{Tl^+}$ in Table 28-2 show no significant differences.   The partial molal entropy of thallium in the amalgams is obtained by standard methods. Q^* in Table 28-2 is the total heat of transfer of the entire salt, that is, $Q_{Tl^+}^* + Q_{X^-}^*$ for the 1-1 salts and $2Q_{Tl^+} + Q_{X^-}^*$ for the 2-1 salts, respectively.

TABLE 28-2. THERMOCELL RESULTS FOR THALLOUS SALTS
Mean temperature 25.2 ± 0.3°C and 0.100 ± 0.001 M Tl^+

Salt	$\bar{s}_{Tl^+}$, cal/deg mole	Q^*, cal/mole	t_{x^-}
TlClO₄.............	33.9	510	0.467
TlNO₃.............	34.0	500	0.481
TlOAc.............	34.0	1840	0.344
Tl₂CO₃............	33.9	2550	0.48
Tl₂SO₄............	34.0	2410	0.532

The effect of concentration on $\bar{s}_{Tl^+}$ is given, to the approximation of an ideal dilute solution, by the term $-R \ln m$.   Table 28-3 shows values for

[1] *Loc. cit.*

cells with $TlClO_4$ electrolyte at several concentrations.   There is good evidence for a concentration dependence of Q^*; hence there is no reason to expect the values of the last column to be constant.   However, the concentration dependence of Q^* in the dilute region might be expected to be about the same for Tl^+ and ClO_4^- since the entropy of polarization of water distant from the ion should not depend on the direction of polarization.   Also we note in Table 28-2 that t_{x^-} is approximately $\frac{1}{2}$ for $TlClO_4$. Consequently the concentration effects of $Q^*_{Tl^+}$ and $Q^*_{x^-}$ in Eq. (28-57) should nearly cancel, and it is not surprising that the values are nearly constant in the next to the last column of Table 28-3.

TABLE 28-3. CONCENTRATION DEPENDENCE OF THERMOCELL EMF
In $\mu v/deg$ for mean temperature 25.2°C

m	$\left(\dfrac{d\varepsilon}{dT}\right)_{in}$	$\left(\dfrac{d\varepsilon}{dT}\right)_{st}$	$\left(\dfrac{d\varepsilon}{dT}\right)_{in} + \dfrac{R}{\mathfrak{F}}\ln m$	$\left(\dfrac{d\varepsilon}{dT}\right)_{st} + \dfrac{R}{\mathfrak{F}}\ln m$
0.050	−509	−567	−251	−309
0.101	−451	−489	−253.5	−292
0.152	−416	−460	−254	−298
0.203	−392	−424	−255	−287
0.256	−369	−400	−252	−283

Further work on thermocells may be expected to yield much interesting information which will constitute a check on theories of electrolyte solutions.   Eventually such theories may yield a reliable basis for separating transported-entropy values into individual-ion entropies and heats of transfer, but there is no adequate basis at present for this division.

29

SURFACE EFFECTS

Heretofore we have explicitly neglected the effects of surfaces or interfaces on the thermodynamic properties of various systems. Frequently surface effects are negligible, but in other cases they may be important and hence must be considered. The first exact treatment of surface thermodynamics, as of so many of our topics, was given by Gibbs.

The boundary between two contiguous phases, which is known as a surface or an interface, is not a mathematical boundary of zero volume and only two-dimensional extension. There is a small but finite thickness of the region in which the properties differ appreciably from the properties in the interior of either phase. Unless the conditions are close to those of the critical point for the disappearance of the interface, there is every reason to believe that the thickness of the surface region is very small—a few molecular layers. Optical properties[1] support this estimate of the thickness of the surface region. We consider initially surfaces between fluid phases and thus avoid the complication of solid surfaces, which differ for different crystal faces of the same substance. The basic equations apply, however, to a solid-fluid interface provided that the crystal face is specified.

Take as our system a region containing a substantial volume of phase I and phase II together with a flat surface of area A_s between the two phases. If we plot the concentration of some component as a function of the distance perpendicular to the interface, a curve such as that in Fig. 29-1 is to be expected. Let us imagine a plane in the region of the surface and at some location such as y' in Fig. 29-1. For the present, the location is arbitrary so long as the plane is parallel to the surface; i.e., it lies at the same point y' throughout the surface region. With this plane as the boundary the volume of phase I is V'_I and of phase II, V'_{II}.

Next we define the surface concentration of each material component as the excess of that quantity in the system over that calculated from the sum of the products of the concentration in each phase times the volume

[1] See N. K. Adam, "The Physics and Chemistry of Surfaces," 3d ed., p. 5, Oxford University Press, New York, 1941.

of each phase. Thus, if the molal concentration of component 1 in phase I is $c_1{}^{\mathrm{I}}$ and in phase II is $c_1{}^{\mathrm{II}}$, then the apparent surface concentration Γ_1' is

$$\Gamma_1' A_s = n_1 - c_1{}^{\mathrm{I}} V_{\mathrm{I}} - c_1{}^{\mathrm{II}} V_{\mathrm{II}} \qquad (29\text{-}1)$$

and similar equations may be written for the other components. We shall presently define thermodynamic quantities in the analogous fashion, but let us first consider certain properties of surface concentrations.

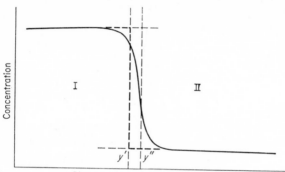

Distance perpendicular to the interface

FIG. 29-1. Actual concentration as a function of distance perpendicular to the surface is given by the solid curve. The surface concentration is the difference in area between the curve and the dashed rectangular path, which implies constant concentration on each side of a mathematical surface at y' (or y'').

Surface Concentration. Are the surface concentrations Γ_i intrinsically uncertain because of an arbitrary location of the dividing plane at y'? Suppose we use, instead, a dividing plane at y''. Now the volume assigned to phase I is increased by $(y'' - y') A_s$, and the volume assigned to phase II is decreased by the same amount. There is no change in the real physical system, however. The change in surface concentration is

$$(\Gamma_i'' - \Gamma_i') = (c_i{}^{\mathrm{II}} - c_i{}^{\mathrm{I}})(y'' - y') \qquad (29\text{-}2)$$

We see that Γ_i does depend on the location of the dividing plane unless $c_i{}^{\mathrm{II}} - c_i{}^{\mathrm{I}}$ is zero.

In a single-component system, where we have a vapor-liquid interface, $c^{\mathrm{liq}} - c^{\mathrm{vap}}$ differs substantially from zero, except near the critical point. But it is possible to choose that dividing plane which makes Γ zero because there is only one component present. Reference to Eq. (29-1) indicates that this location of the plane is the one which makes the two areas of Fig. 29-2 equal.

With two or more components it is not generally possible to make more than one Γ_i zero. While it is a purely arbitrary choice, it is usually preferable to make the surface concentration of the solvent zero, $\Gamma_1 = 0$. Then we expect nonzero values of $\Gamma_{i(1)}$ for all solute components, where the additional subscript indicates the component chosen to define the dividing plane. If in a particular case $c_j^{II} - c_j^{I} = 0$, then we see from Eq. (29-2) that Γ_j is unaffected by the choice of the dividing plane. Except for this special case, all individual Γ_i's are subject to the one arbitrary definition, and we shall usually deal with the series $\Gamma_{i(1)}$, $\Gamma_{1(1)} = 0$.

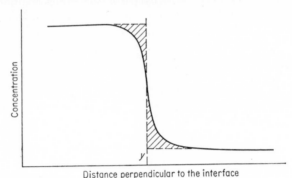

FIG. 29-2. The surface plane at y for zero surface concentration, which requires that the indicated areas be equal.

Surface Free Energy, Entropy, etc. Let us now consider thermodynamic functions for surfaces. Their definition is analogous to that for surface concentration,

$$Y'_s A_s = Y - \Sigma n_i^I \bar{Y}_i^I - \Sigma n_i^{II} \bar{Y}_i^{II} \tag{29-3}$$

where Y is any thermodynamic function, $\bar{Y}_i^I$ and $\bar{Y}_i^{II}$ are the usual partial molal quantities in phases I and II, and n_i^I and n_i^{II} are the numbers of moles in bulk phases I and II, respectively. For single-component systems, where there is no surface concentration of matter, Y_s is unambiguous and represents an additional energy, entropy, etc., associated with the presence of the surface. For multicomponent systems the arbitrary decision about the dividing surface affects surface thermodynamic properties as well as surface concentrations. It will be assumed that the choice $\Gamma_1 = 0$ is made unless otherwise specified.

We note particularly that there is no surface volume. Since all space has been assigned to bulk phases I and II, there is no space remaining to be assigned to the surface.

Alternate definitions are possible which include a surface volume, but the same equations connecting observable quantities are necessarily obtained eventually. Consequently, we believe it is simpler to adopt definitions which eliminate a surface volume at the beginning.

Surface Tension, Interfacial Tension. Consider next a change in shape of our system such that the interfacial area is increased but the number of moles of each constituent remains unchanged. Also the pressure on the exterior boundaries of the system is unchanged.   For this change the increase in free energy of the whole system per unit increase in interfacial area is the *surface tension*, or *interfacial tension*, σ,

$$\sigma = \left(\frac{\partial F}{\partial A_s}\right)_{T,P,n_i} \tag{29-4}$$

The general differential of free energy now contains an additional term, and instead of Eq. (17-9) we have

$$dF = -S\,dT + V\,dP + \sigma\,dA_s + \Sigma\mu_i\,dn_i \tag{29-5}$$

Now consider further the increment of the free energy for the process of increasing the interfacial area by dA_s at constant T and P and total n_i. In view of Eq. (29-4), the definition of surface free energy in Eq. (29-3), and the fact that $\bar{F}_i^{\,I} = \bar{F}_i^{\,II} = \mu_i$,

$$F_s\,dA_s = \sigma\,dA_s - \Sigma\mu_i(dn_i^{\,I} + dn_i^{\,II})$$

The change $dn_i^{\,I} + dn_i^{\,II}$ represents the change in the moles of component i in the bulk phases.   This is just the material added to the surface phase where there is a nonzero surface concentration.   Hence

$$dn_i^{\,I} + dn_i^{\,II} = -\Gamma_i\,dA_s$$

and
$$F_s\,dA_s = \sigma\,dA_s + \Sigma\mu_i\Gamma_i\,dA_s$$

We may now divide by dA_s and obtain

$$\sigma = F_s - \Sigma\mu_i\Gamma_i \tag{29-6}$$

where the same definition of dividing surface must, of course, be used for F_s and the Γ_i.

The differential of Eq. (29-6) is

$$d\sigma = dF_s - \Sigma\mu_i\,d\Gamma_i - \Sigma\Gamma_i\,d\mu_i \tag{29-7}$$

but the differential of F_s for fixed unit surface area from Eq. (29-3), after simplification, is

$$dF_s = -S_s\,dT + \Sigma\mu_i\,d\Gamma_i \tag{29-8}$$

and substitution of Eq. (29-8) into (29-7) yields

$$d\sigma = -S_s\,dT - \Sigma\Gamma_i\,d\mu_i \tag{29-9}$$

This is the basic equation for the surface tension, which was first derived by Gibbs and which will be the basis of much of our further discussion. A completely general treatment beyond this point tends to be unduly complex.   Hence it is more useful to consider several special cases of practical interest, which will also illustrate methods which could be extended to more complex cases as needed.

VAPOR-LIQUID INTERFACES

One-component Systems.   The vapor-liquid interface is the only possible interface between two fluid phases of a one-component system. The term *surface tension of the liquid* is used for σ in this system.

Many surface-tension measurements are made with air present to a total pressure of 1 atm, which introduces additional components to the system.   While the effect of the air is small, it is not necessarily negligible;[1] for example, the true σ for ethyl alcohol is 22.75 ± 0.3 at 20°C, whereas in air the measured value is 22.27 ± 0.1 dynes/cm.   We recall that in a single-component system there is no surface concentration; consequently, Eq. (29-9) simplifies to

$$d\sigma = -S_s \, dT \qquad (29\text{-}10)$$

The heat absorbed per unit of surface formed reversibly at constant amount of liquid and vapor and hence at constant volume is TS_s, where

$$S_s = -\frac{d\sigma}{dT} \qquad (29\text{-}11)$$

If surface area is lost irreversibly at constant pressure and volume, i.e., the work associated with the surface tension is not extracted, then the heat evolved is H_s, which is

$$H_s = F_s + TS_s = \sigma - T \frac{d\sigma}{dT} \qquad (29\text{-}12)$$

Since there is no volume change, $E_s = H_s$ and Eq. (29-12) gives the *total surface energy* as well as the *surface enthalpy*.

The surface tension of various pure liquids shows a similar trend with temperature.   The empirical equation

$$\sigma = \sigma_0 (1 - T_r)^{1\frac{1}{9}} \qquad (29\text{-}13)*$$

is quite successful.[2]   Here T_r is the reduced temperature T/T_c, and σ_0 is an empirical constant which may be regarded as the surface tension of a hypothetical supercooled liquid at 0°K.   The constant σ_0 can be correlated with other properties of the liquid; Appendix 1 includes such a correlation for normal liquids.   Where Eq. (29-13)* holds, the surface entropy is

$$S_s = \frac{11}{9} \frac{\sigma_0}{T_c} (1 - T_r)^{\frac{2}{9}} \qquad (29\text{-}14)*$$

---

[1] The distinction between σ in air and for the pure vapor-liquid interface is clearly made in the excellent section by T. F. Young and W. D. Harkins in the "International Critical Tables," vol. IV, p. 446, McGraw-Hill Book Company, Inc., New York, 1928.

[2] A. Furgeson, *Trans. Faraday Soc.*, **19**, 408 (1923); E. A. Guggenheim, *J. Chem. Phys.*, **13**, 253 (1945); L. Riedel, *Chem.-Ing.-Tech.*, **27**, 209 (1955).

We note that both σ and S_s drop to zero at the critical temperature, as they must. However, while σ decreases rapidly with rising temperature over the full range below T_c, the surface entropy is more nearly constant until the temperature is close to T_c.

Two-component Systems. There are two types of interface between fluid phases in two-component systems which are of interest. First is the vapor-liquid interface analogous to that of the one-component system. Also there may be a liquid-liquid interface between immiscible components, which will be considered later. In the first case temperature and liquid composition are the natural independent variables, and there is one surface concentration. Equation (29-9) becomes

$$d\sigma = -S_s\,dT - \Gamma_2\,d\mu_2 \qquad (29\text{-}15)$$

where we have chosen $\Gamma_1 = 0$ and abbreviated $\Gamma_{2(1)}$ to Γ_2. Also the definition of surface entropy must be consistent with the choice $\Gamma_1 = 0$. It is natural to select x_2 of the liquid as the independent composition variable; hence

$$d\mu_2 = -\bar{s}_2\,dT + \bar{v}_2\,dP + \left(\frac{\partial\mu_2}{\partial x_2}\right)_{T,P} dx_2 \qquad (29\text{-}16)$$

The pressure is determined by the properties of the bulk phases since it is the equilibrium vapor pressure of the system. Hence we may rewrite Eq. (29-16) as

$$d\mu_2 = -\left[\bar{s}_2 - \bar{v}_2\left(\frac{\partial P}{\partial T}\right)_{x_2,\text{sat}}\right] dT + \left(\frac{\partial\mu_2}{x_2}\right)_{T,\text{sat}} dx_2 \qquad (29\text{-}17)$$

and this may be combined with Eq. (29-15) to yield

$$d\sigma = -\left[S_s - \Gamma_2\bar{s}_2 + \Gamma_2\bar{v}_2\left(\frac{\partial P}{\partial T}\right)_{x_2,\text{sat}}\right] dT - \Gamma_2\left(\frac{\partial\mu_2}{\partial x_2}\right)_{T,\text{sat}} dx_2 \quad (29\text{-}18)$$

The partial molal entropy and volume are for the solute in the liquid phase (since x_2 was taken for the liquid).

The quantity in brackets which is $-(\partial\sigma/\partial T)_{x_2}$ may be shown to be the entropy increase per unit increase of surface at constant volume. The first term S_s is the entropy of the surface created, and $\Gamma_2\bar{s}_2$ is the entropy of the Γ_2 moles of component 2 as it previously existed in the bulk liquid. The removal of the Γ_2 moles of component 2 leaves a volume $\Gamma_2\bar{v}_2$ which must be filled with vapor. Since $\partial P/\partial T$ is just $\Delta S/\Delta V$ of vaporization, we see that the third term is the entropy of vaporization required to fill this volume $\Gamma_2\bar{v}_2$ and thus the demonstration is complete.

This argument has shown that the heat absorbed in the reversible increase of surface by unit area at constant volume is $-T(\partial\sigma/\partial T)_{x_2}$ for the two-component system just as for the one-component system. Indeed, a relationship between macroscopic properties of this sort must be independent of the number of components present, as we shall demonstrate presently. Likewise the heat effect of irreversible destruction of vapor-liquid surface area at constant volume is given by $\sigma - T(\partial \sigma / \partial T)$ for a solution as well as for a single component.

The new situation which arises for a two-component system is the possibility of determining a surface concentration from the change of σ with composition. Equation (29-18) yields

$$\Gamma_2 = - \left(\frac{\partial \mu_2}{\partial x_2}\right)_{T,\text{sat}}^{-1} \left(\frac{\partial \sigma}{\partial x_2}\right)_T \tag{29-19}$$

The use of fugacity as the independent variable gives a simpler equation, which is also convenient because the partial pressure may be used as the fugacity at low pressures.

$$\Gamma_2 = - \frac{1}{RT} \left(\frac{\partial \sigma}{\partial \ln f_2}\right)_T \tag{29-20}$$

In an ideal solution the fugacity is proportional to the mole fraction and

$$\Gamma_2 = - \frac{1}{RT} \left(\frac{\partial \sigma}{\partial \ln x_2}\right)_T \qquad \text{ideal solution} \tag{29-21}*$$

From any of the last three equations one sees that a solute which lowers the surface tension is concentrated in the surface, that is, $\Gamma_2 > 0$, while a solute which raises surface tension is less abundant in the surface than in the bulk phase and $\Gamma_2 < 0$. These results are readily understood in terms of molecular behavior.

Let us apply Eq. (29-20) to the data on ethyl alcohol–water solutions following Butler and Wightman.[1] In Table 29-1 the surface tension is given as a function of mole fraction of alcohol, x_2, together with the partial pressure of the alcohol. Figure 29-3 shows the general nature of the surface-tension trend with composition, while Fig. 29-4 gives the graph of σ versus log p_2, whose slope may be used to evaluate Γ_2 in Eq. (29-20).

From our general chemical knowledge we expect that alcohol will concentrate on the surface with the OH groups oriented into the liquid, where they may hydrogen-bond to water molecules, and with the alkyl groups on the surface. The minimum area per surface molecule is about 24 A^2 in the 15 mole per cent solution. This is only a little more than the cross-sectional area per chain in n-paraffin crystals; hence the surface of the water-alcohol solution may have an almost complete coverage of

[1] J. A. V. Butler and A. Wightman, *J. Chem. Soc.*, **1932**, 2089; note also an excellent treatment of the n-butyl alcohol–water system by W. D. Harkins and R. W. Wampler, *J. Am. Chem. Soc.*, **53**, 850 (1931).

TABLE 29-1. THE SURFACE TENSION AND THE SURFACE CONCENTRATION FOR WATER–ETHYL ALCOHOL SOLUTIONS AT 25°C

x_2	σ, dynes/cm	$\log p_2$, mm	$\Gamma_2 \times 10^{-13}$, molecules/cm^2	$\Gamma_2 \times 10^{10}$, moles/cm^2
1.00	21.93	1.771		
0.90	22.59	1.722	14.0	2.32
0.80	23.26	1.679	16.5	2.74
0.70	23.93	1.639	17.5	2.91
0.60	24.67	1.600	20.4	3.39
0.50	25.43	1.565	23.0	3.82
0.40	26.43	1.529	29.5	4.90
0.30	27.60	1.492	33.5	5.56
0.25	28.49	1.467	37.8	6.28
0.20	29.97	1.428	40.3	6.69
0.15	32.20	1.372	42.3	7.02
0.12	34.42	1.316	42.2	7.01
0.10	36.72	1.256	40.7	6.76
0.064	42.13	1.097	37.2	6.18
0.04	47.86	0.908	32.2	5.35
0.02	55.57	0.602	26.8	4.45
0.00	71.97			

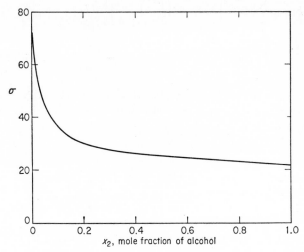

FIG. 29-3. The surface tension of water–ethyl alcohol solutions.

alcohol with most of the ethyl groups perpendicular to the surface. We emphasize, however, that this paragraph of speculation, interesting and probable as it may be, cannot be proved or disproved by thermodynamics.

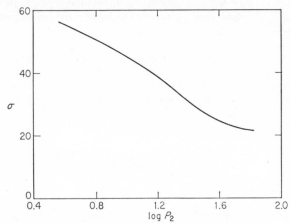

FIG. 29-4. Graph for calculation of Γ_2 for water–ethyl alcohol system.

Multicomponent Systems. The vapor-liquid interface of a system with more than two components has the same properties as have already been discussed. The thermal effects are similar, but the number of surface concentrations is increased. Thus for k components Eq. (29-9) yields

$$d\sigma = -\left\{ S_s - \sum_{i=2}^{k} \Gamma_{i(1)} \left[\bar{S}_i - \bar{V}_i \left(\frac{\partial P}{\partial T} \right)_{x_i,\text{sat}} \right] \right\} dT$$

$$- \sum_{i=2}^{k} \Gamma_{i(1)} RT \, d \ln f_i \quad (29\text{-}22)$$

The temperature derivative of σ gives the entropy quantity in the braces, whose interpretation is analogous to that which was given before. In order to obtain the $k - 1$ surface concentrations, the fugacities (or activities) of the $k - 1$ solute components must be measured in k different solutions with appropriate small differences of composition but at the same temperature. The differences then yield $k - 1$ equations like Eq. (29-22) but with the first term missing since $dT = 0$. These can be solved for the surface concentrations.

AN ALTERNATE APPROACH TO SURFACE PROPERTIES

Before proceeding to consider other examples, let us take an alternate approach to the thermodynamics of flat surfaces. This method deals with observable properties of the whole system together with the surface tension and avoids introducing surface concentrations until later stages. Since we shall wish to consider processes at either constant pressure or constant volume, we need the differential expression for both free energy and work content.   The former was given as Eq. (29-5), and the latter is readily obtained therefrom.

$$dF = -S\,dT + V\,dP + \sigma\,dA_s + \Sigma\,\mu_i\,dn_i \qquad (29\text{-}5)$$
$$dA = -S\,dT - P\,dV + \sigma\,dA_s + \Sigma\,\mu_i\,dn_i \qquad (29\text{-}23)$$

All quantities pertain to the entire system, including bulk phases.

Since each of these expressions is an exact differential, we may use Eq. (3-8) to obtain a series of relationships.   Consider first the entropy

$$\left(\frac{\partial S}{\partial A_s}\right)_{T,P,n_i} = -\left(\frac{\partial \sigma}{\partial T}\right)_{P,A_s,n_i} \qquad (29\text{-}24)$$

$$\left(\frac{\partial S}{\partial A_s}\right)_{T,V,n_i} = -\left(\frac{\partial \sigma}{\partial T}\right)_{V,A_s,n_i} \qquad (29\text{-}25)$$

While these two equations are very similar, the difference between constancy of pressure and volume is important.   For the vapor-liquid problem, one cannot vary temperature at constant pressure and retain both phases; hence, Eq. (29-24) is not applicable.   The use of the constant-volume condition [Eq. (29-25)] yields the results that we have already obtained. $T(\partial S/\partial A_s)_{T,V,n_i}$ is just the heat absorbed for the reversible increase of the surface by unit area at constant temperature, volume, and composition.   For the one-component system, the result is identical with Eq. (29-11).   In the case of two components our earlier discussion was complex, and Eq. (29-25) leads much more directly to the relationship of the heat absorbed to $\partial\sigma/\partial T_{n_i}$, which was obtained eventually from Eq. (29-18).

We may also obtain the earlier result for surface concentration in the two-component vapor-liquid system as follows:

$$\left(\frac{\partial \sigma}{\partial n_2}\right)_{T,V,A_s,n_1} = \left(\frac{\partial \mu_2}{\partial A_s}\right)_{T,V,n_1,n_2} \qquad (29\text{-}26)$$

The escaping tendency of component 2 is affected by the surface area under these conditions only if some material is added to or subtracted from the bulk phases by transfer from or to the surface.   Consequently we consider n_2 as a function of both μ_2 and A_s, and by Eq. (3-3)

$$\left(\frac{\partial \mu_2}{\partial A_s}\right)_{T,V,n_1,n_2} = -\frac{(\partial n_2/\partial A_s)_{\mu_2,T,V,n_1}}{(\partial n_2/\partial \mu_2)_{A_s,T,V,n_1}}$$

The increase of n_2 with surface area at constant μ_2, T, V, and n_1 is just Γ_2

as defined before.   Introduction of these results into Eq. (29-26) with inversion of $\partial n_2 / \partial \mu_2$ yields

$$\left(\frac{\partial \sigma}{\partial n_2}\right)_{T,V,A_s,n_1} = -\Gamma_2 \left(\frac{\partial \mu_2}{\partial n_2}\right)_{A_s,T,V,n_1} \qquad (29\text{-}27)$$

which one may readily show to be equivalent to Eqs. (29-19) and (29-20).

Liquid-Liquid Interface, Two Components.   The case of a liquid-liquid interface between two immiscible components such as benzene and water, although formally similar to the vapor-liquid problem, is quite different in practice.   The composition of each phase is now effectively determined by the equilibrium with the other phase, and the convenient independent variables are temperature and pressure.   While it is possible to apply these conditions to Eq. (29-9), the results are very complex and difficult to interpret.   It seems better to use the alternate approach.

Constant pressure rather than constant volume is the appropriate condition for the temperature derivative, and Eq. (29-24) is now applicable, which gives the entropy absorbed,

$$\left(\frac{\partial S}{\partial A_s}\right)_{T,P,n_i} = -\left(\frac{\partial \sigma}{\partial T}\right)_{P,A_s,n_i} \qquad (29\text{-}24)$$

per unit increase of surface under constant temperature, pressure, and composition.   Since the bulk phases may be made as large as desired, constant composition of the whole system may be interpreted as substantially constant mole fraction in each bulk phase.   It is important to realize, however, that the increase of surface may transfer some material to or from the bulk phases and the surface region.   In the absence of further information, however, it is impossible to determine how much of the entropy increase with surface, $\partial S / \partial A_s$, is associated with material transfer.

The heat effects associated with surface-area changes are related to $T(\partial S / \partial A_s)$ and $(\partial H / \partial A_s)$, as has been discussed before.   For example, the heat evolved on deemulsification at constant T and P is

$$\left(\frac{\partial H}{\partial A_s}\right)_{T,P,n_i} = \sigma - T\left(\frac{\partial \sigma}{\partial T}\right)_{T,P,n_i} \qquad (29\text{-}28)$$

From Eq. (29-5) we may also derive the equation

$$\left(\frac{\partial V}{\partial A_s}\right)_{T,P,n_i} = \left(\frac{\partial \sigma}{\partial P}\right)_{T,A_s,n_i} \qquad (29\text{-}29)$$

This gives the volume increase per unit increase of surface area at constant temperature, pressure, and composition.   While this is a sort of surface volume, it is important to realize that material is also transferred to and from bulk phases and the surface region. Thus a detailed interpretation is not possible without further information. Nevertheless, Eq. (29-29) gives a useful relationship between macroscopic quantities which are potentially observable.

Liquid-Liquid Interface, Three Components. The addition of a third, or solute, component to a system of two immiscible solvents gives a third independent variable, which we may take as n_3 or more conveniently as the concentration of component 3 in one phase or the other. If we hold temperature, pressure, and the amounts of both solvent components constant, we obtain from Eq. (29-5)

$$\left(\frac{\partial \sigma}{\partial n_3}\right)_{T,P,A_s,n_1,n_2} = \left(\frac{\partial \mu_3}{\partial A_s}\right)_{T,P,n_1,n_2,n_3} \tag{29-30}$$

Now we proceed as we did from Eq. (29-26) to Eq. (29-27) for the two-component vapor-liquid case, but omitting the subscripts T, P, n_1, n_2, we obtain first

$$\left(\frac{\partial \mu_3}{\partial A_s}\right)_{n_3} = - \frac{(\partial n_3/\partial A_s)_{\mu_3}}{(\partial n_3/\partial \mu_3)_{A_s}}$$

and then on substitution into Eq. (29-30)

$$\left(\frac{\partial \sigma}{\partial n_3}\right)_{A_s} = - \left(\frac{\partial n_3}{\partial A_s}\right)_{\mu_3} \left(\frac{\partial \mu_3}{\partial n_3}\right)_{A_s} \tag{29-31}$$

The first derivative on the right side of Eq. (29-31) is some sort of surface concentration of component 3. In this system we have no simple way of obtaining $\Gamma_{2(1)}$ (or $\Gamma_{1(2)}$); hence it is difficult to relate $(\partial n_3/\partial A_s)_{\mu_3}$ to $\Gamma_{3(1)}$ or $\Gamma_{3(2)}$. However, if component 3 has a marked effect on the interfacial tension without appreciably changing the mutual solubility of the two solvents, then $(\partial n_3/\partial A_s)_{\mu_3}$ from Eq. (29-31) may be safely interpreted as a surface concentration of component 3 which will approximate $\Gamma_{3(1)}$ or $\Gamma_{3(2)}$.

CURVED INTERFACES

Let us now remove the limitation of planarity of the surface and consider the effects particularly caused by curvature. We shall assume that the surface tension is not affected by surface curvature.[1] This is presumably justified as long as the radius of curvature is large compared with the thickness of the layer within which the properties differ from the bulk phases. We mentioned at the beginning of this chapter that the

---

[1] Treatments considering change of surface tension with curvature have been given by R. C. Tolman, *J. Chem. Phys.*, **17**, 333 (1949); F. O. Koenig, *J. Chem. Phys.*, **18**, 449 (1950); F. P. Buff, *J. Chem. Phys.*, **23**, 419 (1955).

evidence indicates a thickness of only a few molecules, i.e., approximately 10^{-7} cm, under most conditions. Thus curved interfaces with radii of curvature much greater than 10^{-7} cm may be expected to follow the equations to be derived below. If experiments are extended to extremely small droplets, however, deviations may be expected.

Interior Pressure of Bubbles and Drops. A very simple argument shows that the pressure P' within a drop is higher than that outside $P°$. Transfer of a volume dV of liquid from a bulk phase to the interior of the drop requires the work $(P' - P°)\, dV$, which must equal the work of extending the surface $\sigma\, dA_s$. Since $dV = 4\pi r^2\, dr$ and $dA_s = 8\pi r\, dr$, one obtains $dA_s = (2/r)\, dV$ and the Laplace equation[1]

$$P' - P° = \frac{2\sigma}{r} \qquad (29\text{-}32)$$

The excess pressure within a bubble is twice that in a drop of liquid because both the inner and outer surfaces of the thin liquid film contribute separate surface tensions. It is interesting to note that this phenomenon was treated by Kelvin[2] in 1858.

Escaping Tendency from Curved Surfaces. In 1871 Thomson[3] (Lord Kelvin) showed that the vapor pressure of a small droplet will exceed that of a plane surface of the same liquid, while the vapor pressure over a concave surface is decreased. This phenomenon is important in the behavior of finely dispersed volatile material.

We know that increase of pressure raises the fugacity or chemical potential, and we could easily combine this relationship with Eq. (29-32) for the excess pressure inside a drop to obtain the vapor pressure. We shall, however, follow an alternate approach, starting with the general equation for the free energy.

$$dF = -S\, dT + V\, dP + \sigma\, dA_s + \Sigma\mu_i\, dn_i \qquad (29\text{-}5)$$

Our initial applications of this equation were limited to flat surfaces, but the derivation is valid for curved surfaces provided that the surface tension is unchanged by the curvature and provided that the interface within the system does not change the pressure on the exterior. A spherical interface completely within the system satisfies the latter condition. It is also important to note that the chemical potentials in Eq. (29-5) pertain to transfer of matter by a process which does not change

---

[1] This equation was given in 1806 by P. S. de Laplace in his celebrated "Mécanique céleste." The more general form for nonspherical surfaces is $P' - P° = \sigma(1/r_1 + 1/r_2)$, where r_1 and r_2 are the principal radii of curvature.

[2] W. Thomson (Lord Kelvin), *Proc. Roy. Soc. (London)*, **9**, 255 (1858); *Phil. Mag.*, (4)**17**, 61 (1859).

[3] W. Thomson, *Phil. Mag.*, (4)**42**, 448 (1871).

A_s.   Transfer across a planar interface satisfies this criterion; consequently we shall write $\mu_i^{(P)}$ to emphasize the restriction and note that

$$\mu_i^{(P)} = \left(\frac{\partial F}{\partial n_i}\right)_{T,P,A_s,n_j}$$

For a spherical drop, however, the surface area no longer is independently variable but rather is dependent on the volume and therefore the amount of material in the drop.   Consequently, the term $\sigma\, dA_s$ in Eq. (29-5) must be expressed in terms of the various increments of matter, dn_i, and considered in obtaining the true escaping tendency from the drop.

The volume change on addition of increments dn_i moles of the various components is

$$dV = \Sigma \bar{v}_i\, dn_i$$

and, in view of the geometrical properties of a sphere,

$$dA_s = \frac{2\, dV}{r} = \sum \frac{2\bar{v}_i}{r}\, dn_i$$

Combining the last result with Eq. (29-5) yields

$$dF = -S\, dT + V\, dP + \sum \left(\mu_i^{(P)} + \frac{2\bar{v}_i\sigma}{r}\right) dn_i \qquad (29\text{-}33)$$

The escaping tendency of the ith component from the drop is the derivative of the free energy as given in Eq. (29-33) with respect to n_i,

$$\mu_i = \left(\frac{\partial F}{\partial n_i}\right)_{T,P,n_j} = \mu_i^{(P)} + \frac{2\bar{v}_i\sigma}{r}$$

Thus the increased chemical potential for the droplet as compared with the material with a plane surface is

$$\mu_i - \mu_i^{(P)} = \frac{2\bar{v}_i\sigma}{r} \qquad (29\text{-}34)$$

It is convenient to express this result in terms of the fugacity which is approximated by the vapor pressure (unless the pressure is high).

$$\ln \frac{f_i}{f_i^{(P)}} = \frac{2\bar{v}_i\sigma}{rRT} \qquad (29\text{-}35)$$

If the surface is concave toward the vapor, the fugacity is decreased instead of being increased.   Equations (29-34) and (29-35) remain applicable except for the change in sign and the identification of r as a radius of spherical curvature.   It is also possible to consider surfaces with other than spherical curvature, whereupon the factor $2/r$ is replaced by $1/r_1 + 1/r_2$, where r_1 and r_2 are the two radii of curvature.   Thus, for a cylindrical surface, the cylinder radius is r_1 and $r_2 = \infty$.

While these results have been accepted for many years, it is only recently that Thomä[1] and La Mer and Gruen[2] have verified these effects experimentally. Thomä measured the change in vapor pressure for a curved meniscus of isovaleric acid in a fine capillary. In the experiments of La Mer and Gruen uniform-sized droplets of a nonvolatile solvent were produced and then equilibrated with the vapor over a solution of the same solvent and a volatile solute. The droplets rapidly grew to a larger size, which was measured by light-scattering methods. By study of drop size as a function of the fugacity of the solute, it was possible to check Eq. (29-35). The solvent-solute systems tested were dioctyl phthalate–toluene and oleic acid–chloroform.

Measurement of Surface Tension. Most of the methods for the measurement of surface tension depend on the relationship of pressure difference to surface curvature given in Eq. (29-32). In the capillary-rise method the surface is spherical, and the theory is simple provided that the contact angle is known. Methods such as that of measuring the maximum pull to lift a plate or ring out of the surface depend primarily on the tension of flat or nearly flat surfaces, but even these methods are subject to corrections which are required by the curvature of the surface near the plate or ring.

FILMS OF AN INSOLUBLE COMPONENT

If one component of a system is present only on the surface or only in a vapor phase in addition to the surface, then the amount of material on the surface becomes a directly measurable quantity. There are two common examples which have received a great deal of attention. First is the adsorption of a gas on a liquid or solid. Here the vapor pressure is a direct measure of the fugacity of the surface component and its amount may be determined by subtracting the amount remaining as vapor from the total added. The second example is an insoluble and nonvolatile film at a liquid-gas interface. In this case the surface component is insoluble in both bulk phases, but its free energy can be determined by the film pressure, which is the difference in surface tension from that of the pure liquid. While these two examples present very different experimental problems, their interpretation is essentially similar. Let us first relate the different quantities measured in the two cases.

Let us take the fugacity as our measure of escaping tendency. The surface concentration Γ will be used without subscript since it is unambiguous in these cases. Actually the surface area of the solid may be uncertain, but that is a separate problem from the one before us now. The

[1] M. Thomä, *Z. Physik*, **64**, 224 (1930).
[2] V. K. La Mer and R. Gruen, *Trans. Faraday Soc.*, **48**, 410 (1952).

film pressure on a liquid surface may be measured directly by a delicate film balance, which measures the force on a barrier separating a surface film on one side from a clean liquid-gas interface on the other. The force divided by the barrier length gives the film pressure π, which is the difference between the surface tension σ° of the clean liquid and σ of the surface with the film.

$$\pi = \sigma^\circ - \sigma \tag{29-36}$$

Let us start with the following identity,

$$\left(\frac{\partial \ln f}{\partial \Gamma}\right)_T = \left(\frac{\partial \ln f}{\partial \pi}\right)_T \left(\frac{\partial \pi}{\partial \Gamma}\right)_T \tag{29-37}$$

where f and Γ both refer to the film component, and recall the Gibbs adsorption equation, which we take as Eq. (29-15) reduced to isothermal conditions,

$$d\sigma = -\Gamma \, d\mu = -\Gamma RT \, d \ln f$$

or
$$\left(\frac{\partial \ln f}{\partial \pi}\right)_T = \frac{1}{\Gamma RT} \tag{29-38}$$

since $d\pi = -d\sigma$ from Eq. (29-36). The combination of Eqs. (29-37) and (29-38) yields

$$\left(\frac{\partial \pi}{\partial \Gamma}\right)_T = \Gamma RT \left(\frac{\partial \ln f}{\partial \Gamma}\right)_T \tag{29-39}$$

which gives the $\pi - \Gamma$ behavior of an adsorbed film if the $f - \Gamma$ isotherm is available, or vice versa.

Actual adsorption data are commonly reported in terms of a gas volume V at standard conditions where the molal volume is V_M and the area of solid surface Σ. At low pressures $f \cong P$. Also it is customary to deal with the area per mole α, which is just the reciprocal of Γ. In these terms

$$\alpha = \Gamma^{-1} = \frac{V_M \Sigma}{V}$$

and Eq. (29-39) becomes

$$\left(\frac{\partial \pi}{\partial \alpha}\right)_T = -\frac{V^3 RT}{V_M^2 \Sigma^2} \left(\frac{\partial \ln f}{\partial V}\right)_T \tag{29-40}$$

Two-dimensional Gases, Liquids, and Solids. By analogy to three-dimensional phases we might expect surface films to show gaseous behavior at low surface concentrations, where the molecules are far apart, and to show liquidlike and possibly solidlike behavior as the surface molecules come close together. In some respects the film pressure is more analogous to an osmotic pressure than a gas pressure, but the same pressure is predicted on either basis.

There is every theoretical reason to expect a two-dimensional perfect gas to follow the equation

$$\pi \mathfrak{a} = RT \qquad (29\text{-}41)^*$$

and to have other properties analogous to those of a three-dimensional perfect gas. While measurements of very dilute films are particularly difficult, the evidence clearly indicates that this ideal-gas law is approached at low surface concentration on both liquid and solid surfaces.

For the perfect two-dimensional gas on a solid surface, Eqs. (29-40) and (29-41)* yield the isotherm

$$P \cong f = \text{constant} \times V \qquad (29\text{-}42)^*$$

At higher concentrations one or more different surface phases are observed, which have been given various names. It is beyond the scope of our discussion to consider the particular properties of these liquidlike and solidlike phases except to note that some are still quite compressible. The work of Copeland and Harkins[1] provides excellent examples of film-pressure investigations.

Adsorption of Gases.[2]   Our treatment is limited to reversible adsorption processes. Although the adsorption of gases on liquid-vapor interfaces has been studied,[3] much .more attention has been given to adsorption on finely divided solids of large surface area. In this case the surface area is not directly measurable, and while reasonably good indirect estimates are available, it is preferable to avoid unnecessary use of this area in presenting adsorption data and calculations. One does measure the amount of gas adsorbed, which may be recorded in moles n_s or more commonly in volume at STP for a fixed amount of solid. The adsorption equilibrium (or isotherm) then becomes a relationship of the volume adsorbed to the gas pressure.

A typical adsorption isotherm for argon on rutile at 85°K is shown in Fig. 29-5. These results are taken from the particularly extensive study by Morrison, Los, and Drain,[4] to which we shall refer repeatedly. From Fig. 29-5 we see that several hundred cubic centimeters is adsorbed at a very low pressure. There is good reason to believe that this material forms a monomolecular layer. The volume of gas required to complete

[1] L. E. Copeland and W. D. Harkins, *J. Am. Chem. Soc.*, **64**, 1600 (1942); *J. Chem. Phys.*, **10**, 272 (1942).

[2] The thermodynamics of gas adsorption has been treated by several authors. T. L. Hill has given especially complete and careful discussions in *J. Chem. Phys.*, **17**, 520 (1949), **18**, 246 (1950). See also T. L. Hill, P. H. Emmett, and J. G. Joiner, *J. Am. Chem. Soc.*, **73**, 5102 (1951).

[3] For example, see C. L. Cutting and D. C. Jones, *J. Chem. Soc.*, **1955**, p. 4067.

[4] J. A. Morrison, J. M. Los, and L. E. Drain, *Trans. Faraday Soc.*, **47**, 1023 (1951), **48**, 840 (1952); *J. Chem. Phys.*, **19**, 1063 (1951).

the monolayer V_m cannot be obtained from thermodynamics but is inferred from a theory of the detailed mechanism of adsorption.[1] We shall see, however, that a self-consistent interpretation of several phenomena is obtained on this basis.

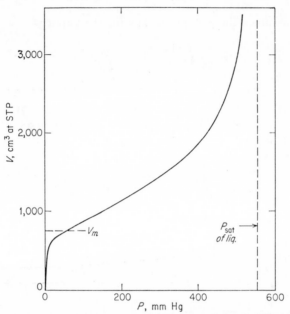

Fig. 29-5. The adsorption equilibrium curve for argon on rutile at 85°K. V_m indicates the complete monolayer according to Brunauer-Emmett-Teller theory.

Although the pressure rises sharply after the monolayer is complete, several additional layers are adsorbed at a pressure appreciably less than the vapor pressure of the bulk liquid.

The heat of adsorption may be measured calorimetrically. While one adds a finite amount of material in a single experiment, the resulting curve for a series of experiments can be differentiated to obtain the *partial molal* (or differential) *heat of adsorption* $\Delta \bar{H}$. This quantity is commonly called the *isosteric heat of adsorption*, since it pertains to a fixed surface coverage, and is given the symbol q_{st}. Data for argon on rutile are shown in Fig. 29-6. The large variation in q_{st} in the range below a monolayer indicates probable nonuniformity of solid surface. One notes that the partial molal heat of adsorption rapidly approaches the heat of vaporization of the liquid after the monolayer is complete.

---

[1] For example, the theory of S. Brunauer, P. H. Emmett, and E. Teller, *J. Am. Chem. Soc.*, **60**, 309 (1938).

The heat of adsorption, like the heat of vaporization of the liquid, may also be obtained from the temperature derivative of the adsorption pressure,

$$\left(\frac{\partial P}{\partial T}\right)_{n_s} = \frac{\Delta \bar{H}}{T \, \Delta v} = \frac{q_{st}}{T \, \Delta v} \tag{29-43}$$

and if one assumes the perfect-gas law for the vapor,

$$\left(\frac{\partial \ln P}{\partial T}\right)_{n_s} = \frac{q_{st}}{RT^2} \tag{29-44}*$$

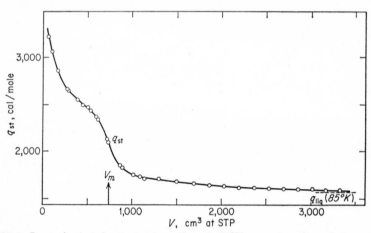

FIG. 29-6. Isosteric (q_{st}) heat of adsorption at 85°K as a function of the volume adsorbed. The dashed line on the right shows the heat of vaporization of the liquid. [*L. E. Drain and J. A. Morrison, Trans. Faraday Soc.,* **48**, 840 (1952), fig. 2.]

Entropy of Adsorbed Gases. The partial molal entropy of an adsorbed material may be obtained from the temperature derivative of the free energy of adsorption together with the entropy of the gas in its standard state,

$$\bar{s}_s = -\frac{\partial}{\partial T} (RT \ln f) + s° \tag{29-45}$$

where it is convenient to express the free energy of adsorption in terms of the equilibrium fugacity of the gas, which is, of course, approximately the gas pressure. If the differentiation in Eq. (29-45) is carried out, the heat of adsorption q_{st} may be identified and one has

$$\bar{s}_s = \frac{q_{st}}{T} - R \ln f + s° \tag{29-46}$$

and we also note that for an ideal gas the last two terms constitute the entropy at the equilibrium pressure and the first term on the right is the partial molal entropy of adsorption from the equilibrium gas.

Drain and Morrison[1] calculated the partial molal entropy of argon on rutile from their adsorption pressures and their calorimetric heats of adsorption and obtained the results shown in Fig. 29-7. Two features

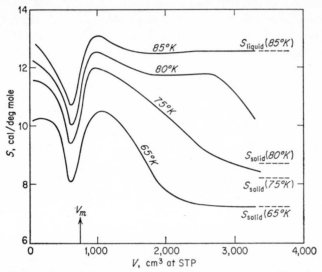

FIG. 29-7. The partial molal entropy of adsorbed argon as a function of the volume adsorbed. [*L. E. Drain and J. A. Morrison, Trans. Faraday Soc.*, **48**, 840 (1952), fig. 4.]

are of particular interest. First is the sharp dip in the partial molal entropy when the monolayer is almost complete. The freedom of motion of the adsorbed atoms is particularly restricted under these conditions because almost all the surface is already filled. The molal entropy at this minimum is not very different from that of solid argon, and one can readily see that an atom in a tightly bound and nearly complete surface layer would be restricted in its motion to about the same extent as in a crystal. It is also interesting to note in Fig. 29-7 that the partial molal entropy for outer layers approaches that of the solid or the liquid in accordance with the temperature.

By sufficiently sensitive calorimetry it is also possible to measure the partial molal heat capacity of adsorbed material. Heat capacities are measured for the bare solid and then for the solid with successively greater amounts of gas adsorbed on it. These data permit the calculation of the change of partial molal entropy with temperature in the usual manner,

$$\mathbf{s}_s(T_2) = \mathbf{s}_s(T_1) + \int_{T_1}^{T_2} \bar{c}_s \, d \ln T \qquad (29\text{-}47)$$

[1] L. E. Drain and J. A. Morrison, *Trans. Faraday Soc.*, **48**, 840 (1952).

By this means Morrison, Los, and Drain calculated the entropy of adsorbed argon at 15°K and extrapolated on to 0°K. They extrapolated along the Einstein curve and found about 0.5 cal/deg mole residual entropy throughout the range from a fraction of a monolayer to nearly two molecular layers. If the extrapolation is made by the Debye theory instead, the entropy at 0°K drops to about 0.2 cal/deg mole, which is zero within experimental error. Until a more adequate basis is available for the extrapolation below 15°K, one can only conclude that little, if any, entropy remains in the adsorbed film of argon on rutile at 0°K.

SURFACE ENERGY AND ENTROPY OF SOLIDS

Since one cannot change the surface area of a solid reversibly, the study of the thermodynamic surface properties of solids is more difficult than, and requires different methods from, such a study of liquids. Also solid surfaces differ according to the crystal face exposed, and there are possibilities of cracks and other irregularities. In spite of these difficulties considerable progress is being made. The heat capacity or heat of solution may be measured for a sample of very large crystals and then for a sample of very small crystals, whereupon the difference yields the surface heat capacity or the surface energy, respectively. Integration of the usual function of the heat capacity yields the entropy, and this in turn allows the calculation of the surface free energy. Jura and Garland[1] applied these methods to magnesium oxide, with the results

$$\frac{S_s}{A_s} = 0.28 \text{ erg/cm}^2 \text{ deg}$$

and $F_s/A_s = \sigma = 1000$ erg cm² at 298°K.

For extremely small solid particles, as for extremely small liquid droplets, the assumption of constant surface properties will fail. In the case of solids one might consider edge effects. In the extreme it may be more convenient to treat a very small solid particle as if it were a large molecule[2] and to characterize the system by the number, size, and shape of the particles rather than by the mass and surface area.

PROBLEMS

29-1. What is the equations for the surface heat capacity C_s implied by Eq. (29-13)* for the surface tension of a normal liquid?

29-2. Calculate the vapor pressure of 0.1-μ (10^{-5}-cm) diameter droplets of water at 25°C. What is the vapor-pressure ratio for these droplets to that of a plane surface of water?

[1] G. Jura and C. W. Garland, *J. Am. Chem. Soc.*, **74**, 6033 (1952).

[2] G. Jura and K. S. Pitzer, *J. Am. Chem. Soc.*, **74**, 6030 (1952).

29-3. What composition of aqueou KCl in droplets of 0.2μ diameter would be in equilibrium with a plane surface of pure water?   Neglect the effect of KCl on the surface tension; see Table A4-2 for the properties of KCl solutions.

29-4. Derive Eq. (29-35) by considering the excess pressure within a small droplet, together with the usual equation for the change of fugacity with pressure.

29-5. Derive an equation for the change with temperature of P/P_{sat} for an adsorption equilibrium (where P_{sat} is the vapor pressure of the bulk liquid).   Sketch curves for $V(STP)$ versus $\log (P/P_{sat})$ for argon on rutile at 85 and $100°K$, based upon the data in Figs. 29-5 and 29-6.

30

SYSTEMS INVOLVING GRAVITATIONAL
OR CENTRIFUGAL FIELDS

In many thermodynamic calculations it can be assumed, as we have usually assumed hitherto, that the thermodynamic state of a substance is determined solely by temperature, pressure, and composition; but there are also many examples in which it is necessary to consider other independent variables. The effects of surfaces were considered in Chapter 29; in this chapter we turn our attention to problems involving centrifugal and gravitational fields.

In treating problems of these more complex types we choose to include in the energy-content function E the energy associated with these additional phenomena. This affects, of course, all the related functions, including particularly the partial molal free energy, which remains a measure of the escaping tendency of the substances present. Thus the equality of $\bar{F}_i$ or a_i throughout the system remains the criterion of equilibrium of distribution of the ith component.

The reader should realize that it is possible and equally correct to retain the symbols E, etc., for the "intrinsic" energy, etc., in the absence of the gravitational or other field and to add the proper additional terms to all equations. Gibbs[1] followed this latter procedure; some recent authors follow one system and some the other.[2]

If 1 mole of substance of molal mass M is situated in a gravitational field, the free energy depends upon the position. If the substance is lifted in a reversible manner, the increase in the free energy of the substance is equal to the work done upon it. All evidence indicates, and we shall assume, that there is no thermal effect and hence no entropy change on reversibly lifting a weight.

$$d_{\mathrm{F}} = Mg\,dh \qquad (30\text{-}1)$$

or
$$\left(\frac{\partial \mathrm{F}}{\partial h}\right)_{P,T,\dots} = Mg \qquad (30\text{-}2)$$

---

[1] J. W. Gibbs, "Collected Works," vol. 1, Thermodynamics, Longmans, Green & Co., Inc., New York, 1928.

[2] Also note that Guggenheim follows Gibbs for gravitational effects but includes electrical energies in his free-energy and electrochemical potentials. E. A. Guggenheim, "Thermodynamics," North Holland Publishing Company, Amsterdam, 1950.

This is the fundamental equation for the change of free energy due to a change in position in a gravitational field, when all other variables which might affect the thermodynamic properties are constant.

Consider a vertical column of a pure liquid in a state of equilibrium. Then the molal free energy must be constant throughout the tube. In the lower part the molal free energy is diminished on account of the influence of gravity, but it is increased by the hydrostatic pressure; and if our equations are correct these two influences must just offset one another. The molal free energy being affected only by the position in the gravitational field and by pressure, we write

$$d_\mathrm{F} = \frac{\partial \mathrm{F}}{\partial h}\, dh + \frac{\partial \mathrm{F}}{\partial P}\, dP = 0 \qquad (30\text{-}3)$$

or by Eqs. (30-2) and (10-16)

$$M g\, dh + \mathrm{v}\, dP = 0 \qquad (30\text{-}4)$$

But the change of pressure with the height is determined by the density ρ, so that

$$\frac{\partial P}{\partial h} = -\rho g \qquad (30\text{-}5)$$

and since ρ is evidently equal to M/v, Eq. (30-4) is satisfied.

It is interesting to consider the variation with height of the pressure of a perfect gas in a gravitational field.   First we rearrange Eq. (30-4),

$$d \ln P = -\frac{M g}{R T}\, dh \qquad (30\text{-}6)^*$$

Then if M, g, and T are constants, one obtains

$$\ln \frac{P''}{P'} = \frac{M g}{R T}\, (h' - h'') \qquad (30\text{-}7)^*$$

In the earth's atmosphere M, g, and T all vary with height; consequently Eq. (30-7)* gives at best a crude approximation to atmospheric pressure over a limited altitude range.

Solutions.   The development of efficient centrifuges has made possible many important investigations of the variation or solution composition in a centrifugal field.   In view of the much greater practical importance of centrifugal as compared with gravitational applications, we note at once that the transformation to centrifugal acceleration is given by

$$g\, dh = -\omega^2 r\, dr \qquad (30\text{-}8)$$

where ω is the angular velocity (ordinarily in radians per second) and r is the distance from the axis of rotation. We shall express all further formulas in terms appropriate to the centrifugal problem; their adaptation to the gravitational case is straightforward.

Again we note that centrifugal energy is a pure work term with no entropy effect (at constant pressure, temperature, and composition). Consequently, in a solution the partial molal free energy varies with radius as follows:

$$\left(\frac{\partial \bar{F}_i}{\partial r}\right)_{P,T,x} = -M_i \omega^2 r \tag{30-9}$$

Now consider a radial column of the solution after it has reached equilibrium. At different radii the pressure and composition may vary, but the partial molal free energy must be constant. So for any constituent

$$d\bar{F}_i = 0 = \left(\frac{\partial \bar{F}_i}{\partial r}\right) dr + \left(\frac{\partial \bar{F}_i}{\partial P}\right) dP + \left(\frac{\partial \bar{F}_i}{\partial x_i}\right) dx_i \tag{30-10}$$

But

$$dP = \rho \omega^2 r \, dr \tag{30-11}$$

where ρ is the density of the solution. Hence, substituting from Eqs. (30-9) and (17-14), we find

$$(\rho \bar{V}_i - M_i)\omega^2 r \, dr + \left(\frac{\partial \bar{F}_i}{\partial x_i}\right) dx_i = 0 \tag{30-12}$$

which relates the radial variation of composition to the molecular weight, partial molal volume, and solution density.

One of the very important uses of the centrifuge has been the determination of the molecular weight of substances composed of very large molecules. Recently centrifuge investigations have considered more subtle properties such as activity coefficients. We shall first integrate Eq. (30-12) under highly simplifying assumptions.

First we assume an ideal solution and that the product $\rho \bar{V}_i$ is a constant independent of the radius. Then, substitution from Eq. (18-3)* yields

$$\frac{dx_i}{x_i} = \frac{(M_i - \rho \bar{V}_i)\omega^2}{RT} r \, dr \tag{30-13}*$$

which upon integration from r', x_i' to r'', x_i'' becomes

$$\ln \frac{x_i''}{x_i'} = \frac{(M_i - \rho \bar{V}_i)\omega^2}{2RT} [(r'')^2 - (r')^2] \tag{30-14}*$$

Another simple case is that of a perfect gaseous solution. In that case it is easier to deal with the partial pressure[1] of each component, $p_i = x_i P$.

---

[1] We shall not use the fugacity function in gravitational or centrifugal problems, although it might be desirable to do so in the future. Presumably one would define f_i to become equal to p_i at the reference height or radius, that is, $r = 0$ or $h = 0$.

Then

$$dF_i = 0 = \left(\frac{\partial F_i}{\partial r}\right) dr + \left(\frac{\partial F_i}{\partial p_i}\right) dp_i$$

$$= -M_i \omega^2 r\, dr + RT\, d \ln p_i \qquad (30\text{-}15)^*$$

$$\ln \frac{p_i''}{p_i'} = \frac{M_i \omega^2}{2RT} [(r'')^2 - (r')^2] \qquad (30\text{-}16)^*$$

This result is equivalent to that from Eq. (30-6)* for a pure perfect gas. As we expect from the properties of perfect gases, the presence of other components had no effect on the behavior of a given component.

In order to investigate the separation of two gaseous components, we subtract Eq. (30-16)* for one component from that for the other

$$\ln \frac{p_2''/p_1''}{p_2'/p_1'} = \frac{(M_2 - M_1)\omega^2}{2RT} [(r'')^2 - (r')^2] \qquad (30\text{-}17)^*$$

It is interesting to note that the separation depends only on the absolute difference in molecular weights—not on the relative difference. Thus it would be as easy to separate by centrifugation heavy isotopes differing by 1 mass unit as to separate He^3 from He^4 (provided that the temperature of the centrifuge remains unchanged).

If the molecular weight of the solute is known, as well as the solution density and partial molal volumes as a function of pressure, then it is possible to derive activities and activity coefficients from equilibrium centrifugal separation data.[1] We shall obtain an equation for the ratio a_β/a_α of the activity of a solute component at compositions x_β, x_α (or m_β, m_α) but at constant P_α and r_α.

Consider the following three processes of transfer of the solute component J and the free-energy change ΔF which corresponds to each:

I:
$$J(r_\beta, P_\beta, x_\beta) \rightarrow J(r_\alpha, P_\beta, x_\beta)$$
$$\Delta F_I = -\int_{r_\beta}^{r_\alpha} M \omega^2 r\, dr$$

II:
$$J(r_\alpha, P_\beta, x_\beta) \rightarrow J(r_\alpha, P_\alpha, x_\beta)$$
$$\Delta F_{II} = \int_{P_\beta}^{P_\alpha} \bar{v}(x_\beta)\, dP$$

III:
$$J(r_\alpha, P_\alpha, x_\beta) \rightarrow J(r_\alpha, P_\alpha, x_\alpha)$$
$$\Delta F_{III} = RT \ln \frac{a_\alpha}{a_\beta}$$

But the sum of all three processes transfers the substance J from position β to position α in the centrifuge at equilibrium; hence

$$\Delta F_I + \Delta F_{II} + \Delta F_{III} = 0$$

---

[1] T. F. Young, K. A. Kraus, and J. S. Johnson, *J. Chem. Phys.*, **22**, 878 (1954); *J. Am. Chem. Soc.*, **76**, 1436 (1954).

The desired quantity is $-\Delta \bar{F}_{III}$. After integrating $\Delta \bar{F}_I$ and reversing the limits of integration in $\Delta \bar{F}_{II}$, we find

$$RT \ln \frac{a_\beta}{a_\alpha} = \frac{M\omega^2}{2} (r_\beta^2 - r_\alpha^2) - \int_{P_\alpha}^{P_\beta} \bar{v}_\beta \, dP \qquad (30\text{-}18)$$

Ordinarily one selects a_α, r_α, etc., to refer to the meniscus of the solution in the centrifuge cell, whereupon $P_\alpha \cong 1$ atm and the activities are directly comparable with those measured by other methods. It should be emphasized that $\bar{v}_\beta$ is the partial molal volume of the solute J at the constant solution composition x_β (or m_β) but under varying pressure from P_α to P_β.

Unless the pressure is measured separately, it must be calculated by integration of Eq. (30-11).

$$P_\beta - P_\alpha = \omega^2 \int_{r_\alpha}^{r_\beta} \rho r \, dr \qquad (30\text{-}19)$$

Here the density ρ must be known as a function of r as it actually exists in the centrifuge at equilibrium.

While it was possible[1] to obtain good activity data for as low a molecular-weight substance as CdI_2, it would appear that this method would have particular advantages for activity studies of very large molecules as soon as their molecular weights can be determined independently with precision.

PROBLEMS

30-1. Show that the substitution of $x_i = p_i/P$ together with other properties of perfect gases reduces Eq. (30-13)* to a form which integrates to Eq. (30-16)*.

30-2. Consider the centrifugal separation of uranium isotopes as UF_6 assumed to be a perfect gas at 100°C. What angular velocity is necessary to obtain enrichment of U^{235} to 1 per cent at $r = 3$ cm if its abundance is 0.7 per cent at $r = 10$ cm?

[1] *Ibid.*

31

SYSTEMS INVOLVING ELECTRIC
OR MAGNETIC FIELDS

While electric fields have entered our discussions of galvanic cells and of electrolyte solutions, the emphasis in these cases was upon the motion and properties of charged ions. At this time we wish to turn our attention to the effect of electric fields on polarizable or dielectric material. It is convenient to consider also the similar, but not identical, effects of magnetic fields on magnetically polarizable material, which then leads to a very important topic—the production of extremely low temperatures by adiabatic demagnetization of paramagnetic materials.

In our general discussion of the thermodynamics of polarizable material in the presence of electric and magnetic fields we shall use mks units and quantities defined in the rational system, although some magnetic field results will be given also in cgs electromagnetic units (emu). Since the discussion in many texts of the work associated with electric and magnetic fields is frequently limited by simplifying assumptions, we derive an expression of general validity in Appendix 8 directly from Maxwell's equations. In our equations we shall use vector notation although in most cases of practical interest all vectors are either parallel or anti-parallel and hence their products are just the products of absolute magnitudes with the proper sign. We shall retain the boldface type for such vectors, in any case, to distinguish electric field $\mathbf{E}$ from energy content E, etc. The resulting equation for the electromagnetic work required to establish a system of fields is

$$w = \int\int \mathbf{E} \cdot d\mathbf{D} \, dV + \int\int \mathbf{H} \cdot d\mathbf{B} \, dV \qquad (31\text{-}1)$$

where the volume integrals are over all space and the field integrals are from zero field to the final state. $\mathbf{E}$ and $\mathbf{H}$ are the electric- and magnetic-field vectors, respectively, while $\mathbf{D}$ and $\mathbf{B}$ are known, respectively, as the electric displacement and the magnetic induction.

In free space $\mathbf{D}$ is proportional to $\mathbf{E}$ and $\mathbf{B}$ to $\mathbf{H}$ with the proportionality constants ϵ_0 and μ_0,[†] called the permittivity and permeability, respectively, of free space.

† Care must be taken not to confuse μ in this usage with the chemical potential.

$$\mathbf{D} = \epsilon_0 \mathbf{E} \qquad \text{and} \qquad \mathbf{B} = \mu_0 \mathbf{H} \qquad (31\text{-}2)$$

The selection of one of these constants is arbitrary, but their product $\epsilon_0 \mu_0$ may be shown to be the reciprocal of the square of the velocity of electromagnetic waves in free space, that is, $\epsilon_0 \mu_0 = c^{-2}$. In the mks system μ_0 is chosen to be $4\pi \times 10^{-7}$ kg m/coulomb² or henry/m. Then ϵ_0 is 8.854×10^{-12} coulomb² sec²/kg m³ or farad/m.

In other media $\mathbf{D}$ and $\mathbf{B}$ are expressed in a formally similar manner, but ϵ and μ are, in general, symmetric tensors of the second rank (or dyadics).[1] However, for isotropic media, i.e., material with uniform properties in all directions, ϵ and μ reduce to scalar quantities but of magnitude different from ϵ_0 and μ_0. Thus

$$\mathbf{D} = \epsilon \mathbf{E} \qquad \text{and} \qquad \mathbf{B} = \mu \mathbf{H} \qquad (31\text{-}3)$$

This same proportionality of $\mathbf{D}$ to $\mathbf{E}$ or $\mathbf{B}$ to $\mathbf{H}$ is also obtained for crystals provided that the field lies along an appropriate axis of the crystal. We shall deal here only with cases where ϵ and μ are simple scalar quantities.

The ratio ϵ/ϵ_0 is the dielectric coefficient, or more commonly, the *dielectric constant* since it is nearly independent of the field except at very high fields. The separate symbol D will be used for ϵ/ϵ_0 only when there is little danger of confusion with the electric displacement $\mathbf{D}$. The similar ratio μ/μ_0 may be described as the magnetic inductive capacity, but it is customary, instead, to use the *magnetic susceptibility* χ, which is defined as

$$\chi = \frac{\mu}{\mu_0} - 1 \qquad (31\text{-}4)$$

or the molal susceptibility $\chi_M = \chi \mathrm{v}$, where v is the molal volume.

The electric and magnetic *polarization* vectors are defined by

$$\mathbf{p} = \mathbf{D} - \epsilon_0 \mathbf{E} \qquad \text{and} \qquad \mathbf{m} = \frac{\mathbf{B}}{\mu_0} - \mathbf{H} \qquad (31\text{-}5)$$

which in an isotropic medium simplify to

$$\mathbf{p} = (\epsilon - \epsilon_0)\mathbf{E} = (D - 1)\epsilon_0 \mathbf{E} \qquad (31\text{-}6)$$

$$\mathbf{m} = \left(\frac{\mu}{\mu_0} - 1\right)\mathbf{H} = \chi \mathbf{H} \qquad (31\text{-}7)$$

---

[1] The components of $\mathbf{D}$ are given in terms of the components of $\mathbf{E}$ by the equations

$$D_x = \epsilon_{xx} E_x + \epsilon_{xy} E_y + \epsilon_{xz} E_z$$
$$D_y = \epsilon_{yx} E_x + \epsilon_{yy} E_y + \epsilon_{yz} E_z$$
$$D_z = \epsilon_{zx} E_x + \epsilon_{zy} E_y + \epsilon_{zz} E_z$$

where $\epsilon_{xy} = \epsilon_{yx}$, etc., but otherwise all the ϵ_{ij}'s may be different and nonzero. The expression for $\mathbf{B}$ in terms of μ_{ij} and $\mathbf{H}$ is similar. In particular cases, crystal symmetry yields simplification provided that the appropriate axes are used. For isotropic media all off-diagonal ϵ_{ij}'s become zero, and $\epsilon_{xx} = \epsilon_{yy} = \epsilon_{zz} = \epsilon$.

For thermodynamic purposes we frequently need, instead of the polarization per unit volume, the polarization of a definite amount of matter, usually 1 mole. The capital letters **P** and **M** will be used for the molal polarization and

$$\mathbf{M} = \chi_M \mathbf{H} \tag{31-8}$$

Under most conditions ϵ, μ, and therefore D and χ are constants for a given material at a particular temperature and pressure. Exceptions occur for ferromagnetic materials and for other materials at extremely high fields and low temperatures. Under such conditions one deals with a differential susceptibility which we shall write explicitly as

$$\left(\frac{\partial \mathbf{m}}{\partial \mathbf{H}}\right)_{P,T} \qquad \text{or} \qquad \text{per mole} \left(\frac{\partial \mathbf{M}}{\partial \mathbf{H}}\right)_{P,T}$$

Additional complications arise for permanent magnetization of ferromagnetic materials where there is a fixed magnetization $\mathbf{M}_0$ in addition to an induced magnetization **M**.

ELECTRIC POLARIZATION

If we assume that the charges generating an electric field are fixed in position, then the change in w associated with the introduction of polarizable material may be associated with the free energy of that substance. An array of fixed charges establishes the electric displacement vector **D** in accordance with the general electromagnetic equation (discussed in Appendix 8)

$$\nabla \cdot \mathbf{D} = \rho \tag{A8-4}$$

where ρ is the charge density. If we limit ourselves to cases where either the polarizable material fills all space or its boundaries are either perpendicular or parallel to the field vector, then **D** is unchanged by the introduction of the dielectric. The change in the field **E** on introduction of dielectric from Eq. (31-3) is

$$\Delta \mathbf{E} = \frac{\mathbf{D}}{\epsilon} - \frac{\mathbf{D}}{\epsilon_0}$$

and from Eq. (31-1) the work of introduction of the dielectric is

$$\Delta w = -\frac{1}{\epsilon_0} \iint \left(1 - \frac{\epsilon_0}{\epsilon}\right) \mathbf{D} \cdot d\mathbf{D} \, dV \tag{31-9}$$

The integration must cover the entire volume of polarizable material, but the integrand is evidently zero elsewhere.

Let us next consider a pair of parallel, uniformly charged plates, one positive and one negative, immersed in a liquid dielectric. If we neglect edge effects, then in the space between the plates **D** is constant and has the magnitude of the surface-charge density, q/a, and the direction perpendicular to the plates. Elsewhere **D** is zero. If the volume of the space between plates is V_c, we have

$$\Delta w = -\frac{V_c}{2\epsilon_0} \int \left(1 - \frac{\epsilon_0}{\epsilon}\right) d(\mathbf{D}^2)$$

Also, by regarding the magnitude of the charges as continuously variable, one may vary **D** similarly and write for an increment in Δw

$$dw = -\frac{V_c}{2\epsilon_0}\left(1 - \frac{\epsilon_0}{\epsilon}\right) d(\mathbf{D}^2) \qquad (31\text{-}10)$$

The differential of the free energy of all the dielectric material is obtained by adding Eq. (31-10) to Eq. (10-12),

$$dF = -S\,dT + V\,dP - \frac{V_c}{2\epsilon_0}\left(1 - \frac{\epsilon_0}{\epsilon}\right) d(\mathbf{D}^2) \qquad (31\text{-}11)$$

The quantity $\epsilon/\epsilon_0 = D$, the dielectric coefficient, which is always greater than 1 and is nearly independent of **D** at low fields. Thus we note that the free energy of polarizable material is reduced when in an electric field. Equation (31-11) is an exact differential; hence we may use Eq. (3-8) to obtain the change of entropy and of volume with the field.

$$\left[\frac{\partial S}{\partial(\mathbf{D}^2)}\right]_{P,T} = -\frac{V_c}{2\epsilon_0}\left[\frac{\partial(\epsilon_0/\epsilon)}{\partial T}\right]_{P,\mathbf{D}} \qquad (31\text{-}12)$$

$$\left[\frac{\partial V}{\partial(\mathbf{D})^2}\right]_{P,T} = \frac{V_c}{2\epsilon_0}\left[\frac{\partial(\epsilon_0/\epsilon)}{\partial P}\right]_{T,\mathbf{D}} \qquad (31\text{-}13)$$

Here P is the pressure on the liquid dielectric outside the capacitor. The derivatives on the right side of these equations are derivatives of the reciprocal of the dielectric constant since $D = \epsilon/\epsilon_0$.

Since ϵ/ϵ_0 usually increases with pressure, Eq. (31-13) indicates a volume decrease in the field. This effect is called electrostriction. The dielectric constant of most substances decreases with rise in temperature; hence we see that the entropy decreases as the material is polarized. This would be expected on a statistical basis for dipolar molecules since the field will tend to reduce the randomness of molecular orientation.

The entropy effect which we just noted implies evolution of heat on polarization in the amount $-T\,dS$. The enthalpy change in the dielectric is given, not by Eq. (31-10), but rather by

$$\left[\frac{\partial H}{\partial(\mathbf{D}^2)}\right]_{P,T} = -\frac{V_c}{2\epsilon_0}\left\{1 - \frac{\epsilon_0}{\epsilon} + T\left[\frac{\partial(\epsilon_0/\epsilon)}{\partial T}\right]\right\} \qquad (31\text{-}14)$$

The electrostrictive effect usually involves much less energy than the thermal effect but must be included in principle to obtain the work content and energy content.

$$\left[\frac{\partial A}{\partial (\mathbf{D}^2)}\right]_{P,T} = -\frac{V_c}{2\epsilon_0}\left\{1 - \frac{\epsilon_0}{\epsilon} + P\left[\frac{\partial(\epsilon_0/\epsilon)}{\partial P}\right]\right\} \tag{31-15}$$

$$\left[\frac{\partial E}{\partial (\mathbf{D}^2)}\right]_{P,T} = -\frac{V_c}{2\epsilon_0}\left\{1 - \frac{\epsilon_0}{\epsilon} + P\left[\frac{\partial(\epsilon_0/\epsilon)}{\partial P}\right] + T\left[\frac{\partial(\epsilon_0/\epsilon)}{\partial T}\right]\right\} \tag{31-16}$$

The last several equations pertain to all the dielectric fluid in between and around the capacitor plates, and, as we have seen, the charging of the plates will tend to force more material between the plates. While the changes in the molal quantities are just v/V_c times the results given above, v varies slightly with $\mathbf{D}$, whereas V_c was a constant.

Let us also write the differential of the molal free energy of material between the plates where the field is described by the vector $\mathbf{D}$,

$$d_\mathrm{F} = -s\,dT + v\,dP - \frac{v}{2\epsilon_0}\left(1 - \frac{\epsilon_0}{\epsilon}\right)d(\mathbf{D}^2) \tag{31-17}$$

where P is now the pressure on the material in the field. If the dielectric is a fluid free to flow into the space between the plates, then F must be the same in and out of the field. This equality of F requires a pressure increase dP on the material in the field which is given by

$$dP = +\frac{1}{2\epsilon_0}\left(1 - \frac{\epsilon_0}{\epsilon}\right)d(\mathbf{D}^2) \tag{31-18}$$

This equation may be integrated if ϵ is known as a function of $\mathbf{D}$, or even more easily if ϵ is constant.

We may also consider a pair of capacitor plates with a sheet of solid isotropic dielectric material between them (Fig. 31-1). If the charged

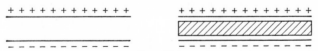

Fig. 31-1

plates are fixed, as before, and there is some free space around the solid dielectric, then the pressure will remain constant. The entropy and volume effects are now obtainable from the equality of cross derivatives of Eq. (31-17).

$$\left[\frac{\partial s}{\partial (\mathbf{D}^2)}\right]_{P,T} = -\frac{v}{2\epsilon_0}\left[\frac{\partial(\epsilon_0/\epsilon)}{\partial T}\right]_{P,\mathbf{D}} + \frac{1}{2\epsilon_0}\left(1 - \frac{\epsilon_0}{\epsilon}\right)\left(\frac{\partial v}{\partial T}\right)_{P,\mathbf{D}} \tag{31-19}$$

$$\left[\frac{\partial v}{\partial (\mathbf{D}^2)}\right]_{P,T} = \frac{v}{2\epsilon_0}\left(\frac{\epsilon_0/\epsilon}{\partial P}\right)_{T,\mathbf{D}} - \frac{1}{2\epsilon_0}\left(1 - \frac{\epsilon_0}{\epsilon}\right)\left(\frac{\partial v}{\partial P}\right)_{T,\mathbf{D}} \tag{31-20}$$

Solutions in Electric Fields. Another interesting possibility is a dielectric solution. Now the chemical potential of each component must remain constant throughout the solution, and the composition between the plates need not be the same as outside. Equation (31-11) may be solved for the derivative of the total free energy with respect to $\mathbf{D}^2$ as a measure of the field strength. Further differentiation with respect to the number of moles of component i between the plates yields the change of chemical potential with $\mathbf{D}^2$.

$$\frac{\partial \mu_i}{\partial (\mathbf{D}^2)} = \frac{V_c}{2\epsilon_0} \frac{\partial (\epsilon_0/\epsilon)}{\partial n_i} \tag{31-21}$$

In order to estimate the effect on composition, we must convert this derivative to one with respect to mole fraction. Let us assume a two-component system, whereupon

$$\frac{\partial \mu_2}{\partial (\mathbf{D}^2)} = \frac{\mathbf{v}^2}{2\bar{\mathbf{v}}_1 \epsilon_0} \frac{\partial (\epsilon_0/\epsilon)}{\partial x_2} \tag{31-22}$$

with $\mathbf{v}$ the molal volume of the solution and $\bar{\mathbf{v}}_1$ the partial molal volume of the other component.

The differential of μ_2 must be zero.

$$d\mu_2 = \frac{\partial \mu_2}{\partial (\mathbf{D}^2)} d(\mathbf{D}^2) + \frac{\partial \mu_2}{\partial \ln x_2} d \ln x_2 = 0$$

Hence
$$\frac{\partial \ln x_2}{\partial (\mathbf{D}^2)} = - \left(\frac{\partial \mu_2}{\partial \ln x_2} \right)^{-1} \left(\frac{\mathbf{v}^2}{2\epsilon_0 \bar{\mathbf{v}}_1} \right) \frac{\partial (\epsilon_0/\epsilon)}{\partial x_2} \tag{31-23}$$

If we assume an ideal solution, this becomes

$$\frac{\partial \ln x_2}{\partial (\mathbf{D}^2)} = - \frac{\mathbf{v}^2}{2\epsilon_0 R T \bar{\mathbf{v}}_1} \frac{\partial (\epsilon_0/\epsilon)}{\partial x_2} \tag{31-24}*$$

Charged Sphere in a Dielectric. An ion in solution is sometimes approximated as a charged sphere in a continuous dielectric medium; hence the latter problem is of interest at this point. The field around a sphere carrying charge q is described by a displacement vector $\mathbf{D}$ of radial orientation and of magnitude

$$\mathbf{D}_r = \frac{q}{4\pi r^2} \tag{31-25}$$

The free energy of polarization of the dielectric is then the integral of Eq. (31-9) from the surface of the ion at radius r_0 to infinity and from charge zero up to ze. In general the dielectric coefficient ϵ/ϵ_0 is a function of the field strength, and one has for 1 mole of ions

$$\Delta F = - \frac{N_0}{\epsilon_0} \int_{r_0}^{\infty} 4\pi r^2 \, dr \int_0^{ze/4\pi r^2} \left(1 - \frac{\epsilon_0}{\epsilon} \right) \mathbf{D} \cdot d\mathbf{D} \tag{31-26}$$

If one assumes a constant value of ϵ/ϵ_0, that is, a dielectric constant D, then one obtains the Born equation

$$\Delta F = -\frac{z^2 e^2 N_0}{8\pi\epsilon_0 r_0}\left(1 - \frac{1}{D}\right) \qquad (31\text{-}27)*$$

The coefficient $e^2 N_0/8\pi\epsilon_0$ has the value 6.95×10^{-5} joule m/mole or 166 kcal A/mole. Thus the predicted ΔF of solvation of an ion of $r_0 = 2$ A is $83z^2(1 - 1/D)$ kcal/mole. While the assumption of a constant value of ϵ/ϵ_0 cannot be very accurate for the high field near an ion in solution, one notes that as long as $\epsilon/\epsilon_0 \gg 1$ the error will not be too serious. Of course, the molecular nature of the solvent dielectric makes it a crude approximation to use a fixed radius r_0; hence results obtained from Eq. (31-27)* should be used with caution.

MAGNETIC POLARIZATION

The basic integral for the magnetic work from Eq. (31-1),

$$w = \int\int \mathbf{H} \cdot d\mathbf{B}\, dV \qquad (31\text{-}28)$$

is identical in form to that for the electrical work, but the generation of a magnetic field is not as simple as the use of one or more electric charges to establish an electric field. The simplest magnetic-field generator is a very long solenoid with uniform windings. If we neglect end effects, the field $\mathbf{H}$ within the magnet is given simply by the number of ampere turns per meter of length. Suppose that we insert a long cylinder of a paramagnetic substance, that is, $\mu > \mu_0$, within the solenoid. We know that the paramagnetic material is attracted into the field and hence that its free-energy content should be lower in the field. But $\mathbf{B} = \mu\mathbf{H}$, and therefore $\mathbf{B}$ is increased by the paramagnetic material, yielding a greater w in Eq. (31-28). The explanation of this apparent contradiction is that additional electrical work must be supplied to keep the solenoid current and thereby $\mathbf{H}$ constant. The magnetic field thus plays a role much like the pressure in volumetric effects.

The additional work associated with 1 mole of magnetically polarizable material in a uniform field is readily obtained from Eqs. (31-5) and (31-28),

$$\Delta w = \mu_0 \int \mathbf{H} \cdot d\mathbf{M} \qquad (31\text{-}29)$$

This work, expressed in differential form, is added to the usual expression for work content to obtain

$$d\mathrm{A} = -\mathrm{s}\, dT - P\, d\mathrm{v} + \mu_0 \mathbf{H} \cdot d\mathbf{M} \qquad (31\text{-}30)$$

The free energy is now defined[1] as

$$\text{F} = \text{A} + P\text{v} - \mu_0 \mathbf{H} \cdot \mathbf{M} \qquad (31\text{-}31)$$

where the final term is added to exclude the electrical work of maintaining a constant magnetic field just as the PV term excludes the volumetric work of maintaining constant pressure. Thus

$$d\text{F} = -\text{s}\,dT + \text{v}\,dP - \mu_0 \mathbf{M} \cdot d\mathbf{H} \qquad (31\text{-}32)$$

The functions E and H are related, as usual, to A and F by the addition of Ts.

$$d\text{E} = T\,ds - P\,dv + \mu_0 \mathbf{H} \cdot d\mathbf{M} \qquad (31\text{-}33)$$

$$d\text{H} = T\,ds + \text{v}\,dP - \mu_0 \mathbf{M} \cdot d\mathbf{H} \qquad (31\text{-}34)$$

We note from the last equation that $d\text{H}$ is the heat absorbed at constant pressure and magnetic field.

Application of Eq. (3-8) to the total differential of the free energy yields

$$\left(\frac{\partial \text{v}}{\partial \mathbf{H}}\right)_{P,T} = -\mu_0 \left(\frac{\partial \mathbf{M}}{\partial P}\right)_{T,\mathbf{H}} \qquad (31\text{-}35)$$

$$\left(\frac{\partial \text{s}}{\partial \mathbf{H}}\right)_{P,T} = \mu_0 \left(\frac{\partial \mathbf{M}}{\partial T}\right)_{P,\mathbf{H}} \qquad (31\text{-}36)$$

In our use of Eqs. (31-33) through (31-38)* we shall be concerned only with examples where the magnetic polarization $\mathbf{M}$ is parallel to the field $\mathbf{H}$; hence each quantity may be regarded as a scalar (rather than a vector). We shall, however, retain the boldface type for $\mathbf{H}$, etc., to distinguish these quantities from others represented by the same letters, e.g., enthalpy H. We have avoided making the further assumption that $\mathbf{M}$ is proportional to $\mathbf{H}$ because exceptions which show saturation of polarization are of importance at low temperatures.

In the more restricted case where $\mathbf{M} = \chi_M \mathbf{H}$, the following approximate equations are readily derived:

$$d\text{F} = -\text{s}\,dT + \text{v}\,dP - \tfrac{1}{2}\mu_0\chi_M\,d(\mathbf{H}^2) \qquad (31\text{-}32')^*$$

$$\left[\frac{\partial \text{v}}{\partial (\mathbf{H}^2)}\right]_{P,T} = -\tfrac{1}{2}\mu_0 \left(\frac{\partial \chi_M}{\partial P}\right)_{T,\mathbf{H}} \qquad (31\text{-}35')^*$$

$$\left[\frac{\partial \text{s}}{\partial (\mathbf{H}^2)}\right]_{P,T} = \tfrac{1}{2}\mu_0 \left(\frac{\partial \chi_M}{\partial T}\right)_{P,\mathbf{H}} \qquad (31\text{-}36')^*$$

The magnetostrictive effect given by Eq. (31-35) or (31-35')* is usually very small and of little interest, but the thermal effect given by Eq. (31-36) or (31-36')* is of major importance.

---

[1] The function $A + PV$ may still be used in a magnetic field but we shall not do so and hence prefer to retain F for the function used.

Cooling by Adiabatic Demagnetization. It was pointed out by Giauque and Debye in 1926 that the thermal effects associated with the magnetization of typical paramagnetic salts at approximately $1°K$ provide a means of producing much lower temperatures. Certain salts of paramagnetic rare earth or transition metal ions such as $Gd_2(SO_4)_3 \cdot 8H_2O$ are appropriate; these materials follow Curie's law to relatively low temperatures.

$$\frac{\mathbf{M}}{\mathbf{H}} = \chi_M = \frac{C_M}{T} \qquad (31\text{-}37)*$$

where C_M is a constant. Hence $\partial \chi / \partial T$ is large and negative at low temperatures, and from Eq. $(31\text{-}36')*$

$$\left[\frac{\partial s}{\partial (\mathbf{H}^2)} \right]_{P,T} = -\frac{\mu_0 C_M}{2T^2} \qquad (31\text{-}38)*$$

We see that there is a large decrease of entropy (and evolution of heat) on application of a strong magnetic field at low temperatures.

After the heat of magnetization has been conducted away, the sample is thermally isolated and the field removed. Adiabatic demagnetization then occurs, with reduction of temperature from an initial value T_1 to a final temperature T_2. In order to treat this process, let us define the molal heat capacity at constant field (analogous to constant pressure),

$$c_H = \left(\frac{\partial \mathbf{H}}{\partial T} \right)_{\mathbf{H}} = T \left(\frac{\partial s}{\partial T} \right)_{\mathbf{H}} \qquad (31\text{-}39)$$

Then the decrease in entropy in isothermal magnetization may be set equal to the entropy decrease on cooling from T_1 to T_2 at zero field.

$$\int_{T_1}^{T_2} c_{\mathbf{H}=0} \, d \ln T = -\frac{\mu_0 \mathbf{H}^2 C_M}{2T_1^2} \qquad (31\text{-}40)*$$

Typical values of C_M are of the order of 5 erg deg/mole oersted[2] in emu (unrationalized)[1] where $\mu_0 = 1$. In a field of 10^4 oersteds and at $1.5°K$ the entropy decrease is 1.1×10^8 ergs/deg mole or 2.6 cal/deg mole. This is a large fraction of the total entropy (excluding nuclear spin) remaining at $1.5°K$; hence a very low temperature is attained on adiabatic demagnetization. Indeed, as nearly all the entropy is removed, the magnetic susceptibility can no longer follow Curie's law and the equations just given are not valid at such a high field.

Let us now review briefly the methods which can be employed without assuming that $\mathbf{M}/\mathbf{H}$ is constant at a given temperature or that Curie's

---

[1] The literature on low-temperature magnetic research uses cgs emu. The unit of energy is the erg, of magnetic-field intensity the oersted. $\mathbf{H}$ in oersteds is (amp turns/m) $\times 4\pi \times 10^{-3}$. Since $\mu_0 = 1$, $\mathbf{B}$ (in gauss) $= \mathbf{H}$ in free space.

law is valid.   Equation (31-36) is appropriate,

$$\left(\frac{\partial s}{\partial \mathbf{H}}\right)_T = \mu_0 \left(\frac{\partial \mathbf{M}}{\partial T}\right)_{\mathbf{H}} \tag{31-36}$$

Thus, if $\mathbf{M}$ is measured as a function of T and $\mathbf{H}$, $(\partial \mathbf{M}/\partial T)_{\mathbf{H}}$ may be obtained and the entropy of isothermal magnetization calculated by integration.

$$s_{\mathbf{H}_f} - s_{\mathbf{H}=0} = \mu_0 \int_{\mathbf{H}=0}^{\mathbf{H}_f} \left(\frac{\partial \mathbf{M}}{\partial T}\right)_{\mathbf{H}} d\mathbf{H} \tag{31-41}$$

Also the magnetic entropy of a sufficiently ideal paramagnetic substance may be calculated by statistical methods which will be given later in this chapter.

After isothermal magnetization at T_1 and thermal isolation the demagnetization is isentropic, and the final temperature T_2 is given by the exact relation equating the two values for the decrease in entropy ΔS below that at T_1 and zero field.

$$\Delta S = \int_{T_1}^{T_2} c_{\mathbf{H}=0}\, d \ln T = \mu_0 \int_{\mathbf{H}=0}^{\mathbf{H}_f} \left(\frac{\partial \mathbf{M}}{\partial T}\right) d\mathbf{H} \tag{31-42}$$

The greatest difficulty in exact analysis of the result is the determination of the heat capacity $c_{\mathbf{H}}$ in the region below 1°K.   The introduction of heat in exactly measured amount is difficult in this temperature range. Giauque and MacDougall[1] used induction heating, Kurti and Simon[2] used gamma-ray heating, while Casimir, de Haas, and de Klerk[3] used heat absorption from an alternating magnetic field due to hysteresis and relaxation in the magnetic material itself.

The thermal measurements of heat capacity serve really to determine the thermodynamic temperature in this region below 1°K, where gas thermometry is impractical.   Measurements are usually interpreted initially in terms of the "magnetic temperature" $T^* = C/\chi$ given by Eq. (31-37)* from the magnetic susceptibility.   A series of demagnetizations from different fields and starting temperatures yields a curve for the entropy as a function of T^* at zero field.   Then an apparent heat capacity on the magnetic temperature scale is obtained by differentiation,

$$C' = \left(\frac{\partial S}{\partial \ln T^*}\right)_{\mathbf{H}=0} \tag{31-43a}$$

Another apparent heat capacity is obtained from the ratio of the increment of heat introduced to the temperature rise on the magnetic scale,

[1] W. F. Giauque and D. P. MacDougall, *J. Am. Chem. Soc.*, **60**, 376 (1938).

[2] N. Kurti and F. Simon, *Phil. Mag.*, **26**, 840 (1938).

[3] H. B. G. Casimir, W. J. de Haas, and D. de Klerk, *Physica*, **6**, 255 (1939).

$$C'' = \left(\frac{\delta q}{\delta T^*}\right)_{H=0},$$

$$(31\text{-}43b)$$

From the second law the thermodynamic temperature is $\delta q_{rev}/\delta S$, and

$$T_{thermo} = T^* \frac{\delta q/\delta T^*}{\partial S/\partial \ln T^*} = T^* \frac{C''}{C'}$$

$$(31\text{-}44)$$

Particularly extensive studies have been made upon chrome alum, $KCr(SO_4)_2 \cdot 12H_2O$, by Bleaney[1] and by de Klerk, Steenland, and Gorter.[2] Their results, which cover different temperature ranges and are based upon different heating methods, nevertheless yield a concordant heat-capacity curve, which is shown in Fig. 31-2. In this region almost all the heat capacity presumably arises from the interaction of the internal crystal fields on the electric and magnetic moments of the Cr^{3+} ions. At temperatures well below $0.01°K$, where the heat capacity has dropped to a very low value, the magnetic moments of the Cr^{3+} ions must have become ordered in some definite pattern, while at $1°K$ and above the magnetic moments are randomly oriented in the absence of a magnetic field.

The preceding discussion concerned the establishment of thermodynamic temperature in the absence of a magnetic field. Let us now consider briefly the situation with a magnetic field present. Under conditions of constant field and pressure, Eq. (31-34) reduces to

$$T = \left(\frac{\partial H}{\partial S}\right)_{H,P}$$

$$(31\text{-}45)$$

where we recall that the enthalpy in the magnetic field was defined to include the term $-\mu_0 H \cdot M$. Giauque and MacDougall[3] showed how magnetic-susceptibility measure-

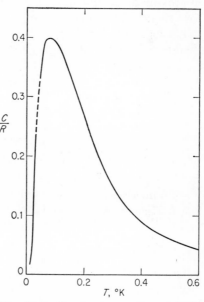

FIG. 31-2. The heat capacity of $KCr(SO_4)_2 \cdot 12H_2O$. The solid curve below $0.03°K$ is from de Klerk, Steenland, and Gorter, while the higher temperature curve is from Bleaney. The dashed curve connects the two sets of experimental results.

[1] B. Bleaney, *Proc. Roy. Soc.* (*London*), **A204**, 216 (1950).

[2] D. de Klerk, M. V. Steenland, and C. J. Gorter, *Physica*, **15**, 649 (1949).

[3] W. F. Giauque and D. P. MacDougall, *J. Am. Chem. Soc.*, **57**, 1175 (1935), **60**, 376 (1938).

ments could be used to evaluate the magnetic contribution to H in Eq. (31-45) and thereby to obtain the thermodynamic temperature.

For many purposes the objective is exact measurements under conditions where the entropy of the substance is reduced to a very low value. This objective is attained as well at high magnetic fields and temperatures somewhat above 1°K as at zero field and a lower temperature. For the former conditions carbon-resistance thermometers have been found to be particularly suitable, and in addition they serve well as heaters.[1]

Theoretical Calculation of Magnetic Polarization. If the effect of a magnetic field on the energy levels of the particles within a substance is known, it is possible to use statistical methods to calculate the magnetic susceptibility. We shall consider a simple ideal model which approximates some real paramagnetic substances but is quite inadequate for others. The same principles are applicable to more complex cases.

Let us assume that the atomic magnetic moment is associated with a quantum number M_J which takes the values $-J$, $-J + 1$, $-J + 2$, . . . , $J - 1$, J. It is found that J is either an integer or an integer plus $\frac{1}{2}$; both cases are consistent with the sequence of M_J values. We also assume that the effect of a magnetic field is to change any particular atomic energy level by the energy difference

$$\epsilon - \epsilon_0 = M_J g \beta \mathbf{B} \qquad (31\text{-}46)*$$

where β is a physical constant known as the Bohr magneton,[2] whose numerical value is 0.9273×10^{-20} erg/gauss or 0.9273×10^{-23} amp m², g is a numerical constant of order of magnitude unity which characterizes the particular system, and $\mathbf{B}$ is the magnetic-field vector. It is of no consequence whether we use the field vector $\mathbf{B}$ or $\mu_0 \mathbf{H}$ in Eq. (31-46)* since the model implies a dilute system with no interaction between atomic units. Under these circumstances the $\mathbf{B}$ or $\mu_0 \mathbf{H}$ measured for a uniform field outside the sample may be used in the formula.

We have purposely avoided specifying other characteristics of the system, which are assumed not to affect the magnetic properties. However, we may note the sort of system approximated by this model. Some constituent particles (atoms or ions) must carry a magnetic dipole moment, and these particles must be sufficiently separated so that their interactions are negligible compared with the interaction of each particle with the field. The magnetic moment may arise from either unpaired electron or nuclear spin or uncompensated orbital angular momentum

---

[1] W. F. Giauque, J. W. Stout, and C. W. Clark, *J. Am. Chem. Soc.*, **60**, 1053 (1938); *Ind. Eng. Chem.*, **28**, 743 (1936).

[2] Quantum mechanics gives the relation $\beta = |e|h/4\pi mc$ in cgs units or $|e|h/4\pi m$ in mks units, where e and m are the charge and mass of the electron, h is Planck's constant, and c is the velocity of light.

of electrons. The nuclear-spin magnetic moments are about three orders of magnitude smaller than those of electrons and hence are usually negligible if electronic effects are being considered. Since magnetic susceptibilities of ideal systems are small in any case, particular interest pertains to relatively concentrated systems rather than gases. Substances such as hydrated salts of rare earth or transition ions have their magnetic particles regularly spaced at distances of 5 to 10 A. In some cases this spacing is sufficient to yield relatively ideal behavior. Thus $Gd_2(SO_4)_3 \cdot 8H_2O$ shows relatively ideal behavior[1] at 1°K, and the more dilute $Gd(PMo_{12}O_{40}) \cdot 30H_2O$ is even more ideal. The diamagnetism of the matter, H_2O, etc., between the ions is so small as to be negligible in an approximate treatment.

Since we have ignored the interaction between atomic magnetic moments, no ordering of the moments will be expected at zero field. Thus our model will not yield a heat capacity in zero field such as that shown for chrome alum in Fig. 31-2.

Comparison of the magnetic term in the free-energy equation (31-32), $-\mu_0 \mathbf{M} \cdot d\mathbf{H}$, with our assumed model for energy levels indicates that the particle magnetic moment is

$$\mathfrak{y} = -M_J g\beta \qquad (31\text{-}47)^*$$

and the mean of these particle moments times N_0 yields the molal magnetization

$$\mathbf{M} = N_0 \bar{\mathfrak{y}} \qquad (31\text{-}48)$$

The arguments in Chapter 27 may be applied to obtain the Boltzmann law for the thermal distribution over these quantum states; hence

$$\bar{\mathfrak{y}} = \frac{\sum\limits_{M_J=-J}^{+J} - M_J g\beta e^{-M_J g\beta \mathbf{B}/kT}}{\sum\limits_{M_J=-J}^{+J} e^{-M_J g\beta \mathbf{B}/kT}} \qquad (31\text{-}49)^*$$

If we define a function,

$$Q = \sum\limits_{M_J=-J}^{+J} e^{-M_J x} \qquad \text{with } x = \frac{g\beta \mathbf{B}}{kT} \qquad (31\text{-}50)$$

we may readily show that

$$\mathbf{M} = N_0 g\beta \frac{d \ln Q}{dx} \qquad (31\text{-}51)^*$$

---

[1] W. F. Giauque, *J. Am. Chem. Soc.*, **49**, 1870 (1927), showed that the magnetic-susceptibility data of $Gd_2(SO_4)_3 \cdot 8H_2O$ fitted this simple theoretical model very well. In this case the difference between the external field and either $\mathbf{B}$ or $\mu_0\mathbf{H}$ within the sample is not trivial, and the agreement with the ideal model is somewhat surprising.

Our task is now reduced to the evaluation of Q, which may be rearranged into the sum of a geometrical progression.

$$Q = e^{Jx} \sum_{m=0}^{2J} e^{-mx} = e^{Jx} \left(\frac{1 - e^{-(2J+1)x}}{1 - e^{-x}} \right)$$
$$= \frac{\sinh\left[(2J+1)x/2\right]}{\sinh\left(x/2\right)} \qquad (31\text{-}52)$$

Then differentiation yields

$$\mathbf{M} = N_0 g\beta \left[\frac{2J+1}{2} \coth \frac{(2J+1)x}{2} - \frac{1}{2} \coth \frac{x}{2} \right] \qquad (31\text{-}53)^*$$

where we recall that $x = g\beta \mathbf{B}/kT$. Figure 31-3 shows a plot of $\mathbf{M}/N_0 g\beta$ versus x for this function. We note the initial proportionality of $\mathbf{M}$ to $\mathbf{B}$

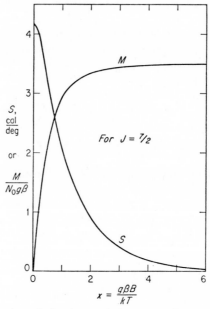

FIG. 31-3. The magnetic polarization and the magnetic entropy as a function of magnetic field for a simple model substance with $J = \frac{7}{2}$. The same numerical scale is used for entropy in cal/deg mole and for the value of $\mathbf{M}/N_0 g\beta$.

and the eventual saturation of $\mathbf{M}$ at the value $JN_0 g\beta$, which corresponds to complete orientation of the molecular magnetic moments.

In the region of low field where $\mathbf{M}$ is proportional to $\mathbf{B}$, Eq. (31-53)* reduces to

$$\mathbf{M} = N_0 g\beta \frac{J(J+1)x}{3} = \frac{N_0 g^2 \beta^2 J(J+1)}{3kT} \mathbf{B} \qquad (31\text{-}54)^*$$

We note the inverse proportionality to temperature, which is the Curie law, and comparison with Eq. (31-37)* gives the Curie constant,

$$C_M = \frac{\mu N_0 g^2 \beta^2 J(J+1)}{3k} \qquad (31\text{-}55)^*$$

where the magnetic permeability $\mu = \mathbf{B}/\mathbf{H}$ is approximately that of free space, i.e., unity in emu but $4\pi \times 10^{-7}$ in mks units.

These results permit the calculation of the entropy and other thermodynamic properties for the ideal magnetic substance which has been assumed. Substitution of Eq. (31-53)* into Eq. (31-36) yields

$$\frac{\partial \mathrm{s}}{\partial \mathbf{H}} = - \frac{\mu_0 x}{T} \frac{d\mathbf{M}}{dx}$$

and integration from zero field to $\mathbf{H}'$ gives, after several steps,

$$\mathrm{s}_{\mathbf{H}'} - \mathrm{s}_{\mathbf{H}=0} = - \frac{\mathbf{B}' \cdot \mathbf{M}'}{T} + R \ln Q' - R \ln (2J+1) \qquad (31\text{-}56)^*$$

where $\mathbf{M}'$ and Q' are obtained by substitution of $\mathbf{H}'$ into Eqs. (31-53)* and (31-52). At the limit of large field the first two terms in Eq. (31-56)* cancel, and one obtains the maximum reduction of entropy by the field, $R \ln (2J+1)$. This is to be expected since at high field all particles are in the single magnetic quantum state with $M_J = -J$, whereas at zero field the particles are randomly distributed over $2J+1$ states. The entropy is also shown in Fig. 31-3 for the case $J = \frac{7}{2}$, which is appropriate for Gd^{3+} ion.

Before concluding this section we wish to emphasize that these results are based on a model which is at best an approximation to any real substance. Even in those cases which show good agreement at moderate temperatures, deviations will occur at very low temperatures, where the small energy interactions between magnetic dipoles are no longer negligible in comparison with kT. Thus our model yields no magnetic heat capacity at zero field, whereas we saw in Fig. 31-2 the large magnetic heat capacity of chrome alum in the 0.01 to 0.6°K region.

The general situation described in Figs. 31-2 and 31-3 shows the limitation on the temperature produced by demagnetization. A moderate field of 10,000 gauss gives $x = 1.0$ at 1.34°K. At this point only about half the magnetic entropy is removed, and adiabatic demagnetization yields a temperature near the peak of the magnetic heat-capacity curve, i.e., about 0.1°K for a typical substance such as chrome alum. The shape of the entropy curve shows that a much larger field is required to attain a significantly lower temperature. We see from Fig. 31-3 that even at $x = 5.0$ or 50,000 gauss at 1.34°K, where $\mathbf{M}$ is within 0.2 per cent of its saturation value, there still remains 0.08 cal/deg of magnetic entropy.

Effect of Sample Shape.   The discussion up to this point has assumed that the field outside the polarizable sample is unchanged by the presence of the sample and that the field in the sample is uniform, with **H** unchanged.   This condition holds for a long cylindrical sample in a solenoid if end effects are neglected.   This is not usually a convenient experimental arrangement.   Other sample systems usually modify the field outside the sample and frequently yield a nonuniform field within the sample. We shall not discuss these complexities in general but shall consider the case of a small ellipsoidal sample in an external field which was uniform before introduction of the sample.   In this case the field in the sample is uniform, but neither **B** nor **H** has its external value.   Instead

$$\mathbf{B} = \mu_0[\mathbf{H}_e + (1 - \alpha)\mathbf{M}] \qquad (31\text{-}57)$$
$$\mathbf{H} = \mathbf{H}_e - \alpha\mathbf{M} \qquad (31\text{-}58)$$

where $\mathbf{H}_e$ is the external field far from the sample and α is the demagnetizing factor[1] of the ellipsoid.   For the most common case, which is a prolate spheroid with the field along the major axis,

$$\alpha = \left(\frac{1}{e^2} - 1\right)\left(\frac{1}{2e}\ln\frac{1 + e}{1 - e} - 1\right) \qquad (31\text{-}59)$$

where the eccentricity e is given from the minor to major axis ratio a/c by

$$e = \left(1 - \frac{a^2}{c^2}\right)^{1/2}$$

A sphere has zero eccentricity, and $\alpha = \frac{1}{3}$, while for a very prolate spheroid α approaches zero.

In order to calculate the work associated with the magnetically polarizable sample, one must integrate Eq. (31-28) over all space and subtract the value of the same integral without the sample.   After several transformations this work term may be reduced to the following integral[2] over the volume of the sample only:

$$\Delta w = \frac{\mu_0}{2}\int dV \int (\mathbf{H}_e \cdot d\mathbf{m} + \mathbf{m} \cdot d\mathbf{H}_e - \mathbf{m} \cdot d\mathbf{H} + \mathbf{H} \cdot d\mathbf{m}) \qquad (31\text{-}60)$$

For the special case of the ellipsoid we substitute Eq. (31-58) and obtain the simple result

$$\Delta w = \mu_0 V \!\int\! \mathbf{H}_e \cdot d\mathbf{m} = \mu_0 \!\int\! \mathbf{H}_e \cdot d\mathbf{M} \qquad (31\text{-}61)$$

where V is the volume of the sample and $\mathbf{M}$ is the total polarization $V\mathbf{m}$.   The important feature is the appearance of $\mathbf{H}_e$ rather than $\mathbf{H}$ and the absence of any ellipsoidal shape factor.   Since it is the external field at a distance from the sample which is readily measured, this is a very convenient formula.

It is convenient to define the free energy in this case as $A + PV - \mu_0\mathbf{H}_e \cdot \mathbf{M}$, whereupon all our earlier formulas apply with $\mathbf{H}_e$ the external field in place of **H**.

[1] J. C. Maxwell, "Treatise on Electricity and Magnetism," 3d ed., vol. 2, p. 69, Clarendon Press, Oxford, 1904; note also that in the unrationalized system M is reduced by the factor $1/4\pi$, while the demagnetizing factor, commonly given the symbol N, becomes $4\pi\alpha$ and the factor $1 - \alpha$ becomes $4\pi - N$.

[2] See J. A. Stratton, "Electromagnetic Theory," McGraw-Hill Book Company, Inc., New York, 1941.   Equation (47) on p. 127 may be transformed into our equation (31-60) by substitution of **B** from Eq. (31-5).

SUMMARY OF THERMODYNAMICS OF
SYSTEMS WITH ADDITIONAL VARIABLES

Chapters 29 to 31 have been devoted to thermodynamic relationships for systems which involve other energy terms in addition to thermal and volumetric energies.   It is valuable to pause a moment and summarize these results.   The change in energy content of the system may be written

$$dE = \Sigma X_i \, dx_i \qquad (31\text{-}62)$$

where each term is the product of an intensive variable or force X_i and the change of an extensive variable x_i.   For the various types of energy one has:

Type of energy	X_i	x_i
Thermal...............................	T	S
Volumetric..............................	P	$-V$
Surface area.	σ	A_s
Length........	$\mathfrak{I}$	l
Gravitational...........................	$g(h - h_0)\dagger$	M
Centrifugal.............................	$-r^2\omega^2/2$	M
Electric charge...........................	$\Delta\mathcal{U}$	q
Material content of component i..........	μ_i	n_i

$\dagger$ More precisely $\displaystyle\int_{h_0}^{h} g \, dh$ if g varies significantly.

We have intentionally omitted electric and magnetic polarization from the table above because of the need to define carefully the treatment of the energy of the fields in the absence of polarizable material.   These problems have been considered in this chapter, and the appropriate energy equations are given.

Since Eq. (31-62) is an exact differential, Eq. (3-8) may be applied to yield

$$\left(\frac{\partial X_i}{\partial x_j}\right)_{x_i,\ldots} = \left(\frac{\partial X_j}{\partial x_i}\right)_{x_j,\ldots} \qquad (31\text{-}63)$$

where x_i and x_j are any pair of variables.   Also, by defining functions such as $E - X_i x_i$ and dealing with their differentials, one may obtain equations of the type

$$\left(\frac{\partial X_i}{\partial X_j}\right)_{x_i,\ldots} = -\left(\frac{\partial x_j}{\partial x_i}\right)_{X_j,\ldots} \qquad (31\text{-}64)$$

and

$$\left(\frac{\partial x_i}{\partial X_j}\right)_{X_i,\ldots} = \left(\frac{\partial x_j}{\partial X_i}\right)_{X_j,\ldots} \qquad (31\text{-}65)$$

Many such equations have been given in these chapters, but others may be derived as desired for any given pair of variables.

It is possible to imagine experiments in which either the extensive or the intensive variable is held constant. Just as either P or V and either T or S may be held constant, so may be either σ or A_s, either gh or M, etc. In some cases it would be extremely difficult to hold a given variable fixed, whereas the other variable is readily controlled. Thus one may readily hold a magnetic field constant but not the magnetic polarization. One may readily prevent flow of electric charge; it is also practical to hold a constant difference in electric potential between wires connected to a pair of electrodes, but there is little to be gained by doing experiments under this condition.

In the case where only thermal and volumetric energies were involved it was useful to define and use the four possible functions E, $H = E + PV$, $A = E - TS$, and $F = E + PV - TS$. Clearly there are many related functions when even a few of the possible additional variables are involved, that is, $E - \sigma A_s$, $E - \sigma A_s - 3l$, $E - \sigma A_s - TS$, etc. But most of these are so rarely if ever of practical use that there is no need to give them names and symbols. Indeed we have found it quite convenient to deal with most of these additional variables without modifying the definitions of E, H, A, and F, except in the single case of the magnetic field, where the term $\mu_0 \mathbf{H} \cdot \mathbf{M}$ was introduced into the definitions of H and F. It seems to us best to avoid the definition of unnecessary new functions but to reserve the possibility of modification of the definition of H and F by the addition of the appropriate product $-X_i x_i$ whenever desired. This modification should be clearly stated, of course, whenever it is made.

PROBLEMS

31-1. Quantum theory and spectroscopic studies give the magnetic properties of Gd^{3+} ion to be $g = 2$ and $J = \frac{7}{2}$. Calculate C_M; also $\mathbf{M}$ at 8000 gauss and 1.433°K for comparison with the value 29,250 in emu reported by Giauque and MacDougall.[1] Note that Eq. (31-53)* must be used to calculate $\mathbf{M}$ unless x is small.

31-2. Calculate the magnetic field necessary to remove 99 per cent of the magnetic entropy for an ideal salt of Gd^{3+} at 1.433°K ($g = 2$, $J = \frac{7}{2}$).

31-3. Derive the equation for the isothermal thermoelastic effect, i.e., the relationship between $(\partial S/\partial 3)_T$ and the coefficient of thermal expansion. Also derive by the method of Eq. (3-3) an equation for the adiabatic thermoelastic effect $(\partial T/\partial 3)_S$. In a measurement of the latter coefficient Rocca and Bever[2] found a ΔT of -0.17°C for a tension increase from zero to 17,700 lb/in.^{2} in nickel at 400°K. Compare this result with the value calculated from the coefficient of expansion and the heat capacity.

[1] W. F. Giauque and D. P. MacDougall, *J. Am. Chem. Soc.*, **60**, 376 (1938).

[2] R. Rocca and M. B. Bever, *Trans. AIME*, **188**, 327 (1950).

32

ESTIMATION OF ENTROPIES AND OTHER THERMODYNAMIC QUANTITIES

We have seen in various typical thermodynamic calculations that enthalpy, entropy, and heat-capacity data are required but that the sensitivity of the final answer to errors in these data varies widely. Also we have remarked that the heat of formation of a vast number of chemical substances is known, at least to the accuracy attained in the work of Thomsen, Berthelot, and others in the latter part of the nineteenth century. While our knowledge of entropy values has increased rapidly since the third law became established and generally applied, nevertheless there are many substances for which the heat of formation is known but the entropy has not been measured. Thus there is particular need to estimate entropy values to an accuracy corresponding to the enthalpies measured by Thomsen. Since an uncertainty of 0.3 to 0.6 kcal/mole is typical of the older heat-of-reaction data, an uncertainty of 1 to 2 cal/deg mole in the entropy has the equivalent effect at room temperature on an equilibrium constant or the free energy of reaction in view of the equation

$$\Delta F = \Delta H - T \Delta S \qquad (15\text{-}2)$$

In order to calculate the free energy at other temperatures, the heat capacities of reactants and products are required, and it is frequently both necessary and feasible to estimate the heat capacity if it has not been measured. The heat capacity of pure substances was considered in Chapter 6 from both theoretical and empirical viewpoints, and the theory of the heat capacity of ideal gases was developed further in Chapter 27. The heat capacities of various types of solutions were considered in Chapters 21 and 25.

Many of the ideas given in the discussions of heat capacity are also applicable to calculations of entropies. Indeed the equation

$$S_{T'} = \int_0^{T'} C_P \, d \ln T \qquad (12\text{-}4)$$

shows the close relationship between the two quantities. But we recall

515

from the rule of Dulong and Petit that the heat capacity of most solid elements is the same at and above room temperature, whereas we know that at lower temperatures the heat capacity of each element falls to low values along a different curve.   Hence the entropies of solid elements are not equal, and we see that estimation of entropy involves factors in addition to those affecting the heat capacity at room temperature and above. If these factors are all known accurately, then the entropy may be calculated by precise statistical methods such as those given in Chapter 27. But in many cases only the major factors affecting the entropy are identified, or the system may be so complex that precise statistical treatment is impractical.   In such cases it is still possible to make a useful estimate of the entropy by a combination of theory and empiricism.   As many important factors as possible are considered theoretically, and then the remaining contributions to the entropy are evaluated by comparison with experimental values for similar substances.

Entropies of Solid Elements.   In Chapter 6 we noted the characteristic behavior of the heat-capacity curve for monatomic solids and that the Debye theory gives a useful approximation.   This theory gives the entropy as a function of T/θ_D; the function is tabulated in Appendix 5. At high temperatures, where $c_v \cong 3R$, the entropy is

$$s = 3R(\ln T - \ln \theta_D + 1.333) \tag{32-1}*$$

Thus the principal factors affecting the entropy are the temperature and those factors which determine θ_D.   We noted in Chapter 6 that θ_D is proportional to a frequency of atomic vibration and therefore is proportional to the square root of an interatomic force constant and inversely proportional to the square root of the atomic mass.   Thus the difference in entropy between two related solid elements at the same high temperature, where both have attained a heat capacity $\cong 3R$, is approximated by

$$s_B - s_A = \tfrac{3}{2}R\left(\ln \frac{M_B}{M_A} - \ln \frac{k_B}{k_A}\right) \tag{32-2}*$$

The atomic weights yield the ratio M_B/M_A at once.   By choosing a comparison element B of known entropy whose hardness or compressibility is about the same as that of the element A, one may assume the force constants to be equal, that is, $k_B \cong k_A$, and the second term drops out, yielding

$$s_B = s_A + \tfrac{3}{2}R \ln \frac{M_B}{M_A} \tag{32-3}*$$

If the element forms molecules which are only weakly bound into a solid lattice, as is the case for I_2 or S_8, then the problem is more complex. The mass is still the predominant factor, but Eq. (32-3)* is accurately applicable only if both the intramolecular and intermolecular forces are equal. Actually the entropies of most elements have been measured, and this discussion is primarily an introduction to more complex solids.

Entropies of Solid Compounds. Kopp proposed the rule that the heat capacity of a solid compound was equal to the sum of the heat capacities of the component elements. If this were true at all temperatures, the corresponding summation rule would apply for the entropy. Actually Kopp's rule is only a rough approximation, but, if used with judgment, it is a very useful method of estimation.

One should keep in mind the factors found to be important in the preceding section—the atomic weight and the force constant restraining atomic motion. It is not usually feasible to obtain numerical values for the force constants; rather one selects reference substances where the force constant should be about the same. The atomic-weight effect can then be taken into account by use of Eq. (32-3)*, which may be generalized for a pair of compounds X and Y differing by the replacement of n atoms of A in X by n atoms of B in Y.

$$s_Y = s_X + \tfrac{3}{2} Rn \ln \frac{M_B}{M_A} \qquad (32\text{-}4)^*$$

If Eq. (32-3)* holds for the elements A and B, it is clear that the entropy additivity rule will yield Eq. (32-4)* for the compounds. However, Eq. (32-4)* may be applied whenever the force constants for the A and B atoms are about the same in the compound even if Eq. (32-3)* would not hold for the elements. For example, chlorine is gaseous and bromine is liquid at room temperature; hence no simple relationship is expected between their entropies. But for the entropy of metal chlorides and bromides Eq. (32-4)* predicts that the bromide entropy will be larger by 2.4 cal/deg mole per halogen in the formula. The values in Table 32-1 show reasonable agreement with this prediction. One may

TABLE 32-1. ENTROPIES OF METAL BROMIDES AND CHLORIDES[†]

$$S_{298}^\circ \text{ for } \frac{1}{n} MX_n(s)$$

M[‡]	X = Cl	X = Br	Difference
Cu^I	20.8	23.0	2.2
Pb^{II}	16.3	19.3	3.0
Hg^{II}	17.3	20.4	3.1
K	19.7	23.0	3.3
Ag	23.0	25.6	2.6
Na	17.3	20.0	2.7

† Values from Appendix 7.
‡ The oxidation state n is indicated by the Roman numeral.

also use the data of Table 32-1 to check Eq. (32-4)* with respect to metal-ion mass by comparing CuCl with AgCl, etc.

An equation of the type of (32-4)* was first given by Latimer[1] in 1921, when he proposed for compounds with the Dulong and Petit heat capacity of $3R$ per atom at 298°K the formula

$$S^{\circ}_{298} = \sum_i (\tfrac{3}{2} R \ln M_i - 0.94) \qquad \text{cal/deg mole}$$

where the sum covers all atoms in the compound. The numerical constant was selected to fit several metal chlorides. More recently Latimer[2] presented extensive tables checking rules of this general type for the entropies of inorganic solids. These tables provide a convenient basis for further estimates.

The melting point may be used under favorable circumstances as an indicator of relative force constants in similar crystals. Thus one may plot against the melting point the deviation of the entropies of a series of compounds from the values predicted by Eq. (32-4)*. Such a graph will aid in estimating the correct entropy of another member of the same series. However, one must be careful to avoid this method for solids with molecular lattices, where molecules with large intramolecular forces may still have very low melting points.

Entropy of Fusion. The entropy of fusion of a monatomic solid is usually in the range 2.0 to 3.5 cal/deg mole. The rare gases have values near 3.4, while most metals have values nearer 2.5. Metallic elements of complex structure such as bismuth have larger entropies of fusion, for example, $\Delta s_{Bi} = 4.8$ cal/deg mole.

Substances with infinite lattices composed of single ions also behave as monatomic solids with molal entropies of fusion in the range of 2.5 to 3.5 cal/deg per ion. However, when there are drastic changes in coordination number in going from the solid to the liquid, the entropy of fusion may be much larger. The occurrence of these abnormally high entropies of fusion can be predicted from the sizes of the anions and cations involved. As an example, the entropy of fusion of $AlCl_3$ is 18.2 cal/deg mole instead of the predicted value of 10 to 14. The size of the aluminum cation is such that it can accommodate a coordination number of 6 in the solid, but the increased thermal agitation of the liquid causes reduction of the coordination number and formation of Al_2Cl_6 molecules. On the other hand CCl_4, which has a molecular lattice, has an entropy of fusion of 2.4 cal/deg mole of CCl_4. Thus each CCl_4 molecule is acting as a single unit.

[1] W. M. Latimer, *J. Am. Chem. Soc.*, **43**, 818 (1921).

[2] W. M. Latimer, "Oxidation Potentials," 2d ed., app. III, Prentice-Hall, Inc., Englewood Cliffs, N.J., 1952.

More complex solids have larger entropies of fusion, and it is difficult to predict the value for a particular solid unless data are available for other solids of similar structure. Even then one must consider whether there are solid-solid transitions below the melting point. It is found that molecular substances of similar structure have approximately equal values of the sum of the entropy of fusion and of any entropies of transition. Thus there are two solid-solid phase transitions in oxygen (see Table 11-1) at which the entropy changes are 0.95 and 4.06 cal/deg mole. The entropy of fusion is 1.95, and the total is 6.96 cal/deg mole. The entropies of fusion of the halogens, which are also diatomic molecules but which have no solid-solid transitions, are Cl_2 8.89, Br_2 9.48, I_2 9.75, all in cal/deg mole.

Entropies of Vaporization. The Trouton[1] rule for entropies of vaporization is probably the oldest and best known of all rules for estimating entropies. It states that the entropy of vaporization at the boiling point (at 1 atm) is approximately constant for various substances. The value 21 cal/deg mole is commonly selected for the Trouton constant. Like the other rules we have been considering, the Trouton rule is valid only if certain other factors are very nearly constant. Thus deviations occur for substances which have hydrogen bonding, for substances boiling at very low or very high temperatures, etc. Hildebrand[2] showed that the entropies of vaporization for a variety of normal liquids were more nearly equal if compared at equal molal volume of vapor instead of equal vapor pressure. But it is found[3] that the entropies of vaporization of the extremely simple and similar substances Ar, Kr, Xe are almost exactly equal when compared at equal reduced temperatures (or equal reduced pressures). In this case the principle of corresponding states holds very precisely, whereas the range of entropies of vaporization at the boiling point is 0.4 cal/deg mole and the range at constant vapor volume is 1.0 cal/deg mole.

So long as one compares similar substances, it makes little difference whether the basis is at constant vapor pressure, constant reduced temperature, or constant vapor volume. One may then assume that, for such similar substances, the entropies of vaporization will be approximately equal.

Entropies of Gases. The entropies of gases with simple molecules can be calculated accurately by methods given in Chapter 27. For more complex molecules, however, the necessary spectroscopic or molecular-structure data may not be available, and estimates are then useful. Sometimes it is best to estimate the missing vibration frequencies or other

[1] F. T. Trouton, *Phil. Mag.*, (5)**18**, 54 (1884).

[2] J. H. Hildebrand, *J. Am. Chem. Soc.*, **37**, 970 (1915), **40**, 45 (1918).

[3] K. S. Pitzer, *J. Chem. Phys.*, **7**, 583 (1939).

molecular parameters; indeed high-frequency vibrations (such as C—H stretching) contribute little to the entropy at moderate temperatures, and their terms can even be omitted without serious error.

There are many regularities in the structure of organic compounds which aid in the estimation of entropies. For example, it is found that the entropy of paraffin isomers is lowered 3.5 cal/deg mole, on the average, for each branch in the carbon skeleton. Also it is easy to calculate certain symmetry factors which may be introduced to improve the accuracy of estimate. On this basis, Pitzer and Scott[1] proposed the formula

$$\text{s} = \text{s}_n + R \ln 2 + R \ln \frac{I}{\sigma_e \sigma_i} - 3.5B \qquad (32\text{-}5)*$$

where s_n is the entropy of the isomeric n-paraffin at the same temperature and pressure, I is the number of isomers included (two for a racemic mixture, otherwise one), σ_e and σ_i are the symmetry numbers for external rotation and for internal rotation of the carbon skeleton, respectively, and B is the number of chain branchings. The $R \ln 2$ corrects for the symmetry number of the n-paraffin. The entropy of the n-paraffins above butane at 298.15°K in the ideal-gas standard state is well represented by the formula

$$\text{s}_n^\circ = 9.31n + 37.07 \text{ cal/deg mole} \qquad (32\text{-}6)*$$

where n is the number of carbon atoms per molecule.

Table 32-2 compares the entropies of the various heptane isomers calculated from Eqs. (32-5)* and (32-6)* with experimental values which have been obtained subsequently.

TABLE 32-2. COMPARISON OF ESTIMATED AND EXPERIMENTAL ENTROPIES OF THE HEPTANES[†]

Values for the ideal-gas standard state at 298.15°K, cal/deg mole

Isomer	B	I	σ_e	σ_i	s_{est}°	s_{exp}°
n-Heptane.....................	0	1	2	1	102.24	102.24†
2-Methylhexane.................	1	1	1	1	100.1	100.35
3-Methylhexane.................	1	2	1	1	101.5	(101.37)
3-Ethylpentane.................	1	1	3	1	97.9	98.30
2,2-Dimethylpentane............	2	1	1	3	94.4	93.85
2,3-Dimethylpentane............	2	2	1	1	98.0	(98.96)
2,4-Dimethylpentane............	2	1	2	1	95.2	94.80
3,3-Dimethylpentane............	2	1	2	1	95.2	(95.53)
2,2,3-Trimethylbutane..........	3	1	1	3	90.9	91.60

† Values in parentheses are calculated by more elaborate statistical methods; others for branched heptanes are experimental values from the U.S. Bureau of Mines, Bartlesville, Okla., which are to be published.

[1] K. S. Pitzer and D. W. Scott, *J. Am. Chem. Soc.*, **63**, 2419 (1941).

Many additional methods appropriate to organic compounds are discussed by Janz in his summarizing monograph.[1]

Electronic Contributions to the Entropy. Most substances have only a single electronic state at low energy and, as a result, no electronic entropy at moderate temperatures. However, substances containing atoms of the transition elements or the rare earths frequently have a considerable number of low-energy electronic states. If the energies of these states are known from spectral data or from theory, the statistical methods of Chapter 27 suffice to calculate the entropy. In other cases the total number n of low-energy states may be known, but not their exact energies. In such cases, the quantity $R \ln n$ gives an upper limit to the possible entropy.

Since such systems with multiple electronic states show special magnetic properties, it is usually possible to use knowledge of the magnetic susceptibility to guide an estimate of actual electronic entropy. A simple system of this sort was discussed in Chapter 31; further consideration is beyond the scope of this book.[2]

Energies of Organic Compounds. Janz[3] also discussed various methods of estimation of free energies and enthalpies of organic compounds, and in view of his excellent summary our section will be brief. Energy estimates are based upon association of energy-of-formation values with structural units which may be taken as bonds or as atoms. In either case some corrections for interactions are needed.

We again take the paraffins as examples. Isomers have the same number of C—C and C—H bonds; so only those changes due to branching of the skeleton need be considered. A very simple empirical formula for the heat of formation at 298.15°K in the ideal-gas state is

$$\Delta H = 10.41 - 4.926n - 1.4b - 4.4b' \qquad \text{kcal/mole} \qquad (32\text{-}7)^*$$

where n is the total number of carbon atoms per molecule, b is the number of carbon atoms bonded to three other carbons, and b' is the number of carbon atoms bonded to four other carbons. The total number of skeletal branches B is $b + 2b'$. Equation (32-7)* is quite successful except when steric hindrance or strain is present in the molecule. The presence of such strain can be judged from molecular models, provided that care is taken to twist single bonds into the proper orientation for the torsional

---

[1] G. J. Janz, "Estimation of Thermodynamic Properties of Organic Compounds," Academic Press, Inc., New York, 1958. See also S. W. Benson and J. H. Buss, *J. Chem. Phys.*, **29**, 546 (1958).

[2] For further discussion of magnetic properties see J. H. Van Vleck, "The Theory of Electric and Magnetic Susceptibilities," Oxford University Press, New York, 1932.

[3] Janz, *op. cit.*

potential minimum, and rough estimates of the amount of strain can be made in favorable cases.[1]

Entropies of Aqueous Solutes Including Ions.   Powell and Latimer[2] considered the entropies of simple solutes in water.   For uncharged solutes they proposed the equation

$$\bar{s}^\circ = \tfrac{3}{2}R \ln M + s_{int} + 10 - 0.22v^\circ \qquad \text{cal/deg mole} \qquad (32\text{-}8)*$$

where M is the molecular weight, s_{int} is the molal entropy of the internal motions (rotation, vibration, electronic) evaluated for the free molecule, and v° is the molal volume in the pure liquid state.   Figure 32-1 shows

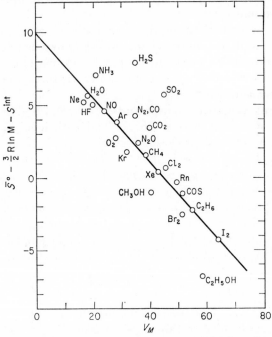

Fig. 32-1. Corrected entropy in aqueous solution vs. molal volume for various neutral molecules.

the correlation of corrected solute entropies with molal volume, which tests Eq. (32-8)*.   We note that the two alcohols fall several cal/deg mole below the line while SO_2 is above, but a large variety of other solutes fit quite well.

The presence of an electrical charge profoundly affects the interaction of a solute with water.   One might hope to get the charge effect by the

[1] A system including only weak hindrances was given by K. S. Pitzer, *J. Chem. Phys.*, **8**, 711 (1940); *Chem. Rev.*, **27**, 39 (1940).

[2] R. E. Powell and W. M. Latimer, *J. Chem. Phys.*, **19**, 1139 (1951).

temperature derivative of the Born equation for the free energy of a charged sphere in a continuous ideal dielectric medium [Eq. (31-27)*]. The appropriate temperature derivative gives the electrostatic entropy of solvation,

$$\Delta \bar{s} = \frac{z^2 e^2 N_0}{8 \pi \epsilon_0 r_0 D^2} \left(\frac{\partial D}{\partial T} \right) \qquad (32\text{-}9)*$$

which, upon substitution of numerical values for water at 25°C, yields

$$\Delta \bar{s} = -9.4 \frac{z^2}{r_0} \qquad \text{cal/deg} \qquad (32\text{-}10)*$$

Here z is the ionic charge and r_0 the radius of the spherical boundary between the ion and the dielectric. Laidler and Pegis[1] have shown that it is possible to fit the observed entropies of hydration of simple ions to the Born formula by adjusting from the arbitrary scale $\bar{s}^o_{H^+} = 0$ to $\bar{s}^o_{H^+} = -5.5$ cal/deg and by either expanding crystal radii 25 per cent or adjusting the coefficient of z^2/r_0 somewhat. In the latter case the empirical formula chosen by Laidler is

$$\bar{s}^o_{298} = -5.5z + \tfrac{3}{2} R \ln M + 10.2 - 11.6 \frac{z^2}{r_u} \qquad \text{cal/deg} \quad (32\text{-}11)*$$

where the term $-5.5z$ converts the results to the conventional scale $\bar{s}^o_{H^+} = 0$ and r_u is Pauling's *univalent ion radius,* i.e., the radius the ion would have if subject only to the compressive forces of single charges.[2]

However, water is not a continuous medium but rather is composed of molecules whose size is comparable with that of the ions. Also the electrically positive side of a water molecule is doubtless quite different from the negative side. These facts indicate that crystal radii can hardly be expected to apply directly to the Born equation and that the corrections to the crystal radii will probably be different for positive and negative ions. Thus the theoretical argument for the particular form of the Born equation chosen by Laidler is not very strong. It is also interesting to note that Powell and Latimer[3] found an empirical equation for simple ion entropies which is of very different form in its term in z/r^2 instead of z^2/r. Their equation is

$$\bar{s}^o = \tfrac{3}{2} R \ln M + 37 - 270 \frac{z}{r_e^2} \qquad \text{cal/deg} \qquad (32\text{-}12)*$$

[1] K. J. Laidler and C. Pegis, *Proc. Roy. Soc. (London),* **A241,** 80 (1957); see also K. J. Laidler, *Can. J. Chem.,* **34,** 1107 (1956).

[2] L. Pauling, "The Nature of the Chemical Bond," 3d ed., p. 514, Cornell University Press, Ithaca, N.Y., 1960.

[3] *Loc. cit.*

where $S°$ is the conventional entropy based upon $S°_{H^+} = 0$ and r_e is an effective radius in angstroms which is 1.0 A greater than the crystal radius for negative ions and 2.0 A greater than that for positive ions.

Equations (32-11)* and (32-12)* fit the known ion entropies about equally well. If both equations agree on the prediction for an unknown case, one will have more confidence in the result than if they disagree.

Several equations have been proposed for the entropies of polyatomic ions. The simplest is that of Connick and Powell,[1] which, for the ion $XO_n(OH)_l^{-z}$, is

$$S°_{298} = 43.5 - 46.5(z - 0.28n) \quad \text{cal/deg} \quad (32\text{-}13)*$$

Note the absence of l, the number of OH groups, in the formula. Also the familiar term $\frac{3}{2}R \ln M$ is missing, which is a weakness since there is full theoretical basis for the mass term.

Couture and Laidler[2] propose the somewhat more complex but theoretically better founded formula for the same class of oxygenated anions,

$$S°_{298} = 5.5|z| + 40.2 + \tfrac{3}{2}R \ln M - 108.8 \frac{z^2}{nr} \quad (32\text{-}14)*$$

where z and n have the same significance as before and r is the X—O bond distance plus 1.40 A. The effect of the symmetry number should be included, as King[3] has shown.

Cobble[4] has considered other sorts of complex ions and gives useful empirical formulas for some of these types. For example, he proposes for the entropy of ionization of a weak acid HOXY, where Y may represent any number of atoms,

$$\Delta S° = 20 - \frac{59}{r_{12}} \quad \text{cal/deg mole}$$

where r_{12} is the O—X bond distance in angstroms.

All these formulas agree qualitatively with the generalization mentioned in Chapter 22 that ionization reactions have substantial negative values of $\Delta S°$ and $\Delta C_P°$. The values -20 and -40 cal/deg mole are typical of a dissociation to singly charged ions in aqueous solution at room temperature. The data for further ionization to more highly charged species are less extensive but indicate even larger negative values. This is a very important factor in this area of chemistry. Thus reactions creating ions must be substantially exothermic to overcome the negative $\Delta S°$ if $\Delta F°$ is to be zero or negative. Conversely, complex-ion formation,

[1] R. E. Connick and R. E. Powell, *J. Chem. Phys.*, **21**, 2206 (1953).
[2] A. M. Couture and K. J. Laidler, *Can. J. Chem.*, **35**, 202 (1957).
[3] E. L. King, *J. Phys. Chem.*, **63**, 1070 (1959).
[4] J. W. Cobble, *J. Chem. Phys.*, **21**, 1443, 1446, 1451 (1953).

which reduces the amount of charged species in solution, may proceed even if substantially endothermic because of the positive $\Delta S°$.

Energies of Inorganic Substances. There are many times when one must be prepared to make rough estimates of the stabilities of various species that might possibly be of chemical importance in a system. These species may often involve oxidation states which are quite unusual under normal conditions but which might well be important under the specific conditions of a given problem. Earlier, when we have had to estimate heat capacities, entropies, or other properties, we have often made assumptions which required a uniform variation of the property across the periodic table. With heats of formation this type of procedure can be somewhat dangerous unless one takes into account the nature of the bonding. This is perhaps well illustrated by Tables 32-3 to 32-5,

TABLE 32-3. ΔH°_{298} FOR $\dfrac{1}{x}$ MF$_x$(s) $= \dfrac{1}{x}$ M(s) $+ \frac{1}{2}$F$_2$(g), kcal/g atom F

LiF	BeF$_2$		
145.7	121 ± 3		
NaF	MgF$_2$	AlF$_3$	
136.3	134	118.8	
KF	CaF$_2$		TiF$_4$
134.5	145.1		98.4
RbF	SrF$_2$		
131.8 ± 2	145.2		
CsF	BaF$_2$		
130.3	143.5		

which present $\Delta H°$ of dissociation per gram atom of nonmetal for the iodides, fluorides, and oxides of a number of the metals in the periodic table. The enthalpies are given for the reaction resulting in the formation of the same number of moles of nonmetal in every instance to ensure that the standard entropy change is relatively independent of the metal involved. Thus the heat of reaction will be the determining factor in predicting trends in the standard free energy of reaction. The variation of the enthalpies per gram atom of halogen is proportional to the variation of the logarithm of the equilibrium partial pressure of halogen resulting from decomposition of the halide to halogen gas and solid metal.

It will be noted that the cesium iodide is the most stable iodide, with stability decreasing as one goes upward or to the right in the periodic table. On the other hand, for the oxides and fluorides there is a ridge of maximum stability from lithium to thorium, with stability falling off on either side. It is clear here that differences of heats of formation between fluorides and iodides would not show uniform behavior across the periodic table.   If one examines the various factors which contribute to the heat of formation of a compound, one can see the reason for this apparently anomalous behavior.   Some of the factors, for example, which come into play are the heats of vaporization of the elements, ionization potentials,

TABLE 32-4. ΔH°_{298} FOR $\frac{1}{x}$ $MI_x(s) = \frac{1}{x} M(s) + \frac{1}{2} I_2(s)$, kcal/g atom I

LiI	BeI$_2$		
64.8	19 $\pm$ 3		
NaI	MgI$_2$	AlI$_3$	
68.8	43.0	24.7	
KI	CaI$_2$		TiI$_4$
78.3	63.9		22.8
RbI	SrI$_2$	YI$_3$	ZrI$_4$
79 $\pm$ 2	68.0	46 $\pm$ 1	33 $\pm$ 2
CsI	BaI$_2$	LaI$_3$	
83.9	72.0	53	

and other properties which vary in different manners across the periodic table so that the combination of these terms does not give a smooth behavior.   Thus it is often of value to break down the heat of formation into simpler quantities which would vary in a simpler way.

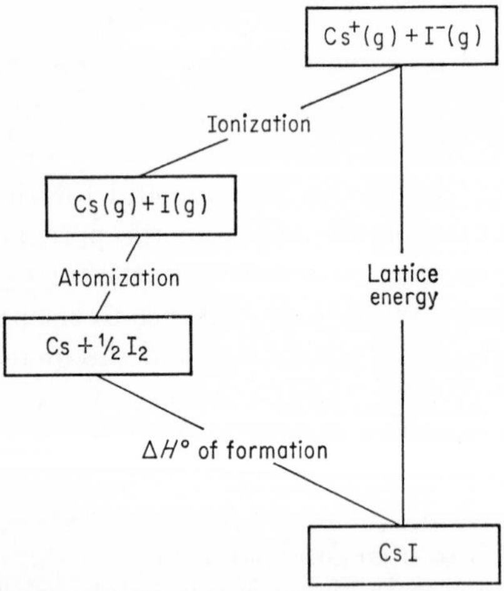

When one is dealing with compounds for which there is a great difference in electronegativity between metal and nonmetal, such that one can expect the compounds to be rather ionic in nature, the use of the Born-Haber cycle can often shed a great deal of light on the factors which influence the variation in the heat of formation. The Born-Haber cycle is illustrated in Fig. 32-2, with CsI as an example.   The over-all process of combination of the elements to the compound is broken up into a number of elementary steps, such as the vaporization of the elements, the atomization of the elements, and subsequent ionization to produce the gaseous ions, which then combine to form the solid lattice.   Let us consider this

FIG. 32-2. Born-Haber cycle.

last step, where the gaseous ions are brought together to their equilibrium distance in the solid lattice. The heat liberated by this step will be determined by the charges of the ions and the distance to which they can approach one another. If we are dealing with the alkali elements, the charges are all the same. For the iodides the interionic distances are largely fixed by the radius of the iodide ion, so that there is little variation in the interionic distance going from lithium iodide to cesium iodide and therefore little variation in the lattice energy—the energy liberated upon combination of the gaseous ions. Since there is little variation in this

TABLE 32-5. ΔH°_{298} FOR $\frac{1}{x} M_y O_x(s) = \frac{y}{x} M(s) + \frac{1}{2}O_2(g)$, kcal/g atom O

Li_2O	BeO	B_2O_3		
142.6	143.1	101.8		
Na_2O	MgO	Al_2O_3	SiO_2	P_2O_5
99.4	143.8	133.5	105.0	72.0
K_2O	CaO	Sc_2O_3	TiO_2	V_2O_5
86.4	151.8	137 ± 3	112.8	74.6
Rb_2O	SrO	Y_2O_3	ZrO_2	Nb_2O_5
80 ± 4	141.1	151.8	130.8	91.0
Cs_2O	BaO	La_2O_3	HfO_2	Ta_2O_5
82 ± 2	133.5	142.9	133.0	97.8
			ThO_2	
			146.7	

step, the variation in the heats of formation must be fixed primarily by the variation in heats of sublimation of the metals and the variation of the ionization potentials of the gaseous metallic atoms. In going from lithium to cesium the energy required to produce gaseous ions decreases rather rapidly and thus accounts for the higher stability of the cesium iodide compared with the lithium iodide.

If one now examines the corresponding steps for the fluorides, one finds that the lattice energies vary greatly because the size of the fluoride ion is about the same size as the cations. Thus there will be a very large percentage change in the ionic distance in going from lithium fluoride to cesium fluoride, and the lithium fluoride will have a much greater lattice energy than the cesium fluoride. This trend in lattice energy is more than enough to offset the variation in the heats of sublimation and ionization potentials of the alkali elements, and thus lithium fluoride is more stable than cesium floride. In terms of this simple model, what seems to be a quite anomalous behavior is readily explained, and the variation of the stability of iodides is correlated to the variation of stability of fluorides.

One can draw some generalizations which can be of aid in predicting heats of formation. One would expect ionic compounds with large anions to show a similar variation across the periodic table, with maximum stability in the lower left-hand corner and stability decreasing as one moves either upward or to the right in the periodic table. Thus one should be able to predict the variation in stability of the perchlorates from lithium to cesium, for example, by taking the same variation as one observes for the iodides. On the other hand, if one wished to predict the variation of stability of the alkali hydroxides, one would examine the data for the fluorides rather than the iodides.

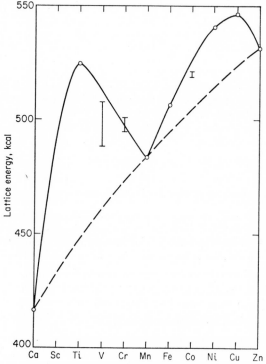

Fɪɢ. 32-3. Lattice energies of transition-metal dichlorides compared with ionic-model values.

For the transition-metal compounds one finds that a simple ionic model is not adequate. This is illustrated by Fig. 32-3, which shows the lattice energies of the transition-metal dichlorides of the third period of the periodic table. In contrast to the expected variation on the basis of an ionic model it can be seen that there is a region of large deviations followed by close agreement for manganous chloride and again large deviations for the transition metals beyond manganese. These anomalies have been explained in terms of the crystal-field theory.[1-5]   The d electrons of the transition-metal ions do not have spherical symmetry except for manganous ion, which has a half-filled shell.   It is possible for some anions or water of hydration to approach more closely without encountering large repulsive forces due to overlap of electron clouds than would be anticipated if the ions had spherical symmetry.   Also there is a stabilization of the gaseous ions with incomplete d shells in an electric field, which is known as the Jahn-Teller effect.   It is possible, from the spectra of the ions in the aqueous solution or in a crystalline lattice, to determine the magnitude of these energies and thus to predict the degree of enhanced stability over that calculated on the basis of an ionic model with spherical ions.   Even when spectral data are lacking, the general shape of the deviation curve is known and with data for one or two of the transition metals one can roughly predict the positions of the other transition-metal elements.

There are yet other ways in which one can deal with deviations from a simple ionic model.   If one plots lattice energies for the alkali halides as a function of the halogen, or as a function of interionic distance, one gets characteristic curves which agree rather well with the calculated curves for an ionic model.   When one goes to the right-hand side of the periodic table, where the cations do not have completed octets, one finds deviations from the calculated curves and these deviations increase as one goes from fluoride to iodide.   In general these deviations are larger the more polarizable the cation or the anion and can be calculated in part from the known polarizabilities of the cations and anions.   When one is dealing with cations with incomplete octets, such as silver, iron, or copper ions, there appear to be additional interactions, which are generally regarded as being due to covalent bonding.   By examining these deviations as a function of position in the periodic table, and therefore of electronic structure of the cations, one has again a guide to the corrections to be applied to a simple ionic model.

Similarly, when one is working with compounds for which there is not a great difference in electronegativity, one finds it very useful to estimate the energies of covalent bonds from Pauling's table[6] of electronegativity or a similar treatment.   Kubaschewski and Evans[7] present correlations

[1] J. S. Griffith and L. E. Orgel, *Quart. Revs. (London)*, **11**, 381–393 (1957).

[2] F. Basolo and R. G. Pearson, "Mechanisms of Inorganic Reactions," chap. 2, John Wiley & Sons, Inc., New York, 1958.

[3] D. S. McClure, *J. Phys. Chem. Solids*, **3**, 311–317 (1957).

[4] N. S. Hush, *Discussions Faraday Soc.*, **26**, 145–156 (1958).

[5] J. S. Griffith, *J. Inorg. & Nuclear Chem.*, **3**, 15–23 (1956).

[6] L. Pauling, "The Nature of the Chemical Bond," 3d ed., p. 91, Cornell University Press, Ithaca, N.Y., 1960.

[7] O. Kubaschewski and E. L. Evans, "Metallurgical Thermochemistry," 3d ed., Pergamon Press, London, 1958.

for intermetallic compounds and semiconductors, which are intermediate between the typical ionic and covalent compounds. They also review many other procedures for the estimation of thermodynamic data.

There are several types of bonding models which one can use when dealing with a specific problem. In general, one gathers together data for compounds which are closely related to the one of interest, and one tests the various methods of estimating bonding energies to determine which methods agree most closely with the known data. Thus one may select a method most likely to give good results for the unknown compound. Procedures of this type will usually allow one to make reasonable estimates of the heats of formation of a very large fraction of compounds for which data are not available.

To determine the importance of various compounds under particular conditions of interest, one first uses the simplest and crudest of these methods of estimating bonding energies, neglects ΔC_P, and uses the roughest rules of estimating entropies of reaction. Once one has limited the possible compounds that need to be considered, more elaborate methods of estimating entropies and heats of formation are employed in an attempt to define more accurately those species of importance under the specific conditions of interest. If these calculations demonstrate that the major species are compounds which have been studied previously and for which quite adequate thermodynamic data exist, then it will be possible to obtain quite accurate results upon applying thermodynamic calculations to the system of interest. If, on the other hand, the preliminary calculations indicate the importance of new species which have not been studied previously, one may be restricted to estimates, but one will not fall into the trap of assuming that only well-known compounds need to be considered and therefore will not make an error of many orders of magnitude.

PROBLEMS

32-1. Compare the entropy of fusion of NH_3 with the sum of entropies of transition and of fusion for PH_3. For data on the latter substance see Stephenson and Giauque,[1] and note that either of two solid forms may be taken as the starting phase at 0°K.

32-2. For ZrC at 298°K, $\Delta H° = -44.4 \pm 1.1$ kcal/g atom of carbon.[2] High-temperature heat-content and entropy data are lacking. Use the data in Appendix 7 for TiC, TiN, and ZrN to calculate $\Delta_F°$ of ZrC at 2000°K. What is the vapor pressure of Zr(g) in equilibrium with ZrC and graphite at 2000°K?

32-3. Low-temperature heat capacities yielded the entropies of RbBr and RbI given in Table A7-3, but no such data are available for RbCl or RbF. By comparison with the data for potassium salts and the predictions of Eq. (32-4)*, estimate $s_{298}°$ for RbCl and RbF.

[1] C. C. Stephenson and W. F. Giauque, *J. Chem. Phys.*, **5**, 149 (1937).

[2] A. D. Mah and B. J. Boyle, *J. Am. Chem. Soc.*, **77**, 6512 (1955).

32-4. Compare the entropy values given in Appendix 7 for KCl and KI with the corresponding values for Cu and Ag compounds. The melting points in degrees Kelvin are as follows:

KCl	CuCl	AgCl	KI	CuI	AgI
1043	703	728	955	861	830

What explanation can be offered for the deviations from Eq. (32-4)*?

32-5. The crystal radii of Cu^+, Ag^+, and Tl^+ are 0.93, 1.21, and 1.59 A, respectively. The entropies of aqueous Ag^+ and Tl^+ are given in Table 25-7. Estimate $\bar{s}^\circ$ of Cu^+(aq) at 25°C.

32-6. Use the entropies of $TiCl_4(g)$, $TiBr_4(g)$, and $TiI_4(g)$ from Tables A7-6 and A7-9 to estimate the entropies of the corresponding zirconium compounds at 298°K. Use the entropy data for $ZrCl_4(s)$ to estimate s°_{298} for $ZrBr_4$ and ZrI_4 solids which have melting points and volatilities close to those for $ZrCl_4$. Combine these estimates to compare Δs°_{298} of sublimation of the three zirconium halides with the experimentally determined values of 45 cal/deg for each from the data of Rahlfs and Fischer.[1] In the Σ-plot treatment of the data, $c_P = 25.5$ cal/deg was used for the gases, and the experimental data for c_P of $ZrCl_4(s)$ was used for all three solids.

32-7. No data are available for $ScCl_2(s)$. What ΔH°_f would be predicted from Fig. 32-2? $\Delta H^\circ = 305$ kcal for the reaction $Sc(s) + Cl_2(g) = Sc^{++}(g) + 2Cl^-(g)$. If ΔH° of $ScCl_3(s)$ is -220 kcal at 25°C, would $ScCl_2(s)$ be stable with respect to disproportionation? Would Mg be capable of reducing $ScCl_3$ to Sc metal?

32-8. Predict the variation of the heats of formation of the $MSeO_4$ compounds from $M = Be$ to $M = Ra$. Predict the trend of solubility in water for these compounds.

---

[1] O. Rahlfs and W. Fischer, *Z. anorg. Chem.*, **211,** 349 (1933).

33

VAPORIZATION PROCESSES

In this chapter we consider primarily vaporization processes where several molecular species are present in the vapor or in the condensed phase or in both. Also examples with two or more condensed phases are frequently of interest. Several molecular species are always involved when two or more components are present, but examples with one-component systems also are common. It is even convenient to use the same methods for the very simplest vaporization process, that of an element such as argon.

No new thermodynamic principles are introduced, nor are any new functions or methods presented; rather we take these problems as examples to illustrate the methods already presented. Also we show how even a limited knowledge of the energy of interaction between molecular species allows one greatly to simplify problems. In order to keep this chapter within reasonable length, most examples are not developed in detail, but many illustrative problems are included at the end.

ONE-COMPONENT SYSTEMS

Treatment for a Single Vapor Species. We first consider the exact treatment of the vaporization of a simple molecular substance to a vapor consisting predominantly of a single molecular species. One may consider this process in terms of the Clapeyron equation (10-22),

$$\frac{dP}{dT} = \frac{\Delta H}{T \, \Delta V} \tag{10-22}$$

where ΔH is the heat of vaporization to the real vapor and ΔV is the increase in volume from the liquid to the real vapor. Both these quantities are affected by gas imperfection. It is possible to apply gas-imperfection corrections to both ΔH and ΔV in a manner which expresses their temperature dependence and eventually to integrate Eq. (10-22) exactly. Since this is cumbersome, it is usually more convenient to consider vaporization as a reaction whose equilibrium constant is the ratio of the vapor fugacity f_g to the liquid activity a_l and hence

$$\frac{d \ln (f_g/a_l)}{dT} = \frac{d}{dT}\left(-\frac{\Delta F^\circ}{RT}\right) = \frac{\Delta H^\circ}{RT^2} \qquad (33\text{-}1)$$

Now all gas-imperfection effects are included in the ratio f_g/P for the gas, and the effect of liquid volume is included in a_l.   Consequently ΔH° is the heat of vaporization to the ideal gas.   The temperature dependence of ΔH° is now just $\Delta C_P^\circ = C_P^\circ(g) - C_P^\circ(l)$, with $C_P^\circ(g)$ the usual heat capacity of the ideal gas.   At low pressures one may take $a_l = 1$ unless very high accuracy is required.

In applying Eq. (33-1) to the vapor-pressure measurements for a particular substance one first converts each pressure measurement to a fugacity of the vapor by one of the methods described in Chapter 16 or Appendix 1.   From the liquid volume one easily calculates a_l at the pressure of the vapor.   Then, if one has equations for the heat capacity of vapor and liquid, one may follow the sigma-plot method illustrated in Eq. (15-38) for the reaction of formation of H_2S from the elements.   An even better treatment is possible if one knows the free-energy functions for the condensed and gaseous phases.   Also from Chapter 15 we recall that

$$-R \ln \frac{f_g}{a_l} = \frac{\Delta F^\circ}{T} = \left(\frac{F^\circ - H_0^\circ}{T}\right)_g - \left(\frac{F^\circ - H_0^\circ}{T}\right)_l + \frac{\Delta H_0^\circ}{T} \qquad (33\text{-}2)$$

Hence a value of ΔH_0° can be calculated from each value of the vapor pressure, and if all component data are correct, the various ΔH_0° values should be the same within experimental error.   It is apparent that $(F^\circ - H_{298}^\circ)/T$ values could be used instead of $(F^\circ - H_0^\circ)/T$ values in an equation analogous to Eq. (33-2), and a value of ΔH_{298}° would then be given by each vapor-pressure value.

Treatment for More than One Vapor Species.   Even the familiar gas imperfection of a normal vapor may be treated as a process of formation of dimeric and polymeric molecular species, but it is not necessary to consider these species in detail for vapors which follow a standard equation of state for a normal fluid.   In other cases, however, it is usually best to consider explicitly the principal species present in the vapor.   We have already discussed in the next to the last section of Chapter 20 the ambiguity which arises between nonideal behavior and dimerization or polymerization when both effects are considered.   We shall not now pursue this matter further but rather shall take ideal behavior of each species present.

Let us consider first bismuth vapor, which in equilibrium with bismuth liquid consists of monatomic bismuth and diatomic bismuth in comparable amounts.   Although one can calculate quite readily the heat capacities of monatomic bismuth and diatomic bismuth from spectroscopic data, one would have to know the ratio of the two species in the saturated vapor as a function of temperature and the heat of dissociation in order to calculate the heat capacity of the equilibrium vapor. Thus the sigma-plot method would not be applicable without some modification. Likewise, one would not be able to calculate the value of $(F° - H_0°)/T$ for the gaseous mixture without knowledge of the ratio, and one could not apply the third-law method. If one can determine, in addition to the total vapor pressure in equilibrium with liquid bismuth, the ratio of the two species, one can evaluate the partial pressures of monatomic and diatomic species and apply the above methods separately to each of the molecular species.

At moderate temperatures it is usually feasible to measure the density of the vapor and calculate its average molecular weight. Also one can frequently measure the concentration of individual gas species by spectroscopic measurements. At very high temperatures, however, these methods become difficult or even impossible, and one must work with vaporization data alone. The mean molecular weight of the vapor may be determined by comparison of the static vapor pressure with the mass evaporated to saturate a known volume of inert gas in a transpiration experiment.

The work of Brewer and Lofgren[1] on cuprous chloride illustrates the last method. At 1119°K the vapor pressure of cuprous chloride is 0.0668 ± 0.008 atm, and 0.212 g was evaporated to saturate 0.0105 mole of argon carrier gas at 1 atm. From these data one calculates that the 0.212 g constituted 7.05×10^{-4} mole of cuprous chloride vapor or that the mean molecular weight was 300 ± 36. This result shows that saturated cuprous chloride vapor at 1119°K has an average composition near Cu_3Cl_3 ($M = 297$), but it does not tell whether the vapor is mostly Cu_3Cl_3 or a mixture of higher and lower polymers of CuCl. A thermodynamic method which does determine the detailed composition will be explained in a later section.

One may consider also the reverse process, whereby one attempts to predict equilibrium conditions starting from thermodynamic data. It is of very great importance to know the possible species that could be of chemical importance, i.e., be present in substantial amounts. One could carry out quite accurate thermodynamic calculations for the process $2As = As_2(g)$. However, the calculated vapor pressures would deviate very greatly from the observed total vapor pressures because the As_4 species is much more important than the As_2 species under most conditions.

[1] L. Brewer and N. L. Lofgren, *J. Am. Chem. Soc.*, **72**, 3038 (1950).

It is thus very important for vaporization processes, as for most processes, to establish the principal net reaction in order to be able to treat the system thermodynamically in an efficient manner. The establishment of the principal net reaction corresponds to fixing the principal species that are involved in the equilibrium. If one is dealing with an unexplored system, there may be a very large number of possible species that one should consider. Also in practical work one may have to consider not only the species which might result from direct vaporization of the material under study but also those species which might result from disproportionation or decomposition reactions, those species which might result from interaction between the containing material and the substance under study, and those species which might result from interaction with the atmosphere present in contact with the material under study. In many instances these reactions with container materials or with the atmosphere are unsuspected or unintended reactions which can indeed be predominant. To ensure that one is not making any gross error, one has to consider every possible gaseous species that can result from the combination of components in the system and to estimate their importance.

When one is starting from initial thermodynamic data in an attempt to predict equilibrium behavior, one must carry out trial calculations on the basis of each of the various possible species. In these preliminary calculations, one can predict that some species are quite unimportant chemically and can be neglected and that others are likely to be of considerable importance and will have to be considered in more detail. Similarly, in the reverse process, when one has experimental vapor-pressure data and wishes to convert them to thermodynamic form, one must again consider the various possible species which could be contributing to the vapor pressure in order to know how to proceed in the treatment of these data.

In order to select the important species, one must be able to estimate in a simple, quick manner the standard heats and standard entropies for every conceivable gaseous species. Fortunately it is possible to make some generalizations that can greatly simplify this task.

Entropy of Vaporization. Consider first the relatively simple situation of a one-component system with the polymeric species A, A_2, A_3, etc., in the vapor phase. The vaporization can be represented by the following equations:

$$A(l) = A(g)$$
$$2A(l) = A_2(g)$$
$$3A(l) = A_3(g)$$
$$\cdot \ \cdot \ \cdot \ \cdot \ \cdot \ \cdot \ \cdot$$

The first generalization that can be made with regard to this series of reactions is that the standard entropy change for all these reactions will be approximately the same. This can be seen by considering the various contributions to the entropy of the reaction. The entropy of the condensed phase is normally small compared with the entropy of the gaseous phase. The major contribution to the entropy of the gaseous molecule is the translational contribution, which will be the same for each of these reactions since we have 1 mole of gaseous species on the right-hand side and the mass effect cancels between gas and liquid. Thus, to the extent to which one can neglect, or can assume cancellation between the gas and the liquid of, the electronic, vibrational, and rotational contributions to entropy, one can take the entropy of vaporization to be the same for all these vaporization reactions. This is the essential basis of Trouton's rule.

These arguments may be applied to an even wider variety of vaporization processes. For example, consider disproportionation reactions such as $2TiBr_3(s) = TiBr_2(s) + TiBr_4(g)$ or $Hg_2Cl_2(s) = Hg(g) + HgCl_2(g)$ or $SiO_2(s) = SiO(g) + O(g)$. If one writes the equilibrium constants for these reactions in terms of the vapor pressure, one finds that the reactions are comparable if they are all written for the formation of 1 mole of gas on the right-hand side, i.e.,

$$2TiBr_3(s) = TiBr_2(s) + TiBr_4(g)$$
$$\tfrac{1}{2}Hg_2Cl_2(s) = \tfrac{1}{2}Hg(g) + \tfrac{1}{2}HgCl_2(g)$$
$$\tfrac{1}{2}SiO_2(s) = \tfrac{1}{2}SiO(g) + \tfrac{1}{2}O(g)$$

On this basis there is the same translational contribution to the entropy of vaporization for the 1 mole of gas formed in each case. Of course the other entropy contributions do not cancel exactly for the condensed phase and the internal motions of gaseous molecules. Nevertheless, this assumption of equal entropy of vaporization per mole of gas produced is useful in the crude estimates one makes in exploring new systems.

If the entropies of vaporization are the same for various processes, the free energies of vaporization will fall in the same order as the heats of the various vaporization processes. Consequently, the species that will predominate in the low-temperature vapor is the species with the lowest heat of vaporization. All the other possible species have higher heats and free energies of vaporization. Also, to the extent that Trouton's rule holds, one can make the generalization that all the minor species which may be present in the low-temperature saturated vapor will increase in importance as the temperature of the saturated system is increased. Let us now consider estimates of heats of vaporization.

Heats of Vaporization. It is not possible to make any generalization comparable with the Trouton rule for entropies about enthalpies of vaporization. One can, however, consider all the knowledge available about binding energies in the condensed and vapor species. A few examples will illustrate the types of considerations which are sometimes useful.

Let us consider first normal liquids in which the intermolecular forces are of the van der Waals type, i.e., London dispersion forces possibly supplemented by some dipole-dipole interactions. The London forces are additive for all intermolecular pairs. A given molecule has possibly 6 to 10 nearest neighbors in the liquid and even 12 in a closest-packed solid. The vaporization of a molecule destroys half this number of inter-molecular interactions (or van der Waals bonds), while dissociation of a gaseous dimer breaks only one such interaction. Hence in a normal liquid one expects the heat of vaporization to the monomer to be 3 to 5 times the dimerization energy (and for a closest-packed solid 6 times).

We refer again to the equations for reactions each producing 1 mole of vapor, i.e.,

$$A(l) = A(g) \qquad \Delta H_1^\circ$$
$$2A(l) = A_2(g) \qquad \Delta H_2^\circ$$

and the dimerization reaction in the gas

$$2A(g) = A_2(g) \qquad \Delta H_D^\circ$$

One may easily show that the heat of vaporization to the dimer is

$$\Delta H_2^\circ = 2\,\Delta H_1^\circ - \Delta H_D^\circ \qquad (33\text{-}3)$$

From the discussion of the normal fluid we estimate ΔH_2° to be $\frac{5}{3}$ to $\frac{9}{5}$ times ΔH_1°. Also, since the free energies of vaporization are expected to differ by the same amounts as the heats, one expects little dimer in the vapor at low temperatures and pressures. The same arguments predict even less trimer or higher polymer. Experimental data on normal fluids are consistent with these predictions.

Since ΔH_2° is larger than ΔH_1° in the normal fluid, the partial pressure of dimer will increase more rapidly with temperature than that of the monomeric species. Thus the vapor becomes a more complex mixture of species at higher temperatures, and near the critical point the vapor of a normal fluid contains many polymeric species.

If the principal binding forces in the condensed phase are saturated with a smaller number of neighbors than for a normal liquid, then one may expect a more complex species in the low-pressure vapor. An extreme example is a monocarboxylic acid, which can form only one strong hydrogen bond per molecule. This bonding may take place to form a dimer with two bonds or a long-chain polymer with one bond per monomer unit. Regardless of which structure is preferred in the liquid, the heat of vaporization to the monomer is only slightly more than half the heat of dissociation of the dimer, and the heat of vaporization to the dimer is considerably smaller than that to the monomer.   In view of the entropy factors given above, one expects and finds that the saturated vapor is largely dimer with some monomer at low temperatures, and the fraction monomer increases with temperature.   Near the critical point, however, one finds a complex mixture of various species, including higher polymers.

Hydrogen fluoride and the simple alcohols provide somewhat related examples.   Like the carboxylic acids there can be but one hydrogen bond per monomer unit, but the geometry in this case does not allow two bonds in a dimer.   Rather, a four-membered or larger ring is required before the average of one strong hydrogen bond per monomer is attained.   In the case of HF, where other attractive forces are small compared with the hydrogen-bonding effect, one finds the vapor to be predominantly $(HF)_6$ and $(HF)_4$ at low temperatures, with the fraction of HF increasing with temperature.[1]   In the alcohols there is much more interaction of the normal van der Waals type, and much less extreme results are noted. Nevertheless, there is appreciable tetramer $(ROH)_4$ in the saturated vapor at the normal boiling point of methanol[2] and ethanol.[3]   In the alcohols this effect was first noticed in the heat capacity of the nearly saturated vapor.   Heating at constant pressure causes dissociation of the polymer, and this leads to an abnormally high heat capacity which is as much as double the expected heat capacity of the monomer species.

Ionic binding yields an intermediate case.   If one makes the crude assumption of rigid spherical ions of equal charge, the binding energies relative to separate ions are readily calculated.   The monomer MX in the gas has an energy $-e^2/d$, where e is the ionic charge and d the interionic distance.   A square dimer of structure

$$
\begin{array}{ccc}
M & — & X \\
| & & | \\
X & — & M
\end{array}
$$

has an energy $-2(e^2/d)(2 - 2^{-1/2})$ or $-2.60(e^2/d)$.   Ionic lattices of the MX type all have energies near $-1.7(e^2/d)$.   From these data we calculate that dimerization energy almost equals the energy of vaporization to the monomer and that ΔH_2° will be roughly $\frac{8}{7} \Delta H_1^\circ$.   These results indicate that the vapor of an ionic salt at low temperatures will be predominantly monomer but with more dimer than is present for a normal fluid.   Except for LiCl, LiBr, and LiI, which have abnormally high proportions of polymers, the alkali halide vapors have 5 to 30 per cent dimer

[1] D. F. Smith, *J. Chem. Phys.*, **28**, 1040 (1958), and references there cited.

[2] W. Weltner and K. S. Pitzer, *J. Am. Chem. Soc.*, **73**, 2606 (1951).

[3] G. M. Barrow, *J. Chem. Phys.*, **20**, 1739 (1952).

at temperatures somewhat below their melting points,[1,2] which is generally consistent with our crude analysis.   Improved calculations for ionic models have been made by Born and Huang,[3] Rittner,[4] O'Konski and Higuchi,[5] Berkowitz,[6] Milne and Cubicciotti,[7] Bauer, Diner, and Porter,[8] and others.

In general one considers the various possible vapor species and the expected binding energies and attempts to predict the relative values of the heats of vaporization for processes forming 1 mole of total vapor species in each case.   It is also important to perform whatever experiments are possible to confirm the estimate of the principal species.   In many cases spectroscopic measurements will confirm the presence or absence of molecular species even though calibration to yield quantitative concentrations of the species is not possible.   Mass spectrometric data are particularly valuable, although their interpretation involves assumptions about the ionization cross sections and stabilities of the positive ions.

VOLATILIZATION BY REACTION WITH A GAS

In many cases additional information about vapor species may be obtained by adding a gas which does not dissolve significantly in the condensed phase.   This allows variation in the activity of the component of interest without affecting the activity of a second component, which is fixed by the presence of the solid or liquid.   The volatilization of copper by HCl gas provides an example of this type.

Earlier in this chapter we noted that saturated cuprous chloride vapor has an average molecular weight close to that of Cu_3Cl_3 at $1119°K$.   In order to determine the particular species present, Brewer and Lofgren[9] studied the unsaturated vapor prepared by equilibrating a mixture of H_2 and HCl with copper.   They showed that all copper chloride vapor species had formulas Cu_xCl_x with equal numbers of copper and chlorine atoms.   Hence one has the set of equations

$$x\text{Cu(s)} + x\text{HCl(g)} = Cu_xCl_x\text{(g)} + \frac{x}{2}\,H_2\text{(g)}$$

[1] G. M. Rothberg, M. Eisenstadt, and P. Kusch, *J. Chem. Phys.*, **30**, 517 (1959), and references there cited.

[2] S. Datz, Ph.D. thesis, Oak Ridge National Laboratory Report 2933 (1960).

[3] M. Born and K. Huang, "Dynamical Theory of Crystal Lattices," Oxford University Press, New York, 1954.

[4] E. S. Rittner, *J. Chem. Phys.*, **19**, 1030–1035 (1951).

[5] C. T. O'Konski and W. I. Higuchi, *J. Chem. Phys.*, **23**, 1175 (1955).

[6] J. Berkowitz, *J. Chem. Phys.*, **29**, 1386 (1958), **32**, 1519 (1960).

[7] T. A. Milne and D. Cubicciotti, *J. Chem. Phys.*, **29**, 846 (1958).

[8] S. H. Bauer, R. M. Diner, and R. F. Porter, *J. Chem. Phys.*, **29**, 991 (1958).

[9] *Loc. cit.*

with the equilibrium constants

$$K_x = \frac{[Cu_xCl_x][H_2]^{x/2}}{[HCl]^x}$$

in which the quantities in brackets are the fugacities (or partial pressures) of the gaseous species. They measured the amount of copper volatilized, which may be related to

$$\sum x[Cu_xCl_x] = K_1 \frac{[HCl]}{[H_2]^{\frac{1}{2}}} + 2K_2 \frac{[HCl]^2}{[H_2]} + 3K_3 \frac{[HCl]^3}{[H_2]^{\frac{3}{2}}} + \cdots \quad (33\text{-}4)$$

Measurements were made at 1309°K with H_2-HCl gas mixtures such that $[HCl]/[H_2]^{\frac{1}{2}}$ varied from 0.0722 to 0.432 atm$^{\frac{1}{2}}$. Each measurement yields an equation of the type (33-4), and simultaneous solution of the various equations for K_1, K_2, K_3, etc., gave the result $K_1 = 6.8 \times 10^{-4}$, $3K_3 = 43 \times 10^{-4}$, and all other K's zero within experimental error.

Less extensive series of measurements were made at other temperatures from 988 to 1340°K, and all results were consistent with the result at 1309°K, indicating just the two species CuCl and Cu_3Cl_3. From these data the heats and entropies of formation as well as the free energies of CuCl and Cu_3Cl_3 could be calculated. It may be shown also that the saturated vapor at 1119°K is nearly all Cu_3Cl_3, in agreement with the result mentioned above.

One can apply the same general entropy reasoning when the vaporization involves interaction with a gas, as was used for a single component. Consider the processes

$$A(s) + \tfrac{1}{2}B_2(g) = AB(g)$$
$$A(s) + \phantom{\tfrac{1}{2}}B_2(g) = AB_2(g)$$
$$A(s) + \tfrac{3}{2}B_2(g) = AB_3(g)$$

If we examine these reactions, we note that the first one corresponds to the production of $\tfrac{1}{2}$ mole of gas. The second corresponds to no net change of moles of gas, and the third corresponds to loss of $\tfrac{1}{2}$ mole of gas. Since the translational contribution to the entropy of the gas is the largest of all the factors contributing to the entropy, we should predict that the entropy change for the first reaction would be positive; the entropy change for the second reaction would be close to zero and the entropy change for the third reaction negative. Typical averages[1] for a series of halide reactions are $\Delta S° = 25$ cal/deg mole for the first reaction, 5 for the second, and -10 for the third. The second reaction usually has a

[1] L. Brewer, L. A. Bromley, P. W. Gilles, and N. L. Lofgren, in L. L. Quill (ed.), "The Chemistry and Metallurgy of Miscellaneous Materials: Thermodynamics," p. 184, McGraw-Hill Book Company, Inc., New York, 1950.

small positive entropy increase rather than a zero entropy change because vibrational and rotational terms do not cancel exactly.

If now we are concerned with endothermic reactions, we can limit the possible reactions that can be of chemical importance. In view of the fact that

$$\Delta F° = \Delta H° - T \Delta S°$$

and with $\Delta H° > 0$, one can see that $\Delta F°$ will always be quite positive if $\Delta S°$ is negative. On the other hand, if $\Delta S°$ is positive, then $\Delta F°$ can become small and even negative if the temperature is increased sufficiently. Reactions where one adds more than 1 mole of B_2 gas to a solid per mole of gaseous product can be important at low temperatures if exothermic, but such reactions become less important as the temperature is increased and at high temperatures one can limit the important processes to those which add either $\frac{1}{2} B_2$ or 1 mole of B_2.

If one has a reaction with atomic gaseous reactants rather than diatomic reactants, then the reaction becomes, for example, $W(s) + O(g) = WO(g)$ or $W(s) + 2O(g) = WO_2(g)$. Here we see that the first reaction corresponds to no change in the number of moles of gas, whereas the second one corresponds to a decrease. Therefore, when one has gaseous atoms reacting with a solid, only the addition of a single gaseous atom per mole of gaseous product will result in an entropy increase and therefore a reaction which can become important at high temperatures even though endothermic.

This behavior is illustrated[1] by Fig. 33-1, where the logarithms of the partial pressures of WO and WO_2 are given as a function of temperature when solid tungsten is equilibrated with oxygen gas at 10^{-3} atm pressure. Above 2700°K one finally reaches the situation where O_2 will be largely dissociated to O. Then the equation of the principal net reaction changes from one involving O_2 molecules to one involving O molecules, and although the WO partial pressure will continue to increase with increased temperature, the WO_2 reaction will have become exothermic and will decrease in importance as one goes to yet higher temperatures. This behavior would be expected for any MX_2 gaseous species which had reached any appreciable chemical importance at the lower temperatures. Of course, if the MX gaseous molecule were sufficiently stable, its formation would also become exothermic at temperatures and pressures at which monatomic oxygen predominates and would also have a maximum partial pressure at the temperature for which X and X_2 molecules are present in equivalent amounts at the given pressure.

---

[1] The figure is based on the data of G. De Maria, R. P. Burns, J. Drowart, and M. G. Inghram, *J. Chem. Phys.*, **32**, 1373 (1960).

A more general statement of this type of reaction could be expressed by the following equation

$$xA(s) + bB_y(g) = A_xB_{by}(g)$$

For such a reaction, $\Delta S° > 0$ only for $b \leq 1$ with any possible values of x.

Let us consider now the relatively simple reactions between oxygen and a metal, like platinum, for which all the gaseous oxides would be produced endothermically $(A = Pt, B = O)$. At temperatures and pressures

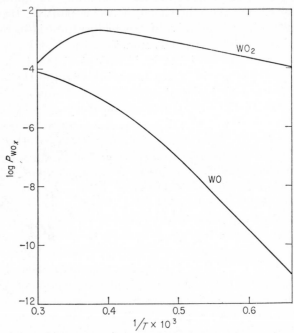

Fig. 33-1. Variation of WO and WO_2 partial pressures with temperatures at constant oxygen pressure of 0.001 atm. The partial pressure of WO_3 (not shown) is even higher than that of WO_2 below 2800°K.

where the O_2 molecule is more important than atomic oxygen $(y = 2)$, one can expect the formation of gaseous platinum oxides with either one or two atoms of oxygen per mole $(b = \frac{1}{2}$ or $1)$, but entropy considerations alone do not place any limit as to the number of Pt atoms per mole. Thus in addition to PtO, one can have Pt_2O, Pt_3O, Pt_4O, etc., as well as PtO_2, Pt_2O_2, Pt_3O_2, etc. However, one would not expect any appreciable contributions from molecules with three or more atoms of oxygen per mole. To establish which species are present, one can study the volatilities as a function of pressure of oxygen. In the moderate-temperature region one finds a proportionality between the vapor pressure of the platinum oxide and the oxygen partial pressure which indicates a major species containing two atoms of oxygen per mole.[1] To establish how many platinum atoms per mole, one would have to alloy the platinum with some other metal to reduce its activity and determine the volatility of the platinum oxide as a function of platinum activity. One would have to guard against the formation of ternary gaseous oxides under such conditions. Alcock and Hooper[2] have used Au-Pt alloys to fix the formula of PtO_2 gas.

At higher temperatures and lower pressures where atomic oxygen predominates, the gaseous species of two atoms of oxygen per mole would be formed exothermically with a decrease in entropy, and the partial pressure would decrease with further increases in temperature at constant oxygen pressure. The species with one atom of oxygen per mole, however, could continue to increase in importance as the temperature increased until one reached the boiling point of platinum, at which temperature the net reaction becomes $Pt(g) + O(g) = PtO(g)$. Now one has a reaction with a decrease in entropy, as there are more moles of gas on the left-hand side of the equation than the right-hand side of the equation, and the reaction has become exothermic; hence further increase in temperature at constant pressure will result in a decrease of the partial pressure of the gaseous platinum oxide.

Another example of a reaction between a gas and a solid to form a new gaseous product would be the reaction of chlorine with chromium trichloride solid, which might produce the endothermic products $CrCl_4$ and $CrCl_5$, but not higher products. Although no solid chlorides of higher oxidation than $CrCl_3$ are known, Doerner[3] has shown that the vapor in equilibrium with $CrCl_3$ solid and Cl_2 gas contains species of higher oxidation number. His results fix $CrCl_4$ gas as the major species. Another example is the vaporization of barium oxide under oxidizing conditions. At temperatures and pressures where O_2 is the main species, solid BaO may be oxidized to form $(BaO)_xO$ or $(BaO)_xO_2$ gas, but from entropy considerations alone one can predict that not more than two atoms of oxygen would react per mole of an endothermic gaseous product.

From bond energy considerations, one would not expect any exothermic reaction between solid BaO and $O_2(g)$ to produce a gaseous product. Thus, of possible gaseous products of appreciable importance, one would not expect any oxidation state higher than that of BaO_3, with BaO_2 and Ba_2O_4 being the next highest possible important species. Although these species have the necessary positive entropy of formation, their enthalpies of formation are so positive that they have not been detected in the BaO-O_2 system. The highest oxidation state observed has been that cor-

[1] L. Brewer, *Chem. Rev.*, **52**, 59 (1953).
[2] C. B. Alcock and G. W. Hooper, *Proc. Roy. Soc. (London)*, **A254**, 551 (1960).
[3] H. A. Doerner, *U.S. Bur. Mines Tech. Paper 577*, 1937.

responding to Ba_2O_3 gas.[1]   The treatment of this system is presented in more detail as a problem at the end of this chapter.

VAPORIZATION OF SOLUTIONS

While we have already considered systems with two or more components in the vapor, the condensed phase was assumed to remain a single component.   Actually, of course, the other vapor components must dissolve to some degree, but if their solubility is small, the activity of the solid or liquid may be assumed to remain unchanged.   We now wish to consider systems where there is wide variation in the activity of the two or more components in the condensed phase.   Systems with two or more condensed phases will also be mentioned.

Many binary systems may be considered by methods discussed in earlier chapters.   In particular, if the vapor composition is known and there is only one vapor species for each component, then the partial pressures and approximate fugacities are readily obtainable.   In addition to ideal or nearly ideal behavior in some systems, one finds in other systems azeotropic points, i.e., constant-boiling compositions.   But our discussion here will concern more diverse types of vaporization processes, where chemical reactions are involved and the vapor molecules may be very different from the molecules or structural units in the solid or liquid. Also we wish to consider the interpretation of evaporation or transpiration experiments where the only information is the composition of the condensed phase and possibly that of the material evaporated.

The system zirconium carbide illustrates some typical problems.   The available evidence indicates that the vapor species are just zirconium atoms and the carbon species, which are primarily C and C_3.   Zirconium and carbon yield a simple binary system with a single intermediate phase ZrC, which has the sodium chloride type of crystal structure.   This ZrC phase extends over a homogeneous composition range which varies with temperature.   However between 1500 and 2500°K the homogeneous range is essentially between the limits $ZrC_{0.55}$ to $ZrC_{0.98}$. $ZrC_{0.55}$ is saturated by Zr metal; thus the partial pressure of Zr in equilibrium with $ZrC_{0.55}$ must be close to the vapor pressure of Zr metal.   At the carbon-rich composition the phase is saturated with essentially pure graphite, and therefore the partial pressure of carbon at this composition must be nearly the same as that of graphite.   From the thermodynamic data for the $ZrC_{0.98}$ composition[2,3] one can calculate the standard free-energy change for the reaction $ZrC_{0.98} = Zr(g) + 0.98C(graphite)$.   At 2000°K,

[1] M. G. Inghram, W. A. Chupka, and R. F. Porter, *J. Chem. Phys.*, **23**, 2159 (1955).

[2] A. D. Mah and B. J. Boyle, *J. Am. Chem. Soc.*, **77**, 6512 (1955).

[3] G. B. Skinner, J. W. Edwards, and H. L. Johnston, *J. Am. Chem. Soc.*, **73**, 174 (1956).

$\Delta F° = 48.3 \pm 3$ kcal. From this one can readily calculate that at 2000°K the zirconium partial pressure in equilibrium with $ZrC_{0.98}$ is 10^{-14} atm, whereas the carbon vapor must be the same as that for pure graphite, which at this temperature consists of $10^{-10.5}$ atm C and $10^{-10.5}$ atm C_3. Thus the vapor in equilibrium with $ZrC_{0.98}$ has a carbon-zirconium ratio of 10^4. Upon vaporization of a $ZrC_{0.98}$ sample, the carbon content of the solid will gradually decrease. Using similar calculations for the zirconium-rich composition, one finds that the vapor phase in equilibrium with $ZrC_{0.55}$ consists of $10^{-8.7}$ atm Zr and $10^{-19.8}$ atm C, or a carbon-zirconium ratio of 10^{-11}. Thus, if $ZrC_{0.55}$ is vaporized, the carbon content will increase. As the carbon content increases, the carbon partial pressure will increase and the zirconium partial pressure will decrease and it is clear that there is some intermediate composition at which the vapor phase will have the same composition as the solid.

If one were to use a method of vapor-pressure determination which removes substantial amounts of material during the process of vaporization, one would not have a clearly defined vapor pressure in general because of the changing composition and pressure as the material is vaporized. However, there are three unique vapor pressures and compositions which can be determined by such methods. One can determine the vapor pressure of the carbon-saturated or the zirconium-saturated composition by adding either excess carbon or excess zirconium to ensure that the system is invariant. A two-component system with two phases in addition to the gaseous phase at a fixed temperature is invariant, and the vapor pressures is uniquely fixed. One could also get a definite vapor pressure by starting with the constant-boiling composition in the ZrC homogeneous-phase region. This is an invariant system even though only two phases are present in the binary system at constant temperature, because of the additional restriction, namely, that the gas and solid (or liquid) have the same composition. This restriction reduces the degrees of freedom of the system and results in an invariant system. On the other hand, if one used a method which yields a measurement with the withdrawal of only a small amount of material, such as the use of a mass spectrometer or other extremely sensitive instrument, then one could determine the partial pressure of each vapor component or even of each vapor species over the full range in composition of the phase. Under such circumstances the composition of the solid phase must be defined.

Some materials do not have a constant-boiling composition. Then the only invariant systems available are those with a second condensed phase. An example is the Ti_2O_3 phase.[1-3] Since the important vapor species is

---

[1] J. Berkowitz, W. A. Chupka, and M. G. Inghram, *J. Phys. Chem.*, **61**, 1569 (1957).

[2] W. C. Groves, M. Hoch, and H. L. Johnston, *J. Phys. Chem.*, **59**, 127 (1955).

[3] P. W. Gilles, private communication.

TiO, the vapor composition has a lower oxygen content than the solid composition over the entire homogeneous range of this phase and it steadily loses titanium as vaporization proceeds until decomposition to Ti_3O_5 commences.   For such a phase, gross evaporization measurements cannot lead to any meaningful result if one has a single solid phase. Unambiguous vapor pressures can be measured only if one adds either solid TiO or Ti_3O_5 to the Ti_2O_3 phase.

After having used entropy estimates to limit the number of gaseous species that need to be considered, one can make generalizations in regard to the heats of sublimation which allow even further restriction of the number of species.   In general this will involve the comparison of the heats of sublimation of the various possible reactions.   When the bonding in the condensed phase and the gaseous phases are of similar type, it is often possible to make broad generalizations about the stabilities of various species.   For example, when the interaction between molecules is of the van der Waals type with London dispersion forces and possibly some dipole-dipole forces in both the condensed phase and the gaseous phase, one can use directly the theory that has been developed in Chapter 16 and Appendix 1 to relate the second virial coefficient to the critical constants or to the vapor-pressure curve for the monomer units.   Likewise we have briefly described the nature of the calculations that can be applied when ionic bonding is assumed.   However, it is beyond the scope of this book to consider these calculations in any additional detail.

PROBLEMS

33-1. Scheffer and Voogd[1] report the following vapor pressures for $Br_2(l)$:

T, °K.............	269.73	271.35	300.19	302.33	355.87	357.93
$\log P_{atm}$........	-1.1397	-1.1107	-0.5118	-0.4682	0.3244	0.3499

The critical pressure and temperature for Br_2 are, respectively, 584°K and 102 atm. Use the Berthelot equation (16-17)* to calculate the fugacity of bromine at each temperature.   Use the heat-capacity equation for $Br_2(g)$ given in Table 6-4 and the molal heat capacity of 17.3 cal/deg for $Br_2(l)$ to make a Σ plot [Eq. (15-37)] of the bromine fugacity data.   Use the resulting ΔH_I and I values to calculate the fugacity of liquid bromine at 500°K.   What is the vapor pressure of liquid bromine at 500°K?   Compare the f/P value calculated for saturated Br_2 vapor at 500°K from Eq. (16-17)* with those from Eq. (A1-7) and from Tables A1-8 and A1-9; the acentric factor may be obtained from these vapor-pressure data.

33-2. Use the free-energy functions of $Br_2(l)$ and $Br_2(g)$ in Table A7-7 together with the molal heat capacity of $Br_2(l)$ given in Problem 33-1 to calculate $(\Delta F° - \Delta H°)/T$ values for the vaporization of $Br_2(l)$ at 298 and 500°K.   Interpolate and extrapolate to obtain values of $(\Delta F° - \Delta H°)/T$ at each temperature of Problem 33-1, and calculate a value of $\Delta H°$ from each vapor-pressure measurement of Scheffer and Voogd.

[1] F. E. C. Scheffer and M. Voogd, *Rec. trav. chim.*, **45,** 214 (1926).

Calculate the fugacity and vapor pressure of liquid bromine at 500°K, and compare with the result of Problem 33-1.

33-3. Data are given in Table A7-2 for Bi and Bi_2 gases. Calculate the equilibrium constant for $Bi_2(g) = 2Bi(g)$ at 1000 and 1500°K. Use the data in Tables A7-1 and A7-2 to calculate the saturated partial pressure of Bi(g) at 1000 and 1500°K. Use the equilibrium constants calculated previously to calculate the ratio of Bi_2 to Bi in the saturated vapor.

33-4. Data for condensed antimony and for Sb and Sb_2 gases may be found in Tables A7-1 and A7-2. Calculate the saturated partial pressures of Sb and Sb_2 at 700 and 1000°K. The actual observed total vapor pressures of antimony at 700 and 1000°K are $10^{-6.89}$ and $10^{-3.06}$ atm, respectively. What explanation can be offered for the discrepancy?

33-5. Kelley and Mah[1] have estimated s_{298}° values for $TiBr_2(s)$ and $TiBr_2(g)$ of 29.5 and 42.2 cal/deg, respectively. Other data may be found in Appendix 7. Compare ΔS_{298}° values for the three reactions

$$2TiBr_3(s) = TiBr_2(s) + TiBr_4(g)$$
$$\tfrac{1}{2}Hg_2Cl_2(s) = \tfrac{1}{2}Hg(g) + \tfrac{1}{2}HgCl_2(g)$$
$$\tfrac{1}{2}SiO_2(s) = \tfrac{1}{2}SiO(g) + \tfrac{1}{2}O(g)$$

with ΔS_{298}° for a simple vaporization reaction such as

$$KF(s) = KF(g)$$

33-6. $BeCl_2$ and Be_2Cl_4 have nearly the same heat of vaporization from the liquid and their relative proportion in the saturated vapor changes little with temperature. Predict what will happen to the ratio of dimer to monomer when the temperature is reduced below the melting point.

33-7. Schäfer and Dohrman[2] observe that reaction between Nb metal and SiO_2 at 800 to 1000°C takes place only if H_2 gas is present. They attribute the chemical transport of silica to the metal to the reaction $SiO_2(s) + H_2(g) = SiO(g) + H_2O(g)$. Use the data of Appendix 7 to determine the partial pressure of SiO(g) in equilibrium with quartz and 10^{-1} atm $H_2(g)$ at 1250°K to check the feasibility of transport. More than 10^{-8} atm SiO would be required.

33-8. Fenton and Garner[3] give for $2HOAc(g) = (HOAc)_2(g)$ the following equilibrium constants:

$T, °K$	K
383	3.72
405	1.329
429	0.479
457	0.168

The total vapor pressure of liquid HOAc at 25°C is 0.0200 atm, and the total pressure[4] of HOAc species above 1.316 M aqueous HOAc is 0.000290 atm and above 2.890 M HOAc is 0.000635 atm. Calculate the proportion of monomer and dimer over pure liquid HOAc at 25°C and over the HOAc solutions.

33-9. Porter and Schoonmaker[5] have determined the equilibrium partial pressures

[1] K. K. Kelley and A. D. Mah, *U.S. Bur. Mines Rept.* 5490, 1959.

[2] H. Schäfer and K. Dohrman, *Z. anorg. u. allgem. Chem.*, **299**, 197 (1959).

[3] T. M. Fenton and W. E. Garner, *J. Chem. Soc.*, **1930**, 694.

[4] W. A. Kaye and G. S. Parks, *J. Chem. Phys.*, **2**, 141 (1934).

[5] R. F. Porter and R. C. Schoonmaker, *J. Phys. Chem.*, **62**, 486 (1958).

of KOH monomer and dimer through use of a mass spectrometer. The results can be summarized by $KOH(l) = KOH(g)$, $\Delta H^\circ_{650} = 39.5 \pm 3$ kcal, and by the reaction $K_2(OH)_2(g) = 2KOH(g)$, $\Delta H^\circ_{650} = 46.5 \pm 5$. Estimate the proportion of the two species at the boiling point of 1600°K if $[K_2O_2H_2]/[KOH] = 2$ in saturated vapor at 650°C.

33-10. Mostowitsch and Pletneff[1] report higher volatility of Au metal in H_2 gas than in O_2, N_2, CO, or CO_2 at 1200 to 1400°C. Farkas[2] also used the transpiration method and observed, in equilibrium with gold and 1 atm H_2, a gold hydride partial pressure at 1400°C of 1.03×10^{-4} atm compared with 1.06×10^{-5} atm Au. On the basis of AuH, these results yield $K = 9.7$ and $\Delta F^\circ_{1673} = 7.55$ kcal. Farkas did not vary the hydrogen pressure to fix the species. Use $s^\circ_{298} = 43.12$ cal/deg mole for Au(g) and $s^\circ_{298} = 50.41$ calculated from spectroscopic data for AuH(g) to determine whether or not the volatility is consistent with the dissociation energy of 69 kcal/mole for AuH obtained from extrapolation of vibrational levels of AuH to convergence.

33-11. Thermodynamic treatment of complex vaporization processes can be quite fruitful even when the necessary auxiliary thermodynamic data are lacking and when the primary observations are qualitative in nature. The zirconium halide systems will be used as an example. Except for the $ZrCl_4(s)$ data in Table A7-3 and the data for elemental zirconium in Tables A7-1 and A7-2, no accurate thermodynamic data are available for the halides of Zr. The methods illustrated in Problem 32-6 can be used to estimate the following free-energy functions:

	$-(F_T - H^\circ_{298})/T$, cal/deg			ΔH°_{298}, kcal
	298°K	500°K	1000°K	
$ZrCl_2(s)$..........	26	28	35	-130
$ZrBr_2(s)$..........	31	34	41	-105
$ZrI_2(s)$..........	34	37	45	-72
$ZrCl_3(s)$..........	35	38	47	-186
$ZrBr_3(s)$..........	44	47	57	-153
$ZrI_3(s)$..........	48	51	62	-104
$ZrBr_4(s)$..........	54	57	...	-195
$ZrI_4(s)$..........	60	63	...	-133
$ZrCl_4(g)$..........	89	92	102	-205
$ZrBr_4(g)$..........	99	102	112	-167
$ZrI_4(g)$..........	105	108	118	-103

Although the free-energy functions can be estimated with a probable error of 3 to 4 cal/deg mole, which is adequate for our present purposes, no satisfactory method of estimating heats of formation of transition-metal compounds of the fourth and fifth periods of the periodic table is known. The heats tabulated above were obtained from qualitative chemical observations in the following manner. The main observations reported in the literature consist of minimum temperatures at which disproportionation reactions take place. Fast[3] and Young[4] have summarized these observa-

---

[1] W. Mostowitsch and W. Pletneff, *J. Russ. Met. Soc.*, **1915**(3), 410–431.

[2] A. Farkas, *Z. physik. Chem.*, **B5**, 467–475 (1929).

[3] J. D. Fast, *Z. anorg. Chem.*, **239**, 145 (1938).

[4] R. C. Young, *J. Am. Chem. Soc.*, **53**, 2148 (1931).

tions as follows: The reaction $2ZrX_3(s) = ZrX_2(s) + ZrX_4(g)$ is reported to take place at temperatures above 600, 580, and 580°K for the chloride, bromide, and iodide, respectively. The reaction $2ZrX_2(s) = Zr(s) + ZrX_4(g)$ is reported to take place at temperatures above 870, 670, and 700°K for the chloride, bromide, and iodide, respectively. For the observation of substantial disproportionation during the period of time of the experiments at the minimum temperature, the partial pressure of escaping ZrX_4 gas would have to be around 10^{-4} atm in apparatus of the size used. If the postulated reactions are correct and if there are no substantial kinetic barriers in the vaporization process, the equilibrium constant for each of these reactions can be taken as 10^{-4} at the minimum temperature. In addition, there are other observations which can be expressed in terms of equilibrium constants or which set limits to equilibrium constants. The decomposition of $ZrI_4(g)$ to metal and atomic iodine on a wire at 1600°K, the nonreduction of $ZrBr_4$ by hydrogen, the reduction of $ZrBr_4$ by aluminum, and similar reactions also set limits to the heats of formation. The solid and gaseous tetrahalides can be related by the accurate vapor-pressure measurements of Rahlfs and Fischer.[1] The heats tabulated above are consistent with the chemical observations. Once the observations have been translated into thermodynamic form, one can test them for self-consistency and thus test the correctness of the interpretation of the data that were used to obtain the enthalpies of formations. Use the above free-energy functions and enthalpies of formation to calculate the temperatures at which the disproportionation reactions have an equilibrium constant of 10^{-4}. By comparison with other halides, one would predict that the ZrX_2 halides would attain partial pressures of 10^{-4} atm at 1000, 900, and 800°K for $ZrCl_2$, $ZrBr_2$, and ZrI_2, respectively. The ZrX_3 partial pressure over the ZrX_3 phases would be 10^{-4} atm at 1050, 1050, and 1000°K for $ZrCl_3$, $ZrBr_3$, and ZrI_3, respectively. Compare the partial pressure of ZrX_4 gas with the partial pressures of ZrX_3 or ZrX_2 in equilibrium with a mixture of ZrX_2 and ZrX_3 or a mixture of Zr and ZrX_2. Is the ZrX_4 molecule the main vaporizing species, as was assumed in interpreting the chemical observations, or is it more likely that the ZrX_2 phases disproportionate to Zr and $ZrX_3(g)$, as has been demonstrated for the titanium system?

33-12. When the vapor pressure of alumina is measured in a tungsten Knudsen cell, calculate whether a reaction between W metal and Al_2O_3 to produce WO_2 or WO_3, gaseous tungsten products, and Al, Al_2O, or AlO, gaseous aluminum species, can be expected to produce a higher vapor pressure in the cell at 2000°K than the reaction $Al_2O_3(s) = 2Al(g) + 3O(g)$. Data will be found in Appendix 7 for all species except $WO_2(g)$ and $WO_3(g)$, for which ΔF°_{2000} of formation from W(s) and $O_2(g)$ may be taken as -3 and -38 kcal, respectively.

33-13. The alumina phase has a narrow composition range from the metal-rich side, which is saturated with aluminum metal, to the oxygen-rich side, saturated with oxygen gas at 1 atm. Calculate the partial pressures of the species Al, Al_2O, AlO, O, and O_2 for the alumina phase at each end of its composition range at 2000°K. What is the sum of the partial pressures of all the gaseous species in equilibrium with alumina at each end of its composition range? What would happen to the composition upon heating each of the limiting compositions at 2000°K under reduced pressure, and what vapor pressure would be measured in a Knudsen-cell experiment which is carried out under reduced pressure? If we neglect the possibility of gaseous aluminum hydroxides, how many moles of alumina would be transported per mole of gas in a transpiration experiment at 2000°K and 1 atm with an equal molal mixture of hydrogen and water as the carrier gas? Compare the vapor pressure of alumina under these conditions with the previously calculated vapor pressures.

[1] O. Rahlfs and W. Fischer, *Z. anorg. Chem.*, **211**, 349 (1933).

33-14. Thompson and Armstrong[1] determined the vapor pressure of BaO by the transpiration method, using air as the carrier gas. At 1530°K, their pressure is 4×10^{-4} atm. The Langmuir and Knudsen type of measurements under reduced pressures have yielded at the same temperature 1.3×10^{-7}, 1.2×10^{-7}, and 1.5×10^{-7} atm.[2-4] Inghram, Chupka, and Porter[4] observed traces of Ba_2O_3 in barium oxide vapor, using a mass spectrometer. In equilibrium with condensed barium oxide and barium vapor at a partial pressure of 2.2×10^{-6} atm at 1664°K, they observed 8×10^{-9} atm Ba_2O_3. Calculate the oxygen partial pressure under such conditions, and calculate the amount of Ba_2O_3 gas to be expected in the presence of 0.2 atm O_2. Is it possible to reconcile the transpiration and Knudsen methods? What simple experiment should Thompson and Armstrong have made to fix the species under their conditions? If the possibility of a barium hydroxide species is not considered, what vapor pressure of barium oxide should be indicated by a transpiration experiment using an equal molal mixture of H_2 and H_2O at 1 atm and 1530°K?

[1] M. D. Thompson and W. G. Armstrong, *Trans. Electrochem. Soc.*, **54**, 85 (1928).

[2] A. Classen and C. F. Veenemans, *Z. Physik*, **80**, 342 (1933).

[3] J. P. Blewett, H. A. Liebhafsky, and E. F. Hennelly, *J. Chem. Phys.*, **7**, 478 (1939).

[4] M. G. Inghram, W. A. Chupka, and R. F. Porter, *J. Chem. Phys.*, **23**, 2159 (1955).

34

MULTICOMPONENT SYSTEMS

The methods of treating binary systems can be directly applied to multicomponent systems with the added complexity of additional variables. In earlier chapters we have touched briefly on the behavior of multicomponent systems such as the distribution of a component between two immiscible phases. In Chapters 19 and 20 it was pointed out that the behavior of a dilute solute in a mixed solvent is the same as that of a dilute solute in a pure solvent. In particular, it was shown that with appropriate choice of components Henry's law is approached for solutes as their concentration approaches zero, whether the solvent is a pure solvent or a solution; this allows the establishment of a solute standard state in a mixed solvent.

In contrast to binary solutions, for which the Gibbs-Duhem equation requires a reduction of activity of one component when it is diluted by another component, the activity of one component of a multicomponent system may either increase or decrease upon addition of another com-

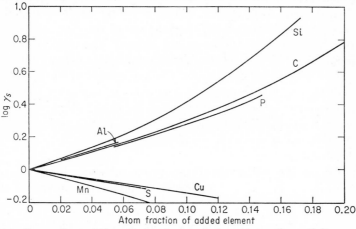

Fig. 34-1. Comparison of the effects of several elements on the activity coefficient of sulfur in molten iron. [*Charles W. Sherman and John Chipman, Trans. AIME,* **194,** 597 (1952), *fig.* 6.]

ponent.   This may be seen by inspection of the Gibbs-Duhem equation:

$$(x_1 d \ln a_1 + x_2 d \ln a_2 + x_3 d \ln a_3 + \cdots = 0)_{P,T} \quad (19\text{-}21)$$

This is also illustrated in Fig. 34-1, where the variation of the activity coefficient of sulfur in molten iron is shown upon addition of various third components.

APPLICATION OF MULTICOMPONENT GIBBS-DUHEM EQUATION

We have seen that for binary solutions the determination of the chemical potential of one component as a function of composition is sufficient to fix the variation of the chemical potential of the other component. From the Gibbs-Duhem equation one might expect that it would be necessary to determine the variations of the chemical potentials of at least two components as a function of composition in order to use the Gibbs-Duhem equation to fix the chemical potential of the third component in a ternary system.   However, Darken[1] has used the principle that a multicomponent solution can be considered as a pure solvent if the ratios of components are kept fixed and has carried out integrations of partial molal properties of a solute along pseudobinary lines corresponding to fixed ratios of the other components.   He has thus shown that it is possible to calculate, for example, the chemical potentials of all other components in the multicomponent system from the experimental determination of the chemical potential of a single component at all compositions.   Darken has shown that the entire free-energy surface may be fixed by integrating the chemical potential of the single component along a number of pseudobinary lines.   The free energy can then be differentiated to obtain the partial molal free energies of the other components.

In an effort to obtain more easily applied equations, Wagner[2] and Schuhmann[3] have developed alternative integrations of the Gibbs-Duhem equation.   All the methods require integrations along pseudobinary lines and require proper choice of components to ensure that Henry's law is approached as a limit.   They all suffer from the inherent difficulty of applying the Gibbs-Duhem equation to very dilute solutions.   McKay[4] has applied similar methods to ternary solutions using molalities for expressing composition.

Gokcen[5] has shown that all the above methods are closely related and may be obtained in a simple manner by applying the properties of exact

[1] L. S. Darken, *J. Am. Chem. Soc.*, **72**, 2909 (1950).

[2] C. Wagner, "Thermodynamics of Alloys," pp. 19–22, Addison-Wesley Publishing Company, Reading, Mass., 1952.

[3] R. Schuhmann, Jr., *Acta Met.*, **3**, 219 (1955).

[4] H. A. C. McKay, *Nature*, **169**, 464 (1952), and *Trans. Faraday Soc.*, **49**, 237 (1953).

[5] N. A. Gokcen, *J. Phys. Chem.*, **64**, 401 (1960).

differentials to either the Gibbs-Duhem equation or the complete differential of F at constant pressure and temperature:

$$dF = \mu_1\, dn_1 + \mu_2\, dn_2 + \mu_3\, dn_3 + \cdots \mu_i\, dn_i \qquad (34\text{-}1)$$

From Eq. (3-8), we obtain

$$\left(\frac{\partial \mu_2}{\partial n_1}\right)_{n_2,n_3,\ldots,n_i} = \left(\frac{\partial \mu_1}{\partial n_2}\right)_{n_1,n_3,\ldots,n_i} \qquad (34\text{-}2)$$

Owing to the symmetry of the subscripts in Eqs. (34-1) and (34-2), Eq. (34-2) may be readily obtained for the other chemical potentials. For example, $(\partial \mu_4/\partial n_1)_{n_2,n_3,n_4,\ldots,n_i}$ is obtained by merely interchanging subscripts 2 and 4 in Eq. (34-2). There are $n-1$ equations like Eq. (34-2) in an n-component system. If μ_1 is known at all compositions of the multicomponent system, each of the $n-1$ partial derivatives of μ_1 can be evaluated at any composition and each of the other chemical potentials can be obtained by integration of each of the $n-1$ equations of the type of Eq. (34-2). Thus all chemical potentials can be evaluated from the complete measurement of a single chemical potential.

It is convenient to express Eq. (34-2) in terms of mole fractions or mole ratios through the introduction of the independent variables

$$s = \frac{n_i}{n_1 + n_3 + \cdots n_r} = \frac{x_i}{1 - x_2} \qquad (34\text{-}3)$$

$$t = \frac{n_i}{n_2 + n_3 + \cdots n_r} = \frac{x_i}{1 - x_1} \qquad (34\text{-}4)$$

with i neither 1 nor 2. For a ternary system Eqs. (34-3) and (34-4) reduce to

$$s = \frac{n_3}{n_1 + n_3} = \frac{x_3}{x_1 + x_3} = \frac{1}{1 + x_1/x_3} \qquad (34\text{-}5)$$

$$t = \frac{n_3}{n_2 + n_3} = \frac{x_3}{x_2 + x_3} = \frac{1}{1 + x_2/x_3} \qquad (34\text{-}6)$$

Partial derivatives of n_1 and n_2 can be expressed in terms of s and t and substituted into Eq. (34-2). For a ternary solution,

$$\left(\frac{\partial s}{\partial n_1}\right)_{n_2,n_3} = -\frac{n_3}{(n_1 + n_3)^2}$$

$$\left(\frac{\partial t}{\partial n_2}\right)_{n_1,n_3} = -\frac{n_3}{(n_2 + n_3)^2}$$

as constant n_2 and n_3 correspond to constant t, and similarly for s,

$$\left(\frac{\partial \mu_2}{\partial s}\right)_t = \frac{t^2}{s^2}\left(\frac{\partial \mu_1}{\partial t}\right)_s = \left(\frac{1 - x_2}{1 - x_1}\right)^2\left(\frac{\partial \mu_1}{\partial t}\right)_s \qquad (34\text{-}7)$$

Use of the multicomponent equations (34-3) and (34-4) in the same manner also yields Eq. (34-7) identically. Equation (34-7) can be given for μ_3 by merely interchanging superscripts 2 and 3 in Eqs. (34-3) to (34-7):

$$\left(\frac{\partial \mu_3}{\partial s_3}\right)_t = \left(\frac{1 - x_3}{1 - x_1}\right)^2 \left(\frac{\partial \mu_1}{\partial t}\right)_{s_3} \tag{34-8}$$

$$s_3 = \frac{n_i}{n_1 + n_2 + n_4 + \cdots + n_r} = \frac{x_i}{1 - x_3} \tag{34-9}$$

where i is now neither 1 nor 3 and the subscript on s designates which chemical potential is being evaluated. Similar equations can be written for the other components.

For a ternary system, integration of Eq. (34-7) for μ_2 yields

$$\mu_2(s_2, t) - \mu_2(s_2 = 1, t) = \int_1^{s_2} \frac{t^2}{s_2{}^2} \left(\frac{\partial \mu_1}{\partial t}\right)_{s_2} ds_2 \qquad \text{(constant } t\text{)} \tag{34-10}$$

In a ternary system, $x_1 = 0$ when $s_2 = 1$ and $\mu_2(s_2 = 1)$ is the chemical potential of component 2 in the binary 2-3 system at a composition corresponding to the constant value of t at which the integration is carried out. The relationships corresponding to constant s_2, s_3, and t can be seen in Fig. 34-2.

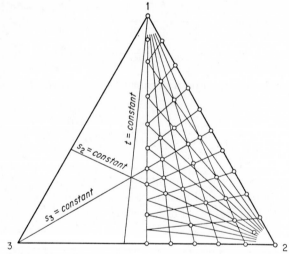

FIG. 34-2. Ternary diagram showing lines corresponding to constant s_2, s_3, and t or constant molal ratios.

All the above equations can be applied to any extensive thermodynamic property with μ being replaced by the corresponding partial molal quantity, for example, $\bar{H}_1$, $\bar{S}_1$, etc. Equation (34-10) can be awkward when dealing with μ and $\bar{S}$, which approach infinity at low mole fractions. It is convenient to express Eq. (34-10) in terms of excess functions such as the excess molal entropy or free energy which were discussed in Chapter 21.

$Y^E = Y - Y^i$, where Y^i is the ideal contribution. The ideal enthalpy is zero; thus Eq. (34-10) in terms of $\bar{H}_1$ and $\bar{H}_2$ is unaltered when expressed in terms of the excess enthalpy. As $\bar{F}_1{}^E = \bar{F}_1 - \bar{F}_1{}^i = RT \ln \gamma_1$, use of excess free energy allows expression of the above equations in terms of activity coefficients. Excess entropy can be similarly expressed. Equation (34-10) becomes

$$\log \gamma_2(s_2, t) - \log \gamma_2(s_2 = 1, t) = \int_1^{s_2} \frac{t^2}{s_2^2} \left(\frac{\partial \log \gamma_1}{\partial t} \right)_{s_2} ds_2$$

$$\text{(constant } t\text{)} \quad (34\text{-}11)$$

At the limit, $s_2 \to 1$, t^2/s_2^2 approaches a constant, and $(\partial \log \gamma_1/\partial t)_{s_2 = 1}$ is finite when the components have been chosen to ensure approach to Henry's law at low concentrations. This integral is well behaved at the limit. The value of $\gamma_2(s_2 = 1, t)$ is merely the value of γ_2 in the binary 2-3 system at the composition corresponding to t and may be taken relative either to the pure 2 standard state or to the solute standard state in solvent 3.

In some cases the availability of data may make it more convenient to carry out the integration of Eq. (34-11) from the limit $s_2 = 0$. It is necessary to demonstrate that the integral approaches a finite value at this limit. As s_2 is reduced to zero, the line $s_2 = $ constant approaches the 1-2 edge of the triangle of Fig. 34-2. The line $s_2 = $ constant will intersect a given $t = $ constant line at larger and larger values of x_1 as s_2 approaches zero. Thus both mole fractions x_2 and x_3 approach zero as s_2 approaches zero at constant t and $t^2/s_2^2 = (1 - x_2)^2/(1 - x_1)^2$ becomes large. However, $\partial \log \gamma_1/\partial t$ becomes very small as s_2 approaches zero and x_1 approaches unity. From the discussion at the end of Chapter 19 on deviations from Henry's law and the succeeding discussion in connection with Eqs. (20-31) and (20-32), we may expect $\log \gamma_1$ to approach proportionality to $(1 - x_1)^2$ at high x_1 for solutions other than those containing electrolytes or high polymer molecules. At high x_1, the integrand of Eq. (34-11) approaches proportionality to

$$\frac{1}{(1-x_1)^2} \left\{ \frac{\partial (1-x_1)^2}{\partial [x_3/(1-x_1)]} \right\}_{x_1/x_3} = \frac{x_1/x_3}{(1-x_1)^2} \left\{ \frac{\partial (1-x_1)^2}{\partial [x_1/(1-x_1)]} \right\}_{x_1/x_3}$$

$$= \frac{x_1/x_3}{(1-x_1)^2} (1-x_1)^2 [-2(1-x_1)] = 2x_1 \left(1 + \frac{x_2}{x_3} \right)$$

Since the integration is carried out at constant t or constant x_2/x_3, the integrand remains finite as s_2 approaches zero.

The evaluation of Eq. (34-11) may be carried out by a series of plots in a manner that can be seen by reference to Fig. (34-2); $\log \gamma_1$ is plotted versus t for various constant values of s_2. Each plot corresponds to taking $\log \gamma_1$ values for compositions corresponding to one of the

$s_2 =$ constant lines which terminate at the 2 vertex of the composition triangle. The tangent lines to the resulting curves yield values of $(\partial \ln \gamma_1/\partial l_2)_{s_2}$. Each of these values is multiplied by the value of $(t/s_2)^2$ at the point of tangency, and the resultant is plotted versus s_2 choosing values corresponding to a given value of t or values corresponding to compositions along one of the lines terminating at the 1 vertex of the composition triangle. Values of log γ_2 are thus obtained along a series of lines of constant t. These lines should be closely enough spaced to allow interpolation to obtain values of γ_2 at any composition.

Values of log γ_3 can be obtained in the same manner using Eqs. (34-11) and (34-4) with subscripts 2 and 3 interchanged and using Eq. (34-9) for s_3. The first set of plots is carried out along $s_3 =$ constant lines, and the integrations are carried out as above along $t =$ constant lines. An economy of effort will result if measurements of γ_1 are made at points of intersection of the lines corresponding to the chosen values of constant s_2, s_3, and t. The right-hand side of Fig. 34-2 shows a set of points that Gokcen recommends.

The above treatment for ternary systems may be carried out for systems of any number of components. The condition of constancy of t and the various s variables as given by Eqs. (34-4) and (34-3) with interchange of subscripts requires that a number of ratios of mole fractions be kept constant. For a quaternary system represented by a tetrahedron, Gokcen has shown that one can take a plane through a vertex that will correspond to compositions with a fixed value for a given mole-fraction ratio. A line in this plane from the vertex can be chosen to correspond to a fixed value of another ratio of mole fractions and corresponds to a constant value of t or one of the s variables. Thus it is possible to choose appropriate planes which allow the evaluation of Eq. (34-11) for the activity coefficient of each of the components. The evaluation of values of activity coefficients along closely spaced lines originating from a vertex is then repeated in a series of closely spaced planes. In this manner, values of the activity coefficient of each component are fixed at a sufficient number of points in composition space to allow interpolation to obtain values for any composition. This procedure of fixing the ratio of mole fractions to reduce the number of variables allows one to carry out the above procedure for systems of any number of variables.

In contrast to Gokcen's[1] method, which is based on Eq. (34-2), Wagner's[2] and Schuhmann's[3] methods are based on the relation

$$\left(\frac{\partial \mu_2}{\partial \mu_1}\right)_{n_2,n_3} = -\left(\frac{\partial n_1}{\partial n_2}\right)_{\mu_1,n_3} \tag{34-12}$$

[1] *Ibid.*
[2] *Loc. cit.*
[3] *Loc. cit.*

This equation may be obtained by applying the properties of perfect differentials to Eq. (34-1). If Eq. (3-7) is written for the perfect differential dZ, Eq. (3-8) results. If dL is expressed as a function of x and y and the operations illustrated by Eqs. (3-1) to (3-3) are applied, the result may be substituted into Eq. (3-8) to yield

$$\left(\frac{\partial x}{\partial y}\right)_L = -\left(\frac{\partial M}{\partial L}\right)_y \tag{34-13}$$

Equation (34-13) may either be applied to Eq. (34-1) to yield Eq. (34-12) or be applied to the Gibbs-Duhem relation in the following form:

$$-d\mu_3 = \frac{n_2}{n_3}\,d\mu_2 + \frac{n_1}{n_3}\,d\mu_1 \tag{34-14}$$

Equation (34-12) may be transformed to mole-fraction form,

$$\left(\frac{\partial \mu_2}{\partial \mu_1}\right)_{n_2/n_3} = -\left[\frac{\partial(x_1/x_3)}{\partial(x_2/x_3)}\right]_{\mu_1} \tag{34-15}$$

or it may be derived directly by expressing Eq. (34-14) in mole-fraction form.

Wagner has used x_1 and the t of Eq. (34-6) as independent variables to express the chemical potentials in a ternary system. x_2 and x_3 are eliminated by carrying out the indicated differentiation of the right-hand side of Eq. (34-15) and recognizing that $d(1/x_3)/d(x_2/x_3) = 1 + d(x_1/x_3)/d(x_2/x_3)$ from the relation $1/x_3 = x_1/x_3 + x_2/x_3 + 1$. Combination of terms yields for the right-hand side of Eq. (34-15) $-x_1/(1 - x_1) - [1/x_3(1 - x_1)]\,\partial x_1/\partial(x_2/x_3)$. From the definition of t, $dt/d(x_2/x_3) = -(1 + x_2/x_3)^{-2}$. Substitution of this result together with $1 - x_1 = x_2 + x_3$ yields

$$\left(\frac{\partial \mu_2}{\partial \mu_1}\right)_{n_2/n_3} = \frac{t}{(1 - x_1)^2}\left(\frac{\partial x_1}{\partial t}\right)_{\mu_1} - \frac{x_1}{1 - x_1}$$

As constant n_2/n_3 corresponds to constant t,

$$\left(\frac{\partial \mu_2}{\partial x_1}\right)_t = \frac{t}{(1 - x_1)^2}\left(\frac{\partial x_1}{\partial t}\right)_{\mu_1}\left(\frac{\partial \mu_1}{\partial x_1}\right)_t - \frac{x_1}{1 - x_1}\left(\frac{\partial \mu_1}{\partial x_1}\right)_t$$

Using the method leading to Eq. (3-3), one sees that

$$\left(\frac{\partial x_1}{\partial t}\right)_{\mu_1}\left(\frac{\partial \mu_1}{\partial x_1}\right)_t = -\left(\frac{\partial \mu_1}{\partial t}\right)_{x_1}$$

$$\left(\frac{\partial \mu_2}{\partial x_1}\right)_t = -\frac{t}{(1 - x_1)^2}\left(\frac{\partial \mu_1}{\partial t}\right)_{x_1} - \frac{x_1}{1 - x_1}\left(\frac{\partial \mu_1}{\partial x_1}\right)_t$$

$$\left(\frac{\partial \log \gamma_2}{\partial x_1}\right)_t = -\frac{t}{(1 - x_1)^2}\left(\frac{\partial \log \gamma_1}{\partial t}\right)_{x_1} - \frac{x_1}{1 - x_1}\left(\frac{\partial \log \gamma_1}{\partial x_1}\right)_t \tag{34-16}$$

The corresponding equation for $\log \gamma_3$ is readily obtained by interchanging subscripts 2 and 3 in Eqs. (34-16) and (34-6). The interchange of subscripts in Eq. (34-6) yields a value of t which is one minus its value when used to evaluate μ_2, and the sign of dt is changed.

Wagner's final result upon integration of Eq. (34-16), with Henry's law being approached as a limit at infinite dilution, is

$$\log \gamma_2(x_1', t) - \log \gamma_2(x_1 = 1, t) = \left[\int_0^1 \frac{\log \gamma_1\,dx_1}{(1 - x_1)^2}\right]_{t=0}$$

$$+ \left\{\int_1^{x_1'}\left[\frac{\log \gamma_1}{(1 - x_1)^2} - t\frac{\partial}{\partial t}\frac{\log \gamma_1}{(1 - x_1)^2}\right]dx_1 - \frac{x_1'\log \gamma_1}{1 - x_1'}\right\}_t \tag{34-17}$$

If the thermodynamic properties of the binary system 2-3 are known, one may obtain higher accuracy by integrating Eq. (34-16) from $x_1 = 0$ rather than from $x_1 = 1$, as we have noted for Eq. (34-11). Tangent lines of a plot of $\log \gamma_1$ versus t at particular values of t yield the integrand of Eq. (34-17). Values of $\log \gamma_2$ from Eq. (34-17) for different values of t should approach the same limiting value for $x_1 \to 1$. Owing to experimental errors, actual data do not yield this equality, but the data may be smoothed to bring about this required consistency. Wagner's method is also readily adapted to analytical evaluation of the integrals.

Schuhmann applies Eq. (34-12) directly. Expressed in terms of activity coefficients, the integrated form is

$$\left[\log \gamma_2'' - \log \gamma_2' = - \int_{\log \gamma_1'}^{\log \gamma_1''} \left(\frac{\partial n_1}{\partial n_2} \right)_{\gamma_1, n_3} d \log \gamma_1 \right]_{n_2/n_3} \quad (34\text{-}18)$$

$(\partial n_1/\partial n_2)_{\gamma_1, n_3}$ can be evaluated as a function of $\log \gamma_1$ by taking the tangents to curves of constant activity coefficient plotted on a composition triangle and reading off the tangent intercepts along the 1-2 composition edge of the triangle. Gokcen[1] has extended Schuhmann's graphical method of tangent intercepts to quaternary systems.

Gokcen[1] has also generalized the Wagner method to systems of any number of components and has illustrated the treatment of quaternary systems. By proper choice of the form of the independent variable t for Wagner's equations, it is possible to obtain equations which are as simple as for a ternary system. The integrations are carried out along lines in appropriate planes, as described for Eq. (34-11). The adaptability of Wagner's method to analytical treatment of the data makes it particularly valuable to systems with many components.

Elliott and Chipman[2] and Mellgren[3] have successfully applied the Darken[4] method to several ternary systems. A concentration cell was used to measure the chemical potential of cadmium for several pseudobinary lines crossing each of the ternary fields in the systems Cd-Pb-Bi, Cd-Pb-Sb, and Cd-Pb-Sn. From the temperature coefficient of the activity of Cd, partial molal enthalpies and entropies of Cd were obtained and all three properties were integrated to obtain integral excess molal free energies, enthalpies, and entropies for the ternary solutions. Figures 34-3 and 34-4 illustrate the molal heat of solution surface and the excess molal entropy of solution surface, respectively, for the Cd-Pb-Bi solution at 500°C. From these results the partial molal quantities for the three

[1] *Loc. cit.*

[2] J. F. Elliott and J. Chipman, *Trans. Faraday Soc.*, **47**, 138 (1951); *J. Am. Chem. Soc.*, **73**, 2682 (1951).

[3] S. Mellgren, *J. Am. Chem. Soc.*, **74**, 5037 (1952).

[4] *Loc. cit.*

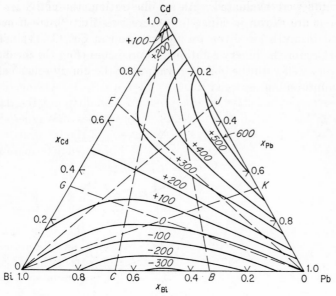

Fig. 34-3. Molal heat of solution surface, Δ$_H$, for the cadmium-lead-bismuth solution at 500°C in 100-cal steps. [*John Chipman and J. F. Elliott, J. Am. Chem. Soc.,* **73**, 2682 (1951), *fig. 6.*]

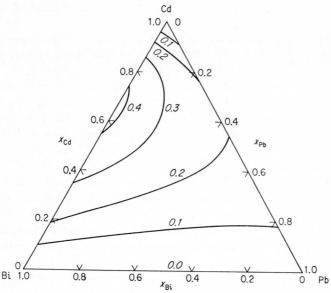

Fig. 34-4. Excess molal entropy of solution surface, Δs^E, for the cadmium-lead-bismuth solution at 500°C in 0.1 cal/deg steps. [*John Chipman and J. F. Elliott, J. Am. Chem. Soc.,* **73**, 2682 (1951), *fig. 7.*]

components were evaluated. As an illustration, the activities of each component are shown in Fig. 34-5, where isoactivity lines of cadmium, lead, and bismuth are given for the solution at 500°C. Values of the free energies for the binary solutions, as predicted from the ternary data, agreed very well with the observed values for the binary solutions of lead and bismuth and of lead and tin.

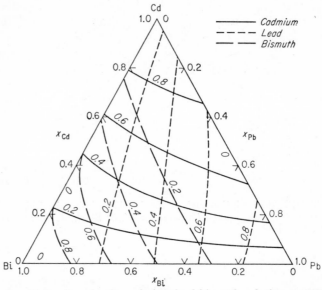

FIG. 34-5. Isoactivity lines in the cadmium-lead-bismuth solution at 500°C. Solid lines, cadmium; short dashes, lead; long dashes, bismuth. [*John Chipman and J. F. Elliott, J. Am. Chem. Soc.*, **73**, 2682 (1951), *fig*. 10.]

In spite of the fact that the binary systems lead-bismuth, lead-antimony, and lead-tin show virtually ideal entropy of mixing, the entropy surfaces in the three ternary systems with cadmium were quite dissimilar. The excess molal entropy of solution was found to be positive for virtually all portions of the ternary diagrams.

Schuhmann[1] has used his method to evaluate the Cd-Pb-Bi data of Elliott and Chipman which had been previously evaluated by Darken's method. He obtained comparable results and concluded that the three methods of Darken, Wagner, and Schuhmann are substantially equivalent in accuracy and in the amount of labor required when relatively complete data are available for a component. Chipman[2] has compared the Wagner and Schuhmann methods for the CaO-SiO$_2$-Al$_2$O$_3$ system and likewise agrees that the methods are almost equivalent when complete data are available and that the Wagner method might have the advantage

[1] *Loc. cit.*

[2] J. Chipman, "Physical Chemistry of Process Metallurgy," Interscience Publishers, Inc., New York, 1960.

under those conditions. However, both Schuhmann and Chipman agree that the Schuhmann method is much superior to the other two methods when the data are not precise or are incomplete and judgment must be used in the evaluation of the data. Because of the simplicity of the manipulations required in the evaluation of Eq. (34-18), one can see more directly the result of any extrapolations or smoothing of the data. With any of the methods, Schuhmann has pointed out that "the calculations for a new system are likely to require a good supply of graph paper, a calculating machine, and a painstaking approach to the study of the data."

The presence of a saturating phase which maintains constant activity of a component in the solution simplifies the integration of the Gibbs-Duhem equation to that of a binary integration. Schuhmann and Ensio[1] and Michal and Schuhmann[2] have carried out such integrations along the Fe and SiO_2 saturation curves in the FeO-Fe_2O_3-SiO_2 system. As might be expected from the designation of constant activity in Eq. (34-12), the presence of any second phase simplifies the Gibbs-Duhem integration since tie lines in a two-phase region are isoactivity lines. The reduction in the number of independent variables when a second phase appears removes the restriction of integrating along pseudobinary lines, and Eq. (34-18) is not restricted to constant n_2/n_3; thus the integration may be readily carried out over the entire two-phase region.

Schuhmann[3] has also pointed out that the Gibbs-Duhem equations require simple relations between the directions of the activity curves of each of the three components at a point in a ternary diagram and derives the result

$$\left(\frac{\partial n_1}{\partial n_2}\right)_{\mu_3, n_3} = -\left(\frac{\partial n_1}{\partial n_3}\right)_{\mu_2, n_2} \left(\frac{\partial n_3}{\partial n_2}\right)_{\mu_1, n_1} \tag{34-19}$$

The tangent to each isoactivity curve will intercept each of the three edges of the composition triangle or their extensions. The intercept with the 1-2 edge of the composition triangle of the tangent to the constant μ_3 curve at a given point yields the term on the left-hand side of Eq. (34-19). The ratio n_1/n_2 given by the intercept corresponds to that composition which can be added to the solution without changing μ_3. The first factor of the right-hand side of Eq. (34-19) is the n_1/n_3 ratio corresponding to the intercept with the 1-3 edge of the composition triangle of the tangent to the constant μ_2 curve. The last factor can be obtained similarly. Schuhmann proved the geometrical relation that the three intercepts corresponding to the three factors of Eq. (34-19) must lie on a straight

[1] R. Schuhmann, Jr., and P. J. Ensio, *Trans. AIME*, **191**, 401–411 (1951); *J. Metals*, May, 1951.

[2] E. J. Michal and R. Schuhmann, Jr., *Trans. AIME*, **194**, 723 (1952); *J. Metals*, July, 1952.

[3] *Loc. cit.*

line.   This provides a simple geometrical procedure for constructing iso-activity curves or checking consistency of isoactivity curves.   From the determination of two of the intercepts of Eq. (34-19) from isoactivity curves of two components, the tangent to the isoactivity curve of the third component at the given point can be immediately drawn.

Gockcen has shown that equations similar to Eq. (34-19) may also be obtained for quaternary systems and has presented other equations which are useful for checking the internal consistency of the calculations.

DILUTE MULTICOMPONENT SYSTEMS

In dealing with a multicomponent solution with most of the components present at low concentration, the representation of thermodynamic properties is simplified.   The iron alloys are an important example of dilute solutions of many components with one major component.   In a very dilute solution one might expect that each component would obey Henry's law and that the Henry's-law constants could be obtained from the binary systems.   While this is true at the limit of zero concentration, it is found that at very low concentration there are already significant interactions between the dilute solutes.   Figure 34-1 illustrates this by representing the variation of the activity coefficient of small amounts of sulfur in iron[1] as various other solutes are added.   Wagner[2] has used a Taylor's-series expansion to represent the logarithm of the activity coefficient of component 2 in a solution with mole fractions x_2, x_3, x_4, etc., of the various solutes.

$$\ln \gamma_2(x_2, x_3, \ldots) = \ln \gamma_2^{\circ} + x_2 k_{22} + x_3 k_{23} + \cdots \qquad (34\text{-}20)^*$$

where the coefficients k_{22}, k_{23}, etc., are defined as $k_{22} = \partial \ln \gamma_2 / \partial x_2$, $k_{23} = \partial \ln \gamma_2 / \partial x_3$, etc.   The derivatives are to be taken for the limiting case of zero concentration of all solutes, and terms involving second and higher derivatives are disregarded.   The first term $x_2 k_{22}$ obviously represents the deviation from Henry's law due to interaction between pairs of molecules of component 2.   Each of the following interaction terms arises from the interaction of pairs of unlike molecules and represents the effect upon the activity coefficient of component 2 as each of the other components is added with fixed mole fraction of component 2.   Thus Eq. (34-20)* can be rewritten

$$\ln \gamma_2(x_2, x_3, \ldots) = \ln \gamma_2^{\circ} + \ln \frac{\gamma_2(x_2,\, x_3 = 0,\, x_4 = 0,\, \ldots)}{\gamma_2^{\circ}}$$

$$+ \ln \frac{\gamma_2(x_2 = 0,\, x_3,\, x_4 = 0,\, \ldots)}{\gamma_2^{\circ}} + \cdots \qquad (34\text{-}21)^*$$

[1] J. Chipman, *J. Iron Steel Inst.* (*London*), **180**, 97 (1955).
[2] *Op. cit.*, p. 51.

Hence the activity coefficient of component 2 in the dilute multicomponent system may be calculated from its value in the system 1-2 and its limiting value in the systems 1-2-3, 1-2-4, etc., as the amount of component 2 approaches 0.

Wagner's equations are equivalent to those used earlier by Chipman and Elliott[1] and others for representing the effect of various solutes upon the activity coefficient of a single solute.

If Eq. (34-2) is expressed in terms of activity coefficients,

$$\frac{\partial \ln \gamma_i}{\partial x_j} = \frac{\partial \ln \gamma_j}{\partial x_i} \qquad (34\text{-}22)$$

At the limit of zero x_i and x_j or over the range of x_i and x_j for which the linear terms of Eq. (34-20)* are valid, Eq. (34-22) yields the relationship

$$k_{ij} = k_{ji} \qquad (34\text{-}23)*$$

There are not many accurate checks of Eq. (34-23)*. Chipman[2] reports $k_{Si,C} = k_{C,Si}$ for mole fractions below 0.016 in iron.

Attempts have been made to relate the interaction coefficients k_{ji} between different solutes to the interaction coefficients k_{ii} and k_{jj} for solutes with themselves. However, as for the binary metallic solutions, where no satisfactory models have been found to relate the solution behavior to the properties of the pure substances for a wide range of metals, no satisfactory models have been found for calculation of the ternary solute interactions. Table 34-1 presents[3] a summary of values of the interaction coefficients k for solutes in molten iron. Equation (34-23)* can be used as a basis for averaging values of k_{ij} and k_{ji}, and

---

[1] J. Chipman and J. F. Elliott, The Thermodynamics of Liquid Metallic Solutions, in "Thermodynamics and Physical Metallurgy," p. 102, American Society for Metals, Cleveland, Ohio, 1950.

[2] J. Chipman, *J. Iron Steel Inst.* (London), **180**, 97 (1955).

[3] *Ibid.*, p. 105, and revision by Chipman (private communication) using the following new sources of data: $k_{H,C}$ from K. T. Kurochkin, P. E. Nizhelskii, and P. V. Umrikhin, *Izvest. Akad. Nauk S.S.S.R., Otdel. Tekh. Nauk*, **1957**, 19; $k_{H,Nb}$ from M. M. Karnaukov and A. N. Morozov, *ibid.*, **1948**, 1845; all C entries from T. Fuwa and J. Chipman, *Trans. Met. Soc. AIME*, **215**, 708 (1959), except $k_{C,C}$ from A. Rist and J. Chipman, "Physical Chemistry of Steelmaking," p. 3, John Wiley & Sons, Inc., New York, 1958; all N entries from R. D. Pehlke and J. F. Elliott, *Trans. Met. Soc. AIME*, in press; all new O entries from T. P. Floridis and J. Chipman, *Trans. Met. Soc. AIME*, **212**, 549 (1958), except $k_{O,C}$ from T. Fuwa and J. Chipman, *Trans. Met. Soc. AIME*, **218**, 887 (1960), and $k_{O,P}$ from D. Dutilloy and J. Chipman, *Trans. Met. Soc. AIME*, **218**, 428 (1960). All S entries from C. W. Sherman and J. Chipman, *Trans. AIME*, **194**, 597 (1952), except $k_{S,Cr}$ from N. Griffing and G. W. Healy, *Trans. Met. Soc. AIME*, **218**, 849 (1960), and $k_{S,Ni}$ from J. A. Cordier and J. Chipman, *Trans. AIME*, **203**, 905 (1955).

values of k_{ji} given in Table 34-1 can be used also for k_{ij}. Thus the values in Table 34-1 allow one to calculate solute interactions for a large variety of solutes simultaneously in dilute liquid-iron solutions.

TABLE 34-1. ATOMIC INTERACTION COEFFICIENTS IN LIQUID IRON, 1600°C

$$\text{Values of } k_{ij} = \frac{\partial \ln \gamma_i}{\partial x_j}$$

Added component j	Component i					
	H	C	N	O	Si	S
C.........	+42?	+11.1	+12?	− 6.4	+10	+6.0
O.........	......	......	+ 3.5	− 13		
Al........	......	+ 6.7	− 0.3	−1300	......	6.5
Si........	3.2	11.2	+ 5.4	− 2?	3.4	7.6
P.........	......	......	......	+ 8.9	11	5.7
S.........	......	......	......	− 12	......	−3.8
V.........	......	− 8.0	−20.5	− 57		
Nb.......	− 1.5	−23				
Ta........	......	......	−25.2			
Cr........	......	− 5.1	− 9.6	− 8.8	......	−4.7
Mo.......	......	− 3.5	− 4.5	1.4		
W.........	......	− 2.3	− 1.1	6.4		
Mn.......	......	− 1.4	− 4.5	0?	0?	−5.7
Co.......	......	2.9	2.7	1.7		
Ni........	......	2.9	2.6	1.4	1.2	0
Cu.......	......	4.2	2.4	− 2.5	......	−3.2
Pt........	......	......	......	3.6		
Au.......	......	......	......	− 3.0		
Sn........	......	0	3.4	0		

Solubility Product of a Slightly Soluble Solute.   If a solvent is saturated with a slightly soluble solute which dissociates according to the net reaction

$$MY(s) = M + Y$$

log $x_M x_Y$ will vary linearly with the excess of either M or Y over the range of mole fractions for which Eq. (34-20)* is a good approximation to the activity coefficients.   This requires that there be no long-range forces and that nearest-neighbor pair interactions predominate.   For the saturated solution,

$$\log x_M x_Y + \log \gamma_M \gamma_Y = K_{SP} \quad (34\text{-}24)$$
$$\log x_M x_Y + (k_{M,M} + k_{Y,M})x_M + (k_{Y,Y} + k_{M,Y})x_Y = K_{SP} \quad (34\text{-}25)^*$$

represent the variation of the product $x_M x_Y$ as the amount of excess M or Y is varied.   Since higher terms beyond the linear terms of Eq. (34-20)* are neglected, $k_{M,Y}$ can be taken equal to $k_{Y,M}$ from Eq. (34-23)*.   The solute standard state is used in this example; a constant term would be added if other standard states were used. If either excess M or Y is added, the mole fraction of the other becomes small and log $x_M x_Y$ varies linearly with the mole fraction of the component in excess. Likewise, it can be shown that the logarithm of the product $x_M x_Y$ will vary linearly upon addition of another solute Z according to

$$\log x_M x_Y = \text{constant} - (k_{M,Z} + k_{Y,Z})x_Z \qquad (34\text{-}26)^*$$

in the dilute-solution range, where higher interaction terms can be neglected within experimental error. The mole-fraction product can either increase or decrease depending upon the absolute sign of the sum of the two interaction coefficients. In general, the logarithm of any equilibrium quotient for an equilibrium in a solution showing the behavior indicated by Eq. (34-20)* will show a linear variation with the mole fractions of any added solutes, whether they appear in the net reaction or not, if the interaction coefficients k are not zero.

The range of application of a linear dependence of the logarithm of an equilibrium constant with mole fraction of added solutes varies widely with the nature of the solution. In most solutions, such behavior is approached as infinite dilution is approached. Quadratic and higher terms appear as the solute mole fractions are increased, but linear behavior can often approximate observations within experimental error up to several mole per cent. In solutions with long-range forces such as electrolyte solutions, the linear behavior is not approached even at infinite dilution. Long-range electrostatic forces also occur in fused salts. However, these do not usually appear in activity-coefficient terms owing to cancellation of the effect between the pure liquids and the solutions, and short-range interactions are often adequate for treating such solutions.

Fused Salts. Flood, Förland, Grjotheim, and coworkers[1] have presented several theoretical derivations and have obtained a variety of experimental data to demonstrate that activity coefficients in fused-salt solutions can be treated by a regular solution equation such as Eq. (21-15)*. Förland[2] has studied the reaction

$$CaCO_3(\text{in melt}) = CaO(s) + CO_2(g)$$

Figure 34-6 shows $RT \log (P_{CO_2}/x_{Ca^{++}})$ as a function of the amount of $CaCO_3$ dissolved in an equimolal mixture of Na_2CO_3 and K_2CO_3 liquids.

[1] H. Flood, T. Förland, and K. Grjotheim, *Z. anorg. allgem. Chem.*, **276**, 19 (1954); H. Flood and A. Muan, *Acta Chem. Scand.*, **4**, 364 (1950); H. Flood, T. Förland, and K. Motzfeldt, *Acta Chem. Scand.*, **6**, 257 (1952); H. Flood, T. Förland, and A. Nesland, *Acta Chem. Scand.*, **5**, 1193 (1951).

[2] T. Förland, *J. Phys. Chem.*, **59**, 152 (1955).

$x_{Ca^{++}}$ is the cation fraction of calcium ions, and its use corresponds to the Temkin[1] entropy of mixing, with alkali ions and calcium ions randomly distributed over the cation positions. The abscissa $(x'_{Na,K})^2$ is obtained from the equation

$$x'_{Na,K} = \frac{\frac{1}{2}(n_{Na} + n_{K})}{n_{Ca} + \frac{1}{2}(n_{Na} + n_{K})}$$

which is based on an energy of mixing expressed in terms of characteristic cation-cation pair interactions. There is no term due to cation-anion interaction since no change in coordination number is assumed. Figure 34-6 shows that the predicted dependence upon the square of the mole

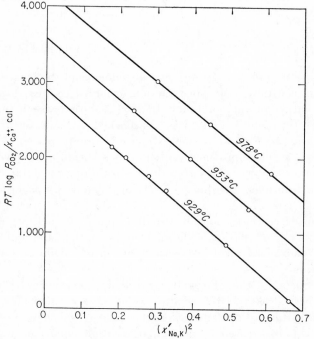

FIG. 34-6. Solubility of $CaCO_3$ as a function of CO_2 pressure and alkali composition. [*T. Förland, J. Phys. Chem.*, **59**, 152 (1955), *fig. 4*.]

fraction holds well over a considerable range. From the average of similar plots for the $CaCO_3$-Na_2CO_3 and $CaCO_3$-K_2CO_3 systems, the slope of the line for the $CaCO_3$-$(NaK)CO_3$ system was estimated to be -4040 cal compared with the observed -4190 cal.

A similar treatment with fixed concentrations of reactants in solution and variation of the ratio of K to Na would predict a linear dependence

[1] M. Temkin, *Acta Physica chim. U.R.S.S.*, **20**, 411 (1945).

of the logarithm of equilibrium constant upon mole fraction of one of the alkalies. Flood and Muan[1] studied the reaction

$$M_2Cr_2O_7(\text{in melt}) = M_2CrO_4(\text{in melt}) + \tfrac{1}{2}Cr_2O_3(s) + \tfrac{3}{4}O_2(g)$$

with the metal ion M^+ consisting of various proportions of K^+, Na^+, and Tl^+. Their entropy results were in good agreement with a model requiring random mixing of the two anions and random distribution of the cations among the cation sites, and they obtained only a small deviation from a linear variation of the logarithm of the equilibrium constants with cation fraction, as shown in Fig. 34-7.

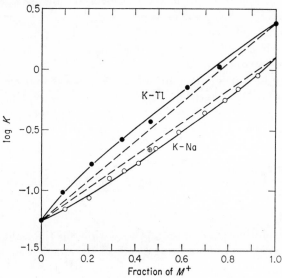

Fig. 34-7. Variation of log K with fraction of cation. [*H. Flood and A. Muan, Acta Chem. Scand.*, **4**, 364 (1950), *fig. 2.*]

AQUEOUS ELECTROLYTES

Multicomponent systems involving aqueous solutions have been of considerable interest, and a large body of data has been accumulated. As has been pointed out in the previous section, one can obtain activity coefficients of all the components of a solution if the activity coefficient of a single component is obtained as a function of all concentrations. A valuable source of data on activity coefficients in mixed electrolyte solutions is the cell $H_2|\ HX(m_2),\ MX(m_3)\ |AgX$, Ag. In Chapter 24, the calculation of activity coefficients from potential measurements for this type of cell has been presented for binary solutions. The same procedure

[1] *Loc. cit.*

is applicable here, with the total molality of X^- being given by $m_2 + m_3$ and with $\mathcal{E}°$ available from measurements for pure acid solutions. Güntelberg[1] showed that such a cell yields accurate $\gamma_\pm$ values for HCl in the presence of other chlorides. Åkerlöf[2] applied the method to sulfates, and the extensive work of Harned[3] and his collaborators yielded a large body of data for chlorides and bromides.

Additional methods which determine the activity of a solute component include the measurement of the solubility of a salt in mixed electrolytes. The activity of the water may also be measured, and we shall consider the treatment of isopiestic data presently.

We shall consider for simplicity 1-1 electrolytes almost exclusively, but generalization to other valence types involves no new principles. By considering mixing processes at constant ionic strength, all terms related to the Debye-Hückel limiting law are eliminated.

For a ternary solution of water, HX, and MX with constant total molality m and a varying molality of salt m_3, it was found empirically in many cases that the logarithm of the activity coefficient of HX, log γ_2, varied linearly with m_3 or with $f_3 = m_3/m$, the fraction of the total molality due to the salt. This relation is called Harned's rule.

$$\log \gamma_2 = \log \gamma_{2(0)} - \alpha_{23}m_3 \qquad (34\text{-}27a)^*$$

where the term on the left side is log $\gamma_\pm$ for HX at a molality

$$m_2 = (1 - f_3)m$$

in a ternary solution of total molality m, while on the right side $\gamma_{2(0)}$ is the activity coefficient of a pure HX solution of molality m, and α_{23} is an empirical coefficient which may be a function of the total molality m. The corresponding equation for the salt MX is

$$\log \gamma_3 = \log \gamma_{3(0)} - \alpha_{32}m_2 \qquad (34\text{-}27b)^*$$

Harned[4] has summarized the data for the HCl-NaCl system up to 3 M and between 0 and 50°C. Equation $(34\text{-}27a)^*$ is an accurate representation of the data, and Eq. $(34\text{-}27b)^*$ holds well for γ_{NaCl} between 20 and 40°C. The variation of α_{ij} with molality is small but significant; for example, α_{23} is 0.0299 at 1 M and 0.0290 at 2 M and 30°C. At lower temperatures for HCl-NaCl and for other systems there are departures from Eq. $(34\text{-}27b)^*$ such that the dependence on m_2 is no longer linear. It is then convenient to add further terms with empirical coefficients such

[1] E. Güntelberg, *Z. physik. Chem.*, **123**, 199 (1926).

[2] G. Åkerlöf, *J. Am. Chem. Soc.*, **48**, 1160 (1926).

[3] H. S. Harned, *J. Phys. Chem.*, **63**, 1299 (1959).

[4] *Ibid.*

that the general expressions are

$$\log \gamma_2 = \log \gamma_{2(0)} - \alpha_{23} m_3 - \beta_{23} m_3{}^2 - \delta_{23} m_3{}^3 - \cdots \quad (34\text{-}28a)$$
$$\log \gamma_3 = \log \gamma_{3(0)} - \alpha_{32} m_2 - \beta_{32} m_2{}^2 - \delta_{32} m_2{}^3 - \cdots \quad (34\text{-}28b)$$

The coefficients α_{ij}, β_{ij}, δ_{ij}, etc., may be functions of the total molality m but are independent of the fraction of each electrolyte present. The coefficients in one equation are related to those in the other equation together with certain properties of the pure salt solutions. The method by which one set of coefficients may be calculated from the other set will be given.

Although we shall illustrate most of the relationships with Eqs. (34-28) and analytical methods, the same basic equations and paths of integration could be employed with graphical methods.

Dilute Solutions. In Chapter 23 we have already presented the Bronsted-Guggenheim equation for the activity coefficient of an aqueous electrolyte in the presence of other electrolytes. For a mixture of 1-1 salts with a low total molality m, Eq. (23-38)* for the MX-NY mixture (NY = 2, MX = 3) yields

$$\log \gamma_2 = -A_\gamma \frac{m^{\frac{1}{2}}}{1 + m^{\frac{1}{2}}} + m_2 B_{\text{NY}} + \tfrac{1}{2} m_3 (B_{\text{NX}} + B_{\text{MY}}) \quad (34\text{-}29a)^*$$

$$\log \gamma_3 = -A_\gamma \frac{m^{\frac{1}{2}}}{1 + m^{\frac{1}{2}}} + m_3 B_{\text{MX}} + \tfrac{1}{2} m_2 (B_{\text{NX}} + B_{\text{MY}}) \quad (34\text{-}29b)^*$$

where B_{NY}, etc., are the specific ion interaction constants. It is interesting to note that $\log \gamma_2$ for a solution with $m_2 = 0$, $m_3 = m$ is equal to $\log \gamma_3$ when $m_3 = 0$, $m_2 = m$. Thus the activity coefficient of a trace of NY in MX solution is the same as that for a trace of MX in NY solution if the total molality is the same in each case.

If one now subtracts from Eqs. (34-29)* the corresponding functions for the pure electrolyte solutions, one readily verifies that Harned's rule is obeyed with the coefficients

$$\alpha_{23} = B_{\text{NY}} - \tfrac{1}{2}(B_{\text{MY}} + B_{\text{NX}}) \quad (34\text{-}30a)^*$$
$$\alpha_{32} = B_{\text{MX}} - \tfrac{1}{2}(B_{\text{MY}} + B_{\text{NX}}) \quad (34\text{-}30b)^*$$

and β_{23}, etc., are all zero.

The difference in the Harned coefficients is seen by Eq. (22-31)* to be related to the difference in osmotic coefficients for the two pure electrolyte solutions:

$$\alpha_{32} - \alpha_{23} = B_{\text{MX}} - B_{\text{NY}} = \frac{2(\phi_3 - \phi_2)}{2.303 m} \quad (34\text{-}31)^*$$

The sum of the Harned coefficients,

$$\alpha_{23} + \alpha_{32} = B_{\text{MX}} + B_{\text{NY}} - B_{\text{MY}} - B_{\text{NX}}$$

is not zero in the general case.   But in the case where the two electrolytes have one common ion, $M = N$ or $X = Y$, then $\alpha_{23} + \alpha_{32} = 0$, and one finds the very simple relationship

$$\alpha_{23} = -\alpha_{32} = \frac{\phi_3 - \phi_2}{2.303m} = \frac{1}{2}(B_{NY} - B_{MX}) \qquad (34\text{-}32)*$$

Güntelberg[1] measured γ_{HCl} from cells of the type mentioned above for aqueous mixtures of HCl with LiCl, NaCl, KCl, and CsCl at total molality 0.1.   These results are compared in Table 34-2 with values calculated from the ΔB values from Table A4-3.   Since the ΔB_{MX} of Table A4-3 are always $B_{MX} - B_{KCl}$, these ΔB values may be used in the equations given above as if they were B values and the B_{KCl} will cancel unless KCl is an actual constituent.

The agreement between calculated and experimental values in Table 34-2 is good for NaCl and KCl.   The differences are probably not much more than experimental error in the other cases.   Thus the use of the Bronsted-Guggenheim equations and constants will probably yield good results for mixed electrolytes at ionic strengths less than 0.1 M.

TABLE 34-2. ACTIVITY COEFFICIENT OF HCl IN AQUEOUS HCl-MCl
SOLUTIONS OF 0.1 TOTAL MOLALITY

MCl	α_{23} [Eq. (34-32)]	α_{23} (expt.)
LiCl............	0.02	0.013
NaCl..........	0.049	0 043
KCl...........	0.074	0 077
CsCl..........	0.117	0 143

Osmotic Coefficient.   The Gibbs-Duhem equation may be applied to determine the restrictions on the variation of the activity or of the osmotic coefficient of water with fraction of salt when Eqs. (34-28) are valid.   If we express the Gibbs-Duhem equation in terms of total molality m and fraction of each of the salts, f_2, f_3, etc., of a multicomponent system,

$$\frac{1000}{M_1} d \ln a_1 + \nu_2 m f_2\, d \ln mf_2\gamma_2 + \nu_3 m f_3\, d \ln f_3 m \gamma_3 + \cdots = 0$$

The osmotic coefficient for a mixed salt solution is defined by

$$\frac{1000}{M_1} \ln a_1 = -\left(\sum_i \nu_i m_i\right) \phi \qquad (34\text{-}33)$$

---

[1] E. Güntelberg, *loc. cit.;* "Studier over Elektrolyt-Activiteter," G. E. C. Gads Forlag, Copenhagen, 1938.

where the sum gives the total molality of all solute species.

$$-d[(m\phi)(\nu_2 f_2 + \nu_3 f_3 + \cdots)] + \nu_2 m f_2 \, d \ln m f_2 \gamma_2 + \cdots = 0$$

If ν is the same for all the salts and m is held constant, $m \sum_i \nu_i f_i = \nu m$

cancels out. Also $\sum_i f_i \, d \ln f_i = 0$, and

$$d\phi = f_2 \, d \ln \gamma_2 + f_3 \, d \ln \gamma_3 + \cdots \qquad (34\text{-}34)$$

From Eqs. (34-28) we obtain for $d \ln \gamma_2$, with a similar expression for $d \ln \gamma_3$, etc.,

$$\frac{1}{2.303} d \ln \gamma_2 = -(m\alpha_{23} + 2m^2 f_3 \beta_{23} + 3m^3 f_3^2 \delta_{23} + \cdots) \, df_3$$

If there are only two solute components, $df_3 = -df_2$, and

$$\frac{1}{2.303} d\phi = m[\alpha_{23} f_2 - \alpha_{32} f_3 + 2m f_2 f_3 (\beta_{23} - \beta_{32})$$
$$+ 3m^2 f_2 f_3 (f_3 \delta_{23} - f_2 \delta_{32}) + \cdots] \, df_2$$

Substitution of $f_3 = 1 - f_2$ and integration between zero and f_2 yields

$$\phi_{f_2} - \phi_{f_2=0} = 2.303m \left\{ -\alpha_{32} f_2 + \left[\frac{\alpha_{23} + \alpha_{32}}{2} + m(\beta_{23} - \beta_{32}) \right. \right.$$
$$\left. + \tfrac{3}{2} m^2 \delta_{23} + \cdots \right] f_2^2 - [\tfrac{2}{3} m (\beta_{23} - \beta_{32})$$
$$\left. + m^2 (2\delta_{23} + \delta_{32}) + \cdots] f_2^3 + \tfrac{3}{4} m^2 (\delta_{23} + \delta_{32}) f_2^4 + \cdots \right\} \qquad (34\text{-}35)$$

Even when $\beta_{23} - \beta_{32} = 0$ and δ_{23}, δ_{32} and all higher terms are zero, the variation of ϕ with f_2 is quadratic unless $\alpha_{23} = -\alpha_{32}$. The substitution $f_2 = 1$ yields a general relationship between the difference of the osmotic coefficients of the two pure salt solutions:

$$\phi_2 - \phi_3 = 2.303m \left[\tfrac{1}{2}(\alpha_{23} - \alpha_{32}) + \frac{m}{3}(\beta_{23} - \beta_{32}) \right.$$
$$\left. + \frac{m^2}{4}(\delta_{23} - \delta_{32}) + \cdots \right] \qquad (34\text{-}36)$$

This result reduces to Eq. (34-31)* when the β, δ, . . . are omitted. The equation corresponding to Eq. (34-35) for the other solute component may be obtained by interchange of subscripts 2 and 3 throughout.

Total Free Energy of Mixing. While the total free energy of mixing is not usually directly measurable, the heat of mixing is frequently measured. It is convenient to calculate the total free energy of mixing in terms of the activity-coefficient expressions and then to take the temperature derivative to obtain the heat. The free-energy change on mixing is just the sum of the changes in chemical potential for the various components. Again we take 1-1 electrolytes and mix solutions of molality m to obtain a final amount of solution containing 1 kg of solvent and m_2, m_3 moles of solutes, respectively, with $m = m_2 + m_3$:

$$\Delta F_{\mathrm{mix}} = 2RT\left(-m\phi + m_2\phi_2 + m_3\phi_3 + m_2 \ln \frac{m_2\gamma_2}{m\gamma_{2(0)}} + m_3 \ln \frac{m_3\gamma_3}{m\gamma_{3(0)}}\right)$$

At infinite dilution where all ϕ's and γ's are unity this reduces to

$$\Delta F_{\mathrm{mix}} = 2RT(m_2 \ln f_2 + m_3 \ln f_3)$$

with $f_2 = m_2/m$ and $f_3 = m_3/m$ as before. This is just the free energy of mixing of ideal solutions and may be subtracted to yield the excess free energy of mixing:

$$\Delta F_{\mathrm{mix}}^E = 2mRT\left(-\phi + f_2\phi_2 + f_3\phi_3 + f_2 \ln \frac{\gamma_2}{\gamma_{2(0)}} + f_3 \ln \frac{\gamma_3}{\gamma_{3(0)}}\right) \quad (34\text{-}37)$$

Now the expressions of Eqs. (34-28) and (34-35) may be substituted and after simplification

$$\Delta F_{\mathrm{mix}}^E = -2.303RTm_2m_3[\alpha_{23} + \alpha_{32} + 2(m_3\beta_{23} + m_2\beta_{32}) \\ + \tfrac{2}{3}(\beta_{23} - \beta_{32})(m_2 - m_3) + \cdots] \quad (34\text{-}38)$$

We noted previously that the difference $\alpha_{23} - \alpha_{32}$ was related solely to the difference between the solutions of pure-component electrolytes, and it is now interesting to note that the sum $\alpha_{23} + \alpha_{32}$ is the leading term in the expression for the excess free energy of mixing. It is apparent that solutions following Eq. (34-32)* would have zero $\Delta F_{\mathrm{mix}}^E$.

Calculation of γ_3 from γ_2. McKay[1] applied the cross-differentiation equation analogous to Eqs. (23-33) and (34-2), which for mean activity coefficients is

$$\nu_3 \left(\frac{\partial \ln \gamma_3}{\partial m_2}\right)_{m_3} = \nu_2 \left(\frac{\partial \ln \gamma_2}{\partial m_3}\right)_{m_2} \quad (34\text{-}39)$$

to obtain α_{32} from α_{23}. We apply the same procedure to the more general equation (34-28) but retain the restriction to binary electrolytes of equal valence. In differentiating the expressions of Eqs. (34-28) it is convenient to note that, if Y is a function of total molality but not of fractional solute composition,

$$\left(\frac{\partial Y}{\partial m_2}\right)_{m_3} = \left[\frac{\partial Y}{\partial (m_2 + m_3)}\right]_{m_3} = \frac{dY}{dm}$$

[1] H. A. C. McKay, *Trans. Faraday Soc.*, **51**, 903 (1955); *Nature*, **169**, 464 (1952).

In this and following equations we shall ignore possible dependence of functions on pressure, temperature, and such variables and shall write total derivatives if there is no other dependence on composition variables. Then

$$\frac{\partial}{\partial m}(\alpha_{32}m_2 + \beta_{32}m_2{}^2 + \cdots)_{m_3} = \frac{d}{dm}\left(\log\frac{\gamma_{3(0)}}{\gamma_{2(0)}}\right) + \alpha_{23}$$

$$+ 2\beta_{23}m_3 + m_3\frac{d\alpha_{23}}{dm} + m_3{}^2\frac{d\beta_{23}}{dm} + \cdots \quad (34\text{-}40)$$

Integration between $m = m_3'$ (that is, $m_2 = 0$) and $m = m'$, holding m_3 constant at m_3' and with $m_2' + m_3' = m'$, yields

$$(\alpha_{32} + \beta_{32}m_2' + \cdots)m_2' = \left[\log\frac{\gamma_{3(0)}}{\gamma_{2(0)}}\right]_{m=m_3'}^{m=m'}$$

$$+ \int_{m_3'}^{m'}(\alpha_{23} + 2\beta_{23}m_3' + \cdots)\,dm + [m_3'\alpha_{23} + (m_3')^2\beta_{23} + \cdots]_{m_3'}^{m'}$$

$$(34\text{-}41)$$

The value of $\alpha_{32} + \beta_{32}m' + \cdots$ is obtained if $m_3' \to 0$ and $m_2' = m'$:

$$(\alpha_{32} + \beta_{32}m' + \cdots)m' = \log\frac{\gamma_{3(0)}}{\gamma_{2(0)}} + \int_0^{m'}\alpha_{23}\,dm \quad (34\text{-}42)$$

where $\gamma_{3(0)}$ and $\gamma_{2(0)}$ for the pure salts and α_{32}, β_{32}, etc., are values for $m = m'$. The value of α_{32} is obtained if $m_2 \to 0$ and $m_3 = m$ in Eq. (34-40):

$$\alpha_{32} = \frac{d}{dm}\left(\log\frac{\gamma_{3(0)}}{\gamma_{2(0)}}\right) + \alpha_{23} + m\frac{d\alpha_{23}}{dm} + 2\beta_{23}m$$

$$+ m^2\frac{d\beta_{23}}{dm} + \cdots \quad (34\text{-}43)$$

Since α_{32}, β_{32}, etc., may be functions of the total molality, their evaluation in general would require the use of a large range of values of m_2' and m' in Eq. (34-41) and its limiting forms Eqs. (34-42) and (34-43). By means of an adequate number of simultaneous equations at fixed m' but varying m_2', one can solve for α_{32}, β_{32}, etc., for that value of m'. A more elegant method of obtaining α_{32}, β_{32}, etc., will be presented.

Harned[1] treated the data for the system HCl-NaCl (components 2 and 3, respectively) by the Eq. (34-41) method; in this case

$$\beta_{23} = \delta_{23} \cdots = 0$$

[1] H. S. Harned, *J. Phys. Chem.*, **63**, 1299 (1959).

within experimental error.   The data cover the range 0 to 50°C and 1 to 3 M total electrolyte.   The quantity $(\alpha_{32} + \beta_{32}m_2 + \cdots)$ varies with total molality and with temperature.

Equation (34-40) may be employed to calculate γ_3 from data on γ_2 by another method which is in some respects more elegant, although it does not appear to have been used frequently.   If the differentiations indicated in Eq. (34-40) are carried out and the substitutions $m_2 = f_2m$ and $m_3 = (1 - f_2)m$ are made, one obtains

$$\alpha_{32} + mf_2\frac{d\alpha_{32}}{dm} + 2mf_2\beta_{32} + m^2f_2{}^2\frac{d\beta_{32}}{dm} + \cdots = \frac{d}{dm}\left(\log\frac{\gamma_{3(0)}}{\gamma_{2(0)}}\right)$$

$$+ \alpha_{23} + m(1 - f_2)\frac{d\alpha_{23}}{dm} + 2m(1 - f_2)\beta_{23} + m^2(1 - f_2)^2\frac{d\beta_{23}}{dm} + \cdots$$

This equation must be satisfied for all values of f_2; hence we transpose all terms to the left side, collect the terms according to powers of f_2, and equate to zero the coefficient of each power of f_2:

$$\alpha_{32} - \frac{d}{dm}\left(\log\frac{\gamma_{3(0)}}{\gamma_{2(0)}}\right) - \alpha_{23} - m\frac{d\alpha_{23}}{dm} - 2m\beta_{23}$$

$$- m^2\frac{d\beta_{23}}{dm} - \cdots = 0 \quad (34\text{-}44a)$$

$$m\left(\frac{d\alpha_{32}}{dm} + \frac{d\alpha_{23}}{dm}\right) + 2m(\beta_{32} + \beta_{23}) + 2m^2\frac{d\beta_{23}}{dm} + \cdots = 0 \quad (34\text{-}44b)$$

$$m^2\left(\frac{d\beta_{32}}{dm} - \frac{d\beta_{23}}{dm}\right) + 3m^2(\delta_{32} - \delta_{23}) - 3m^3\frac{d\delta_{23}}{dm} + \cdots = 0 \quad (34\text{-}44c)$$

The first of these equations is just the equivalent of Eq. (34-43) which determines α_{32}.   The following equations may be solved successively for β_{32}, δ_{32}, etc.   This provides an elegant method of determining these coefficients if the necessary derivatives of the empirical quantities are available.

The data summarized by Harned for the HCl-NaCl system at 25°C will be used to illustrate the use of Eqs. (34-44).   The value of the derivative of $\log(\gamma_{3(0)}/\gamma_{2(0)})$ is readily obtained from Table A4-3.   The remaining data are the α_{23} values from the cell electromotive-force measurements[1] and $d\alpha_{23}/dm$ derived from these values together with the result that β_{23}, δ_{23}, etc., are all zero.   The calculation is summarized in Table 34-3, where the results show that β_{32} is zero within experimental error.   Similar calculations show that β_{32} is zero within experimental error from 20 to 40°C and allow one to conclude that this system follows Harned's rule for both solute components in this range.

[1] *Ibid.*

TABLE 34-3. CALCULATION OF γ_3 FROM γ_2 IN MIXED ELECTROLYTE SOLUTIONS

Equation (34-44a) yields:

	1 M	2 M	3 M
$\dfrac{d}{dm}\left(\log \dfrac{\gamma_{NaCl}}{\gamma_{HCl}}\right)$...........	-0.0896	-0.0873	-0.0856
α_{23}......................	0.0315	0.0308	0.0300
$m\left(\dfrac{d\alpha_{23}}{dm}\right)$...............	-0.0007	-0.0015	-0.0024
α_{32}......................	-0.0588	-0.0580	-0.0580

Equation (34-44b) reduces to $\beta_{32} = -\dfrac{1}{2}\left(\dfrac{d\alpha_{32}}{dm} + \dfrac{d\alpha_{23}}{dm}\right)$ and yields:

	1–2 M	2–3 M
β_{32}............	0.0000	0.0004

Isopiestic Measurements for Multicomponent Salt Solutions.
Just as it was possible to obtain the thermodynamic properties of the
other components from the measurement of the activity of HCl in solutions, the measurement of the vapor pressure of the solvent can also
yield the activity coefficients of the solutes. The equations of Darken,
Wagner, Schuhmann, and Gokcen which were reviewed earlier in this
chapter can be used for this purpose. McKay and Perring[1] obtained
related equations which are particularly suited for isopiestic measurements on mixed electrolyte solutions. Ternary solutions of constant
fraction of the two salts $f_3 = n_3/(n_2 + n_3) = m_3/m$ are particularly
adapted for isopiestic studies. In Eq. (34-12), subscript 1 is replaced by
subscript w to distinguish the component water from the salt components,
and the equation is expressed in terms of the variables m and f_3, it being
noted that

$$\left(\frac{\partial n_w}{\partial n_2}\right)_{a_w, n_3} = \left[\frac{\partial(n_w/n_3)}{\partial(n_2/n_3)}\right]_{a_w} = \frac{1000}{M_w}\left[\frac{\partial(1/m_3)}{\partial(1/f_3)}\right]_{a_w}$$
$$= \frac{1000}{M_w}\frac{1}{m}\left(1 + \frac{f_3}{m}\frac{dm}{df_3}\right)_{a_w}$$

Equation (34-12) becomes

$$-\left(\frac{\partial \ln \gamma_2^2 m_2^2}{\partial \ln a_w}\right)_{f_3} = \frac{1000}{M_w}\frac{1}{m}\left(1 + \frac{d \ln m}{d \ln f_3}\right)_{a_w}$$

[1] H. A. C. McKay and J. K. Perring, *Trans. Faraday Soc.*, **49**, 163 (1953).

where γ_2 is $\gamma_\pm$ for a 1-1 salt in a solution with fraction f_3 of salt 3. Integration is carried out at constant f.

$$\frac{2M_w}{1000} \ln m_2\gamma_2 = -\int_0^{\ln a_w{}'} \left[\frac{1}{m^2}\left(\frac{\partial m}{\partial \ln f_3}\right)_{a_w} + \frac{1}{m}\right] d \ln a_w$$

It is not convenient to evaluate this integral to the limit of $m = 0$ and $\ln a_w = 0$, but a more useful expression is obtained if a pure solution of 2 is equilibrated to the same water activity as the mixed salt solution of total molality m and $m_2 = f_2 m$. The molality of the pure salt solution in isopiestic equilibrium with the mixed salt solution is designated as $^i m_2$. The Gibbs-Duhem equation for the pure salt solution yields

$$\frac{2M_w}{1000} \ln {}^i m_2 \gamma_{2(0)} = -\int_0^{\ln a_w{}'} \frac{d \ln a_w}{{}^i m_2}$$

The preceding equation may be subtracted from this equation to yield

$$\frac{2M_w}{1000} \ln \frac{{}^i m_2 \gamma_{2(0)}}{m_2 \gamma_2} = \int_0^{\ln a_w{}'} \left[\frac{1}{m^2}\left(\frac{\partial m}{\partial \ln f_3}\right)_{a_w} + \frac{1}{m} - \frac{1}{{}^i m_2}\right] d \ln a_w \quad (34\text{-}45)$$

The quantity $(\partial m/\partial \ln f_3)_{a_w}$ is determined as a function of f_3 and a_w by equilibrating a large number of dishes in a desiccator containing different ratios of the two salts. The equilibration is repeated after successive water additions. Values of m for solutions of different f_3 in isopiestic equilibrium at a given a_w are plotted versus $\ln f_3$ to determine the slope $(\partial m/\partial \ln f_3)_{a_w}$ at the particular value of f_3 for which the integration of Eq. (34-45) is to be performed. This is repeated for each a_w to obtain sufficient values of $(\partial m/\partial \ln f_3)_{a_w}$ for a given f_3 to allow evaluation of the integral of Eq. (34-45).

Isopiestic data have been reported for a number of salt mixtures, but very few data have been taken so as to be easily adapted to the McKay-Perring method.

Review of Data for Aqueous Mixed Salt Solutions. Harned and Owen[1] and Robinson and Stokes[2] have reviewed the data for mixed salt solutions. Harned's rule [Eqs. (34-27)*] is a good approximation for many mixed salt solutions at fixed total molality. Many systems show noticeable deviations for one or both salts requiring higher terms as in Eqs. (34-28), but the error due to neglect of the deviation is not usually large. The deviations from Harned's rule appear to be worst in the range 0.1 to 1 M and become smaller at both lower and higher molalities. For mix-

[1] H. S. Harned and B. B. Owen, "The Physical Chemistry of Electrolytic Solutions," 3d ed., Reinhold Publishing Corporation, New York, 1958.

[2] R. A. Robinson and R. H. Stokes, "Electrolyte Solutions," 2d ed., Butterworth & Co. (Publishers) Ltd., London, 1959.

tures of salts with ions of different charge, Harned's rule can still be applied if fixed total ionic strength is maintained.

Even with close adherence to Harned's rule, one may find a range of behavior. The thermodynamic restrictions consistent with Harned's rule, Eq. (34-44b), only fix the sum $\alpha_{23} + \alpha_{32}$ as independent of molality. Both α_{23} and α_{32} may vary with molality and need not show a simple relationship to one another. The osmotic coefficient of water may vary linearly (when $\alpha_{23} = -\alpha_{32}$) or as a quadratic function of f, the fraction of salt.

The simplest behavior is that predicted by the Bronsted-Guggenheim equation (34-32)* for dilute solutions of salts with a common ion, namely, $\alpha_{23} = -\alpha_{32}$, $B_{23} = \log (\gamma_{2(0)}/\gamma_{3(0)}) = \alpha_{23} - \alpha_{32} = 2\alpha_{23}$ and a linear variation of ϕ, the osmotic coefficient of water, with f_2. B_{23} is the difference for salts 2 and 3 of the $B^{\cdot}$ values of Eq. (22-38) at low concentration and may be obtained from the tables of Appendix 4 from the relation $B_{23} = \Delta B_2^{\cdot} - \Delta B_3^{\cdot}$. With such behavior the properties of the mixed salt solution can be predicted from those of the pure salt solutions. Because of the large experimental uncertainty of the α_{23} and α_{32} values at low molalities, it is difficult to test the Bronsted-Guggenheim equation directly in mixed salt solutions by the methods described above, although one test was given in Table 34-2 and the extensive literature on the variation of the solubility of slightly soluble salts upon addition of other salts was the original basis for Bronsted's theory of specific-ion interaction and provides confirmation for Eq. (34-32)* in dilute solutions. The freezing-point measurements of salt mixtures by Scatchard and Prentiss[1] provide additional confirmation of Bronsted's theory of specific-ion interaction.

From the magnitude of triple-ion interactions, as given by $dB^{\cdot}/dm$ or C of Eq. (22-40) for pure salt solutions, one would expect the variation with molality of the α terms to be negligible below 0.1 M except for work of extreme accuracy; also the β terms should be negligible. Between 0.1 and 1 M, one would not expect Eq. (34-32)*, based on pair interactions only, to be satisfactory in view of the significant change of $B^{\cdot}$ with molality for pure salt solutions. Thus it is not surprising that, with few exceptions, none of the predictions of Eq. (34-32)* are found to hold well between 0.1 and 1 M. However, surprisingly enough, better agreement is found above 1 M in that a $B_{23}^{\cdot}$ calculated from $B^{\cdot}$ values for such concentrations and $\alpha_{23} - \alpha_{32}$ do appear to approach one another as the molality is increased in many systems. This behavior is shown in Fig. 34-8, where $B_{23}^{\cdot}$ and $\alpha_{23} - \alpha_{32}$ are both plotted versus m for HX-MX mixtures, where 2 represents HX and 3 represents MX. We have noted in Chapter 22 that $B^{\cdot}$ of Eq. (22-38), and therefore $B_{23}^{\cdot}$, becomes almost

---

[1] G. Scatchard and S. S. Prentiss, *J. Am. Chem. Soc.*, **56**, 2320 (1934); G. Scatchard, *Chem. Rev.*, **19**, 309 (1936).

independent of molality at high molality.   Likewise $\alpha_{23} + \alpha_{32}$ approaches constancy as the molality is increased, thus indicating better adherence to Eqs. (34-27)*.   It would seem that the improvement in the performance of Eq. (34-32)* as the molality is increased might be attributed to the shrinkage of the ionic atmosphere.   Thus at high molalities in aqueous solutions, as in fused salts, nearest-neighbor interactions become predominant.

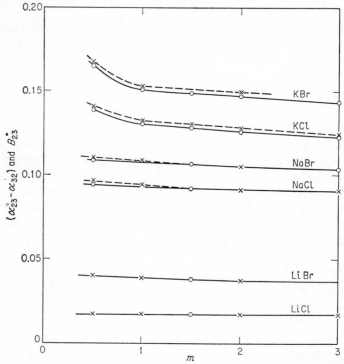

FIG. 34-8. Comparison of $\alpha_{23} - \alpha_{32}$ and $B_{23}^{\cdot}$.   x, $B_{23}^{\cdot}$; o, $\alpha_{23} - \alpha_{32}$.   (*H. S. Harned and B. B. Owen, "The Physical Chemistry of Electrolytic Solutions," 3d ed., fig. 14-6-3, Reinhold Publishing Corporation, New York, 1958.*)

In the region of intermediate molalities, where Eq. (34-32)* is not a good representation of the behavior of mixed salt solutions, it is necessary to make measurements on the mixed salt solutions to obtain values of α_{23} or α_{32}.   For such solutions, Eq. (34-27a), for example, is conveniently expressed as

$$\log \gamma_2 = \log \gamma_{\text{KCl}} + (\Delta B_2^{\cdot} - \alpha_{23} f_3) m \qquad (34\text{-}46)$$

where $\Delta B_2^{\cdot}$ is tabulated in Appendix 4 along with values of $\log \gamma_{\text{KCl}}$ as a function of m.   At low molalities, where Eq. (34-32)* serves as a good approximation, α_{23} may also be expressed in terms of $B^{\cdot}$ or $\Delta B^{\cdot}$ values.

For the example of a salt mixture with a common anion, Eq. (34-46) then reduces to

$$\log \gamma_2 = \log \gamma_{KCl} + [\Delta B_2^{\cdot} - \frac{1}{2}(\Delta B_2^{\cdot} - \Delta B_3^{\cdot})f_3]m \qquad (34\text{-}47)^*$$

where both $\Delta B^{\cdot}$ values can be obtained from Table A4-3. As $\Delta B^{\cdot}$ varies somewhat with molality at all molalities, the $\Delta B^{\cdot}$ should be taken for each salt at a molality equal to the total molality of the mixed solution. Equations corresponding to Eqs. (34-46) and (34-47)* can be obtained for the other component with 3 and 2 interchanged, and similar equations can be given for salt mixtures with ions of different charge using ionic strength as the variable in place of molality. Equations of the type of Eq. (34-47)* provide a means of estimating mixed-salt-solution behavior when only data for the pure salts are available, and may be used as strictly empirical equations at the higher molalities.[1] The type of accuracy that can be expected is illustrated in Table 34-4, where the activity coefficient of one of the components, calculated from Eq. (34-47)*

TABLE 34-4. ACTIVITY COEFFICIENT OF MX AT LOW MOLALITY IN A SOLUTION OF NY AT MOLALITY m†

NY	MX	$m = 0.5$		$m = 1$		$m = 3$	
		Calculated	Observed	Calculated	Observed	Calculated	Observed
CsCl	HCl	0.66	0.67	0.66	0.64	0.79	0.67
KCl	HCl	0.70	0.71	0.70	0.72	0.87	0.86
NaCl	HCl	0.72	0.73	0.73	0.75	0.97	1.07
LiCl	NaCl	0.71	....	0.71	0.72	0.91	0.91
CsCl	KCl	0.63	0.63	0.58	0.58	0.52	0.53

† The HCl data are from H. S. Harned and B. B. Owen, "The Physical Chemistry of Electrolytic Solutions," 3d ed., Reinhold Publishing Corporation, New York, 1958. The salt-mixture data are from R. A. Robinson, *Trans. Faraday Soc.*, **49**, 1147 (1953), and R. A. Robinson and C. A. Lim, *Trans. Faraday Soc.*, **49**, 1144 (1953). The calculated values are based on Eq. (34-47)*, which uses only data for pure salt solutions.

using the $\Delta B^{\cdot}$ values of Appendix 4, is compared with the experimental value over a range of molalities for a variety of mixtures. In each instance the activity coefficient is given for salt 2 at a small value of f_2 or close to unit value of f_3 to make the test of Eq. (34-47)* as severe as possible. Thus, for the HCl example, the HCl molality is 0.01 M and Eq. (34-47)* reduces to

$$\log \gamma_{2(f_3=1)} = \log \gamma_{KCl} + \frac{1}{2}(\Delta B_2^{\cdot} + \Delta B_3^{\cdot})m$$

According to this result $\gamma_{2(f_3=1)}$ is the geometric mean of γ_3 and γ_2 in pure solutions of molality m.

[1] E. Glueckauf, *Nature*, **163**, 414 (1949).

It is common procedure in equilibrium or kinetic studies to maintain constant ionic strength when the composition of aqueous electrolyte solutions is being varied to determine the dependence of rates or equilibrium constants upon the reactant concentrations. For example, when one wishes to determine the order of dependence upon hydrogen ion, the variation of hydrogen ion is accompanied by variation of an alkali ion so as to maintain constant ionic strength. When this is done, it is often the practice to flood the solution with excess electrolyte so that the percentage replacement is small. Thus, when the hydrogen ion concentration is being varied between 0 and 0.1 M, the total ionic strength might be kept at 3 M. Application of Harned's rule shows that this procedure is not necessarily effective. To the extent that the α_{23} is independent of molality one will obtain the same percentage change in the activity coefficient of the acid when 0.1 mole/liter of hydrogen ion is replaced by 0.1 mole/liter sodium ion, whether the total ionic strength is 0.1 or whether it is 3 M. If $\alpha_{23} = 0.1$, replacement of 0.1 mole of HX by 0.1 mole of MX in 1000 g water results in a 3 per cent change in $\gamma_\pm$, while replacement of 1 mole results in a 26 per cent change in $\gamma_\pm$. If it is desired to keep the hydrogen-ion activity coefficient as constant as possible, it would be more desirable to choose an alkali ion which makes α_{23} as small as possible. When α values are not known, Eq. (34-47)* or the more generalized form based on Eq. (23-38)* may be used as empirical equations to predict a replacement for the variable ion that will minimize change in the activity coefficient.

Solubility of Salts in Multicomponent Solutions. The solubility of salt MX in solution of salt NY may be expressed through use of Eq. (34-46) as

$$\log K_{SP} = \log (m_M m_X) + 2 \log \gamma_{KCl} + 2(\Delta B'_{MX} - \alpha f_{NY})m \quad (34\text{-}48)$$

with γ_{KCl} and $\Delta B'_{MX}$ at molality m. If Harned's rule is valid, the single parameter α can be determined from a solubility measurement at single value of m if K_{SP} and the activity coefficient of pure MX are both known. If γ_{MX} and $(\Delta B'_{MX})$ are not known for pure MX solutions up to the molality of interest, for example, when m is much larger than the solubility and if f is close to unity, $\Delta B'_{MX} - \alpha$ may be evaluated as a single parameter. From solubilities at several values of m, a plot of $\log (m_M m_X) + 2 \log \gamma_{KCl}$ versus m yields $\Delta B'_{MX} - \alpha$ from the slope and K_{SP} from the intercept. To illustrate Eq. (34-48) for a 1-1 salt, Fig. 34-9 shows a plot of the data[1,2] for the solubility of AgOAc in KNO_3 and $NaNO_3$ solutions at 25°C. The data fall on straight lines for solutions with the fraction of

---

[1] F. H. MacDougall, *J. Am. Chem. Soc.*, **52**, 1390 (1930).

[2] F. H. MacDougall and J. Rehner, Jr., *J. Am. Chem. Soc.*, **56**, 368 (1934).

nitrate greater than 0.88.   The slopes -0.014 and -0.026 fall remarkably close to the averages -0.007 and -0.024 of the ΔB values at 1 M for AgNO$_3$ and for KOAc and NaOAc, respectively.

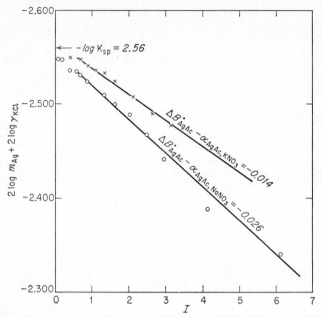

Fig. 34-9. Determination of solubility product of AgOAc at 25°C.

Heats of Mixing of Electrolyte Solutions.   In addition to their intrinsic interest, heats of mixing of aqueous salt solutions are valuable for fixing the temperature coefficient of activity coefficients.   Young, Wu, and Krawetz[1] have reported heats of mixing for 20 pairs of salt solutions at a total molality of 1 M.   The heats of mixing were quadratic functions of the fraction of salt 2, f_2.   McKay[2] pointed out that the Harned rule for activity coefficients would require a quadratic dependence of the heat of mixing upon f_2.

Consider the mixing of a solution containing m_2 moles of solute 2 at molality m with a solution containing m_3 moles of solute 3 at molality m to yield a mixed solution with 1 kg of solvent.   Again we take 1-1 electrolytes and assume, of course, no chemical reaction on mixing.   Since the ideal heat of mixing is zero, the heat of mixing is obtained from the appropriate temperature derivative of the excess free energy of mixing,

[1] T. F. Young, Y. C. Wu, and A. A. Krawetz, *Discussions Faraday Soc.*, **24**, 37, 77 (1957).

[2] H. A. C. McKay, *Discussions Faraday Soc.*, **24**, 76 (1957).

which was given in Eqs. (34-37) and (34-38).   If we choose Eq. (34-38), we obtain

$$\Delta H_{mix} = T^2 \frac{\partial(-\Delta F^E_{mix}/T)}{\partial T}$$

$$= 2.303RT^2 m_2 m_3 \frac{\partial}{\partial T} [\alpha_{23} + \alpha_{32} + 2(m_3\beta_{23} + m_2\beta_{32})$$

$$+ \tfrac{2}{3}(\beta_{23} - \beta_{32})(m_2 - m_3) + \cdots] \quad (34\text{-}49)$$

In the case of an equimolal mixture, $m_2 = m_3 = m/2$, and

$$\Delta H_{mix} = \frac{2.303RT^2 m^2}{4} \frac{\partial}{\partial T} [\alpha_{23} + \alpha_{32} + m(\beta_{23} + \beta_{32}) + \cdots] \quad (34\text{-}50)$$

This equation reduces, on omission of the β terms, to that given by McKay.   Table 34-5 presents values determined calorimetrically by Young, Wu, and Krawetz for equimolal mixing of 1 M solutions at 25°C.

TABLE 34-5. ΔH_{mix} OF EQUIMOLAL 1 M SOLUTIONS AT 25°C

Solution	ΔH_{mix}, cal/mole
HCl-LiCl	13
HCl-NaCl	32
HCl-KCl	−3.5
HCl-CsCl	−34
LiCl-NaCl	21
LiCl-KCl	−16.1
LiCl-CsCl	−49
NaCl-KCl	−9.5
NaCl-CsCl	−9
KCl-CsCl	1.7
NaCl-NaBr	0.8
KCl-KBr	0.8
LiCl-LiBr	0.8
KCl-KNO₃	0.34
LiBr-KBr	−17.1
LiBr-KCl	−31.75
LiCl-KBr	+0.17

We noted before that, in the case of mixing of solutions with a common ion, a zero excess free energy of mixing is predicted by the Bronsted theory.   Likewise a zero heat of mixing is predicted.   If we take HCl-NaCl as an example, the H^+ ions are equally surrounded by Cl^- ions before and after mixing.   Also, the Bronsted theory predicts no specific interaction of H^+ and Na^+ ions that is different from the interaction of H^+ ions or Na^+ ions with one another.

Bronsted's theory would be expected to be restricted to dilute solutions where pair interactions predominate over higher-order interactions; thus it is surprising to find the heat of mixing of 1 M solutions so small compared with the specific-ion interaction enthalpies, $2(2.303)RT^2(dB^{\cdot}/dT)$, which may be obtained from Tables A4-2b and A4-4a. Typical values given in cal kg/mole^2 at 25°C are:

HCl	LiCl	NaCl	KCl	LiBr	KBr
−301	−147	532	520	−127	653

A more direct test is illustrated in Table 34-6, where the temperature coefficients of $-\alpha_{23}$ and α_{32} calculated for equimolal solutions of HCl(2) and NaCl(3) from the data given by Harned[1] and Harned and Owen[2] are compared with $-\frac{1}{2}d(B_2^{\cdot} - B_3^{\cdot})/dT$ calculated from the values for pure salt solutions in Table A4-4c and with the $\partial(\alpha_{32} + \alpha_{23})/\partial T$ value for $m = 1$ calculated from the calorimetric data in Table 34-5 by use of Eq. (34-50). The units are kg/mole deg.

TABLE 34-6. COMPARISON OF TEMPERATURE COEFFICIENTS OF $B_2^{\cdot} - B_3^{\cdot}$, α_{23}, α_{32}, AND $\alpha_{23} + \alpha_{32}$, FOR HCl-NaCl AT 25°C, KG/MOLE DEG

Molality	$-\dfrac{1}{2}\dfrac{d(B_2^{\cdot} - B_3^{\cdot})}{dT} \times 10^4$	$-\dfrac{d\alpha_{23}}{dT} \times 10^4$	$\dfrac{d\alpha_{32}}{dT} \times 10^4$	$\dfrac{d(\alpha_{23} + \alpha_{32})}{dT} \times 10^4$
0.5	5.4	4 ± 1		
1	5.1	3.6 ± 1	6 ± 1	3.1
2	4.7	3.5 ± 0.5	4 ± 1	
3	4.3	4 ± 1	2 ± 2	

It will be noted from Table A4-5b that $d^2(B_2^{\cdot} - B_3^{\cdot})/dT^2$ varies from 2.6×10^{-5} at 0.5 M to 1.74×10^{-5} kg/mole deg^2 at 2 M for HCl-NaCl. Thus $d(B_2^{\cdot} - B_3^{\cdot})/dT$ rapidly approaches zero as the temperature is increased and should become zero between 65 and 75°C. In a parallel manner the temperature coefficient of $\alpha_{23} + \alpha_{32}$ goes to zero as the temperature is increased and for 2 to 3 M solutions is close to zero at 25°C.

In general, the results of Young, Wu, and Krawetz given in Table 34-5 indicate that the heats of mixing of large ions are small but that the heats of mixing of small cations with other cations can be quite appreciable. The closeness of the heats of mixing of Li and K ions in chloride and bromide solutions indicates that the heat of mixing of cations may be approximately independent of the anion, but the difference between the LiCl-KBr and LiBr-KCl values points out that the anion does have some influence. They also find that the heat of mixing of two cations in the presence of two anions is the weighted arithmetic mean of the heats of

[1] H. S. Harned, *J. Phys. Chem.*, **63**, 1299 (1959).

[2] H. S. Harned and B. B. Owen, "The Physical Chemistry of Electrolytic Solutions," 3d ed., Reinhold Publishing Corporation, New York, 1958.

mixing of the same cations in the presence of the single anions for the types of salts that they have studied.   These rules are very useful in allowing one to predict heats of mixing for a variety of salts containing ions common to the salts listed in Table 34-5.   Böttcher,[1] Young, Wu, and Krawetz,[2] and Friedman[3] have reviewed the implications of the empirical rules listed above.

AQUEOUS MIXTURES OF ELECTROLYTES AND NONELECTROLYTES

The phenomenon of salting out of a nonelectrolyte or the increase in activity of a nonelectrolyte in aqueous solution upon addition of a salt is well known.   Thus the vapor pressures of most organic solutes in an aqueous solution are increased upon addition of salt, although a few solutes such as hydrogen peroxide and hydrogen cyanide are salted in. When dealing with relatively nonsoluble solutes, a common method of determining the variation of activity of nonelectrolytes is by variation of the solubility upon addition of salt.   When the solubility of water and of the electrolyte in the nonelectrolyte phase is small, the activity of the nonelectrolyte remains substantially constant.   Measurements of the solubilities of large numbers of nonelectrolytes as a function of salt concentration have been carried out on gaseous, liquid, and solid nonelectrolytes.   A number of empirical equations for representing the data have been proposed.   The most commonly used is that of Setchenow[4] with S_0 and S, the solubilities in pure water and in salt solution of molality m_2, respectively,

$$\log \gamma_3 = \log \frac{S_0}{S} = k_{32}^{\cdot} m_2 \qquad (34\text{-}51)$$

Randall and Failey[5] reviewed the extensive data on salting out of nonelectrolytes.   Their summary indicates that Eq. (34-51) with a constant value of $k^{\cdot}$ is not strictly valid but that it is a good first approximation. The superscript $\cdot$ on $k^{\cdot}$ indicates that $k^{\cdot}$ is a function of m and will be dropped if k is constant or is a limiting value.

Debye and McAulay[6] and Debye[7] have shown theoretically that the

[1] C. J. F. Böttcher, *Discussions Faraday Soc.*, **24,** 78 (1957).

[2] *Op. cit.*, pp. 77, 80.

[3] H. L. Friedman, *Discussions Faraday Soc.*, **24,** 74 (1957).

[4] M. Setchenow, *Ann. chim. phys.*, (6)**25,** 226 (1892).

[5] M. Randall and C. F. Failey, *Chem. Rev.*, **4,** 271 (1927); see also F. A. Long and W. F. McDevit, *Chem. Rev.*, **51,** 119 (1952).

[6] P. Debye and J. McAulay, *Physik. Z.*, **26,** 22 (1925).

[7] P. Debye, *Z. physik. Chem.*, **130,** 55 (1927).

order of magnitude of the salting-out effects can be accounted for on the basis of coulombic forces alone. Their theories are based upon the change of the dielectric constant of the solvent upon addition of the nonelectrolyte. If the nonelectrolyte decreases the dielectric constant of the aqueous solution, the energy of the nonelectrolyte molecules in the vicinity of an ion is increased and the concentration of the nonelectrolyte in the vicinity of the ions is reduced. On the other hand, a nonelectrolyte which increases the dielectric constant of the solution is concentrated in the vicinity of the ions and is therefore salted in. The theories predict as a limiting behavior a linear variation of the logarithm of the activity coefficient of the nonelectrolyte with salt concentration in agreement with the empirical equation (34-51). From Eq. (34-39) it is clear that the effect of salt upon the activity coefficient of a nonelectrolyte is directly related to the effect of nonelectrolyte upon the activity coefficient of the salt. Using a method similar to that of Debye and McAulay, Born[1] had much earlier derived an equation for the activity coefficient of an electrolyte as a function of the dielectric constant of the solution due to the addition of a nonelectrolyte.

The magnitude of the salting-out and salting-in effects predicted by theory are of the right order of magnitude, but in view of the approximations made in the derivation the theoretical results cannot be considered more than limiting laws. The theory predicts the effect of varying ionic radius of the electrolyte, but the order of the effect does not agree with experimental results. Gross[2] has extended the theory of Debye, and McDevit and Long[3] have attempted to account for some of the deviation in terms of additional interactions. Scatchard[4] has reviewed the status of the theory and has pointed out that there are a number of interactions which will contribute terms of the same order as the coulombic term calculated by the theories of Debye, or Debye and McAulay. The various interactions which contribute to the salting-out or salting-in effects also play a part in the activity coefficient of electrolyte solutions, and the B· of Eq. (22-38) is of the same form as the k· term in Eq. (34-51) and involves similar interactions, at least in part.

There are some generalizations which can be made in regard to the interaction of electrolyte and nonelectrolyte solutes. Generally the values of k· in Eq. (34-51) show the same order from salt to salt for a large range of nonelectrolytes, although minor changes in the order are often observed. To illustrate the trends, Table 34-7 presents values of k· at

[1] M. Born, *Z. Physik*, **1**, 45 (1920).

[2] P. Gross, *Monatsh. Chem.*, **53**, 4445 (1929), and **55**, 287 (1930).

[3] W. F. McDevit and F. A. Long, *J. Am. Chem. Soc.*, **74**, 1773 (1952).

[4] G. Scatchard, *Trans. Faraday Soc.*, **23**, 454 (1927).

TABLE 34-7. VALUES OF $k^{\cdot}$ FOR SALTING OUT OF NONELECTROLYTES AT 25°C
AND APPROXIMATELY 1 M SALT, KG MOLE
DA = $(CH_3)_2COHCH_2COCH_3$; MO = $(CH_3)_2C{=}CHCOCH_3$

Salt	Ar	C_6H_5COOH	DA	MO	C_2H_5OAc
NaOAc............	....	....	....	0.180	
KOAc.............	....	....	....	0.17	
NaCl.............	0.058	0.191	0.139	0.174	0.166
KCl..............	0.061	0.152	0.118	0.149	0 143
MgSO$_4$...........	....	....	0.105	......	0.127
NaBr.............	....	....	0.109	0.12	0.119
KBr..............	... ...	....	0.090	0.102	0.105
LiCl.............	0.037	....	0.077	......	0.088
BaCl$_2$............	0.070	0.100	....	......	0.080
NaNO$_3$...........	0.058	0.064	....	0.08	0.074
KNO$_3$............	....	0.025	....	0.08	0.060
CaCl$_2$............	0.018	....	....		
MgCl$_2$............	0.012	....	0.060		
NaI.............	....	....	0.041	0.024	0.024
KI...............	....	....	0.034	......	0.037
NaClO$_4$..........	....	....	....	0.022	
HClO$_4$...........	−0.022	....	....	<0	

25°C for argon[1] gas, benzoic acid,[2] diacetone alcohol[3] [$(CH_3)_2COHCH_2$-$COCH_3$], mesityl oxide[4,5] [$(CH_3)_2C{=}CHCOCH_3$], and ethyl acetate[2] for approximately 1 M salt solutions. These data are based on determinations of vapor pressures, solubilities, and distribution between immiscible liquids. From Eq. (34-39), we can readily relate the salting out of a nonelectrolyte as given by Eq. (34-51) with constant k to the reverse effect of a nonelectrolyte on the activity coefficient of an electrolyte,

$$\log \gamma_3 = k_{32}m_2 \qquad (34\text{-}51)^*$$
$$\nu_2 \log \gamma_2 = k_{23}m_3 \qquad k_{32} = k_{23} \qquad (34\text{-}52)^*$$

Haugen and Friedman have verified the equality of k_{32} and k_{23} for aqueous solutions of nitromethane with KClO$_4$ and with CsClO$_4$. To illustrate the effect of nitromethane, the solubility of CsClO$_4$ is 22 per cent smaller in 1 M nitromethane than in pure water.[6]

When $k^{\cdot}$ is not a constant, Eq. (34-39) must be applied with $k^{\cdot}$ expressed as a function of molality. A typical example of the variation of

[1] G. Åkerlof, *J. Am. Chem. Soc.*, **57**, 1196 (1935).
[2] M. Randall and C. F. Failey, *Chem. Revs.*, **4**, 271 (1927).
[3] G. Åkerlof, *J. Am. Chem. Soc.*, **51**, 984 (1929).
[4] H. Lucas, D. Pressman, and L. Brewer, *J. Am. Chem. Soc.*, **64**, 1117 (1942).
[5] L. Brewer, T. R. Simonson, and L. K. J. Tong, *J. Phys. Chem.*, **65**, 420 (1961).
[6] G. R. Haugen and H. L. Friedman, *J. Phys. Chem.*, **60**, 1363 (1956).

$k\cdot$ as a function of both m_2 and m_3 is the behavior of mesityl oxide, $(CH_3)_2CH{=}CHCOCH_3$ or MO, in aqueous NaCl solutions.[1] Figure 34-10 shows the variation of $k_{MO,NaCl}$ of Eq. (34-51) at 20 and 25°C with m_{NaCl} at constant mesityl oxide concentration. A variation of 0.004 in k causes a 1 per cent change in γ around 1 M.

Figure 34-10 shows that the partial molal enthalpy of transfer of mesityl oxide from water to NaCl solutions is less than the experimental error, which is around 100 cal/mole for 1 M solutions. On the other

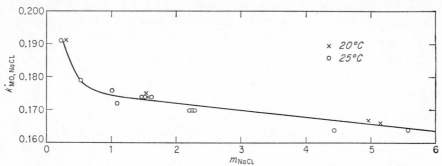

FIG. 34-10. Mesityl oxide–sodium chloride interaction coefficient in water.

hand, for salts with small k values such as $NaNO_3$ and $NaClO_4$, as much as 1.4 kcal/mole is absorbed upon transfer of mesityl oxide from water to 1 M salt solutions. These data were obtained by equilibrating mesityl oxide through the vapor phase until pure water and salt solutions had essentially the same mesityl oxide partial pressures. If $k\cdot$ were independent of the mesityl oxide concentration, e.g., if Henry's law held rigorously, the ratio of mesityl oxide molalities would be independent of the absolute molality of mesityl oxide. From the observed variation of the ratio upon varying the molality, the deviation from Henry's law was determined and can be expressed as

$$\log \gamma_{MO} = k_{MO,MO}m_{MO} = -0.30m_{MO} \qquad (34\text{-}53)$$

where γ_{MO} is the activity coefficient of mesityl oxide at 25°C relative to the solute standard state of mesityl oxide in pure water. No variation of $k_{MO,MO}$ with molality of salt could be detected within experimental error up to 1.5 M NaCl. The measurement of $k_{MO,MO}$ could not be extended because of the reduced solubility of mesityl oxide at higher salt molalities. Equation (34-53) corresponds to a 23 per cent deviation from Henry's law for a saturated (0.304 M) mesityl oxide solution at 25°C.

Dissociation Constant of Weak Acids. In the section on cells without liquid junctions in Chapter 24 the method of Guggenheim and

[1] Brewer, Simonson, and Tong, *loc. cit.*

Turgeon[1] for extrapolation of buffered-cell data to obtain the ionization constant of a weak acid was reviewed. We are now prepared to consider these four-component systems in more detail and to discuss the ambiguity inherent in the extrapolation to infinite dilution, which was briefly mentioned in Chapter 24.

The evaluation of Eq. (24-22) for a buffered cell involves a correction for the hydrogen ion and acetate ion liberated by the ionization of the acid and the determination of the activity-coefficient quotient. The hydrogen-ion molality is frequently evaluated by Eq. (24-23)* by using for K_a not the thermodynamic-equilibrium constant but the molality quotient based on activity coefficients calculated from the limiting Debye-Hückel law. We shall follow a similar procedure, using the first term of Eq. (23-38) to calculate $\gamma_{\pm}^{St}$ values from which

$$K_a^{St} = K_a(\gamma_{H^+X^-}^{St})^{-2}$$

where K_a is the thermodynamic-dissociation constant for acid HX. However, we shall extend our treatment to a higher degree of approximation through use of all the terms of Eq. (23-38) as well as the electrolyte-nonelectrolyte terms of Eqs. (34-51) and (34-52)*. In an equal molal buffer mixture with $m_1 = m_2 = m_3 = m$, after expansion of logarithmic terms in power series and neglecting m_H^2 terms,

$$K_a^{St} - m_H = 2.303 m m_H(-k_{MCl,HX} - k_{MX,HX} + k_{H^+X^-,HX} - k_{HX,HX}$$
$$+ B_{H^+X^-} + B_{HCl} + 2B_{MX}) \quad (34\text{-}54)*$$

where the B terms are the Bronsted specific-ion interaction terms of Eq. (23-38) and the k terms are the electrolyte-nonelectrolyte interaction terms of Eqs. (34-51)* and (34-52)*. The subscripts refer to the interacting species, with the charge of the ions being indicated only for H^+X^- to distinguish it from un-ionized HX. We shall modify the previous treatment of Chapter 24 to redefine

$$\mathcal{E}' = \mathcal{E} + \frac{RT}{\mathcal{F}} \ln \frac{m(m - K_a^{St})}{m + K_a^{St}}$$

Substitution of $\mathcal{E}$ into Eq. (24-22) and expansion of $\ln[(m - K_a^{St})(m + m_H)/(m + K_a^{St})(m - m_H)]$ in a power series with neglect of third and higher powers yields

$$\mathcal{E}' = \mathcal{E}^\circ - \frac{RT}{\mathcal{F}} \ln K_a - \frac{RT}{\mathcal{F}} \ln \frac{\gamma_{Cl}\gamma_{HX}}{\gamma_X} + \frac{RT}{\mathcal{F}} \frac{2}{m} (m_H - K_a^{St}) \quad (34\text{-}55)*$$

The last term of Eq. (34-55)* is given by Eq. (34-54)*. If the activity coefficients of Eq. (34-55)* are expressed in terms of the B and k terms of Eqs. (23-38), (34-51)*, and (34-52)*, we obtain the following equations:

[1] E. A. Guggenheim and J. C. Turgeon, *Trans. Faraday Soc.*, **51**, 747 (1955).

$$-\frac{1}{2.303}\ln\frac{\gamma_{\text{Cl}}\gamma_{\text{HX}}}{\gamma_{\text{X}}} = -m(2B_{\text{MCl}} - 2B_{\text{MX}} + 2k_{\text{MCl,HX}} + k_{\text{HX,HX}})$$

$$-m_{\text{H}}(B_{\text{HCl}} - B_{\text{H}^+\text{X}^-} + k_{\text{H}^+\text{X}^-,\text{HX}} + k_{\text{MX,HX}} - k_{\text{MCl,HX}} - k_{\text{HX,HX}}) \quad (34\text{-}56)^*$$

$$\varepsilon' = \varepsilon° - 2.303\frac{RT}{\mathfrak{F}}(\log K_a + bm + cm_{\text{H}}) \quad (34\text{-}57)^*$$

$$b = 2B_{\text{MCl}} - 2B_{\text{MX}} + 2k_{\text{MCl,HX}} + k_{\text{HX,HX}}$$

$$c = 3B_{\text{HCl}} + 4B_{\text{MX}} + 3k_{\text{MCl,HX}} - k_{\text{MX,HX}} - 3k_{\text{HX,HX}} + B_{\text{H}^+\text{X}^-} + 3k_{\text{H}^+\text{X}^-,\text{HX}}$$

If m_{H} is small compared with m, a series of measurements with m values well below 0.1 M, where Eq. (23-38) is a good approximation, can be represented by

$$\varepsilon' = \varepsilon° - \frac{2.303RT}{\mathfrak{F}}(\log K_a + bm) \quad (34\text{-}58)^*$$

One may then evaluate K_a from the intercept of the best straight line through the plot of ε' versus m, or, with sufficient data, one may predict b and the slope of the straight line. Thus measurements at different buffer ratios along with equations corresponding to Eq. (34-57)* will yield values of the interaction coefficients appearing in the b term of Eq. (34-57)*, or values of the B and k terms may be known from freezing-point depressions, solubility, salting out, and other types of measurements.

For NaCl-NaOAc-HOAc buffers, we obtain from Table A4-3

$$B_{\text{NaCl}} - B_{\text{NaOAc}} = \Delta B_{\text{NaCl}} - \Delta B_{\text{NaOAc}} = -0.10$$

From the salting out of HOAc by NaCl as given by the distribution data of Sugden,[1] $k_{\text{HOAc,NaCl}} = 0.06$; also $k_{\text{HOAc,HOAc}} = -0.03$ (see page 303). Thus

$$\varepsilon' = \varepsilon° - \frac{2.303RT}{\mathfrak{F}}(\log K_a - 0.11m) \quad (34\text{-}59)^*$$

from which one can obtain $\log K_a = -4.757$ from the Fig. 34-11 plot of ε' versus m from the data of Harned and Ehlers.[2] Hamer[3] has reviewed the extrapolation of cell data to obtain values of K_a. If one does not fix the value of the b term from the values of B and k given by other experiments and arbitrarily assigns a value of b, as is often done, one obtains a range of K_a values depending upon the choice of b. There need be no such ambiguity in K_a, as b is a quantity which can be experimentally determined.

[1] J. N. Sugden, *J Chem. Soc.*, **128**, 174 (1926).

[2] H. S. Harned and R. W. Ehlers, *J. Am. Chem. Soc.*, **54**, 1350 (1932).

[3] W. J. Hamer, "The Structure of Electrolytic Solutions." p. 236, John Wiley & Sons, Inc., New York, 1959.

However, if one attempts to obtain data at lower values of m or for stronger acids such that the contribution of the m_H terms of Eq. (34-57)* cannot be neglected, an ambiguity in the extrapolation to K_a does arise. Even when the contribution of the m_H term cannot be neglected, m_H is

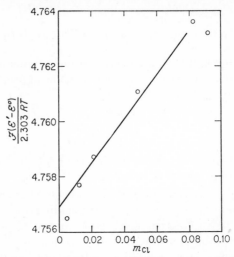

FIG. 34-11. Extrapolation of buffered-cell data to value of K_{HOAc} at 25°C.

very close to K_a and does not vary rapidly with m as long as the ionization of HX is not very extensive. Thus Eq. (34-57)* becomes

$$\mathcal{E}' = \mathcal{E}^\circ - \frac{2.303RT}{\mathfrak{F}} \ (\log K_a + cK_a + bm) \qquad (34\text{-}60)^*$$

When K_a is large enough so that cK_a is appreciable compared with $\log K_a$, the linear plot of $\mathcal{E}'$ versus m will extrapolate to an intercept at zero m which will not yield $\log K_a$ but will yield $\log K_a$ plus a constant. One can evaluate K_a from knowledge of the B and k terms that contribute to c. For acetic acid the cK_a term is approximately 10^{-5} and does not contribute appreciably. Thus the fixing of the b term of Eq. (34-60) is sufficient to allow an unambiguous extrapolation for acetic acid. But for stronger acids the cm_H term becomes much more important. Thus Hamer found that the experimental data gave a range of 0.0056 in $\log K_a$ of HCOOH and a range of 0.549 in $\log K_a$ of HSO_4^- as the b term was arbitrarily varied. If b and c are calculated from the ΔB data of Appendix 4 and from salting-out coefficients, the range of K_a values is greatly reduced. However, there is some remaining ambiguity. The first five B and k terms of the c term of Eq. (34-57)* are experimentally determinable. The last two terms, B_{H^+,X^-} and $k_{H^+X^-,HX}$, cannot be uniquely separated from the H^+X^- interactions which are taken into account by the dissociation constant K_a. Although there are a number of possible choices, we would suggest that the H^+X^- interaction be divided between the dissociation constant of the chemical treatment and the B and k terms of Bronsted's interaction treatment such that $B_{H^+X^-} + 3k_{H^+X^-,HX}$ be taken equal to zero. Thus the c term of Eqs. (34-57)* and (34-60)* reduces to

$$c = 3B_{HCl} + 4B_{MX} - 3k_{MCl,HX} - k_{MX,HX} - 3k_{HX,HX}$$

and can be experimentally determined with no fundamental ambiguity. This procedure uniquely fixes the value of K_a even for relatively strong acids if one has data at low enough molalities to allow neglect of the variation of the B and k terms with molality.

Unbuffered Cells. The cell $H_2|HCl(m_2)$, $A(m_3)|AgCl-Ag$ has been used to study the effect of a wide variety of nonelectrolytes, A, upon the activity coefficient of HCl. Harned and Owen[1] review the data in terms of Eq. (34-61) for $\log \gamma_\pm$ of HCl relative to the pure-water-solute standard state.

$$\log \gamma_{I,m} = \log \gamma_{0,m} + \log \frac{\gamma_{I,m}}{\gamma_{0,m}} \qquad (34\text{-}61)$$

where the first subscript indicates the ionic strength of the solution and the second indicates the molality of nonelectrolyte. The first term is called the total medium effect. The term $\log \gamma_{0,m}$ is called the primary medium effect and represents the effect of the nonelectrolyte upon very low molalities of HCl. The secondary medium effect is given by $\log (\gamma_{I,m}/\gamma_{0,m})$, which corresponds to the activity coefficient of HCl at a given I and m relative to the solute standard state of HCl in the mixed solvent. The primary[1] medium effects at 25°C for 20 weight per cent solutions of methanol and dioxane are $\log \gamma = 0.1148$ and $\log \gamma = 0.1636$, respectively. At low molalities of HCl, the secondary medium effect appears to follow the expected variation of the limiting Debye-Hückel slope with variation of dielectric constant. Harned and Owen review such data in considerable detail.

When the nonelectrolyte is a weak acid such as acetic acid, the treatment of the data is complicated by the ionization of the acid,

$$m_{H^+} \cong m_2 + Q\frac{m_3}{m_2} \qquad (34\text{-}62)^*$$

where Q is the quotient $m_{H^+}m_{X^-}/m_{HX}$ for the given solution and is a function of both I and m_3.

$$\varepsilon = \varepsilon^\circ - \frac{RT}{\mathfrak{F}}\left[\ln m_2\left(m_2 + Q\frac{m_3}{m_2}\right) + 2\ln \gamma_{I,m}\right] \qquad (34\text{-}63)^*$$

[1] Harned and Owen, *op. cit.*

Unless HX is a quite strong acid or m_3/m_2 is large, the terms in Eqs. (34-62)* and (34-63)* are small enough so that Q may be approximated by $K_a/(\gamma^{St})^2$. Substitution of Q and the experimental ε values for the various solutions into Eq. (34-63)* yields values of $\gamma_{I,m}$. As a second approximation corresponding to $\gamma_X = \gamma_{Cl}$, one may use $Q = K_a/\gamma_{I,m}^2$ with

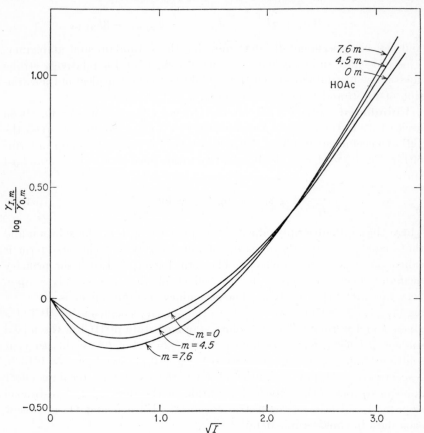

FIG. 34-12. Secondary medium effect for HCl in aqueous HOAc at 25°C.

values of $\gamma_{I,m}^2$ from the first calculation. If the m_H/m_2 is much above unity, further approximations for Q may be necessary taking into account differences between γ_X and γ_{Cl} and also considering γ_{HX}.

Figure 34-12 presents the secondary medium effect, log $(\gamma_{I,m}/\gamma_{0,m})$ for HCl in aqueous acetic acid solutions of molality m at 25°C. The values given have been obtained by recalculation of the data of Owen.[1] The buffered-cell data of Harned and Owen[2] were useful for approximating Q,

[1] B. B. Owen, *J. Am. Chem. Soc.*, **54**, 1759 (1932).

[2] H. S. Harned and B. B. Owen, *J. Am. Chem. Soc.*, **52**, 5079 (1930).

particularly at high HOAc.   The HCl concentrations have been changed
to moles per kilogram of solvent (water and acetic acid).   In Fig. 34-13 is
presented the primary medium effect of HOAc on HCl in solutions up to
pure HOAc.   Molality is not prac-
tical for such a range and the mole
fraction of HOAc is used.   The 4.5
and 7.6 M HOAc solutions of Fig.
34-12 correspond to 0.1 and 0.2 mole
fraction of HOAc in the pure solvent.

   The secondary medium effect
differs little between pure water
and $x_{HOAc} = 0.3$, especially for ionic
strengths above 3 M.   However, the
primary and therefore the total medium effect causes a tenfold increase
in $\gamma_\pm$ of HCl between pure water
and $x_{HOAc} = 0.3$ and a 900-fold increase on going to pure HOAc.   The
data at high I and high x_{HOAc} are
based on vapor-pressure measurements of HCl over HOAc solutions.[1]
These data gave total medium effects
which were joined onto the data of

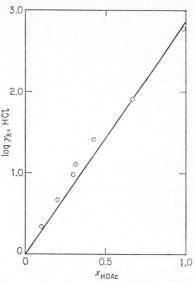

Fig. 34-13. Primary medium effect for HCl in aqueous HOAc at 25°C.

Owen by extrapolation of the small secondary medium effect.   The
straight line of Fig. 34-13 corresponds to the equivalent of Eq. (34-51)*
expressed in mole fraction,

$$\log \gamma_{HCl} = k x_{HOAc} \tag{34-64}*$$

In many instances, as in the present example, this equation holds surprisingly well over a wide range.

PROBLEMS

34-1. Use the data of Appendix 7 for Mn_3C, C, and Mn at 1500°K to calculate the
Mn activity relative to pure Mn in a solution of liquid iron saturated with graphite,
$x_C = 0.173$, and Mn_3C.   Use the k_{Mn_3C} value in Table 34-1, and assume Eq. (34-20)*
and a perfect solution for Fe and Mn to calculate the Mn mole fraction in this solution.   What is the partial pressure of Mn over this solution?   What is the partial
pressure of CO(g) in equilibrium with a mixture of Mn_3C, MnO, and Mn at 1440°K?
What is the H_2O/H_2 ratio in equilibrium with Mn and MnO at 1440°K?

34-2. Duke and Iverson[2] have determined the solubilities of various metal chromates
in fused potassium nitrate–sodium nitrate eutectic mixtures as a function of added

[1] V. Cupr, *Rec. trav. chim.*, **47,** 55 (1928).

[2] F. R. Duke and M. L. Iverson, *J. Phys. Chem.*, **62,** 417 (1958).

sodium chloride. The solubilities of cadmium chromate in the KNO_3-$NaNO_3$ eutectic melt as a function of added sodium chloride are given as follows:

NaCl, m	CdCrO$_4$, $m \times 10^2$	
	250°	300°
0.0	0.59	0.8
0.1	1.2	1.7
0.2	1.7	2.6
0.3	2.3	3.5
0.4	2.9	4.4
0.5	3.5	5.3

Calculate $\gamma_{\pm}$ for cadmium chromate in 0.5 M NaCl. Use Eq. (34-20)* in molality form to calculate values of $k_{CdCrO_4,NaCl}$ between 0.1 and 0.5 M NaCl. Is an equation of the type of Eq. (34-20)* adequate with just a linear term in m?

34-3. When excess silver oxide and silver chloride solids are shaken with a solution of potassium hydroxide and potassium chloride, the hydroxide ion–chloride ion ratio at equilibrium is larger than 100. Use Eqs. (34-29a)* and (34-29b)* to show that $\log (m_{OH}m_{Cl})$ would be expected to decrease linearly with ionic strength below 0.1 M with a slope of $-\tfrac{1}{2}\Delta B_{KOH}$.

34-4. If a series of buffered cells with $m_{NaCl} = 2m_{NaX} = 2m_{HX}$ are prepared for the determination of K_{HX}, show that the b term of Eq. (34-57)* becomes

$$b = 3B_{MCl} - 3B_{MX} + 3k_{MCl,HX} + k_{HX,HX}$$

34-5. Harned and Ehlers[1] report the following data for the cell $H_2(g)|HOAc(m_1)$, $NaOAc(m_2)$, $NaCl(m_3)|AgCl = Ag$ at 20°C, where $\varepsilon° = 0.2255$ volt:

m_1	m_2	m_3	ε
0.004779	0.004599	0.004896	0.63580
0.012035	0.011582	0.012426	0.61241
0.021006	0.020216	0.021516	0.59840
0.04922	0.04737	0.05042	0.57699
0.08101	0.07796	0.08297	0.56456
0.09056	0.08716	0.09276	0.56171

Calculate the hydrogen-ion concentration, the ionic strength, and K^{St} by successive approximations, starting with $K_a = 1.74 \times 10^{-5}$. Calculate

$$\varepsilon' = \varepsilon + \frac{RT}{\mathcal{F}} \ln \frac{m_3(m_1 - K^{St})}{(m_2 + K^{St})}$$

and plot $\varepsilon' - \varepsilon°$ versus m_3. Will this new value of K_a require any recalculation of ε'?

[1] H. S. Harned and R. W. Ehlers, *J. Am. Chem. Soc.*, **54**, 1350 (1932).

35

HYDROGEN AND HELIUM
AT LOW TEMPERATURES

Both hydrogen and helium show special effects at low temperatures which are, for the most part, unique to the various isotopes of these two elements. The underlying cause in all cases is to be found in quantum mechanics, but we shall not pretend to discuss that aspect in any considerable detail.[1] However, it is of interest to note that none of these peculiar phenomena show any disagreement with thermodynamic principles. Here again we are impressed by the general validity of thermodynamics with respect to any system containing many particles.

Hydrogen is also of interest for its general significance, and the low-temperature properties of both elements are important in view of their value as refrigerants.

HYDROGEN

The unique properties of hydrogen arise from the fact that the moment of inertia of the molecule is smaller than that of the other familiar diatomic molecules by a factor of about 10 or more. The rotational energy levels are given by the combination of Eqs. (27-17) and (27-18),

$$\epsilon_J = \frac{h^2 J(J+1)}{8\pi^2 I} \tag{35-1}$$

where J is the quantum number and I the moment of inertia. The successive ϵ_J values are spaced so far apart for hydrogen that this spacing is much larger than kT at low temperatures of practical interest. In addition there are the three isotopes, stable H^1 (or H) and H^2 (or D) and radioactive H^3 (or T), each of which has nuclear spin. Two or more nuclei with spin in the same molecule will have their angular momenta

---

[1] For an elementary discussion of the relevant quantum theory see K. S. Pitzer, "Quantum Chemistry," secs. 9b, 10j, and 11g, Prentice-Hall, Inc., Englewood Cliffs, N.J., 1953; see also C. J. Gorter (ed.), "Progress in Low Temperature Physics," vol. I, chaps. I, II, and V, North Holland Publishing Company, Amsterdam, 1955.

coupled together in a quantized fashion.   The $\frac{1}{2}$-unit spins of H^1 couple parallel or antiparallel.   It is found that only rarely will this coupling be changed unless a magnetic catalyst is present.   Consequently H_2 comprises two species, depending on the relative coupling, and the amount of each species will be relatively permanent and subject to only very slow change in the absence of an appropriate catalyst.

In Chapter 27 we noted that a particular symmetrical diatomic molecule shows only the rotational energy levels corresponding to even values or to odd values of J, but not both.   It is found that the nuclear-spin coupling determines which set of J values is permitted.   For most substances the macroscopic properties of the species with even and with odd J values are indistinguishable, and these characteristics are detectable only in the spectra and in other microscopic properties.   However, for hydrogen we noted that the rotational energy spacing was large, and therefore we expect a substantial difference in various thermodynamic properties for the two species.

Ortho and Para Hydrogen.   With the development of quantum mechanics and the emergence of the ideas just discussed, several investigators sought to show that these two species of hydrogen existed and that their relative amounts could be changed.   In 1928 Giauque and Johnston[1] held gaseous H_2 at 85°K for 6 months and on subsequent liquefaction found a detectable difference in vapor pressure.   Bonhoeffer and Harteck[2] and independently Eucken and Hiller[3] in 1929 successfully catalyzed the conversion and measured the difference in thermal conductivity, heat capacity, etc., of the two species of H_2.   The species with even J values, which has antiparallel nuclear-spin orientation, is called *para*hydrogen, and the other species with parallel spins and odd J values is called *ortho*-hydrogen.   Since the lowest rotational energy level has $J = 0$, the para species becomes the stable form at low temperature.   It is found, however, that the parallel spin coupling is three times as probable as the antiparallel coupling;[4] hence ortho is three times as abundant as para at equilibrium at high temperatures.   This mixture of $\frac{1}{4}$ para and $\frac{3}{4}$ ortho is known as normal hydrogen.

The thermodynamic functions c_P°, s°, $H^\circ - H_0^\circ$, and $(F^\circ - H_0^\circ)/T$ may be calculated[5] for each species of hydrogen by substituting the appropriate

[1] W. F. Giauque and H. L. Johnston, *J. Am. Chem. Soc.*, **50,** 3221 (1928).

[2] K. F. Bonhoeffer and P. Harteck, *Naturwiss.*, **17,** 182 (1929); *Sitzber. preuss. Akad. Wiss.*, **1929,** 103.

[3] A. Eucken and K. Hiller, *Z. physik. Chem.*, **B4,** 142 (1929).

[4] The 3:1 probability ratio arises from the number of nuclear-spin states.   The parallel coupling of two spins, each of $\frac{1}{2}$ unit, yields a net spin of 1, which may have a component special in a given axis of $+1$, 0, or -1 and hence 3 states, whereas a net spin of zero can only have a component of zero.

[5] W. F. Giauque, *J. Am. Chem. Soc.*, **52,** 4816 (1930).

spectroscopic data into the formulas of Chapter 27. At low temperatures it is easier to sum Eq. (27-20) over the rotational states directly than it is to correct adequately the integral form in Eq. (27-21). The resulting heat capacities are shown in Fig. 35-1, which shows also the experimental values of Clusius and Hiller[1] for a sample of about 95 per cent para-H_2. The curve for n-H_2 is just $\frac{3}{4}c_P^{\circ}$(ortho) $+ \frac{1}{4}c_P^{\circ}$(para), and it is well confirmed by experiment.

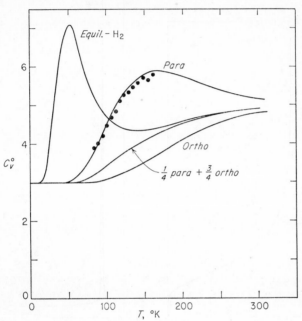

Fig. 35-1. The heat capacity of hydrogen gas. The experimental points shown are for a composition approximately 95 per cent para.

One may also consider the case of H_2 in the presence of an ortho-para conversion catalyst and calculate the properties of an equilibrium mixture by applying the usual formulas for chemical equilibria. In this simple case of two species of the same formula, that is para-H_2 = ortho-H_2, the equilibrium constant is just

$$K = e^{-\Delta \epsilon_0/kT} \frac{Q(\text{ortho})}{Q(\text{para})} \tag{35-2}$$

where $\Delta\epsilon_0 = \epsilon_0(\text{ortho}) - \epsilon_0(\text{para}) = h^2/4\pi^2 I = 337$ cal/mole and the Q's are the partition functions defined by Eq. (27-6). The equilibrium composition is shown in Fig. 35-2.

The heat capacity of the equilibrium mixture includes a term for the heat of conversion multiplied by the change of composition. Likewise,

[1] K. Clusius and K. Hiller, *Z. physik. Chem.*, **B4**, 158 (1929).

the entropy of ortho-para mixtures contains an entropy-of-mixing term. An alternate method is to treat equilibrium H_2 as a single species in Eq. (27-6) and the subsequent formulas of Chapter 27, but with an additional factor of 3 in all g_i values for the odd J states.   The two methods yield the same result, of course, and this is also shown in Fig. 35-1.

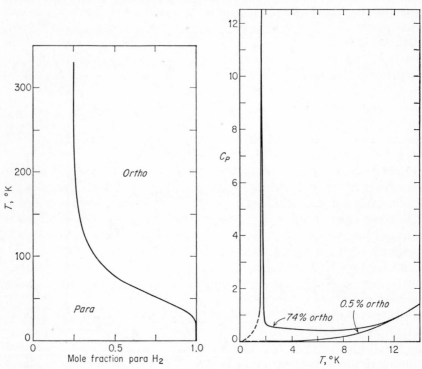

FIG. 35-2. The equilibrium ortho-para composition of hydrogen.

FIG. 35-3. The heat capacity of solid hydrogen.

Various properties of solid, liquid, and gaseous parahydrogen have been measured and compared with the corresponding properties of normal hydrogen.   These results, obtained before 1935, were excellently summarized by Farkas.[1] Woolley, Scott, and Brickwedde[2] reviewed some of these properties in 1948.   We shall not discuss many of these differences, but we do wish to consider the interesting problem of the entropy of solid hydrogen.

[1] A. Farkas, "Orthohydrogen, Parahydrogen, and Heavy Hydrogen," Cambridge University Press, New York, 1935.
[2] H. W. Woolley, R. B. Scott, and F. G. Brickwedde, *J. Research Natl. Bureau Standards*, **41**, 379 (1948).

The heat capacity of solid hydrogen has been measured by several investigators.[1] The individual measurements are quite concordant and yield the curves shown in Fig. 35-3, together with others for intermediate compositions. It is clear that above 11 or 12°K the heat capacity is little affected by the ortho-para composition but that drastic differences appear at lower temperatures. These properties are explained on the basis that the solid at 12 to 14°K consists of freely rotating molecules but that interactions between molecules become significant (as compared with kT) at lower temperatures. At 12°K parahydrogen molecules are almost all in the $J = 0$ rotational state; hence the heat capacity of solid parahydrogen arises only from the linear vibration of the molecules in the lattice. At 12°K the ortho molecules are randomly distributed over nine states which result from the three possible components of the unit nuclear spin and three components of the unit molecular rotational angular momentum on a given axis. The second factor is the usual $2J + 1$ multiplicity of all rotational energy levels and the first factor is the threefold multiplicity of all orthohydrogen states. Hill and Ricketson calculated the extra entropy of the anomalous heat capacity of normal H_2 (in excess of the curve for para-H_2) and obtained a value close to $R \ln 3$ per mole of ortho-H_2. It is presumed that the threefold multiplicity of nuclear-spin orientation remains at 1°K but that the molecular rotational multiplicity has been lost.

In Chapter 12 it was explained that the extrapolation to 0°K of experimental heat-capacity values yields *practical entropy values* which exclude the multiplicity of nuclear-spin orientation (and that of isotope mixing). In the case of the proton with spin $\frac{1}{2}$, the entropy of spin orientation is $R \ln 2$ per proton or $2R \ln 2$ for the two protons in H_2. Consequently, the practical entropy of hydrogen for use in combination with third-law values for other substances is $2R \ln 2 = 2.75$ cal/deg mole less than that calculated from the spectroscopic data with consideration of the ortho and para species. While the low-temperature thermal data on solid hydrogen do not yield as precise an entropy for the gas as that calculated from spectroscopic data, nevertheless it is interesting to consider the results obtained by extrapolation of the heat capacity of the solid to 0°K. An extrapolation from 12°K would seem very plausible for normal H_2 if one were to ignore the data at lower temperatures; the extrapolation would follow close to the curve for para-H_2. This yields an entropy which is smaller than the correct practical value by approximately $\frac{3}{4}R \ln 3 = 1.54$ cal/deg mole. But if one considers the measurements

[1] F. Simon and E. Lange, *Z. Physik*, **15**, 312 (1923); Clusius and Hiller, *loc. cit.*; F. Simon, K. Mendelssohn, and M. Ruhemann, *Naturwiss.*, **18**, 34 (1930); *Z. physik. Chem.*, **B15**, 121 (1931); R. W. Hill and B. W. A. Ricketson, *Phil. Mag.*, (7)**45**, 277 (1954).

down to 1°K and extrapolates from that point, the correct practical entropy is obtained.   The remaining entropy of nuclear-spin orientation, $2R \ln 2$, may be classified into the entropy of mixing of the ortho and para species, given by Eq. (18-8)* for an ideal solution,

$$\Delta S_{\text{mix}} = -R(\tfrac{1}{4} \ln \tfrac{1}{4} + \tfrac{3}{4} \ln \tfrac{3}{4})$$

together with the nuclear-spin entropy of the ortho species, which is $\tfrac{3}{4}R \ln 3$ per mole of normal H_2.   The sum of these two terms is $2R \ln 2$.

Heavy Hydrogen Species.   While the abundance of deuterium in natural hydrogen is too low appreciably to affect its properties, it is possible to separate deuterium and in addition to prepare relatively pure gases of HD and of D_2 molecules.   More recently the third isotope of hydrogen has been prepared in significant amounts, but its radioactivity would hamper low-temperature studies of its properties.   The deuteron D (that is, H^2) has unit nuclear spin, while T (or H^3) has $\tfrac{1}{2}$-unit spin like the proton.

The properties of D_2 may be calculated by methods analogous to those used for H_2 provided that appropriate adjustment is made for the different mass and spin of the deuteron.   There are ortho and para species as before.   The situation for T_2 is even more similar to that for H_2 since only the mass is different.

The unsymmetrical molecule HD presents some new aspects, and it is interesting to consider these briefly.   In this case all J values are allowed in the rotational energy-level pattern.   For "practical" calculations recognizing different isotopes one wishes an entropy which still excludes nuclear-spin effects, and this is readily obtained from the equations of Chapter 27 by simply ignoring nuclear spin at all points.   The thermal properties of HD at low temperatures were measured by Clusius, Popp, and Frank.[1]   An extrapolation of the heat capacity of solid HD to 0°K in the usual fashion also yields this practical entropy value.

The zero entropy state (excluding nuclear spin) of HD at low temperatures might be thought to imply that the H and D atoms had formed an ordered array, but this seems unlikely.   The properties of the various species of H_2 are best understood on the assumption that the molecules rotate freely in the solid at 11 to 14°K.   In this case HD will be in its lowest rotational state which is the single $J = 0$ state.   Consequently an extrapolation from this range, as made by Clusius, Popp, and Frank, implies free rotation even at 0°K but no rotational entropy because all the molecules are in a single quantum state.

Crystals of other diatomic molecules have properties indicating that rotation has ceased at these temperatures.   Therefore, a molecule such as $Cl^{35}Cl^{37}$ has a residual

---

[1] K. Clusius, L. Popp, and A. Frank, *Physica*, **4,** 1105 (1937).

entropy $R \ln 2$ because of random end-for-end orientation like that found for CO and N_2O (see Chapter 12).   It would be interesting to study solid HD under high pressure, where the rotation might be suppressed.

HELIUM

Helium is a relatively unexciting substance at ordinary temperatures, but below 2.2°K its behavior provides many surprises.   We shall do little more than mention some of these unique properties; other books[1] and reviews must be consulted for details.   The abundant isotope of helium, He^4, shows relatively normal behavior down to 2.18°K, where it undergoes a transition to a second liquid state known as He II.   There is no latent heat of transition—only a region of high heat capacity.   He II, although still mechanically fluid, approaches zero entropy as the temperature approaches 0°K.   The flow and heat-transport properties of He II are quite abnormal, but we shall not consider these.

The rare isotope of helium, He^3, which has recently become available from the radioactive decay of artificially produced H^3, also remains liquid to exceedingly low temperatures.   Whereas He^4 has no nuclear spin, He^3 has $\frac{1}{2}$ unit of spin and a spin entropy of

$$R \ln 2 = 1.38 \text{ cal/deg mole}$$

at any ordinary temperature.   The total entropy[2] of liquid He^3 is only about 0.3 cal/deg mole at the lowest temperature of measurement, 0.1°K; hence much of the nuclear-spin entropy has already been lost.   Extrapolation of the heat-capacity curve for liquid He^3 to 0°K yields an entropy value agreeing with the calculated value for the vapor.   But the difference should be remembered that for other substances the nuclear-spin entropy is not included in the value obtained by integration of the experimental heat capacity together with extrapolation from 0.1 to 0.0°K, whereas for He^3 the nuclear-spin entropy is included.

Both He^3 and He^4 may be solidified under pressure of approximately 25 to 30 atm at 0.5°K.   The solid-liquid diagram in Fig. 35-4 shows that the equilibrium pressure of He^4 remains nearly constant up to 2°K and then rises sharply, but the pressure for He^3 rises steadily above 0.5°K.

The determination of temperature in the region below about 3°K is an interesting thermodynamic problem since gas thermometry becomes very

[1] W. H. Keesom, "Helium," Elsevier Press, Inc., Houston, Tex., 1942; also chapters on helium in the volumes edited by C. J. Gorter (ed.), "Progress in Low Temperature Physics," North Holland Publishing Company, Amsterdam, vol. I, 1955, vol. II, 1957.

[2] D. F. Brewer, A. K. Speedhar, H. C. Kramers, and J. G. Daunt, *Phys. Rev.*, **110**, 282 (1958).

difficult.   The use of the third law in connection with the measured heat
capacity, heat of vaporization, and vapor pressure of liquid helium provides an interesting basis for temperature determination.   A thermodynamic vapor-pressure equation is calculated from the other data and is

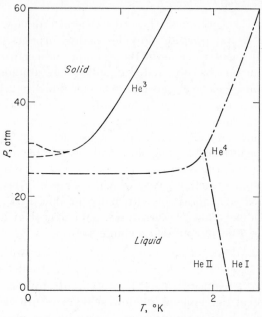

FIG. 35-4. The solid-liquid phase diagram for He³ (solid and broken lines) and for He⁴
(dot-dash lines).   The He I to II transition in liquid He⁴ is also shown.

then used to determine temperature from vapor-pressure measurements.
The vapor-liquid equilibrium requires, of course, $\Delta F = 0$ or that

$$\ln f = \frac{F(\text{sat l})}{RT} - \frac{F^{\circ}(g)}{RT} \qquad (35\text{-}3)$$

where f, the fugacity, can be obtained from the vapor pressure by making
corrections for gas imperfection as indicated in Chapter 16.   If we take
the heat content of the liquid at 0°K and zero pressure $H_0(l)$ as our reference point, then the heat content of the gas is greater by the heat of
vaporization, which we shall indicate by ΔH_0° at 0°K.   The standard free
energy of the helium gas is given by Eq. (27-3), which, in these terms,
becomes

$$\frac{F^{\circ}(g)}{RT} = \frac{\Delta H_0^{\circ}}{RT} - \frac{5}{2}\ln T - i \qquad (35\text{-}4)$$

where the constant i (sometimes called the *chemical constant*) has the
value $\frac{3}{2}\ln M - (7.283/R) = -1.584$ for He⁴.

The free energy of the saturated liquid is obtained by integrating the equation

$$d\text{F} = -\text{s}\, dT + \text{v}\, dP \tag{10-12}$$

from 0°K, where $\text{F}_0 = \text{H}_0(\text{l})$ to the temperature at which the vapor pressure is measured. Combination of these equations yields at T'' and P'

$$\ln f_{T'} = \frac{-\Delta\text{H}_0^\circ}{RT''} + \frac{5}{2}\ln T' - \frac{1}{RT}\left(\int_0^{T'} \text{s}_l\, dT - \int_0^{P'} \text{v}_l\, dP\right) + i \tag{35-5}$$

The entropy of the liquid s_l is obtained by integration of the experimental heat-capacity curve, and the volume of the liquid v_l is measured also. The remaining constant ΔH_0° may be obtained by measuring the heat of vaporization at some temperature and calculating the correction to 0°K or by fitting the vapor-pressure curve at a relatively high temperature which is known by gas thermometry.

This temperature determination is, in practice, not independent of other methods but rather an important supplement. If gas imperfection can be measured, as is necessary near 4°K, then gas thermometry is also quite feasible. At lower temperatures the vapor pressure decreases rapidly, and gas thermometry becomes more difficult, but the vapor approaches perfect-gas behavior. At still lower temperatures, below 1°K, the vapor pressure becomes too small to measure accurately, and the magnetic properties of certain materials provide the basis of thermometry, as described in Chapter 31. The entropy of the liquid in Eq. (35-5) is obtained by integration from 0°K to the temperature of interest, and this calculation requires the use of the temperature scale below 1°K as well as a self-consistent use of the temperature on the helium vapor-pressure scale in the region above 1°K.

The same method could be applied to He³ or, for that matter, to any other substance. In the case of He³ it would be necessary to include nuclear-spin entropy consistently for both gas and liquid.

PROBLEM

35-1. Calculate all the energy levels of H_2 within 5000 cm⁻¹ above the lowest level. Indicate the multiplicity g_i of each level; remember that all states of ortho-H_2, that is, with odd J, have triple multiplicity from nuclear spin. Obtain Q and Q' at 500°K by direct summation of Eqs. (27-6) and (27-11). Calculate s° and $-(\text{F}° - \text{H}_0°)/T$ at 500°K; remember to subtract $R \ln 4$ to eliminate the nuclear-spin entropy, which is excluded on the practical scale.

Appendix 1

PROPERTIES OF NORMAL FLUIDS

In Chapters 10 and 16, thermodynamic principles are applied to fluids deviating from ideal-gas behavior. Also, the continuity of gas and liquid states is noted. While the properties of many substances have been measured over broad ranges of pressure and temperature, far more substances have not been studied and a system of prediction is to be desired. The very brief discussion of predictive methods in Chapter 16 is extended here. Molecular theory is used to guide forms of equations in some cases, but the final results are based upon the observed properties of numerous substances. It is impractical to repeat here the numerous comparisons with observed data—for that the reader must refer to the original papers cited below.

The prediction of the volumetric and thermodynamic properties of pure fluids has been the subject of many studies since 1860. Although the underlying principles in terms of intermolecular forces are now well understood, the calculation or even the quantitative empirical representation of the resulting macroscopic properties has proved to be unusually difficult. The failure to find a really satisfactory analytical equation of state arises, in the authors' opinion, not from any lack of skill in selecting the appropriate combination of common mathematical functions, but rather from the nonexistence of any suitable function. Consequently, it is best, in effect, to let nature generate mathematical functions and to develop an equation in which the minimum number of such functions are combined with other precisely defined parameters to yield the desired volumetric and thermodynamic properties for a broad class of pure substances.

A correlation based on these ideas has been developed and presented in full in two series of papers.[1,2] Statistical theory[3] shows that a group of substances will conform to the principle of corresponding states only if their intermolecular potentials are identical except for distance and energy-scale factors characteristic of each substance. Also, their intermolecular motion must be classical; i.e., quantum effects must be negligible. The only group of substances which may be expected to conform to these criteria are the heavier rare gases Ar, Kr, and Xe, which do conform accurately to corresponding-states behavior. These are called *simple fluids*. Various types of molecular shapes and molecular dipole moments might be expected to cause different deviations from the macroscopic properties of simple fluids. It is found, however,

[1] L. Riedel, *Chem.-Ing. Tech.*, **26**, 83, 259, 679 (1954), **27**, 209, 475 (1955), **28**, 557 (1956).

[2] K. S. Pitzer, D. Z. Lippmann, R. F. Curl, Jr., C. M. Huggins, and D. E. Petersen, *J. Am. Chem. Soc.*, **77**, 3427, 3433 (1955), **79**, 2369 (1957); *Ind. Eng. Chem.*, **50**, 265 (1958).

[3] K. S. Pitzer, *J. Chem. Phys.*, **7**, 583 (1939).

that the reduced theoretical second virial coefficients for a wide variety of these molecular types fall into a single family of curves which may be characterized by a single parameter. The only exception noted is that of molecules with large dipole moments, although there are unusual types of abnormality that have not been tested. The molecules falling into this single family are just those commonly called normal liquids or fluids. Normal fluids are defined more precisely below.

Acentric Factor. The theory thus suggests an extension of the corresponding-states correlation involving a third parameter.[1] The slope of the reduced vapor-pressure curve is the most sensitive property upon which to base the third parameter, and it has the additional advantage that vapor pressures are readily measured with high accuracy. An arbitrary but convenient definition[2] is based upon the reduced vapor pressure at a point well removed from the critical point and takes the form

$$\omega = -\log\frac{P_s}{P_c} - 1.000 \tag{A1-1}$$

where P_s = vapor pressure at $T_r = 0.700$

The form is chosen to make $\omega = 0$ for the simple fluids Ar, Kr, and Xe, with simple spherical molecules. Other normal fluids have small positive values of ω. The name *acentric factor* was adopted to indicate that the factor measures the deviation of the intermolecular potential function from that of the simple spherical molecules.

Any property of the fluid, in reduced or dimensionless form, is assumed to be given by a function of the three variables: reduced pressure, reduced temperature, and acentric factor. For example, the compressibility factor z may be written

$$z = z(P_r, T_r, \omega) \tag{A1-2}$$

It is found that a linear equation in ω is usually adequate.

$$z = z^{(0)}(P_r, T_r) + \omega z^{(1)}(P_r, T_r) \tag{A1-3}$$

Analytical representation of the functions $z^{(0)}$ and $z^{(1)}$, however, is usually impractical, and they are tabulated in most cases.

Vapor Pressure, Heat and Entropy of Vaporization. Where two phases exist in equilibrium, the vapor pressure and other properties are given as functions of two variables, T_r and ω.

$$\log P_r = (\log P_r)^{(0)} + \omega \left(\frac{\partial \log P_r}{\partial \omega}\right)_T \tag{A1-4}$$

These two functions are tabulated in Table A1-1. From the definition of the acentric factor $(\log P_r)^{(0)} = -1.000$, and $(\partial \log P_r / \partial \omega)_T = -1.000$ at $T_r = 0.7$.

Also included in Table A1-1 are values of the compressibility-factor functions for the vapor and for the liquid. These are to be used in Eq. (A1-3). The values for z for the liquid are not very accurate at the lower temperatures, where they are very small; a better source of volumetric data for liquids is Table A1-16. By combination of the compressibility-factor data with the temperature derivative of the vapor

---

[1] A third parameter was first suggested by Nernst in 1907 and by various authors since. See papers by Riedel, *op. cit.*, and by Pitzer et al., *op. cit.*, for references and for the advantages of the present system over others.

[2] The parameter α_k chosen by Riedel, *op. cit.*, is equivalent but is different in detail of definition. $\alpha_k = 5.808 + 4.93\omega$.

TABLE A1-1. DATA FOR COEXISTING LIQUID AND VAPOR PHASES

T_r	$-(\log P_r)^{(0)}$	$-\left(\dfrac{\partial \log P_r}{\partial \omega}\right)_T$	Vaporization†		Vapor		Liquid	
			$\Delta S^{(0)}$	$\Delta S^{(1)}$	$z^{(0)}$	$z^{(1)}$	$z^{(0)}$	$z^{(1)}$
1.00	0.000	0.000	0.00	0.00	0.291	−0.080	0.291	−0.080
0.99	0.025	0.021	2.57	2.83	0.43	−0.030	0.202	−0.090
0.98	0.050	0.042	3.38	3.91	0.47	0.000	0.179	−0.093
0.97	0.076	0.064	4.00	4.72	0.51	+0.020	0.162	−0.095
0.96	0.102	0.086	4.52	5.39	0.54	0.035	0.148	−0.085
0.95	0.129	0.109	5.00	5.96	0.565	0.045	0.136	−0.095
0.94	0.156	0.133	5.44	6.51	0.59	0.055	0.125	−0.094
0.92	0.212	0.180	6.23	7.54	0.63	0.075	0.108	−0.092
0.90	0.270	0.230	6.95	8.53	0.67	0.095	0.0925	−0.087
0.88	0.330	0.285	7.58	9.39	0.70	0.110	0.0790	−0.080
0.86	0.391	0.345	8.19	10.3	0.73	0.125	0.0680	−0.075
0.84	0.455	0.405	8.79	11.2	0.756	0.135	0.0585	−0.068
0.82	0.522	0.475	9.37	12.1	0.781	0.140	0.0498	−0.062
0.80	0.592	0.545	9.97	13.0	0.804	0.144	0.0422	−0.057
0.78	0.665	0.620	10.57	13.9	0.826	0.144	0.0360	−0.053
0.76	0.742	0.705	11.20	14.9	0.846	0.142	0.0300	−0.048
0.74	0.823	0.800	11.84	16.0	0.864	0.137	0.0250	−0.043
0.72	0.909	0.895	12.49	17.0	0.881	0.131	0.0210	−0.037
0.70	1.000	1.00	13.19	18.1	0.897	0.122	0.0172	−0.032
0.68	1.096	1.12	13.89	19.3	0.911	0.113	0.0138	−0.027
0.66	1.198	1.25	14.62	20.5	0.922	0.104	0.0111	−0.022
0.64	1.308	1.39	15.36	21.8	0.932	0.097	0.0088	−0.018
0.62	1.426	1.54	16.12	23.2	0.940	0.090	0.0068	−0.015
0.60	1.552	1.70	16.92	24.6	0.947	0.083	0.0052	−0.012
0.58	1.688	1.88	17.74	26.2	0.953	0.077	0.0039	−0.009
0.56	1.834	2.08	18.64	27.8	0.959	0.070	0.0028	−0.007

† Strictly there is a small $\Delta S^{(2)}$ which is almost always negligible.   See K. S. Pitzer et al., *J. Am. Chem. Soc.*, **77**, 3439 (1955).

pressure, the heat and entropy of vaporization may be calculated.   The Clapeyron equation rearranges into

$$\Delta S = \frac{R \, \Delta z}{T_r} \frac{\partial \ln P}{\partial (1/T)_r} \tag{A1-5}$$

where Δz is the change in compressibility factor on vaporization.   The values of $\Delta S^{(0)}$ and $\Delta S^{(1)}$ in Table A1-1 come from this source.   The heat of vaporization is, of course, $T \, \Delta S$.

The full vapor-pressure functions of Table A1-1 make it possible to evaluate the acentric factor from any vapor-pressure value at a temperature well below the critical point.

608   *Thermodynamics*

Second Virial Coefficient. In Chapter 16 the Berthelot equation for the second virial coefficient was discussed, and its limited accuracy was apparent from Fig. 16-1. The acentric factor makes a substantially improved equation possible. A series of inverse powers of T_r is a convenient form. The constants were adjusted[1] to fit both volumetric and thermal data on numerous normal gases.

$$Pv = RT + {}_BP + \cdots$$
$$\frac{{}_BP_c}{RT_c} = (0.1445 + 0.073\omega) - (0.330 - 0.46\omega)T_r^{-1} - (0.1385 + 0.50\omega)T_r^{-2}$$
$$- (0.0121 + 0.097\omega)T_r^{-3} - 0.0073\omega T_r^{-8} \quad (A1\text{-}6)^*$$

The selection of the term in T_r^{-8} is, of course, somewhat arbitrary, but a very high power is required to fit the low-temperature thermal data. Figure A1-1 compares the curves of Eq. (A1-6)* for a few ω values with experimental data for appropriate substances.

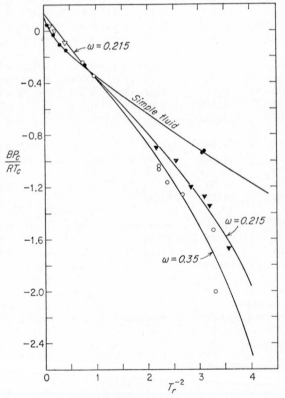

FIG. A1-1. A comparison of calculated and experimental values of the second virial coefficient. Solid circles are Ar, Kr, or Xe points, and the line through these points is that of Eq. (A1-6)* with $\omega = 0$. The open triangles are CO_2, and solid triangles are benzene points. These are to be compared with the line for $\omega = 0.215$. The open circles are points for *n*-heptane ($\omega = 0.35$). [*K. S. Pitzer and R. F. Curl, Jr., J. Am. Chem. Soc.*, **79**, 2369 (1957), *fig.* 1.]

[1] K. S. Pitzer and R. F. Curl, Jr., *J. Am. Chem. Soc.*, **79**, 2369 (1957).

Volumetric Data at High Pressure.   The volumetric data for a number of substances were interpolated to a series of even reduced pressures and temperatures and plotted against the acentric factor.   Figure A1-2 shows a sample of these plots at $P_r = 3.0$ and a series of reduced temperatures.   It is seen that the linear dependence on acentric factor which was assumed in Eq. (A1-3) is a good approximation.

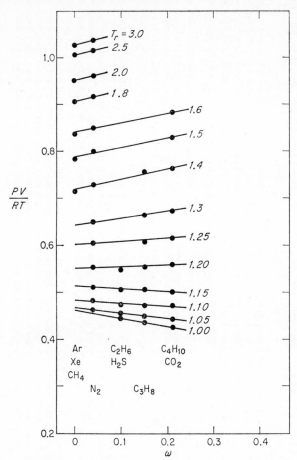

FIG. A1-2. The compressibility factor as a function of the acentric factor for $P_r = 3.0$ and the indicated values of T_r.   The substances are indicated below the columns of points.   [*K. S. Pitzer, D. Z. Lippmann, R. F. Curl, Jr., C. M. Huggins, and D. E. Petersen, J. Am. Chem. Soc.*, **77**, 3433 (1955), *fig. 2.*]

The intercepts on the $\omega = 0$ axis constitute the function $z^{(0)}$ and the slopes the function $z^{(1)}$.   The values from the plots were checked for smoothness of variation with P_r and T_r and adjusted within experimental error if necessary.   The resulting functions are given in Tables A1-2 and A1-5.   Additional values near the two-phase region and the critical point are given in Tables A1-3, A1-4, A1-6, and A1-7.

Fugacity, Enthalpy, and Entropy.   The fugacity and other thermodynamic functions may be calculated from the equation of state.   For the low-pressure region,

TABLE A1-2. VALUES OF $z^{(0)}$ FOR COMPRESSIBILITY-FACTOR CALCULATION†

P_r

T_r	0.2	0.4	0.6	0.8	1.0	1.2	1.4	1.6	1.8	2.0	2.2	2.4	2.6	2.8	3.0	3.2	3.4	3.6	3.8	4.0	4.5	5.0	6.0	7.0	8.0	9.0
0.80	0.851	0.066	0.100	0.133	0.164	0.192	0.225	0.258	0.287	0.318	0.347	0.376	0.405	0.433	0.461	0.490	0.519	0.547	0.576	0.605	0.675	0.746	0.883	1.017	1.15	1.28
0.85	0.882	0.067	0.101	0.134	0.165	0.194	0.226	0.258	0.287	0.316	0.345	0.374	0.403	0.431	0.459	0.487	0.515	0.542	0.569	0.597	0.663	0.730	0.861	0.990	1.115	1.24
0.90	0.904	0.778	0.102	0.135	0.167	0.198	0.229	0.258	0.288	0.316	0.345	0.373	0.402	0.430	0.458	0.485	0.512	0.538	0.565	0.591	0.655	0.718	0.842	0.966	1.089	1.21
0.95	0.920	0.819	0.697	0.145	0.176	0.205	0.235	0.262	0.292	0.321	0.347	0.375	0.403	0.430	0.457	0.484	0.510	0.536	0.561	0.587	0.647	0.709	0.828	0.947	1.066	1.185
1.00	0.932	0.849	0.756	0.638	0.291	0.231	0.250	0.278	0.304	0.329	0.356	0.381	0.407	0.433	0.458	0.484	0.509	0.534	0.557	0.582	0.642	0.702	0.819	0.932	1.048	1.166
1.05	0.942	0.874	0.800	0.714	0.609	0.470	0.341	0.320	0.332	0.350	0.372	0.393	0.417	0.441	0.466	0.489	0.512	0.535	0.557	0.580	0.639	0.700	0.814	0.923	1.032	1.147
1.10	0.950	0.893	0.833	0.767	0.691	0.607	0.512	0.442	0.408	0.402	0.405	0.420	0.440	0.462	0.484	0.504	0.525	0.547	0.567	0.589	0.643	0.699	0.810	0.916	1.019	1.129
1.15	0.958	0.908	0.858	0.805	0.746	0.684	0.620	0.562	0.514	0.484	0.477	0.478	0.485	0.498	0.513	0.529	0.546	0.563	0.581	0.600	0.651	0.705	0.809	0.911	1.008	1.113
1.20	0.963	0.921	0.879	0.835	0.788	0.737	0.690	0.640	0.598	0.568	0.553	0.545	0.544	0.548	0.554	0.563	0.574	0.587	0.601	0.618	0.664	0.714	0.810	0.907	1.000	1.100
1.25	0.968	0.930	0.896	0.858	0.820	0.778	0.740	0.702	0.664	0.636	0.618	0.606	0.599	0.597	0.598	0.602	0.609	0.618	0.629	0.643	0.682	0.726	0.816	0.907	0.994	1.088
1.30	0.971	0.940	0.909	0.878	0.846	0.811	0.780	0.749	0.718	0.691	0.671	0.657	0.649	0.644	0.642	0.642	0.645	0.651	0.659	0.668	0.701	0.740	0.824	0.910	0.992	1.078
1.4	0.977	0.952	0.929	0.908	0.883	0.859	0.838	0.817	0.795	0.777	0.759	0.745	0.734	0.725	0.720	0.718	0.718	0.722	0.727	0.734	0.754	0.781	0.844	0.921	0.994	1.071
1.5	0.982	0.963	0.945	0.927	0.909	0.892	0.875	0.859	0.844	0.831	0.819	0.808	0.800	0.794	0.790	0.785	0.784	0.784	0.786	0.790	0.805	0.826	0.877	0.934	1.000	1.070
1.6	0.985	0.971	0.957	0.944	0.930	0.917	0.904	0.893	0.882	0.872	0.863	0.855	0.848	0.843	0.840	0.836	0.834	0.833	0.834	0.835	0.844	0.860	0.904	0.953	1.010	1.075
1.7	0.988	0.977	0.966	0.956	0.946	0.936	0.926	0.919	0.911	0.903	0.896	0.889	0.883	0.879	0.875	0.873	0.872	0.872	0.873	0.874	0.882	0.895	0.930	0.972	1.023	1.082
1.8	0.991	0.982	0.974	0.966	0.958	0.950	0.944	0.937	0.931	0.926	0.921	0.916	0.913	0.910	0.908	0.907	0.906	0.906	0.907	0.908	0.914	0.925	0.955	0.993	1.039	1.091
1.9	0.993	0.986	0.980	0.974	0.968	0.962	0.958	0.952	0.948	0.944	0.940	0.936	0.933	0.931	0.930	0.929	0.929	0.930	0.932	0.934	0.941	0.950	0.976	1.010	1.051	1.097
2.0	0.995	0.989	0.984	0.979	0.975	0.971	0.968	0.964	0.961	0.959	0.956	0.953	0.953	0.953	0.952	0.952	0.953	0.954	0.954	0.956	0.962	0.972	0.996	1.027	1.064	1.106
2.5	1.000	0.999	0.999	0.998	0.998	0.998	0.998	0.997	0.999	1.000	1.001	1.001	1.002	1.004	1.006	1.008	1.009	1.012	1.014	1.018	1.026	1.035	1.055	1.079	1.105	1.136
3.0	1.001	1.002	1.003	1.004	1.005	1.007	1.008	1.010	1.012	1.014	1.016	1.019	1.022	1.025	1.028	1.030	1.033	1.036	1.038	1.041	1.049	1.058	1.077	1.10	1.124	1.150
3.5	1.002	1.004	1.006	1.008	1.011	1.013	1.015	1.018	1.020	1.022	1.024	1.027	1.030	1.033	1.036	1.039	1.042	1.045	1.048	1.051	1.058	1.067	1.086	1.105	1.126	1.148
4.0	1.003	1.005	1.008	1.010	1.013	1.015	1.017	1.020	1.022	1.024	1.026	1.029	1.032	1.035	1.038	1.041	1.044	1.047	1.050	1.053	1.060	1.068	1.086	1.104	1.124	1.143

† K. S. Pitzer, D. Z. Lippmann, R. F. Curl, Jr., C. M. Huggins, and D. E. Petersen, *J. Am. Chem. Soc.*, **77**, 3427 (1955), table II. See Tables A1-3 and A1-4 for additional data in the region enclosed by dashed lines.

TABLE A1-3. VALUES OF $z^{(0)}$ NEAR THE TWO-PHASE REGION†

T_r	P_r						
	0.4	0.5	0.6	0.7	0.8	0.9	1.0
0.90	0.778	0.701	0.102	0.118	0.135	0.151	0.167
0.91	0.787	0.715	0.104	0.120	0.136	0.152	0.168
0.92	0.796	0.728	0.650	0.122	0.138	0.153	0.169
0.93	0.805	0.740	0.666	0.124	0.140	0.155	0.170
0.94	0.812	0.751	0.681	0.125	0.142	0.157	0.173
0.95	0.819	0.762	0.697	0.612	0.145	0.160	0.176
0.96	0.826	0.772	0.711	0.632	0.149	0.164	0.180
0.97	0.832	0.782	0.724	0.652	0.56	0.170	0.186
0.98	0.838	0.791	0.735	0.669	0.591	0.177	0.193
0.99	0.844	0.800	0.746	0.685	0.616	0.514	0.205
1.00	0.849	0.807	0.757	0.699	0.638	0.554	0.291
1.01	0.854	0.813	0.767	0.713	0.654	0.583	0.476
1.02	0.860	0.820	0.776	0.726	0.672	0.608	0.525
1.03	0.865	0.826	0.784	0.737	0.687	0.630	0.558
1.04	0.870	0.833	0.793	0.748	0.701	0.648	0.586
1 05	0.874	0.838	0.800	0.758	0.714	0.665	0.609

† K. S. Pitzer, D. Z. Lippmann, R. F. Curl, Jr., C. M. Huggins, and D. E. Petersen, *J. Am. Chem. Soc.*, **77**, 3427 (1955), table III*A*.

where the second virial coefficient gives an adequate measure of the deviation from the ideal-gas state, Eq. (A1-6)* may be substituted into the appropriate thermodynamic equations. In these equations H°, s°, c_P° refer to the properties of the standard state, i.e., the hypothetical ideal gas at 1 atm.

$$\log \frac{f}{P} = \frac{P_r}{2.303} [(0.1445 + 0.073\omega)T_r^{-1} - (0.330 - 0.46\omega)T_r^{-2}$$
$$- (0.1385 + 0.50\omega)T_r^{-3} - (0.0121 + 0.097\omega)T_r^{-4}$$
$$- 0.0073\omega T_r^{-9}] \quad (A1\text{-}7)$$

$$\frac{\mathrm{H} - \mathrm{H}^\circ}{RT_c} = P_r[(0.1445 + 0.073\omega) - (0.660 - 0.92\omega)T_r^{-1} - (0.4155 + 1.50\omega)T_r^{-2}$$
$$- (0.0484 + 0.388\omega)T_r^{-3} - 0.0657\omega T_r^{-8}] \quad (A1\text{-}8)$$

$$\frac{\mathrm{s} - \mathrm{s}^\circ}{R} = -\ln P + P_r[-(0.330 - 0.46\omega)T_r^{-2} - (0.2770 + 1.00\omega)T_r^{-3}$$
$$- (0.0363 + 0.29\omega)T_r^{-4} - 0.0584T_r^{-9}] \quad (A1\text{-}9)$$

$$\frac{c_P - c_P^\circ}{R} = P_r[(0.660 - 0.92\omega)T_r^{-2} + (0.831 + 3.0\omega)T_r^{-3}$$
$$+ (0.145 + 1.16\omega)T_r^{-4} + 0.526\omega T_r^{-9}] \quad (A1\text{-}10)$$

These equations are valuable in calculating gas-imperfection corrections for gases at low pressures. Figure A1-3 compares a few data for gas heat capacity with the curves of Eq. (A1-10). Since the heat capacity involves the second derivative of the volume, this is a very sensitive test.

T_r	P_r										
	1.0	1.1	1.2	1.3	1.4	1.5	1.6	1.7	1.8	1.9	2.0
0.98	0.193	0.204	0.217	0.230	0.244	0.257	0.270	0.284	0.298	0.313	0.324
0.99	0.205	0.210	0.223	0.235	0.247	0.260	0.273	0.287	0.301	0.315	0.327
1.00	0.291	0.220	0.231	0.241	0.250	0.265	0.278	0.290	0.304	0.317	0.329
1.01	0.476	0.283	0.243	0.248	0.259	0.271	0.283	0.294	0.307	0.319	0.332
1.02	0.525	0.402	0.273	0.260	0.270	0.278	0.291	0.300	0.311	0.323	0.335
1.03	0.558	0.466	0.34	0.29	0.283	0.288	0.297	0.306	0.316	0.328	0.339
1.04	0.586	0.509	0.41	0.33	0.307	0.302	0.307	0.314	0.324	0.334	0.343
1.05	0.609	0.543	0.470	0.375	0.341	0.324	0.320	0.323	0.332	0.341	0.350
1.06	0.628	0.572	0.505	0.423	0.370	0.349	0.336	0.333	0.343	0.348	0.358
1.07	0.645	0.597	0.534	0.468	0.408	0.379	0.358	0.349	0.356	0.358	0.367
1.08	0.663	0.618	0.562	0.504	0.445	0.412	0.385	0.373	0.370	0.369	0.375
1.09	0.677	0.636	0.587	0.535	0.480	0.443	0.412	0.396	0.387	0.383	0.387
1.10	0.691	0.652	0.607	0.561	0.512	0.473	0.442	0.422	0.408	0.400	0.402
1.11	0.703	0.667	0.625	0.584	0.538	0.502	0.469	0.448	0.428	0.418	0.417
1.13	0.726	0.693	0.658	0.621	0.584	0.549	0.520	0.494	0.472	0.456	0.450
1.15	0.746	0.715	0.684	0.652	0.620	0.589	0.562	0.536	0.514	0.495	0.484

† K. S. Pitzer, D. Z. Lippmann, R. F. Curl, Jr., C. M. Huggins, and D. E. Petersen, *J. Am. Chem. Soc.*, **77**, 3427 (1955), table III*B*.

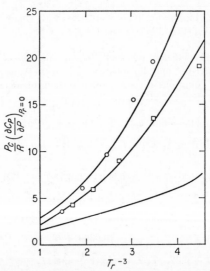

FIG. A1-3. A comparison of calculated and experimental values of the pressure derivative of the heat capacity [see Eq. (A1-10)]. The circles and upper curve are for *n*-heptane ($\omega = 0.35$). The squares and intermediate curve are for benzene ($\omega = 0.215$). The bottom curve is for $\omega = 0$, but no experimental data are available for comparison. [*K. S. Pitzer and R. F. Curl, Jr., J. Am. Chem. Soc.*, **79**, 2369 (1957), *fig.* 2.]

TABLE A1-5. VALUES OF $z^{(1)}$ FOR COMPRESSIBILITY-FACTOR CALCULATION†

T_r	P_r																				
	0.2	0.4	0.6	0.8	1.0	1.2	1.4	1.6	1.8	2.0	2.2	2.4	2.6	2.8	3.0	4.0	5.0	6.0	7.0	8.0	9.0
0.80	-0.095	-0.028	-0.044	-0.058	-0.07	-0.08	-0.10	-0.11	-0.12	-0.13	-0.14	-0.15	-0.16	-0.17	-0.18	-0.23	-0.26	-0.29	-0.32	-0.35	-0.37
0.85	-0.067	-0.031	-0.049	-0.064	-0.08	-0.09	-0.11	-0.12	-0.13	-0.14	-0.15	-0.16	-0.17	-0.18	-0.18	-0.22	-0.25	-0.28	-0.31	-0.34	-0.36
0.90	-0.042	-0.09	-0.053	-0.068	-0.085	-0.10	-0.11	-0.12	-0.13	-0.14	-0.15	-0.16	-0.17	-0.17	-0.18	-0.21	-0.24	-0.27	-0.30	-0.32	-0.35
0.95	-0.025	-0.050	-0.10	-0.072	-0.091	-0.10	-0.11	-0.12	-0.12	-0.13	-0.14	-0.15	-0.15	-0.16	-0.17	-0.20	-0.22	-0.25	-0.28	-0.31	-0.34
1.00	-0.012	-0.016	-0.020	-0.05	-0.080	-0.090	-0.099	-0.108	-0.115	-0.123	-0.13	-0.13	-0.14	-0.14	-0.15	-0.17	-0.20	-0.23	-0.26	-0.30	-0.33
1.05	0.000	+0.001	+0.005	+0.015	+0.02	+0.01	-0.01	-0.04	-0.06	-0.07	-0.08	-0.09	-0.10	-0.10	-0.11	-0.14	-0.17	-0.20	-0.24	-0.28	-0.31
1.10	+0.002	0.008	0.016	0.030	0.055	0.082	+0.11	+0.082	+0.035	0.000	-0.02	-0.03	-0.05	-0.06	-0.07	-0.10	-0.13	-0.16	-0.21	-0.25	-0.28
1.15	0.004	0.012	0.012	0.040	0.064	0.093	0.12	0.140	0.136	+0.100	+0.07	+0.04	+0.02	0.00	-0.01	-0.04	-0.08	-0.12	-0.16	-0.20	-0.24
1.20	0.009	0.018	0.028	0.044	0.069	0.10	0.13	0.16	0.17	0.17	0.16	0.14	0.12	+0.09	+0.07	0.00	-0.04	-0.08	-0.12	-0.16	-0.19
1.25	0.011	0.023	0.036	0.050	0.069	0.10	0.13	0.16	0.18	0.19	0.19	0.18	0.16	0.14	0.12	+0.05	0.00	-0.03	-0.07	-0.11	-0.13
1.30	0.013	0.027	0.041	0.055	0.072	0.10	0.13	0.16	0.18	0.20	0.20	0.20	0.20	0.19	0.18	0.10	+0.04	0.00	-0.04	-0.07	-0.09
1.4	0.016	0.032	0.049	0.065	0.082	0.10	0.13	0.16	0.18	0.19	0.20	0.21	0.21	0.21	0.20	0.15	0.11	+0.07	+0.04	+0.01	-0.01
1.5	0.017	0.035	0.052	0.070	0.088	0.10	0.13	0.15	0.17	0.18	0.20	0.20	0.21	0.21	0.21	0.20	0.17	0.14	0.11	0.09	+0.07
1.6	0.018	0.036	0.054	0.07	0.08	0.10	0.12	0.14	0.16	0.17	0.18	0.19	0.20	0.20	0.20	0.22	0.21	0.19	0.17	0.15	0.14
1.7	0.018	0.036	0.054	0.07	0.09	0.10	0.11	0.13	0.15	0.16	0.17	0.18	0.19	0.20	0.21	0.24	0.25	0.26	0.25	0.24	0.22
1.8	0.018	0.036	0.054	0.07	0.09	0.10	0.11	0.13	0.15	0.16	0.17	0.18	0.19	0.20	0.21	0.26	0.29	0.31	0.32	0.32	0.30
1.9	0.018	0.035	0.05	0.07	0.09	0.10	0.11	0.13	0.15	0.16	0.17	0.18	0.19	0.20	0.21	0.26	0.30	0.35	0.38	0.40	0.40
2.0	0.016	0.031	0.05	0.07	0.08	0.10	0.11	0.13	0.14	0.15	0.16	0.17	0.19	0.20	0.20	0.26	0.30	0.35	0.40	0.43	0.45
2.5	0.01	0.02	0.04	0.05	0.07	0.08	0.10	0.11	0.10	0.13	0.15	0.16	0.18	0.19	0.20	0.25	0.30	0.35	0.40	0.45	0.50
3.0	0.01	0.02	0.03	0.05	0.06	0.07	0.08	0.09	0.10	0.11	0.13	0.14	0.15	0.16	0.17	0.23	0.28	0.34	0.38	0.45	0.50
3.5	0.01	0.02	0.03	0.04	0.05	0.06	0.07	0.08	0.08	0.09	0.10	0.11	0.12	0.13	0.14	0.19	0.24	0.28	0.33	0.38	0.42
4.0	0.01	0.02	0.02	0.03	0.04	0.05	0.06	0.06	0.07	0.08	0.09	0.10	0.10	0.11	0.12	0.16	0.20	0.23	0.27	0.31	0.35

† K. S. Pitzer, D. Z. Lippmann, R. F. Curl, Jr., C. M. Huggins, and D. E. Petersen, *J. Am. Chem. Soc.*, 77, 3427 (1955), table IV.
See Tables A1-6 and A1-7 for additional data in the region enclosed by dashed lines.

TABLE A1-6. VALUES OF $z^{(1)}$ NEAR THE TWO-PHASE REGION[†]

T_r	P_r			
	0.4	0.6	0.8	1.0
0.90	−0.09	−0.053	−0.068	−0.085
0.91	−0.08	−0.053	−0.069	−0.087
0.92	−0.072	−0.18	−0.070	−0.089
0.93	−0.066	−0.15	−0.071	−0.090
0.94	−0.058	−0.12	−0.072	−0.091
0.95	−0.050	−0.10	−0.072	−0.091
0.96	−0.042	−0.08	−0.072	−0.091
0.97	−0.035	−0.065	−0.14	−0.091
0.98	−0.027	−0.050	−0.11	−0.090
0.99	−0.021	−0.033	−0.08	−0.087
1.00	−0.016	−0.020	−0.05	−0.080
1.01	−0.012	−0.012	−0.02	−0.02
1.02	−0.008	−0.006	0.00	−0.01
1.03	−0.005	−0.001	+0.005	0.00
1.04	−0.002	+0.002	+0.010	+0.01
1.05	+0.001	+0.005	+0.015	+0.02

[†] K. S. Pitzer, D. Z. Lippmann, R. F. Curl, Jr., C. M. Huggins, and D. E. Petersen, *J. Am. Chem. Soc.*, **77,** 3427 (1955), table V*A*.

At higher pressures the second virial coefficient does not suffice, and we must use the general equation of state defined in Eq. (A1-3) and the tables. The thermodynamic properties are obtained from the usual equations. In terms of the compressibility factor they are

$$\ln \frac{f}{P} = \int_0^{P_r'} \frac{z-1}{P_r} dP_r \tag{A1-11}$$

$$\frac{H^\circ - H}{RT_c} = T_r^2 \int_0^{P_r'} \left(\frac{1}{P_r}\right)\left(\frac{\partial z}{\partial T_r}\right) dP_r \tag{A1-12}$$

$$\frac{S^\circ - S}{R} = \frac{1}{T_r} \frac{H^\circ - H}{RT_c} + \ln \frac{f}{P} + \ln P \tag{A1-13}$$

In Eqs. (A1-11) and (A1-12) the resulting fugacity and enthalpy pertain to the reduced pressure P_r' which is the upper limit of integration. The superscript °, as usual, denotes the ideal-gas standard state. The integration and differentiation must be carried out by graphical and numerical methods, but this causes no fundamental difficulty. A more important feature arises when the path of integration crosses the condensation curve. As ω increases, the condensation occurs at a lower pressure for the same temperature. As a result there arises a special contribution to $[\ln (f/P)]^{(1)}$

TABLE A1-7. VALUES OF $z^{(1)}$ IN THE CRITICAL REGION[†]

T_r	P_r					
	1.0	1.2	1.4	1.6	1.8	2.0
0.98	−0.090	−0.099	−0.109	−0.118	−0.125	−0.130
0.99	−0.087	−0.095	−0.104	−0.114	−0.121	−0.127
1.00	−0.080	−0.090	−0.099	−0.108	−0.115	−0.123
1.01	−0.02	−0.080	−0.091	−0.102	−0.10	−0.100
1.02	−0.01	−0.065	−0.082	−0.095	−0.09	−0.09
1.03	0.00	−0.047	−0.068	−0.085	−0.08	−0.09
1.04	+0.01	−0.025	−0.050	−0.073	−0.07	−0.08
1.05	+0.02	+0.01	−0.01	−0.04	−0.06	−0.07
1.06	+0.03	+0.06	+0.07	−0.02	−0.05	−0.073
1.07	+0.04	+0.08	+0.09	0.000	−0.038	−0.059
1.08	+0.047	+0.08	+0.10	+0.030	−0.015	−0.041
1.09	+0.050	+0.08	+0.11	+0.056	+0.012	−0.022
1.10	+0.055	+0.082	+0.11	+0.082	+0.035	0.000
1.11	+0.057	+0.085	+0.12	+0.099	+0.062	+0.020
1.13	+0.062	+0.089	+0.12	+0.123	+0.105	+0.060
1.15	+0.064	+0.093	+0.122	+0.140	+0.136	+0.100

[†] K. S. Pitzer, D. Z. Lippmann, R. F. Curl, Jr., C. M. Huggins, and D. E. Petersen, *J. Am. Chem. Soc.*, **77**, 3427 (1955), table V*B*.

as follows:

$$\Delta \left(\ln \frac{f}{P} \right)^{(1)} = \left(\left\{ \frac{\partial [\ln (f/P)]^{(0)}}{\partial P_r} \right\}_{\text{gas}} - \left\{ \frac{\partial [\ln (f/P)]^{(0)}}{\partial P_r} \right\}_{\text{liq}} \right) \left(\frac{\partial P_r}{\partial \omega} \right)_{T,\text{sat}} \quad \text{(A1-14)}$$

Corresponding contributions arise for the enthalpy and entropy.

The resulting functions are given in Tables A1-8 to A1-15. Because of the lower vapor pressure for liquids of high acentric factor at a given temperature, it is sometimes necessary to extrapolate the functions on the liquid side. This extrapolation should be taken with respect to pressure at constant temperature. The complete equations, which have the form of Eq. (A1-3), are as follows:

$$\log \frac{f}{P} = \left(\log \frac{f}{P} \right)^{(0)} + \omega \left(\log \frac{f}{P} \right)^{(1)} \quad \text{(A1-15)}$$

$$\frac{\text{H}^\circ - \text{H}}{RT_c} = \left(\frac{\text{H}^\circ - \text{H}}{RT_c} \right)^{(0)} + \omega \left(\frac{\text{H}^\circ - \text{H}}{RT_c} \right)^{(1)} \quad \text{(A1-16)}$$

$$\frac{\text{s}^\circ - \text{s}}{R} = \left(\frac{\text{s}^\circ - \text{s}}{R} \right)^{(0)} + \omega \left(\frac{\text{s}^\circ - \text{s}}{R} \right)^{(1)} + R \ln P \quad \text{(A1-17)}$$

Figure A1-4 shows a comparison of the calculated enthalpy difference from the ideal gas with experimental values from Joule-Thomson expansion and other methods.

Tables A1-11 and A1-13 give more closely spaced values for the enthalpy functions in the critical region.   Simple graphs are still desirable for accurate interpolation here. The fugacity functions are interpolated easily in the critical region.   The entropy is, of course, related to the fugacity and enthalpy.   General entropy tables are given for convenience, but in the critical region the entropy should be calculated from the detailed enthalpy tables and the fugacity.

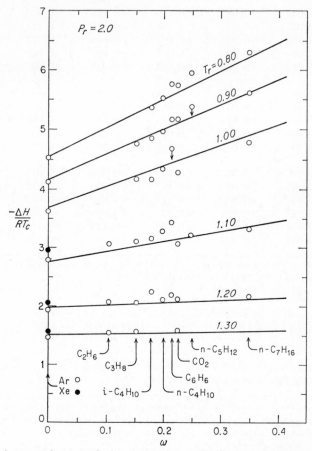

Fig. A1-4. A comparison of calculated and observed values of the enthalpy difference from the ideal gas at the same pressure and temperature.   All values are for reduced pressure of 2.0 and at the reduced temperatures indicated.   The acentric factor is a constant for a particular substance; consequently the points lie in vertical columns as labeled.   [R. F. Curl, Jr., and K. S. Pitzer, Ind. Eng. Chem., **50,** 265 (1958), fig. 3.]

Internal-energy and work-content values can be obtained by appropriate combination of the functions tabulated.   Also temperature derivatives of the enthalpy will yield rough heat-capacity values.

Liquid Density.   This completes the results for the gas phase and the compressed liquid.   The properties of the liquid at low pressure are especially accessible to measurement and have particular interest.   Strictly, we consider the liquid under its own

TABLE A1-8. VALUES OF [LOG (f/P)][(a)]

T_r	P_r												
	0.2	0.4	0.6	0.8	1.0	1.2	1.4	1.6	1.8	2.0	2.2	2.4	2.6
0.80	-0.060	-0.262	-0.425	-0.535	-0.618	-0.683	-0.736	-0.780	-0.817	-0.849	-0.877	-0.901	-0.922
0.85	-0.046	-0.120	-0.281	-0.392	-0.474	-0.539	-0.592	-0.636	-0.673	-0.705	-0.733	-0.757	-0.779
0.90	-0.042₅	-0.087₅	-0.163	-0.273	-0.356	-0.421	-0.474	-0.517	-0.554	-0.587	-0.614	-0.639	-0.680
0.95	-0.033	-0.070	-0.112	-0.173	-0.255	-0.319	-0.372	-0.415	-0.452	-0.483	-0.511	-0.535	-0.557
1.00	-0.028	-0.059	-0.094	-0.131	-0.175	-0.237	-0.287	-0.330	-0.367	-0.398	-0.425	-0.449	-0.470
1.05	-0.024	-0.051	-0.079	-0.109	-0.142	-0.178	-0.218	-0.257	-0.292	-0.322	-0.349	-0.372	-0.393
1.10	-0.021	-0.044	-0.067	-0.093	-0.120	-0.147	-0.177	-0.207	-0.237	-0.264	-0.289	-0.311	-0.331
1.15	-0.018	-0.037	-0.058	-0.079	-0.101	-0.123	-0.146	-0.170	-0.194	-0.217	-0.238	-0.258	-0.276
1.20	-0.016	-0.032	-0.050	-0.067	-0.086	-0.104	-0.124	-0.143	-0.163	-0.182	-0.200	-0.217	-0.233
1.25	-0.014	-0.029	-0.044	-0.059	-0.075	-0.091	-0.107	-0.123	-0.139	-0.155	-0.171	-0.186	-0.199
1.3	-0.012	-0.025	-0.038	-0.051	-0.065	-0.078	-0.092	-0.106	-0.119	-0.133	-0.146	-0.159	-0.171
1.4	-0.010	-0.021	-0.031	-0.041	-0.052	-0.062	-0.072	-0.082	-0.092	-0.102	-0.111	-0.120	-0.130
1.5	-0.008	-0.016	-0.024	-0.032	-0.040	-0.047	-0.055	-0.063	-0.070	-0.078	-0.085	-0.092	-0.099
1.6	-0.007	-0.013	-0.019	-0.026	-0.032	-0.038	-0.044	-0.050	-0.056	-0.062	-0.067	-0.072	-0.077
1.7	-0.005	-0.010	-0.015	-0.020	-0.025	-0.030	-0.034	-0.039	-0.043	-0.047	-0.051	-0.056	-0.059
1.8	-0.004	-0.008	-0.012	-0.015	-0.019	-0.022	-0.026	-0.030	-0.033	-0.036	-0.039	-0.042	-0.045
1.9	-0.003	-0.006	-0.009	-0.012	-0.015	-0.018	-0.020	-0.023	-0.025	-0.028	-0.030	-0.033	-0.035
2.0	-0.002	-0.004	-0.007	-0.009	-0.011	-0.013	-0.015	-0.017	-0.019	-0.021	-0.023	-0.025	-0.026
2.5	0.000	0.000	0.000	0.000	-0.001	-0.001	-0.001	-0.001	-0.001	-0.001	-0.001	-0.001	-0.001
3.0	0.000	+0.001	+0.001	+0.002	+0.002	+0.003	+0.003	+0.004	+0.004	+0.005	+0.005	+0.006	+0.007
3.5	+0.001	0.002	0.003	0.003	0.004	0.005	0.006	0.007	0.008	0.009	0.010	0.011	0.012
4.0	0.001	0.002	0.003	0.005	0.006	0.007	0.008	0.009	0.010	0.011	0.012	0.013	0.014

TABLE A1-8. VALUES OF $[\text{Log } (f/P)]^{(0)}$ *(Continued)*

P_r

T_r	2.8	3.0	3.2	3.4	3.6	3.8	4.0	4.5	5.0	6.0	7.0	8.0	9.0
0.80	-0.941	-0.957	-0.972	-0.985	-0.997	-1.007	-1.016	-1.035	-1.048	-1.064	-1.067	-1.063	-1.052
0.85	-0.797	-0.814	-0.829	-0.842	-0.854	-0.864	-0.874	-0.893	-0.907	-0.924	-0.929	-0.926	-0.917
0.90	-0.679	-0.696	-0.710	-0.724	-0.736	-0.746	-0.756	-0.775	-0.789	-0.807	-0.814	-0.813	-0.805
0.95	-0.575	-0.592	-0.607	-0.621	-0.632	-0.643	-0.652	-0.672	-0.687	-0.706	-0.713	-0.713	-0.707
1.00	-0.489	-0.505	-0.520	-0.534	-0.545	-0.556	-0.566	-0.586	-0.601	-0.620	-0.629	-0.630	-0.624
1.05	-0.411	-0.428	-0.442	-0.455	-0.467	-0.478	-0.488	-0.508	-0.523	-0.543	-0.552	-0.553	-0.549
1.10	-0.348	-0.364	-0.378	-0.391	-0.403	-0.413	-0.422	-0.442	-0.457	-0.477	-0.487	-0.489	-0.486
1.15	-0.293	-0.307	-0.321	-0.333	-0.344	-0.354	-0.363	-0.383	-0.397	-0.417	-0.427	-0.429	-0.426
1.20	-0.247	-0.261	-0.273	-0.285	-0.295	-0.305	-0.314	-0.332	-0.346	-0.366	-0.375	-0.378	-0.376
1.25	-0.212	-0.224	-0.236	-0.246	-0.256	-0.264	-0.273	-0.290	-0.304	-0.322	-0.331	-0.334	-0.332
1.3	-0.182	-0.193	-0.203	-0.212	-0.221	-0.229	-0.237	-0.253	-0.266	-0.283	-0.292	-0.295	-0.294
1.4	-0.138	-0.146	-0.154	-0.162	-0.169	-0.175	-0.181	-0.194	-0.205	-0.220	-0.228	-0.231	-0.229
1.5	-0.104	-0.112	-0.117	-0.124	-0.129	-0.134	-0.139	-0.149	-0.158	-0.170	-0.176	-0.178	-0.176
1.6	-0.082	-0.087	-0.092	-0.096	-0.100	-0.104	-0.108	-0.116	-0.123	-0.132	-0.137	-0.138	-0.136
1.7	-0.063	-0.067	-0.071	-0.074	-0.077	-0.080	-0.083	-0.089	-0.094	-0.101	-0.105	-0.105	-0.102
1.8	-0.048	-0.051	-0.053	-0.056	-0.058	-0.060	-0.063	-0.067	-0.071	-0.076	-0.078	-0.077	-0.074
1.9	-0.037	-0.039	-0.041	-0.043	-0.045	-0.046	-0.048	-0.051	-0.054	-0.057	-0.057	-0.055	-0.051
2.0	-0.028	-0.029	-0.031	-0.032	-0.033	-0.034	-0.035	-0.037	-0.039	-0.040	-0.039	-0.038	-0.034
2.5	-0.001	-0.001	-0.001	-0.001	0.000	0.000	0.000	+0.001	+0.003	+0.006	+0.011	+0.016	+0.022
3.0	+0.007	+0.008	+0.009	+0.010	+0.011	+0.012	+0.012	0.015	0.017	0.023	0.028	0.035	0.042
3.5	0.013	0.014	0.015	0.016	0.017	0.018	0.020	0.022	0.025	0.031	0.038	0.044	0.051
4.0	0.015	0.016	0.017	0.019	0.020	0.021	0.022	0.025	0.028	0.034	0.040	0.047	0.054

P_r

T_r	0.2	0.4	0.6	0.8	1.0	1.2	1.4	1.6	1.8	2.0	2.2	2.4	2.6	2.8	3.0	4.0	5.0	6.0	7.0	8.0	9.0
0.80	−0.04	−0.47	−0.48	−0.48	−0.48	−0.49	−0.50	−0.50	−0.51	−0.51	−0.52	−0.52	−0.53	−0.53	−0.54	−0.56	−0.59	−0.61	−0.63	−0.65	−0.67
0.85	−0.03	−0.31	−0.31	−0.32	−0.33	−0.33	−0.34	−0.35	−0.35	−0.36	−0.37	−0.37	−0.38	−0.38	−0.39	−0.41	−0.44	−0.46	−0.48	−0.50	−0.51
0.90	−0.02	−0.04	−0.18	−0.20	−0.20	−0.21	−0.21	−0.22	−0.23	−0.23	−0.24	−0.24	−0.25	−0.26	−0.26	−0.29	−0.31	−0.33	−0.35	−0.36	−0.38
0.95	−0.01	−0.02	−0.03	−0.09	−0.10	−0.11	−0.12	−0.12	−0.13	−0.13	−0.14	−0.15	−0.15	−0.16	−0.16	−0.18	−0.20	−0.22	−0.24	−0.26	−0.27
1.00	−0.01	−0.01	−0.01	−0.02	−0.03	−0.03	−0.04	−0.05	−0.05	−0.06	−0.06	−0.07	−0.07	−0.08	−0.08	−0.10	−0.12	−0.13	−0.15	−0.17	−0.18
1.05	0.00	0.00	0.00	0.00	+0.01	+0.01	+0.01	+0.01	0.00	0.00	0.00	0.00	−0.01	−0.01	−0.01	−0.03	−0.05	−0.06	−0.07	−0.09	−0.11
1.10	0.00	0.00	0.00	+0.01	0.01	0.02	0.02	0.03	+0.03	+0.03	+0.03	+0.03	+0.03	+0.03	+0.03	+0.02	0.00	−0.01	−0.02	−0.03	−0.05
1.15	0.00	0.00	0.00	0.01	0.02	0.02	0.03	0.04	0.04	0.05	0.05	0.05	0.06	0.06	0.06	0.05	+0.05	+0.04	+0.02	+0.01	0.00
1.20	0.00	+0.01	+0.01	0.01	0.02	0.03	0.04	0.05	0.05	0.06	0.07	0.07	0.08	0.08	0.08	0.09	0.09	+0.08	+0.07	+0.07	+0.06
1.25	0.00	0.01	0.01	0.02	0.03	0.03	0.04	0.05	0.06	0.07	0.07	0.08	0.09	0.09	0.10	0.11	0.11	0.11	0.10	0.10	0.09
1.3	+0.01	0.01	0.02	0.03	0.03	0.04	0.04	0.05	0.06	0.07	0.08	0.08	0.09	0.10	0.10	0.12	0.13	0.13	0.13	0.12	0.12
1.4	0.01	0.01	0.02	0.03	0.04	0.04	0.05	0.06	0.07	0.08	0.08	0.09	0.10	0.11	0.11	0.13	0.15	0.15	0.16	0.16	0.16
1.5	0.01	0.02	0.02	0.03	0.04	0.05	0.05	0.06	0.06	0.07	0.08	0.08	0.09	0.10	0.11	0.13	0.15	0.16	0.17	0.17	0.18
1.6	0.01	0.02	0.02	0.02	0.03	0.04	0.05	0.05	0.06	0.07	0.08	0.08	0.09	0.10	0.11	0.14	0.16	0.18	0.19	0.20	0.21
1.7	0.01	0.02	0.02	0.03	0.04	0.05	0.05	0.05	0.06	0.07	0.08	0.08	0.09	0.10	0.11	0.14	0.16	0.18	0.20	0.21	0.23
1.8	0.01	0.02	0.02	0.03	0.04	0.05	0.05	0.06	0.06	0.07	0.08	0.08	0.09	0.10	0.11	0.14	0.16	0.19	0.21	0.23	0.24
1.9	0.01	0.02	0.02	0.03	0.04	0.05	0.05	0.06	0.06	0.07	0.08	0.09	0.09	0.10	0.11	0.14	0.16	0.19	0.21	0.23	0.25
2.0	0.01	0.01	0.02	0.03	0.04	0.05	0.05	0.06	0.07	0.07	0.08	0.09	0.09	0.09	0.11	0.13	0.16	0.19	0.21	0.23	0.26
2.5	0.01	0.01	0.02	0.02	0.03	0.04	0.04	0.05	0.06	0.06	0.07	0.07	0.08	0.08	0.09	0.12	0.14	0.17	0.19	0.22	0.24
3.0	0.00	0.01	0.01	0.02	0.02	0.03	0.04	0.04	0.05	0.05	0.05	0.06	0.06	0.07	0.07	0.10	0.12	0.15	0.17	0.20	0.22
3.5	0.00	0.01	0.01	0.02	0.02	0.02	0.03	0.03	0.04	0.04	0.04	0.04	0.05	0.06	0.06	0.08	0.10	0.13	0.15	0.17	0.19
4.0	0.00	0.01	0.01	0.02	0.02	0.02	0.02	0.03	0.03	0.03	0.04	0.04	0.04	0.05	0.05	0.07	0.09	0.10	0.12	0.14	0.15

TABLE A1-10. VALUES OF $\left(\dfrac{H^\circ - H}{RT_c}\right)^{(0)}$

P_r

T_r	0.2	0.4	0.6	0.8	1.0	1.2	1.4	1.6	1.8	2.0	2.2	2.4	2.6	2.8	3.0	3.2	3.4	3.6	3.8	4.0	4.5	5.0	6.0	7.0	8.0	9.0
0.80	0.37	4.52	4.52	4.52	4.52	4.52	4.52	4.52	4.52	4.52	4.52	4.52	4.52	4.52	4.52	4.51	4.51	4.50	4.50	4.49	4.46	4.43	4.38	4.34	4.29	4.24
0.85	0.32	4.35	4.35	4.35	4.34	4.34	4.33	4.32	4.32	4.32	4.32	4.32	4.32	4.32	4.32	4.32	4.31	4.31	4.31	4.30	4.28	4.24	4.20	4.16	4.10	4.04
0.90	0.27	0.60	4.06	4.10	4.14	4.14	4.15	4.15	4.14	4.13	4.13	4.12	4.12	4.12	4.11	4.10	4.10	4.10	4.10	4.10	4.09	4.05	3.99	3.93	3.92	3.85
0.95	0.23	0.52	0.86	3.69	3.80	3.85	3.87	3.88	3.89	3.90	3.90	3.90	3.90	3.90	3.90	3.90	3.90	3.89	3.89	3.89	3.85	3.85	3.84	3.81	3.76	3.71
1.00	0.21	0.45	0.76	1.15	2.3₅	3.09	3.32	3.44	3.52	3.57	3.60	3.63	3.65	3.68	3.70	3.71	3.71	3.70	3.70	3.70	3.69	3.68	3.67	3.64	3.63	3.57
1.05	0.19	0.40	0.64	0.95	1.35	1.94	2.54	2.86	3.07	3.21	3.30	3.36	3.39	3.42	3.43	3.46	3.47	3.49	3.50	3.51	3.51	3.50	3.49	3.48	3.46	3.45
1.10	0.17	0.36	0.57	0.82	1.10	1.44	1.83	2.25	2.55	2.75	2.89	3.00	3.08	3.15	3.20	3.24	3.27	3.29	3.30	3.32	3.34	3.34	3.34	3.34	3.33	3.32
1.15	0.14	0.30	0.49	0.70	0.93	1.19	1.48	1.78	2.07	2.33	2.52	2.67	2.78	2.86	2.93	2.98	3.02	3.05	3.09	3.12	3.17	3.18	3.18	3.19	3.19	3.20
1.20	0.13	0.27	0.44	0.63	0.83	1.03	1.25	1.49	1.73	1.95	2.13	2.30	2.44	2.56	2.66	2.73	2.78	2.82	2.87	2.91	2.99	3.02	3.05	3.07	3.07	3.08
1.25	0.12	0.25	0.39	0.56	0.73	0.91	1.09	1.29	1.50	1.70	1.87	2.03	2.17	2.29	2.39	2.48	2.55	2.61	2.67	2.72	2.82	2.87	2.92	2.92	2.95	2.98
1.3	0.11	0.23	0.36	0.50	0.66	0.81	0.97	1.14	1.32	1.49	1.64	1.79	1.93	2.05	2.16	2.24	2.32	2.39	2.45	2.52	2.63	2.72	2.79	2.81	2.84	2.88
1.4	0.09	0.19	0.31	0.42	0.54	0.67	0.80	0.94	1.08	1.23	1.36	1.47	1.59	1.70	1.79	1.88	1.96	2.04	2.11	2.18	2.32	2.42	2.53	2.58	2.62	2.65
1.5	0.09	0.18	0.29	0.39	0.49	0.59	0.70	0.80	0.93	1.04	1.15	1.26	1.36	1.45	1.53	1.61	1.68	1.75	1.82	1.88	2.01	2.12	2.25	2.33	2.39	2.41
1.6	0.09	0.18	0.27	0.36	0.45	0.54	0.62	0.71	0.81	0.91	1.00	1.09	1.18	1.26	1.33	1.39	1.44	1.51	1.57	1.64	1.78	1.87	2.01	2.12	2.19	2.21
1.7	0.08	0.16	0.25	0.33	0.41	0.48	0.56	0.64	0.71	0.80	0.87	0.95	1.02	1.08	1.15	1.21	1.27	1.33	1.39	1.45	1.57	1.66	1.79	1.91	2.00	2.03
1.8	0.07	0.15	0.23	0.30	0.37	0.44	0.51	0.58	0.64	0.71	0.78	0.84	0.90	0.95	1.01	1.06	1.12	1.18	1.24	1.31	1.42	1.50	1.62	1.74	1.83	1.86
1.9	0.06	0.13	0.19	0.26	0.33	0.40	0.46	0.51	0.57	0.63	0.68	0.73	0.78	0.82	0.87	0.92	0.97	1.02	1.08	1.13	1.23	1.32	1.44	1.54	1.63	1.67
2.0	0.06	0.12	0.18	0.24	0.30	0.36	0.42	0.46	0.51	0.55	0.59	0.64	0.69	0.74	0.78	0.82	0.85	0.89	0.93	0.97	1.07	1.14	1.25	1.33	1.43	1.46
2.5	0.04	0.08	0.12	0.16	0.19	0.22	0.25	0.28	0.31	0.34	0.37	0.40	0.43	0.45	0.47	0.50	0.52	0.54	0.56	0.58	0.63	0.67	0.75	0.81	0.87	0.91
3.0	0.03	0.05	0.07	0.09	0.11	0.14	0.16	0.18	0.20	0.22	0.24	0.26	0.28	0.30	0.31	0.33	0.34	0.36	0.37	0.38	0.41	0.44	0.50	0.53	0.56	0.58
3.5	0.02	0.04	0.05	0.06	0.07	0.09	0.10	0.11	0.12	0.13	0.15	0.16	0.17	0.18	0.19	0.20	0.21	0.22	0.23	0.23	0.25	0.26	0.28	0.30	0.30	0.29
4.0	0.01	0.02	0.03	0.04	0.04	0.05	0.06	0.07	0.08	0.08	0.09	0.10	0.10	0.11	0.11	0.11	0.12	0.12	0.12	0.13	0.13	0.14	0.14	0.14	0.13	0.12

TABLE A1-11. VALUES OF $\left(\dfrac{H^\circ - H}{RT_c}\right)^{(0)}$ IN THE CRITICAL REGION

T_r	P_r												
	0.8	0.9	1.0	1.1	1.2	1.3	1.4	1.5	1.6	1.7	1.8	1.9	2.0
0.98	1.25	3.2	3.3₅	3.43	3.50	3.55	3.59	3.63	3.65	3.67	3.68	3.70	3.71
0.99	1.19	1.6	3.1	3.20	3.31	3.39	3.46	3.51	3.55	3.58	3.61	3.63	3.64
1.00	1.15	1.5	2.3₅	2.9	3.09	3.23	3.32	3.39	3.44	3.48	3.52	3.54	3.57
1.01	1.11	1.38	1.8	2.5	2.86	3.05	3.18	3.27	3.34	3.39	3.44	3.47	3.50
1.02	1.07	1.31	1.6	2.1	2.59	2.85	3.02	3.14	3.23	3.30	3.36	3.40	3.44
1.03	1.03	1.24	1.52	1.88	2.33	2.63	2.87	3.02	3.12	3.21	3.27	3.32	3.36
1.04	0.99	1.18	1.43	1.74	2.12	2.47	2.71	2.87	2.99	3.09	3.17	3.23	3.28
1.05	0.95	1.13	1.35	1.62	1.94	2.27	2.54	2.73	2.86	2.98	3.07	3.14	3.21
1.06	0.92	1.08	1.29	1.53	1.80	2.09	2.37	2.58	2.74	2.88	2.98	3.06	3.13
1.07	0.89	1.04	1.23	1.45	1.69	1.96	2.22	2.43	2.61	2.76	2.88	2.97	3.04
1.08	0.87	1.02	1.18	1.37	1.59	1.83	2.07	2.29	2.49	2.64	2.77	2.86	2.95
1.09	0.84	0.99	1.14	1.33	1.51	1.72	1.94	2.16	2.37	2.54	2.66	2.77	2.86
1.10	0.82	0.96	1.10	1.26	1.44	1.64	1.83	2.05	2.25	2.42	2.55	2.66	2.75
1.11	0.80	0.93	1.06	1.21	1.38	1.55	1.74	1.94	2.15	2.31	2.45	2.57	2.66
1.13	0.74	0.87	1.00	1.14	1.28	1.44	1.59	1.77	1.95	2.11	2.25	2.38	2.50
1.15	0.70	0.82	0.93	1.06	1.19	1.33	1.48	1.62	1.78	1.92	2.07	2.20	2.33

vapor pressure, but usually the properties under atmospheric pressure are not significantly different. These results are based upon the correlations of Riedel.[1] For the ratio of the density of the liquid to that of the critical point, the equation is

$$\frac{d}{d_c} = 1 + 0.85(1 - T_r) + (1 - T_r)^{1/3}(1.89 + 0.91\omega) \qquad \text{(A1-18)}$$

If we substitute $T_r = 0$, we find a hypothetical density d_0 for the liquid at 0°K. It turns out that the ratio of the actual density to d_0 is effectively independent of the acentric factor at temperatures lower than $T_r = 0.8$. The resulting values are listed in Table A1-16. This table together with Eq. (A1-18) summarizes the correlation of liquid density. They allow prediction of the density variation over the full range of temperature from a single measured value. Thus the coefficient of expansion is determined and may be obtained by differentiating Eq. (A1-18).

Surface Tension, a Criterion of a Normal Fluid. Another important property of the liquid is the surface tension. It is well represented by the equation

$$\frac{\sigma}{\sigma_0} = (1 - T_r)^{11/9} \qquad \text{(A1-19)}$$

where σ_0 is a hypothetical surface tension at 0°K. The constant σ_0 may be predicted

[1] L. Riedel, *op. cit.*

Table A1-12. Values of $\left(\dfrac{H^\circ - H}{RT_c}\right)^{(1)}$

T_r	P_r 0.2	0.4	0.6	0.8	1.0	1.2	1.4	1.6	1.8	2.0	2.2	2.4	2.6	2.8	3.0	4.0	5.0	6.0	7.0	8.0	9.0
0.80	0.44	5.05	5.04	5.02	4.98	4.97	4.94	4.93	4.90	4.88	4.87	4.86	4.84	4.83	4.82	4.83	4.85	4.86	4.88	4.90	4.91
0.85	0.37	4.74	4.74	4.70	4.67	4.65	4.63	4.63	4.63	4.60	4.57	4.57	4.56	4.56	4.56	4.57	4.61	4.63	4.65	4.68	4.71
0.90	0.31	0.71	4.33	4.32	4.31	4.31	4.30	4.30	4.30	4.30	4.29	4.28	4.29	4.30	4.30	4.36	4.39	4.43	4.43	4.50	4.53
0.95	0.25	0.55	1.01	3.83	3.80	3.81	3.84	3.84	3.87	3.87	3.88	3.90	3.90	3.93	3.93	4.04	4.12	4.19	4.23	4.28	4.32
1.00	0.20	0.41	0.68	0.95	2.66	3.17	3.27	3.33	3.38	3.41	3.45	3.49	3.53	3.58	3.61	3.76	3.88	3.97	4.02	4.07	4.10
1.05	0.14	0.27	0.40	0.54	0.68	1.22	1.77	2.19	2.45	2.59	2.71	2.79	2.88	2.94	2.99	3.26	3.48	3.57	3.66	3.75	3.81
1.10	0.12	0.22	0.29	0.36	0.42	0.52	0.69	0.92	1.32	1.71	1.97	2.11	2.24	2.30	2.39	2.74	3.01	3.17	3.30	3.42	3.53
1.15	0.09	0.17	0.23	0.28	0.32	0.36	0.39	0.45	0.58	0.81	1.08	1.32	1.52	1.67	1.78	2.24	2.52	2.73	2.93	3.09	3.23
1.20	0.08	0.14	0.20	0.24	0.28	0.29	0.29	0.32	0.37	0.47	0.59	0.74	0.91	1.08	1.21	1.70	1.98	2.21	2.41	2.59	2.79
1.25	0.06	0.13	0.18	0.21	0.25	0.25	0.26	0.28	0.28	0.30	0.36	0.43	0.51	0.63	0.76	1.26	1.56	1.79	1.98	2.15	2.41
1.3	0.03	0.06	0.10	0.13	0.17	0.19	0.24	0.23	0.24	0.25	0.27	0.32	0.39	0.46	0.53	0.85	1.11	1.43	1.60	1.83	2.10
1.4	0.02	0.04	0.06	0.08	0.10	0.11	0.11	0.12	0.12	0.12	0.14	0.16	0.19	0.22	0.25	0.45	0.73	0.98	1.15	1.40	1.66
1.5	0.01	0.01	0.02	0.03	0.04	0.04	0.04	0.04	0.04	0.04	0.05	0.06	0.07	0.08	0.09	0.20	0.47	0.68	0.86	1.10	1.36
1.6	0.00	0.00	0.00	0.00	0.00	0.00	0.00	0.00	−0.01	−0.01	−0.01	−0.01	−0.02	−0.02	−0.02	0.05	0.28	0.45	0.64	0.88	1.10
1.7	−0.01	−0.01	−0.02	−0.03	−0.04	−0.04	−0.05	−0.05	−0.06	−0.06	−0.07	−0.08	−0.09	−0.10	−0.10	−0.07	0.11	0.25	0.43	0.63	0.85
1.8	−0.01	−0.02	−0.03	−0.04	−0.06	−0.07	−0.08	−0.09	−0.10	−0.10	−0.11	−0.12	−0.13	−0.14	−0.16	−0.16	−0.03	0.08	0.23	0.42	0.61
1.9	−0.02	−0.03	−0.05	−0.07	−0.09	−0.11	−0.13	−0.15	−0.17	−0.19	−0.20	−0.20	−0.21	−0.21	−0.21	−0.21	−0.15	−0.07	0.06	0.22	0.39
2.0	−0.02	−0.04	−0.06	−0.08	−0.10	−0.13	−0.16	−0.19	−0.22	−0.25	−0.25	−0.25	−0.26	−0.26	−0.26	−0.28	−0.26	−0.19	−0.10	0.03	0.18
2.5	−0.03	−0.06	−0.10	−0.13	−0.17	−0.20	−0.23	−0.26	−0.29	−0.32	−0.35	−0.39	−0.43	−0.45	−0.48	−0.53	−0.55	−0.57	−0.59	−0.63	−0.61
3.0	−0.04	−0.08	−0.13	−0.17	−0.21	−0.25	−0.29	−0.33	−0.37	−0.41	−0.45	−0.49	−0.53	−0.57	−0.60	−0.74	−0.85	−0.95	−1.05	−1.14	−1.21
3.5	−0.04	−0.08	−0.13	−0.17	−0.21	−0.25	−0.29	−0.34	−0.37	−0.42	−0.47	−0.51	−0.55	−0.59	−0.63	−0.83	−1.02	−1.19	−1.37	−1.50	−1.67
4.0	−0.04	−0.08	−0.13	−0.17	−0.21	−0.25	−0.29	−0.34	−0.37	−0.42	−0.46	−0.50	−0.54	−0.58	−0.62	−0.84	−1.03	−1.25	−1.47	−1.66	−1.86

TABLE A1-13. VALUES OF $\left(\dfrac{H^\circ - H}{RT_c}\right)^{(1)}$ IN THE CRITICAL REGION

T_r	P_r						
	0.8	1.0	1.2	1.4	1.6	1.8	2.0
0.98	1.15	3.28	3.39	3.48	3.55	3.57	3.59
0.99	1.0	3.05	3.28	3.39	3.44	3.47	3.51
1.00	0.95	2.66	3.17	3.27	3.33	3.38	3.41
1.01	0.84	1.47	2.97	3.11	3.18	3.24	3.29
1.02	0.75	1.19	2.61	2.92	3.00	3.08	3.14
1.03	0.68	0.99	2.10	2.67	2.82	2.91	3.00
1.04	0.60	0.82	1.60	2.30	2.55	2.66	2.80
1.05	0.54	0.68	1.22	1.77	2.19	2.45	2.59
1.06	0.49	0.61	0.96	1.40	1.89	2.18	2.42
1.07	0.44	0.57	0.80	1.12	1.58	1.95	2.25
1.08	0.42	0.51	0.69	0.95	1.35	1.71	2.07
1.09	0.38	0.46	0.59	0.83	1.10	1.52	1.91
1.10	0.36	0.42	0.52	0.69	0.92	1.32	1.71
1.11	0.33	0.40	0.47	0.61	0.80	1.08	1.47
1.13	0.29	0.34	0.40	0.47	0.59	0.78	1.04
1.15	0.28	0.32	0.36	0.39	0.45	0.58	0.81

from the equation

$$\frac{\sigma_0 V_0^{2/3}}{T_c} = 1.86 + 1.18\omega \qquad \text{(dynes/cm)(cc/mole)}^{2/3}\text{(}^\circ\text{K)} \qquad \text{(A1-20)}$$

The quantity V_0 is the hypothetical molal volume at 0°K as calculated from d_0. Given a density at a known reduced temperature, one may obtain d/d_0 from Table A1-16; then from the definitions one has

$$V_0 = \frac{M}{d}\frac{d}{d_0} \qquad \text{(A1-21)}$$

A test of Eq. (A1-20) is shown in Fig. A1-5. We note that typical normal liquids agree with the predicted line quite well, whereas the values of $\sigma_0 V_0^{2/3}/T_c$ for hydrogen-bonding liquids are as low as 50 per cent of that predicted. Thus the surface tension represents a sensitive test for a normal liquid. A deviation of over 5 per cent from Eq. (A1-20) appears to indicate significant abnormality. This relationship for σ_0 is recommended as the operational test of a normal liquid.

Estimating Critical Constants. We have already noted that the vapor pressure need not be known at exactly the temperature of definition in order to obtain the acentric factor since the entire vapor-pressure function is given in Table A1-1. Similarly, other observations may be substituted for the critical temperature and pressure which we have assumed to be known. Riedel has proposed several systems for predicting T_c and P_c, of which the following seems most useful. The underlying equations are just those presented above; they are now recast for convenience in another form.

Table A1-14. Values of $\left(\dfrac{s° - s}{R}\right)^{(0)}$

P_r

T_r	0.2	0.4	0.6	0.8	1.0	1.2	1.4	1.6	1.8	2.0	2.2	2.4	2.6	2.8	3.0	3.2	3.4	3.6	3.8	4.0	4.5	5.0	6.0	7.0	8.0	9.0
0.80	0.33	5.04	4.66	4.41	4.23	4.08	3.94	3.85	3.76	3.69	3.63	3.56	3.53	3.48	3.44	3.39	3.36	3.33	3.30	3.26	3.19	3.12	3.01	2.96	2.91	
0.85	0.27	4.85	4.47	4.21	4.01	3.86	3.73	3.62	3.53	3.46	3.39	3.34	3.29	3.25	3.21	3.18	3.13	3.11	3.08	3.05	2.98	2.90	2.82	2.76	2.70	
0.90	0.20	0.47	4.13	3.92	3.78	3.63	3.52	3.42	3.32	3.23	3.18	3.11	3.06	3.01	2.97	2.92	2.89	2.86	2.83	2.81	2.76	2.68	2.61	2.56	2.48	2.42
0.95	0.17	0.39	0.64	3.48	3.41	3.32	3.22	3.13	3.05	2.99	2.93	2.87	2.82	2.78	2.74	2.70	2.67	2.64	2.61	2.58	2.57	2.48	2.43	2.37	2.31	2.29
1.00	0.14	0.31	0.54	0.85	1.95	2.55	2.66	2.68	2.68	2.65	2.62	2.60	2.57	2.55	2.54	2.51	2.48	2.44	2.42	2.40	2.34	2.30	2.24	2.19	2.14	2.13
1.05	0.12	0.27	0.43	0.66	0.96	1.44	1.91	2.13	2.25	2.31	2.31	2.34	2.32	2.31	2.30	2.28	2.26	2.25	2.23	2.22	2.17	2.13	2.08	2.05	2.02	2.02
1.10	0.11	0.23	0.36	0.53	0.73	0.97	1.25	1.57	1.77	1.89	1.96	2.01	2.04	2.06	2.07	2.07	2.07	2.06	2.05	2.05	2.02	1.98	1.94	1.91	1.89	1.90
1.15	0.08	0.17	0.30	0.43	0.57	0.75	0.95	1.16	1.36	1.53	1.64	1.73	1.81	1.82	1.84	1.85	1.86	1.86	1.87	1.88	1.88	1.85	1.81	1.79	1.78	1.80
1.20	0.07	0.15	0.25	0.37	0.49	0.62	0.76	0.91	1.07	1.21	1.32	1.42	1.50	1.57	1.62	1.64	1.66	1.67	1.69	1.70	1.72	1.72	1.70	1.69	1.68	1.70
1.25	0.07	0.14	0.21	0.31	0.41	0.52	0.62	0.75	0.88	1.00	1.10	1.20	1.28	1.34	1.39	1.44	1.47	1.50	1.53	1.55	1.59	1.60	1.59	1.58	1.59	1.62
1.30	0.06	0.12	0.19	0.27	0.36	0.44	0.53	0.63	0.74	0.84	0.92	1.01	1.09	1.15	1.22	1.25	1.29	1.33	1.35	1.39	1.44	1.48	1.49	1.48	1.51	1.54
1.4	0.04	0.09	0.15	0.21	0.26	0.34	0.41	0.49	0.56	0.64	0.71	0.77	0.84	0.89	0.94	0.99	1.03	1.07	1.11	1.14	1.21	1.26	1.30	1.32	1.34	1.37
1.5	0.04	0.09	0.14	0.19	0.23	0.28	0.34	0.39	0.45	0.51	0.57	0.63	0.68	0.72	0.76	0.81	0.83	0.87	0.91	0.93	0.99	1.05	1.11	1.15	1.18	1.20
1.6	0.04	0.08	0.12	0.17	0.21	0.25	0.29	0.33	0.38	0.43	0.47	0.51	0.56	0.60	0.63	0.66	0.68	0.71	0.74	0.78	0.84	0.89	0.95	1.01	1.05	1.07
1.7	0.03	0.07	0.11	0.15	0.18	0.21	0.25	0.29	0.32	0.36	0.39	0.43	0.46	0.49	0.52	0.55	0.58	0.61	0.64	0.66	0.72	0.76	0.82	0.88	0.94	0.96
1.8	0.03	0.06	0.09	0.13	0.16	0.19	0.22	0.26	0.28	0.31	0.34	0.37	0.39	0.42	0.44	0.47	0.49	0.52	0.55	0.58	0.63	0.67	0.73	0.79	0.84	0.86
1.9	0.03	0.05	0.08	0.11	0.14	0.17	0.19	0.22	0.24	0.27	0.28	0.31	0.33	0.35	0.37	0.39	0.41	0.43	0.46	0.48	0.53	0.57	0.63	0.68	0.73	0.76
2.0	0.03	0.05	0.08	0.10	0.13	0.15	0.18	0.19	0.21	0.23	0.24	0.27	0.29	0.31	0.33	0.34	0.35	0.37	0.39	0.41	0.45	0.48	0.54	0.58	0.63	0.65
2.5	0.02	0.03	0.05	0.06	0.07	0.09	0.10	0.12	0.13	0.14	0.15	0.16	0.17	0.18	0.19	0.20	0.21	0.22	0.23	0.23	0.25	0.27	0.31	0.35	0.37	0.41
3.0	0.01	0.02	0.03	0.04	0.04	0.05	0.06	0.07	0.08	0.08	0.09	0.10	0.11	0.12	0.12	0.13	0.14	0.15	0.16	0.16	0.18	0.19	0.22	0.24	0.27	0.29
3.5	0.01	0.02	0.02	0.03	0.03	0.04	0.04	0.05	0.05	0.06	0.07	0.08	0.08	0.09	0.09	0.10	0.10	0.11	0.11	0.11	0.12	0.13	0.15	0.17	0.19	0.20
4.0	0.005	0.01	0.01	0.02	0.02	0.03	0.03	0.04	0.04	0.05	0.05	0.05	0.06	0.06	0.06	0.07	0.07	0.07	0.08	0.08	0.09	0.10	0.11	0.13	0.14	0.15

TABLE A1-15. VALUES OF $\left(\dfrac{s^\circ - s}{R}\right)^{(1)}$

T_r											P_r										
	0.2	0.4	0.6	0.8	1.0	1.2	1.4	1.6	1.8	2.0	2.2	2.4	2.6	2.8	3.0	4.0	5.0	6.0	7.0	8.0	9.0
0.80	0.46	5.23	5.20	5.16	5.12	5.08	5.03	5.00	4.96	4.92	4.89	4.86	4.83	4.81	4.79	4.74	4.71	4.68	4.65	4.62	4.60
0.85	0.37	4.86	4.82	4.79	4.74	4.70	4.66	4.63	4.61	4.58	4.54	4.52	4.50	4.48	4.46	4.43	4.41	4.39	4.37	4.36	4.35
0.90	0.30	0.70	4.38	4.36	4.33	4.31	4.29	4.27	4.26	4.24	4.22	4.20	4.19	4.19	4.18	4.18	4.17	4.17	4.16	4.16	4.15
0.95	0.24	0.53	0.98	3.81	3.77	3.76	3.76	3.76	3.76	3.76	3.76	3.76	3.76	3.77	3.77	3.83	3.87	3.90	3.90	3.91	3.92
1.00	0.19	0.39	0.65	0.90	2.60	3.09	3.18	3.23	3.26	3.28	3.31	3.34	3.37	3.41	3.42	3.53	3.61	3.66	3.67	3.68	3.68
1.05	0.13	0.26	0.38	0.51	0.67	1.18	1.71	2.11	2.34	2.47	2.58	2.65	2.72	2.77	2.82	3.03	3.20	3.26	3.32	3.36	3.39
1.10	0.11	0.20	0.27	0.35	0.41	0.51	0.68	0.91	1.27	1.63	1.86	1.99	2.11	2.16	2.23	2.53	2.75	2.86	2.96	3.04	3.10
1.15	0.09	0.16	0.22	0.28	0.32	0.36	0.41	0.48	0.60	0.81	1.06	1.28	1.45	1.58	1.68	2.07	2.30	2.46	2.60	2.71	2.81
1.20	0.08	0.14	0.20	0.24	0.28	0.31	0.33	0.38	0.44	0.53	0.65	0.79	0.94	1.09	1.21	1.62	1.85	2.03	2.18	2.31	2.38
1.25	0.06	0.12	0.17	0.22	0.26	0.28	0.30	0.33	0.35	0.39	0.46	0.53	0.61	0.71	0.83	1.25	1.50	1.68	1.82	1.95	2.02
1.3	0.04	0.07	0.10	0.13	0.16	0.18	0.20	0.22	0.24	0.26	0.31	0.36	0.41	0.46	0.51	0.77	0.98	1.23	1.36	1.53	1.74
1.4	0.03	0.05	0.07	0.09	0.11	0.13	0.14	0.15	0.16	0.17	0.20	0.22	0.24	0.26	0.29	0.45	0.67	0.85	0.98	1.16	1.31
1.5	0.02	0.03	0.05	0.06	0.07	0.07	0.08	0.09	0.09	0.10	0.11	0.12	0.13	0.15	0.17	0.26	0.46	0.61	0.74	0.90	1.09
1.6	0.01	0.0	0.03	0.04	0.04	0.04	0.05	0.05	0.06	0.06	0.07	0.08	0.08	0.09	0.10	0.17	0.33	0.46	0.59	0.74	0.90
1.7	0.00	0.01	0.01	0.02	0.02	0.02	0.02	0.03	0.03	0.03	0.04	0.04	0.04	0.05	0.05	0.10	0.22	0.33	0.45	0.59	0.73
1.8	0.00	0.00	0.01	0.01	0.01	0.01	0.01	0.01	0.01	0.01	0.01	0.01	0.01	0.02	0.02	0.05	0.14	0.23	0.34	0.46	0.58
1.9	0.00	0.00	−0.01	−0.01	−0.01	−0.01	−0.02	−0.02	−0.03	−0.03	−0.02	−0.02	−0.04	−0.01	0	0.03	0.08	0.15	0.24	0.35	0.46
2.0	0.00	0.00	−0.01	−0.01	−0.01	−0.02	−0.03	−0.04	−0.05	−0.06	−0.05	−0.05	−0.09	−0.04	−0.03	−0.01	0.03	0.09	0.16	0.25	0.35
2.5	−0.01	−0.02	−0.02	−0.03	−0.04	−0.04	−0.05	−0.06	−0.06	−0.07	−0.08	−0.09	−0.11	−0.10	−0.10	−0.09	−0.08	−0.06	−0.05	−0.03	0
3.0	−0.01	−0.02	−0.03	−0.04	−0.05	−0.06	−0.07	−0.08	−0.09	−0.09	−0.10	−0.11	−0.10	−0.12	−0.13	−0.15	−0.16	−0.17	−0.18	−0.18	−0.18
3.5	−0.01	−0.02	−0.02	−0.03	−0.04	−0.05	−0.06	−0.07	−0.08	−0.08	−0.09	−0.09	−0.10	−0.11	−0.12	−0.16	−0.19	−0.21	−0.24	−0.26	−0.29
4.0	−0.01	−0.01	−0.02	−0.02	−0.03	−0.04	−0.05	−0.06	−0.07	−0.08	−0.08	−0.09	−0.10	−0.10	−0.11	−0.14	−0.17	−0.21	−0.25	−0.28	−0.32

To estimate T_c, P_c, and ω for a fluid, three pieces of information are required. These are its liquid density at some temperature near or below the boiling point, the normal boiling point T_{760}, and the temperature T_{100} at which the vapor pressure is 100 mm. T_{100} may be calculated from a vapor-pressure equation for the substance.

The density is used to calculate a characteristic molal volume for the substance. We use the V_0 defined in Eq. (A1-21).

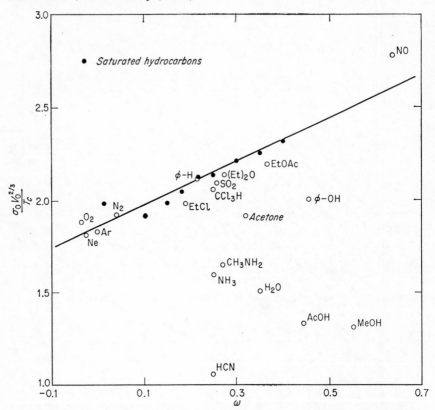

FIG. A1-5. A criterion for a normal fluid. The reduced surface-tension parameter $\sigma_0 V_0^{2/3}/T_c$ as a function of the acentric factor ω. Points for normal fluids fall within 5 per cent of the line of Eq. (A1-20). [*R. F. Curl, Jr., and K. S. Pitzer, Ind. Eng. Chem.*, **50**, 265 (1958), *fig.* 4.]

The calculation of a set of critical constants is a rapidly converging sequence of approximations. One starts with the assumption that $T_c = T_{760}/0.65$ and calculates an approximate value of T_r at which the liquid density is known. One then obtains from Table A1-16 the values of d/d_0 at this reduced temperature and the volume $V_0 = M(d/d_0)/d$, where M is the molecular weight. Next the values of $\log(T_{760}/V_0)$ and T_{100}/T_{760} are calculated and their intersection found in Fig. A1-6. In the first expression T_{760} is in degrees Kelvin and V_0 in cubic centimeters per mole. This gives a new estimate of T_{760}/T_c and a value of ω.

This estimate of T_{760}/T_c is used to calculate a new value of T_r for the temperature at which the density is measured. This T_r is used to look up d/d_0. Then $\log(T_{760}/V_0)$

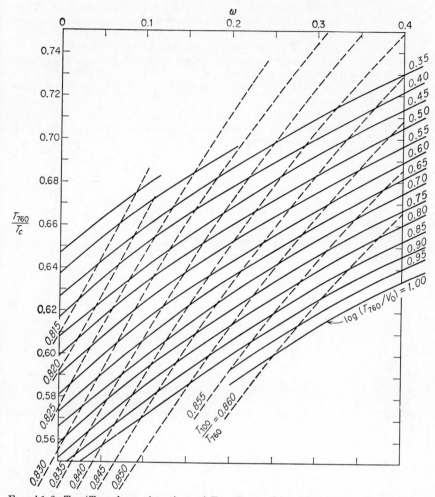

FIG. A1-6. T_{760}/T_c and ω as functions of T_{100}/T_{760} and log (T_{760}/V_0). The lines running steeply upward from left to right are constant values of T_{100}/T_{760}. The lines running upward less steeply from left to right are constant values of log (T_{760}/V_0) with T_{760} in degrees Kelvin and V_0 in cubic centimeters per mole. By finding the point given by the values of T_{100}/T_{760} and log (T_{760}/V_0) the values of T_{760}/T_c and ω may be read from the ordinate and abscissa, respectively. [*R. F. Curl, Jr., and K. S. Pitzer, Ind. Eng. Chem.,* **50,** 265 (1958), *fig. 5.*]

is recalculated, and a new estimate of T_{760}/T_c and ω is found. This process is continued until no further change is found.

T_c is given by $T_c = (T_{760}/T_c)^{-1}T_{760}$. P_c is then found by using the calculated values of T_c and ω and the equation log $P_c = (\log P_c)^{(0)} + \omega(\log P_c)^{(1)}$, together with Table A1-17, which is merely a convenient rearrangement of the vapor-pressure equation of Table A1-1. Riedel calculated the critical parameters of 26 substances. The rms percentage error for these substances for $T_c = 0.8$ per cent, for $P_c = 3.3$ per cent.

TABLE A1-16. VALUES OF d/d_0

T_r	0	0.01	0.02	0.03	0.04	0.05	0.06	0.07	0.08	0.09	0.10
0.3	0.875	0.870	0.866	0.862	0.858	0.853	0.849	0.844	0.840	0.835	0.830
0.4	0.830	0.826	0.821	0.816	0.812	0.807	0.802	0.797	0.792	0.787	0.782
0.5	0.782	0.777	0.772	0.767	0.762	0.757	0.752	0.747	0.742	0.737	0.731
0.6	0.731	0.726	0.720	0.715	0.709	0.703	0.698	0.692	0.686	0.680	0.674
0.7	0.674	0.668	0.662	0.656	0.649	0.642	0.636	0.629	0.622	0.613	0.608

TABLE A1-17. DEPENDENCE OF P_c ON T_{760}/T_c AND ω, ATM

$$\log P_c = (\log P_c)^{(0)} + \omega(\log P_c)^{(1)}$$

T_{760}/T_c	$(\log P_c)^{(0)}$	$(\log P_c)^{(1)}$
0.76	0.742	0.705
0.74	0.823	0.800
0.72	0.909	0.895
0.70	1.000	1.00
0.68	1.096	1.12
0.66	1.198	1.25
0.64	1.308	1.39
0.62	1.426	1.54
0.60	1.552	1.70
0.58	1.688	1.88
0.56	1.834	2.08

Examples. The calculation of the critical constant of toluene is given as an example of this procedure:

$$T_{760} = 383.8°K \qquad T_{100} = 325.1°K$$
$$d = 0.866 \text{ g/cc at } 293.2°K$$
$$\frac{T_{100}}{T_{760}} = 0.847$$

Assume $T_{760} = 0.65T_c$

$T_r = 0.65(^{293.2}\!/_{383.8}) = 0.496$ for the density value

From Table A1-16,

$$\log \frac{T_{760}}{V_0} = \log \frac{383.8 \times 0.866}{93 \times 0.784} = 0.659$$

From Fig. A1-6,

$$\frac{T_{760}}{T_c} = 0.643 \qquad \omega = 0.233$$

$T_r = 0.643(^{293.2}\!/_{383.8}) = 0.491$ for the density value

From Table A1-16,

$$\log \frac{T_{760}}{V_0} = \log \frac{383.8 \times 0.866}{93 \times 0.786} = 0.658$$

There is no change in this approximation; therefore

$$\omega = 0.233$$
$$T_c = {}^{383.8}\!/_{0.643} = 597°K$$

From Table A1-17, at $T_r = 0.643$,

$$\log P_c = 1.291 + \omega(1.37)$$
$$= 1.610$$
$$P_c = 40.8 \text{ atm}$$

In addition let us calculate V_c.

$$z_c = z_c^{(0)} + \omega z_c^{(1)}$$
$$= 0.291 + \omega(-0.080)$$
$$= 0.272$$
$$V_c = \frac{z_c R T_c}{P_c} = \frac{0.272 \times 0.082 \times 597}{40.8}$$
$$= 0.326 \text{ liter/mole}$$

	T_c, °K	P_c, atm	V_c, liters/mole
Estimate............	597	40.8	0.326
Experimental.........	594.0	40	0.32

We may continue the example by predicting the properties of toluene at 600°K and 80 atm. We take $\omega = 0.233$ and the experimental critical properties.

$$T_r = {}^{600}\!/_{594} = 1.01$$
$$P_r = {}^{80}\!/_{40} = 2.00$$
$$z = 0.332 + 0.233(-0.100) = 0.309$$
$$V = \frac{RT}{P} = (82.0 \times 600)/(80 \times 0.309) = 1990 \text{ cc/mole}$$

$$\log \frac{f}{P} = -0.382 - 0.233(-0.05) = -0.370$$

$$f = 80 \times 0.427 = 34.1 \text{ atm}$$

$$\frac{\text{H}° - \text{H}}{RT_c} = 3.50 + 0.233 \times 3.29 = 4.27$$

$$\text{H}° - \text{H} = 1.987 \times 594 \times 4.27 = 5040 \text{ cal}$$

$$\text{s}° - \text{s} = 1.987[{}^{4.27}\!/_{1.01} + 2.303(-0.370)] + R \ln 80$$
$$= 6.7 + 8.7 = 15.4 \text{ cal/deg}$$

Appendix 2

PROPERTIES OF GASEOUS SOLUTIONS

In Chapter 21 we remarked about the difficulties of predicting properties of gaseous solutions. One approach which has shown promise is based upon the theory of corresponding states and the pseudocritical concept of Kay.[1] An intrinsic limitation on this system is the fact that the pure components seldom follow the law of corresponding states strictly. Brown[2] showed that the deviations of the CO-CH$_4$ system from solution theory of this type differed substantially depending on the choice of CO or CH$_4$ as the reference substance. Recently Pitzer and Hultgren[3] applied to solutions the extended theory based upon the acentric factor which had been applied to pure fluids and which is described in Appendix 1.

This procedure removes the contradiction which is inherent in a corresponding-states treatment of a solution of two components which do not themselves follow the principle of corresponding states. The acentric-factor theory is not exact, but it was found to reduce the magnitude of deviation in pure substances by about a factor of 10 from that of simple corresponding states theory. At the present level of accuracy of solution studies, this residual uncertainty in treatment of pure substance properties is unimportant.

Determination of Pseudocritical Constants. In the acentric-factor theory the compressibility factor is given by the equation

$$\frac{PV}{RT} = z = z^{(0)} + \omega z^{(1)} \qquad \text{(A1-3)}$$

where $z^{(0)}$ and $z^{(1)}$ are functions of the reduced temperature T_r and the reduced pressure P_r which are given in Tables A1-2 to A1-7. The acentric factor ω is defined by Eq. (A1-1) and methods of determination are discussed in Appendix 1.

We seek to determine an acentric factor and pseudocritical constants for a solution such that its volumetric behavior is given by Eq. (A1-3). Since solutions cannot be expected to conform exactly with even the acentric-factor extension of the corresponding-states theory for pure substances, there is necessarily some arbitrariness in the selection of pseudocritical constants. The most characteristic and sensitive single-phase region is that just above the critical point in both temperature and pressure. The two-phase region is expanded for solutions, as compared with pure substances; consequently the range available for comparison of single-phase properties begins somewhat above the point $T_r = 1$, $P_r = 1$. For typical systems the isotherms from $T_r = 1.15$ to 1.4 are most useful. These isotherms have minima in the compressibility

[1] W. B. Kay, *Ind. Eng. Chem.*, **28**, 1014 (1936).

[2] W. B. Brown, *Phil. Trans. Roy. Soc. London*, **A250**, 175 (1957).

[3] K. S. Pitzer and G. O. Hultgren, *J. Am. Chem. Soc.*, **80**, 4793 (1958).

factor at reduced pressures between 2.3 and 3.3, and data over a pressure range of about a factor of 2 each way from the minimum are usually available.   A technique of determining these pseudocritical constants is given in Pitzer and Hultgren's paper.[1]

The resulting constants are presented in Table A2-1.   The deviation in z between the observed and theoretical values is less than 1 per cent in most regions.   Comparisons are made at temperatures and pressures well outside the region used in determining the pseudocritical constants.   In an extreme comparison of this type the second virial coefficient of 40 per cent n-butane, 60 per cent methane is predicted from the pseudocritical constants fitted in the high-pressure region and Eq. (A1-6) for second virial coefficient.   The calculated result differs from experiment by not over 6 cc/mole in the range 378 to 510°K.   The resulting error in the compressibility factor at atmospheric pressure would be 0.02 per cent.

The precision of determination of the pseudocritical constants depends, of course, on the precision of agreement with the theoretical curves   The uncertainty in ω for the systems in Table A2-1 ranges from 0.02 to 0.1, while that in T_c or P_c ranges from 0.2 to 1 per cent.

TABLE A2-1. PSEUDOCRITICAL CONSTANTS OF SOLUTIONS[†]
Values for 50 mole per cent

System	Ref.	T_{cm}, °K	P_{cm}, atm	ω_m
CH_4-C_2H_6	a	254	48.5	0.05
CH_4-C_3H_8	b	287	37.9	0.12
CH_4-iso-C_4H_{10}	c	309	40.3	0.10
CH_4-n-C_4H_{10}	d	324	42.2	0.17
C_3H_8-iso-C_5H_{12}	e	415	37.3	0.20
C_3H_8-C_6H_6	f	464	45.2	0.20
N_2-C_2H_6	g	213	41.4	0.05
CH_4-CO_2	h	242	57.6	0.15
CO_2-C_2H_6	i	292	55.3	0.10
CO_2-C_3H_8	j	322	49.5	0.14
CO_2-n-C_4H_{10}	k	352	46.1	0.20
CH_4-H_2S	l	269	63.2	0.05

[†] K. S. Pitzer and G. O. Hultgren, *J. Am. Chem. Soc.*, **80**, 4793 (1958), table I.

[a] B. H. Sage and W. N. Lacey, *Ind. Eng. Chem.*, **31**, 1497 (1939); the 70°F isotherms appear to be inaccurate; constants in table I are based on higher-temperature data.

[b] H. H. Reamer, B. H. Sage, and W. N. Lacey, *ibid.*, **42**, 534 (1950).

[c] R. H. Olds, B. H. Sage, and W. N. Lacey, *ibid.*, **34**, 1008 (1942).

[d] H. H. Reamer, K. J. Kaapi, B. H. Sage, and W. N. Lacey, *ibid.*, **39**, 206 (1947)

[e] W. E. Vaughan and F. C. Collins, *ibid.*, **34**, 885 (1942).

[f] J. W. Glanville, B. H. Sage, and W. N. Lacey, *ibid.*, **42**, 508 (1950).

[g] H. H. Reamer, F. T. Selleck, B. H. Sage, and W. N. Lacey, *ibid.*, **44**, 198 (1952).

[h] H. H. Reamer, R. H Olds, B. H. Sage, and W. N. Lacey, *ibid.*, **36**, 88 (1944).

[i] H. H. Reamer, R. H. Olds, B. H. Sage, and W. N. Lacey, *ibid.*, **37**, 688 (1945).

[j] H. H. Reamer, B. H. Sage, and W. N. Lacey, *ibid.*, **43**, 2515 (1951).

[k] R. H. Olds, H. H. Reamer, B. H. Sage, and W. N. Lacey, *ibid.*, **41**, 475 (1949).

[l] H. H. Reamer, B. H. Sage, and W. N. Lacey, *ibid.*, **43**, 976 (1951).

[1] *Ibid.*

It is remarkable that pairs of components with such different volatility as methane and *n*-butane or nitrogen and ethane yield solutions which behave nearly like pure substances in the single-phase region.   These systems appear to be near the boundary, however, because the system methane–*n*-pentane deviates significantly.   Experimental isotherms fitted near their minima in *z* fall about 2 per cent below the theoretical curves at a pressure one-half or twice that of the minimum.

Composition Dependence of Pseudocritical Constants.   The prima-facie extension of Kay's postulates to the acentric-factor system is the assumption that ω as well as T_c and P_c are linear functions of the mole fraction.   Table A2-2 presents the tests of this assumption.   The fourth, sixth, and eighth columns contain the quantities which are zero if the critical temperature, the critical pressure, and the acentric factor, respectively, are linear functions of the mole fraction.   Three systems, propane-isopentane, propane-benzene, and nitrogen-ethane, appear to fit since none of their deviations exceeds the limit of uncertainty.   The remaining systems deviate by significant amounts and in a few cases by large amounts.

TABLE A2-2. RELATIONSHIP OF PSEUDOCRITICAL CONSTANTS OF 50 MOLE PER CENT SOLUTIONS TO PROPERTIES OF COMPONENTS[†]

System		Critical temperature		Critical pressure		Acentric factor		
Component 1	Component 2	$\dfrac{T_{c2}}{T_{c1}}$	$\dfrac{2T_{cm}}{T_{c1}+T_{c2}}-1$	$\dfrac{P_{c2}}{P_{c1}}$	$\dfrac{2P_{cm}}{P_{c1}+P_{c2}}-1$	$\omega_2-\omega_1$	$\omega_m-\dfrac{\omega_1+\omega_2}{2}$	
CH_4	C_2H_6	1.60	0.02	1.07	0.02	0.09	−0.01	
CH_4	C_3H_8	1.94	0.02	0.92	0.03	0.14	+0.04	
CH_4	*iso*-C_4H_{10}	2.14	0.03	0.79	−0.01	0.17	0.00	
CH_4	*n*-C_4H_{10}	2.23	0.05	0.82	+0.01	0.19	0.06	
C_3H_8	*iso*-C_5H_{12}	1.25	0.00	0.78	0.00	0.06	0.02	
C_3H_8	C_6H_6	1.52	0.00	1.16	0.00	0.06	0.02	
N_2	C_2H_6	2.42	−0.01	1.39	0.00	0.06	−0.02	
CH_4	CO_2	1.60	−0.02	1.59	−0.03	0.21	+0.03	
CO_2	C_2H_6	1.00	−0.04	0.67	−0.09	−0.12	−0.06	
CO_2	C_3H_8	1.21	−0.05	0.58	−0.14	−0.07	−0.05	
CO_2	*n*-C_4H_{10}	1.40	−0.03	0.52	−0.16	−0.01	−0.02	
CH_4	H_2S	1.96	−0.05	1.94	−0.06	0.09	−0.01	

† K. S. Pitzer and G. O. Hultgren, *J. Am. Chem. Soc.*, **80**, 4793 (1958), table II.

In cases where there is deviation from linear dependence on the mole fraction, a quadratic term suffices to yield agreement.   Thus the following equations are obtained for the pseudocritical constants of a solution of mole fraction x_1 and x_2:

$$T_{cx} = x_1 T_{c1} + x_2 T_{c2} + 2x_1 x_2 (2T_{cm} - T_{c1} - T_{c2}) \qquad (A2\text{-}1)$$

$$P_{cx} = x_1 P_{c1} + x_2 P_{c2} + 2x_1 x_2 (2P_{cm} - P_{c1} - P_{c2}) \qquad (A2\text{-}2)$$

$$\omega_x = x_1 \omega_1 + x_2 \omega_2 + 2x_1 x_2 (2\omega_m - \omega_1 - \omega_2) \qquad (A2\text{-}3)$$

The T_c's and P_c's refer to the critical temperature and pressure throughout.   The quantities with subscript 1 or 2 pertain to the pure components, those with subscript x to the mixture of composition x_1, x_2, while those with subscript m apply to the equimolal mixtures.   The relationship of the quantities in parentheses to those in Table A2-2 is self-evident.

Two conclusions may be drawn from the results in Table A2-2.   First, the acentric factor of the solution is, within the limits of uncertainty, a linear function of the acentric factors of the components.   Second, the deviation from linear dependence of the pseudocritical temperature and pressure exceeds the limit of uncertainty in many cases but does not appear to follow any simple formula based on differences of properties of pure components.

Calculation of Fugacities.   The system defined by Eqs. (A1-3) and (A2-1) to (A2-3) provides a basis for the calculation of various thermodynamic properties of solutions.   The last term in Eq. (A2-3) is dropped in view of the results already discussed.   The fugacity is probably of greatest interest because it determines both phase equilibria and chemical-reaction equilibria.   The derivation of fugacities from a pseudocritical corresponding-states system of equations has been given by Gamson and Watson.[1]   The method is straightforward, and we shall merely give the resulting equations for the two components.

$$\log \frac{f_1}{x_1 P} = \left(\log \frac{f}{P} \right)_x + x_2 Y \tag{A2-4}$$

$$\log \frac{f_2}{x_2 P} = \left(\log \frac{f}{P} \right)_x - x_1 Y \tag{A2-5}$$

$$Y = \left(\frac{\mathrm{H}^\circ - \mathrm{H}}{RT_c} \right)_x \frac{T_{c2} - T_{c1} + 2(1 - 2x_2)(2T_{cm} - T_{c1} - T_{c2})}{2.303 T}$$
$$+ \frac{(z_x - 1)[P_{c2} - P_{c1} + 2(1 - 2x_2)(2P_{cm} - P_{c1} - P_{c2})]}{2.303 P_c}$$
$$- \left(\log \frac{f}{P} \right)^{(1)} (\omega_2 - \omega_1) \tag{A2-6}$$

All the quantities appearing in Eqs. (A2-4) to (A2-6) have been defined or tabulated before.   The subscript x indicates that the quantity is to be calculated for the reduced temperature, pressure, and the acentric factor of the solution of mole fraction x_1, x_2.   The compressibility-factor functions are given in Tables A1-2 through A1-7, while the fugacity and enthalpy functions are given in Tables A1-8 through A1-13.   In the low-pressure region Eqs. (A1-6)* to (A1-8) give values for all the functions.

In the two-phase region the fugacity of each component must be the same for each phase.   The calculated fugacities for four sets of experimental equilibrium compositions in the propane-isopentane system[2] are presented in Table A2-3 as an example

TABLE A2-3.   FUGACITIES IN THE PROPANE-ISOPENTANE SYSTEM†
Component 1 propane, component 2 isopentane; fugacities, atm

T, °K	P, atm	Liquid			Vapor		
		x_1	f_1	f_2	x_1	f_1	f_2
348.2	15.0	0.509	10.6	2.0	0.822	10.6	1.8
373.2	20.0	0.414	12.1	4.0	0.713	12.2	3.7
373.2	39.0	0.909	25.8	0.87	0.951	25.9	0.74
398.2	37.0	0.577	20.9	5.0	0.735	21.6	4.7

† K. S. Pitzer and G. O. Hultgren, *J. Am. Chem. Soc.*, **80**, 4793 (1959), table III.

[1] B. W. Gamson and K. M. Watson, *Natl. Petrol. News*, **36**, R623 (1944); see also J. Jaffe, *Ind. Eng. Chem.*, **40**, 738 (1948).

[2] W. E. Vaughan and F. C. Collins, *Ind. Eng. Chem.*, **34**, 885 (1942).

and test of this method.   The agreement for propane fugacity is good, and that for isopentane is reasonably satisfactory in view of the fact that no experimental data in or near the two-phase region were used in evaluating the pseudocritical constants.

The extension of this method to other thermodynamic functions or to more than two components is quite straightforward.   The principal limitation for predictive purposes is the need for the coefficients of the quadratic terms in Eqs. (A2-1) and (A2-2) for the pseudocritical temperature and pressure.   While these quadratic terms are negligible for a few systems, they are of considerable importance for some others. The number of systems studied is not large enough to justify broad conclusions on an empirical basis.   It would be desirable to have these quadratic-term coefficients for additional systems.   Nevertheless, consideration of the various factors affecting intermolecular forces should allow the experienced scientist to estimate values of $2T_{cm} - T_{c1} - T_{c2}$ and $2P_{cm} - P_{c1} - P_{c2}$ for systems similar to those listed in Table A2-2 with sufficient accuracy for many purposes.

It should be noted that in this study the maximum ratio of critical temperatures of components is 2.4, that of critical pressures is 2.0, and the maximum ratio of critical volumes implied is about 3.   The authors have reason to believe that the behavior of mixed gases may become more complex when the components differ from one another to a greater degree.   Thus the present methods should be used with caution outside the range defined above.   Nevertheless the present method should yield useful predictions of considerable accuracy for a large variety of solution systems.

Appendix 3

TRANSLATIONAL ENTROPY
OF AN IDEAL GAS

In Chapters 8 and 12 the basic relationship of entropy and probability was given,

$$S = k \ln W \qquad \text{(A3-1)}$$

where k is the Boltzmann constant, R/N_0, and W is the total number of individual quantum states which are included in the thermodynamic state of entropy S. While a number of simple applications of statistics to entropy calculations are given in Chapters 21, 23, and 27, the calculation of the absolute entropy of translational motion for an ideal gas requires a somewhat more complex treatment. The statistical calculation is outlined below, starting from an assumed basic quantum-mechanical equation for the particle in a box.

As an introductory digression we note that a surprisingly crude calculation of W suffices to yield as accurate a value of S as that of the most precise experiments. Thus an accuracy of measurement of entropy to 0.001 cal/mole deg is only rarely attainable. The corresponding error in probability is

$$\delta \ln W = {}^{0.001}\!/_{1.98} \times 6 \times 10^{23} = 3 \times 10^{20}$$

Thus one only needs to get the decimal point in W to within about 10^{20} places from the correct location to have attained high accuracy in the calculated entropy.

The adequacy of the various approximations to be made below has been analyzed. We shall not consider these questions here, however, but shall note only that, in view of the result of the preceding paragraph, high precision in W or $\ln W$ is not required.

The quantum mechanics of a particle moving freely within a cubical box[1] yields the following equation for the energies of the quantum states,

$$\epsilon = \frac{h^2(n_1{}^2 + n_2{}^2 + n_3{}^2)}{8ml^2} \qquad \text{(A3-2)}$$

where m is the mass of the particle, l is the edge length of the box, h is Planck's constant, and n_1, n_2, n_3 are three quantum numbers each of which can have any positive integral values: 1, 2, 3, etc. We shall be interested in the statistical distribution of these states with respect to energy. For this purpose it is convenient to consider a three-dimensional cartesian-coordinate space with integral spacing of points in each direction. Each point in the octant of all positive coordinates then represents a quantum state, and the energy of the state is proportional to the square of the distance

[1] This result is given in various elementary texts on quantum mechanics, for example, L. Pauling and E. B. Wilson, Jr., "Introduction to Quantum Mechanics," McGraw-Hill Book Company, Inc., New York, 1935.

from the origin.   Thus

$$r^2 = n_1{}^2 + n_2{}^2 + n_3{}^2 \tag{A3-3}$$

$$\epsilon = r^2 \frac{h^2}{8ml^2} \tag{A3-4}$$

We shall need to have the number of quantum states g_i which have, within narrow limits, the same energy.   This is readily seen to be the volume of the octant of a spherical shell of radius r and thickness δr.

$$g_i = \frac{\pi}{2}\, r^2\, \delta r \tag{A3-5}$$

We must now obtain an expression for the total number of quantum states for 1 mole of particles at a given temperature.   It is actually more convenient to calculate this number of quantum states W for a fixed energy and to consider later the relationship between energy and temperature.

The particles are assumed not to interact with one another; hence the energy for an N-particle system is just the sum of the energies of the N one-particle states.   Each different set of one-particle quantum numbers yields a different N-particle quantum state.   In order to make the problem tractable, we segregate the N particles into groups of N_i particles each of which has the same energy ϵ_i.

QUANTUM STATES FOR SYSTEM OF TWO PARTICLES
AND FOUR ONE-PARTICLE STATES
One-particle state

1	2	3	4
Type 1			
A	B		
B	A		
A		B	
B		A	
A			B
B			A
	A	B	
	B	A	
	A		B
	B		A
		A	B
		B	A
Type 2			
AB			
	AB		
		AB	
			AB

Now we must obtain the number of individual quantum states which a set of N_i particles of a given energy may have if there are g_i single-particle states of that energy. Let us first consider a very simple example of two particles and four single-particle states in the accompanying diagram.   The two-particle states fall into two groups:

those with only one particle in any one-particle state and those with both particles in the same one-particle state. In our example there are 12 states of the first type and 4 of the second type, or 16 in all. In a more general case with g_i single-particle states but still $N_i = 2$, the result would be a total of $g_i{}^2$ states of which $g_i{}^2 - g_i$ would be of the first type and g_i would be of the second type.

Atoms and molecules of the same sort, however, are not labeled A, B, etc., but rather are completely indistinguishable from one another. There is only one two-particle quantum state with one particle in state 1 and one in state 2. Thus the first two states in the diagram are really the same state; likewise the third and fourth are a single state, etc. Thus the actual number of states of the first type is $\frac{1}{2}(g_i{}^2 - g_i)$. For reasons which we shall not discuss the second type of state exists for some atoms or molecules and not for others. Thus the total number is either $\frac{1}{2}(g_i{}^2 + g_i)$ or $\frac{1}{2}(g_i{}^2 - g_i)$. We readily see, however, that, if g_i is a large number, the second term is negligible and the result is $g_i{}^2/2$.

Now we must generalize this result for N_i particles. The total number of states of both types, if the particles were distinguishable, would clearly be $g_i{}^{N_i}$. The number of rearrangements of N_i particles is $N_i!$; consequently the desired result is

$$W_i = \frac{g_i{}^{N_i}}{N_i!} \tag{A3-6}$$

This result is valid only in the approximation $g_i \gg N_i$, where states of the second type have negligible effect.

Actually we shall use the logarithm of the probability, and it may be simplified by the use of Stirling's approximation[1] for a factorial as follows:

$$\begin{aligned}
\ln W_i &= N_i \ln g_i - \ln N_i! \\
&= N_i \ln g_i - N_i \ln N_i + N_i \\
&= N_i \left(1 + \ln \frac{g_i}{N_i} \right)
\end{aligned} \tag{A3-7}$$

The probability of the complete array of N particles is[2]

$$\begin{aligned}
\ln W &= \sum \ln W_i \\
&= \sum N_i \left(1 + \ln \frac{g_i}{N_i} \right)
\end{aligned} \tag{A3-8}$$

where this sum, and those to follow, cover all values of i. The distribution of the particles over the various groups of states will follow the Boltzmann law. The reader may accept this result and proceed directly to Eq. (A3-10); however, we shall next give a statistical derivation of the Boltzmann law.

The most probable distribution of N_i values will be that which maximizes $\ln W$ subject to the conditions of a fixed total number of particles and fixed total energy. This maximum is found by the use of Lagrange's method of undetermined multipliers. We maximize the function

$$\ln W - \alpha \Sigma N_i - \beta \Sigma \epsilon_i N_i$$

[1] See first footnote of Chapter 21.

[2] For another derivation of Eq. (A3-8) see the discussion leading to eq. (7.31) on p. 114 of "Quantum Chemistry" by K. S. Pitzer, Prentice-Hall, Inc., Englewood Cliffs, N.J., 1953.

where α and β are the undetermined multipliers.   The partial derivative with respect to each N_i vanishes.

$$\frac{\partial}{\partial N_i}\left[\sum N_i\left(1 - \alpha - \beta\epsilon_i + \ln\frac{g_i}{N_i}\right)\right] = 0$$

$$-\alpha - \beta\epsilon_i + \ln\frac{g_i}{N_i} = 0$$

and there results

$$N_i = g_i e^{-\alpha}e^{-\beta\epsilon_i} \qquad (A3\text{-}9)$$

Since there are N particles in all and the energy is fixed, α and β can be evaluated from the two equations

$$N = \Sigma N_i = e^{-\alpha}\Sigma g_i e^{-\beta\epsilon_i}$$
$$E = \Sigma\epsilon_i N_i = e^{-\alpha}\Sigma g_i\epsilon_i e^{-\beta\epsilon_i}$$

Since β controls the total energy by the relative distribution over high and low energy states, it is clearly a sort of temperature.   We shall presently evaluate the energy of translation for an ideal gas which will serve to determine β.   In the meantime we shall avoid rewriting formulas by stating that the relationship is $\beta = (kT)^{-1}$.   Equation (A3-9) is then the desired result.

The Boltzmann distribution law may be written in the form

$$N_i = e^{-\alpha}g_i e^{-\epsilon_i/kT} \qquad (A3\text{-}10)$$

which is the same as Eq. (27-4) except that the constant n_0 is here written as $e^{-\alpha}$.

Let us evaluate α by the insertion of the values of g_i and ϵ_i from Eqs. (A3-4) and (A3-5) into (A3-10).   We sum over all particles,

$$\Sigma N_i = N = e^{-\alpha}\Sigma g_i e^{-\epsilon_i/kT}$$

$$e^{\alpha} = \frac{\pi}{2N}\int_0^{\infty} r^2\,dr\,e^{-r^2 h^2/8ml^2 kT}$$

$$= \frac{l^3}{N}\left(\frac{2\pi mkT}{h^2}\right)^{3/2}$$

Here the sum is replaced by an integral, and we also note that l^3 is the volume of the box V.   Thus

$$\alpha = \ln\frac{V}{N} + \frac{3}{2}\ln\frac{2\pi mkT}{h^2} \qquad (A3\text{-}11)$$

Also the energy may be calculated as follows:

$$E = \Sigma N_i\epsilon_i$$
$$= e^{-\alpha}\frac{\pi h^2}{16ml^2}\int_0^{\infty} r^4\,dr\,e^{-r^2 h^2/8ml^2 kT}$$
$$= \tfrac{3}{2}NkT \qquad (A3\text{-}12)$$

This is the familiar result.

We may now calculate the logarithm of the total probability from Eq. (A3-8) and the Boltzmann law.

$$\ln\frac{g_i}{N_i} = \alpha + \frac{\epsilon_i}{kT}$$

$$\ln W = \sum N_i\left(1 + \alpha + \frac{\epsilon_i}{kT}\right)$$

But from Eq. (A3-12) this becomes

$$\ln W = N(\alpha + \tfrac{5}{2}) \qquad (A3\text{-}13)$$

and we may now calculate the entropy for 1 mole of gas from Eq. (A3-1) with the substitution of α from Eq. (A3-11).

$$s = R(\alpha + \tfrac{5}{2})$$
$$= R\left(\frac{5}{2} + \ln\frac{V}{N_0} + \frac{3}{2}\ln\frac{2\pi mkT}{h^2}\right) \tag{A3-14}$$

To convert this to the usual form of the Sackur-Tetrode equation, one substitutes the atomic or molecular weight $M = N_0 m$, whence

$$s = R(\ln V + \tfrac{3}{2}\ln T + \tfrac{3}{2}\ln M) + \mathfrak{c} \tag{A3-15}$$
$$\mathfrak{c} = R\left(\frac{5}{2} + \frac{3}{2}\ln\frac{2\pi k}{h^2} - \frac{5}{2}\ln N_0\right) \tag{A3-16}$$

This result was given without proof in Chapter 27.

As a final comment we note that we calculated only the number of quantum states represented by the most probable distribution of particles N_i over the possible energies ϵ_i.   All the other states of energy E constitute a very large number, but the addition of this number to the value of W obtained above has a negligible effect on the calculated entropy.   Similar refinements, such as allowing fluctuations in E about its most probable value, also have no significant effect on the entropy of a macroscopic amount of gas.

Appendix 4

DATA FOR AQUEOUS
ELECTROLYTE SOLUTIONS

In Chapter 23, the Debye-Hückel limiting law for electrolyte solutions was presented and the limiting variation of activity coefficient, enthalpy, and heat capacity with molality was given in Eqs. (23-20)*, (23-25)*, and (23-27)* in terms of the constants A_γ, A_H, and A_J. These constants can be calculated from fundamental constants and data on the dielectric constant and volume of the solvent as a function of temperature as given in Eqs. (23-21), (23-26), and (23-28).

The dielectric constants of water at 1 atm determined between 0 and 100°C by Malmberg and Maryott[1] and the density of water given by N. E. Dorsey[2] yields the values of the Debye-Hückel constants for 1 atm tabulated in Table A4-1a at 10° intervals between 0 and 100°C. The dielectric constants evaluated by Franck[3] for water at a density of 1 g/cc yield the values of the constant-volume Debye-Hückel constants tabulated at 100° intervals between 0 and 800°C in Table A4-1b.

TABLE A4-1a. DEBYE-HÜCKEL LIMITING CONSTANTS FOR WATER AT 1 ATM

t, °C	A_γ	A_H, cal/mole	A_J, cal/deg mole
0	0.492	464	7.5
10	0.499	545	8.7
20	0.507	637	9.9
25	0.511	688	10.4
30	0.517	743	11.0
40	0.524	856	12.1
50	0.534	982	13
60	0.545	1122	15
70	0.556	1277	16
80	0.569	1450	18
90	0.582	1643	21
100	0.596	1865	23

[1] G. C. Malmberg and A. A. Maryott, *J. Research Natl. Bur. Standards,* **56,** 1 (1956).
[2] N. E. Dorsey, "Water-substance," Reinhold Publishing Corporation, New York, 1940.
[3] E. U. Franck, *Z. physik. Chem.,* **8,** 107 (1956).

TABLE A4-1*b*. CONSTANT-VOLUME DEBYE-HÜCKEL LIMITING CONSTANTS
FOR WATER AT 1 G/CC

t, °C	A_γ	A_E, cal/mole	A_{C_V}, cal/deg mole
0	0.49	450	1
100	0.53	510	0.5
200	0.56	530	0.2
300	0.58	550	0.2
400	0.60	570	0.3
500	0.61	600	0.5
600	0 63	650	0.7
800	0.64	800	1

The constant-volume Debye-Hückel constants in Table A4-1*b* are given by the same Eqs. (23-21), (23-26), and (23-28) that are used to obtain the constant-pressure values, except that all partial differentiations are to be carried out at constant volume rather than at constant pressure.   Equation (23-26) yields a limiting constant for energies of dilution, A_E, when evaluated for constant-volume conditions, rather than A_H, the heat-content term obtained for constant-pressure conditions.   Likewise, Eq. (23-28) yields A_{C_V}.   The equations are considerably simplified when volume is constant, as $\alpha/3$ of Eq. (23-26) is zero and the four terms of Eq. (23-28) that contain volume derivatives are also zero.

To aid in the use of the Debye-Hückel limiting equations as well as the extended terms such as those suggested by Bronsted and Guggenheim, the equations for activity coefficient, osmotic coefficient, and the corresponding partial molal and integral enthalpy and heat-capacity equations are summarized here.   The function $\sigma(m^{1/2})$, which appears in many of these equations, is given in Eq. (23-40).   Two useful relations involving $\sigma(m^{1/2})$ are

$$\sigma(m^{1/2}) = \frac{3}{m^{3/2}} \int \frac{m \, dm^{1/2}}{(1 + m^{1/2})^2} = \frac{3}{2m^{3/2}} \int \frac{m^{1/2} \, dm}{(1 + m^{1/2})^2}$$

$$\sigma(m^{1/2}) + \tfrac{2}{3}m \frac{d\sigma(m^{1/2})}{dm} = (1 + m^{1/2})^{-2}$$

$\sigma(x)$ is tabulated as a function of x by Harned and Owen,[1] and interpolated values of $\sigma(m^{1/2})$ at a few even values of m are given in Table A4-2*b*.

Although all data in this appendix are expressed in terms of molality, it is sometimes of interest to use mole fraction x or moles per liter c.   The solute standard states and the activity coefficients will be different for data expressed in each of the units.   The relationships are[2]

$$x_\pm = \frac{m_\pm}{\nu m + 1000/M_1} = \frac{c_\pm}{\nu c + (1000d - cM_2)/M_1}$$

$$\log \gamma_x = \log \gamma_m + \log \frac{m_\pm}{x_\pm} \frac{M_1}{1000} = \log \gamma_c + \log \frac{c_\pm}{x_\pm} \frac{M_1}{1000 d_0}$$

[1] H. S. Harned and B. B. Owen, "The Physical Chemistry of Electrolytic Solutions," Reinhold Publishing Corporation, New York, table (B-4-1), 2d ed., 1950, or table (5-2-*b*), 3d ed., 1958.

[2] H. S. Harned and B. B. Owen, "The Physical Chemistry of Electrolytic Solutions," 3d ed., p. 11, Reinhold Publishing Corporation, New York, 1958.

where γ_x, γ_m, and γ_c are the $\gamma_\pm$ values of salt 2 with respect to the solute standard state in solvent 1, d is the density of the solution, and d_0 is the density of the pure solvent.

The following series of equations for solutions of single electrolytes is more complete in some instances than the corresponding equations given in the text of Chapters 22, 23, and 25 in that the equations in the text are usually presented for 1-1 electrolytes for simplicity, while the equations presented here are applicable to multicharged ions and, in addition, usually indicate more of the extended terms.   Equation numbers are given here to indicate the location of the equations in the text even when the equation given in the text is not as general as that given here.   In some cases the text equations are more general in that they apply to mixed electrolytes. z_+ and z_- are the charges of the positive and negative ions, and ν_+ and ν_- are the number of each type of ion produced from a molecule of salt. $\nu = \nu_+ + \nu_-$, and $\nu_+ z_+ = -\nu_- z_-$.

$$\log \gamma_\pm = -A_\gamma |z_+ z_-| \frac{I^{\frac{1}{2}}}{1 + I^{\frac{1}{2}}} + \frac{2\nu_+\nu_-}{\nu} Bm \left(1 + \frac{C}{B} m + \cdots \right) \qquad \text{(23-38) (A4-1)}$$

$$dm(1 - \phi) = -m \, d \ln \gamma \qquad \text{(22-29)}$$

$$1 - \phi = \frac{2.303}{3} A_\gamma |z_+ z_-| I^{\frac{1}{2}} \sigma(I^{\frac{1}{2}}) - 2.303 Bm \frac{\nu_+\nu_-}{\nu} \left(1 + \frac{4C}{3B} m + \cdots \right)$$
$$\text{(23-39) (A4-2)}$$

The total free energy of a solution relative to the pure water and solute standard states is given for a solution containing 1000 g water by

$$F - F^\circ = m(\bar{\mathrm{F}}_2 - \bar{\mathrm{F}}_2^\circ) + \frac{1000(\bar{\mathrm{F}}_1 - \mathrm{F}_1^\circ)}{M_1} \qquad \text{(17-32)}$$

$$\frac{F - F^\circ}{\nu RT} = -m\phi + 2.303m \log m_\pm \gamma_\pm$$

Using Eq. (15-7) to obtain enthalpies, we find that

$$H - H^\circ = m \,{}^\phi L = L \qquad \text{(17-22) (25-5) (A4-3)}$$

$${}^\phi L = \frac{\nu}{2} A_H |z_+ z_-| I^{\frac{1}{2}} \left[(1 + I^{\frac{1}{2}})^{-1} - \frac{\sigma(I^{\frac{1}{2}})}{3} \right]$$
$$- 2.303 RT^2 \frac{dB}{dT} m\nu_+\nu_- \left(1 + \frac{2}{3}\frac{dC/dT}{dB/dT} m + \cdots \right) \qquad \text{(25-15)* (A4-4)}$$

$$\frac{\bar{L}_1}{m} = \frac{-(\nu/2) A_H |z_+ z_-| I^{\frac{1}{2}} \sigma(I^{\frac{1}{2}})}{3(1000/M_1)}$$
$$+ \frac{2.303 RT^2}{1000/M_1} \frac{dB}{dT} m\nu_+\nu_- \left(1 + \frac{4}{3}\frac{dC/dT}{dB/dT} m + \cdots \right) \qquad \text{(25-14)* (A4-5)}$$

$$\bar{L}_2 = \frac{\nu}{2} A_H |z_+ z_-| \frac{I^{\frac{1}{2}}}{1 + I^{\frac{1}{2}}}$$
$$- 2 \times 2.303 RT^2 \frac{dB}{dT} m\nu_+\nu_- \left(1 + \frac{dC/dT}{dB/dT} m + \cdots \right) \qquad \text{(25-17)* (A4-6)}$$

The heat-capacity equations are readily obtained by differentiation with respect to temperature from Eqs. (25-15)*, (25-14)*, and (25-17)*.   The C or higher terms are not given in Eqs. (A4-7) and (A4-8); in Eq. (A4-9) the higher terms just yield $B^{\cdot}$ as indicated.

$${}^\phi C_P - \bar{C}_{P2}^\circ = \frac{\nu}{2} A_J |z_+ z_-| I^{\frac{1}{2}} \left[(1 + I^{\frac{1}{2}})^{-1} - \frac{\sigma(I^{\frac{1}{2}})}{3} \right]$$
$$- 2.303 RT^2 m\nu_+\nu_- \left(\frac{2}{T}\frac{dB}{dT} + \frac{d^2B}{dT^2}\right) + \cdots \qquad \text{(25-2)* (A4-7)}$$

$$\frac{\mathfrak{J}_1}{m} = \frac{\bar{c}_{P1} - \bar{c}_{P1}^{\circ}}{m} = \frac{-(\nu/2)\sigma(I^{\frac{1}{2}})}{3(1000/M_1)} A_J |z_+ z_-| I^{\frac{1}{2}}$$

$$+ 2.303 R T^2 m \nu_+ \nu_- \left(\frac{2}{T} \frac{dB}{dT} + \frac{d^2 B}{dT^2} \right) + \cdots \quad \text{(A4-8)}$$

$$\mathfrak{J}_2 = \bar{c}_{P2} - \bar{c}_{P2}^{\circ} = \frac{\nu}{2} A_J |z_+ z_-| \frac{I^{\frac{1}{2}}}{1 + I^{\frac{1}{2}}}$$

$$- 2 \times 2.303 R T^2 m \nu_+ \nu_- \left(\frac{2}{T} \frac{dB^{\cdot}}{dT} + \frac{d^2 B^{\cdot}}{dT^2} \right) \quad \text{(A4-9)}$$

The equations for the integral and partial molal heat contents and heat capacities
are consistent with the first and second temperature derivatives of the free energy
with respect to temperature, and it can also be confirmed that they satisfy the general

TABLE A4-2a. THERMODYNAMIC PROPERTIES OF AQUEOUS KCl AT 25°C

m	ϕ	$-\log \gamma_\pm$	$\gamma_\pm$	$^\phi L$, cal/mole	$\bar{L}_2$, cal/mole	$^\phi C_P$, cal/deg mole	$\mathfrak{J}_2$, cal/deg mole
0.0005	0.9916	0.0111	0.9747	10	15	−27.6	0.2
0.001	0.9883	0.0156	0.9648	14	21	−27.5	0.3
0.002	0.9838	0.0217	0.951	19	28	−27.4	0.5
0.01	0.9672	0.0455	0.901	38	54	−27.0	1.0
0.02	0.9569	0.0613	0.868	50	70	−26.7	1.5
0.03	0.9501	0.0725	0.846	60	79	−26.5	1.8
0.04	0.9451	0.0812	0.829	65	82	−26.3	2.1
0.05	0.9411	0.0885	0.816	69	87	−26.1	2.4
0.06	0.9377	0.0948	0.804	72	89	−25.9	2.7
0.07	0.9350	0.1004	0.794	75	90	−25.7	3.0
0.08	0.9326	0.1053	0.785	77	91	−25.6	3.3
0.09	0.9304	0.1098	0.777	79	91	−25.45	3.6
0.1	0.9286	0.1140	0.769	81	92	−25.35	3.9
0.2	0.9150	0.1439	0.718	81	72	−24.00	6.0
0.3	0.9083	0.1631	0.687	76	44	−22.90	7.7
0.4	0.9037	0.1771	0.665	63	12	−22.05	9.0
0.5	0.9008	0.1880	0.649	48	−21	−21.30	10.15
0.6	0.8995	0.1966	0.636	33	−55	−20.55	11.25
0.7	0.8989	0.2037	0.626	18	−85	−19.9	12.3
0.8	0.8989	0.2097	0.617	3	−115	−19.25	13.25
0.9	0.8990	0.2150	0.610	−12	−146	−18.65	14.2
1.0	0.8993	0.2195	0.603	−26	−176	−18.10	15.05
1.5	0.9047	0.2352	0.582	−99	−316	−15.60	19.0
2.0	0.9144	0.2424	0.572	−169	−446	−13.45	22.35
3.0	0.9387	0.2454	0.568	−300	−648	−9.85	27.85
4.0	0.9668	0.2394	0.576	−405	−782		
5.0	0.9962	0.2292	0.590		−870		

partial molal equations

$$Y = n_1 \bar{Y}_1 + n_2 \bar{Y}_2 \qquad (17\text{-}32)$$

$$0 = n_1\, d\bar{Y}_1 + n_2\, d\bar{Y}_2 \qquad (17\text{-}34)$$

$$\bar{Y}_2 = \frac{d(m\,{}^\phi Y)}{dm} = m\,\frac{d\,{}^\phi Y}{dm} + {}^\phi Y \qquad (17\text{-}23)$$

In Tables A4-2a and A4-2b, the thermodynamic properties of aqueous KCl solutions at 25°C are presented. In Chapter 22, the evaluation of the KCl activity-coefficient data at low molalities was presented in Fig. 22-9. The constants of Table A4-1a were used. The measurements from concentration cells with transference by Hornibrook,

TABLE A4-2b. THERMODYNAMIC PROPERTIES OF AQUEOUS KCl AT 25°C

m	$\sigma(m^{1/2})$	$B^{\cdot}$	$2 \times 2.303\,RT^2\dfrac{dB^{\cdot}}{dT}$, cal kg/mole^2	$\dfrac{dB^{\cdot}}{dT} \times 10^4$	$-\dfrac{d^2B^{\cdot}}{dT^2} \times 10^5$
0.0005	0.9673	0.10	760	9.3	2
0.001	0.9544	0.10	760	9.3	2
0.002	0.9364	0.10	760	9.3	2
0.005	0.9023	0.10	760	9.3	2
0.01	0.8662	0.10	760	9.3	2
0.02	0.8194	0.10	760	9.3	2
0.03	0.7854	0.10	760	9.3	2
0.04	0.7588	0.095	760	9.3	2
0.05	0.7362	0.093	760	9.3	2
0.06	0.7173	0.092	760	9.3	2
0.07	0.7003	0.090	750	9.2	2
0.08	0.6851	0.089	750	9.2	2
0.09	0.6712	0.088	740	9.1	2
0.1	0.6585	0.087	740	9.10	2
0.2	......	0.069	705	8.67	2.3
0.3	......	0.059	667	8.20	2.2
0.4	......	0.051	638	7.84	2.0
0.5	......	0.047	612	7.52	1.9
0.6	......	0.044	592	7.28	1.9
0.7	......	0.0414	569	6.99	1.8
0.8	......	0.0394	550	6.76	1.7
0.9	......	0.0374	534	6.56	1.7
1.0	......	0.0359	520	6.39	1.63
1.5	......	0.0307	463	5.69	1.47
2	......	0.0284	425	5.22	1.35
3	......	0.0261	361	4.44	1.17
4	......	0.0253	310	3.81	
5	......	0.0248	269	3.31	

TABLE A4-3. $\Delta B^{\cdot}$ VALUES FOR AQUEOUS ELECTROLYTES
$$\log \gamma_{\pm} = \log \gamma_{KCl} + m \, \Delta B^{\cdot}$$

m	$\log \gamma_{KCl}$	$\Delta B^{\cdot} = B^{\cdot}_{MX} - B^{\cdot}_{KCl}$						
		HCl	HBr	HI	HClO$_4$	HNO$_3$	LiOH	LiCl
0.1	-0.1140	0.148	0.20	0.25	0.17	0.12	-0.27	0.11
0.5	-0.1880	0.134	0.170	0.221	0.146	0.09	-0.09	0.11
1.0	-0.2195	0.1275	0.160	0.201	0.134	0.079	-0.06	0.108
2.0	-0.2424	0.1232	0.155	0.186	0.132	0.071	-0.034	0.103
3.0	-0.2454	0.1216	0.1564	0.183	0.135	0.068	-0.027	0.1026
4.0	-0.2394	0.1213	......	.....	0.1390	.....	-0.025	0.1045
5.0	-0.2292	0.1211	......	.....	0.1441	.....	....	0.1067

m	$\Delta B^{\cdot} = B^{\cdot}_{MX} - B^{\cdot}_{KCl}$							
	LiBr	LiI	LiClO$_4$	LiNO$_3$	LiOAc	LiOTs†	NaOH	NaF
0.1	0.14	0.24	0.22	0.10	0.07	0.01	-0.03	-0.03
0.5	0.13	0.204	0.19	0.097	0.064	0.013	$+0.05$	-0.023
1.0	0.1238	0.177	0.167	0.090	0.056	0.009	0.05	-0.023
2.0	0.1242	0.1601	0.152	0.082	0.052	-0.002	0.046	
3.0	0.1241	0.1595	0.1477	0.0766	0.0487	-0.003	0.046	
4.0	0.1292	......	0.1441	0.0725	0.0453	-0.0021	0.048	
5.0	0.1334	......	......	0.0691	......	-0.0005	0.052	

m	$\Delta B^{\cdot} = B^{\cdot}_{MX} - B^{\cdot}_{KCl}$							
	NaCl	NaBr	NaI	NaClO$_3$	NaClO$_4$	NaBrO$_3$	NaNO$_3$	NaHCO$_2$
0.1	0.049	0.065	0.10	0.00	0.03	-0.08	-0.05	0.045
0.5	0.043	0.062	0.094	-0.007	0.025	-0.063	-0.044	0.047
1.0	0.0365	0.056	0.086	-0.012	0.018	-0.057	-0.042	0.039
2.0	0.0335	0.0529	0.078	-0.014	0.013	-0.052	-0.0395	0.030
3.0	0.0330	0.0515	0.0762	-0.015	0.0103	-0.048	-0.0382	0.0254
4.0	0.0332	0.0517	0.0752	-0.015	0.0089	......	-0.0376	0.0215
5.0	0.0340	.....	......	......	0.0081	......	-0.0370	

† OTs = toluenesulfonate.

TABLE A4-3. $\Delta B^{\cdot}$ VALUES FOR AQUEOUS ELECTROLYTES (*Continued*)

m	$\Delta B^{\cdot} = B^{\cdot}_{MX} - B^{\cdot}_{KCl}$							
	NaOAc	NaOPr†	NaOBu‡	NaOHep§	NaH malonate	NaH succinate	NaH adipate	NaOTs¶
0.1	0.15	0.15	0.16	0.17	−0.03	−0.03	0.03	−0.03
0.5	0.12	0.14	0.16	0.16	−0.031	−0.019	0.028	−0.03
1.0	0.10	0.125	0.157	0.03	−0.0305	−0.018	0.023	−0.038
2.0	0.087	0.113	0.137	−0.08	−0.0266	−0.014	.....	−0.048
3.0	0.080	0.103	0.117	−0.090	−0.0255	−0.012	.....	−0.0503
4.0	0.074	.....	0.099	−0.084	−0.0251	−0.0103	.....	−0.0488
5.0	.....	.....	.....	−0.077	−0.0246	−0.0088		

m	$\Delta B^{\cdot} = B^{\cdot}_{MX} - B^{\cdot}_{KCl}$							
	NaCNS	NaH$_2$PO$_4$	NaH$_2$AsO$_4$	KOH	KF	KBr	KI	KClO$_3$
0.1	0.09	−0.14	−0.02	0.03	0.03	0.01	0.045	−0.12
0.5	0.08	−0.13	−0.053	0.08	0.028	0.011	0.035	−0.116
1.0	0.071	−0.112	−0.064	0.085	0.028	0.009	0.029	−0.118
2.0	0.057	−0.095	....	0.089	0.030	0.0075	0.023	
3.0	0.0518	−0.084	....	0.089	0.031	0.0065	0.0197	
4.0	0.0479	−0.074	....	0.089	0.0326	0.0057	0.0167	
5.0	......	−0.066	....	0.090	......	0.0050	0.0139	

m	$\Delta B^{\cdot} = B^{\cdot}_{MX} - B^{\cdot}_{KCl}$							
	KBrO$_3$	KNO$_3$	KOAc	KH malonate	KH succinate	KH adipate	KOTs¶	KCNS
0.1	−0.14	−0.18	0.14	−0.06	−0.05	0.01	−0.05	−0.007
0.5	−0.14	−0.15	0.127	−0.054	−0.041	0.007	−0.06	−0.004
1.0	.....	−0.135	0.113	−0.058	−0.038	0.0036	−0.077	−0.004
2.0	.....	−0.118	0.1004	−0.0523	−0.033	......	−0.086	−0.0065
3.0	.....	−0.1084	0.0936	−0.0482	−0.0283	......	−0.085	−0.0081
4.0	.....	−0.1010	0.0864	−0.0462	−0.0253	......	−0.082	−0.0094
5.0	.....	....	......	−0.0448	−0.0251	......	....	−0.0105

† OPr = propionate.
‡ OBu = butyrate.
§ OHep = heptylate.
¶ OTs = toluenesulfonate.

TABLE A4-3. $\Delta B^{\cdot}$ VALUES FOR AQUEOUS ELECTROLYTES (*Continued*)

m	$\Delta B^{\cdot} = B^{\cdot}_{MX} - B^{\cdot}_{KCl}$						
	KH_2PO_4	KH_2AsO_4	NH_4Cl	NH_4NO_3	RbCl	RbBr	RbI
0.1	−0.225	−0.11	0.00	−0.17	−0.03	−0.04	−0.05
0.5	−0.178	−0.12	0.000	−0.09	−0.02	−0.023	−0.027
1.0	−0.157	−0.115	−0.001	−0.08	−0.015	−0.019	−0.021
2.0	−0.127	......	−0.0011	−0.07	−0.0105	−0.0145	−0.0157
3.0	......	......	−0.0021	−0.063	−0.0087	−0.0130	−0.0136
4.0	......	......	−0.0033	−0.060	−0.0076	−0.0126	−0.0124
5.0	......	......	−0.0044	−0.058	−0.0069	−0.0120	−0.0116

m	$\Delta B^{\cdot} = B^{\cdot}_{MX} - B^{\cdot}_{KCl}$						
	$RbNO_3$	RbOAc	CsOH	CsCl	CsBr	CsI	$CsNO_3$
0.1	−0.21	0.14	0.22	−0.085	−0.09	−0.10	−0.215
0.5	−0.169	0.13	0.13	−0.061	−0.064	−0.072	−0.179
1.0	−0.148	0.118	0.114	−0.046	−0.050	−0.055	−0.156
2.0	−0.126	0.1059	....	−0.0316	−0.0358	−0.0435	−0.13
3.0	−0.1151	0.0988	.....	−0.0251	−0.0292	−0.0396	
4.0	−0.1067	0.0926	.....	−0.0215	−0.0253		
5.0	−0.0997	......	.....	−0.0191	−0.0231		

m	$\Delta B^{\cdot} = B^{\cdot}_{MX} - B^{\cdot}_{KCl}$				
	CsOAc	$AgNO_3$	$TlClO_4$	$TlNO_3$	TlOAc
0.1	0.15	−0.21	−0.23	−0.40	−0.11
0.5	0.14	−0.166	−0.18	−0.28	−0.084
1.0	0.122	−0.149	......	......	−0.069
2.0	0.110	−0.129	......	......	−0.055
3.0	0.101	−0.1179	......	......	−0.0492
4.0	0.097	−0.1097	......	......	−0.0465
5.0	.....	−0.1028	......	......	−0.0445

Janz, and Gordon[1] and by Shedlovsky and MacInnes[2] were the most important in fixing the value of B in Eq. (22-38).   In addition, the NaCl data were evaluated in a similar manner.   Finally, the relation between the NaCl and KCl measurements through the isopiestic data, as well as evaluation of the cell data by taking differences between the NaCl and KCl data, was used to fix $B_{NaCl} - B_{KCl}$.   Thus the data in Tables A4-2a and A4-2b represent a summary of the best values not only for KCl but

[1] W. J. Hornibrook, G. J. Janz, and A. R. Gordon, *J. Am. Chem. Soc.*, **64**, 513 (1942).
[2] T. Shedlovsky and D. A. MacInnes, *J. Am. Chem. Soc.*, **61**, 200 (1939).

also for NaCl. Above 0.1 M, the osmotic and activity coefficients tabulated by Robinson and Stokes[1] were corrected to agree with the value chosen at 0.1 M. Thus 0.0004 was added to all $- \log \gamma_{\pm}$ values. Also 0.0020 was added to all ϕ values.

The enthalpy data have been treated in a similar manner by using the integral heats tabulated by the Bureau of Standards[2] and by Harned and Owen.[3] The values below 0.1 M are based on the difference between the KCl and NaCl data and the NaCl measurements of Gulbransen and Robinson.[4] The apparent and partial molal enthalpy data tabulated by Harned and Owen above 0.1 M were used unaltered. The heat-capacity data of Randall and Rossini[5] were treated as described in Chapter 25. In Table A4-2b, the values of $B^{.}$ and its derivatives were obtained by applying Eqs. (22-38), (A4-6), and (A4-9) to the values of Table A4-2a. Since $B^{.}m$ is dimensionless, $B^{.}$ has the dimensions of the reciprocal of moles per 1000 g water. The temperature derivatives of $B^{.}$ are expressed in similar dimensions, with the corresponding

TABLE A4-4a. RELATIVE PARTIAL MOLAL ENTHALPIES

m	$\bar{L}_{KCl}$	$\dfrac{\bar{L}_{MX} - \bar{L}_{KCl}}{m}$, cal kg/mole^2								
		HCl	LiCl	LiBr	NaCl	NaBr	NaOH	KF	KBr	HBr
0	0	1240	860	780	90	-100	500	700	-300	700
0.1	91	1110	820	750	110	-130	360	640	-210	720
0.2	72	1005	810	755	90	-70	305	535	-165	750
0.3	44	960	783	743	60	-43	260	487	-143	
0.4	12	928	765	733	40	-25	233	463	-135	
0.5	-21	902	746	720	22	-18	220	448	-136	718
0.6	-55	883	732	710	12	-18	207	440	-135	
0.7	-85	861	710	693	0	-21	194	429	-136	
0.8	-115	844	694	674	-6	-26	185	419	-136	
0.9	-146	833	681	660	-11	-30	177	413	-136	
1.0	-176	821	667	647	-12	-30	172	407	-133	676
1.2	-233	801	644	623	-16	-39	163	394	-131	
1.5	-316	779	620	597	-18	-44	164	380	-127	
2.0	-446	751	595	571	-10	-46	180	360	-116	
2.5	-558	731	577	552	1	-38	194	340	-106	
3.0	-648	711	564	536	7	-34	208	322	-103	
4.0	-782	(670)	539	505	24	-29	241	310	-100	
5.0	-870	....	523	483	43	-22	...	303	-95	

[1] R. A. Robinson and R. H. Stokes, "Electrolyte Solutions," 2d ed., Butterworth & Co. (Publishers) Ltd., London, 1959.

[2] *Natl. Bur. Standards (U.S.) Circ.* 500, 1952.

[3] H. S. Harned and B. B. Owen, "The Physical Chemistry of Electrolytic Solutions," 2d ed., Reinhold Publishing Corporation, New York, 1950.

[4] E. A. Gulbransen and A. L. Robinson, *J. Am. Chem. Soc.*, **56**, 2637 (1934).

[5] M. Randall and F. D. Rossini, *J. Am. Chem. Soc.*, **51**, 323 (1929).

TABLE A4-4*b*. RELATIVE PARTIAL MOLAL ENTHALPIES
AT MOLALITIES BELOW 0.1 M

Salt	$\dfrac{\bar{L}_{MX} - \bar{L}_{KCl}}{m}$, cal	$-\dfrac{d\Delta B}{dT} \times 10^3$
LiI	700	0.9
LiClO$_4$	1600	2.0
LiNO$_3$	640	0.79
LiOH	1300	1.6
NaF	700	0.9
NaI	80	0.10
NaClO$_3$	−760	−0.93
NaBrO$_3$	−1220	−1.5
NaIO$_3$	−2800	−3.4
NaNO$_3$	−940	−1.2
NaOAc	1360	1.67
NaCNS	−540	−0.66
KI	−680	−0.84
KClO$_4$	−4000	−4.9
KClO$_3$	−2200	−2.7
KIO$_3$	−2700	−3.3
KNO$_3$	−2500	−3.1
KOAc	1200	1.5
KCNS	−980	−1.2
KBrO$_3$	−1100	−0.14
RbF	−260	−0.32
RbCl	−360	−0.44
RbBr	−520	−0.64
RbNO$_3$	−2700	−3.3
CsF	−600	−0.7
CsCl	−660	−0.81
CsBr	−880	−1.08
CsI	−920	−1.13
CsNO$_3$	−3000	−3.7
NH$_4$Cl	0	0.0
NH$_4$NO$_3$	−2600	−3.2
HI	800	0.9
AgF	−2360	−2.9

T^{-1} and T^{-2} for the first and second derivatives. The fourth column of Table A4-2*b*, labeled $2 \times 2.3030RT^2(dB^{\cdot}/dT)$, was obtained by subtracting $A_H m^{1/2}/(1 + m^{1/2})$ from $\bar{L}_2$, which is tabulated in Table A4-2*a*.

Values of the various thermodynamic quantities for KCl can be obtained at intermediate molalities by direct interpolation, or they may be obtained by interpolation of the values of $B^{\cdot}$ or its derivatives and evaluation of $\sigma(m^{1/2})$ and $(1 - m^{1/2})^{-1}$ to obtain the Debye-Hückel term.

The CaCl$_2$ data in Table A4-7 were obtained in the manner described for KCl by using the cell measurements of McLeod and Gordon[1] and the isopiestic measurements

[1] H. G. McLeod and A. R. Gordon, *J. Am. Chem. Soc.*, **68**, 58 (1946).

of Stokes[1] as tabulated by Robinson and Stokes.[2]   The original heat-of-dilution data of Lange and Streeck[3] were treated in the manner illustrated in Chapter 25 for NaCl solutions.   Above 0.1 M, the enthalpy values are based on the older and less accurate heats of dilution which are summarized by the National Bureau of Standards[4] and are uncertain by as much as 10 to 20 cal/mole.   The heat-capacity values are based on the measurements of Rutskov[5] and of Richards and Dole.[6]   Richards and Dole

TABLE A4-4c

$$\frac{d\Delta B^{\cdot}}{dT} \times 10^4 = -1.229 \times 10^{-2} \frac{\bar{L}_{MX} - \bar{L}_{KCl}}{m}$$

m	$\left(10^4 \dfrac{dB_{KCl}^{\cdot}}{dT}\right)$	HCl	LiCl	LiBr	NaCl	NaBr	NaOH	KF	KBr	HBr
0	+9.3	−15.2	−10.6	−9.6	−1.1	+0.49	−6.9	−8.6	+3.7	−4.9
0.1	9.10	−13.6	−10.1	−9.2	−1.4	1.60	−4.4	−7.9	2.6	−8.9
0.2	8.67	−12.4	−10.0	−9.3	−1.1	0.86	−3.75	−6.6	2.0	−9.2
0.3	8.20	−11.8	−9.6	−9.13	−0.74	0.53	−3.20	−6.0	1.8	
0.4	7.84	−11.4	−9.40	−9.01	−0.49	0.31	−2.86	−5.7	1.76	
0.5	7.52	−11.1	−9.17	−8.85	−0.27	0.22	−2.70	−5.5	1.67	−8.8
0.6	7.28	−10.9	−9.00	−8.73	−0.15	0.22	−2.54	−5.4	1.66	
0.7	6.99	−10.6	−8.70	−8.52	0.00	0.26	−2.38	−5.27	1.67	
0.8	6.76	−10.4	−8.53	−8.28	+0.07	0.32	−2.27	−5.15	1.67	
0.9	6.56	−10.2	−8.37	−8.11	0.14	0.37	−2.18	−5.08	1.67	
1.0	6.39	−10.1	−8.20	−7.95	0.15	0.37	−2.11	−5.00	1.63	−8.3
1.2	6.06	−9.85	−7.92	−7.66	0.20	0.48	−2.00	−4.84	1.61	
1.5	5.69	−9.57	−7.62	−7.34	0.22	0.54	−2.02	−4.67	1.56	
2.0	5.22	−9.23	−7.31	−7.02	+0.12	0.56	−2.21	−4.42	1.43	
2.5	4.82	−8.98	−7.09	−6.78	−0.01	0.48	−2.38	−4.18	1.30	
3.0	4.44	−8.74	−6.93	−6.59	−0.09	0.42	−2.56	−3.96	1.27	
3.5	4.11	−8.47	−6.76	−6.38	−0.17	0.39	−2.34	−3.85	1.24	
4.0	3.81	....	−6.62	−6.21	−0.30	0.36	−2.96	−3.81	1.23	
4.5	3.55	....	−6.43	−6.05	−0.39	0.32	......	−3.75	1.20	
5.0	+3.31	....	−6.43	−5.94	−0.53	+0.27	......	−3.72	+1.17	

present heat-of-dilution data at 20 and 25° from which values of $dB^{\cdot}/dT$ were obtained for the purpose of joining the heat-capacity data to the limiting Debye-Hückel slope by the method illustrated in Chapter 25 for NaCl.   The values of $\bar{J}_2$ in Table A4-7a are uncertain by about 10 per cent.   The $^{\phi}C_P$ values are uncertain by 2 to 3 cal/deg mole, although values are given to 0.1 unit at low molalities, where differences are known accurately.

[1] R. H. Stokes, *Trans. Faraday Soc.*, **41**, 637 (1945).

[2] Robinson and Stokes, *op. cit.*

[3] E. Lange and H. Streeck, *Z. physik. Chem.*, **152A**, 1 (1931).

[4] *Natl. Bur. Standards (U.S.) Circ.* 500, 1952.

[5] A. P. Rutskov, *Zhur. Priklad. Khim.*, **21**, 8 (1948).

[6] T. W. Richards and M. Dole, *J. Am. Chem. Soc.*, **51**, 794 (1929).

$(\bar{J}_{MX} - \bar{J}_{KCl})/m$, cal kg/mole² deg[1]

m	$\bar{J}_{KCl}$	NaCl	NaBr	NaI	NaNO₃	KBr	KI	KNO₃	LiCl	NH₄NO₃	HCl	HBr	NaOH	KOH
0	0	14.5	14	24	26	2	8	24	−13	....	−26	−24	5	
0.05	2.4	13	13	26	25	1	13	29	−12.5	....	−20	....		
0.1	3.9	10.5	11.5	21	25	2	10.5	24.5	−12.5	....	−36	−10.5	4.5	
0.2	6.0	9.5	9.3	16.5	22	1.5	9	20.8	−14.8	....	−18	....	4.5	
0.5	10.15	6.2	6.8	11.3	17.7	1.1	7.2	15.9	−8.9	....	−8.7	−9.1	3.3	−42
1.0	15.05	4.75	5.3	7.95	14.25	1.0	5.6	12.5	−7.0	2.35	−7.05	−7.35	2.27	−15.2
2.0	22.35	3.33	3.63	....	....	0.60	....	....	−5.5	1.01	−5.6	....	1.58	−3.48
2.5	25.25	3.0	3.10	....	....	0.31	....	....	....	0.6	....	....	1.18	
3.0	27.85	....	2.69	....	....	0.11	....	....	....	0.23	....	....		

The $(2\nu_+\nu_-/\nu)$ $\Delta B^.$ values of Tables A4-3 and A4-8 are based on the tabulation of Robinson and Stokes with changes in $\gamma_\pm$ up to 1 per cent due to new extrapolations to the solute standard state. There is room for additional improvement in some of the extrapolations. The tabulated values were obtained from

$$\log \frac{\gamma_\pm}{\gamma_{std}} = m\left(\frac{2\nu_+\nu_-}{\nu}\, \Delta B^.\right) \tag{A4-3}$$

with KCl as the standard for 1-1 salts and $CaCl_2$ as the standard for 2-1 and 1-2 salts.

The values of $(\mathsf{L}_{MX} - \mathsf{L}_{KCl})/m$ and of $-d\Delta B^./dT$ for molalities below 0.1 M given in Tables A4-4a to A4-4c and A4-9 were obtained by application of Eq. (A4-4) for $^\phi L$ to the integral heats of dilution tabulated by the National Bureau of Standards,[1]

TABLE A4-5b. VALUES OF $\dfrac{d^2\Delta B^.}{dT^2} \times 10^5$ AT 25°C

m	$\dfrac{d^2B^._{KCl}}{dT^2} \times 10^5$	$\dfrac{d^2\Delta B^.}{dT^2} \times 10^5$										
		HCl	LiCl	NaCl	NaBr	NaOH	KBr	HBr	NaI	KI	NaNO₃	KNO₃
0.05	-2	4	2.2	-1.5	-1.7	-0.2	-0.5	2.5	-3.1	-2.2	-3.9	-5.5
0.1	-2	5.3	2.2	-1.2	-1.5	-0.25	-0.4	1.9	-2.6	-2		
0.2	-2.3	3.1	2.5	-1.1	-1.2	-0.3	-0.3	...	-2.0	....	-3	-5
0.5	-1.9	1.82	1.70	-0.74	-0.85	-0.2	-0.25	1.7	-1.4			
1.0	-1.63	1.54	1.41	-0.59	-0.67	-0.14	-0.23	1.5	-1.0			
2.0	-1.35	1.31	1.17	-0.42	-0.49	-0.04	-0.17					
2.5	-1.2	....	....	-0.37	-0.41	$+0.02$	-0.125					
3.0	-1.17	....	....	......	-0.36	......	-0.099					

TABLE A4-6. PARTIAL MOLAL HEAT CAPACITIES
AT INFINITE DILUTION AT 25°C

Salt	$\bar{c}^\circ_{P2}$, cal/deg	Salt	$\bar{c}^\circ_{P2}$, cal/deg
LiCl	-15.8	RbI	-40
LiBr	-17.0	CsCl	-48.7
LiI	-17.1	CsBr	-49.9
LiNO₃	-3.6	CsI	-50
NaCl	-22.1	NH₄Cl	-13.1
NaBr	-23.3	NH₄Br	-14.3
NaI	-23.4	NH₄I	-14.4
NaNO₃	-9.9	NH₄NO₃	-0.9
KCl	-27.7	HCl	-30
KBr	-28.9	HBr	-31
KI	-29.0	HI	-31
KNO₃	-15.5	HNO₃	-18
RbCl	-38.7	NaOH	-19
RbBr	-39.9	KOH	-25

[1] *Natl. Bur. Standards (U.S.) Circ.* 500, 1952.

TABLE A4-7a. THERMODYNAMIC PROPERTIES OF AQUEOUS $CaCl_2$ AT 25°C

m	ϕ	$\log \gamma_\pm$	$\gamma_\pm$	ϕL, cal/mole	$\bar{L}_2$, cal/mole	ϕC_P, cal/deg mole	$\bar{J}_2$, cal/deg mole
0.0001	0.9869	−0.017	0.962	23	35	−68.2	0.5
0.0005	0.9719	−0.037	0.918	51	76	−67.7	1.2
0.001	0.9615	−0.052	0.887	71	104	−67.4	1.6
0.005	0.9250	−0.106	0.783	148	212	−66	3.5
0.01	0.9048	−0.140	0.724	199	280	−65	4.8
0.02	0.884	−0.180	0.661	261	360	−64	6.4
0.04	0.867	−0.2237	0.597	338	460	−62.5	8.6
0.05	0.861	−0.2408	0.574	367	490	−62	9.5
0.07	0.857	−0.2624	0.547	415	550	−61	10.8
0.1	0.854	−0.2857	0.518	465	620	−60	12.5
0.2	0.862	−0.3261	0.472	560	690	−56	17
0.3	0.876	−0.3420	0.455	610	760	−54	20
0.4	0.894	−0.3487	0.448	650	850	−51	24
0.5	0.917	−0.3487	0.448	700	940	−49	27
0.6	0.940	−0.3439	0.453	750	1020	−48	31
0.7	0.963	−0.3372	0.460	790	1090	−46	33
0.8	0.988	−0.3279	0.470	830	1140	−44	36
0.9	1.017	−0.3152	0.484	870	1200	−42	38
1.0	1.046	−0.3010	0.500	900	1250	−41	40
1.2	1.107	−0.2684	0.539	970	1400	−39	44
1.4	1.171	−0.2314	0.587	1040	1500	−36	48
1.6	1.237	−0.1911	0.644	1100	1600	−34	51
1.8	1.305	−0.1475	0.712	1160	1700	−32	54
2.0	1.376	−0.1013	0.792	1220	1800	−30	57
2.5	1.568	+0.0265	1.063	1370	....	−25	64
3.0	1.779	0.1711	1.483	....	....	−21	70
3.5	1.981	0.3177	2.078	...	....	−17	75
4.0	2.182	0.4675	2.934	....	....	−14	81
4.5	2.383	0.6201	4.17	....	....	−10	86
5.0	2.574	0.7701	5.89	....	....	−7	91
5.5	2.743	0.9128	8.18	....	....	−4	95
6.0	2.891	1.0457	11 11	....	....	−2	99
6.5	3.003	1.1623	14.53	....	....	+1	103
7.0	3.081	1 2620	18 28	....	....	4	107
7.5	3.127	1.3450	22.13	....	....	6	
8.0	3.151	1.4153	26.02				
8.5	3 165	1 4786	30.1				
9.0	3.171	1.5340	34.2				
9.5	3.171	1.5855	38.5				
10.0	3.169	1.6335	43.0				

TABLE A4-7b. THERMODYNAMIC PROPERTIES OF AQUEOUS $CaCl_2$ AT 25°C

m	$\frac{4}{3}B\dot{}$	$4 \times 2.303RT^2\frac{dB\dot{}}{dT}$, cal kg/mole²	$\frac{dB\dot{}}{dT} \times 10^4$	$\frac{d^2B\dot{}}{dT^2} \times 10^5$
0.001	1.0	2600	16	−2
0.005	1.0	2600	16	−2
0.01	1.0	2500	15	−2
0.02	0.99	2200	14	−2
0.04	0.98	1800	11	−1 5
0.05	0.89	1600	10	−1.4
0.07	0.84	1400	9	−1.4
0.1	0.76	1100	7	−1.4
0.2	0.600	1050	6	−1.4
0.3	0.5185	820	5	−1.4
0.4	0.4639	680	4	−1.4
0.5	0.4278	390	2	−1 4
0.6	0.4027	270	1.7	−1 4
0.7	0.3821	190	1.2	−1.4
0.8	0.3665	140	0.9	−1.3
0.9	0.3558	90	0.6	−1.3
1.0	0.3469	50	0.4	−1.3
1.2	0.3341	−30	−0.2	−1.2
1.4	0.3254	−80	−0.5	−1.1
1.6	0.3191	−110	−0.7	−1.1
1.8	0.3150	−140	−0.9	−1.1
2.0	0.3122	−170	−1	−1.0
2.5	0.3101			
3.0	0.3125			
3.5	0.3139			
4.0	0.3151			
4.5	0.3163			
5.0	0.3165			
5.5	0.3151			
6.0	0.3121			
6.5	0.3070			
7.0	0.3001			
7.5	0.3127			
8.0	0.2830			
8.5	0.2743			
9.0	0.2657			
9.5	0.2575			
10.0	0.2498			

TABLE A4-8 $\frac{4}{3} \Delta B^{\cdot}$ VALUES FOR AQUEOUS 2-1 AND 1-2 ELECTROLYTES
$$\log \gamma_{\pm} = \log \gamma_{CaCl_2} + m \times \tfrac{4}{3} \Delta B^{\cdot}$$

m	$\log \gamma_{CaCl_2}$	$\frac{4}{3} \Delta B^{\cdot} = \frac{4}{3}B_{MX_2}^{\cdot} - \frac{4}{3}B_{CaCl_2}^{\cdot}$					
		$MgCl_2$	$MgBr_2$	MgI_2	$Mg(ClO_4)_2$	$Mg(NO_3)_2$	$Mg(OAc)_2$
0.1	$-0\ 2857$	0.08	0.20	0.35	0.40	0.04	-0.53
0 5	$-0\ 3487$	0 06	0.16	0.25	0 29	0.04	$-0\ 25$
1.0	$-0\ 3010$	0.056	0.155	0.238	0 26	0.030	-0.202
2.0	-0.1013	0.0614	0.1517	0.236	0.254	0.0114	-0.186
3.0	$+0.1711$	0.0647	0.1505	0.2382	0.2585	-0.0034	-0.195
4.0	0.4675	0.0689	0.1530	0.2454	0 2620	-0.0135	-0.2045
5.0	0.7701	0.0747	0.1575	0.2552	......	-0.0188	
6.0	1.0457						

m	$\frac{4}{3} \Delta B^{\cdot} = \frac{4}{3}B_{MX_2}^{\cdot} - \frac{4}{3}B_{CaCl_2}^{\cdot}$						
	$CaBr_2$	CaI_2	$Ca(ClO_4)_2$	$Ca(NO_3)_2$	$SrCl_2$	$SrBr_2$	SrI_2
0.1	0.12	0.21	0.24	-0.26	-0.03	0.07	0.21
0.5	0.078	0.17	0.19	-0.18	-0.029	0.036	0.14
1.0	0.076	0.158	0.165	-0.17	-0.032	0.0294	0.126
2.0	0.075	0.152	0.154	-0.179	-0.0346	0.0291	0.121
3.0	0.0773	. ...	0.149	-0.196	-0.0388		
4.0	0 0825	.....	0.1395	-0.2067	-0.0420		
5.0	0 0991	.....	0.1299	-0.2125			
6.0	0.1167	....	0.1252	-0.2117			

m	$\frac{4}{3} \Delta B^{\cdot} = \frac{4}{3}B_{MX_2}^{\cdot} - \frac{4}{3}B_{CaCl_2}^{\cdot}$						
	$Sr(ClO_4)_2$	$Sr(NO_3)_2$	$BaCl_2$	$BaBr_2$	BaI_2	$Ba(ClO_4)_2$	$Ba(NO_3)_2$
0.1	0.11	-0.35	-0.09	-0.01	0.15	0.05	-0.80
0.5	0 11	-0.269	-0.094	-0.019	0.124	0.03	-0.55
1.0	0.11	-0.260	-0.096	-0.024	0.109	0.01	
2.0	0.095	-0.267	-0.111	-0.039	0.092	-0.02	
3.0	0.081	-0.278	....	....	.....	$-0\ 05$	
4.0	0.063	-0.286	....	....	.....	-0.07	
5.0	0.0474	....	....	....	.....	-0.088	
6.0	0.0371						

TABLE A4-8., $\frac{4}{3}\Delta B^{\cdot}$ VALUES FOR AQUEOUS 2-1 AND
1-2 ELECTROLYTES (*Continued*)

m	$\frac{4}{3}\Delta B^{\cdot} = \frac{4}{3}B^{\cdot}_{MX_2} - \frac{4}{3}B^{\cdot}_{CaCl_2}$						
	$Ba(OAc)_2$	$MnCl_2$	$FeCl_2$	$CoCl_2$	$CoBr_2$	CoI_2	$Co(NO_3)_2$
0.1	-0.50	0.00	0.017	0.04	0.18	0.34	0.02
0.5	-0.20	-0.012	0.0069	0.026	0.14	0.25	0.00
1.0	-0.175	-0.017	0.0069	0.026	0.135	0.246	-0.006
2.0	-0.204	-0.036	0.0015	0.0178	0.133	0.232	-0.0179
3.0	-0.2427	-0.0662	......	-0.0025	0.1192	0.2327	-0.0319
4.0	-0.271	-0.0935	......	-0.0302	0.1025	0.2246	-0.0425
5.0	....	-0.1155	......	....	0.0823	0.2016	-0.0496
6.0	....	-0.1283	......	....	......	0.1583	

m	$\frac{4}{3}\Delta B^{\cdot} = \frac{4}{3}B^{\cdot}_{MX} - \frac{4}{3}B^{\cdot}_{CaCl_2}$						
	$NiCl_2$	$CuCl_2$	$Cu(NO_3)_2$	$ZnCl_2$	$ZnBr_2$	ZnI_2	$Zn(ClO_4)_2$
0.1	0.04	-0.07	-0.05	0.00	0.24	0.43	0.40
0.5	0.03	-0.071	-0.042	-0.11	0.114	0.26	0.275
1.0	0.03	-0.077	-0.040	-0.166	0.043	0.198	0.260
2.0	0.029	-0.114	-0.057	-0.218	-0.071	0.053	0.2647
3.0	0.0191	-0.151	-0.072	-0.237	-0.132	-0.0424	0.2729
4.0	0.0010	-0.177	-0.0815	-0.2444	-0.161	-0.0936	0.2778
5.0	-0.0198	-0.195	-0.0917	-0.2437	-0.176	-0.1215	
6.0	....	-0.2025	-0.0951	-0.2371	-0.1795	-0.1340	

m	$\frac{4}{3}\Delta B^{\cdot} = \frac{4}{3}B^{\cdot}_{MX_2} - \frac{4}{3}B^{\cdot}_{CaCl_2}$						
	$Zn(NO_3)_2$	$CdCl_2$	$CdBr_2$	CdI_2	$Cd(NO_3)_2$	$Pb(ClO_4)_2$	$Pb(NO_3)_2$
0.1	0.10	-3.57	-4.35	-6.88	-0.02	0.06	-1.07
0.5	0.043	-1.30	-1.52	$-2.15'$	-0.04	0.032	-0.66
1.0	0.028	-0.874	-0.985	-1.30	-0.06	0.0195	-0.538
2.0	0.006	-0.626	-0.67	-0.82	-0.092	0.0045	-0.460
3.0	-0.013	-0.542	-0.562	-0.64	-0.118	-0.0097	
4.0	-0.0264	-0.495	-0.506	......	....	-0.0222	
5.0	-0.0367	-0.465	....	......	....	-0.0325	
6.0	-0.0402	-0.438	....	......	....	-0.0370	
7.0	....	....	....	......	....	-0.0333	
8.0	....	....	....	......	....	-0.0254	
10.0	....	....	....	......	....	-0.0101	

TABLE A4-8. $\frac{4}{3} \Delta B^{\cdot}$ VALUES FOR AQUEOUS 2-1 AND
1-2 ELECTROLYTES (*Continued*)

m	$\frac{4}{3} \Delta B^{\cdot} = \frac{4}{3} B^{\cdot}_{MX_2} - \frac{4}{3} B^{\cdot}_{CaCl_2}$						
	UO_2Cl_2	$UO_2(ClO_4)_2$	$UO_2(NO_3)_2$	Li_2SO_4	Na_2SO_4	Na_2CrO_4	$Na_2S_2O_3$
0.1	0.17	0.67	0.20	−0.35	−0.59	−0.34	−0.46
0.5	0.116	0.46	0.15	−0.276	−0.44	−0.30	−0.35
1.0	0.089	0.428	0.133	−0.247	−0.389	−0.282	−0.32
2.0	0.0435	0.429	0.093	−0.2345	−0.355	−0.270	−0.297
3.0	0.0050	0.433	0.043	−0.2343	−0.343	−0.261	−0.288
4.0	......	0.4306	−0.0114	....	−0.332	−0.250	−0.279
5.0	......	0.4179	−0.0583				
6.0	......	0.401	−0.090				

m	$\frac{4}{3} \Delta B^{\cdot} = \frac{4}{3} B^{\cdot}_{MX_2} - \frac{4}{3} B^{\cdot}_{CaCl_2}$						
	Na_2CO_3	Na_2HPO_4	Na_2HAsO_4	Na_2Fu†	Na_2Ma‡	K_2SO_4	K_2CrO_4
0.1	−0.46	−0.45	−0.26	−0.44	−0.84	−0.75	−0.46
0.5	−0.31	−0.444	−0.32	−0.25	−0.44	−0.469	−0.35
1.0	−0.277	−0.409	−0.322	−0.192	−0.339	−0.40	−0.319
2.0	−0.29	....	....	−0.180	−0.286	....	−0.298
3.0	....	....	....	....	−0.277	....	−0.295
4.0	....	....	....	....	....	....	−0.289

m	$\frac{4}{3} \Delta B^{\cdot} = \frac{4}{3} B^{\cdot}_{MX_2} - \frac{4}{3} B^{\cdot}_{CaCl_2}$				
	K_2HPO_4	K_2HAsO_4	$(NH_4)_2SO_4$	Rb_2SO_4	Cs_2SO_4
0.1	−0.43	−0.15	−0.88	−0.52	−0.48
0.5	−0.38	−0.22	−0.51	−0.39	−0.36
1.0	−0.347	−0.23	−0.423	−0.35	−0.319
2.0	....	......	−0.370	−0.32	−0.299
3.0	....	......	−0.3586		
4.0	....	......	−0.3495		
5.0	....	......	−0.344		
6.0	....	......	−0.336		

† Fu = fumarate.
‡ Ma = maleate.

TABLE A4-9. RELATIVE PARTIAL MOLAL ENTHALPIES AT 0.01 AND 0.05 M

Salt	$\dfrac{\bar{L}_{MX_2} - \bar{L}_{CaCl_2}}{m}$, cal kg/mole^2		$-\dfrac{d\Delta B^{\cdot}}{dT} \times 10^4$	
	0.01 M	0.05 M	0.01 M	0.05 M
Li$_2$SO$_4$.............	3,400	1000	21	6
Na$_2$SO$_4$...........	−5,100	−4900	−32	−30
K$_2$SO$_4$.............	−5,500	−4600	−34	−28
Rb$_2$SO$_4$...........	−6,600	−5400	−41	−33
Cs$_2$SO$_4$............	−10,000	−7000	−60	−40
MgCl$_2$............	420	470	2.6	2.9
MgBr$_2$............	−340	−200	−2	−1
Mg(NO$_3$)$_2$.........	−1,800	−1600	−11	−10
CaBr$_2$.............	−2,000	−1000	−12	−6
Ca(NO$_3$)$_2$..........	−6,900	−4700	−42	−29
SrCl$_2$.............	−370	−540	−2.3	−3.3
SrBr$_2$.............	−2,900	−1900	−18	−12
Sr(NO$_3$)$_2$..........	−10,000	−7000	−60	−40
BaCl$_2$.............	−600	−700	−3.7	−4.3
BaBr$_2$.............	−3,000	−2000	−20	−12

by Harned and Owen,[1] and by Lange and Streeck.[2] The values for molalities above 0.1 M were obtained from the $\bar{L}$ values given by Harned and Owen.

The heat-capacity data of Tables A4-5a and A4-6 were obtained as indicated in Chapter 25. The values of $d^2\Delta B^{\cdot}/dT^2$ of Table A4-5b were obtained by the relationship

$$\frac{d^2\Delta B^{\cdot}}{dT^2} = \left(\frac{2}{T} \frac{d\Delta B^{\cdot}}{dT} + \frac{d^2\Delta B^{\cdot}}{dT^2} \right) - \frac{2}{T} \frac{d\Delta B^{\cdot}}{dT}$$

$$= \frac{-1}{2 \times 2.303 R T^2} \left(\frac{\bar{J}_{MX} - \bar{J}_{KCl}}{m} - \frac{2}{T} \frac{\bar{L}_{MX} - \bar{L}_{KCl}}{m} \right)$$

[1] H. S. Harned and B. B. Owen, "The Physical Chemistry of Electrolytic Solutions," 3d ed., Reinhold Publishing Corporation, New York, 1958.

[2] E. Lange and H. Streeck, *Z. physik. Chem.*, **152A**, 1 (1931), and **157A**, 1 (1931).

Appendix 5

DEBYE FUNCTIONS FOR THE THERMODYNAMIC PROPERTIES OF SOLIDS

The chief features of the derivation of the Debye heat capacity theory are outlined very briefly, after which the final equations and tables of values are given.

The vibrations of continuous solid media are distributed with a density of modes of oscillation proportional to the square of the frequency. Thus the number of modes of frequency between ν and $\nu + d\nu$ may be written

$$\rho(\nu)\, d\nu = c\nu^2\, d\nu \tag{A5-1}$$

where c is a constant. We know, however, that in a solid composed of N atoms there are $3N$ total modes of vibration. This follows because the interaction of the atoms can only rearrange the modes and frequencies of motion but cannot change the total number. Debye assumes that Eq. (A5-1) holds up to a certain maximum frequency of vibration ν_D which may be evaluated by calculating the total number of modes.

$$3N = \int_0^{\nu_D} c\nu^2\, d\nu = \frac{c\nu_D^3}{3} \tag{A5-2}$$

where
$$c = \frac{9N}{\nu_D^3}$$

Since the contribution of each single vibrational mode of frequency ν is well known [Eqs. (27-39) to (27-42)], one need only integrate over the Debye frequency distribution. The Debye characteristic temperature is defined by $\theta_D = h\nu_D/k$, where h and k are the Planck and Boltzmann constants, respectively. Also let $u = h\nu/kT$. Then the frequency distribution becomes

$$\rho(\nu)\, d\nu = 9N \left(\frac{T}{\theta_D}\right)^3 u^2\, du \tag{A5-3}$$

and the thermodynamic functions are

$$\frac{C_v}{3R} = 3 \left(\frac{T}{\theta_D}\right)^3 \int_0^{\theta_D/T} \frac{u^4 e^u\, du}{(e^u - 1)^2} \tag{A5-4}$$

$$\frac{E - E_0}{3RT} = 3 \left(\frac{T}{\theta_D}\right)^3 \int_0^{\theta_D/T} \frac{u^3\, du}{e^u - 1} \tag{A5-5}$$

$$-\frac{A - E_0}{3RT} = 3 \left(\frac{T}{\theta_D}\right)^3 \int_0^{\theta_D/T} \ln\,(1 - e^{-u}) u^2\, du \tag{A5-6}$$

Table A5-1. Debye Heat-capacity Function, $c_v/3R$, as a Function of θ_D/T

When $\dfrac{\theta_D}{T} \geq 16$, $\dfrac{c_v}{3R} = 77.927 \left(\dfrac{T}{\theta_D}\right)^3$

θ_D/T	0.0	0.1	0.2	0.3	0.4	0.5	0.6	0.7	0.8	0.9	1.0
0.0	1.0000	0.9995	0.9980	0.9955	0.9920	0.9876	0.9822	0.9759	0.9687	0.9606	0.9517
1.0	0.9517	0.9420	0.9315	0.9203	0.9085	0.8960	0.8828	0.8692	0.8550	0.8404	0.8254
2.0	0.8254	0.8100	0.7943	0.7784	0.7622	0.7459	0.7294	0.7128	0.6961	0.6794	0.6628
3.0	0.6628	0.6461	0.6296	0.6132	0.5968	0.5807	0.5647	0.5490	0.5334	0.5181	0.5031
4.0	0.5031	0.4883	0.4738	0.4595	0.4456	0.4320	0.4187	0.4057	0.3930	0.3807	0.3686
5.0	0.3686	0.3569	0.3455	0.3345	0.3237	0.3133	0.3031	0.2933	0.2838	0.2745	0.2656
6.0	0.2656	0.2569	0.2486	0.2405	0.2326	0.2251	0.2177	0.2107	0.2038	0.1972	0.1909
7.0	0.1909	0.1847	0.1788	0.1730	0.1675	0.1622	0.1570	0.1521	0.1473	0.1426	0.1382
8.0	0.1382	0.1339	0.1297	0.1257	0.1219	0.1182	0.1146	0.1111	0.1078	0.1046	0.1015
9.0	0.1015	0.09847	0.09558	0.09280	0.09011	0.08751	0.08500	0.08259	0.08025	0.07800	0.07582
10.0	0.07582	0.07372	0.07169	0.06973	0.06783	0.06600	0.06424	0.06253	0.06087	0.05928	0.05773
11.0	0.05773	0.05624	0.05479	0.05339	0.05204	0.05073	0.04946	0.04823	0.04705	0.04590	0.04478
12.0	0.04478	0.04370	0.04265	0.04164	0.04066	0.03970	0.03878	0.03788	0.03701	0.03617	0.03535
13.0	0.03535	0.03455	0.03378	0.03303	0.03230	0.03160	0.03091	0.03024	0.02959	0.02896	0.02835
14.0	0.02835	0.02776	0.02718	0.02661	0.02607	0.02553	0.02501	0.02451	0.02402	0.02354	0.02307
15.0	0.02307	0.02262	0.02218	0.02174	0.02132	0.02092	0.02052	0.02013	0.01975	0.01938	0.01902

TABLE A5-2. DEBYE FUNCTION FOR ENERGY CONTENT, $(E - E_0)/3RT$, AS A FUNCTION OF θ_D/T

When $\dfrac{\theta_D}{T} \geq 16$, $\dfrac{E - E_0}{3RT} = 19.482 \left(\dfrac{T}{\theta_D}\right)^3$

θ_D/T	0.0	0.1	0.2	0.3	0.4	0.5	0.6	0.7	0.8	0.9	1.0
0.0	1.0000	0.9630	0.9270	0.8920	0.8580	0.8250	0.7929	0.7619	0.7318	0.7026	0.6744
1.0	0.6744	0.6471	0.6208	0.5954	0.5708	0.5471	0.5243	0.5023	0.4811	0.4607	0.4411
2.0	0.4411	0.4223	0.4042	0.3868	0.3701	0.3541	0.3388	0.3241	0.3100	0.2965	0.2836
3.0	0.2836	0.2712	0.2594	0.2481	0.2373	0.2269	0.2170	0.2076	0.1986	0.1900	0.1817
4.0	0.1817	0.1739	0.1664	0.1592	0.1524	0.1459	0.1397	0.1338	0.1281	0.1227	0.1176
5.0	0.1176	0.1127	0.1080	0.1036	0.09930	0.09524	0.09137	0.08768	0.08415	0.08079	0.07758
6.0	0.07758	0.07452	0.07160	0.06881	0.06615	0.06360	0.06118	0.05886	0.05664	0.05453	0.05251
7.0	0.05251	0.05057	0.04873	0.04696	0.04527	0.04366	0.04211	0.04063	0.03921	0.03786	0.03656
8.0	0.03656	0.03532	0.03413	0.03298	0.03189	0.03084	0.02983	0.02887	0.02794	0.02705	0.02620
9.0	0.02620	0.02538	0.02459	0.02384	0.02311	0.02241	0.02174	0.02109	0.02047	0.01987	0.01930
10.0	0.01930	0.01874	0.01821	0.01769	0.01720	0.01672	0.01626	0.01581	0.01538	0.01497	0.01457
11.0	0.01457	0.01418	0.01381	0.01345	0.01311	0.01277	0.01245	0.01213	0.01183	0.01153	0.01125
12.0	0.01125	0.01098	0.01071	0.01045	0.01020	0.00996	0.00973	0.00950	0.00928	0.00907	0.00886
13.0	0.00886	0.00866	0.00846	0.00827	0.00809	0.00791	0.00774	0.00757	0.00741	0.00725	0.00710
14.0	0.00710	0.00695	0.00680	0.00666	0.00652	0.00639	0.00626	0.00613	0.00601	0.00589	0.00577
15.0	0.00577	0.00566	0.00555	0.00544	0.00533	0.00523	0.00513	0.00503	0.00494	0.00485	0.00476

TABLE A5-3. DEBYE FUNCTION FOR WORK CONTENT, $(A - E_0)/3RT$, AS A FUNCTION OF θ_D/T

When $\dfrac{\theta_D}{T} \geq 16$, $-\dfrac{A - E_0}{3RT} = 6.494 \left(\dfrac{T}{\theta_D}\right)^3$

θ_D/T	0.0	0.1	0.2	0.3	0.4	0.5	0.6	0.7	0.8	0.9	1.0
0.0	∞	2.6732	2.0168	1.6476	1.3956	1.2077	1.0602	0.9403	0.8405	0.7560	0.6835
1.0	0.6835	0.6205	0.5653	0.5166	0.4734	0.4349	0.4003	0.3692	0.3410	0.3156	0.2925
2.0	0.2925	0.2714	0.2522	0.2346	0.2185	0.2037	0.1901	0.1776	0.1661	0.1554	0.1456
3.0	0.1456	0.1365	0.1281	0.1203	0.1130	0.1063	0.1000	0.09423	0.08882	0.08377	0.07906
4.0	0.07906	0.07467	0.07057	0.06674	0.06316	0.05981	0.05667	0.05373	0.05097	0.04839	0.04596
5.0	0.04596	0.04368	0.04154	0.03952	0.03763	0.03584	0.03416	0.03258	0.03108	0.02967	0.02834
6.0	0.02834	0.02709	0.02590	0.02477	0.02371	0.02271	0.02175	0.02085	0.02000	0.01918	0.01841
7.0	0.01841	0.01768	0.01699	0.01633	0.01570	0.01510	0.01454	0.01400	0.01348	0.01299	0.01252
8.0	0.01252	0.01208	0.01165	0.01124	0.01085	0.01048	0.01013	0.00979	0.00946	0.00915	0.00886
9.0	0.00886	0.00857	0.00830	0.00804	0.00779	0.00755	0.00731	0.00709	0.00688	0.00667	0.00648
10.0	0.00648	0.00629	0.00611	0.00593	0.00576	0.00560	0.00544	0.00529	0.00515	0.00501	0.00487
11.0	0.00487	0.00474	0.00462	0.00450	0.00438	0.00427	0.00416	0.00405	0.00395	0.00385	0.00376
12.0	0.00376	0.00366	0.00357	0.00349	0.00340	0.00332	0.00325	0.00317	0.00310	0.00302	0.00296
13.0	0.00296	0.00289	0.00282	0.00276	0.00270	0.00264	0.00258	0.00253	0.00247	0.00242	0.00237
14.0	0.00237	0.00232	0.00227	0.00222	0.00217	0.00213	0.00209	0.00204	0.00200	0.00196	0.00192
15.0	0.00192	0.00189	0.00185	0.00181	0.00178	0.00174	0.00171	0.00168	0.00165	0.00162	0.00159

Table A5-4. Debye Entropy Function s/3R as a Function of θ_D/T

When $\frac{\theta_D}{T} \geq 16$, $\frac{s}{3R} = 25.976 \left(\frac{T}{\theta_D}\right)^3$

θ_D/T	0.0	0.1	0.2	0.3	0.4	0.5	0.6	0.7	0.8	0.9	1.0
0.0	∞	3.6362	2.9438	2.5396	2.2536	2.0327	1.8531	1.7022	1.5723	1.4587	1.3579
1.0	1.3579	1.2676	1.1861	1.1120	1.0442	0.9820	0.9246	0.8714	0.8222	0.7763	0.7336
2.0	0.7336	0.6937	0.6564	0.6214	0.5886	0.5578	0.5289	0.5017	0.4761	0.4519	0.4292
3.0	0.4292	0.4077	0.3875	0.3683	0.3503	0.3332	0.3171	0.3018	0.2874	0.2737	0.2608
4.0	0.2608	0.2486	0.2370	0.2260	0.2156	0.2057	0.1964	0.1875	0.1791	0.1711	0.1636
5.0	0.1636	0.1564	0.1496	0.1431	0.1369	0.1311	0.1255	0.1203	0.1152	0.1105	0.1059
6.0	0.1059	0.1016	0.09750	0.09358	0.08986	0.08631	0.08293	0.07971	0.07664	0.07371	0.07092
7.0	0.07092	0.06826	0.06572	0.06329	0.06097	0.05876	0.05665	0.05463	0.05270	0.05085	0.04908
8.0	0.04908	0.04739	0.04578	0.04423	0.04274	0.04132	0.03996	0.03866	0.03741	0.03621	0.03506
9.0	0.03506	0.03395	0.03289	0.03187	0.03090	0.02996	0.02905	0.02818	0.02735	0.02655	0.02577
10.0	0.02577	0.02503	0.02431	0.02362	0.02296	0.02232	0.02170	0.02111	0.02053	0.01998	0.01944
11.0	0.01944	0.01893	0.01843	0.01795	0.01749	0.01704	0.01660	0.01618	0.01578	0.01539	0.01501
12.0	0.01501	0.01464	0.01428	0.01394	0.01361	0.01328	0.01297	0.01267	0.01237	0.01209	0.01181
13.0	0.01181	0.01155	0.01129	0.01103	0.01079	0.01055	0.01032	0.01010	0.00988	0.00967	0.00946
14.0	0.00946	0.00926	0.00907	0.00888	0.00870	0.00852	0.00834	0.00818	0.00801	0.00785	0.00770
15.0	0.00770	0.00754	0.00740	0.00725	0.00711	0.00697	0.00684	0.00671	0.00659	0.00646	0.00634

663

At low pressure H $\cong$ E, and F $\cong$ A; hence the equations may be used also in good approximation for those more commonly used functions.   The entropy is given by (E − A)/T.   The numerical values for the functions as calculated by Beattie[1] and Shomate[2] are tabulated.

At low temperatures the upper limit of the integral in Eqs. (A5-4) to (A5-6) may be approximated by infinity.   This yields the equations under the table titles, which are valid at large θ_D/T.

[1] J. A. Beattie, *J. Math. Phys.*, **6,** 1 (1926).
[2] C. H. Shomate, private communication.

Appendix 6

TABULAR SUMMARY OF
THERMODYNAMIC FORMULAS

In this appendix we give first a list of the most widely used formulas, together with the equation numbers, which will locate their derivation in the text. The text should be consulted for the conditions for validity of each equation. In a second portion we give a shorthand system by which additional formulas may be obtained for any desired partial derivative.

Differential energy expressions:

$$dE = T\,dS - P\,dV \qquad \text{(10-7)}$$
$$dH = T\,dS + V\,dP \qquad \text{(10-8)}$$
$$dA = -S\,dT - P\,dV \qquad \text{(10-11)}$$
$$dF = -S\,dT + V\,dP \qquad \text{(10-12)}$$

Maxwell relations:

$$\left(\frac{\partial T}{\partial V}\right)_S = -\left(\frac{\partial P}{\partial S}\right)_V \qquad \text{(A6-1)[1]}$$

$$\left(\frac{\partial T}{\partial P}\right)_S = \left(\frac{\partial V}{\partial S}\right)_P \qquad \text{(A6-2)[1]}$$

$$\left(\frac{\partial S}{\partial V}\right)_T = \left(\frac{\partial P}{\partial T}\right)_V \qquad \text{(10-14)}$$

$$\left(\frac{\partial S}{\partial P}\right)_T = -\left(\frac{\partial V}{\partial T}\right)_P \qquad \text{(10-17)}$$

Energy-function derivatives:

$$\left(\frac{\partial E}{\partial S}\right)_V = \left(\frac{\partial H}{\partial S}\right)_P = T \qquad \text{(A6-3)[1]}$$

$$\left(\frac{\partial E}{\partial V}\right)_S = \left(\frac{\partial A}{\partial V}\right)_T = -P \qquad \text{(A6-4)}$$

$$\left(\frac{\partial H}{\partial P}\right)_S = \left(\frac{\partial F}{\partial P}\right)_T = V \qquad \text{(10-16)}$$

$$\left(\frac{\partial F}{\partial T}\right)_P = \left(\frac{\partial A}{\partial T}\right)_V = -S \qquad \text{(10-15)}$$

Heat-capacity relations·

$$C_V = \left(\frac{\partial E}{\partial T}\right)_V = T\left(\frac{\partial S}{\partial T}\right)_V \qquad \text{(4-11) (10-3)}$$

$$C_P = \left(\frac{\partial H}{\partial T}\right)_P = T\left(\frac{\partial S}{\partial T}\right)_P \qquad \text{(4-15) (10-4)}$$

[1] Derived from Eqs. (10-7) and (10-8).

$$C_P - C_V = -T \left(\frac{\partial V}{\partial T}\right)_P^2 \left(\frac{\partial P}{\partial V}\right)_T \qquad (10\text{-}27)$$

$$C_P - C_{\text{sat}} = T \left(\frac{\partial V}{\partial T}\right)_P \left(\frac{\partial P}{\partial T}\right)_{\text{sat}} \qquad (6\text{-}12)$$

$$\left(\frac{\partial C_V}{\partial V}\right)_T = T \left(\frac{\partial^2 P}{\partial T^2}\right)_V \qquad (16\text{-}39)$$

$$\left(\frac{\partial C_P}{\partial P}\right)_T = -T \left(\frac{\partial^2 V}{\partial T^2}\right)_P \qquad (16\text{-}23)$$

Effect of P or V on H or E:

$$\left(\frac{\partial H}{\partial P}\right)_T = V - T \left(\frac{\partial V}{\partial T}\right)_P \qquad (10\text{-}24)$$

$$\left(\frac{\partial E}{\partial V}\right)_T = T \left(\frac{\partial P}{\partial T}\right)_V - P \qquad (10\text{-}23)$$

Temperature effect on $\Delta F/T = -R \ln K$:

$$\left[\frac{\partial(\Delta F/T)}{\partial T}\right]_P = -R \frac{\partial \ln K}{\partial T} = -\frac{\Delta H}{T^2} \qquad (15\text{-}7)$$

Partial molal quantities, Y any extensive quantity:

$$\bar{Y}_1 = \left(\frac{\partial Y}{\partial n_1}\right)_{P,T,n_2,n_3,\ldots} \qquad (17\text{-}29)$$

$$Y = n_1 \bar{Y}_1 + n_2 \bar{Y}_2 + \cdots \qquad (17\text{-}32)$$

$$x_1 \left(\frac{\partial \bar{Y}_1}{\partial x_1}\right) + x_2 \left(\frac{\partial \bar{Y}_2}{\partial x_1}\right) + \cdots = 0 \qquad (17\text{-}38)$$

$$\left(\frac{\partial \bar{Y}_i}{\partial n_j}\right) = \frac{\partial^2 Y}{\partial n_i \, \partial n_j} = \left(\frac{\partial \bar{Y}_j}{\partial n_i}\right) \qquad (17\text{-}39)$$

Several schemes of shorthand notation for a wide variety of thermodynamic formulas have been proposed. Use may be made of mathematical functions known as Jacobians, which are defined as follows:

$$\frac{\partial(x,y)}{\partial(\alpha,\beta)} = \begin{vmatrix} \left(\dfrac{\partial x}{\partial \alpha}\right)_\beta & \left(\dfrac{\partial y}{\partial \alpha}\right)_\beta \\[2ex] \left(\dfrac{\partial x}{\partial \beta}\right)_\alpha & \left(\dfrac{\partial y}{\partial \beta}\right)_\alpha \end{vmatrix} = \left(\frac{\partial x}{\partial \alpha}\right)_\beta \left(\frac{\partial y}{\partial \beta}\right)_\alpha - \left(\frac{\partial x}{\partial \beta}\right)_\alpha \left(\frac{\partial y}{\partial \alpha}\right)_\beta \qquad (A6\text{-}5)$$

If α and β remain unchanged throughout, as will be the case here, a shorter notation may be used.

$$J(x,y) = \frac{\partial(x,y)}{\partial(\alpha,\beta)} \qquad (A6\text{-}6)$$

The following properties can be easily established by reference to the basic definition of Jacobians:

$$J(y,x) = -J(x,y) \qquad (A6\text{-}7)$$

$$\left(\frac{\partial y}{\partial x}\right)_z = \frac{J(y,z)}{J(x,z)} \qquad (A6\text{-}8)$$

$$J(\alpha,\beta) = 1 \qquad (A6\text{-}9)$$

We choose $T = \alpha$ and $P = \beta$. Also we shall follow Bridgman,[1] who chose to express the results in terms of T, P, and S and the three derivatives which are most readily capable of experimental measurement, namely, $(\partial V/\partial T)_P$, $(\partial V/\partial P)_T$, and $C_P = (\partial H/\partial T)_P$. Since all results will arise from the ratio of two Jacobians to yield a partial derivative, following Eq. (A6-8), we shall use a notation immediately suggestive of this property, $(\partial y)_z = J(y,z)$. Then Eq. (A6-8) becomes

$$\left(\frac{\partial y}{\partial x}\right)_z = \frac{(\partial y)_z}{(\partial x)_z} \tag{A6-10}$$

The summary follows:

$$(\partial T)_P = -(\partial P)_T = 1$$

$$(\partial V)_P = -(\partial P)_V = \left(\frac{\partial V}{\partial T}\right)_P$$

$$(\partial S)_P = -(\partial P)_S = \frac{C_P}{T}$$

$$(\partial E)_P = -(\partial P)_E = C_P - P\left(\frac{\partial V}{\partial T}\right)_P$$

$$(\partial H)_P = -(\partial P)_H = C_P$$

$$(\partial F)_P = -(\partial P)_F = -S$$

$$(\partial A)_P = -(\partial P)_A = -\left[S + P\left(\frac{\partial V}{\partial T}\right)_P\right]$$

$$(\partial V)_T = -(\partial T)_V = -\left(\frac{\partial V}{\partial P}\right)_T$$

$$(\partial S)_T = -(\partial T)_S = \left(\frac{\partial V}{\partial T}\right)_P$$

$$(\partial E)_T = -(\partial T)_E = T\left(\frac{\partial V}{\partial T}\right)_P + P\left(\frac{\partial V}{\partial P}\right)_T$$

$$(\partial H)_T = -(\partial T)_H = -V + T\left(\frac{\partial V}{\partial T}\right)_P$$

$$(\partial F)_T = -(\partial T)_F = -V$$

$$(\partial A)_T = -(\partial T)_A = P\left(\frac{\partial V}{\partial P}\right)_T$$

$$(\partial S)_V = -(\partial V)_S = \frac{1}{T}\left[C_P\left(\frac{\partial V}{\partial P}\right)_T + T\left(\frac{\partial V}{\partial T}\right)_P^2\right]$$

$$(\partial E)_V = -(\partial V)_E = C_P\left(\frac{\partial V}{\partial P}\right)_T + T\left(\frac{\partial V}{\partial T}\right)_P^2$$

$$(\partial H)_V = -(\partial V)_H = C_P\left(\frac{\partial V}{\partial P}\right)_T + T\left(\frac{\partial V}{\partial T}\right)_P^2 - V\left(\frac{\partial V}{\partial T}\right)_P$$

$$(\partial F)_V = -(\partial V)_F = -\left[V\left(\frac{\partial V}{\partial T}\right)_P + S\left(\frac{\partial V}{\partial P}\right)_T\right]$$

$$(\partial A)_V = -(\partial V)_A = -S\left(\frac{\partial V}{\partial P}\right)_T$$

$$(\partial E)_S = -(\partial S)_E = \frac{P}{T}\left[C_P\left(\frac{\partial V}{\partial P}\right)_T + T\left(\frac{\partial V}{\partial T}\right)_P^2\right]$$

$$(\partial H)_S = -(\partial S)_H = -\frac{VC_P}{T}$$

$$(\partial F)_S = -(\partial S)_F = -\frac{1}{T}\left[VC_P - ST\left(\frac{\partial V}{\partial T}\right)_P\right]$$

[1] P. W. Bridgman, *Phys. Rev.*, **3**, 273 (1914); A. N. Shaw, *Phil. Trans. Roy. Soc. London*, **A334**, 299 (1935).

$$(\partial A)_S = -(\partial S)_A = \frac{1}{T}\left\{P\left[C_P\left(\frac{\partial V}{\partial P}\right)_T + T\left(\frac{\partial V}{\partial T}\right)_P^2\right] + ST\left(\frac{\partial V}{\partial T}\right)_P\right\}$$

$$(\partial H)_E = -(\partial E)_H = -V\left[C_P - P\left(\frac{\partial V}{\partial T}\right)_P\right] - P\left[C_P\left(\frac{\partial V}{\partial P}\right)_T + T\left(\frac{\partial V}{\partial T}\right)_P^2\right]$$

$$(\partial F)_E = -(\partial E)_F = -V\left[C_P - P\left(\frac{\partial V}{\partial T}\right)_P\right] + S\left[T\left(\frac{\partial V}{\partial T}\right)_P + P\left(\frac{\partial V}{\partial P}\right)_T\right]$$

$$(\partial A)_E = -(\partial E)_A = P\left[(C_P + S)\left(\frac{\partial V}{\partial P}\right)_T + T\left(\frac{\partial V}{\partial T}\right)_P^2\right] + ST\left(\frac{\partial V}{\partial T}\right)_P$$

$$(\partial F)_H = -(\partial H)_F = -V(C_P + S) + TS\left(\frac{\partial V}{\partial T}\right)_P$$

$$(\partial A)_H = -(\partial H)_A = -\left[S + P\left(\frac{\partial V}{\partial T}\right)_P\right]\left[V - T\left(\frac{\partial V}{\partial T}\right)_P\right] + PC_P\left(\frac{\partial V}{\partial P}\right)_T$$

$$(\partial A)_F = -(\partial F)_A = -S\left[V + P\left(\frac{\partial V}{\partial P}\right)_T\right] - PV\left(\frac{\partial V}{\partial T}\right)_P$$

Appendix 7

TABLES OF THERMODYNAMIC PROPERTIES

Introduction

At the end of Chapter 15 a list was given of compilations from which one can obtain a wide variety of thermodynamic data needed for calculations. The following tables present representative thermodynamic data which are likely to be needed frequently in typical calculations. For the most part, these tables are restricted to data of high accuracy which are not likely to be appreciably altered by subsequent determinations. However, there will be instances of important compounds where most of the data are of high accuracy, but perhaps the high-temperature heat-capacity data, for example, may be lacking. In such instances it has been necessary to include estimated data to complete the tabulation. Such estimated values are given in parentheses.

It is customary in thermodynamic tabulations to give more significant figures than are warranted by the absolute accuracy when differences between thermodynamic quantities are known very accurately. Thus if high-temperature c_P° values are known accurately, $(F^\circ - H_{298}^\circ)/T$ values above 298.15°K are given to 0.01 cal/deg mole even when the value at 298.15°K can be given only to 0.1 cal/deg mole to indicate that the increments above 298.15°K are known to 0.01 cal/deg mole.

The quantities presented in the following tables consist of three types. The major part of the tables is devoted to free-energy functions based either on H_0° or on $H_{298.15}^\circ$. The use of these free-energy functions has been discussed in Chapter 15. For those substances where the free-energy functions are based on 298°K, values of the enthalpy of formation ΔH_{298}° are also tabulated. This is the enthalpy change on formation of the indicated substance from the elements in their standard reference states. The standard reference state for an element is the stable form of the element at 1 atm pressure and the temperature of interest, for example, 25°C. Thus, with a few exceptions such as hydrogen, oxygen, nitrogen, fluorine, and chlorine, which are gaseous diatomic molecules, and mercury and bromine, which are liquid, all the other elements which are included in the following tables are solids in their standard reference states at 25°C. Where two possible forms exist, such as for sulfur, the more stable form, rhombic sulfur, is the standard reference state except for the use of white phosphorus. Except for the elements that are gaseous at 25°C and are retained in the gaseous state for expressing ΔH_0° for the formation of a compound, ΔH_0° is based on the solid state of the elements at 0°K. In the tabulations of free-energy functions for condensed substances, the state of the condensed substances is either solid or liquid depending upon whether the indicated temperatures are above or below the melting point. The symbol m is inserted to indicate that melting took place between the tabulated temperatures. When gaseous substances are specifically indicated, all the data presented at all temperatures are for the gaseous state of the substance whether or not the gaseous state is the stable form at 1 atm.

The data are tabulated at rather wide temperature intervals, and interpolations are required for values at intermediate temperatures. It is very commonly found that the function $(\Delta F^\circ - \Delta H^\circ_{298})/T$ or the corresponding function based on $0°K$ varies more slowly with temperature than the free-energy functions of the individual substances. Thus it is usually expedient to calculate the value of $(\Delta F^\circ - \Delta H^\circ_{298})/T$ at the even temperatures given in the tables and then to interpolate to obtain directly values of the $(\Delta F^\circ - \Delta H^\circ_{298})/T$ function at the desired temperature. If one of the elements undergoes a transition, such as the melting of sulfur or the vaporization of bromine, the rate of change with temperature of $(\Delta F^\circ - \Delta H^\circ_{298})/T$ for the formation of a compound will show a discontinuity at the transition point, which makes interpolation difficult. Under those circumstances it is often expedient to use the gaseous standard state even at room temperature for elements such as iodine, bromine, and sulfur so that the free-energy functions for the formation of compounds will be expressed in terms of the same elemental form at all temperatures. This is easily done with the following tables since free-energy functions are given not only for the condensed states of these elements but also for the gaseous states. The heats of formation of compounds at 298.15°K are given with respect to the solid states of the element for sulfur and iodine and the liquid state for bromine, but the heats of sublimation and vaporization are given for these elements so that one can convert the heats of formation with respect to the condensed states of the elements to heats of formation with respect to the gaseous states.

References for Tables

Table A7-1. Condensed Elements. These tables, which present the free-energy functions based on H°_{298} for the elements in the solid and liquid states, are based primarily on the tabulations of Stull and Sinke,[1] with corrections where the new revisions of the *U.S. Bureau of Mines Bulletins* 476 and 477 indicate new data.[2]

Table A7-2. Gaseous Elements. As the new tabulation of Kelley and King[2] does not differ substantially from the values given by Stull and Sinke[1] for gaseous elements, the values given here are taken directly from Stull and Sinke except for revisions of some of the heats of formation of gaseous species. In particular the following references are cited as sources of newer data upon which revisions are based:

Au—Hultgren[3] has evaluated recent vapor-pressure data.

B—Evans, Wagman, and Prosen[4] list 129.4, 133, and 138.9 kcal/mole as reported values of ΔH°_0 for the sublimation of boron and take 134 kcal as the average. A new determination of 135 kcal[5] is in agreement with the average.

Bi and Bi$_2$—Brackett and Brewer[6] have applied the third-law treatment to the available data to obtain the heats of sublimation given in Table A7-2.

[1] D. R. Stull and G. C. Sinke, Thermodynamic Properties of the Elements, *Advances in Chem. Ser.* 18, 1956.

[2] K. K. Kelley, Contributions to the Data on Theoretical Metallurgy. XIII. High-temperature Heat-content, Heat-capacity, and Entropy Data for the Elements and Inorganic Compounds, *U.S. Bur. Mines Bull.* 584, 1960; K. K. Kelley and E. G. King, Contributions to the Data on Theoretical Metallurgy. XIV. Entropies of Inorganic Substances, *U.S. Bur. Mines Bull.* 592, in press.

[3] See bibliography listing 6 at end of Chapter 15.

[4] W. H. Evans, D. D. Wagman, and E. J. Prosen, *Natl. Bur. Standards (U.S.) Rept.* 6252, Dec. 15, 1958.

[5] P. W. Gilles and H. E. Robson, paper presented at American Chemical Society meeting, Boston, April, 1959.

[6] E. Brackett and L. Brewer, The Heat of Dissociation of Bi$_2$, UCRL-3712, Mar. 8, 1957. See also R. F. Porter and C. W. Spencer, *J. Chem. Phys.*, **32**, 943 (1960).

TABLE A7-1. SOLID AND LIQUID ELEMENTS
Free energies based on H°_{298}

	$-(F^\circ - H^\circ_{298})/T$, cal/deg					$H^\circ_{298} - H^\circ_0$,
	298.15°K	500°K	1000°K	1500°K	2000°K	kcal
Ag.............	10.20	10.90	m 13.40 m	15.87	(17.99)	1.373
Al.............	6.77	7.45	m 10.20	(13.14)	(15.22)	1.094
As.............	8.4	9.08	.....	.....	.....	1.226
Au.............	11.32	12.02	14.51	16.77	(18.87)	1.434
B.............	1.40	1.76	(3.36)	(4.88)	(6.21)	0.292
Ba.............	16.0	16.79	m (19.5)	(22.6)	.....	(1.76)
Be.............	2.28	2.78	4.82	6.62 m	(8.55)	0.468
Bi.............	13.5	14.23	m 19.19	22.3	.....	1.536
Br₂.............	36.4	.....	.....	.....	.....	3.24
C.............	1.37	1.67	3.04	4.37	5.51	0.251
Ca.............	9.95	10.69	13.59	m (16.53)	.....	1.38
Cd.............	12.37	13.09	m 16.75			1.49
Co.............	7.18	7.88	10.62	13.10	m 15.48	1.146
Cr.............	5.68	6.34	8.84	11.01	12.86	0.973
Cs.............	20.16	21.70	25.23	.....	.....	1.859
Cu.............	7.97	8.64	11.09	m 13.33	(15.47)	1.201
Fe.............	6.49	7.21	10.07	12.87	m 15.24	1.07
Gd.............	15.8	(16.6)	(19.6)	(21.9)	m (24.3)	2.172
Hf.............	10.91	(11.61)	(14.13)	(16.17)	(17.82)	1.448
Hg.............	18.17	18.92	.....	.....	.....	2.23
I₂.............	27.76	.....	.....	.....	.....	3.154
K.............	15.34	m 16.74	20.2	.....	.....	1.695
Li.............	6.73	m 7.60	11.00	13.35	.....	1.092
Mg.............	7.81	8.51	m 11.28	.....	.....	1.195
Mn.............	7.64	8.41	11.28	14.00	m 16.94	1.194
Mo.............	6.84	7.50	9.92	11.89	(13.5)	1.092
Na.............	12.24	m 13.49	16.96	.....	.....	1.531
Ni.............	7.14	7.90	10.77	13.09	m 15.31	1.144
Pb.............	15.49	16.23	m 19.68	(22.18)	(24.00)	1.644
Pt.............	9.95	10.66	13.21	15.29	16.98	1.384
Rb.............	18.1	m 19.6	.....	.....	.....	1.79
S.............	7.62	m 8.59	.....	.....	.....	1.053
Sb.............	10.92	11.61	m 14.66	18.34	.....	1.226
Se.............	10.14	m 10.93	.....	.....	.....	1.319
Si.............	4.51	5.09	7.30	9.12	m 11.7	0.769
Sn.............	12.29	13.07	m 17.55	20.34	(22.35)	1.507
Ta.............	9.91	10.62	13.13	15.13	16.71	1.358
Te.............	11.88	12.62	m 17.11	.....	.....	1.463
Th.............	12.76	13.53	16.50	19.16	21.56	1.556
Ti.............	7.30	8.01	10.58	12.95	m 14.9	1.15
Tl.............	15.35	16.09	m 19.67	(22.23)	.....	1.632
U.............	12.03	12.81	15.98	m 19.42	(22.4)	1.559
V.............	7.01	7.69	10.23	12.35	14.14	1.122
W.............	8.00	8.68	11.10	13.03	14.59	1.216
Zn.............	9.95	10.66	m 14.06	.....	.....	1.349
Zr.............	9.32	10.05	12.75	(15.2)	(17.1)	1.313

TABLE A7-2. GASEOUS ELEMENTS
Free energies based on H_{298}°

	$-(F^\circ - H_{298}^\circ)/T$, cal/deg					$H_{298}^\circ - H_0^\circ$, kcal	ΔH_{298}°, kcal
	298.15°K	500°K	1000°K	1500°K	2000°K		
Ag........	41.32	41.89	43.85	45.37	46.56	1.481	68.4
Al........	39.30	39.88	41.87	43.40	44.59	1.654	77.5 ± 1.5
Au........	43.12	43.69	45.65	47.17	48.35	1.481	87.3
B.........	36.65	37.22	39.18	40.70	41.89	1.51	135 ± 4
Ba........	40.67	41.23	43.20	44.72	45.94	1.481	42.5 ± 0.5
Be........	32.55	33.11	35.08	36.59	37.78	1.481	77.9
Bi........	44.67	45.24	47.20	48.72	49.90	1.481	49.5 ± 1
Bi_2........	65.40	66.40	69.92	72.65	74.77	2.453	52.5 ± 1
Br........	41.81	42.37	44.34	45.88	47.11	1.481	26.76
Br_2........	58.65	59.64	63.13	65.87	68.00	2.325	7.45
C.........	37.76	38.33	40.30	41.81	43.00	1.56	170.9
Ca........	36.99	37.56	39.53	41.04	42.22	1.481	42.2
Cd........	40.07	40.64	42.60	44.11	45.30	1.481	26.75
Cl........	39.46	40.06	42.19	43.83	45.09	1.499	28.94
Cl_2........	53.29	54.24	57.63	60.32	62.43	2.194	0
Co........	42.88	43.53	45.90	47.79	49.28	1.52	101.6
Cr........	41.64	42.20	44.17	45.69	46.89	1.481	95.0 ± 1
Cs........	41.94	42.51	44.48	45.99	47.18	1.481	18.67
Cu........	39.74	40.31	42.28	43.79	44.98	1.481	81.1
F.........	37.92	38.53	40.62	42.21	43.43	1.558	18.9 ± 1
F_2........	48.45	49.33	52.55	55.16	57.22	2.11	0
Fe........	43.11	43.81	46.14	47.88	49.19	1.637	99.5
Gd........	46.42	47.16	49.69	51.57	53.01	1.825	82.5 ± 5
H.........	27.39	27.96	29.92	31.44	32.63	1.481	52.09
H_2........	31.21	32.00	34.76	36.95	38.69	2.024	0
Hf........	44.64	45.22	47.27	49.01	50.47	1.481	168 ± 10
Hg........	41.79	42.36	44.33	45.84	47.03	1.481	14.66
I.........	43.18	43.75	45.72	47.24	48.42	1.481	25.52
I_2........	62.28	63.29	66.82	69.58	71.74	2.418	14.92
K.........	38.30	38.87	40.83	42.34	43.53	1.481	21.45
Li........	33.14	33.71	35.68	37.19	38.38	1.481	38.44
Mg........	35.51	36.07	38.04	39.55	40.74	1.481	35.6
Mn........	41.49	42.06	44.03	45.54	46.73	1.481	67.2
Mo........	43.46	44.03	45.99	47.51	48.70	1.481	157.5
N.........	36.61	37.18	39.15	40.66	41.85	1.481	112.98
N_2........	45.77	46.57	49.39	51.67	53.51	2.072	0
Na........	36.71	37.28	39.25	40.76	41.95	1.481	25.9
Ni........	43.52	44.16	46.45	48.25	49.64	1.631	102.8
O.........	38.47	39.06	41.07	42.61	43.81	1.607	59.55
O_2........	49.01	49.83	52.78	55.19	57.15	2.075	0
Pb........	41.89	42.46	44.42	45.95	47.15	1.481	46.8 ± 1
Pt........	45.96	46.68	49.17	51.00	52.35	1.572	135.2
Rb........	40.63	41.20	43.16	44.68	45.86	1.481	19.5 ± 1

TABLE A7-2. GASEOUS ELEMENTS (*Continued*)

	$-(F° - H°_{298})/T$, cal/deg					$H°_{298} - H°_0$, kcal	$\Delta H°_{298}$, kcal
	298.15°K	500°K	1000°K	1500°K	2000°K		
S..........	40.09	40.72	42.88	44.49	45.73	1.591	66.4 ± 2
S₂..........	54.51	55.42	58.72	61.35	63.43	2.141	30.84
Sb..........	43.06	43.63	45.59	47.11	48.30	1.481	62 ± 7
Sb₂.........	60.90	61.89	65.38	68.10	70.21	2.363	54.9
Se..........	42.21	42.78	44.82	46.46	47.77	1.481	49.4 ± 1
Se₂.........	60.23	61.21	64.67	67.39	69.52	2.57	34.12 ± 1
Si..........	40.12	40.71	42.74	44.29	45.49	1.805	108 ± 3
Sn..........	40.24	40.85	43.28	45.43	47.13	1.485	72.0
Ta..........	44.24	44.82	46.95	48.78	50.33	1.482	186.8
Te..........	43.64	44.21	46.18	47.71	48.93	1.481	46 ± 2
Te₂.........	64.10	65.10	68.60	71.32	73.43	2.379	39.6 ± 1
Th..........	40.46	43.03	46.51	49.80	51.35	1.483	136.6
Ti..........	43.07	43.72	45.84	47.44	48.70	1.802	112.7 ± 1
Tl..........	43.23	43.79	45.76	47.28	48.48	1.481	43.0 ± 1
U..........	47.73	48.37	50.61	52.34	53.73	1.553	125 ± 3
V..........	43.55	44.23	46.55	48.38	49.79	1.89	122.75 ± 5
W..........	41.55	42.16	44.77	47.26	49.27	1.486	200.0 ± 1
Zn..........	38.45	39.02	40.98	42.50	43.69	1.481	31.18
Zr..........	43.32	44.06	46.62	48.56	50.08	1.629	146 ± 1

Fe and Ni—The heats of sublimation in Table A7-2 are based on the recent work of Morris and Zellars and Morris, Zellars, Payne, and Kipp[1] and the confirming work of Bell, Davis, King, and Routly.[2]

Pt—The free-energy functions are from Hultgren,[3] and the enthalpy of sublimation is from Dreger and Margrave.[4]

S—Brewer[5] and Marsden[6] review the recent data which relate to the dissociation energy of S₂.

Sb, Se, and Te—The values given by Stull and Sinke have been reappraised in the light of the observations of Goldfinger, Ackerman, and Jeunehomme.[7]

Si—The value given is from Davis, Anthrop, and Searcy.[8]

[1] J. P. Morris and G. R. Zellars, *J. Metals*, **8**, 1086 (1956); J. P. Morris, G. R. Zellars, S. L. Payne, and R. L. Kipp, *U.S. Bur. Mines Rept. Invest.* 5364, 1957.

[2] G. D. Bell, M. H. Davis, R. B. King, and P. M. Routly, *Astrophys. J.*, **127**, 775 (1958).

[3] See bibliography listing 6 at end of Chapter 15.

[4] L. H. Dreger and J. L. Margrave, *J. Phys. Chem.*, **64**, 1323 (1960).

[5] L. Brewer, *J. Chem. Phys.*, **31**, 1143 (1959).

[6] D. G. H. Marsden, *J. Chem. Phys.*, **31**, 1144 (1959).

[7] P. Goldfinger, M. Ackerman, and M. Jeunehomme, Final Technical Report Contract AF 61(052)-19, Brussels, January, 1959.

[8] S. G. Davis, D. F. Anthrop, and A. W. Searcy, *J. Chem. Phys.*, **34**, 659 (1961).

TABLE A7-3. SOLID AND LIQUID HALIDES
Free energies based on H_{298}°

	$-(F^\circ - H_{298}^\circ)/T$, cal/deg				ΔH_{298}°, kcal
	298.15°K	500°K	1000°K	1500°K	
AlF$_3$	15.89	18.08	26.67	33.7	-356.3 ± 1
AlCl$_3$	26.58 m	30.5	(50.1)	.....	-168.6 ± 0.5
AgCl	23.00	24.49 m	31.25	(37)	-30.36
AgBr	25.6	27.13 m	34.05	(41)	-23.8
AgI	27.6	29.3 m	(37)	.....	-14.9
BaF$_2$	23.03	25.1	32.7	(38)	-286.9
BiCl$_3$	45.7	(49) m	.....	.....	-90.6
CaF$_2$	16.46	18.4	25.7	31.8	-290.2
CaCl$_2$	27.2	29.3	36.6 m	44.6	-190.0
CdCl$_2$	27.6	29.6 m	38.4	.....	-93.0
CdBr$_2$	34.9	(36.9) m	(45)	.....	-75.8 ± 1
CdI$_2$	38	(40) m	(48)	.....	-48.4 ± 1
CoCl$_2$	25.4	27.7	36.4 m	.....	-75 ± 2
CuCl	20.8	22.2 m	29.3	.....	-32.6
CuBr	23.0	(24) m	(31)	.....	-25.3
CuI	23.1	(24) m	(31)	.....	-16.4
CrF$_3$	22.44	(24)	(34) m	(46)	-266 ± 5
CrCl$_2$	27.7	(29.7)	(36.7) m	(46)	-97 ± 3
CrCl$_3$	29.4	(31)	(42) m	(51)	-132 ± 5
FeCl$_2$	28.7	30.85 m	39.1	(49.0)	-81.86
FeBr$_2$	33.6	36 m	(45)	.....	-60.0
FeI$_2$	36.8	39	(48)	.....	-30 ± 2
FeCl$_3$	35	37.8 m	.....	.....	-95.7
GdCl$_3$	34.9	(37) m	(49)	.....	-240.1
HfCl$_4$	45.6	48.95 m	.....	.....	-237 ± 2
Hg$_2$Cl$_2$	46.2	(48)	.....	.....	-63.3
Hg$_2$Br$_2$	52.4	54	.....	.....	-48.7
HgCl$_2$	34.5	36.5 m	.....	.....	-53.4 ± 1
HgBr$_2$	40.7	(42.7) m	.....	.....	-40.7
HgI$_2$	42.4	44.9 m	.....	.....	-25.3
KF	15.91	17.27	22.29 m	(27.9)	-134.5
KCl	19.72	21.1	26.34 m	(32.7)	-104.18
KBr	22.96	24.4	29.68 m	(36.0)	-93.73
KI	25.2	26.8 m	(32.3)	(38.6)	-78.31
LiF	8.52	9.73	14.39 m	(19.84)	-145.7 ± 2
LiCl	14.17	15.52 m	21.25	(27.12)	-96 ± 2
MgF$_2$	13.68	15.51	22.3	27.9	-268
MgCl$_2$	21.40	23.41 m	30.92	40.43	-153.2
MnF$_2$	22.25	24.3	(31.1) m	(41)	-190 ± 5
MnCl$_2$	28.0	30.1 m	38.3	(47.6)	-115.19
NaF	12.26	13.57	18.42 m	23.39	-136.3
NaCl	17.33	18.72	23.85 m	29.92	-98.2
NaBr	(20.8)	(22.3)	(27.4) m	(33.8)	-86.0

TABLE A7-3. SOLID AND LIQUID HALIDES (*Continued*)

	$-(\text{F}° - \text{H}°_{298})/T$, cal/deg				$\Delta\text{H}°_{298}$, kcal
	298.15°K	500°K	1000°K	1500°K	
NaI................	23.55	25.0 m	(30.8)	(37.0)	−68.8
NiF₂..............	17.59	(19)	(26) m	.....	−156 ± 2
NiCl₂..............	23.33	25.34	32.69 m	40.55	−72.97
PbCl₂..............	32.5	34.6 m	44.4	.....	−85.7
PbBr₂..............	38.6	(40.8) m	(50.2)	.....	−66.3
PbI₂........	42.2	44.4 m	55.95	.....	−41.6
RbBr..............	26.0	(27.0) m	(32.77)	(38.88)	−93.5
RbI..............	28.2	(29.7) m	(35.11)	(41.22)	−79.0
SbCl₃..............	44.5 m	(50.8)	.....	.....	−91.4
TiCl₄(l)............	60.3	64.55	.....	.....	−192.3
TlF..............	22.0	(22) m	(31)	.....	−74 ± 5
TlCl..............	26.59	28.04 m	34.92	.....	−48.79
TlBr..............	29.5	30.9 m	37.64	.....	−41.9
TlI................	31.0	(32.7) m	(41)	.....	−29.6
UF₄..............	36.25	39	52	m (61)	−443 ± 5
VCl₂..............	23.2	25.2	32.5	m (41)	(−102) ± 10
VCl₃..............	31.3	33.9	43.5	.....	(−137) ± 10
ZnCl₂..............	26.64	(28.8) m	(42.5)	(51.2)	−99.6
ZnBr₂..............	32.5	(34.5) m	(47.5)	(57)	−78.4 ± 1
ZnI₂..............	35.0	(37.0) m	(49)	(58)	−49.8
ZrCl₄..............	44.5	47.84 m	.....	.....	−233 ± 2

Th—Darnell, McCollum, and Milne[1] give values of the free-energy functions and the enthalpy of sublimation. Values of the free-energy function at lower temperatures were obtained from Margrave.[2]

U—The value given is based on an appraisal of all available measurements, along with recent observations[3] which indicate that some of the previous Knudsen-cell measurements have yielded too high a vapor pressure owing to porosity of the lids.

Tables A7-3 to A7-6. Condensed Halides, Oxides, Carbides, Nitrides, and Gaseous Halides. The free-energy functions of these tables were largely derived from the values of $\text{S}°_{298}$, $\text{H}° - \text{H}°_{298}$, and $\text{S}° - \text{S}°_{298}$ tabulated by Kelley and King.[4] The $\Delta\text{H}°_{298}$

[1] A. J. Darnell, W. A. McCollum, and T. A. Milne, *J. Phys. Chem.*, **64,** 341 (1960).

[2] See bibliography listing 13 at end of Chapter 15.

[3] J. S. Hubbs, private communication; G. De Maria, R. P. Burns, J. Drowart, and M. G. Inghram, *J. Chem. Phys.*, **32,** 1373 (1960).

[4] See bibliography listing 4 at the end of Chapter 15. In addition to the three major compilations listed there, a number of *U.S. Bureau of Mines Reports of Investigations* and journal publications under the authorship of K. K. Kelley, S. S. Todd, J. P. Coughlin, R. L. Orr, K. R. Bonnickson, G. L. Humphrey, R. Barany, E. G. King, A. D. Mah, and other members of the staff of the U.S. Bureau of Mines at Berkeley, Calif., provided additional $\Delta H°$ data.

TABLE A7-4. SOLID OXIDES, SULFIDES, AND RELATED COMPOUNDS
Free energies based on H_{298}°

	$-(F^{\circ} - H_{298}^{\circ})/T$, cal/deg					ΔH_{298}°, kcal
	298.15°K	500°K	1000°K	1500°K	2000°K	
Ag_2O	29.1	30.99	......	......	......	-7.2
Ag_2S	33.5	35.93	45.24 m	......	......	-7.6
Al_2O_3	12.18	14.61	24.46	32.86	39.67	-400.4
B_2O_3	12.91	14.88	m 25.7	35.8	(43.4)	-305.3
BaO	16.8	18.11	23.00	26.98	(30.2)	-133.5
BeO	3.37	4.19	7.76	(11.0)	(13.7)	-143.1
$CaCO_3$	22.2	24.69	34.47	......	......	-288.45
$Ca(NO_3)_2$	46.2	50.67	m	......	......	-224.0
CaO	9.5	10.73	15.26	18.94	21.9	-151.79
$Ca(OH)_2$	19.93	22.48	32.17	......	......	-235.65
$CaSO_4$	25.5	28.38	39.76	50.5	......	-342.42
$CaSiO_3$	19.6	22.18	32.21	40.68 m	......	-382.9
$Ca_2SiO_4(\beta)$	30.5	34.25	48.9	61.42	72.04	-543.7
$Ca_2SiO_4(\gamma)$	28.8	32.47	46.72	......	......	-546.2
Ca_3SiO_5	40.3	45.36	64.9	81.36	......	-692.3
$CaTiSiO_5$	30.90	35.08	51.15	64.63	77.77	-613.6
$CaTiO_3$	22.40	25.38	36.49	45.56	52.88	-396.9
$Ca_3Ti_2O_7$	56.10	(63.29)	(90.04)	(111.86)	(129.39)	-944.1
Cr_2O_3	19.4	22.51	33.62	42.61	49.75	-272.7
CoO	12.66	14.10	19.18	23.21	26.47	-57.1
Co_3O_4	24.5	28.18	42.77	......	......	-204 ± 4
Cu_2O	22.4	24.33	31.48	(37.2) m	......	-40.4
CuO	10.19	11.43	16.09	(20.07) m	......	-37.6
Cu_2S	28.9	31.85	41.51	51.66 m	......	-19.0
$Fe_{0.947}O$	13.74	17.43	20.34	24.31 m	28.35	-63.8
Fe_3O_4	35.0	39.56	58.97	73.21 m	(86)	-267.8
Fe_2SiO_4	34.7	38.57	53.72 m	66.62	......	-345.7
Fe_2O_3	20 90	24.00	36.43	46.91	......	-196.8
$FeTiO_3$	25.30	28.21	39.22	48.38 m	58.45	-295.55
FeS	14.41	16.32	23.15 m	28.09	33.16	-22.72
FeS_2	12.7	14.55	21.32	......	......	-42.52
HfO_2	14.18	15.90	22.65	28.06	32.54	-266.05
$KMg_3AlSi_3O_{10}F_2$	75.9	85.97	125.29	165.22 m	......	-1497.0 ± 1
Li_2O	8.97	10.60	17.12	(22.84)	......	-142.4
MgO	6.55	7 63	11.70	15.12	(17.9)	-143.77
$Mg(OH)_2$	15.09	17.21	......	......	......	-220.97
$MgSiO_3$	16.22	18.63	28.12	36.26 m	(43.8)	-362.4
Mg_2SiO_4	22.75	26.24	40.10	51.90	(61.6)	-512.6
$MgTi_2O_5$	33.20	36.51	54.08	68.05	79.61	-599.8
Mg_2TiO_4	27.51	31.27	45.94	58.47 m	68.7	-517.45
$MgTiO_3$	17.82	20.61	31.27	40.15 m	47.4	-375.9
MnO	14.27	15.53	20.15	23.93	27.0	-92.05

TABLE A7-4. SOLID OXIDES, SULFIDES, AND RELATED COMPOUNDS (*Continued*)

	$-(F° - H°_{298})/T$, cal/deg					$\Delta H°_{298}$, kcal
	298.15°K	500°K	1000°K	1500°K	2000°K	
Mn$_3$O$_4$.........	35.5	39.68	55.21	68.2 m	......	−331.4
Mn$_2$O$_3$.........	26.4	29.24	40.13	49.5	......	−228.4
MnO$_2$..........	13.68	15.31	21.64	......	......	−124.45
MnS............	18.7	20.06	24.95	28.85 m 32.33		−49.5
MnSiO$_3$........	21.3	23.89	33.98	42.46 m	......	−307.87
MoO$_3$..........	18.58	20.75	29.23 m 40.1		......	−178.1
Na$_2$CO$_3$........	32.5	35.63	48.78 m 61.9		......	−270.3
NaNO$_3$.........	27.85	30.71	m 46.45	......	......	−111.54
Na$_2$O..........	18.2	20.2	(27.5) m (35.4)	(42.1)		−99.4
NaO$_2$..........	27.7	(29.3)	(36.7)	......	......	−62.0
Na$_2$SO$_4$*.......	35.73(V)	39.74(III)	57.53(I)	72.70(l)	84.69	−330.9
Nb$_2$O$_5$.........	32.8	36.61	51.07	63.08 m 74.36		−455.2
NiO...........	9.08	10.40	15.53	19.51	22.7	−57.3
PbO y.........	16.1	17.39	22.28 m (27.87)		......	−52.05
PbO r.........	15.6	16.96	(22.0)	......	......	−52.45
PbS...........	21.8	23.1	28.2	......	......	−22.54
SiO$_2$ q.........	10.00	11.36	17.00	21.84 m 25.83		−209.9
SiO$_2$ cr.........	10.20	11.56	17.32	22.11	25.98	−209.6
SiO$_2$ tr.........	10.4	11.82	17.48	22.23 m 26.08		−209.4
SnS...........	18.4	19.75	24.83 m	......	......	−24.3
SnS$_2$..........	20.9	22.84	29.82			
Ta$_2$O$_5$..........	34.2	38.04	52.86	65.25	75.4	−488.8
TiO...........	8.31	9.52	14.12	18.12	21.6	−124.15
Ti$_2$O$_3$..........	18.83	21.83	34.11	44.25	52.38	−363.4
Ti$_3$O$_5$..........	30.9	36.61	57.19	72.64	84.74	−587.65
TiO$_2$ rut........	12.04	13.76	20.09	25.25	29.4	−225.75
TiO$_2$ an........	11.93	13.64	19.99	(25.2)		
TiS$_2$...........	18.73	20.70	27.81			
UO$_2$..........	18.63	20.52	27.69	33.62	38.4	−259.2
UO$_3$...........	23.58	25.9	34.6	......	......	−291.0
V$_2$O$_3$...........	23.5	26.55	38.10	47.64	55.55	−296
VO$_2$...........	12.25	14.96	22.67	28.56 m 33.9		−171
V$_2$O$_5$...........	31.3	35.35	m 51.41	69.23	82.0	−373
ZnO...........	10.43	11.63	16.10	19.77	22.76	−83.25
ZnS...........	13.8	15.10	19.96	......	......	−48.5
ZrO$_2$...........	12.12	13.77	20.09	25.35	29.76	−261.5

* Roman numerals in parentheses refer to crystalline modifications designated by F. C. Kracek and R. E. Gibson, *J. Phys. Chem.*, **34**, 188 (1930).

TABLE A7-5. SOLID CARBIDES AND NITRIDES
Free energies based on H_{298}°

	$-(F^\circ - H_{298}^\circ)/T$, cal/deg					ΔH_{298}°, kcal
	298.15°K	500°K	1000°K	1500°K	2000°K	
B$_4$C............	6.47	8.32	16.47	23.90	30.14	-12.7 ± 3
BN.............	3.67	4.31	7.23	10.0	.....	-60.3
CaC$_2$...........	16.8	18.59	25.72	31.42	35.8	-14.1
Cr$_{23}$C$_6$..........	145.8	164.1	234.15	293.53	.....	-98.3
Cr$_7$C$_3$............	48.0	50.96	62.61	72.49	81.10	-42.5
Cr$_3$C$_2$............	20.42	23.38	35.03	44.91	53.52	-21.0
Fe$_3$C............	24.2	27.27	38.26	46.9	.....	5.0
Mn$_3$C..........	23.6	26.34	36.72	45.81	.....	$-1 \quad \pm 3$
Mo$_3$Si...........	25.4	27.99	37.44	45.14	.....	-23.5 ± 4
SiC(hex)........	3.94	4.78	8.41	11.60	14.25	-15 ± 2
SiC(cub)........	3.97	4.82	8.45	11.63	14.27	-15 ± 2
TiC.............	5.79	6.85	11.03	14.54	17.38	-44.1
TiN.............	7.24	8.37	12.71	16.32	19.24	-80.7
ZrN.............	9.29	10.45	14.85	18.52	21.47	-87.3

values came mainly from the publications of the U.S. National Bureau of Standards[1] and of the U.S. Bureau of Mines[2] at Berkeley and from Latimer.[3]

A few additional sources of data will be noted. Free-energy functions for the condensed alkali halides and heats of vaporization of alkali halides are based on a compilation by Brewer and Brackett.[4] The data for all Rb and Cs compounds given by *U.S. Bureau of Standards Circular* 500 have been corrected to be consistent with $\Delta \bar{H}^\circ$ of Cs$^+$ determined by Friedman and Kahlweit[5] and to correct for an error in $\bar{S}^\circ$ of Rb$^+$. To all ΔH° and ΔF° values, -0.5 kcal was added per gram atom of Rb and -3.4 kcal was added per gram atom of Cs. Recent work also indicates that the ΔH° values for Na and K compounds should be more negative by as much as 0.4 kcal/mole. These last corrections have not been applied because the thermodynamic data for Na and K compounds are so interrelated with one another and with data for other elements that a complete systematic reevaluation would be required before any changes could be made without danger of introducing serious inconsistencies. If the proposed unified atomic-weight scale is adopted, the changes resulting from the new atomic weights as well as from changes in fundamental constants will require recalculation of all the accurate thermodynamic data. A revision of *Circular* 500 might then reflect not only the changes in the constants but also more recent data which have accumulated since 1950.

The data of Schäfer and Oehler[6] have been used to fix the entropy of solid FeCl$_3$.

[1] See bibliography listing 2 at the end of Chapter 15. In addition to the supplements in the *Journal of Research of the National Bureau of Standards, Natl. Bur. Standards (U.S.) Repts.* 6252, 1958; 6484, 1959; and 6297, 1959, were very valuable.

[2] See footnote 4 on page 675.

[3] See bibliography listing 7 at the end of Chapter 15.

[4] L. Brewer and E. Brackett, paper in preparation.

[5] H. L. Friedman and M. Kahlweit, *J. Am. Chem. Soc.*, **78**, 4243 (1956).

[6] H. Schäfer and E. Oehler, *Z. anorg. allgem. Chem.*, **271**, 206 (1953).

TABLE A7-6. GASEOUS HALIDES
Free energies based on H_{298}°

	$-(F^{\circ} - H_{298}^{\circ})T$, cal/deg					ΔH_{298}°, kcal
	298.15°K	500°K	1000°K	1500°K	2000°K	
AlBr	57.2	58.17	61.6	64.4	66.4	0 ± 5
AlI	59.1	60.1	63.6	66.4	68.4	15 ± 5
AsF₃	69.08	70.96	77.99	......	88.2	-218.3
AsCl₃	78.3	80.4	87.9	......	98.5	-71.5
AsI₃	92.7	94.7	102.7	......	113.4	8
BiF	58.2	59.15	62.54	......	67.3	7
BiCl	60.9	61.89	65.36	......	70.2	10.7
BiBr	63.6	64.6	68.1	......	88.1	12.7
CsF	58.03	59.02	62.47	65.17	67.27	-82
CsBr	63.99	......	68.52	71.26	73.36	-51.1
CsI	66.01	67.02	70.57	73.30	75.41	-38.2
CuF	54.1	55.02	58.36	......	63.1	44
CuCl	56.5	57.46	60.9	......	65.7	32
CuBr	59.2	60.18	63.66	......	68.5	38
CuI	61.1	62.1	65.6	......	70.4	62
HgF	59.3	60.23	63.66	......	68.43	14
HgCl	62.1	63.09	66.57	......	71.4	19
HgBr	65.0	66.0	69.5	......	74.4	23
HgI	67.1	68.1	71.6	......	68.9	33.0
HgCl₂	70.3	71.9	77.6	......	......	-33.5 ± 1
HgBr₂	76.5	78.15	83.94	......	......	-20.7 ± 1
HgI₂	80.3	82.0	87.8	......	......	-3.5 ± 1
KF	54.19	55.16	58.61	61.30	63.40	-77.2
KI	61.78	62.78	66.30	69.03	71.16	-29.8
MgF₂	55.92	57.30	62.52	66.75	70.12	-182.9
NaF	51.98	52.93	56.36	59.05	61.14	-69.2
PbF	59.6	60.55	63.94	......	68.71	-9 ± 9
PbCl	62.3	63.29	66.77	......	71.59	$+4 \pm 7$
PbBr	65.0	66.00	69.51	......	74.35	16 ± 9
PbI	66.8	67.8	71.3	......	76.2	26 ± 9
RbF	56.61	57.57	61.03	63.73	65.84	-77.7
RbBr	62.36	......	66.87	69.61	71.74	-44.1
RbI	64.33	......	68.85	71.59	73.72	-31.7
SiCl₄	79.20	81.76	91.18	98.71	104.66	-150 ± 1
SnCl₄	87.2	89.94	99.7	......	......	-117.9 ± 1
SnBr₄	98.2	101.04	111.05	......	......	-78.2 ± 1
SnI₄	106.6	109.5	119.6			
TiBr₄	94.9	97.67	107.56	114.3	121.4	-131.1
TiI₄	102.6	105.54	115.81	123.8	130.0	-70.2
ZnF	55.6	56.52	59.85	......	57.13	
ZnCl	58.3	59.27	62.73	......	67.53	10 ± 5
ZnBr	60.8	61.8	65.3	......	70.15	

TABLE A7-7. ELEMENTS
Free energies based on H_0°

	$-(F^\circ - H_0^\circ)/T$, cal/deg					$H_{298}^\circ - H_0^\circ$, kcal	ΔH_0°, kcal
	298.15°K	500°K	1000°K	1500°K	2000°K		
Br(g)	36.84	39.41	42.85	44.89	46.36	1.481	26.90
Br₂(g)	50.85	54.99	60.80	64.31	66.83	2.325	8.37
Br₂(l)	25.5	.....	.....	.....	.....	3.240	0
C(g)	32.52	35.20	38.73	40.77	42.21	1.56	169.6
C₂(g)	39.14	43.68	49.80	53.33	55.86	2.530	196 ± 6
C₃(g)	42.91	47.18	53.85	58.33	61.75	2.319	188 ± 5
C(graphite)	0.53	1.16	2.78	4.19	5.38	0.251	0
Cl(g)	34.43	37.06	40.69	42.83	44.34	1.499	28.54
Cl₂(g)	45.93	49.85	55.43	58.85	61.34	2.194	0
F(g)	32.69	35.41	39.06	41.16	42.64	1.558	18.4 ± 1
F₂(g)	41.37	45.10	50.44	53.74	56.17	2.110	0
H(g)	22.42	24.99	28.44	30.45	31.88	1.481	51.62
H₂(g)	24.42	27.95	32.74	35.59	37.67	2.024	0
I(g)	38.22	40.78	44.23	46.24	47.68	1.481	25.61
I₂(g)	54.18	58.46	64.40	67.96	70.52	2.148	15.66
I₂(s)	17.18	.....	.....	.....	.....	3.154	0
N(g)	31.65	34.22	37.66	39.67	41.10	1.481	112.54
N₂(g)	38.82	42.42	47.31	50.28	52.48	2.072	0
O(g)	33.08	35.84	39.46	41.54	43.00	1.607	58.98
O₂(g)	42.06	45.68	50.70	53.81	56.10	2.07	0
O₃(g)	48.79	53.27	60.17	64.70	68.0	2.475	34.6
P(white)	5.51	m 8.09	.....	.....	.....	1.28	0
S(g)	34.75	37.54	41.28	43.43	44.93	1.591	66 ± 2
S₂(g)	47.33	51.13	56.57	59.92	62.36	2.141	30.81
S₈(g)	77.44	92.06	115.68	.....	.....	7.547	25.23
S(rh)	4.09	m 6.48	.....	.....	.....	1.053	0

Shirley[1] gives the entropy of LiCl. Burns, Osborne, and Westrum[2] give the entropy of UF_4. The entropies of $CdCl_2$ and CdI_2 are based on data of Itskevich and Strelkov.[3] The heats of formation reported by Gross, Hayman, and Levi[4] have been very useful in fixing the data for AlF_3, MgF_2, TiF_4, and $ZrCl_4$. They have also determined data for other halides of Ti, V, Nb, and Ta, which are not given in these tables because of lack of entropy data. The heats of formation given for SiC are from Searcy.[5] The ΔH° values for $NiCl_2$ and TlCl are from Busey and Giauque[6] and Bartky and Giauque.[7]

[1] D. A. Shirley, *J. Am. Chem. Soc.*, **82**, 384 (1960).

[2] H. Burns, D. W. Osborne, and E. F. Westrum, Jr., *J. Chem. Phys.*, **33**, 387 (1960).

[3] E. S. Itskevich and P. G. Strelkov, *Zhur. Fiz. Khim.*, **33**, 1575 (1959).

[4] P. Gross, C. Hayman, and D. L. Levi, *Trans. Faraday Soc.*, **50**, 477 (1954), **53**, 1285, 1601 (1957), and private communication.

[5] S. G. Davis, D. F. Anthrop, and A. W. Searcy, *J. Chem. Phys.*, **34**, 659 (1961).

[6] R. H. Busey and W. F. Giauque, *J. Am. Chem. Soc.*, **75**, 1791 (1953).

[7] I. R. Bartky and W. F. Giauque, *J. Am. Chem. Soc.*, **81**, 4169 (1959).

$\Delta H°$ of SnS is from Richards.[1] Data for Mn and its compounds come from Mah.[2] All the data for $Ca(OH)_2$ are from Hatton, Hildenbrand, Sinke, and Stull.[3]

Coughlin[4] gives the high-temperature enthalpy and entropy values for Na_2SO_4 based on $H_{298}^°$ of $Na_2SO_4(V)$. These values were revised to agree with recent measurements of Christensen as reported by Kelley.[5]

Table A7-7. Elements. This table presents for convenience the data for a few elements on the basis of $H_0^°$. Most of the compounds for which the available free-energy functions are based on the enthalpy at $0°K$ are compounds of these elements. The presentation of these data as well as the presentation of the heats of formation of the gaseous forms of these elements from the solid form allows one to use either the condensed or the gaseous standard states of these elements. The free-energy functions for the gases are taken from *Series* III of the National Bureau of Standards and supplements in the *Journal of Research of the National Bureau of Standards*. The free-energy function for solid iodine is that given by Shirley and Giauque.[6] The free-energy functions for bromine and sulfur are those chosen in the King and Kelley compilation, and those for C_2 are from Altman.[7] The C_3 free-energy functions are based on the molecular constants given by Pitzer and Clementi.[8] The references for the $\Delta H_0^°$ values are the same as those of Table A7-2.

Table A7-8. Gaseous Carbon Compounds. Most free-energy functions calculated for gaseous substances are reported on the basis of $H_0^°$. Since $H_{298}^° - H_0^°$ is also given in such calculations, the results can be readily converted to the basis of $H_{298}^°$ if desired. These values for carbon compounds have been taken largely from the American Petroleum Institute[9] tabulations and *Circular* 500 of the National Bureau of Standards with additional values from the following sources:

COS and CS_2—The free-energy functions are from Brewer.[10]

Chlorinated hydrocarbons—The free-energy functions are from Gelles and Pitzer.[11]

CF and CF_4—All data were from Haar and Beckett[12] and Potocki and Mann.[13]

CH_3OH—The free-energy functions are from Ivash, Li, and Pitzer.[14]

CH_2O—Zeise[15] gives the free-energy functions.

CH_3SH—The data are from Barrow and Pitzer.[16]

$COCl_2$, C_2H_5OH, and HCOOH—The free-energy functions were calculated from

[1] A. W. Richards, *Trans. Faraday Soc.*, **51**, 1193 (1955).

[2] A. D. Mah, *U.S. Bur. Mines Rept. Invest.* 5600, 1960.

[3] W. E. Hatton, D. L. Hildenbrand, G. C. Sinke, and D. R. Stull, *J. Am. Chem. Soc.*, **81**, 5028 (1959).

[4] J. P. Coughlin, *J. Am. Chem. Soc.*, **77**, 868 (1955).

[5] K. K. Kelley, *U.S. Bur. Mines Bull.* 584, 1960.

[6] D. A. Shirley and W. F. Giauque, *J. Am. Chem. Soc.*, **81**, 4778 (1959).

[7] R. L. Altman, *J. Chem. Phys.*, **32**, 615 (1960).

[8] K. S. Pitzer and E. Clementi, *J. Am. Chem. Soc.*, **81**, 4477 (1959). The free-energy functions were extended to lower temperatures by Dr. R. Altman.

[9] See bibliography listing 1 at the end of Chapter 15.

[10] L. Brewer, Paper 5, in L. L. Quill (ed.), "The Chemistry and Metallurgy of Miscellaneous Material: Thermodynamics," vol. 19B, National Nuclear Energy Series, McGraw-Hill Book Company, Inc., New York, 1950.

[11] E. Gelles and K. S. Pitzer, *J. Am. Chem. Soc.*, **75**, 5259 (1953).

[12] L. Haar and C. W. Beckett, *Natl. Bur. Standards (U.S.) Rept.* 1164, 1951.

[13] R. M. Potocki and D. E. Mann, *Natl. Bur. Standards (U.S.) Rept.* 1439, 1952.

[14] E. V. Ivash, J. C. M. Li, and K. S. Pitzer, *J. Chem. Phys.*, **23**, 1814 (1955).

[15] H. Zeise, *Z. Elektrochem.*, **48**, 673 (1942).

[16] G. M. Barrow and K. S. Pitzer, *Ind. Eng. Chem.*, **41**, 2737 (1945).

TABLE A7-8. GASEOUS CARBON COMPOUNDS
Free energies based on H_0°

	$-(F - H_0^\circ)/T$, cal/deg					ΔH_{298}°, kcal	$H_{298}^\circ - H_0^\circ$, kcal	ΔH_0°, kcal
	298°K	500°K	1000°K	1500°K	2000°K			
CO........	40.25	43.86	48.77	51.78	53.99	−26.416	2.073	−27.202
CO₂.......	43.56	47.67	54.11	58.48	61.85	−94.052	2.238	−93.969
COS......	47.39	51.79	58.75	63.40	66.9	−32.8	2.37	−32.8 ± 2
CS₂.......	48.28	53.04	60.51	65.44	69.1	27.55	2.55	27.39 ± 2
CH₄.......	36.46	40.75	47.65	52.84	57.1	−17.89	2.397	−15.99
CH₃Cl.....	47.45	52.06	59.78	65.54	.....	−19.6	2.489	−17.7
CH₂Cl₂....	55.08	60.36	69.58	76.17	.....	−21	2.834	−19
CHCl₃.....	59.29	65.81	76.78	84.36	.....	−24	3.390	−23
CCl₄.......	60.15	68.12	81.41	89.96	.....	−25.5	4.111	−25
COCl₂.....	57.50	63.33	72.79	79.13	83.92	−52.47	3.075	−52.06
CF........	43.62	47.38	52.60	55.81	58.17	74	2.170	73 ± 7
CF₄.......	52.27	58.24	68.83	76.40	82.5	−218	3.043	−216.6
CH₃OH....	48.13	53.14	61.58	.....	.....	−48.08	2.731	−45.47
CH₃SH....	51.18	56.66	65.92	.....	.....	−3.7	2.90	−1.2
CH₂O.....	44.25	48.54	55.11	59.81	63.58	−27.7	2.393	−26.8
HCOOH...	50.72	55.60	63.99	70.17	75.14	−90.39	2.601	−88.65
HCN......	40.82	44.85	51.01	55.15	58.31	31.2	2.21	31.3
CN.......	41.46	45.05	49.99	53.00	55.22	109.8	2.073	109 ± 3
C₂H₂......	39.98	44.51	52.01	57.23	61.33	54.19	2.392	54.33
C₂H₄......	43.98	48.74	57.29	63.94	69.46	12.50	2.525	14.52
C₂H₆......	45.27	50.77	61.11	69.46	.....	−20.24	2.856	−16.52
C₂H₅OH ..	56.20	62.82	75.28	85.15	.....	−56.625	3.39	−52.41
C₂H₅SH...	58.64	65.76	79.07	.....	.....	−11.04	3.617	−7.0
(CH₃)₂S. .	55.71	63.04	76.47	.....	.....	−8.98	3.76	−5.11
CH₃CHO..	52.85	58.67	69.03	.....	.....	−39.67	3.070	−37.15
CH₃COOH.	56.50	63.24	75.92	85.35	.....	−103.8	3.30	−100.5
CH₃CN. ..	48.54	54.00	63.58	.....	.....	21.0	2.877	23.7
CH₃NC....	48.75	54.4	.....	.....	.....	35.9	2.996	38.5
C₃H₆. . . .	52.95	59.32	71.57	81.43	.....	4.88	3.237	8.47
C₃H₈.	52.73	59.81	74.10	85.86	.....	−24.82	3.512	−19.48
(CH₃)₂CO..	57.45	65.03	79.22	90.54	....	−51.72	3.889	−47.74
n-C₄H₁₀....	58.54	67.91	86.60	101.95	... .	−30.15	4.645	−23.67
iso-C₄H₁₀ ..	56.08	64.95	83.38	98.64	....	−32.15	4.276	−25.30
n-C₅H₁₂....	64.52	75.94	98.87	117.72	....	−35.00	3.146	−27.23
iso-C₅H₁₂ ..	64.36	75.28	97.96	116.78	...	−36.92	2.888	−28.81
neo-C₅H₁₂..	56.36	67.04	89.90	108.91	....	−39.67	2.574	−31.30
C₆H₆......	52.93	60.24	76.57	90.45	.....	19.82	3.401	24.00
cy-C₆H₁₀...	60.29	69.43	90.48	108.68	.....	−1.70	4.168	5.76
cy-C₆H₁₂. .	57.07	66.39	88.74	108.8	.....	−29.43	4.237	−20.01

TABLE A7-9. GASEOUS HALIDES
Free energies based on H_0°

	$-(F - H_0^\circ)/T$, cal/deg					ΔH_{298}°, kcal	$H_{298}^\circ - H_0^\circ$, kcal	ΔH_0°, kcal
	298.15°K	500°K	1000°K	1500°K	2000°K			
AlF....	44.27	48.04	53.43	56.76	59.18	-61 ± 5	2.125	-61
AlCl....	46.99	50.96	56.62	60.07	62.57	-11.3	2.228	-11.3
AlCl$_3$...	62.0	69.3	80.6	87.8	93.1	-137.1	3.958	-136.7 ± 1.5
Al$_2$Cl$_6$..	81.74	96.79	120.78	136.25	147.67	-308.76	7.852	-307.84 ± 0.8
BF.....	40.92	44.55	49.64	52.79	55.11	-42.9	2.078	-43.6 ± 5
BCl....	43.84	47.59	52.95	56.27	58.69	46.2	2.118	45.5 ± 5
BBr....	46.54	50.36	55.84	59.20	61.65	64.2	2.150	64
BF$_3$....	51.38	56.63	65.32	71.32	75.93	-270	2.784	-269.3 ± 2
BCl$_3$....	58.04	64.39	74.56	81.25	86.25	-97.1	3.362	-96.88
BeF....	42.17	45.82	50.95	54.14	56.49	4	2.082	3.3 ± 30
BeCl....	44.89	48.64	54.00	57.32	59.74	37	2.118	36.4 ± 30
BeF$_2$..	44.81	48.93	55.60	60.17	63.65	-192.1	2.226	-191.7 ± 5
BeCl$_2$..	48.95	53.66	61.21	66.18	69.91	-84	2.506	-84 ± 6
BrF$_5$...	61.43	70.41	85.95	96.47	104.40	-120	4.475	-118 ± 10
Cl$_2$O....	54.52	59.49	67.04	71.91	......	18.1	2.719	18.61
ClO....	47.04	50.79	56.14	59.46	61.88	24.2	2.115	24.2
ClO$_2$....	52.79	57.48	64.65	69.33	......	25.0	2.577	25.59
ClF....	44.92	48.70	54.11	57.45	59.88	-13.42	2.129	-13.4
ClF$_3$....	56.34	62.60	72.81	79.55	84.59	-38.79	3.262	-37.9
CsCl....	53.06	57.35	63.32	66.90	69.48	-58.6	2.424	-58.1
F$_2$O....	50.25	55.02	62.35	67.09	......	7.6	2.606	8.14 ± 2
HF.....	34.62	38.19	43.00	45.87	47.95	-64.2	2.055	-64.2
HCl....	37.72	41.31	46.16	49.08	51.23	-22.063	2.065	-22.019
HBr....	40.53	44.12	48.99	51.95	54.13	-8.66	2.067	-8.1
HI.....	42.40	45.99	50.90	53.90	56.11	6.2	2.069	6.7
HClO...	48.23	52.58	59.00	63.13	64.4	....	2.442	
IF.....	49.13	53.01	58.55	61.94	......	-22.5	2.175	-22
IF$_5$.....	63.32	72.38	87.94	98.45	......	-195.1	4.565	-192.8
IF$_7$.....	64.23	75.74	96.44	......	......	-224.2	5.585	-220.8
KCl....	49.17	53.31	59.24	62.79	65.35	-50.9	2.330	-50.4
KBr....	51.78	56.06	62.03	65.61	68.18	-42.7	2.421	-41.8
LiF....	40.74	44.48	49.81	53.12	55.54	-80.2	2.110	-80.2
LiCl....	43.58	47.43	52.95	56.34	58.80	-44.5	2.163	-44.5
LiBr....	46.25	50.15	55.72	59.14	61.63	-35.0	2.189	-34.5
LiI.....	48.04	52.00	57.65	61.10	63.55	-20.6	2.220	-20.2
MgF...	45.63	49.44	54.89	58.25	60.69	-20	2.143	-20 ± 20
MgCl..	48.23	52.21	57.88	61.33	63.83	2	2.236	2 ± 20
MgCl$_2$	51.53	57.07	65.60	70.98	74.94	-100.8	2.975	-100.4 ± 0.6
MoF$_6$..	62.09	72.77	......	......	......	-382		
NF$_3$....	52.78	58.20	67.32	73.58	78.35	-29.7	2.829	-28.3 ± 2
NaCl...	47.18	51.25	57.05	60.56	63.09	-42.7	2.295	-42.4
NaBr...	49.75	53.91	59.77	63.30	65.86	-34.0	2.342	-33.2
NaI....	51.51	55.72	61.64	65.20	67.75	-21.3	2.379	-20.6

TABLE A7-9. GASEOUS HALIDES (*Continued*)
Free energies based on H_0°

	$-(F - H_0^\circ)/T$, cal/deg					ΔH_{298}°, kcal	$H_{298}^\circ - H_0^\circ$, kcal	ΔH_0°, kcal
	298.15°K	500°K	1000°K	1500°K	2000°K			
PF$_3$....	54.82	60.83	70.37	76.92	81.8	....	2.92	
PCl$_3$....	61.66	68.87	80.07	......	......	−66.6	3.84	−65.9
PBr$_3$...	68.89	76.82	......	......	......	−35.9	4.25	−34.0
PCl$_5$....	66.76	76.30	91.56	......	......	−88.7	5.22	−87.2
POCl$_3$..	63.45	71.57	84.76	93.1	99.6	−141.5	4.23	−140.1
PSCl$_3$...	64.77	73.21	86.81	......	....	....	4.35	
RbCl...	51.79	56.04	62.00	65.57	68.14	−52.3	2.406	−51.8
ReF$_6$...	64.25	75.08	......	......	......	−273		
SiF$_4$....	55.03	62.37	74.40	82.6	88.9	−372.5	3.658	−371.2
S$_2$Cl$_2$...	65.55	73.14	84.74	......	......	−5.70	4.082	−5.48
SCl$_2$....	57.57	63.21	71.55	......	......	....	3.070	
SOCl$_2$...	61.65	68.33	78.96	85.80	90.88	−50.5	3.563	−49.8
SF$_6$.....	56.21	64.67	80.55	92.02	100.88	−289	4.001	−285.6
SeF$_6$....	59.27	68.98	......	......	......	−246		
TeF$_6$...	62.51	73.04	......	......	......	−315		
TiCl$_4$...	67.11	76.74	91.54	100.88	107.73	−182.4	5.119	−182.0 ± 0.7
UF$_6$....	68.92	81.25	101.37	......	......			
WF$_6$....	63.52	74.31	......	......	......	−416		

the molecular constants.[1] ΔH° of HCOOH(g) was derived from Sinke's value for the liquid.[2] The enthalpy of formation of COCl$_2$ was calculated from equilibrium data for the dissociation to CO + Cl$_2$.

HCN—The free-energy functions are from Bradley, Haar, and Friedman.[3]

CN—The free-energy functions were calculated from the entropy and enthalpy values of Kelley.[4] The enthalpy of formation is that given by Brewer, Hicks, and Krikorian.[5]

C$_2$H$_5$SH and (CH$_3$)$_2$S—The values are from McCullough et al.[6]

CH$_3$CHO and CH$_3$COOH—The data are from Pitzer and Weltner[7] and from Weltner.[8] Evans and Skinner[9] give ΔH° of CH$_3$COOH(l).

[1] The constants for COCl$_2$, C$_2$H$_5$OH, and HCOOH were obtained from E. Catalano and K. S. Pitzer, *J. Am. Chem. Soc.*, **80**, 1054 (1958), G. M. Barrow, *J. Chem. Phys.*, **20**, 1739 (1952), and R. C. Millikan and K. S. Pitzer, *J. Chem. Phys.*, **27**, 1305 (1957), respectively. The authors wish to thank Dr. Robert Altman for help with the calculations.

[2] G. C. Sinke, *J. Phys. Chem.*, **63**, 2063 (1959).

[3] J. C. Bradley, L. Haar, and A. S. Friedman, *J. Research Natl. Bur. Standards*, **56**, 197 (1956).

[4] See bibliography listing 4 at end of Chapter 15.

[5] L. Brewer, W. Hicks, and O. Krikorian, paper in press.

[6] J. P. McCullough et al., *J. Am. Chem. Soc.*, **79**, 561 (1957).

[7] K. S. Pitzer and W. Weltner, Jr., *J. Am. Chem. Soc.*, **71**, 2842 (1949).

[8] W. Weltner, *J. Am. Chem. Soc.*, **77**, 3941 (1955).

[9] F. W. Evans and H. A. Skinner, *Trans. Faraday Soc.*, **55**, 260 (1959).

TABLE A7-10. GASEOUS OXIDES, HYDRIDES, AND RELATED COMPOUNDS
Free energies based on H_0°

	$-(F - H_0^\circ)/T$, cal/deg					ΔH_{298}°, kcal	$H_{298}^\circ - H_0^\circ$, kcal	ΔH_0°, kcal
	298°K	500°K	1000°K	1500°K	2000°K			
Al_2O	53.0	57.7	64.9	69.6	73.0	-39.4 ± 5	2.582	-38.8 ± 5
AlO	45.13	48.83	54.11	57.38	59.77	16.9	2.101	16.9 ± 8
BH	34.12	37.71	42.61	45.61	47.83	117.76	2.065	117 ± 5
BO	41.66	45.26	50.21	53.26	55.51	5.75	2.073	5 ± 10
B_2O_2	47.7	53.2	62.2	68.4	73.2	-107.8	2.862	-108 ± 3
B_2O_3	54.89	60.28	69.66	76.49	81.92	-210.1	2.829	-209.5 ± 2
BS	44.66	48.31	53.48	56.68	59.03	....	2.085	
BN	43.75	47.37	52.42	55.54	57.85	154.75	2.076	154 ± 3
BeO	40.24	43.87	48.92	52.06	54.37	30.44	2.077	29.87
HO	36.82	40.48	45.39	48.30	50.41	10.06	2.106	10 ± 0.3
H_2O	37.17	41.29	47.01	50.60	53.32	-57.798	2.368	-57.107
H_2O_2	46.95	51.72	59.15	64.28	.....	-32.53	2.59	-31.04
HS	39.51	43.27	48.42	51.50	53.74	32	2.172	32 ± 3
H_2S	41.17	45.34	51.29	55.16	58.1	-4.815	2.385	-4.12
H_2Se	44.30	48.50	54.54	.....	.....	20.5	2.388	21.5
H_2Te	46.61	51	.....	.....	.....	36.9	2.409	38.0
NH_3	37.99	42.28	48.63	53.03	56.56	-11.04	2.37	-9.37
NO	42.98	46.76	51.86	54.96	57.24	21.60	2.194	21.48
N_2O	44.89	49.11	55.76	60.27	.....	19.49	2.291	20.31
NO_2	49.19	53.60	60.23	64.58	67.88	8.091	2.465	8.68
PH_3	42.10	46.47	53.34	58.9	63.0	2.21	2.42	4.10
SO	46.03	49.69	54.87	58.09	60.43	1.4	2.088	1.4 ± 1
SO_2	50.82	55.38	62.28	66.82	70.2	-70.96	2.519	-70.36
SO_3	51.89	57.14	66.08	72.40	77.1	-94.45	2.77	-93.06
SiO	43.56	47.21	52.35	55.54	57.88	-21.5	2.083	-21.8

CH_3CN and CH_3NC—Ewell and Bourland[1] give the free-energy functions.

$(CH_3)_2CO$—All the data are from Pennington and Kobe.[2]

Tables A7-9 and A7-10. References to papers from which free-energy functions and $H_{298}^\circ - H_0^\circ$ values were obtained may be found in the comprehensive bibliographies by Kelley and Kelley and King.[3] ΔH° values for gaseous alkali halides are from Brewer and Brackett.[4] The MoF_6 and WF_6 data are from Myers and Brady.[5] The data for gaseous compounds of Be, B, Mg, and Al are from recent *National Bureau of Standards Reports*[6] except for B_2O_2 and B_2O_3, which are from White et al.[7]

[1] R. H. Ewell and J. F. Bourland, *J. Chem. Phys.*, **8**, 635 (1940).

[2] R. E. Pennington and K. A. Kobe, *J. Am. Chem. Soc.*, **79**, 300 (1957).

[3] See bibliography listing 4 at the end of Chapter 15.

[4] L. Brewer and E. Brackett, forthcoming paper.

[5] O. E. Myers and A. P. Brady, *J. Phys. Chem.*, **64**, 591 (1960).

[6] *Natl. Bur. Standards (U S.) Repts.* 6252, 1958; 6484, 1959; and 6297, 1959.

[7] D White, P. N. Walsh, D. E. Mann, and A. Sommer, RF Project 691, *Tech. Rept.* 4, July, 1959.

The $\Delta H°$ value of $TiCl_4$ is from Johnson, Nelson, and Prosen.[1] Durie and Ramsay[2] give the $\Delta H°$ of ClO. Neale and Williams[3] give the $\Delta H°$ of PCl_3, and Stevenson and Yost[4] give the data for PCl_5. Enthalpies of formation for AlF, AlCl, and SF_6 are given by Gross, Hayman, and Levi.[5] Any other $\Delta H°$ values were obtained from *Circular* 500 of the National Bureau of Standards, although changes were necessary in some instances to ensure consistency with related values which have been changed since the completion of *Circular* 500.

In a number of instances where published tables of free-energy functions did not extend to 2000°K, Stull[6] and Gordon[7] kindly supplied extensions of the tables.

[1] W. H. Johnson, R. A. Nelson, and E. J. Prosen, *J. Research Natl. Bur. Standards*, **62,** 49 (1959).

[2] R. A. Durie and D. A. Ramsay, *Can. J. Phys.*, **36,** 35 (1958).

[3] E. Neale and L. T. D. Williams, *J. Chem. Soc.*, **1954,** 2156.

[4] D. P. Stevenson and D. M. Yost, *J. Chem. Phys.*, **9,** 403 (1941); the more recent paper of J. K. Wilmhurst and H. J. Bernstein [*J. Chem. Phys.*, **27,** 661 (1957)] contains an error.

[5] P. Gross, C. Hayman, and D. L. Levi, *Trans. Faraday Soc.*, **50,** 477 (1954), **53,** 1285, 1601 (1957), and private communication.

[6] D. R. Stull, Thermal Laboratory, The Dow Chemical Co., Midland, Mich., private communication, and G. C. Sinke, *Rept.* AR-1S-59, April, 1959.

[7] John S. Gordon, private communication and *Rept.* RMD 210-E3, Thiokol Chemical Corp., Reaction Motors Div., Denville, N.J., 1960.

Appendix 8

ELECTROMAGNETIC WORK

In this appendix we derive a general expression for electromagnetic work from Maxwell's equations.[1] The applications of the result to systems of physical and chemical interest are considered in Chapter 31. The mks units and the rational system will be used. Definitions of terms and a discussion of polarizable matter are given in Chapter 31 and will not be repeated here.

In the rational system Maxwell's equations are

$$\nabla \times \mathbf{E} + \frac{\partial \mathbf{B}}{\partial t} = 0 \tag{A8-1}$$

$$\nabla \times \mathbf{H} - \frac{\partial \mathbf{D}}{\partial t} = \mathbf{J} \tag{A8-2}$$

In these equations $\mathbf{E}$ and $\mathbf{H}$ are vectors commonly called the intensities of the electric and magnetic fields, respectively, while $\mathbf{D}$ and $\mathbf{B}$ are known, respectively, as the electric displacement and magnetic induction. $\mathbf{J}$ is the vector current density, and the density of electric charge (a scalar) will be given the symbol ρ.

The assumption of conservation of electric charge yields the continuity equation

$$\nabla \cdot \mathbf{J} + \frac{\partial \rho}{\partial t} = 0$$

From this equation, the fact that the divergence of the curl of any vector vanishes, and the assumption that all the fields may be reduced to zero, it is possible to derive the two additional equations

$$\nabla \cdot \mathbf{B} = 0 \tag{A8-3}$$
$$\nabla \cdot \mathbf{D} = \rho \tag{A8-4}$$

These equations are frequently listed with Eqs. (A8-1) and (A8-2) as if they were separate postulates. We have seen, however, that, if one makes the assumptions noted above, these are not independent relations.

The electric and magnetic fields may be expressed as functions of a scalar potential ψ and a vector potential $\mathbf{A}$. There is an arbitrary function to be chosen in the complete definition of these potentials in a general case. But in our considerations we shall use only the scalar electric potential and then only in stationary-field cases, whereupon the vector potential and the arbitrary function drop out and one has

$$\mathbf{E} = -\nabla \psi \tag{A8-5}$$

[1] J. A. Stratton, "Electromagnetic Theory," McGraw-Hill Book Company, Inc., New York, 1941, is an excellent reference and one wherein electromagnetic work is discussed in rigorous terms.

In an isotropic medium where $\mathbf{D} = \epsilon\mathbf{E}$ with ϵ a constant, combination of Eqs. (A8-4) and (A8-5) yields the Poisson equation[1]

$$\nabla^2\psi = -\frac{\rho}{\epsilon} \qquad (A8\text{-}6)$$

For thermodynamic purposes we need an expression for the electromagnetic work. Consider a cylindrical volume element parallel to the current-density vector $\mathbf{J}$ of length dl and cross-section area da. The flow of electric charge through this element is $\mathbf{J}\,da\,dt$. The electric-potential difference $d\psi = -\mathbf{E}_J\,dl$, where $\mathbf{E}_J$ is the component of $\mathbf{E}$ in the direction of $\mathbf{J}$. Consequently, the electrical work absorbed in time dt in this volume element $dV = da\,dl$ is

$$dw = -\mathbf{E}\cdot\mathbf{J}\,dV\,dt \qquad (A8\text{-}7)$$

and the total work absorbed is the integral of this quantity over all space and over the time interval of interest.

Next we need an expression for $\mathbf{E}\cdot\mathbf{J}$ in terms of all of field vectors. Take the scalar product of Eq. (A8-2) with $\mathbf{E}$, and subtract from it the corresponding product of Eq. (A8-1) with $\mathbf{H}$.

$$\mathbf{E}\cdot\mathbf{J} = -\mathbf{E}\cdot\frac{\partial\mathbf{D}}{\partial t} - \mathbf{H}\cdot\frac{\partial\mathbf{B}}{\partial t} + (\mathbf{E}\cdot\nabla\times\mathbf{H} - \mathbf{H}\cdot\nabla\times\mathbf{E}) \qquad (A8\text{-}8)$$

The final term in parentheses may be shown to yield a zero integral by first making the vector transformation

$$\mathbf{E}\cdot\nabla\times\mathbf{H} - \mathbf{H}\cdot\nabla\times\mathbf{E} = \nabla\cdot(\mathbf{H}\times\mathbf{E})$$

and then noting that the volume integral may be transformed to a surface integral,

$$\int_V \nabla\cdot(\mathbf{H}\times\mathbf{E})\,dV = \int_S (\mathbf{H}\times\mathbf{E})\cdot d\mathbf{S}$$

In systems of interest we may assume that both $\mathbf{H}$ and $\mathbf{E}$ go to zero at the boundary, which may be at infinity, and by choosing the boundary of the system in this way the surface integral vanishes. Substitution of the remainder of Eq. (A8-8) into Eq. (A8-7) yields

$$dw = \left(\mathbf{E}\cdot\frac{\partial\mathbf{D}}{\partial t} + \mathbf{H}\cdot\frac{\partial\mathbf{B}}{\partial t}\right)dt\,dV \qquad (A8\text{-}9)$$

and if we assume that the field is built up from an initial value of zero everywhere,

$$w = \iint\mathbf{E}\cdot d\mathbf{D}\,dV + \iint\mathbf{H}\cdot d\mathbf{B}\,dV \qquad (A8\text{-}10)$$

The first integral is clearly the electrostatic work and the second integral the magnetic work.

The work content A of the entire system may be obtained by adding the expression for electrical and magnetic work from Eq. (A8-10) to the usual expression $E - TS$. However, the entire system commonly contains magnets or capacitor plates, wires, containers, and other components besides the sample we wish to study. A completely general consideration of all interactions is very difficult, but good approximations can be made for several cases of practical interest. These are discussed in Chapter 31.

---

[1] In Chapter 23 the Poisson equation was given in unrationalized form and for esu, whereupon the right side becomes $-4\pi\rho/D$. This form is conventional in the literature on Debye-Hückel theory.

Appendix 9

SYMBOLS†

Italic capital letters

1. Extensive thermodynamic quantities

A	Work content (ψ of Gibbs or Helmholtz free energy)
B, C, D	Virial coefficients for gases
C	Heat capacity
E	Energy content
F	Free energy (Gibbs potential ζ in absence of gravitation, electric, etc., fields)
H	Enthalpy (heat content)
J	Relative heat capacity
L	Relative enthalpy
Q^*	Heat of transfer
S	Entropy
V	Volume
Y	Any thermodynamic function

2. Other quantities

A	Area; moment of inertia
A_γ, A_H	Debye-Hückel coefficients
A_{12}	Nonideal-solution parameter
B	Spectroscopic rotational constant; aqueous ion-interaction parameter
C_M	Curie constant (magnetic)
D	Determinant of inertia; dielectric constant; spectroscopic rotational stretching constant
I	Moment of inertia; constant of integration; ionic strength
J	Rotational quantum number; flux
K	Equilibrium constant; quantum number
L	Rate constant
M	Molecular weight
M_J	Quantum number
N	Number of molecules
N_0	Avogadro's number
P	Pressure
Q	Partition function
R	Gas constant; isopiestic ratio
T	Temperature

† Symbols and usages confined to a single section of text are omitted here.

V	Potential energy in a molecule
W	Multiplicity (statistical)
X	Potential difference in Eq. (28-9)

Small Roman capitals (molal thermodynamic quantities)

A	Molal work content
B, C, D	Molal virial coefficients
C	Molal heat capacity
E	Molal energy content
F	Molal free energy (Gibbs)
H	Molal enthalpy (heat content)
J	Relative molal heat capacity
L	Relative molal enthalpy
Q*	Molal heat of transfer
S	Molal entropy
V	Molal volume
S̄, F̄, etc.	Partial molal quantities
S̃	Transported molal entropy

Script capital letters

$\mathcal{A}$	Surface area per mole
$\mathcal{E}$	Galvanic-cell electromotive force
$\mathcal{F}$	Faraday constant
$\mathcal{P}$	Relative probability (statistical)
$\mathcal{U}$	Electric or electrode potential
$\mathcal{Z}$	Coordination number

Boldface capital letters (vectors)

A	Electromagnetic potential
B	Magnetic induction
D	Electric displacement
E	Electric field
H	Magnetic field
J	Current density
M	Molal magnetic polarization
P	Molal electric polarization

Lower-case italic

a, b, c	Constants in heat-capacity equations
a	Activity
b	Volume parameter, van der Waals equation, etc.
c	Concentration; velocity of light
d	Density
e	Charge on the electron; eccentricity of spheroid
f	Fugacity
g	Acceleration of gravity; magnetic-moment factor; number of quantum states of equal energy
h	Planck constant; height
k	Boltzmann constant; Henry's-law constant; and interaction coefficient of Eq. (34-20)*

l	Length
m	Molality; mass of single atom
n	Number of moles; quantum number
p	Partial pressure; vapor pressure of a single component
q	Quantity of heat absorbed
r	Radius; mole ratio
t	Time; transference number; temperature, °C
u	Statistical quantity for molecular vibration
v	Vibrational quantum number
w	Quantity of work absorbed; nonideal-solution parameter
x	Mole fraction; spectroscopic anharmonicity constant
y	Statistical quantity for molecular rotation
z	Compressibility factor; composition measure of Eq. (21-23)
z_+, z_-	Charge on a positive, negative ion
x, y, z	Linear coordinates

Boldface lower case (vectors)

$\mathbf{m}$	Magnetic polarization per unit volume
$\mathbf{p}$	Electric polarization per unit volume

Greek letters

α	Coefficient of thermal expansion; degree of ionization; spectroscopic rotation-vibration interaction constant
β	Bohr magneton; β_{MX}, aqueous ion-interaction parameter of Eq. (23-34)* for ions M and X; β_{23}, salt interaction coefficient of Eq. (34-28)
γ	Activity coefficient (stoichiometric)
Γ	Surface concentration; activity of a reference state at pressure other than 1 atm
ϵ	Molecular energy level; electric permittivity
ϵ_0	Electric permittivity of free space
θ	Freezing-point depression
κ	Debye-Hückel ion atmosphere parameter; compressibility
λ	Molal freezing-point lowering
Λ	Equivalent conductivity
μ	Chemical potential; Joule-Thomson coefficient; magnetic permeability
μ_0	Magnetic permeability of free space
ν	Frequency; number of ions per molecule of electrolyte
π	Osmotic pressure; film pressure; 3.14159
ρ	Density; electric charge density
σ	Surface tension; symmetry number; D-H osmotic function [Eq. (23-40)]
Σ	Area of solid surface; sum in Eq. (15-38)
ϕ	Osmotic coefficient; volume fraction
χ	Magnetic susceptibility
χ_M	Molal magnetic susceptibility
ψ	Electric potential
ω	Vibration frequency in reciprocal centimeters; acentric factor; angular velocity

Superscripts, other than mathematical exponents and prime marks

$^\circ$	Standard state
$*$	Ideal-gas state

.	Used with aqueous ion-interaction parameter B and k to indicate dependence on molality
E	Excess over value for ideal solution
i	Ideal solution
(0), (1)	Functions in acentric-factor theory for fluids
ϕ	Apparent molal quantity, ϕV, ϕH, etc.

Subscripts

The absolute temperature, that is, 0 for 0°K, 298 for 298.15°K

1, 2	To represent components in a solution (if a solvent is defined, it is component 1)
c	Critical
D	Debye, as in θ_D
f	Reaction of formation from the elements; also fusion
I	Constant of integration, ΔH_I
irr	Contribution from irreversibility, as S_{irr}
P	Constant pressure
r	Reduced, i.e., ratio to value for critical state
rev	Quantity for reversible process
s	Surface quantity
t.p.	Triple-point property
V	Constant volume
vap	Vaporization
w	To represent the component water as in a_w
$\pm$	Mean quantity for ions of an electrolyte
23	Coefficient for component 2 as a result of interaction by component 3, as in k_{23}

Appendix 10

PHYSICAL CONSTANTS AND NUMERICAL FACTORS†

Absolute temperature of ice point (0°C)..................... 273.150°K
Absolute temperature of triple point of H_2O, by definition...... 273.16000°K
Thermochemical calorie, by definition...................... 4.1840 joules
Standard gravity, by definition.......................... 980.665 cm/sec^2
Atmosphere (by definition 760 mm Hg at s.c.).............. 1,013,250 dynes/cm^2

$$[Pv]_{0°C}^{P=0} = R(273.150) = 22,414.6 \text{ cc atm/mole}$$
$$= 2271.16 \text{ joules/mole}$$
$$R = 1.98726 \text{ cal/deg mole}$$
$$= 8.31470 \text{ joules/deg mole}$$
$$= 82.0597 \text{ cc atm/deg mole}$$
$$\mathfrak{F} = 96,493.5 \text{ coulombs/equiv or joules/volt equiv}$$
$$= 23,062.3 \text{ cal/volt equiv}$$
$$N_0 = 6.0232 \times 10^{23} \text{ mole}^{-1}$$
$$k = 1.38045 \times 10^{-16} \text{ erg/deg}$$
$$e = 4.8029 \times 10^{-10} \text{ esu}$$
$$= 1.60206 \times 10^{-19} \text{ coulomb}$$
$$h = 6.6252 \times 10^{-27} \text{ erg sec}$$
$$c = 2.99793 \times 10^{10} \text{ cm/sec}$$
$$\ln 10 = 2.302585$$
$$R \ln x = 4.5758 \log_{10} x \text{ cal/deg mole}$$
$$298.15 R \ln x = 1364.3 \log_{10} x \text{ cal/mole}$$
$$hc/k = c_2 = 1.4388 \text{ cm deg}$$

English Units

1 Btu = 252.16 cal = 1055.04 joules
1 Btu/lb = 0.55592 cal/g = 2.3260 joules/g
1 Btu/(lb)(°F) = 1.00065 cal/g deg
$\qquad\qquad$ = 4.1867 joules/g deg
14.696 lb/in.^{2} = 1 atm
2.2046 lb = 1 kg
1 ft^3 = 28,317 cc
1.8°F = 1°C
R = 1.9860 Btu/(°F)(lb mole)
$\quad$ = 10.732 ft^3(lb/in.^{2})/(°F)(lb mole)

† See end of Chapter 4 for discussion and references; the mole is based on 16 g of natural oxygen.

694

ENERGY CONVERSION FACTORS

	cal/mole	joule/mole	cc atm/mole	cm^{-1}	electron volt	erg/molecule
1 cal/mole	1	4.1840	41.292	0.34974	4.3361×10^{-5}	6.9465×10^{-17}
1 joule/mole	0.23901	1	9.8692	0.08359	1.0364×10^{-5}	1.6602×10^{-17}
1 cc atm/mole	0.024218	0.10133	1	8.470×10^{-3}	1.0501×10^{-6}	1.6823×10^{-18}
1 cm^{-1}	2.8593	11.963	118.07	1	1.2398×10^{-4}	1.9862×10^{-16}
1 electron volt	23,062	96,493	9.523×10^5	8065.7	1	1.6020×10^{-12}
1 erg/molecule	1.4396×10^{16}	6.023×10^{16}	5.944×10^{17}	5.0348×10^{15}	6.2422×10^{11}	1

ANSWERS TO SELECTED PROBLEMS

3-1. (a) $(\partial v/\partial T)_P = R/P + d_B/dT$; $(\partial v/\partial P)_T = -RT/P^2$

4-2. (a) $w = -730$, $\Delta E = 9000$, $\Delta H = 9730$ cal; (b) $w = 0$

4-3. 446 cal/g of lead

5-2. $-_BRT/(V - _B)^2$ **5-3.** $q = 2777$ cal

6-1. $c_V^\circ = 23.1$ cal/deg   **6-2.** 6.85 cal/deg mole

6-3. Calculate $c_V = 0.123$ and 7.89 cal/deg at 25 and 100°K

6-4. 2870 and 8510 cal at 1000 and 2000°K

6-8. 6880 kcal, using analytical expression, and 6860 kcal, using tabulated enthalpies

6-10. 3110°K

7-1. 0.00858 cal/deg   **7-2.** 0.874 cal/deg

7-3. $\Sigma \Delta S = 0.17$ cal/deg for combined system

9-3. 0.546   **9-4.** 225°K

9-5. 2250 cal   **9-6.** 5×10^5 cm/sec

10-3. $q = 266$, and $\Delta H = 296$ cal; $\Delta S = 1.02$ cal/deg

10-8. $C_V = 7.2 \times 10^{-10}$ cal/deg cc 1.807×10^{-10} cal./deg c.c

10-10. -0.00746 deg/atm   **10-11.** -5.6%

11-3. 4790 cal of work   **11-5.** 12.6 cal/deg mole

11-7. 71°K

11-8. $\Sigma \Delta S = 22$ cal/deg; 5500 cal of work

13-2. 0.44 cal   **13-3.** 0.031 atm

14-2. 3.1×10^{-3} atm   **14-4.** 0.035 atm

15-2. $d\varepsilon/dT = 3.39 \times 10^{-4}$ volt/deg

15-6. $\Delta F_{298} = -254$ cal; $\Delta F_{298}^\circ = 2440$ cal

15-7. $-22,740$ cal   **15-8.** $\Delta F_{1538}^\circ = -2675$ cal

15-11. 0.18%, 5.8%, and 49%

15-12. 0.54 atm H_2, 0.26 atm S_2, 0.16 atm H_2S, 0.04 atm HS

15-13. $\Delta F_{273}^\circ = 2210$ cal   **15-14.** 1.55% dissociation

15-15. 0.97 atm Cl_2, 0.03 atm CCl_4, $10^{-3.4}$ atm Cl

16-4. $\ln f = \ln RT/(v - b) - 2a/RTv + b/(v - b)$

16-9. 1897 atm

17-1. $x_2 = 0.0177$ **17-3.** 6.258, 0.1013

17-4. 20.2 cc   **17-6.** 16.3 cc

17-7. 102.9 cc   **17-8.** $\bar{v}_2 = 85.71$ cc at $x_2 = 0.35007$

17-9. $\bar{v}_2 = 28.46$ cc at 0.3 M

18-1. 2.92   **18-2.** 4.88°C

18-3. $RT \ln 2$, $2RT \ln 2$ **18-5.** 24 cal

19-4. $-103.3°$/mole, $-1.860°$/mole   **19-8.** 13.3 mm Hg

20-1. 1.075   **20-2.** 0.92

20-3. -1275 cal   **20-5.** 0.252

20-7. 0.0390 volt

20-10. $\gamma = 83$ for aniline, 3.7 for water

20-11. Liquid is 9.2% *n*-pentane, $\Gamma_{iso} = 1.607$

21-2. $p = 0.562 p^\circ_{C_6H_6} + 0.445 p^\circ_{C_2H_4Cl_2}$

21-3. $x(SnI_4) = 0.0026$ **21-4.** $w = 1140$ cal/mole

21-5. Ag-Cr system; first approximation 3.5% and 91% Ag; exact solution 3.3% and 74% Ag

22-1. 1.15   **22-2.** 1560 cal/mole   **22-3.** $m = 0.40$

22-4. For $Al_2(SO_4)_3$ at molality m, $a_\pm = 2^{2/5}3^{3/5}\,m$, $a_2 = 108\;m^5$; in Na_2SO_4, $AlCl_3$ solution, $a_\pm = [(m_{Na_2SO_4})^3(m_{AlCl_3})^2]^{1/5}$

22-6. $\varepsilon = 0.114$ volt   **22-7.** $P - P^\circ = 6.27$ atm

22-10. $\gamma_{NaOAc} = 0.74$ at 1 M

23-2. 8.68 cal/deg mole for limiting law at 25°C

24-1. -0.143 volt   **24-2.** 10^3, 10^{-32}

25-3. 175 cal

25-6. $KCl(s) = KCl(1\ M)$, $\Delta \bar{H} = 3.94$ kcal/mole

25-7. $(dm_{sat}/dT) = 6.0 \times 10^{-3}$ mole/kg deg

25-9. 0.018%/deg

26-2. $\gamma_\pm = 0.804$ at 0.1 M, $\gamma_\pm = 0.751$ at 0.6 M

26-3. $\gamma = 0.94$ for HOAc   **26-4** $\phi = 1.066$

26-6. If $\bar{J}_2 = -2.8$ cal/deg, $\gamma_\pm = 0.774$ at 25°C

27-2. $Q' = 6.77$, $s_{int} = 5.79$ cal/deg mole

27-4. $s^\circ = 58.9 \pm 0.5$ cal/deg mole   **27-5.** 23% Na_2

29-2. Pressure ratio 1.021   **29-3.** 3.1 M

30-2. $\omega = 0.9 \times 10^4$ radian/sec

31-2. 62,000 gauss

32-2. $10^{-12.9}$ atm

33-3. $p_{Bi} = 10^{-4.7}$ atm, $[Bi_2]/[Bi] = 1.0$, both at 1000°K

33-7. $p_{SiO} = 10^{-6.3}$ atm; transport will occur

33-8. 78%, 17%, and 28% dimer   **33-9.** $[K_2O_2H_2]/[KOH] = 0.1$

33-11. ZrX_4 is principal gaseous species for all systems

33-12. WO_2 and WO_3 are unimportant compared with Al

33-13.

	Al_2O_3(Al sat)	Al_2O_3(Knudsen)	Al_2O_3(H_2O/H_2)	Al_2O_3(O_2 sat)
Al	$10^{-2.05}$	$10^{-8.16}$	$10^{-10.25}$	$10^{-15.5}$
O	$10^{-12.18}$	$10^{-8.10}$	$10^{-6.70}$	$10^{-3.18}$
O_2	$10^{-18.0}$	$10^{-9.85}$	$10^{-7.04}$	10^0
Al_2O	$10^{-1.3}$	$10^{-9.42}$	$10^{-13.3}$	$10^{-19.3}$
AlO	$10^{-7.1}$	$10^{-9.16}$	$10^{-9.85}$	$10^{-11.6}$
Total	$10^{-1.2}$	$10^{-7.8}$	$10^{-6.7}$ (or $10^{-9.7}$ excluding O and O_2)	1 or $10^{-11.6}$

33-14. At 1664°K, $p_{O_2} = 10^{-15.3}$ atm; with 0.2 atm O_2, $p_{Ba_2O_3} = 10^{-0.9}$ atm. At 1530°K in H_2O/H_2, $p_{BaO} = 10^{-6.9}$ atm and $p_{Ba_2O_3} = 10^{-8.6}$ atm.

NAME INDEX

Abegg, R., 414
Ackerman, M., 673
Ackermann, Th., 378, 403, 408
Adam, N. K., 470
Adams, L. H., 319, 410, 411
Agar, J. N., 460, 461, 465, 468
Ahlberg, J. E., 125
Åkerlöf, G., 315, 328, 568, 586
Alcock, C. B., 543
Altman, R. L., 681, 684
Amagat, E. H., 294
Andrews, T., 9, 193
Angstadt, R. T., 360
Anthrop, D. F., 673
Armstrong, W. G., 550
Arrhenius, S. A., 301, 332
Aston, J. G., 372
Ayres, F. D., 389

Barany, R., 675
Barrow, G. M., 538, 681, 684
Barrow, R. F., 178, 179
Bartky, I. R., 680
Bartoli, A., 111
Basolo, F., 529
Bates, R. G., 315, 350, 358
Bates, S. J., 312
Bauer, S. H., 539
Bawn, C. E. H., 293
Bearden, J. A., 41
Bearman, R. J., 452
Beattie, J. A., 664
Beckett, C. W., 181, 681
Becktold, M. F., 320
Beebe, C. W., 288
Bell, G. D., 673
Benedict, M., 195
Benedict, W. S., 181
Benson, S. W., 521
Berger, W., 190
Berkowitz, J., 539, 545

Bernstein, H. J., 686
Berthelot, D., 187
Berthelot, M., 67, 159, 515
Bever, M. B., 514
Bichowsky, F. R., 174
Bird, R. B., 101
Bisson, C. S., 383, 386, 387, 389
Bjerrum, N., 263, 321, 332, 333, 337, 341
Black, J., 3, 29
Bleaney, B., 507
Blewett, J. P., 550
Bockris, J. O'M., 182, 349
Boltzmann, L., 54, 91, 112, 130, 335, 421
Bonhoeffer, K. F., 596
Bonnickson, K. R., 675
Born, M., 55, 98, 539, 585
Böttcher, C. J. F., 584
Bourland, J. F., 685
Bousfield, W. R., 320
Bower, V. E., 315, 358
Boyle, B. J., 530, 544
Brackett, E., 670, 678, 685
Bradley, J. C., 684
Brady, A. P., 685
Brann, B. F., 362
Bray, W., 415
Breck, W. G., 465, 468
Brewer, D. F., 601
Brewer, L., 180, 534, 539, 540, 543, 586,
    587, 670, 673, 678, 681, 684, 685
Brickwedde, F. G., 598
Bridgman, P. W., 36, 667
Briggs, L. J., 195
Bromley, L. A., 180, 540
Bronsted, J. N., 4, 318, 334, 344, 345, 415
Brown, A. S., 362
Brown, I., 288
Brown, O. L. I., 320
Brown, P. G. M., 410, 411
Brown, W. B., 630
Brunauer, S., 487
Buckdahl, H. A., 98

SUBJECT INDEX

Absolute temperature, 33, 44, 83

Absolute zero, thermodynamic properties at, 130

Acentric factor, gaseous solutions, 630, 632
- normal fluids, 606, 609
- theory, 606–630

Acetic acid, dissociation constant, 302–306, 364–366, 587–591
- freezing-point lowering of water, 300
- reaction with H_2, 270

Acetone solution, carbon bisulfide, 212, 215, 286
- chloroform, 216

Acids, weak, dissociation, 302–306, 364–366, 587–591

Activity and activity coefficient, accuracy, 319, 325, 326
- Bronsted-Guggenheim equation, 318–319, 346, 366, 376, 390, 569, 577, 579, 588, 642
- calculation, from activity of other component, 260–267, 322–323, 410, 412–413, 551–564, 572–577, 586
- from distribution of solute between two solvents, 255–256, 586–587
- when solute forms compounds with solvent, 272
- centrifuge measurements, 494–496
- concentration, 326, 641
- correlations with thermal data, 394–398
- Debye-Hückel theory, 337–340
- definition, 243, 246, 250
- diffusion measurements, 325
- dimensions, 251
- dissociating solutes, 273
- electrolytes, 309–330
  - mixed, 567–570, 572–575, 578–580, 591–593
  - tables of data, 317, 643–647, 653–657
- electromotive force, 256–260, 314, 359–361, 567

Activity and activity coefficient, equilibrium, 267, 309, 311
- freezing-point-lowering measurements, 406, 410–414
- fugacity, 242–248, 251–255
- galvanic cell, 256–260, 314–319, 359–361
- infinite dilution, 246, 275, 318
- ions, individual, 309–310
- isopiestic measurements, 298, 320–323, 411, 575
- mean activity coefficient, 310, 333
- multicomponent systems, 551–593
- osmotic coefficient, 263, 321–324, 339, 344, 347, 406
- osmotic pressure measurements, 323–325
- partial molal free energy, 243, 260, 315
- phase diagrams, calculations from, 414
- polymer solutions, 294, 324
- pressure coefficient, 243, 246, 249, 251
- quotient, 269
- reference state, 249–251, 270, 321
- regular solutions, 284–289
- standard state, 244–249, 271–278, 308, 311, 316
- tabulation methods, 326–330, 652
- temperature coefficient, 392
- vapor-pressure measurements, 246, 252–255, 312, 319

Actual process (*see* Process, irreversible)

Adiabatic calorimeter, 62

Adiabatic demagnetization, 505–508, 511

Adiabatic expansion, 36, 46, 108

Adiabatic gas compressor, 99

Adiabatic process, 36, 46, 48, 108

Adsorption, gases, 486–490
- isosteric heat, 487
- partial molal entropy, 488–490
- partial molal heat capacity, 489
- surface films, 485

Adsorption isotherm, 486–488